Geoinformatics and Data Mining in Earth Sciences

Geoinformatics and Data Mining in Earth Sciences

Editors

Alexei Gvishiani
Boris Dzeboev

MDPI

Basel • Beijing • Wuhan • Barcelona • Belgrade • Novi Sad • Cluj • Manchester

Editors
Alexei Gvishiani
Geophysical Center of the Russian Academy of Sciences (GC RAS)
Moscow
Russia

Boris Dzeboev
Geophysical Center of the Russian Academy of Sciences (GC RAS)
Moscow
Russia

Editorial Office
MDPI AG
Grosspeteranlage 5
4052 Basel, Switzerland

This is a reprint of articles from the Topical Collection published online in the open access journal *Applied Sciences* (ISSN 2076-3417) (available at: https://www.mdpi.com/journal/applsci/topical_collections/Geoinformatics_Datamining).

For citation purposes, cite each article independently as indicated on the article page online and as indicated below:

Lastname, A.A.; Lastname, B.B. Article Title. *Journal Name* **Year**, *Volume Number*, Page Range.

ISBN 978-3-7258-1867-9 (Hbk)
ISBN 978-3-7258-1868-6 (PDF)
doi.org/10.3390/books978-3-7258-1868-6

Contents

About the Editors ix

Alexei D. Gvishiani, Inessa A. Vorobieva, Peter N. Shebalin, Boris A. Dzeboev, Boris V. Dzeranov and Anna A. Skorkina
Integrated Earthquake Catalog of the Eastern Sector of the Russian Arctic
Reprinted from: *Appl. Sci.* **2022**, *12*, 5010, doi:10.3390/app12105010 1

Inessa A. Vorobieva, Alexei D. Gvishiani, Peter N. Shebalin, Boris A. Dzeboev, Boris V. Dzeranov, Anna A. Skorkina, et al.
Integrated Earthquake Catalog II: The Western Sector of the Russian Arctic
Reprinted from: *Appl. Sci.* **2023**, *13*, 7084, doi:10.3390/ app13127084 25

Inessa A. Vorobieva, Alexei D. Gvishiani, Peter N. Shebalin, Boris A. Dzeboev, Boris V. Dzeranov, Natalia A. Sergeeva, et al.
Integrated Earthquake Catalog III: Gakkel Ridge, Knipovich Ridge, and Svalbard Archipelago
Reprinted from: *Appl. Sci.* **2023**, *13*, 12422, doi:10.3390/app132212422 49

Alexei D. Gvishiani, Boris A. Dzeboev, Boris V. Dzeranov, Ernest O. Kedrov, Anna A. Skorkina and Izabella M. Nikitina
Strong Earthquake-Prone Areas in the Eastern Sector of the Arctic Zone of the Russian Federation
Reprinted from: *Appl. Sci.* **2022**, *12*, 11990, doi:10.3390/app122311990 79

Leopold I. Lobkovsky, Irina S. Vladimirova, Yurii V. Gabsatarov and Dmitry A. Alekseev
Keyboard Model of Seismic Cycle of Great Earthquakes in Subduction Zones: Simulation Results and Further Generalization
Reprinted from: *Appl. Sci.* **2021**, *11*, 9350, doi:10.3390/app11199350 95

Alexandra Moshou, Antonios Konstantaras, Panagiotis Argyrakis, Nikolaos S. Petrakis, Theodoros N. Kapetanakis and Ioannis O. Vardiambasis
Data Management and Processing in Seismology: An Application of Big Data Analysis for the Doublet Earthquake of 2021, 03 March, Elassona, Central Greece
Reprinted from: *Appl. Sci.* **2022**, *12*, 7446, doi:10.3390/app12157446 115

Ioannis Karampinis, Lazaros Iliadis and Athanasios Karabinis
Investigation of Structural Seismic VulnerabilityUsing Machine Learning on Rapid Visual Screening
Reprinted from: *Appl. Sci.* **2024**, *14*, 5350, doi:10.3390/app14125350 139

Stepan Ustinov, Alexey Ostapchuk, Alexey Svecherevskiy, Alexey Usachev, Grigorii Gridin, Antonina Grigor'eva and Igor Nafigin
Prospects of Geoinformatics in Analyzing Spatial Heterogeneities of Microstructural Properties of a Tectonic Fault
Reprinted from: *Appl. Sci.* **2022**, *12*, 2864, doi:10.3390/app12062864 154

Alexei D. Gvishiani, Igor N. Rozenberg, Anatoly A. Soloviev, Andrey G. Kostianoy, Sofia A. Gvozdik, Ilya V. Serykh, et al.
Electronic Atlas of Climatic Changes in the Western Russian Arctic in 1950–2021 as Geoinformatic Support of Railway Development
Reprinted from: *Appl. Sci.* **2023**, *13*, 5278, doi:10.3390/app13095278 177

Alexey Lyubushin
Identification of Areas of Anomalous Tremor of the Earth's Surface on the Japanese Islands According to GPS Data
Reprinted from: *Appl. Sci.* **2022**, *12*, 7297, doi:10.3390/app12147297 . . . . **214**

Alexey Lyubushin
Singular Points of the Tremor of the Earth's Surface
Reprinted from: *Appl. Sci.* **2023**, *13*, 10060, doi:10.3390/app131810060 . . . . **227**

Vladislav Zaalishvili, Olga Burdzieva, Aleksandr Kanukov and Tamaz Zaks
Eco-Geophysical and Geoecological Factors in Assessing the State of the Geological Environment Based on the Analysis of Spatial Databases of the Territory of the Republic of North Ossetia–Alania
Reprinted from: *Appl. Sci.* **2022**, *12*, 2644, doi:10.3390/app12052644 . . . . **238**

Dmitry Melkov, Vladislav Zaalishvili, Olga Burdzieva and Aleksandr Kanukov
Temporal and Spatial Geophysical Data Analysis in the Issues of Natural Hazards and Risk Assessment (in Example of North Ossetia, Russia)
Reprinted from: *Appl. Sci.* **2022**, *12*, 2790, doi:10.3390/app12062790 . . . . **256**

Sergei M. Agayan, Ilya V. Losev, Ivan O. Belov, Victor N. Tatarinov, Alexander I. Manevich and Maksim A. Pasishnichenko
Dynamic Activity Index for Feature Engineering of Geodynamic Data for Safe Underground Isolation of High-Level Radioactive Waste
Reprinted from: *Appl. Sci.* **2022**, *12*, 2010, doi:10.3390/app12042010 . . . . **278**

Sergey Agayan, Shamil Bogoutdinov, Dmitriy Kamaev, Vladimir Kaftan, Maxim Osipov and Victor Tatarinov
Theoretical Framework for Determination of Linear Structures in Multidimensional Geodynamic Data Arrays
Reprinted from: *Appl. Sci.* **2021**, *11*, 11606, doi:10.3390/app112411606 . . . . **295**

Krzysztof Karsznia, Edward Osada and Zbigniew Muszyński
Real-Time Adjustment and Spatial Data Integration Algorithms Combining Total Station and GNSS Surveys with an Earth Gravity Model
Reprinted from: *Appl. Sci.* **2023**, *13*, 9380, doi:10.3390/app13169380 . . . . **319**

Yabin Tao, Han Du, Ruixin Zhang, Jianzhao Feng and Zhiyun Deng
Experiment Study on Mechanical Evolution Characteristics of Coal and Rock under Three-Dimensional Triaxial Stress
Reprinted from: *Appl. Sci.* **2022**, *12*, 2445, doi:10.3390/app12052445 . . . . **338**

Mohamed Hafedh Hamza and Mohamed Chmit
GIS-Based Planning and Web/3D Web GIS Applications for the Analysis and Management of MV/LV Electrical Networks (A Case Study in Tunisia)
Reprinted from: *Appl. Sci.* **2022**, *12*, 2554, doi:10.3390/app12052554 . . . . **353**

Jian Chen, Jianpeng Chen, Xiangrong She, Jian Mao and Gang Chen
Deep Contrast Learning Approach for Address Semantic Matching
Reprinted from: *Appl. Sci.* **2021**, *11*, 7608, doi:10.3390/app11167608 . . . . **373**

Bandar Fuad Khashoggi and Abdulkader Murad
Use of 2SFCA Method to Identify and Analyze Spatial Access Disparities to Healthcare in Jeddah, Saudi Arabia
Reprinted from: *Appl. Sci.* **2021**, *11*, 9537, doi:10.3390/app11209537 . . . . **389**

Jinyu Guo and Lizhen Wang
A Spatial Fuzzy Co-Location Pattern Mining Method Based on Interval Type-2 Fuzzy Sets
Reprinted from: *Appl. Sci.* **2022**, *12*, 6259, doi:10.3390/app12126259 . . . . **411**

Liuchang Xu, Ruichen Mao, Chengkun Zhang, Yuanyuan Wang, Xinyu Zheng, Xingyu Xue and Fang Xia
Deep Transfer Learning Model for Semantic Address Matching
Reprinted from: *Appl. Sci.* **2022**, *12*, 10110, doi:10.3390/app121910110 . . . . **441**

About the Editors

Alexei Gvishiani

Alexei Gvishiani—Chief Scientist of the Geophysical Center of the Russian Academy of Sciences (RAS), Professor, Full member of RAS. Well-known scientist in the fields of mathematical geophysics, geomagnetism, geoinformatics, systems analysis and Big Data in geophysical studies. A. Gvishiani is the Full Member of the European Academy of Sciences, an Honorary Scholar of the International Institute for Applied Systems Analysis (Austria), Chairman of the RAS Scientific Council of the Arctic and Antarctic Studies, Chairman of the National CODATA Committee, Chairman of the National Geophysical Committee of RAS, Deputy Chairman of the RAS Committee for Systems Analysis, Member of the Scientific and Technical Council at the Russian University of Transport, Deputy Chairman of the RAS Scientific Council "Environment and Transport", Member of the Scientific Publishing Council of RAS, Member of the Interdepartmental Coordinating Council for International Scientific and Technical Cooperation at the RAS Presidium, Member of the Russian Geographical Society, Professor of the Lomonosov Moscow State University and the Paris and Strasbourg Institutes of Physics of the Earth, Chief Editor of the "Russian Journal of Earth Sciences". He was awarded with the following state awards: Order of Honor, Order of Friendship, Medal "300th Anniversary of the Russian Academy of Sciences", Medal of Peter Gubonin of the Ministry of Transport of the Russian Federation, Certificate of Honor of the State Corporation for Space Activities "Roscosmos", "Honorary Worker of Science and High Technologies of the Russian Federation", Medal "850th Anniversary of Moscow". He is a laureate of the 2024 Boris Golitsyn Prize of RAS.

Boris Dzeboev

Boris Dzeboev—Deputy Director for Science at the Geophysical Center of the Russian Academy of Sciences (RAS), Doctor of Science in physics and mathematics, and well-known scientist in the fields of mathematical methods of artificial intelligence in geosciences, seismic hazard and seismic risk assessment, and geoinformatics. He is a Member of the RAS Scientific Council of the Arctic and Antarctic Studies, Deputy Chief Editor of the "Russian Journal of Earth Sciences", and a Member of the editorial board of the "Journal of Volcanology and Seismology". He was also awarded with a medal for the "300th Anniversary of the Russian Academy of Sciences".

applied sciences

MDPI

Article

Integrated Earthquake Catalog of the Eastern Sector of the Russian Arctic

Alexei D. Gvishiani [1,2], Inessa A. Vorobieva [1,3], Peter N. Shebalin [1,3], Boris A. Dzeboev [1,*], Boris V. Dzeranov [1] and Anna A. Skorkina [1,2,3]

1 Geophysical Center of the Russian Academy of Sciences (GC RAS), 119296 Moscow, Russia; adg@wdcb.ru (A.D.G.); vorobiev@mitp.ru (I.A.V.); p.n.shebalin@gmail.com (P.N.S.); b.dzeranov@gcras.ru (B.V.D.); anna@mitp.ru (A.A.S.)
2 Schmidt Institute of Physics of the Earth of the Russian Academy of Sciences (IPE RAS), 119296 Moscow, Russia
3 Institute of Earthquake Prediction Theory and Mathematical Geophysics of the Russian Academy of Sciences (IEPT RAS), 117997 Moscow, Russia
* Correspondence: b.dzeboev@gcras.ru; Tel.: +7-495-930-05-46

Abstract: The objective of this study was to create a representative earthquake catalog for the Eastern Sector of the Arctic zone of the Russian Federation that combines all available data from Russian and international seismological agencies, with magnitude reduction to a uniform scale. The article describes the catalog compilation algorithm, as well as formalized procedures for removing duplicates and choosing the optimal magnitude scale. Due to different network configurations and record processing methods, different agencies may register/miss different events. This results in the absence of some events in different earthquake catalogs. Therefore, merging the data of various seismological agencies will provide the most complete catalog for the studied region. When merging catalogs, the problem of identifying duplicates (records related to the same seismic event) necessarily arises. An additional difficulty arises when distinguishing between aftershocks and duplicates since both are events that are close in space and time. To solve this problem, we used a modified nearest neighbor method developed earlier by the authors. The modified version, which is focused on identifying duplicates and distinguishing between duplicates and aftershocks, uses a probabilistic metric in the network error space to determine the epicenters and times of seismic events. In the present paper, a comparison and regression analysis of the different magnitude types of the integrated catalog is carried out, and based on the obtained ratios, the magnitude estimates are unified.

Keywords: merging catalogs; earthquake; clustering algorithm; Arctic region; magnitude unification; duplicate events

Citation: Gvishiani, A.D.; Vorobieva, I.A.; Shebalin, P.N.; Dzeboev, B.A.; Dzeranov, B.V.; Skorkina, A.A. Integrated Earthquake Catalog of the Eastern Sector of the Russian Arctic. *Appl. Sci.* **2022**, *12*, 5010. https://doi.org/10.3390/app12105010

Academic Editor: Amadeo Benavent-Climent

Received: 21 March 2022
Accepted: 13 May 2022
Published: 16 May 2022

1. Introduction

As is known, the Eastern Sector of the Arctic zone of the Russian Federation (AZRF) is a seismically active region [1–4]. Rather strong earthquakes can occur within its limits. For example, only in the last two decades, events with magnitudes of $M \geq 6$ have occurred there: the Olyutorsk earthquake, with $M = 7.6$, on 20 April 2006; the Ilin-Tas (Abyi) earthquake, with $M = 6.6$, on February 14 2013; the earthquake near the border of Kamchatka and Chukotka, with $M = 6.4$, on 9 January 2020; and others [5–10]. The analysis of the seismic regime of the Arctic territories of Russia and the construction of seismic hazard maps [11–25] are topical problems today. The solution to these problems is impossible without the creation of a representative instrumental earthquake catalog [26]. The importance of these problems for the study and development of the Arctic is emphasized by the increasing level of industrial development in the region.

Earthquake studies in the Eastern part of the AZRF started not so long ago. The authors of the *New Catalog of Strong Earthquakes in the U.S.S.R. from ancient times through 1977* [27], Nikolay Shebalin and Nadezhda Kondorskaya, made the first step in these studies in the late 1970s. Intensive earthquake catalog projects concerning the AZRF were implemented in the early 2000s by V.I. Ulomov [22], V.S. Imaev, and L.P. Imaeva, B.M. Koz'min, et al. [5,6]. The seismic zonation map of the AZRF (as a part of the general seismic zoning map of Russian Federation) was created in recent years.

At the same time, there still remains a need in the AZRF representative Eastern Sector catalog which combines data from available Russian and international sources with the magnitude reduction to a uniform scale. This paper describes the results of the study on such catalog creation.

Nowadays, the seismic monitoring of the Russian Arctic is carried out by regional branches of the Geophysical Survey of the Russian Academy of Sciences (GS RAS) (http://www.gsras.ru/new/eng/catalog/, (accessed on 20 March 2022)). In the Eastern part of the Russian Arctic, this work is carried out by the Yakutsk, Magadan, and, partially, the Kamchatka Branches of the GS RAS. In addition, detailed information on earthquakes can be found in the global catalog of the International Seismological Center (ISC) (http://www.isc.ac.uk/isc-ehb/search/catalogue/, (accessed on 20 March 2022)), which combines the data from several global and national seismological networks. It has to be noted that a comparative analysis of the catalogs showed that for the Eastern Sector of the Russian Arctic, the ISC catalog does not contain many events that are presented in the regional catalogs of the GS RAS. This is explained, among other things, by the fact that the survey reports bulletins to the ISC for events starting at a certain magnitude threshold. For this reason, the information on low-magnitude seismicity is mainly contained in regional catalogs.

It should be also noted that due to the different configurations of seismic networks and methods for processing records, some agencies may skip earthquakes recorded by other networks. Thus, merging earthquake catalogs is a method for improving the completeness and representativeness of seismic events in the final catalog [28–32].

When merging catalogs, the problem of identifying duplicates arises. The main difficulty is discrimination [33] between aftershocks and duplicates since both of them are similar events in space and time. This problem is analogous to the discrimination between aftershocks and independent seismic events. In [34], an algorithm for merging two earthquake catalogs was developed, the main task of which was to identify the resulting duplicates and separate them from the aftershocks. The algorithm is based on the author's modification of the nearest neighbor method [35,36] for duplicate identification. It is based on the fact that, unlike aftershocks, duplicates do not have a causal relationship. The algorithm establishes a correspondence between the events from two catalogs, after which the classification of earthquakes into unique and duplicates is performed using the Euclidean metric. The sequential application of the algorithm automates the integration of any number of earthquake catalogs. The developed algorithm efficiency was demonstrated in [34] using the example of merging the ComCat Advanced National Seismic System and the Japan Meteorological Agency catalogs for the aftershock sequence of the 2011 Tohoku earthquake. In this paper, a unified earthquake catalog is created for the Eastern Sector of the Arctic zone of the Russian Federation. For this purpose, the following main issues are solved:

- The sequential merging of three regional catalogs of the GS RAS and the ISC catalog, which implies the identification of duplicate events in the border areas of responsibility of the different networks; and
- The unification of magnitude estimates in the integrated catalog by constructing regression relationships for the different types of magnitude/energy class due to the exact association of data from the different catalogs related to the same event.

2. Materials and Methods

The studied region represents a geographical area limited by the following coordinates: 60° N, 100° E; 77° N, 100° E; 77° N, 165° W; 57.5° N, 165° W; 57.5° N, 138° E; and 60° N, 138° E (dashed line in Figure 1). All of the following four existing earthquake catalogs for the period 1962–2020 were considered as the initial data (Tables 1 and 2):

1. The regional catalog of Yakutia from the annual journals *Earthquakes in the USSR* (1962–1991), *Earthquakes in Northern Eurasia* (1992–2014), and *Earthquakes in Russia* (2015–2019) (GS RAS) (hereinafter YAK);
2. The regional catalog of the northeast of Russia from the annual journals *Earthquakes in the USSR* (1968–1991), *Earthquakes in Northern Eurasia* (1992–2014), and *Earthquakes of Russia* (2015–2019) (GS RAS) (hereinafter NER);
3. The regional catalog of earthquakes in Kamchatka of the Kamchatka Branch of the GS RAS, 1962–2019 (hereinafter KAM); and
4. The ISC 1962–2020 catalog, which is a composite catalog containing data from many world and Russian agencies.

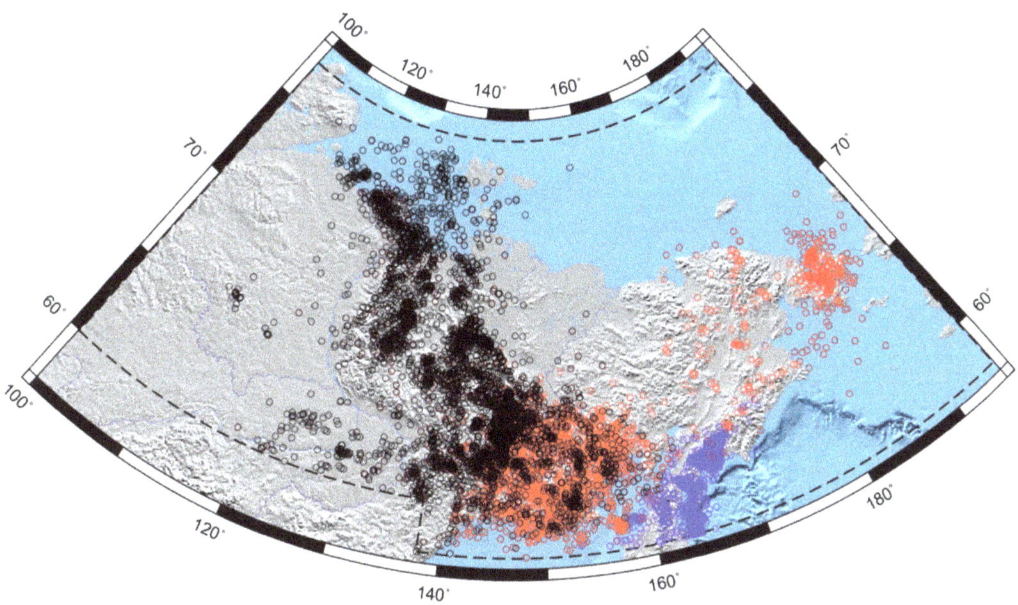

Figure 1. Studied region. The circles are the earthquake epicenters from the YAK (black), NER (red), and KAM (blue) catalogs.

Table 1. Data of Russian agencies.

Catalog	Period	Number of Earthquakes with Energy Classes and/or Magnitudes	Number of Earthquakes with Unknown Energy Classes and Magnitudes
YAK	1962, 1968–2019	6600	46
NER	1968–2019	7668	1
KAM	1962–2019	4498	0

Table 2. ISC catalog statistics from 1962–2020.

Agency Abbreviation	Agency	Number of Earthquakes with Energy Classes and/or Magnitudes *
AEIC	Alaska Earthquake Information Center, USA	184
ANDRE	USSR	16
ANF	USArray Array Network Facility, USA	2
BJI	China Earthquake Networks Center, China	1
BYKL	Baykal Regional Seismological Centre, GS SB RAS, Russia	4
DNAG	USA	13
EIDC	Experimental (GSETT3) International Data Center, USA	22
GCMT	The Global CMT Project, USA	1
IDC	International Data Centre, CTBTO, Austria	123
ISC	International Seismological Centre, United Kingdom	1507
KRSC	Kamchatka Branch of the Geophysical Survey of the RAS, Russia	2684
MATSS	USSR	1400
MOS	Geophysical Survey of Russian Academy of Sciences, Russia	26
MSUGS	Michigan State University, Department of Geological Sciences, USA	2585
NEIC	National Earthquake Information Center, USA	192
NEIS	National Earthquake Information Service, USA	1
NERS	North Eastern Regional Seismological Centre, GS RAS, Russia	4688
NKSZ	USSR	8
SBDV	USSR	107
SYKES	Sykes Catalogue of earthquakes 1950 onwards	2
USCGS	United States Coast and Geodetic Survey, NEIC, USA	1
WASN	USA	328
YARS	Yakutiya Regional Seismological Center, GS SB RAS, Russia	2256
ZEMSU	USSR	1884
	Total:	18,035

* The ISC catalog contains 6441 events with unknown energy classes and magnitudes.

It has to be noted that the technique developed in [34] allows the pairwise merging of earthquake catalogs. Thus, any two chosen catalogs are merged in the first step. Then, another catalog is merged with them, and so on. At the same time, we emphasize that the list of events in the integrated catalog weakly depend on the sequence of pairwise merging. It was shown in [34] that the merging procedure is symmetric. In other words, when two catalogs are merged, the same events are selected as duplicates, regardless of the catalogs' merging sequence. The only difference will be which version of the earthquake record (from which input catalog) falls into the merged catalog.

We believe that earthquake identification based on global network data is the most reliable. A subset of these events from the ISC catalog is the core—the main catalog to which other catalogs will be added. The core only contains information about strong and moderate earthquakes in the region because weak earthquakes are not registered by global networks. Further, it is logical to add data from local networks, which provide information about weak earthquakes in the region. In the final merge step, we use the data from the ISC catalog that was not included in the core. Thus, to merge catalogs (Tables 1 and 2), the following sequence for sources of the initial data was chosen:

1. Earthquakes from the ISC global catalog (the abbreviation of the ISC and GCMT agency in Table 2) with the magnitudes $M_W{}^{GCMT}$ and/or mb^{ISC} are the core (hereinafter CORE) (1393 events);
2. Earthquakes from Russian catalogs with local estimates for the magnitude of weak events. In the intersection zones, preference is given to the data from the catalog of Yakutia (Table 1);
3. Other earthquakes from the ISC (abbreviation of the ISC agency in Table 2, without the magnitude data $M_W{}^{GCMT}$ or mb^{ISC}), as well as data from other agencies in the ISC catalog (16,642 events). This selection from the ISC catalog will be further denoted by ISC_Other.

To discriminate and remove duplicates resulting from the merging of catalogs, we apply the modified nearest neighbor method and the Euclidean metric in the space of the variance in the definitions of seismic event parameters by different networks [34]. We apply a basic three-parameter model that takes into account the differences in time and the coordinates of the epicenter, the effectiveness of which was shown in [34]. We do not analyze the difference in depth because for a significant number of events, information on the hypocentral depth is not presented in the original catalogs or a standard value of 10 km is given. The magnitude is also excluded from consideration because different catalogs use different types of magnitude.

At the input, there are two catalogs: main Catalog 1 and additional Catalog 2. We believe that neither Catalog 1 nor Catalog 2 contain duplicates within themselves, since the modern automatic processing of seismic records almost completely eliminates technical errors. The problem is to find records in Catalogs 1 and 2 that will correspond to the same seismic events (duplicates) and divide Catalog 2 into events that have duplicates in Catalog 1 and unique events.

A modification of the nearest neighbor method is based on the assumption that duplicates form pairs in which the events must belong to different source catalogs. As a result of applying the modified nearest neighbor method, a set of pairs of potential duplicates is formed. We consider the events of the additional Catalog 2, with the value of the neighborhood function as less than the threshold one, as duplicates. The rest of the Catalog 2 events are declared unique and added to Catalog 1. Further, any number of catalogs can be sequentially added.

The choice of the proximity function is based on a probabilistic model. We assume that the difference in earthquake detection by different networks is a random variable with a normal distribution and zero mean for each of the parameters:

$$f(DT) = \frac{1}{\sigma_T\sqrt{2\pi}} \exp\left(-\frac{DT^2}{2\sigma_T{}^2}\right),$$

$$f(DX) = \frac{1}{\sigma_X\sqrt{2\pi}} \exp\left(-\frac{DX^2}{2\sigma_X{}^2}\right),$$

$$f(DY) = \frac{1}{\sigma_Y\sqrt{2\pi}} \exp\left(-\frac{DY^2}{2\sigma_Y{}^2}\right).$$

Here DT, DX, and DY are the differences in time, longitude, and latitude, respectively, between different determinations of a seismic event, and σ_T, σ_X, and σ_Y are the corresponding standard deviations. If we assume that all errors are independent, then the duplicate probability density will be the product of the error probabilities for all parameters. This will be the multivariate normal distribution, as follows:

$$f(DT, DX, DY) = \frac{1}{\sigma_T\sigma_X\sigma_Y(2\pi)^{\frac{3}{2}}} \cdot \exp\left(-\left(\frac{DT^2}{2\sigma_T{}^2} + \frac{DX^2}{2\sigma_X{}^2} + \frac{DY^2}{2\sigma_Y{}^2}\right)\right).$$

Thus, we naturally arrive at the Euclidean metric:

$$Ro = \sqrt{\frac{DT^2}{\sigma_T{}^2} + \frac{DX^2}{\sigma_X{}^2} + \frac{DY^2}{\sigma_Y{}^2}} \tag{1}$$

The preliminary identification of duplicates is done with the standard metric parameters in (1): $\sigma_{0T} = 0.05$ min and $\sigma_{0X} = \sigma_{0Y} = 15$ km. The initial values of the parameters have little effect on the identification of duplicates; however, they significantly affect the value of the duplicate probability and the estimate of the percentage of errors. At this stage, we check that each of the parameters follows a normal distribution, and we refine the values of the standard deviations σ_T, σ_X, and σ_Y. After that, the final identification of duplicates is performed. The choice of the optimal metric threshold for identifying duplicates and the estimation of the percentage of errors will be explained in detail below.

Before proceeding with the merging process of the four catalogs, we checked each of them for internal duplicates. For this reason, we built the distribution of metric (1) for the nearest events within each catalog (Figure 2). The analysis was performed with the metric parameters $\sigma_T = 0.05$ min and $\sigma_X = \sigma_Y = 15$ km. As a result of the analysis, no events with the same time and epicenter coordinates were found in any of the four catalogs. For such events, $Ro = 0$, and, thereafter, we will call them absolute duplicates. Statistical analysis also did not reveal anomalous groups of nearest events. The duplicates are characterized by the value $Ro < 10$. The value $Ro = 10$ corresponds to a distance of 150 km or a time interval of 0.5 min. From our experience, we know that duplicates have smaller differences in instrumental catalogs, and the number of such nearest events within each of the catalogs is very small. These are mainly the early aftershocks of the Olyutorsk earthquake, with $M = 7.6$, on 20 April 2006, and the Ilin-Tas earthquake, with $M = 6.6$, on 14 February 2013. There is no reason to consider such events as duplicates since the early aftershocks can occur at very small distances and time intervals. Thus, the necessary condition for applying the method in [34] is met.

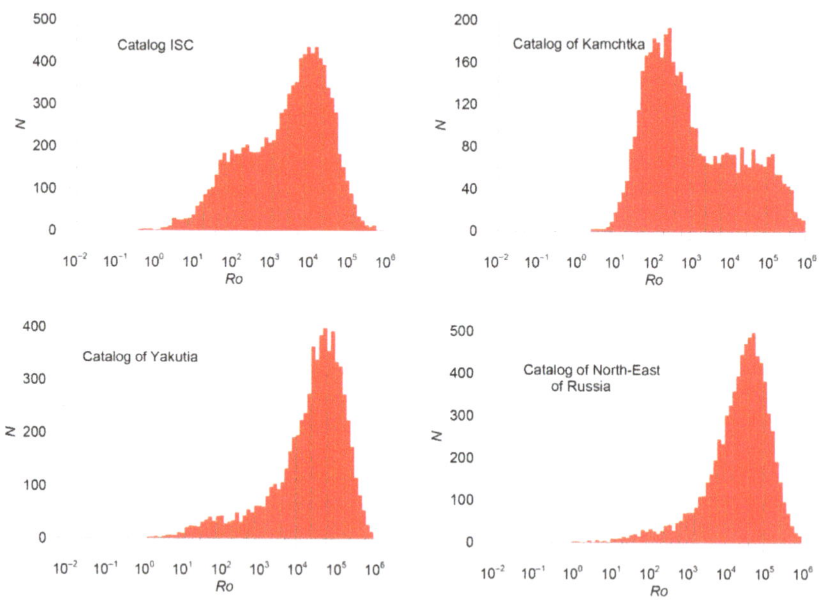

Figure 2. Distribution of metric (1) for events within the source catalogs (Tables 1 and 2). The catalogs are indicated on the histograms.

3. Results

3.1. Merging Catalogs

At the initial stage, the regional data from the catalogs of the Russian agencies YAK, NER, and KAM were merged and then combined with ISC data. The earthquakes with unknown magnitudes/classes were not included in the merging. Below is a sequence of the stages of merging the catalogs.

3.1.1. Stage 1. Merging YAK and NER

YAK was considered as the main catalog and NER as the additional one. The preliminary analysis of duplicates was performed with the distribution parameters $\sigma_T = 0.05$ min and $\sigma_X = \sigma_Y = 15$ km. As a result, 1834 absolute duplicates and about 370 potential duplicates (events with a small metric) were identified. The preliminary threshold was determined by the minimum distribution of the metric. Absolute duplicates were not used to determine the dispersions (Figure 3). It was verified that each of the parameters followed a normal distribution and that the mean was small compared to the standard deviation for all three parameters (*DT*, *DX*, and *DY*). It was also verified that the variance was almost independent of the event magnitude and time (Figure 3).

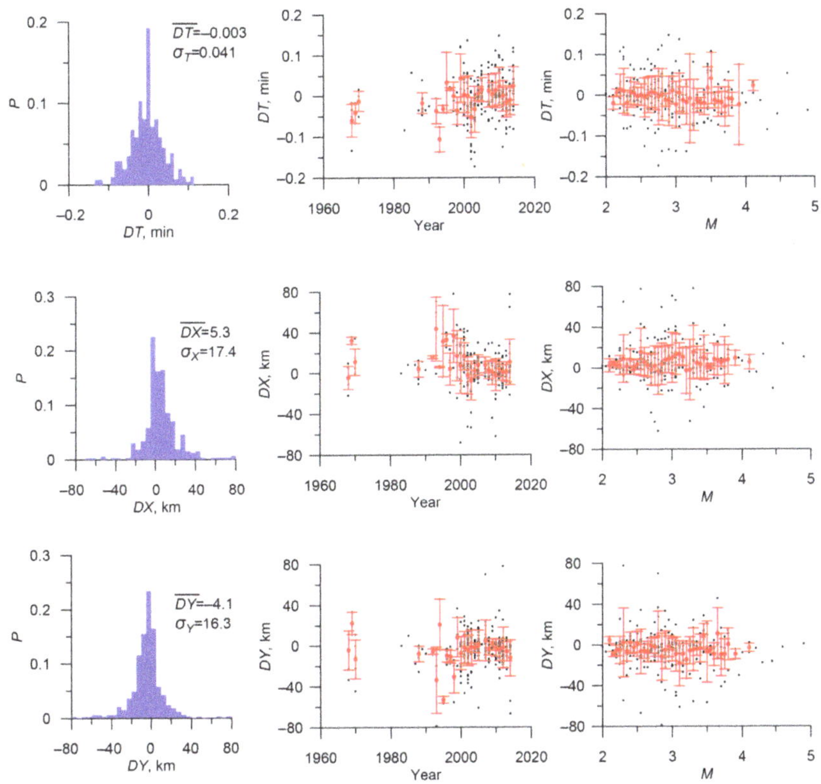

Figure 3. Distributions of *DT*, *DX*, and *DY* for the nearest events from the YAK and NER catalogs, and the dependence of the standard deviations σ_T, σ_X, and σ_Y and the mean values $\overline{DT}$, $\overline{DX}$, and $\overline{DY}$ on the time and magnitude of the events. The red dots and bars are the population mean values and standard deviations, respectively.

The final duplicate analysis was performed with the parameters $\sigma_T = 0.041$ min, $\sigma_X = 17.4$ km, and $\sigma_Y = 16.3$ km. The metric values between the nearest events of the YAK catalog were also calculated. This made it possible to estimate the probability that the duplicate was chosen incorrectly due to the high density of earthquakes. Figure 4a shows the distributions of metric (1) for the YAK/NER pairs and the same metric for the YAK/YAK earthquakes (the algorithm for calculating the metric is the same as for two different catalogs, but only the comparison of the earthquake with itself is excluded). A group of anomalously close YAK/NER events is identified well. The optimization of the threshold value of the metric is illustrated in Figure 4b. The red line is the probability of missing a duplicate in the 3D normal distribution model (error of the first kind) and the blue line is the probability of a false duplicate (error of the second kind), which is defined as the ratio of the number of YAK/YAK pairs for a given value of the metric *Ro* to the number of events in the YAK catalog.

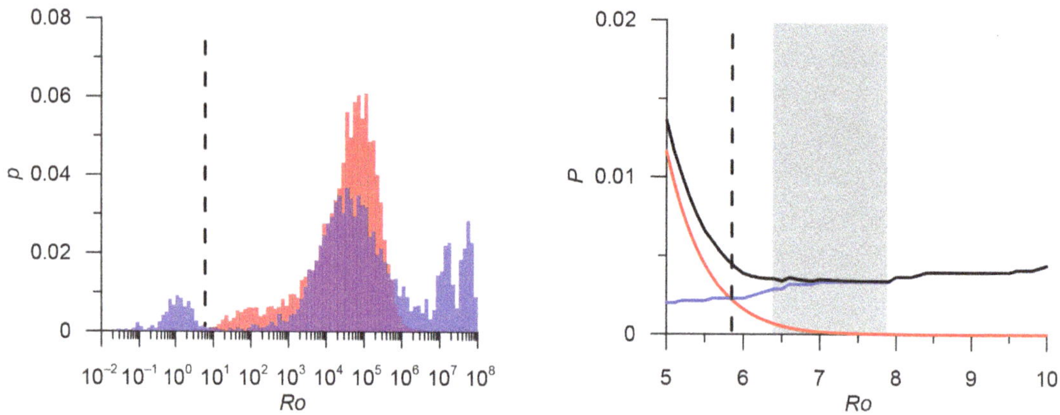

(**a**) (**b**)

Figure 4. (**a**) Comparison of the metric distribution for YAK/NER pairs (blue histogram) and the same metric for YAK/YAK earthquakes (red histogram). (**b**) Threshold optimization: the red line shows the probability of missing a duplicate in the model with metric (1), the blue line is the probability of a false duplicate (see text), the black line is the total probability of errors of the first and second kind, the dashed line *Ro* = 5.8 corresponds to an equal number of errors of the first and of the second kind (the number of false duplicates is equal to the number of missed duplicates), the estimate of the total number of errors is approximately 0.5%, and the gray bar shows the range of values for the metric *Ro* = 6.3 ÷ 7.9, minimizing the total number of errors (approximately 0.4%).

An equal number of errors of the first and second kind is achieved at *Ro* = 5.8. In the NER catalog, only 15 earthquakes have a distance to the nearest neighbor of *Ro* < 5.8. This made it possible to estimate the probability that the duplicate was chosen incorrectly due to the high density of earthquakes. The upper estimate of the probability of false duplicates *P* = 15/6600 = 0.0023 is approximately 0.25%. At *Ro* = 7.9, the number of such earthquakes increases to 26, which corresponds to a probability of 0.4% (see the blue line in Figure 4b).

The choice of the metric threshold for identifying duplicates depends on the objective of further research of the integrated catalog. If it is important to ensure that duplicates are removed, then a higher *Ro* threshold is preferable. If it is important to keep the integral characteristics of the catalog, then the *Ro* threshold that ensures the equality of errors of the first and second kind is preferable.

We chose the threshold *Ro* = 5.8. In this case, in addition to 1834 absolute duplicates, 319 more duplicates were identified. In total, 5515 unique events were identified in the NER catalog in the study area. These events were added to the YAK catalog, and thus a merged YAK_NER catalog containing 12,115 events was obtained. Figure 5 shows the spatio–temporal structure of the YAK/NER duplicates and the naturally grouped events in the YAK catalog. The metric values for earthquakes in the YAK catalog are significantly larger than those for the YAK/NER duplicates. The metric (1) level lines provide a close-to-optimal separation of duplicates and naturally grouped events (the lower cluster of black dots).

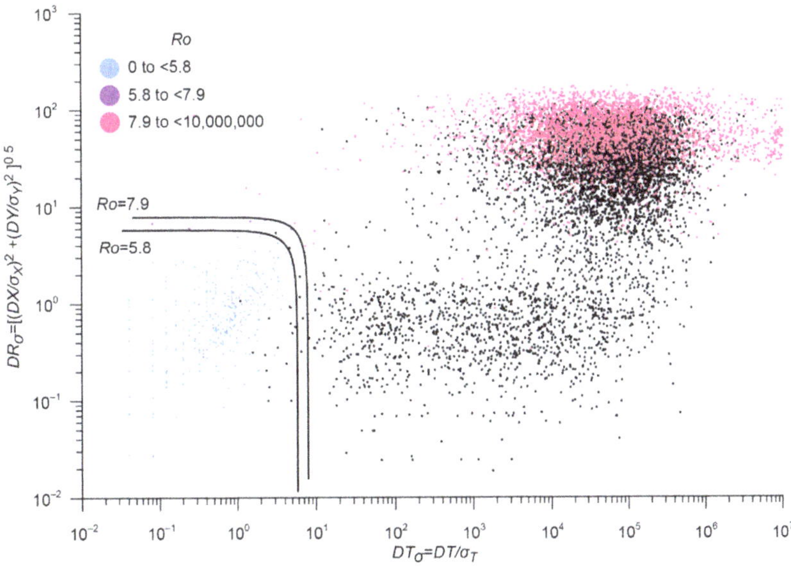

Figure 5. Distribution of normalized *DT* and *DR* and metric level lines (1). The colored dots are YAK/NER pairs and the black dots are the distances between YAK/YAK events in metric (1). The metric levels *Ro* = 5.8 and *Ro* = 7.9 are shown by black lines, and absolute duplicates are not shown.

3.1.2. Stage 2. Merging YAK_NER and KAM into the RUS Catalog

The catalog YAK_NER, obtained in the previous step, was taken as the main one, with KAM as an additional catalog. The preliminary analysis of the duplicates was performed with the distribution parameters determined for the NER and YAK catalogs: $\sigma_T = 0.041$ min, $\sigma_X = 17.4$ km, and $\sigma_Y = 16.3$ km. Twenty-eight potential duplicates were identified (Figure 6), which is not enough to determine the variances. For this reason, the metric parameters defined for the YAK and NER catalogs were used. With *Ro* = 5.8, the KAM catalog contains 26 duplicates and 4472 unique events that have been added to the YAK_NER catalog. The merged RUS catalog obtained in this way contains 16,587 events.

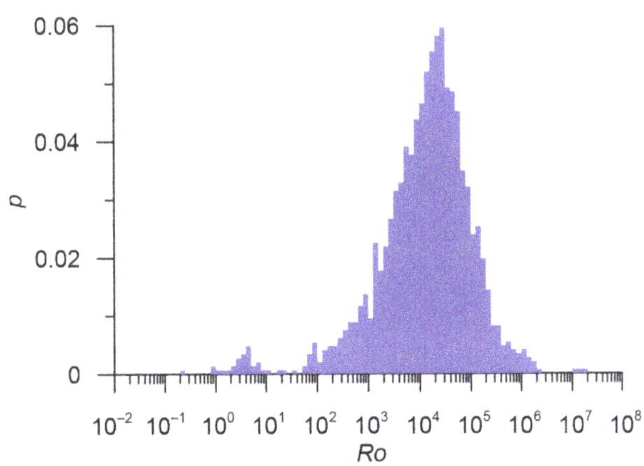

Figure 6. Metric distributions for the YAK_NER and KAM pairs.

3.1.3. Stage 3. Merging RUS and Data from the ISC_Other Catalog

The ISC catalog contains a large amount of data from Russian agencies (Table 2). Accordingly, at this stage of the merging procedure, a large number of duplicates, including absolute ones, are expected.

When merging, the catalog RUS obtained at *Stage 2* was taken as the main one and ISC_Other was taken as the additional. The resulting catalog will be designated RUS_ISC. The preliminary analysis of duplicates was performed with the standard distribution parameters $\sigma_T = 0.05$ min, $\sigma_X = 15$ km, and $\sigma_Y = 15$ km (Figure 7).

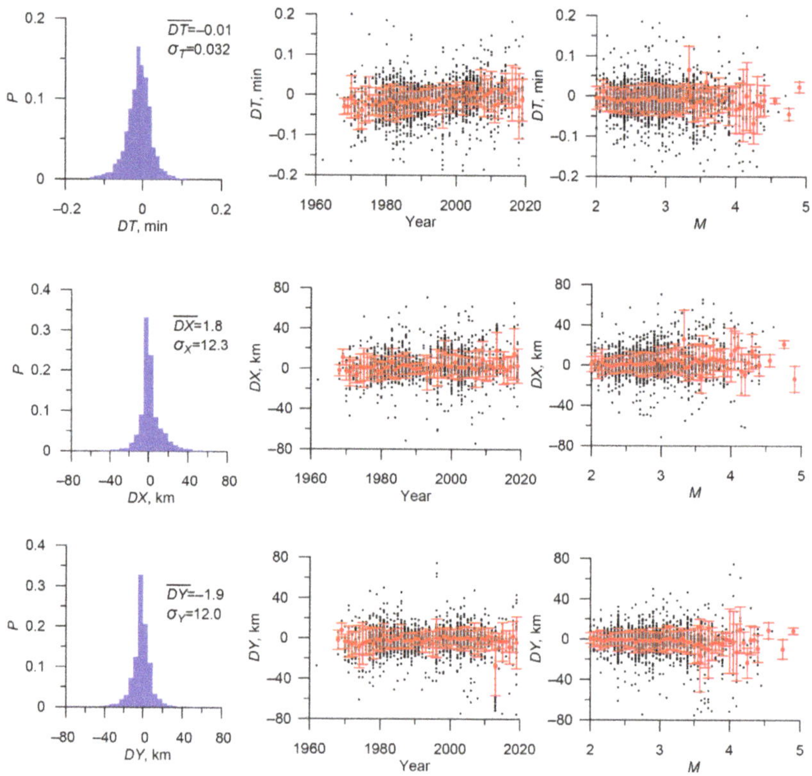

Figure 7. Distributions of *DT, DX*, and *DY* for the nearest events from the RUS and ISC_Other catalogs, and the dependence of the standard deviations σ_T, σ_X, and σ_Y and the mean values $\overline{DT}$, $\overline{DX}$, *and* $\overline{DY}$ on the time and magnitude of the events. The red dots and bars are the population mean values and standard deviations, respectively.

More than 10,000 potential duplicates have been identified, about 5000 of which have the same times and/or epicenters. Such pairs represent the same registration of events by the networks of the GS RAS, which are included in the Russian catalogs and the ISC catalog. They were excluded to determine the variances. We have verified that each of the parameters follows a normal distribution and that the mean is small compared to the standard deviation for all three parameters (*DT, DX*, and *DY*) (Figure 7). It was also verified that the variance is almost independent of the event magnitude and time. The final analysis of duplicates was performed with the parameters $\sigma_T = 0.032$ min, $\sigma_X = 12.3$ km, and $\sigma_Y = 12.0$ km. We calculated the metric values between events of the RUS catalog. This made it possible to estimate the probability that the duplicate was determined incorrectly due to the high density of earthquakes.

We chose the threshold *Ro* = 6.0 (Figure 8). In this case, in addition to 4802 absolute duplicates, another 5706 potential duplicates are identified. Many pairs of earthquakes have the same time or the same coordinates of the epicenter, and 73 of such pairs have large metric values of *Ro* > 6. An analysis of these pairs indicates that the records in the RUS and ISC catalogs differ in one digit. Most likely, these are technical errors of the era of manual information entry, which were corrected when compiling the catalog *Earthquakes of Northern Eurasia*. These events are considered to be duplicates and they are not included in the integrated catalog, despite the large values of the metric.

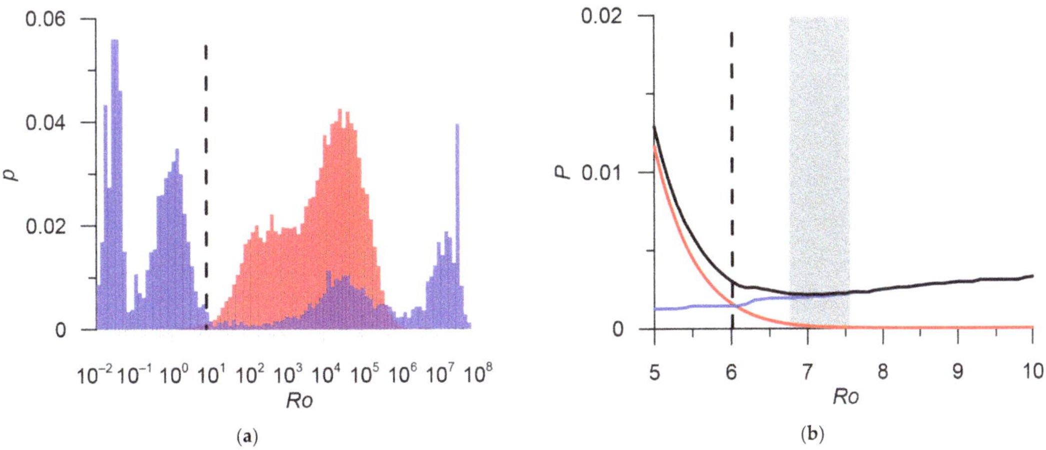

Figure 8. (**a**) Comparison of the metric distribution for RUS/ISC_Other pairs (blue histogram) and the same metric for RUS/RUS earthquakes (red histogram). (**b**) Threshold optimization: the red line is the probability of missing a duplicate in the model with metric (1), the blue line is the probability of a false duplicate, the black line is the total probability of errors of the first and second kind, the dashed line *Ro* = 6.0 corresponds to an equal number of errors of the first and second kind (number of false duplicates is equal to the number of missed duplicates), the estimate of the total number of errors is approximately 0.3%, and the gray bar shows the range of values for the metric *Ro* = 6.7 ÷ 7.6, minimizing the total number of errors (approximately 0.2%).

Figure 9 shows the spatio–temporal structure of the duplicates in RUS/ISC_Other and the naturally grouped events in the RUS catalog. The metric values for the earthquakes in the RUS catalog are significantly larger than for the RUS/ISC_Other duplicates. The metric level lines *Ro* = 6 and *Ro* = 7.6 provide close-to-optimal separation of the duplicates and naturally grouped events (the lower cluster of black dots). In total, for the studied territory there are 6411 unique events in the ISC_Other catalog. These events have been added to the RUS catalog. The merged catalog RUS_ISC contains 22,998 events.

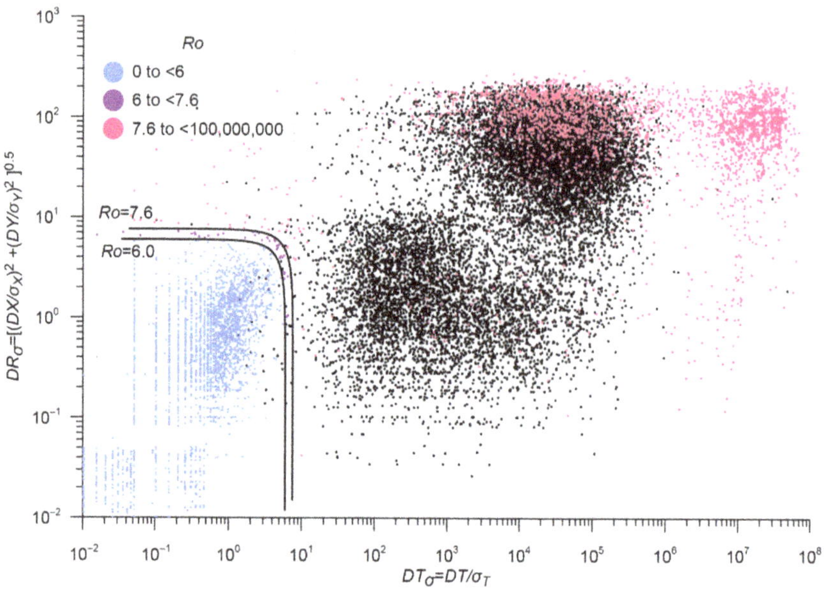

Figure 9. Distribution of normalized *DT* and *DR*, and metric level lines (1). The colored dots are the RUS/ISC_Other pairs and the black dots are the distances between the RUS/RUS events in metric (1). The metric levels *Ro* = 6 and *Ro* = 7.6 are shown by the black lines, and absolute duplicates are not shown.

3.1.4. Stage 4. Merging RUS_ISC and CORE

As the main catalog selected, CORE includes events from the ISC catalog with the magnitudes M_W^{GCMT} or mb^{ISC}. As an additional catalog, we consider RUS_ISC, obtained at the previous stage. The preliminary analysis of the duplicates was performed with the standard catalog distribution parameters $\sigma_T = 0.05$ min, $\sigma_X = 15$ km, and $\sigma_Y = 15$ km. Approximately 1000 duplicates were identified and used to determine the variances. It was verified that each of the parameters follows a normal distribution and that the mean is small compared to the standard deviation for all three parameters (*DT*, *DX*, and *DY*). It was also verified that the variance is almost independent of the event magnitude and time (Figure 10).

The final analysis of the duplicates was performed with the parameters $\sigma_T = 0.044$ min and $\sigma_X = \sigma_Y = 18.3$ km. The metric values between the events of the CORE catalog were also calculated. This made it possible to estimate the probability that the duplicate was chosen incorrectly due to the high density of earthquakes.

The value *Ro* = 5.9 was chosen as a threshold (Figure 11). In this case, 1011 duplicates are detected. In total for the studied territory, there are 21,987 unique events in the RUS_ISC catalog. These events were added to the CORE catalog, and a combined ARCTIC catalog containing 23,370 events was obtained. Figure 12 shows the space–time structure of the CORE/RUS_ISC duplicates and the naturally grouped events in the CORE catalog. The metric level lines *Ro* = 5.9 and *Ro* = 8.4 provide close-to-optimal separation of the duplicates and naturally grouped events (the lower cluster of black dots).

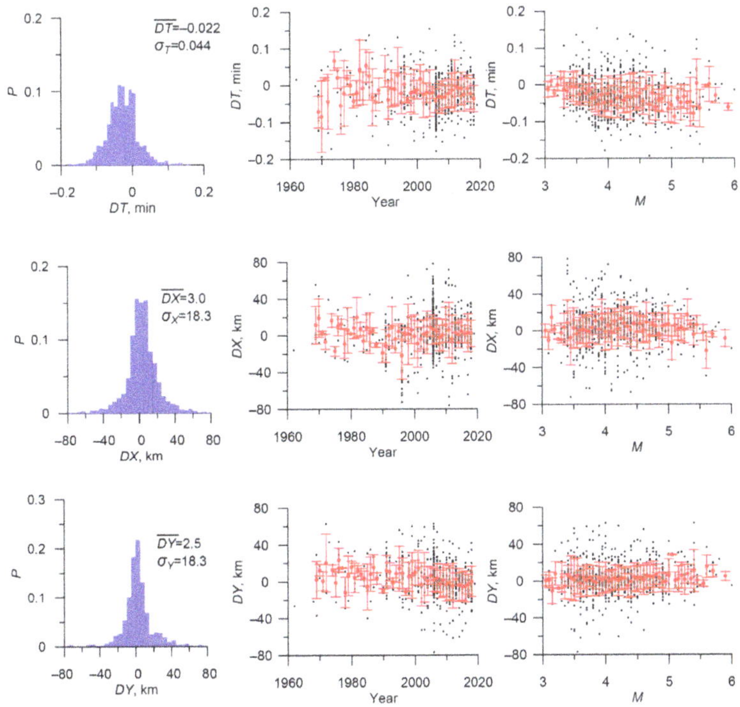

Figure 10. Distributions of *DT*, *DX*, and *DY* for the nearest events from the CORE/RUS_ISC catalogs, and the dependence of the standard deviations σ_T, σ_X, and σ_Y and the mean values $\overline{DT}$, $\overline{DX}$, and $\overline{DY}$ on the time and magnitude of the events. The red dots and bars are the population mean values and standard deviations, respectively.

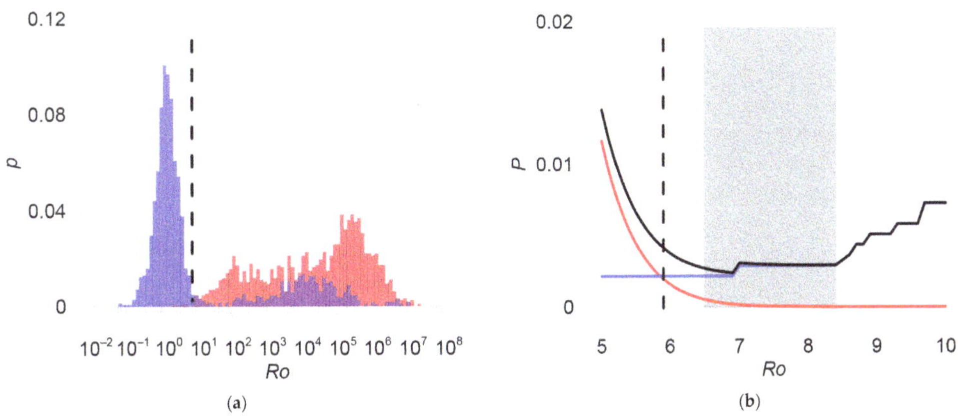

Figure 11. (**a**) Comparison of the distribution of the metric for the CORE/RUS_ISC pairs (blue histogram) and the same metric for the CORE/CORE earthquakes (red histogram). (**b**) Threshold optimization: the red line is the probability of missing a duplicate in the model with metric (1), the blue line is the probability of a false duplicate, the black line is the total probability of errors of the first and second kind, the dashed line *Ro* = 5.9 corresponds to an equal number of errors of the first and second kind (number of false duplicates is equal to the number of missed duplicates), the estimate of the total number of errors is approximately 0.4%, and the gray bar shows the range of values of the metric *Ro* = 6.5 ÷ 8.4, minimizing the total number of errors (approximately 0.3%).

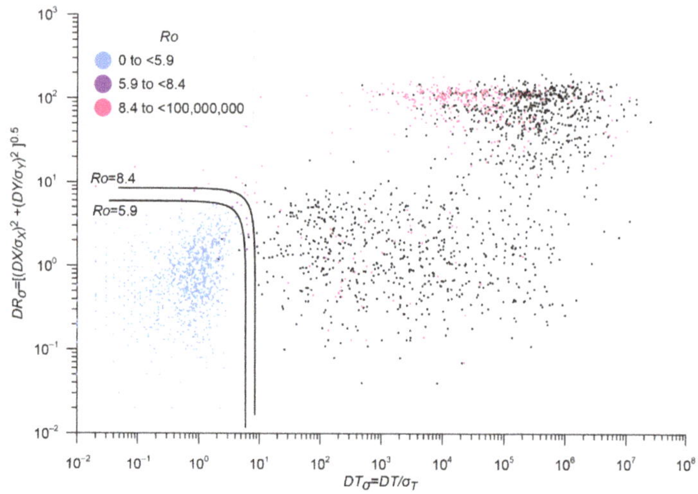

Figure 12. Distribution of normalized *DT* and *DR* and metric level lines (1). The colored dots are the CORE/RUS_ISC pairs and the black dots are the distances between the CORE/CORE events in metric (1). The metric levels *Ro* = 5.9 and *Ro* = 8.4 are shown by lines.

3.1.5. Stage 5. Exclusion of Explosions

At the final stage, we check the catalog for any type of explosion. The information about the 12 explosions in the ISC catalog is given in the ISC bulletins. In addition, 104 events are labeled "exp" or "exp?" in the NER and YAK catalogs from the annual journal *Earthquakes in Northern Eurasia*. In the annual journal *Earthquakes in Russia*, explosions are excluded by the authors. We merged all explosions into the EXP catalog and performed duplicate analysis.

We choose EXP as the main catalog and ARCTIC, which was obtained as a result of combining ISC and Russian data, as the additional one. The duplicate analysis was performed with the standard distribution parameters $\sigma_T = 0.05$ min and $\sigma_X = \sigma_Y = 15$ km. All 116 events of the EXP catalog are absolute duplicates of the events from the ARCTIC catalog. After the explosions were removed, the final integrated catalog of the Eastern Sector of the Arctic zone of the Russian Federation, E_ARCTIC, contains 23,254 events. The assembly scheme, statistics, and parameters for excluding duplicates are presented in Table 3.

Table 3. Scheme and compilation parameters of the integrated catalog.

Stage	Main Catalog	Additional Catalog	Metric Parameters σ_T min, σ_X km, and σ_Y km	Threshold Value of the Metric	Estimation of the Number of Errors	Number of Duplicates	Merged Catalog
1	Catalog of Yakutia YAK 6600 events	Catalog of the Northeast of Russia NER 7668 events	0.041; 17.4; 16.3	5.8	0.5%	2153	YAK_NER 12,115 events
2	YAK_NER 12,115 events	Earthquake catalog of the Kamchatka Branch of the GS RAS KAM 4498 events	0.041; 17.4; 16.3	5.8 *	-	26 *	RUS 16,587 events
3	RUS 16,587 events	ISC, events of various agencies ISC_Other 16,642 events	0.032; 12.3; 12.0	6.0	0.3%	10,231	RUS_ISC 22,998 events
4	CORE 1383 events	RUS_ISC 22,998 events	0.044; 18.3; 18.3	5.9	0.4%	1011	ARCTIC 23,370 events
5	Exclusion of explosions EXP 116 events	ARCTIC 23,370 events	0.05; 15.0; 15.0	-	0%	116 **	E_ARCTIC 23,254 events

* A small number of duplicates does not allow estimating the number of errors and optimizing the metric threshold.
** All duplicates are absolute.

3.2. Magnitudes in the Integrated Catalog of the Eastern Sector of the Russian Arctic

The integrated catalog of the Eastern Sector of the Russian Arctic contains 23,254 events that have different types of magnitude estimates determined by different agencies (Table 4). It is necessary to unify them (bring them to the reference scale of magnitudes).

Table 4. Magnitudes in the unified catalog of the Eastern Sector of the Russian Arctic.

Agency	Type of Magnitude	Priority	Number of Events	Formula for Magnitude in the Integrated Catalog	Figure	M_{min}–M_{max}. Initial Magnitude Scale	Note
GCMT	M_W	1	105	$M = M_W{}^{GCMT}$		4.7–7.6	
ISC	mb	2	1287	$M = mb^{ISC}$	Figure 13a	3.0–5.9	
ISC	M_S	1	4	$M = M_S{}^{ISC}$	Figure 13b	5.7–7.5	Strong events before 1976
YAK, NER, agencies of Russia and the USSR from ISC	K_{PS}	3	16,301	$M = 0.5\ K_{PS} - 1.6$	Figure 14a,b	0.6–14.0	Information about energy classes is given in the ISC bulletins
KAM, KRSC	K_S	3	4050	$M = 0.5K_S - 0.75$	Figure 14c	3.0–13.1	
NEIC, NEIS	mb	4	27	$M = mb^{NEIC} - 0.2$	Figure 15a	3.5–4.9	
MOS	mb	4	16	$M = mb^{MOS} - 0.2$	Figure 15b	4.0–4.8	
EIDC	mb	4	24	$M = mb^{EIDC} + 0.2$	Figure 15c	3.0–4.3	
IDC	mb	4	107	$M = mb^{IDC} + 0.2$	Figure 15d	2.9–4.4	
YARS	ML	4	357	$M = ML^{YARS} + 0.6$	Figure 16a	0.5–3.0	Unreliable correlation
YARS	MSV	4	95	$M = MSV^{YARS} + 0.2$	Figure 16b	0.0–2.1	Unreliable correlation
AEIC	ML	4	351	$M = ML^{AEIC}$	Figure 17a	2.2–4.2	
MSUGS	M	4	24	$M = M^{MSUGS} + 0.1$	Figure 18a	0.1–4.6	
USCGS	mb	4	1	$M = mb^{USCGS}$	Figure 18b	4.1	Unreliable correlation
YARS	M	4	104	$M = M^{YARS} + 0.1$		3.2–3.3	Indirect correlation with energy class. The magnitude M^{YARS} represents a conversion from the energy class K_S according to the formula of Rautian $M^{YARS} = (K_S - 4)/1.8$. For M[3.2–3.3] up to rounding, this is a shift of 0.1.
NERS	M	4	24	$M = M^{NERS} + 0.2$		2.3–2.5	Indirect correlation with energy class. The magnitude M^{NERS} represents a conversion from the energy class K_{PS} according to the formula of Rautian $M^{NERS} = (K_{PS} - 4)/1.8$. For M[2.3–2.5] up to rounding, this is a shift of 0.2.
NEIC	ML	4	13	$M = ML^{NEIC} - 0.1$	Figure 17b	2.5–4.2	Unreliably used indirect correlation ML^{AEIC}
NEIC	$mbLg$	4	2	$M = mbLg^{NEIC} + 0.1$	Figure 17c	2.6–3.0	Unreliably used indirect correlation ML^{AEIC}
LAO	M	4	2	$M = M^{LAO}$	Figure 19a	4.0	Very unreliable correlation
ZEMSU	M	4	2	$M = M^{ZEMSU}$	Figure 19b	3.4–4.5	Very unreliable correlation
MOS	M	4	1	$M = M^{MOS} + 0.1$	Figure 19c	5.0	Very unreliable correlation
NEIC	M	4	6	$M = M^{NEIC}$		2.5–4.9	Very unreliable correlation. Only three events with two magnitudes were found, $M^{NEIC} = mb^{ISC}$.
ANF	ML	4	2	$M = ML^{ANF} - 1$		4.2–4.3	Very unreliable correlation. Found only two events with two magnitudes, $ML^{ANF} >> mb^{ISC}$.
DNAG	M	4	14	$M = M^{DNAG}$		2.5–4.4	Correlation not established
WASN	M	4	328	$M = M^{WASN}$		0.1–4.4	Correlation not established
ZEMSU	MPV	4	1	$M = MPV^{ZEMSU}$		4.5	Correlation not established
YARS	MU	4	2	$M = MU\_YARS$		1.7–2.1	Correlation not established
OTT	ML	4	1	$M = ML^{OTT}$		3.9	Correlation not established
PAL	M	4	1	$M = M^{PAL}$		4.7	Correlation not established
BJI	mb	4	1	$M = mb^{BJI}$		4.8	Correlation not established
EIDC	ML	4	1	$M = ML^{EIDC}$		2.8	Correlation not established
Total			23,254				

At present, the only physical magnitude scale is the seismic moment-based magnitude M_W, which is preferable when analyzing estimates of different magnitude scales [37,38]. However, when moving from large to small magnitudes (from global estimates to regional ones), discrepancies in M_W estimates are observed everywhere at $M < 5.0$ [39,40]. It should be noted that the Eastern Arctic region was not considered in [39].

In the present work, we use only global estimates of the M_W^{GCMT} magnitude. If the global estimate of moment magnitude is unknown, we prefer the magnitude mb^{ISC}, which is used by ISC in its practice to obtain "quasi-M_W" estimates in the range $M < 5.0$ [41].

A feature of the catalogs used is the presence of estimates of energy classes, and not magnitudes. Theoretically, the energy class estimate proposed by T.G. Rautian [42] was assumed to be the same physical parameter as the radiated seismic energy, or moment magnitude, which was presented later [43]. However, it was shown in [40] that, in practice, the Rautian energy classes are rather a magnitude characteristic (with its saturation) than a physical one. It should also be noted that there are two approaches for index in the energy class abbreviation. It can be "K_R", "K_F", or "K_S" for Rautian, Fedotov, or Solov'ev, respectively, or "K_P", "K_S", or "K_{PS}" for the wave type, which were used for calculation. In this paper, we stand for the second approach to emphasize the difference in energy class scales. Therefore, the estimates expressed in energy classes were converted to M_W using regression relations. In the studied territory, the number of earthquakes with the known magnitude M_W^{GCMT} is small, so regressions with the magnitude mb^{ISC} are built, which is well aligned with M_W^{GCMT} [41].

It is necessary to notice that about 7% of the events in the integrated catalog have other types of magnitudes. If an event has several magnitudes, then preference was given to those for which it is possible to construct a correlation with magnitude mb^{ISC}. In few cases, when there were no pairs to determine direct correlations to mb^{ISC}, we used indirect correlations with other magnitudes, and we consider these correlations unreliable (indicated in the "Note" column of Table 4). The 95% confidence intervals are constructed by the Grapher Golden Software built-in tool (https://www.goldensoftware.com/products/grapher, (accessed on 20 March 2022)).

Thus, we adhere to the following priority when choosing the optimal magnitude estimate:

1. M_W^{GCMT} or M_S^{ISC} for strong earthquakes before 1976;
2. mb^{ISC};
3. Magnitude by energy class; and
4. Other magnitudes.

Statistics on magnitudes in the integrated catalog are given in Table 4.

Figure 13a shows the correlation between the M_W^{GCMT} and mb^{ISC} magnitudes in the studied territory ($mb = 0.99M_W + 0.03$). Earthquakes with an M_W of <6.0 were used to construct the best linear approximation. For stronger earthquakes, the magnitude mb saturates and becomes smaller than M_W. The magnitude $M_S \approx M_W$ is used for earthquakes with an M_W of $\geq$6.0. For weaker earthquakes, $M_S < M_W$ (Figure 13b) is used, which generally agrees with previously obtained correlations [41].

There are 105 events in the integrated catalog with the magnitude M_W^{GCMT}. During the considered period, 15 earthquakes with an M of >6.0 occurred. Two strong earthquakes occurred in 1969 (mb = 6.4 and M_S = 7.5) and 1971 (mb = 6.0 and M_S = 7.0). For these earthquakes, the M_S estimate is preferred (Figure 13b, Table 5). For the remaining 13 strong earthquakes, the magnitude M_W^{GCMT} is known.

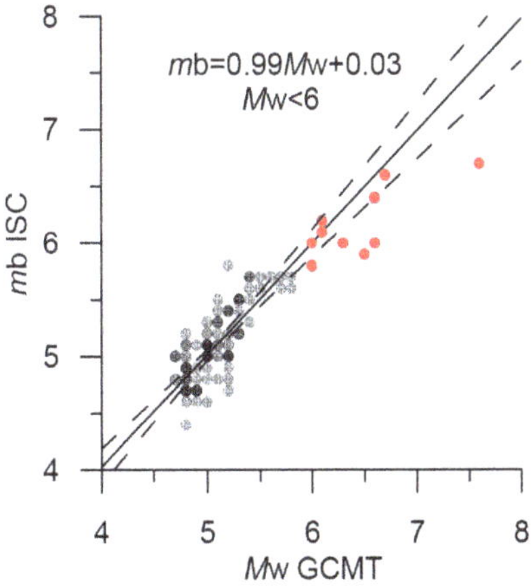

(a)

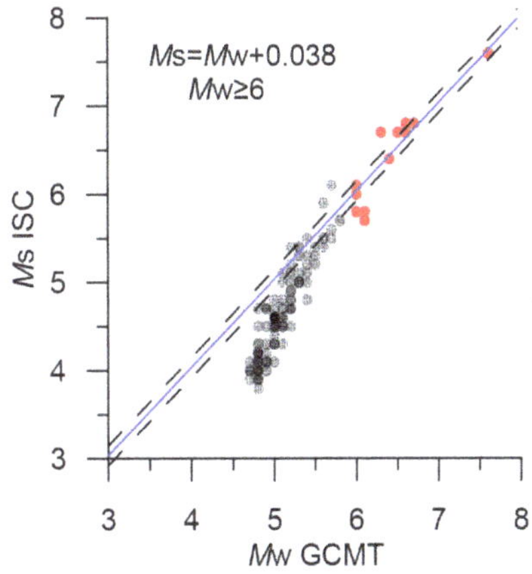

(b)

Figure 13. Correlation ratios of GCMT and ISC magnitudes: (**a**) $M_W{}^{GCMT}$ and mb^{ISC} for $M_W < 6.0$ and (**b**) $M_W{}^{GCMT}$ and $M_S{}^{ISC}$ for $M_W \geq 6.0$. The dashed lines show 95% confidence intervals. The events with an M_W of ≥6.0 are highlighted in red.

Table 5. Strong earthquakes in the Eastern Arctic.

	Date	TIME	Lat	Lon	Dep	Mag	
1	22.11.1969	23:09:38	57.67	163.51	25.6	7.5	$M_S{}^{ISC}$
2	18.05.1971	22:44:41	63.93	145.96	1.5	7.0	$M_S{}^{ISC}$
3	08.03.1991	11:36:31	60.83	167.08	16.5	6.6	$M_W{}^{GCMT}$
4	24.10.1996	19:31:55	66.92	−173.04	22.2	6.0	$M_W{}^{GCMT}$
5	20.04.2006	23:25:02	60.88	167.05	23.9	7.6	$M_W{}^{GCMT}$
6	21.04.2006	4:32:44	60.45	165.96	14.6	6.1	$M_W{}^{GCMT}$
7	21.04.2006	11:14:16	61.30	167.75	22.8	6.0	$M_W{}^{GCMT}$
8	29.04.2006	16:58:06	60.45	167.62	10.9	6.6	$M_W{}^{GCMT}$
9	22.05.2006	11:11:59	60.73	165.81	13.9	6.6	$M_W{}^{GCMT}$
10	22.06.2008	23:56:30	67.70	141.39	18.8	6.1	$M_W{}^{GCMT}$
11	30.04.2010	23:11:43	60.46	−177.91	14.7	6.5	$M_W{}^{GCMT}$
12	30.04.2010	23:16:29	60.48	−177.60	18.3	6.3	$M_W{}^{GCMT}$
13	24.06.2012	3:15:01	57.50	163.41	16	6.0	$M_W{}^{GCMT}$
14	14.02.2013	13:13:52	67.52	142.70	8.9	6.7	$M_W{}^{GCMT}$
15	09.01.2020	8:38:08	62.36	171.06	10	6.4	$M_W{}^{GCMT}$

To determine the magnitude by energy class, the correlation with the magnitude mb^{ISC} was used (Figure 14). The Yakutsk and Northeastern branches of the GS RAS estimate the Rautian K_{PS} class [42], while the Kamchatka Branch estimates the Fedotov K_S class [44]. Earthquakes in Kamchatka were selected only in the studied territory, north of latitude 57.5°.

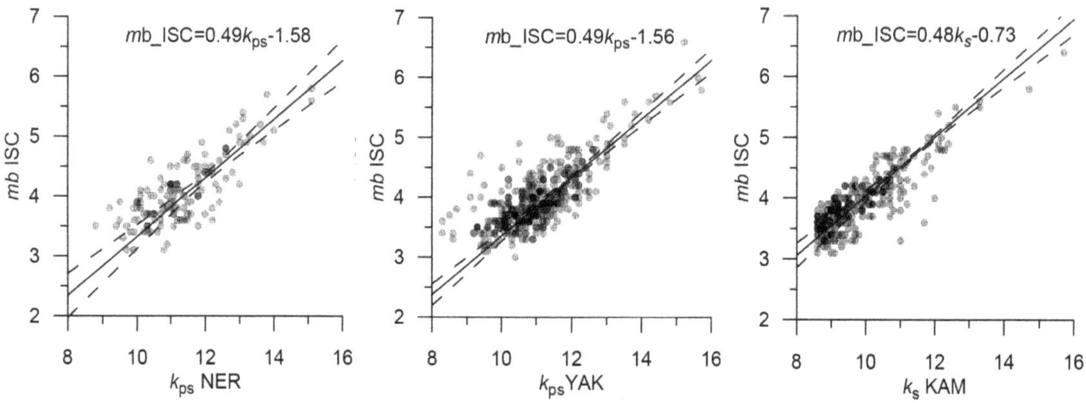

(a) (b) (c)

Figure 14. Correlation ratios of the energy classes K_{PS} and K_S and the magnitude mb^{ISC}. (**a**) Northeast, K_{PS}; (**b**) Yakutia, K_{PS}; and (**c**) Kamchatka, K_S. The dashed lines show 95% confidence intervals.

The magnitude ratio of mb^{ISC} and K_{PS} is the same in Yakutia and in the northeast. The ratio for K_S in Kamchatka is noticeably different since in the formula of energy classes according to Fedotov [43,44], $K_S = 2lgA_{peak} + f(r)$, while in the classes according to Rautian [42,43], $K_{PS} = 1.8lgA_{peak} + f(r)$, where A_{peak} is a peak amplitude of an S-wave or the sum of peak amplitudes of P- and S-waves, and $f(r)$ is an attenuation function.

In the unified catalog, 1551 earthquakes have other types of magnitudes. We try to give estimates in M_W, which is an absolute scale in the first approximation. Therefore, a shift-type transformation $M = M$ + constant was used, corresponding to the approach of [45], which assumed that one day, relative logarithmic magnitude estimates would be converted to absolute energy estimates by adding a constant ("Since the scale is logarithmic, any future reduction to an absolute scale can be accomplished by adding a constant to the scale numbers"). Taking into account that such ratios were obtained for limited ranges of magnitudes, and, in particular, magnitude ranges within $M < 5$, we consider that the assumption of the absence of nonlinear effects can be applied.

The number of earthquakes with the magnitude $M_W{}^{GCMT}$ is small; therefore, to construct correlations, we used all earthquakes from the ISC catalog for which the magnitude mb^{ISC} and the studied magnitude are known. Reliable correlations with mb^{ISC} are determined for 617 earthquakes. For 609 events, unreliable correlations are determined. This is due either to a small number of events or to the use of an indirect correlation with other magnitudes.

No correlations are defined for 349 events, and 324 of these are reported by the WASN agency (Table 4). Correlations of different magnitudes are shown in Figures 15–19.

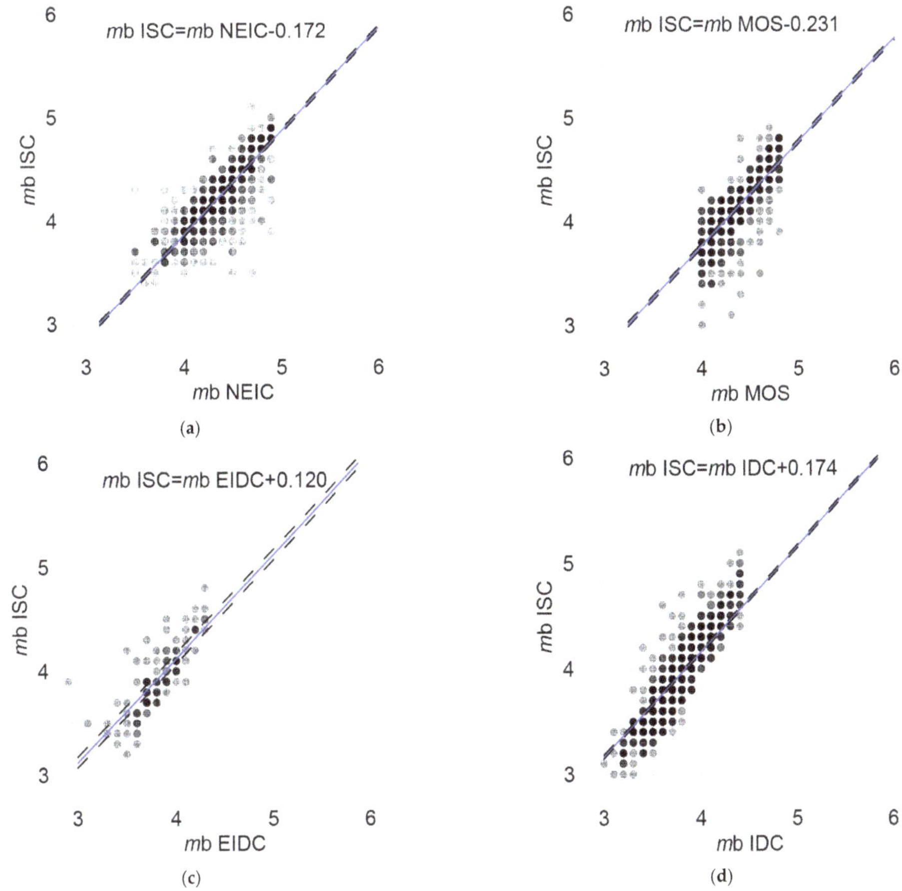

Figure 15. (**a**) NEIC; (**b**) MOS; (**c**) EIDC; and (**d**) IDC. Correlations of the "shift" type are determined in the interval of the magnitudes indicated in Table 4. The dashed lines show 95% confidence intervals.

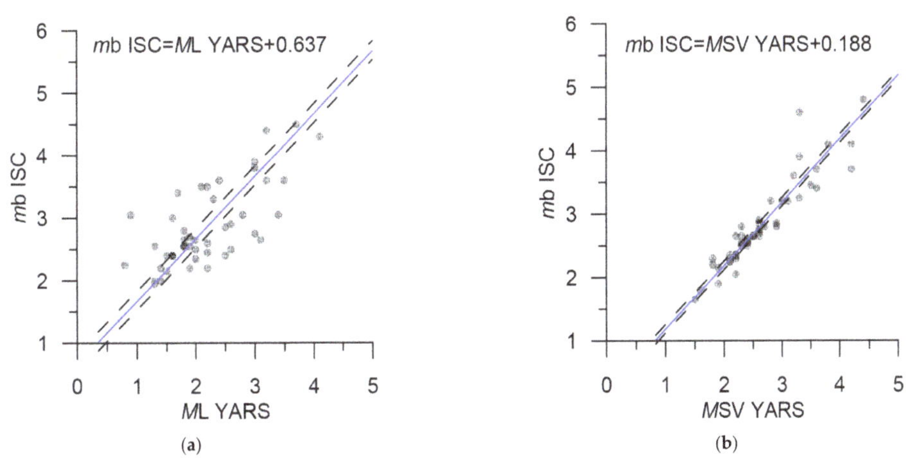

Figure 16. Correlation ratios "shift" of the magnitudes YARS and mb^{ISC}. (**a**) ML^{YARS}. (**b**) MSV^{YARS}. The dashed lines show 95% confidence intervals.

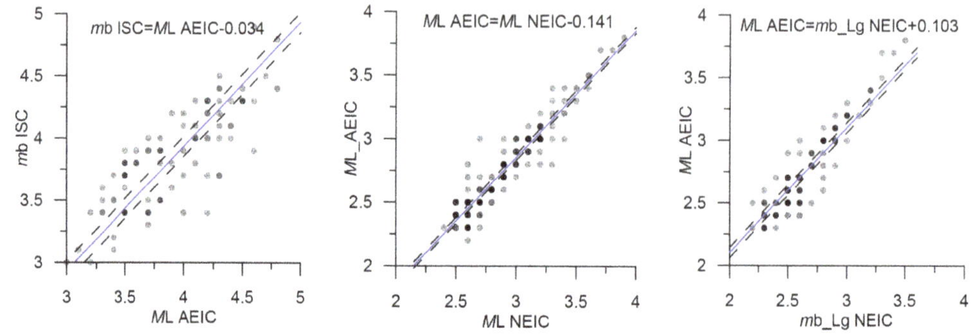

(**a**) (**b**) (**c**)

Figure 17. Correlation ratios "shift" of (**a**) the ML^{AEIC} and mb^{ISC} magnitudes; indirect relationships; (**b**) ML^{AEIC} and ML^{NEIC}; and (**c**) ML^{AEIC} and $mbLG^{NEIC}$. The dashed lines show 95% confidence intervals.

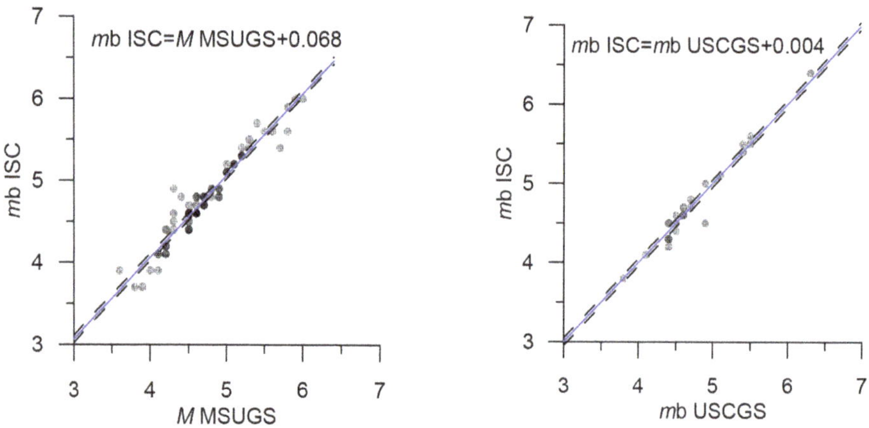

(**a**) (**b**)

Figure 18. Correlation ratios of the magnitudes MSUGS, USCGS, and mb^{ISC}. (**a**). M^{MSUGS}. (**b**) mb^{USCGS}. The dashed lines show 95% confidence intervals.

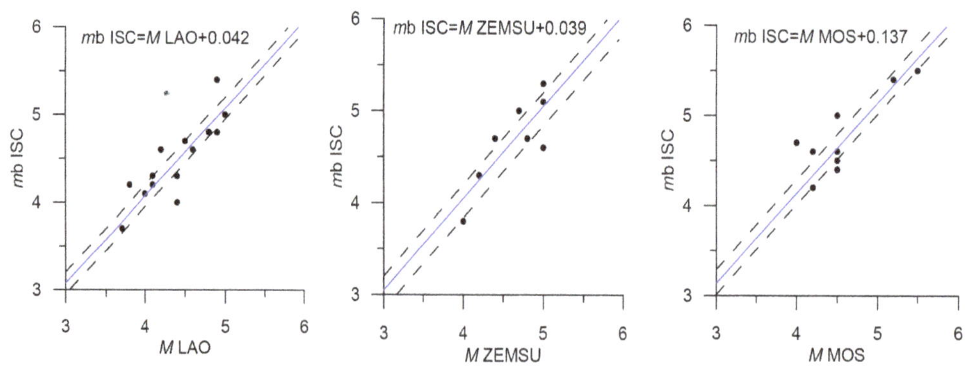

(**a**) (**b**) (**c**)

Figure 19. Poorly defined "shift" correlations with the magnitude mb^{ISC}. (**a**) M^{LAO}; (**b**) M^{ZEMSU}; and (**c**) M^{MOS}. The dashed lines show 95% confidence intervals.

4. Conclusions

Based on the generalization and integration of data from the various networks that serve the Eastern Sector of the Arctic zone of the Russian Federation, the most complete and representative earthquake catalog has been compiled. The catalog contains information on 23,254 seismic events for the period 1962–2020, of which 7781 events are from ISC and 15,473 events are from the Russian catalogs of GS RAS. Such a detailed and universal catalog for the whole Eastern part of the AZRF never existed before. Before 1968, the catalog contained quite a small number of events. In 2020, the catalog contained the events only from the ISC because the GS RAS catalog was not completed yet. The integrated catalog is to be updated accordingly once the former catalog is available.

The correlation of the magnitude types in the catalog was analyzed for various seismic networks. Based on the relations obtained, the unification of the magnitude estimates was carried out. For the most earthquakes, the quasi-M_W magnitude is calculated by converting the energy class using the original regression relationships (Table 4). The distribution of event magnitudes over time and magnitude–frequency graphs are shown in Figure 20. The integrated catalog completeness is quite heterogeneous. A detailed analysis of the changes in the level of registration in space and time is a big work that goes beyond the scope of the present study. We plan to conduct this work in the future.

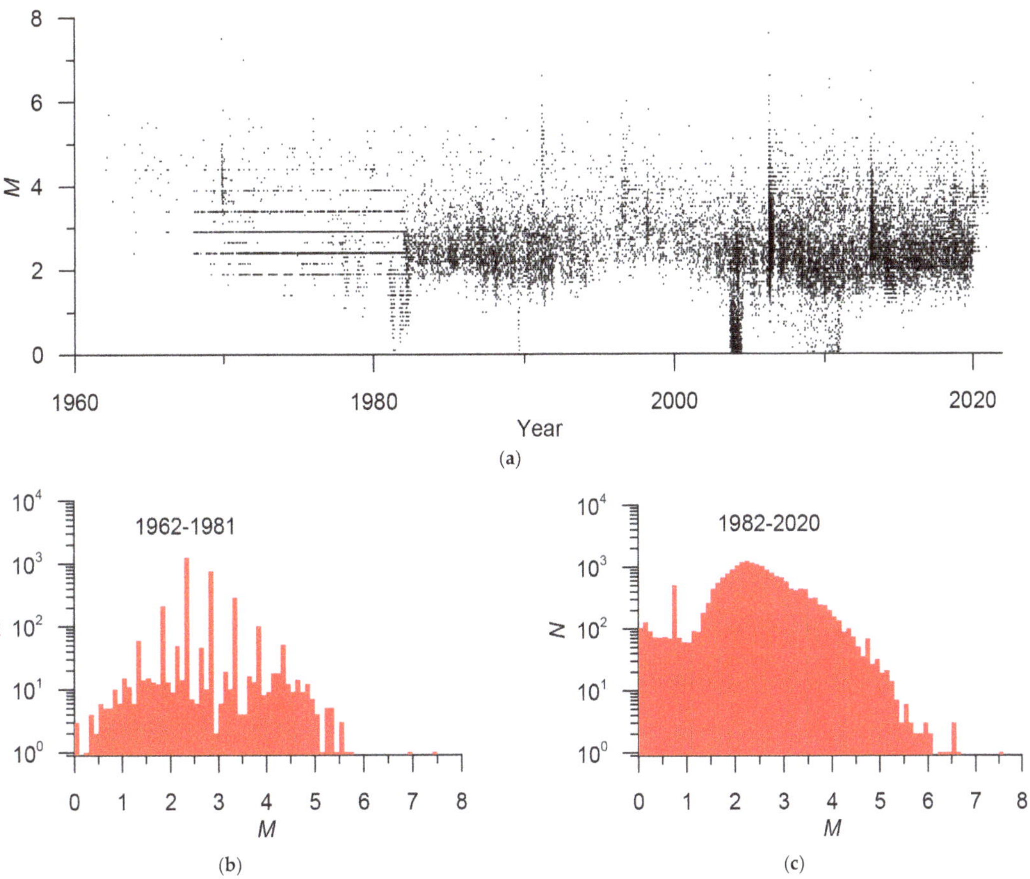

Figure 20. (**a**) Distribution of the magnitude of events in time. (**b**,**c**) Magnitude–frequency graphs before 1982 (**b**) and after 1982 (**c**). Before 1982, the energy classes were integers, and so the magnitudes are in increments of 0.5.

The creation of the unified magnitude-based integrated earthquake catalog realized in this paper opens new prospects in earthquake studies in the Arctic region. Further development of the Russian Arctic seismic zonation and systems analysis of strong earthquake-prone areas are among them.

The map of earthquake epicenters of the integrated catalog is shown in Figure 21. The catalog developed in this article is made available to the public on the website of the World Data Center for Solid Earth Physics, Moscow, at http://www.wdcb.ru/arctic_antarctic/arctic_seism.html, (accessed on 20 March 2022).

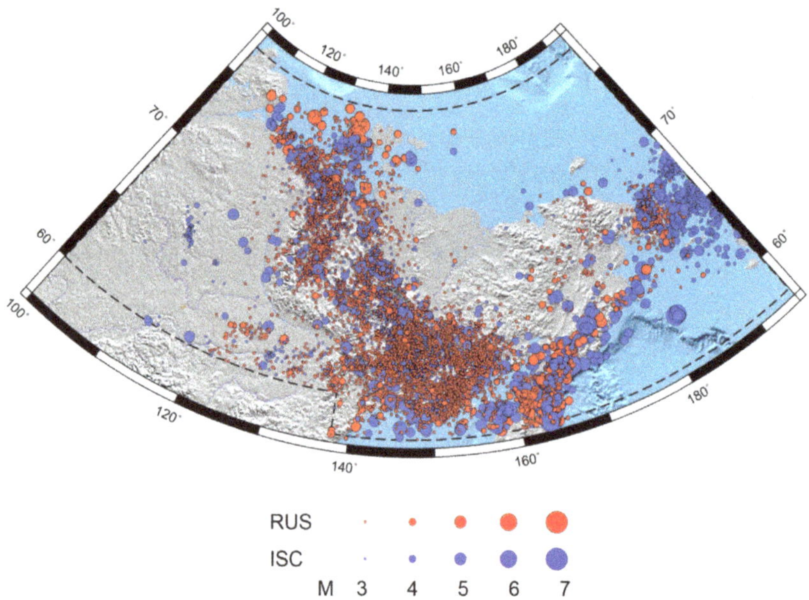

Figure 21. Map of earthquake epicenters of the integrated catalog of the Eastern Sector of the Arctic zone of the Russian Federation. The red and blue circles are the epicenters from the catalogs of GS RAS and ISC, respectively.

Author Contributions: Conceptualization, A.D.G., I.A.V. and P.N.S.; data curation, P.N.S., B.A.D. and A.A.S.; formal analysis, I.A.V.; investigation, P.N.S., B.A.D., B.V.D. and A.A.S.; methodology, I.A.V.; resources, B.A.D.; software, I.A.V.; validation, I.A.V. and A.A.S.; visualization, I.A.V. and B.V.D.; writing—original draft, A.D.G., I.A.V., P.N.S., B.A.D., B.V.D. and A.A.S.; writing—review and editing, A.D.G., I.A.V., P.N.S., B.A.D., B.V.D. and A.A.S. All authors have read and agreed to the published version of the manuscript.

Funding: The reported study was funded by the Russian Science Foundation, project number 21-77-30010, System analysis of geophysical process dynamics in the Russian Arctic and their impact on the development and operation of the railway infrastructure.

Institutional Review Board Statement: Not applicable.

Informed Consent Statement: Not applicable.

Data Availability Statement: Not applicable.

Acknowledgments: This work employed data provided by the Shared Research Facility, the Analytical Geomagnetic Data Center of the Geophysical Center of RAS (http://ckp.gcras.ru/ accessed on: 24 February 2022).

Conflicts of Interest: The authors declare no conflict of interest.

References

1. Imaeva, L.P.; Imaev, V.S.; Koz'min, B.M. Dynamics of the Zones of Strong Earthquake Epicenters in the Arctic–Asian Seismic Belt. *Geosciences* **2019**, *9*, 168. [CrossRef]
2. Daragan-Sushchova, L.A.; Petrov, O.V.; Sobolev, N.N.; Daragan-Sushchov, Y.I.; Grin'ko, L.R.; Petrovskaya, N.A. Geology and tectonics of the northeast Russian Arctic region, based on seismic data. *Geotectonics* **2015**, *49*, 469–484. [CrossRef]
3. Kanao, M.; Suvorov, V.; Toda, S.; Tsuboi, S. Seismicity, structure and tectonics in the Arctic region. *Geosci. Front.* **2015**, *6*, 665–677. [CrossRef]
4. Skorkina, A.A. Scaling of two corner frequencies of source spectra for earthquakes of the Bering fault. *Russ. J. Earth Sci.* **2020**, *20*, ES2001. [CrossRef]
5. Imaev, V.S.; Imaeva, L.P.; Koz'min, B.M. Strong Ulakhan-Chistay earthquake (Ms = 5.7) January 20, 2013 in the zone of influence Ulakhan fault system in North East Russia. *Vestn. St. Petersburg Univer. Earth Sci.* **2020**, *65*, 740–759. (In Russian) [CrossRef]
6. Shibaev, S.V.; Kozmin, B.M.; Imaev, V.S.; Imaeva, L.P.; Petrov, A.F.; Starkova, N.N. The February 14, 2013 Ilin-Tas (Abyi) earthquake (*Mw* = 6.7), Northeast Yakutia. *Russ. J. Seismol.* **2020**, *2*, 92–102. [CrossRef]
7. Chebrov, V.N. The Olyutorskii earthquake of April 20, 2006: Organizing surveys, observations, problems, and results. *J. Volcanol. Seismol.* **2010**, *4*, 75–78. [CrossRef]
8. Rogozhin, E.A.; Ovsyuchenko, A.N.; Marakhanov, A.V.; Novikov, S.S. A geological study of the epicentral area of the April 20(21), 2006 Olyutorskii earthquake. *J. Volcanol. Seismol.* **2010**, *4*, 79–86. [CrossRef]
9. Lander, A.V.; Levina, V.I.; Ivanova, E.I. The earthquake history of the Koryak Upland and the aftershock process of the M_W7.6 April 20(21), 2006 Olyutorskii earthquake. *J. Volcanol. Seismol.* **2010**, *4*, 87–100. [CrossRef]
10. Dzeboev, B.A.; Agayan, S.M.; Zharkikh, Y.I.; Krasnoperov, R.I.; Barykina, Y.V. Strongest Earthquake-Prone Areas in Kamchatka. *Izv. Phys. Solid Earth* **2018**, *54*, 284–291. [CrossRef]
11. Dzeboev, B.A.; Gvishiani, A.D.; Belov, I.O.; Agayan, S.M.; Tatarinov, V.N.; Barykina, Y.V. Strong Earthquake-Prone Areas Recognition Based on an Algorithm with a Single Pure Training Class: I. Altai-Sayan-Baikal Region, M $\geq$ 6.0. *Izv. Phys. Solid Earth* **2019**, *55*, 563–575. [CrossRef]
12. Dzeboev, B.A.; Karapetyan, J.K.; Aronov, G.A.; Dzeranov, B.V.; Kudin, D.V.; Karapetyan, R.K.; Vavilin, E.V. FCAZ-recognition based on declustered earthquake catalogs. *Russ. J. Earth Sci.* **2020**, *20*, ES6010. [CrossRef]
13. Gorshkov, A.I.; Soloviev, A.A. Recognition of earthquake-prone areas in the Altai-Sayan-Baikal region based on the morphostructural zoning. *Russ. J. Earth Sci.* **2021**, *21*, ES1005. [CrossRef]
14. Gvishiani, A.; Dobrovolsky, M.; Agayan, S.; Dzeboev, B. Fuzzy-based clustering of epicenters and strong earthquake-prone areas. *Environ. Eng. Manag. J.* **2013**, *12*, 1–10.
15. Gvishiani, A.D.; Dzeboev, B.A.; Agayan, S.M. FCAZm intelligent recognition system for locating areas prone to strong earthquakes in the Andean and Caucasian Mountain belts. *Izv. Phys. Solid Earth* **2016**, *52*, 461–491. [CrossRef]
16. Gvishiani, A.D.; Dzeboev, B.A.; Sergeyeva, N.A.; Belov, I.O.; Rybkina, A.I. Significant Earthquake-Prone Areas in the Altai–Sayan Region. *Izv. Phys. Solid Earth* **2018**, *54*, 406–414. [CrossRef]
17. Gvishiani, A.D.; Soloviev, A.A.; Dzeboev, B.A. Problem of Recognition of Strong-Earthquake-Prone Areas: A State-of-the-Art Review. *Izv. Phys. Solid Earth* **2020**, *56*, 1–23. [CrossRef]
18. Levin, B.W.; Kim, C.U.; Solovjev, V.N. A seismic hazard assessment and the results of detailed seismic zoning for urban territories of Sakhalin Island. *Russ. J. Pac. Geol.* **2013**, *7*, 455–464. [CrossRef]
19. Lunina, O.V.; Gladkov, A.S.; Gladkov, A.A. Systematization of active faults for the assessment of the seismic hazard. *Russ. J. Pac. Geol.* **2012**, *6*, 42–51. [CrossRef]
20. Magrin, A.; Peresan, A.; Kronrod, T.; Vaccari, F.; Panza, G.F. Neo-deterministic seismic hazard assessment and earthquake occurrence rate. *Eng. Geol.* **2017**, *229*, 95–109. [CrossRef]
21. Nekrasova, A.; Kossobokov, V.G.; Parvez, I.A.; Tao, X. Seismic hazard and risk assessment based on the unified scaling law for earthquakes. *Acta Geod. Et Geophys.* **2015**, *50*, 21–37. [CrossRef]
22. Ulomov, V.I. Seismic hazard of Northern Eurasia. *Ann. Di Geofis.* **1999**, *42*, 1023–1038. [CrossRef]
23. Zaalishvili, V.; Chernov, Y.K. Methodology of detailed assessment of the seismic hazard of the Republic of North Ossetia-Alania. *Open Constr. Build. Technol. J.* **2018**, *12*, 309–318. [CrossRef]
24. Zaalishvili, V.B.; Rogozhin, E.A. Assessment of seismic hazard of territory on basis of modern methods of detailed zoning and seismic microzonation. *Open Constr. Build. Technol. J.* **2011**, *5*, 30–40. [CrossRef]
25. Zavyalov, A.D.; Peretokin, S.A.; Danilova, T.I.; Medvedeva, N.S.; Akatova, K.N. General Seismic Zoning: From Maps GSZ-97 to GSZ-2016 and New-Generation Maps in the Parameters of Physical Characteristics. *Seism. Instr.* **2019**, *55*, 445–463. [CrossRef]
26. Beauval, C.; Yepes, H.; Palacios, P.; Segovia, M.; Alvarado, A.; Font, Y.; Aguilar, J.; Troncoso, L.; Vaca, S. An Earthquake Catalog for Seismic Hazard Assessment in Ecuador. *Bull. Seismol. Soc. Am.* **2013**, *103*, 773–786. [CrossRef]
27. Kondorskaya, N.V.; Shebalin, N.V. (Eds.) *New Catalog of Strong Earthquakes in the USSR from Ancient Times through 1977*; Report SE-31; World Data Center A for Solid Earth Geophysics: Boulder, CO, USA, 1982; 608p.
28. Shebalin, P.N. Compilation of earthquake catalogs as a task of clustering analysis with learning. *Dokl. Akad. Nauk. SSSR* **1987**, *292*, 1083–1086.

29. Markušić, S.; Gülerce, Z.; Kuka, N.; Duni, L.; Ivančić, I.; Radovanović, S.; Glavatović, B.; Milutinović, Z.; Akkar, S.; Kovačević, S.; et al. An updated and unified earthquake catalogue for the Western Balkan Region. *Bull. Earthq. Eng.* **2016**, *14*, 321–343. [CrossRef]
30. Sawires, R.; Santoyo, M.A.; Peláez, J.A.; Corona Fernández, R.D. An updated and unified earthquake catalog from 1787 to 2018 for seismic hazard assessment studies in Mexico. *Sci. Data* **2019**, *6*, 241. [CrossRef]
31. Braclawska, A.; Idziak, A. Unification of data from various seismic catalogues to study seismic activity in the Carpathians Mountain arc. *Open Geosci.* **2019**, *11*, 837–842. [CrossRef]
32. Long, F.; Jiang, C.; Qi, Y.; Liu, Z.; Fu, Y. A joint probabilistic approach for merging earthquake catalogs of two neighboring seismic networks: An example of the 2014 Ludian sequence catalog. *Acta Geophys. Sin.* **2018**, *61*, 2815–2827. [CrossRef]
33. Muller, S.; Garda, P.; Muller, J.-D.; Cansi, Y. Seismic events discrimination by neuro-fuzzy merging of signal and catalogue features. *Phys. Chem. Earth Part. A Solid Earth Geod.* **1999**, *24*, 201–206. [CrossRef]
34. Vorobieva, I.; Gvishiani, A.; Dzeboev, B.; Dzeranov, B.; Barykina, Y.; Antipova, A. Nearest Neighbor Method for Discriminating Aftershocks and Duplicates When Merging Earthquake Catalogs. *Front. Earth Sci. SI Nat. Clust. Earthq. Process. Var. Scales Lab. Exp. Large Earthq.* **2022**, *10*, 820277. [CrossRef]
35. Zaliapin, I.; Ben-Zion, Y. Earthquake clusters in southern California. I: Identification and stability. *J. Geophys. Res. Solid Earth* **2013**, *118*, 2847–2864. [CrossRef]
36. Zaliapin, I.; Ben-Zion, Y. A global classification and characterization of earthquake clusters. *Geophys. J. Int.* **2016**, *207*, 608–634. [CrossRef]
37. Kanamori, H. The Energy Release in Great Earthquakes. *J. Geophys. Res.* **1977**, *82*, 2981–2987. [CrossRef]
38. Hanks, T.C.; Kanamori, H. A Moment Magnitude Scale. *J. Geophys. Res. Solid Earth* **1979**, *84*, 2348–2350. [CrossRef]
39. Di Giacomo, D.; Harris, J.; Storchak, D.A. Complementing regional moment magnitudes to GCMT: A perspective from the rebuilt International Seismological Centre Bulletin. *Earth Syst. Sci. Data* **2021**, *13*, 1957–1985. [CrossRef]
40. Abubakirov, I.R.; Gusev, A.A.; Guseva, E.M.; Pavlov, V.M.; Skorkina, A.A. Mass determination of moment magnitudes Mw and establishing the relationship between Mw and ML for moderate and small Kamchatka earthquakes. *Izv. Phys. Solid Earth* **2018**, *54*, 33–47. [CrossRef]
41. Di Giacomo, D.; Bondár, I.; Storchak, D.A.; Engdahl, E.R.; Bormann, P.; Harris, J. ISC-GEM: Global Instrumental Earthquake Catalogue (1900–2009), III. Re-computed MS and mb, proxy MW, final magnitude composition and completeness assessment. *Phys. Earth Planet. Inter.* **2015**, *239*, 33–47. [CrossRef]
42. Rautian, T.G. Energy of earthquakes. In *Methods for the Detailed Study of Seismicity*; The USSR Academy of Sciences: Moscow, Russia, 1960; pp. 75–114. (In Russian)
43. Rautian, T.G.; Khalturin, V.I.; Fujita, K.; Mackey, K.G.; Kendall, A.D. Origins and methodology of the Russian energy K-class system and its relationship to magnitude scales. *Seismol. Res. Lett.* **2007**, *78*, 579–590. [CrossRef]
44. Fedotov, S.A. *Energy Classification of the Kuril-Kamchatka Earthquakes and the Problem of Magnitudes*; Nauka: Moscow, Russia, 1972; p. 117. (In Russian)
45. Richter, C.F. An instrumental earthquake magnitude scale. *Bull. Seismol. Soc. Am.* **1935**, *25*, 1–32. [CrossRef]

applied sciences

MDPI

Article

Integrated Earthquake Catalog II: The Western Sector of the Russian Arctic

Inessa A. Vorobieva [1,2], Alexei D. Gvishiani [1,3], Peter N. Shebalin [1,2], Boris A. Dzeboev [1,*], Boris V. Dzeranov [1], Anna A. Skorkina [2], Natalia A. Sergeeva [1] and Natalia A. Fomenko [1]

1 Geophysical Center of the Russian Academy of Sciences (GC RAS), 119296 Moscow, Russia; vorobiev@mitp.ru (I.A.V.); adg@wdcb.ru (A.D.G.); p.n.shebalin@gmail.com (P.N.S.); b.dzeranov@gcras.ru (B.V.D.); n.sergeyeva@gcras.ru (N.A.S.); n.fomenko@gcras.ru (N.A.F.)

2 Institute of Earthquake Prediction Theory and Mathematical Geophysics of the Russian Academy of Sciences (IEPT RAS), 117997 Moscow, Russia; anna@mitp.ru

3 Schmidt Institute of Physics of the Earth of the Russian Academy of Sciences (IPE RAS), 119296 Moscow, Russia

* Correspondence: b.dzeboev@gcras.ru; Tel.: +7-495-930-05-46

Abstract: The article is a continuation of the research on creating the most complete and representative earthquake catalogs by combining all available data from regional, national, and international seismological agencies and reducing magnitudes to a uniform scale. The task of identifying and removing duplicates that arise during the merging process is solved using the authors' modification of the nearest neighbor method. It is evident that the intelligent merging of different earthquake catalogs for the same territory will improve the completeness and representativeness of events in the final integrated catalog. In this article, the earthquake catalog of the western sector of the Arctic zone of the Russian Federation (AZRF) covering the period 1962–2022 was created by merging three regional Russian catalogs and the ISC catalog. The ratio of magnitude types in the catalog for different seismic networks was analyzed, and magnitude estimates were unified based on the obtained ratios. For analyzing seismic activity in the western AZRF, it is recommended to use earthquakes from the period 1998–2020 when the catalog was significantly cleaned from explosions and other events of the "non-earthquake" type.

Keywords: merging catalogs; earthquake; Russian Arctic; magnitude unification; duplicate events; seismic networks

Citation: Vorobieva, I.A.; Gvishiani, A.D.; Shebalin, P.N.; Dzeboev, B.A.; Dzeranov, B.V.; Skorkina, A.A.; Sergeeva, N.A.; Fomenko, N.A. Integrated Earthquake Catalog II: The Western Sector of the Russian Arctic. *Appl. Sci.* **2023**, *13*, 7084. https://doi.org/10.3390/app13127084

Academic Editor: Nicholas Vassiliou Sarlis

Received: 23 May 2023
Revised: 7 June 2023
Accepted: 9 June 2023
Published: 13 June 2023

1. Introduction

The paper [1] demonstrated that some seismic agencies may miss events detected by other agencies. This was illustrated using the JMA and ANSS catalogs for the aftershock sequence of the Tohoku earthquake (11 March 2011, $Mw = 9.0$) [2] recorded between 1 March and 31 December 2011. Based on the methodology developed by the authors [1], more than 700 events were identified in the ANSS catalog that were absent in JMA. Among the missed events were several aftershocks with $M > 6.0$ that occurred in the first hours after the main shock [3].

It is obvious that the intellectual merging of different earthquake catalogs for the same territory will make it possible to collect together the most complete information about all seismic events that have occurred. However, it should be noted that in the process of such merging, duplicates will arise, which, similarly to aftershocks [4], are events close in space and time, and there will be a task of their identification. The developed algorithm for merging catalogs with the identification and removal of duplicates is based on the nearest neighbor method [5,6] modified in [1]. The algorithm is based on the knowledge that duplicates do not have the causal relationship that exists for aftershocks. The application of the algorithm allows for the sequential merging of any number of earthquake catalogs [1].

In the paper [7], the earthquake catalog of the eastern sector of the Arctic zone of the Russian Federation (AZRF) was created by combining three regional catalogs (Yakutia, Northeast Russia, and Kamchatka) of the Geophysical Survey of the Russian Academy of Sciences (GS RAS) and the International Seismological Centre (ISC) with a unified magnitude scale, which is made available to the public. To confirm the feasibility of merging catalogs and the effectiveness of the algorithm, it should be noted that the obtained catalog [7] includes 23,254 events (1962–2020), of which 7781 are from ISC and 15,473 are from Russian catalogs. This article is a logical continuation of the study carried out in [7]. It is aimed at the creation of the integrated catalog of the western sector of the AZRF with unified magnitude estimates.

The western sector of the AZRF that is considered in the present study is a seismically active region [8,9] and includes the Arkhangelsk region, the Republic of Karelia, the Kola Peninsula, the western part of the Taymyr Peninsula, archipelagoes of Franz Josef Land, Novaya Zemlya, Severnaya Zemlya, and the basin of the White, Barents, and Kara Seas. In [10], information is provided on 10 earthquakes with $M \geq 5.0$ that occurred in the region starting from the 17th century. It should be noted that the epicenters of four events are located in the continental part of the region, while the others are significantly further north in the Barents–Kara Basin. The strongest earthquake known in the western sector of the AZRF occurred in the latter region (14 October 1908, $M = 6.6$) [10,11]. For events that occurred in the pre-instrumental period of seismic observations, the coordinates of the epicenters and magnitude estimates have a significant uncertainty [12]. However, in recent years, a number of studies have been published in which the parameters of historical events have been redefined, including the considered region [11,13,14].

Many researchers note the modern seismicity of the Barents–Kara Basin in the transition zone from continent to ocean and the Novaya Zemlya Archipelago, where an earthquake with $mb = 4.8$ was registered on 11 October 2010 [15]. Seismic activity is observed on the Kola Peninsula and in the Karelia region.

The Severnaya Zemlya Archipelago and the northern part of the Taymyr Peninsula are located within the Kara Plate. At the same time, only the northeast of the peninsula is characterized by seismic activity, where most of the earthquakes registered during the instrumental period are associated with the mouth of the Khatanga Bay [14]. It should be noted that earthquakes that occurred in the northeast of the peninsula were included in the integrated catalog of the eastern sector of the AZRF [7] and are not considered in this article.

The White Sea Basin, including its waters and adjacent territory, is one of the most fragmented and mobile regions within the East European Platform. The most active tectonic structure here is the Kandalaksha Graben. The maximum intensity of paleo-earthquakes in the western part of Kandalaksha Bay (the bay of the White Sea in the south of the Kola Peninsula) is IX, in the central part of the sea, it is VIII, and on the western coast of the White Sea and in Onega Bay, it is VII–VIII. The White Sea Basin has a high seismic potential with a rare recurrence of strong earthquakes [16].

The full-scale study of the seismic regime of the western sector of the AZRF began in the late 1970s as part of the creation of the catalog [10]. Subsequently, for several decades, researchers noted the insufficient development of the seismic observation network in the region, which allowed the registration of events only with $M \geq 5.0$ [17,18]. Over the past two decades, significant contributions to the development of seismic monitoring in the west of the AZRF have been made by the unique scientific facilities of Federal Center for Integrated Arctic Research, Russia, Arkhangelsk (FCIAR) [19–22]. Based on unique seismic data obtained by FCIAR, the structure of the lithosphere and upper mantle has been specified, regional hodographs have been constructed [23], and parameters of earthquakes that occurred in marine waters during the period from 1908 to 2020 have been refined. As a result, an updated catalog containing data on 125 earthquakes has been obtained [24].

On the maps of the General Seismic Zoning GSZ-97A, GSZ-2015A, and GSZ-2016A, the region is classified as zones with macroseismic intensity estimates V–VI [25–30]. In recent years, a seismic zoning map of the AZRF has been created [31].

Currently, the Geophysical Survey of the Russian Academy of Sciences (GS RAS) carries out the seismic monitoring of the entire territory of the Russian Arctic (http://www.gsras.ru/new/eng/catalog/, accessed on 1 June 2023). The western sector of the AZRF falls within the responsibility zone of the Kola and Arkhangelsk branches as well as the Central Department of GS RAS. Detailed information on earthquakes in the region is presented in the global ISC catalog (http://www.isc.ac.uk/isc-ehb/search/catalogue/, accessed on 1 June 2023). It should be noted that the analysis of earthquake catalogs of GS RAS branches and ISC showed that information about weak earthquakes is mainly contained in regional catalogs.

To create a unified (integrated) earthquake catalog of the western sector of the AZRF, the following main tasks were solved:

- The sequential merging of two regional catalogs of GS RAS, the ISC catalog, and the catalog from [24], identifying duplicates that arise during merging and excluding explosions and other anthropogenic events from the final catalog;
- The unification of magnitude estimates in the integrated catalog by constructing correlation ratios for different types of magnitude/energy class for the same events.

2. Materials and Methods

The studied region covers a geographic area bounded by the coordinates: 60° N–84° N, 25° E–100° E (Figure 1). Five earthquake catalogs are used as input data (Table 1):

1. The Arctic catalog from the annual journals *Earthquakes in the USSR* 1962–1991, *Earthquakes in Northern Eurasia* 1992–2017, and *Earthquakes in Russia* 2018–2020 (hereinafter ARC);
2. The merged earthquake catalog of the territory of the East European platform and its nearest surroundings from the annual journals *Earthquakes in the USSR* 1962–1991, *Earthquakes in Northern Eurasia* 1992–2017, and *Earthquakes in Russia* 2018–2020 (hereinafter VEP);
3. The ISC 1962–2022 catalog, which is a composite and contains data from many world and also Russian agencies (Table 2);
4. The catalog *Seismicity of the western sector of the Russian Arctic* [24] (hereinafter Morozov) 1962–2020;
5. The merged catalog of seismic events in the territory of the Russian Federation from the annual journals *Earthquakes in Russia* 2003–2020. Only data on events that are non-earthquakes are used (hereinafter ER_EXP).

As can be seen from Figure 1, a large number of earthquake epicenters are located in the southwestern part of the studied region on the Kola Peninsula and Karelia as well as in the northern part of the region on the shelf boundary of the Kara and Barents Seas. The peculiarity of the western sector of the AZRF is the large number of seismic events that are not tectonic earthquakes (explosions, rock bursts, and other anthropogenic events) [32–34].

All the catalogs used (Table 1) are divided into "Earthquakes" (the nature of the event is "earthquake" or "possibly earthquake") and other events. Events with unknown magnitude/class are excluded from the "Earthquakes" catalogs.

The methodology of the sequential merging of earthquake catalogs with the identification of duplicates is described in detail in [1,7]. A Euclidean metric (1) is used to find pairs of nearest events from two catalogs. After that, the classification of events into unique and duplicates is carried out using the threshold value of the metric. A basic three-parameter model is used, where the metrics for assessing the proximity between earthquakes take into account the time difference DT, and epicenter difference in the longitude DX and latitude DY. Depth information is not used, since the depth of hypocenters is often unknown or given as a fixed value for many events. Magnitudes are also not used, since earthquakes have magnitude estimates of different types from various agencies.

$$Ro = \sqrt{\frac{DT^2}{\sigma_T{}^2} + \frac{DX^2}{\sigma_X{}^2} + \frac{DY^2}{\sigma_Y{}^2}}, \quad (1)$$

where σ_T, σ_X , and σ_Y are the standard deviations of time, longitude, and latitude differences between the closest events from two source catalogs.

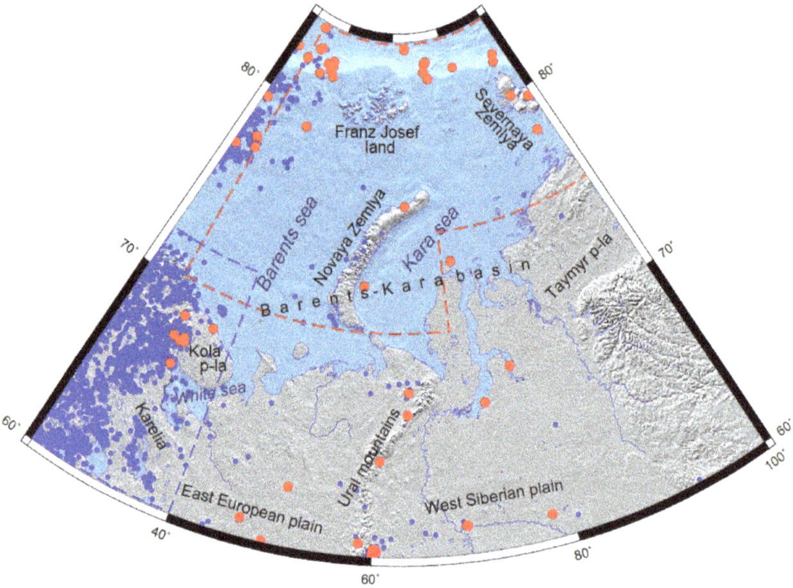

Figure 1. The studied region with earthquake epicenters from the ISC catalog (blue and red dots). Events with magnitude definitions *mb*_ISC and *mb*_NEIC are marked in red. The red outline is the region covered by the Morozov catalog [24]. The blue outline is the sub-region of the Kola Peninsula and Karelia.

Table 1. Input catalogs.

Catalog	Period	Number of Events	Number of Earthquakes with Energy Classes and/or Magnitudes	Number of Non-Earthquakes
ARC	1965–2020	175	175	0
VEP	1987–2020	971	742	170
ISC	1962–2022	105,656	4418	100,112
Morozov	1962–2020	118 *	118	0
ER_EXP	2003–2020	14,990	-	14,990

* 116 events from the main catalog are included (6 events before 1962 and 3 events outside the studied region are excluded) as well as 2 events from the additional catalog that were relocated outside the studied area considered in [24] but are included in the region studied in the present paper.

Before the merging process, each of the source catalogs (Table 1) was checked for internal duplicates. Statistical analysis did not reveal any anomalous groups of close events (Figure 2). It should be noted that duplicates typically have a metric value (1) $Ro < 10$ [1,7]. The number of such close events within each catalog is small, and there are no statistical reasons to consider such events as duplicates. Analysis (Figure 2) was performed with metric parameters $\sigma_T = 0.05$ min, $\sigma_X = \sigma_Y = 15$ km.

Table 2. Statistics of ISC catalog.

Agency Abbreviation	Agency	With Magnitude
BER	University of Bergen, Norway	316
CSEM	Centre Sismologique Euro-Méditerranéen, France	36
EIDC	Experimental (GSETT3) International Data Center, U.S.A.	729
FCIAR	Federal Center for Integrated Arctic Research, Russia	13
HEL	Institute of Seismology, University of Helsinki, Finland	1390
HFS	Hagfors Observatory, Sweden	55
IDC	International Data Centre, CTBTO, Austria	356
IEPN	Institute of Environmental Problems of the North, Russian Academy of Sciences, Russia	6
ISC	International Seismological Centre, United Kingdom	495
KOLA	Kola Regional Seismic Centre, GS RAS, Russia	433
LVSN	Latvian Seismic Network, Latvia	124
MIRAS	Mining Institute of the Ural Branch of the Russian Academy of Sciences, Russia	211
MOS	Geophysical Survey of Russian Academy of Sciences, Russia	3
NAO	Stiftelsen NORSAR, Norway	171
NEIC	National Earthquake Information Center, USA	15
NNC	National Nuclear Center, Kazakhstan	3
UPP	University of Uppsala, Sweden	62
	TOTAL:	**4418**

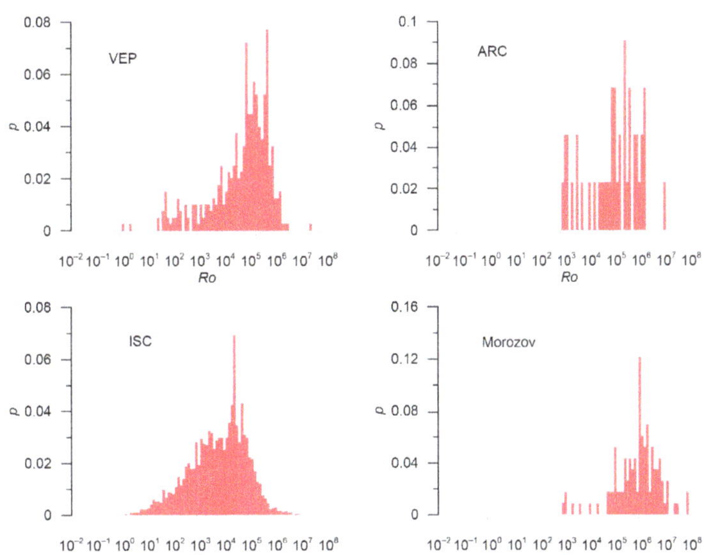

Figure 2. Distribution of the metric for events within the source earthquake catalogs (Table 1). The catalog name is indicated on the histogram.

3. Results

3.1. Integrated Catalog of the Western Sector of the AZRF

As the results of [7] show, in the eastern sector of the AZRF, most earthquakes are registered by the Russian regional seismic networks of GS RAS. As shown in Table 2, in the western sector of the AZRF, a significant majority of events are registered by European seismic networks, whose data are collected in the composite ISC catalog. The number of earthquake records in the ISC catalog is approximately five times greater than the total number of events in the GS RAS catalogs. In addition, a new catalog was recently presented in [24], which covers a substantial part of the studied region. In this catalog, earthquakes are relocated based on the analysis and merging of all available seismic bulletins from Russian and European seismic networks using modern velocity models. However, data from [22] are not used, since they lack estimates of earthquake magnitudes.

Thus, in the task of merging earthquake catalogs, the following priority sources of earthquake data were established in the western sector of the AZRF (Table 1):

1. Earthquakes from the Morozov catalog (118 events);
2. Earthquakes from the ISC catalog (4418 events);
3. Earthquakes from Russian catalogs ARC and VEP (175 + 742 events), with preference given to data from the ARC catalog in overlapping areas.

As mentioned above, when merging earthquake catalogs in the western sector of the AZRF, the task of excluding duplicates, i.e., records in different catalogs related to the same seismic event, as well as the task of excluding explosions and other events that are not earthquakes, were solved. The assembly diagram of the integrated catalog is shown in Figure 3, and the numerical parameters for merging are given in Table 3.

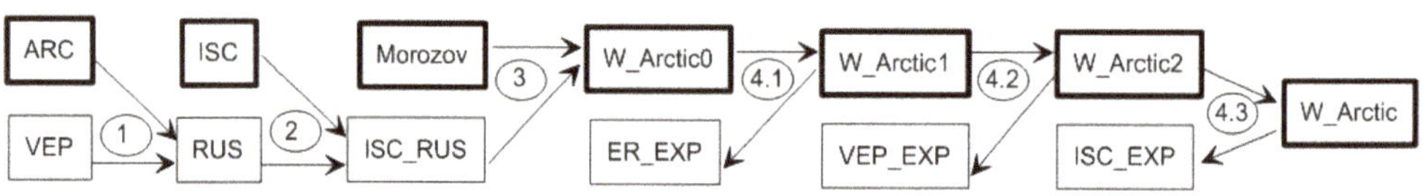

Figure 3. Assembly diagram of the integrated catalog of the western sector of the AZRF. At each stage, the main catalog is marked with a bold frame. The stage numbers are given in ellipses and correspond to Table 3.

3.1.1. Stage 1. Merging the ARC and VEP Catalogs

The main catalog was considered ARC, while VEP was the supplementary one. Duplicate analysis was performed with standard distribution parameters $\sigma_T = 0.05$ min, $\sigma_X = \sigma_Y = 15$ km. At the threshold value $Ro = 10$, three duplicates were identified, which was insufficient for optimizing metric parameters and estimating possible errors (Figure 4). The merged RUS catalog contains 920 events.

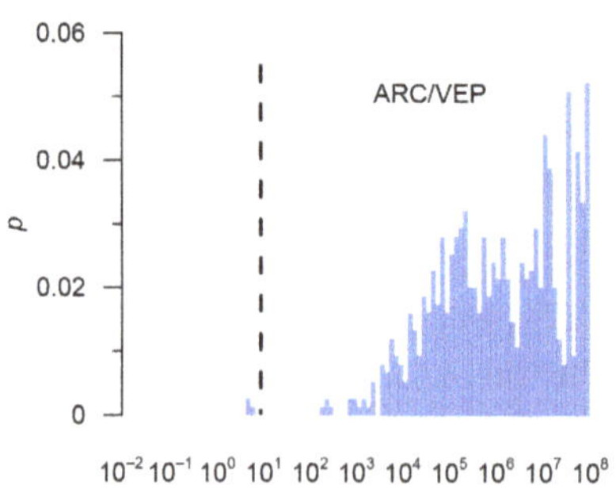

Figure 4. Metric distribution for ARC/VEP pairs. The threshold metric value $Ro = 10$ is shown by a dashed line.

Table 3. Scheme and compilation parameters of the integrated catalog.

Stage	Main Catalog	Additional Catalog	Metric Parameters σ_T min, σ_X km, σ_Y km	Threshold Value of the Metric	Estimation of the Number of Errors	Number of Duplicates	Merged Catalog
1	**ARC 175 events**	**VEP 748 events**	0.05; 15.0; 15.0	10 *	-	3 *	**RUS 920 events**
2	**ISC 4418 events**	**RUS 920 events**	0.035; 16.3; 16.6	13	0.2%	282	**ISC_RUS 5056 events**
3	**Morozov 118 events**	**ISC_RUS 5056 events**	0.05; 15.0; 15.0	20		108	**W_ARCTIC0 5066**
4.1	**W_ARCTIC0 5066**	**ER_EXP 14,490 events**	0.05; 15.0; 15.0	10	-	284	**W_ARCTIC1 4782 events**
4.2	**W_ARCTIC1 4782 events**	**VEP_EXP 170 events**	0.05; 15.0; 15.0	10	-	4	**W_ARCTIC2 4778 events**
4.3	**W_ARCTIC2 4778 events**	**ISC_EXP 100,112 events**	0.05; 15.0; 15.0	10	-	92	**W_ARCTIC 4686 events**

* The number of duplicates is not enough to optimize the metric parameters.

3.1.2. Stage 2. Merging the ISC and RUS Catalogs

The main catalog was considered ISC, while RUS was the supplementary one. Preliminary duplicate analysis was performed with standard distribution parameters $\sigma_T = 0.05$ min, $\sigma_X = \sigma_Y = 15$ km. At the threshold value *Ro* = 10, 263 potential duplicates were identified, of which six were absolute duplicates. Absolute duplicates were not used to determine dispersions (Figure 5).

It can be observed that each of the *DT*, *DX*, and *DY* parameters follows a normal distribution, and the mean is small compared to the standard deviation for all three parameters. Additionally, Figure 5 demonstrates that dispersion is almost independent of event magnitude and time.

The final analysis of duplicates was performed with the parameters $\sigma_T = 0.035$ min, $\sigma_X = 16.3$ km, and $\sigma_Y = 16.6$ km (Figure 6). The metric values were also calculated between the nearest events in the RUS catalog. The RUS catalog contains only two earthquakes with a distance to the nearest neighbor *Ro* < 35 (the algorithm for calculating the metric is the same as for two different catalogs, except for excluding the comparison of an earthquake with itself). This allowed us to estimate the probability that a duplicate was selected incorrectly due to high earthquake density: $p = 2/884 = 0.0022$, with an upper estimate of false duplicate probability at approximately 0.22%.

The selection of a metric threshold for identifying duplicates depends on the task of further study of the merged catalog. We chose a threshold that minimizes the total number of errors. Analysis of the spatial–temporal structure of the distribution for nearest neighbors in the ISC and RUS catalogs (Figure 7) indicates that a threshold of *Ro* = 20 provides a close-to-optimal separation of duplicates and naturally grouped events. In this case, 282 duplicates were identified. In total, there are 638 unique events in the RUS catalog for the studied region. These events were added to the ISC catalog, resulting in a merged ISC_RUS catalog that includes 5056 events.

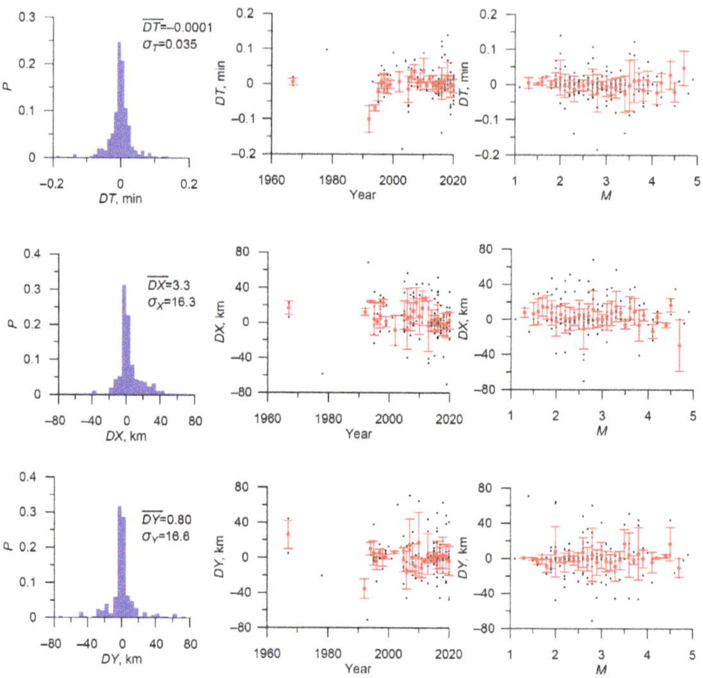

Figure 5. Distributions of metric parameters *DT*, *DX*, and *DY*, and the dependence of dispersion and mean values on time and magnitude of events for merging the ISC and RUS catalogs.

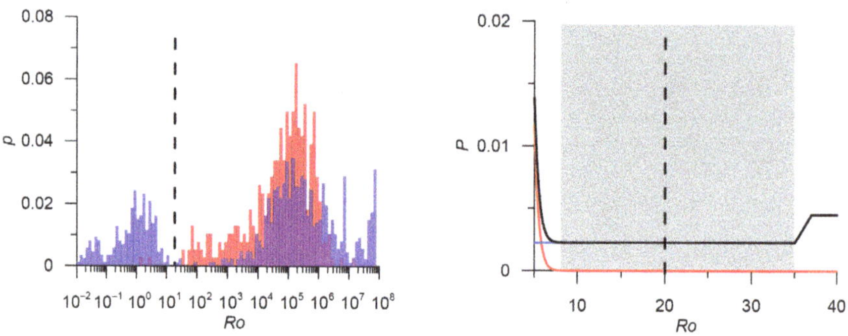

(**a**) (**b**)

Figure 6. (**a**) Comparison of the metric distribution ISC/RUS pairs (transparent blue histogram) and the same metric for RUS/RUS earthquakes (transparent red histogram); the intersection of histograms appears purple; (**b**) Threshold optimization: the red line shows the probability of missing a duplicate in the model with metric (1), the blue line shows the probability of a false duplicate, and the black line shows the total probability of first and second type errors. The gray band shows the range metric values *Ro* = 8 ÷ 35, which minimize the total number of errors, approximately 0.2%. The threshold value *Ro* = 20 is shown by a dashed line.

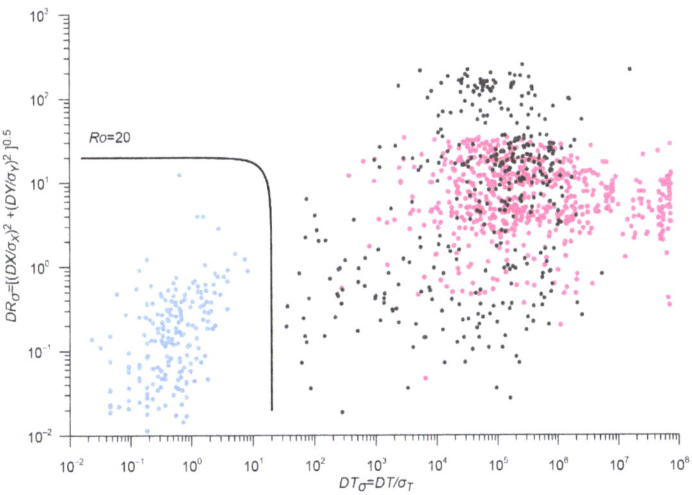

Figure 7. Distribution of normalized *DT* and *DR* and contour line of metric (1). Colored dots represent ISC/RUS pairs (blue dots are duplicates, pink dots are unique events), and black dots are distances between RUS/RUS events in metric (1). The values of the metric for earthquakes in the RUS catalog are significantly larger than for ISC/RUS duplicates (blue dots). The contour line of metric (1) *Ro* = 20 provides a close to optimal separation of duplicates and naturally grouped events. Absolute duplicates are not shown.

3.1.3. Stage 3. Merging Morozov and ISC_RUS Catalogs

The main catalog is Morozov, the additional one is ISC_RUS obtained in the previous step. Duplicate analysis was performed with standard distribution parameters $\sigma_T = 0.05$ min, $\sigma_X = \sigma_Y = 15$ km. Figure 8 shows the distribution of metric (1). At *Ro* = 20, the ISC_RUS catalog contains 108 duplicates and 4948 unique events that were added to the Morozov catalog. The resulting catalog W_ARCTIC0 contains 5066 events.

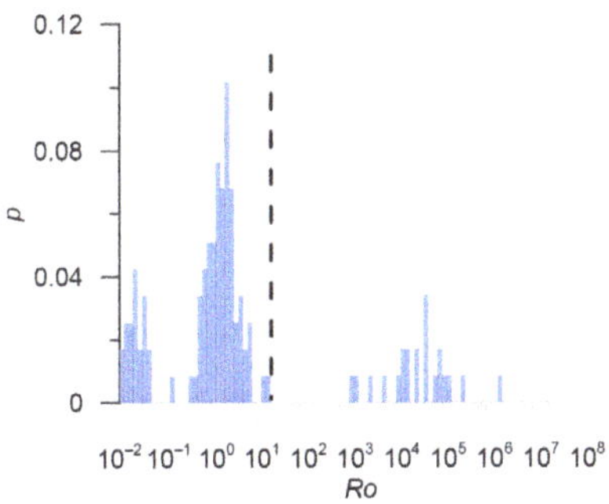

Figure 8. Distributions of the metric for the Morozov/ISC_RUS event pairs. The threshold value of metric *Ro* = 20 is indicated by a dashed line.

3.1.4. Stage 4. Exclusion of Explosions and Other Anthropogenic Events

The preliminary exclusion of explosions and other anthropogenic events was performed based on the information on the nature of the events provided in the source catalogs. However, events marked as "non-earthquake" in one catalog may not be marked as such in other catalogs. We considered an event as a "non-earthquake" if it was marked as such in at least one of the input catalogs. In addition to the information from the ISC and VEP catalogs, data from the "Earthquakes of Russia" ER_EXP merged catalog (Table 1) were used.

Explosions and other anthropogenic events presented in the ER_EXP, VEP_EXP, and ISC_EXP catalogs (Tables 1 and 3) were sequentially removed from the W_ARCTIC0 merged catalog obtained in the previous stage. Duplicate analysis was performed with standard distribution parameters $\sigma_T = 0.05$ min, $\sigma_X = \sigma_Y = 15$ km, and a threshold value of metric $Ro = 10$. A total of 380 duplicates were detected (Figure 9). These events were excluded from the W_ARCTIC0 catalog. The final integrated earthquake catalog, W_ARCTIC, for the western sector of the AZRF contains 4686 events. Table 3 provides statistics and parameters used at each step for excluding non-earthquake duplicates.

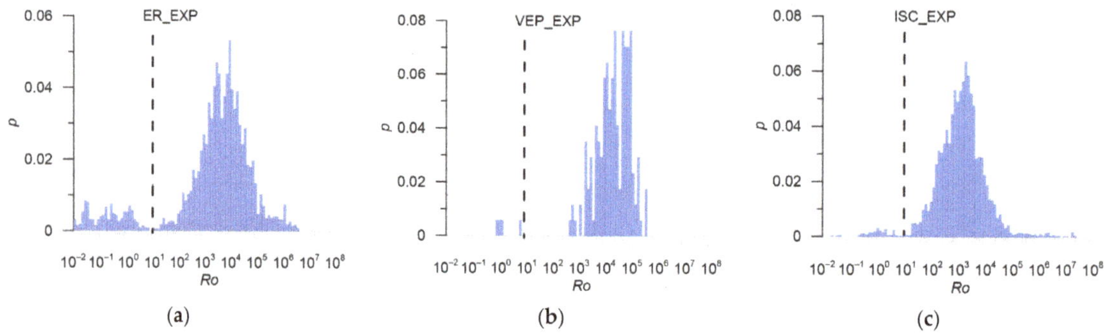

Figure 9. Distributions of metric for the pairs: (**a**) W_ARCTIC0/ER_EXP; (**b**) W_ARCTIC1/VEP_EXP; (**c**) W_ARCTIC2/ISC_EXP. The threshold value of metric, $Ro = 10$ is indicated by a dashed line.

The results of the analysis of the final W_ARCTIC catalog suggest that explosions were not completely excluded. The histogram of the number of events (Figure 10a) shows an abnormally large number of events in 1994–1997 and 2021–2022. The spatial and spatiotemporal distribution of events (Figure 10b,c) also indicates an incomplete removal of explosions during these years. In this sense, the period of 1989–1993 also raises some suspicions. We were unable to find reliable sources of information on explosions during these periods.

In the area of the Severouralsk bauxite deposit (coordinates 60° N, 60° E) in the Northern Urals, there is a significant cluster of 57 events. According to [32], this area experiences a large number of technological explosions and rock bursts (hundreds per year). Therefore, we assume that most of the events in this cluster are not earthquakes, although this is not indicated in the catalogs used (Table 1). All 46 events that occurred in 2021–2022 are highly likely to be explosions (the information about explosions has not yet been published for this period). The other 11 events, if not explosions or rock bursts, are likely to be triggered earthquakes. Based on this, a decision was made to exclude events from the catalog that are part of the identified Urals cluster. The final integrated W_ARCTIC catalog contains 4629 events.

3.2. *Magnitudes in the Integrated Catalog of the Western Sector of the AZRF*

The western sector of the AZRF includes several seismically active zones with different types of magnitudes. In addition, it is assumed that the ratios between magnitudes may differ in different seismically active zones. Magnitude unification was carried out in two sub-regions: the Kola Peninsula and Karelia (60° N–72° N, 25° E–40° E) and the Kara and Barents Seas shelf. Earthquakes located within the East European and West Siberian platforms were considered together with events from the Kara and Barents Seas shelf.

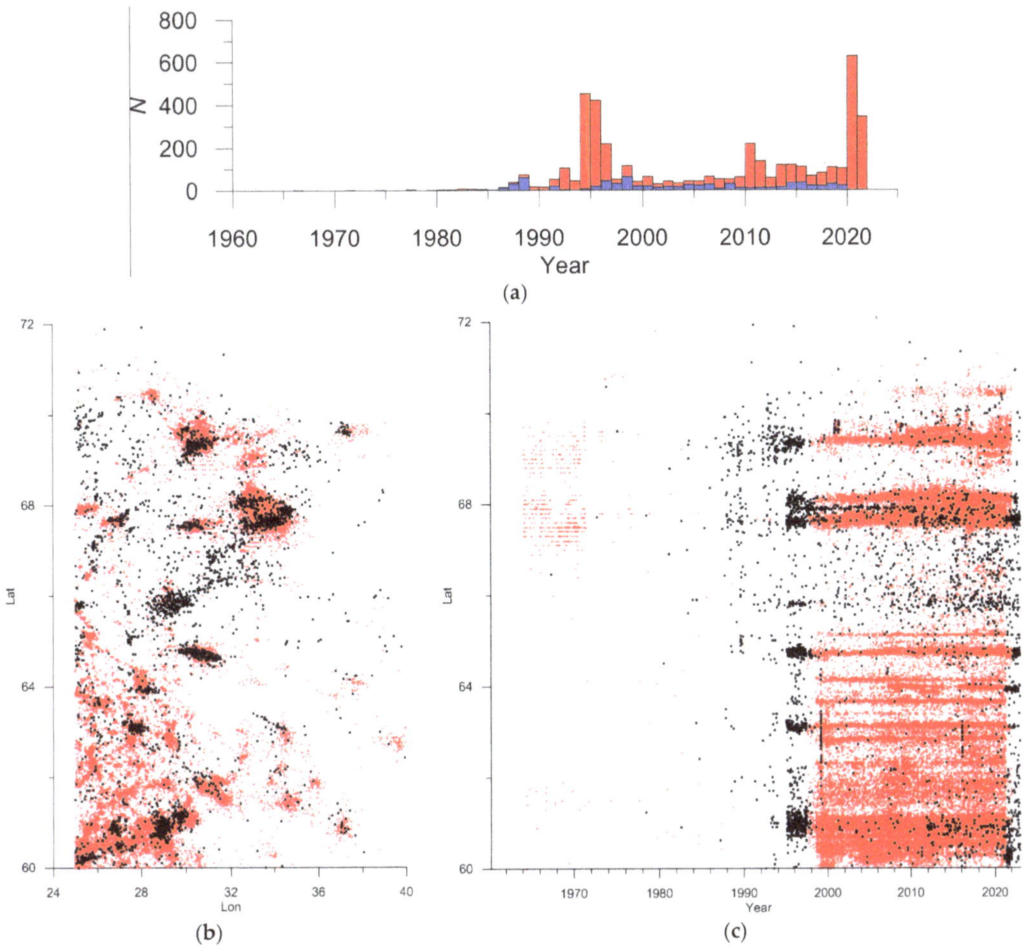

Figure 10. Analysis of the W_ARCTIC catalog: (**a**) Red histogram—number of events per year (blue histogram shows a similar distribution for the VEP catalog); (**b**) Spatial distribution of epicenters; (**c**) Space–time distribution of events. Red dots represent explosions from the ISC_EXP catalog, while black dots represent earthquakes from the merged W_ARCTIC catalog in the Kola Peninsula and Karelia (60° N–72° N, 25° E–40° E).

3.2.1. The Kola Peninsula and Karelia

There is no earthquake in the Kola Peninsula and Karelia with a determined Mw^{GCMT} magnitude, while only seven events have mb^{ISC} and mb^{NEIC} magnitude estimates (Figure 1). Due to these reasons, direct magnitude unification to the moment magnitude scale is impossible. Most events have local ML magnitudes from various agencies. ML^{KOLA} was chosen as the base local magnitude because the local network of the Kola branch of GS RAS is located directly in the studied area. However, most earthquakes have local ML^{HEL} magnitudes, allowing for more reliable correlations with other magnitude types (Figure 11). Thus, the unified local magnitude ML is determined by a regression relationship with ML^{HEL}, which is approximately 0.3 less than ML^{KOLA} (Figure 11a).

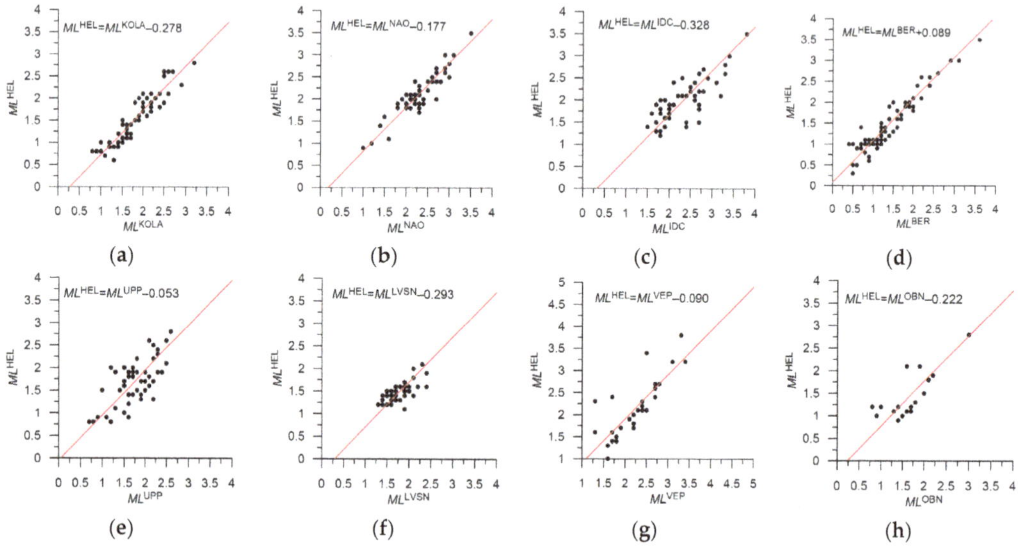

Figure 11. Shift-type correlation ratio for local magnitudes from various agencies in the Kola Peninsula and Karelia sub-region. Magnitude types are specified on the figures. All eight relationships appear reliable.

Out of 104 events with $MPSP^{VEP}$ magnitude, only two have ML^{HEL} and mb^{ISC} magnitudes. To refine the magnitude relationships, events outside the region (Svalbard) were used (Figure 12). Two events from the studied region fit well into the regression for Svalbard events. For this reason, we used the obtained relationships for earthquakes in the Kola Peninsula and Karelia. All events with $MPSP^{VEP}$ magnitude occurred in 1987–1989. It was previously suggested that not all explosions and other non-earthquake events were removed from the catalog for this period. Therefore, it is possible that these events are not tectonic earthquakes. In addition, most events have $MPSP^{VEP}$ < 3.5 magnitudes; i.e., they are outside the range for which relationships are established. Therefore, they were considered highly unreliable.

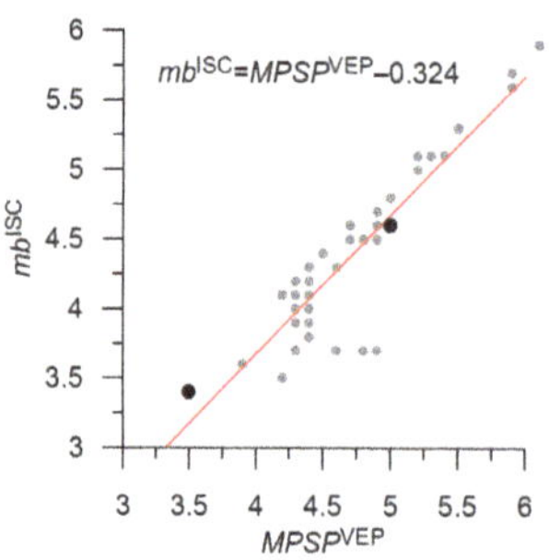

Figure 12. Shift-type correlation ratio for $MPSP^{VEP}$ magnitude. Magnitude types are specified on the figure. Gray dots are events in Svalbard; black dots are two events in the shelf sub-region.

Only MD^{HEL} and MD^{BER} duration magnitudes have 166 events. This type of magnitude is often used for records without clear arrival (explosions, landslides, etc.). For example, the ML and MD relationship is sometimes used to identify landslides in regional catalogs [35]. Events with this type of magnitude are only present in the catalog during the period 1988–1997 when explosions are presumably not removed. It should be noted that ML^{EIDC} and mb^{HFS} magnitudes appear in the catalog only during this period, for which indirect correlations through MD^{BER} were used for unification (Figure 13). Thus, it was assumed that most of these events are not earthquakes.

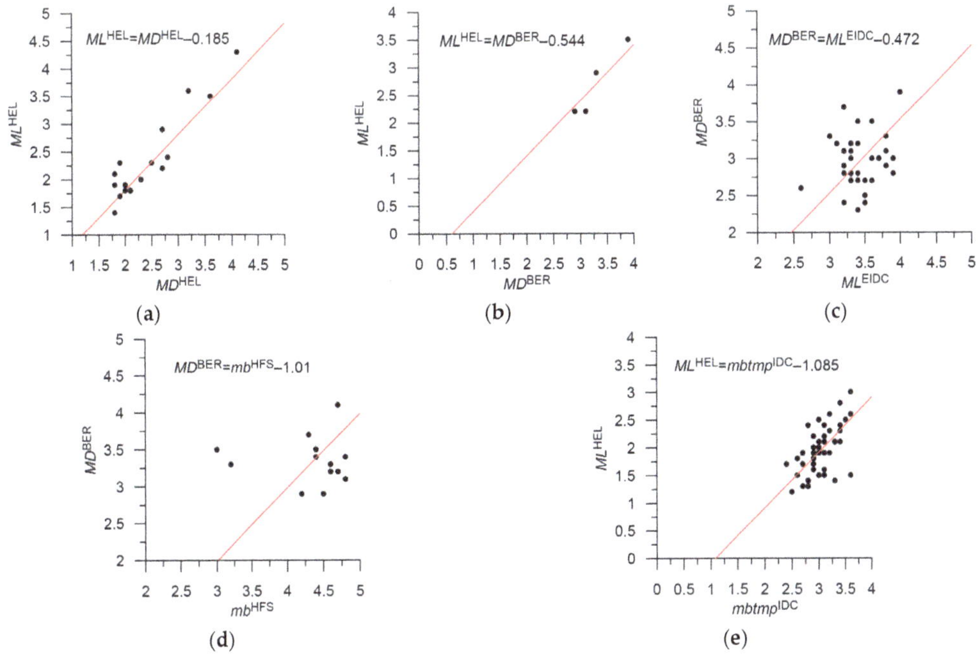

Figure 13. Shift-type correlation ratios for magnitudes of various agencies in the sub-region of the Kola Peninsula and Karelia. Magnitude types are specified on the figures. Only the MD^{HEL} (**a**) and $mbtmp^{IDC}$ (**e**) relationships appear reliable out of 5 ratios.

The regression ratios could not be established for eight events. A relationship was determined between the unified local magnitude ML and mb^{ISC}, mb^{NEIC}: $ML = mb - 0.08$, i.e., they are very close. In fact, the difference lies within the 95% confidence interval, and they can be considered equal. There are only five events with both types of magnitudes, but they fit very well on a straight line (Figure 14). It should be noted that most events in the sub-region have a unified magnitude $ML < 3.3$, which is outside the range for which $ML \approx mb$. Therefore, the relationship is unreliable.

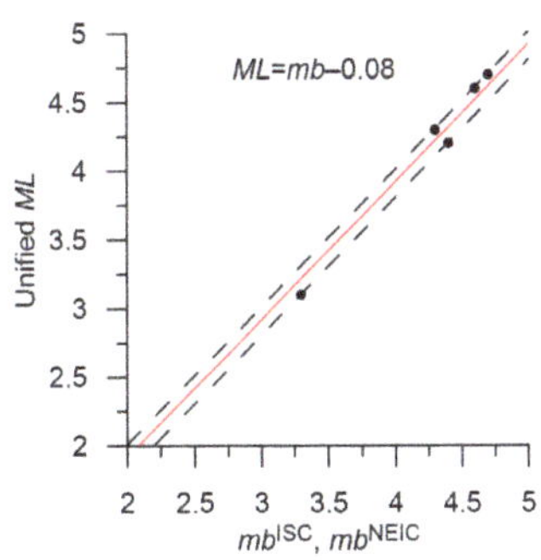

Figure 14. Shift-type correlation ratio between the unified magnitude ML and mb^{ISC}, mb^{NEIC} in the sub-region of the Kola Peninsula and Karelia. Magnitude types are specified on the figure. The dashed lines show the 95% confidence interval.

Statistics and formulas for converting magnitudes are presented in Table 4.

Table 4. Unified magnitude in the catalog of the western sector of the Russian Arctic: sub-region Kola–Karelia.

Agency	Type of Magnitude	Priority	Number of Events	Magnitude in the Integrated Catalog	Figure	Mmin—Mmax. Initial Magnitude Scale	Note
ISC	mb	1	4	$mb = mb^{ISC}$		3.3–4.8	
NEIC, NEIS	mb	1	3	$mb = mb^{NEIC}$		4.3–4.7	
KOLA	ML	2	398	$ML = ML^{KOLA}$		0.3–2.7	
HEL	ML	3	1554	$ML = ML^{HEL} + 0.3$	Figure 11a	0.0–3.6	
NAO	ML	3	140	$ML = ML^{NAO} + 0.1$	Figure 11b	1.1–4.5	
IDC	ML	3	309	$ML = ML^{IDC}$	Figure 11c	0.8–3.4	
BER	ML	3	27	$ML = ML^{BER} + 0.4$	Figure 11d	0.3–3.4	
UPP	ML	3	42	$ML = ML^{UPP} + 0.2$	Figure 11e	1.5–3.2	
LVSN	ML	3	76	$ML = ML^{LVSM}$	Figure 11f	0.8–2.4	
VEP	ML	3	319	$ML = ML^{VEP} + 0.2$	Figure 11g	0.1–4.0	
OBN	ML	3	4	$ML = ML^{OBN} + 0.1$	Figure 11h	0.1–1.0	
HEL	MD	4	16	$ML = MD^{HEL} + 0.1$	Figure 13a	1.6–2.6	Probably non-earthquakes
BER	MD	4	150	$ML = MD^{BER} - 0.2$	Figure 13b	2.0–4.1	Probably non-earthquakes
EIDC	ML	4	1001	$ML = ML^{EIDC} - 0.7$	Figure 13c	1.1–4.4	Indirect correlation through MD^{BER}. Probably non-earthquakes
HFS	mb	4	52	$ML = mb^{HFS} - 1.2$	Figure 13d	3.0–5.3	Indirect correlation through MD^{BER}. Probably non-earthquakes
IDC	$mbtmp$	4	2	$ML = mbtmp^{IDC} - 0.8$	Figure 13e	2.8–3.4	
VEP	$MPSP$	4	102	$mb = MPSP^{VEP} - 0.3$	Figure 12	1.0–4.9	See note to Figure 12. Probably non-earthquakes
NAO	mb	5	2	$M = mb^{NAO}$		3.1–3.2	Not determined
EIDC	mb	5	1	$M = mb^{EIDC}$		3.6	Not determined
BER	M	5	3	$M = M^{BER}$		2.5–4.0	Not determined
HEL	M	5	2	$M = M^{HEL}$		1.8	Not determined
			4207				

3.2.2. The Kara and Barents Seas Shelf

In the sub-region of the Kara and Barents Seas shelf, there are no events with magnitude Mw^{GCMT}. Thus, a direct unification of magnitude to the moment scale is impossible. Magnitudes mb^{ISC} and mb^{NEIC} were determined for 45 events (Figure 1). There are also a small number of determinations for magnitudes mb from other agencies. Most events have a local magnitude ML from various agencies, predominantly ML^{BER}. For some earthquakes, magnitudes MD^{BER} and other types determined by the GS RAS are available. The unified magnitude mb was determined by regression ratios with mb^{ISC} (Figure 15). The unified local magnitude ML was determined by a regression relationship with ML^{BER} (Figure 16). In some cases, it was necessary to use indirect correlations—relationships are unreliable (Figure 17). Regression relationships could not be established for six events.

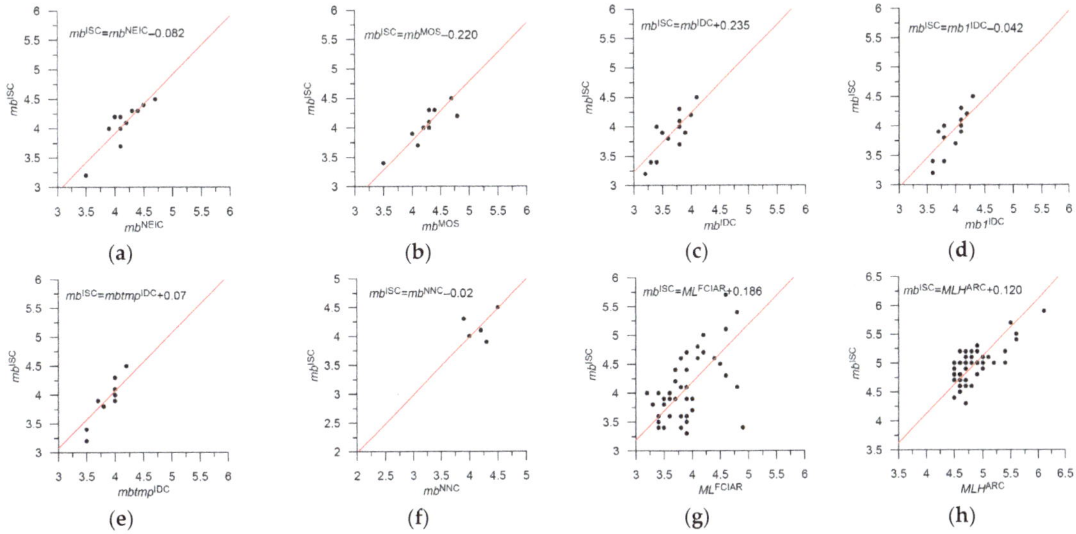

Figure 15. Shift-type correlation ratios for magnitudes of various agencies with the magnitude mb^{ISC} in the sub-region of the Kara and Barents Seas shelf. Magnitude types are specified on the figures.

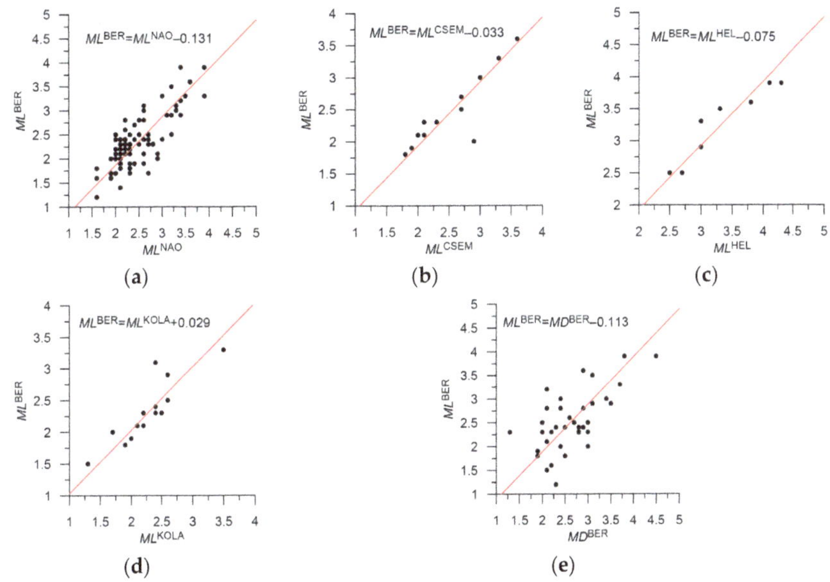

Figure 16. Shift-type correlation ratios for magnitudes of various agencies with the magnitude ML^{BER} in the sub-region of the Kara and Barents Seas shelf. Magnitude types are specified on the figures.

There are 10 events with energy class measures k on the Franz Josef Land archipelago. Eight of them occurred in the end of 1983 or the beginning of 1984 and have the same epicenter coordinates. The coordinates for this period are given with an accuracy of 0.1°. These events are not included in the Morozov catalog. Whether these events are a swarm of natural earthquakes or technogenic events is an open question, so we do not exclude them from the integrated catalog.

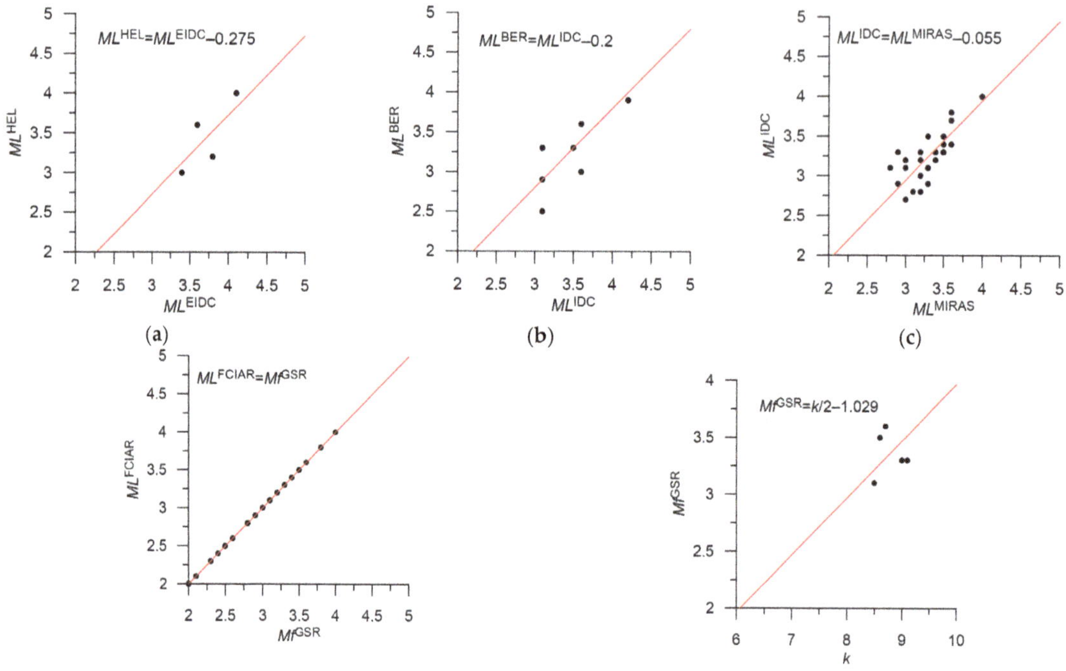

Figure 17. Shift-type correlation ratios for magnitudes of different agencies that were used to calculate indirect correlations in the sub-region of the Kara and Barents Seas shelf. Magnitude types are specified on the figures.

A relationship was determined between the unified local magnitude *ML* and *mb*. The similarity of a linear relationship is observed for earthquakes with $ML > 3.0$. There are only 10 such events (Figure 18). At lower magnitudes, the relationship breaks down. The unified local magnitudes *ML* and *mb* are very close. In fact, the difference is less than 0.1, which is within the 95% confidence interval. Thus, they can be considered equal.

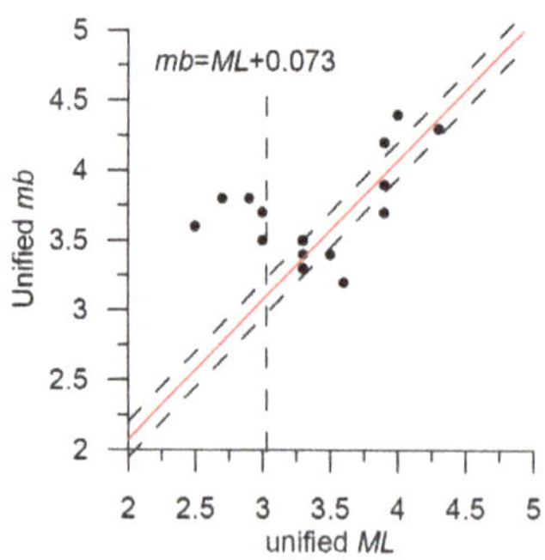

Figure 18. Shift-type correlation ratio between unified magnitudes *ML* and *mb* ($ML > 3.0$) in the sub-region of the Kara and Barents Seas shelf. Magnitude types are specified on the figure. The dashed lines show the 95% confidence interval.

Statistics and formulas for magnitude conversion are presented in Table 5.

Table 5. Magnitude in the integrated catalog of the western sector of the Russian Arctic: sub-region Barents and Kara Seas shelf.

Agency	Type of Magnitude	Priority	Number of Events	Magnitude in the Integrated Catalog	Figure	Mmin—Mmax. Initial Magnitude Scale	Note
ISC	*mb*	1	42	$mb = mb^{ISC}$	-	3.2–4.8	
NEIC, NEIS	*mb*	1	3	$mb = mb^{NEIC} - 0.1$	Figure 15a	4.2–4.3	
MOS	*mb*	2	3	$mb = mb^{MOS} - 0.2$	Figure 15b	3.6–3.9	
IDC	*mb*	3	12	$mb = mb^{IDC} + 0.2$	Figure 15c	2.8–3.7	
IDC	*mb1*	3	4	$mb = mb1^{IDC}$	Figure 15d	3.5–3.8	
IDC	*mbtmp*	3	2	$mb = mbtmp^{IDC} + 0.1$	Figure 15e	3.6–3.7	
NNC	*mb*	3	2	$mb = mb^{NNC}$	Figure 15f	3.5–4.0	
BER	*ML*	4	140	$ML = ML^{BER}$	-	0.9–3.5	
NAO	*ML*	4	45	$ML = ML^{NAO} - 0.1$	Figure 16a	1.7–3.7	
CSEM	*ML*	4	3	$ML = ML^{CSEM}$	Figure 16b	2.9–3.8	
HEL	*ML*	4	10	$ML = ML^{HEL} - 0.1$	Figure 16c	1.6–4.0	
KOLA	*ML*	4	12	$ML = ML^{KOLA}$	Figure 16d	1.5–2.3	
FCIAR	*ML*	4	92	$mb = ML^{FCIAR} + 0.2$	Figure 15g	1.3–3.6	
BER	*MD*	4	11	$ML = MD^{BER} - 0.1$	Figure 16e	2.2–4.0	
ARC	*MLH*	4	1	$mb = MLH^{ARC} + 0.1$	Figure 15h	4.7	
MIRAS	*ML*	5	4	$ML = ML^{MIRAS} - 0.3$	Figure 17b,c	2.6–3.1	Indirect corr thru ML^{IDC}
VEP	*MPSP*	5	8	$mb = MPSP^{VEP} - 0.3$	Figure 13	2.9–4.8	See note to Figure 12
EIDC	*ML*	5	7	$ML = ML^{EIDC} - 0.3$	Figure 17a Figure 16c	3.2–4.1	Indirect corr thru ML^{HEL}
GSR	*Mf*	5	5	$mb = Mf^{GSR} + 0.2$	Figure 17d Figure 15g	2.0–3.3	In the region under study $Mf^{GSR} = ML^{FCIAR}$
GSR	*Klass*	5	10	$mb = k/2 - 1.2$	Figure 17e	8.5–12	See note to Figure 17. Probably non-earthquakes
NAO	*mb*	5	4	$M = mb^{NAO}$		3.4–4.4	Not determined
UPP	*ML*	5	1	$M = ML^{UPP}$	-	4.1	Not determined
LAO	*M*	5	1	$M = M^{LAO}$	-	3.7	Not determined
Total			**4022**				

3.3. Statistics of the Integrated Catalog for Two Sub-Regions

The integrated catalog of the western sector of the AZRF contains information on 4629 events. In turn, for the period 1998–2020, when all "non-earthquakes" were removed with a high degree of reliability, the catalog contains 2126 records (Figure 19). Detailed statistics for the catalog for the period 1998–2020 are presented in Table 6. Figure 19 shows that the registration level in the western part of the region is significantly better than in its central and eastern parts.

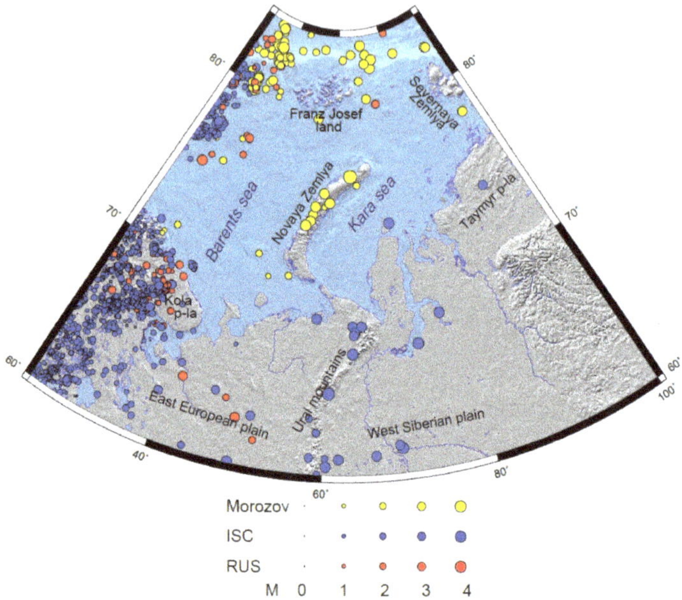

Figure 19. Map of earthquake epicenters of the integrated catalog for the western sector of the AZRF for the period 1998–2020. Yellow dots show events from Morozov catalog, blue dots show events from the ISC catalog, and red dots show events from Russian catalogs. Preferred location of event corresponds to the priority of data sources: 1 Morozov, 2 ISC, 3 RUS.

Table 6. Statistics of the integrated catalog.

Time Period, Catalog	*N* Total	*N* from *ISC*	*N* from GS RAS, Morozov	*Mc*	*N*, *M* ≥ *Mc*	*M*max
1962–2022						
E_Arctic	4629	4072 (88%)	558 (12%)			4.8
Kola–Karelia	4207	3754 (89%)	454 (11%)			4.8
The Kara and Barents shelf	422	256 (61%)	166 (39%)			4.8
1998–2020						
E_Arctic	2126	1715 (81%)	411 (19%)	3.5	50	4.7
Kola–Karelia	1715	1506 (88%)	209 (12%)	1.7	561	4.6
The Kara and Barents shelf	410	209 (51%)	201 (49%)	3.5	44	4.7

Figures 20–23 show event distributions over time and magnitude as well as differential magnitude–frequency graphs for various time periods for the sub-regions of the Kola Peninsula and Karelia, the Kara, and Barents Seas shelf.

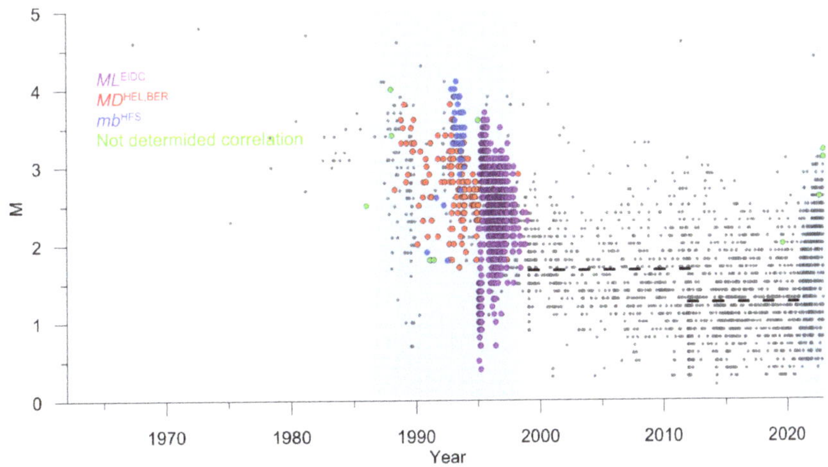

Figure 20. Sub-region of the Kola Peninsula and Karelia. Event distribution over time and magnitude. Blue indicates periods unfiltered for explosions/rock bursts. Colored dots show events with unreliable magnitudes. The overwhelming majority of such events belong to periods with unfiltered explosions. The dashed line shows preliminary estimates of completeness magnitude *Mc*.

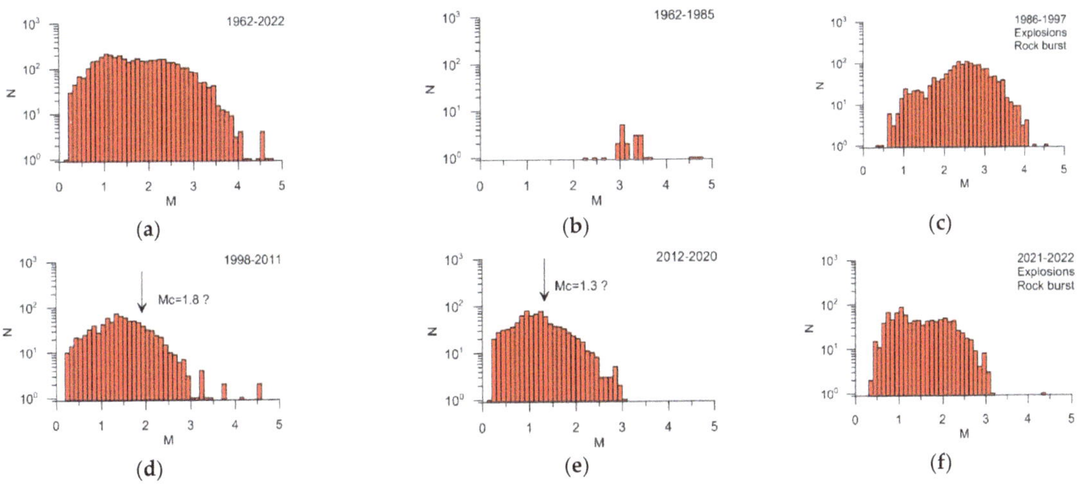

Figure 21. Sub-region of the Kola Peninsula and Karelia. Differential magnitude-frequency graphs for different time periods and preliminary estimates of completeness magnitude *Mc*.

Until 1980, only a few seismic events were registered in the sub-region of the Kola Peninsula and Karelia. The period 1998–2020 can be used to determine seismicity parameters.

Preliminary estimates-$Mc = 1.8$ in 1998–2011, $Mc = 1.3$ in 2012–2020.

Until 1995, the catalog contains few events in the sub-region of the Kara and Barents Seas shelf with $Mc = 4.5$. To determine seismicity parameters, the period 1998–2020 can be used. Preliminary estimates suggest $Mc = 3.5$ for 1995–2020. The increase in the number of small events is related to the western part of the shelf near Svalbard, where the registration level is much better, and the preliminary estimate $Mc = 2.3$.

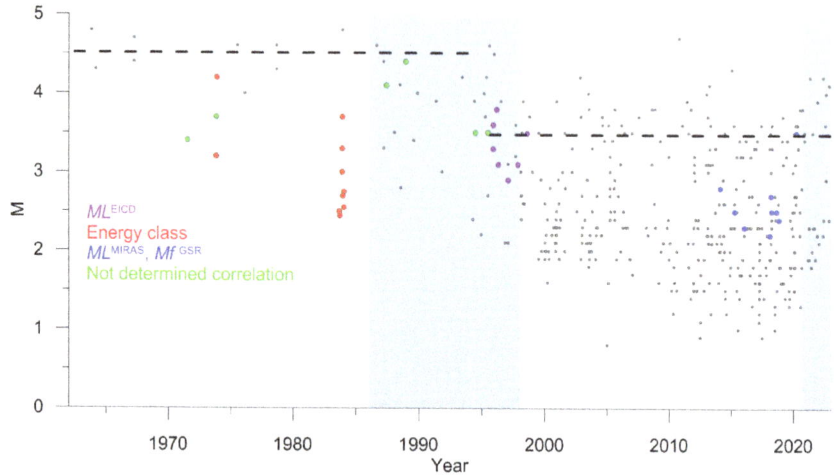

Figure 22. Sub-region of the Kara and Barents Seas shelf. Distribution of events by time and magnitude. The blue color highlights periods not cleared from explosions/rock bursts. Colored dots show events with unreliable magnitudes. The dashed line is a preliminary estimate of completeness magnitude *Mc*.

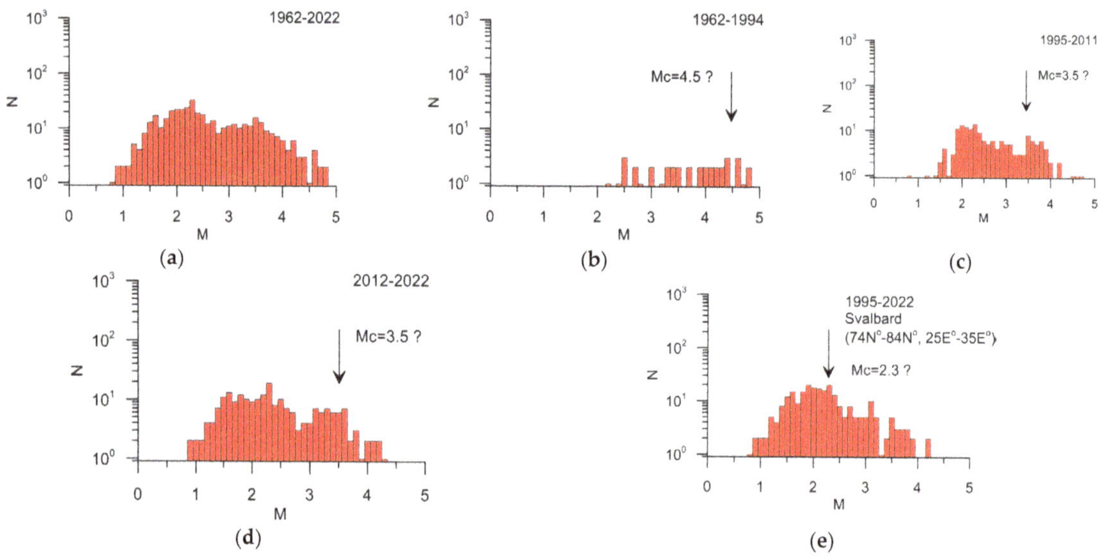

Figure 23. Sub-region of the Kara and Barents Seas shelf: (**a**–**d**) represent differential magnitude-frequency graphs in different periods and preliminary estimates of completeness magnitude. (**e**) shows the magnitude-frequency graph in the vicinity of the Svalbard Archipelago.

Figure 24 shows a map of spatial variations in the completeness magnitude *Mc* in the integrated catalog for the period 1998–2020, which was constructed using a multi-scale method [36]. The volume of registered seismicity is sufficient for mapping only in the western part of the region on the Kola Peninsula and in Karelia as well as in the vicinity of the Svalbard Archipelago. In the rest of the region, only a regional value of *Mc* = 3.5 can be determined. Preliminary estimates of *Mc* = 1.8 on the Kola Peninsula and in Karelia and *Mc* = 2.3 in the vicinity of the Svalbard Archipelago were confirmed.

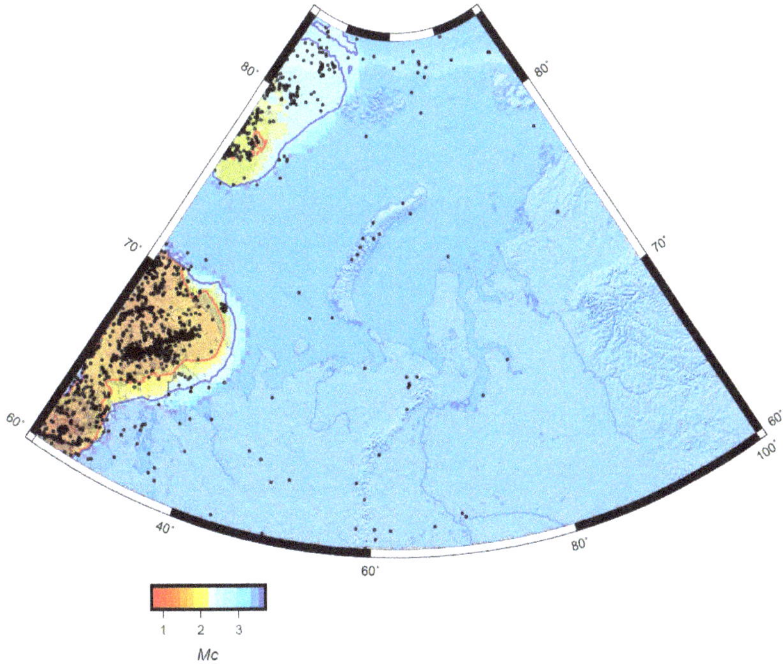

Figure 24. Integrated catalog. Map of spatial variations in the completeness magnitude *Mc* for the period 1998–2020. The black dots show the earthquake epicenters, the red line is the level *Mc* = 1.8, the blue line is *Mc* = 2.3.

4. Conclusions

This paper describes step by step the procedure for compiling the integrated earthquake catalog for the western sector of the AZRF with a unified magnitude scale (Figure 19). The unification was carried out based on the analysis of ratios of magnitude types in the source catalogs. It should be noted that the obtained catalog, in contrast to the original catalogs, is characterized by the completeness and representativeness of events. The unified catalog is made available to the public at: http://www.wdcb.ru/arctic_antarctic/arctic_seism.html, (accessed on 1 June 2023). The integrated earthquake catalog obtained in this paper is intended for a wide range of researchers involved in both the study of the seismic regime of the Arctic and, in general, the seismic hazard assessment [37–45].

Based on the results obtained in this study, the authors consider it possible to formulate the following conclusions:

1. Merging data from the GS RAS, Morozov, and the ISC catalogs significantly increased the number of registered events in the western sector of the AZRF (Tables 3 and 6);
2. The modified nearest neighbor method [1] efficiently identifies records in different catalogs that correspond to the same seismic event (Figures 4–8, Table 3). The error rate for duplicate identification in the studied region was found to be 0.2%;
3. Information on explosions and other events that are not tectonic earthquakes, presented in the catalogs of GS RAS and ISC, is incomplete. Approximately 300 such events from the GS RAS catalog (ER_EXP in Table 1) are classified as earthquakes in the ISC catalog, and conversely, about 100 explosions from the ISC catalog are classified as earthquakes in the GS RAS catalogs (Figure 9, Table 3);

4. Explosions before 1998 and after 2020 have not been removed (Figure 10). Explosions after 2020 may be removed with the arrival of new data–GS RAS catalogs and reviewed ISC catalog. Identifying and removing explosions that occurred before 1998 is an extremely difficult task beyond the scope of the mathematical methods used in this study;
5. In general, there is no sufficient guarantee that "non-earthquakes" have been completely removed from the combined catalog during the period of 1998–2020. However, there are reasons to believe that the number of such events is small and they will not have a significant impact on regional seismicity statistics;
6. In the western sector of the AZRF, there are no Mw^{GCMT} determinations. Therefore, a direct unification of magnitude to moment magnitude scale was impossible. However, there were a sufficient number of mb^{ISC} determinations, which are used in ISC practice to obtain proxy-Mw estimates for earthquakes with $M < 5.0$ [46]. In the eastern sector of the AZRF, the $Mw^{GCMT} = mb^{ISC}$ ratio is directly confirmed [7]. All magnitude scales were aligned with mb^{ISC} (Figures 14 and 18), and it was hypothesized that the estimates made could be used as proxy-Mw in the western sector of the AZRF for regionally strong earthquakes with $M \geq 3.5$, for the overwhelming majority of which mb^{ISC} and mb^{NEIC} determinations are known. Interpreting weaker magnitudes as proxy-Mw is certainly controversial, since they fall outside the range for which regression relationships are constructed.
7. Unlike the eastern sector of the AZRF, where the overwhelming majority of events had only three types of magnitude estimates (Mw^{GCMT}, mb^{ISC}, and energy class k), the diversity of magnitude estimates for weak earthquakes in the western sector is very high. This reduces the reliability of determining a unified magnitude scale. For some earthquakes, regression relationships are not determined or extremely unreliable. It should be noted that the number of such events is insignificant during the period of 1998–2020 (Figures 20 and 22);
8. The regional magnitude of complete registration $Mc = 3.5$ during the period of 1998–2020. The registration level is significantly better in the Kola Peninsula and Karelia, where $Mc = 1.8$, and in the vicinity of the Svalbard archipelago, where $Mc = 2.3$ (Figures 21, 23 and 24);
9. The quality of the catalog after 1998 is significantly better. Therefore, for statistical analysis of seismicity parameters, such as the slope of the magnitude–frequency graph (b-value) and the level of seismic activity, we recommend using the catalog for the period of 1998–2020. However, data prior to 1998 may also be useful for other studies, for example, for studying possible sources of regionally strong earthquakes in the basins of the Barents and Kara Seas, where, for natural reasons, the presence of technogenic events is unlikely.

Author Contributions: Conceptualization, A.D.G., I.A.V. and P.N.S.; Data curation, P.N.S., B.A.D., N.A.S. and A.A.S.; Formal analysis, I.A.V.; Investigation, P.N.S., B.A.D., B.V.D. and A.A.S.; Methodology, I.A.V.; Resources, B.A.D. and N.A.S.; Software, I.A.V.; Validation, I.A.V., A.D.G., P.N.S., B.A.D. and B.V.D.; Visualization, I.A.V., B.V.D. and N.A.F.; Writing—original draft, A.D.G., I.A.V., P.N.S., B.A.D., B.V.D., N.A.S., A.A.S. and N.A.F.; Writing—review and editing, A.D.G., I.A.V., P.N.S., B.A.D., B.V.D., N.A.S., A.A.S. and N.A.F. All authors have read and agreed to the published version of the manuscript.

Funding: The reported study was funded by the Russian Science Foundation, project number 21-77-30010 «System analysis of geophysical process dynamics in the Russian Arctic and their impact on the development and operation of the railway infrastructure».

Institutional Review Board Statement: Not applicable.

Informed Consent Statement: Not applicable.

Data Availability Statement: Not applicable.

Acknowledgments: This work employed data provided by the Shared Research Facility «Analytical Geomagnetic Data Center» of the Geophysical Center of RAS (http://ckp.gcras.ru/ accessed on 22 May 2023).

Conflicts of Interest: The authors declare no conflict of interest.

References

1. Vorobieva, I.; Gvishiani, A.; Dzeboev, B.; Dzeranov, B.; Barykina, Y.; Antipova, A. Nearest Neighbor Method for Discriminating Aftershocks and Duplicates when Merging Earthquake Catalogs. *Front. Earth Sci.* **2022**, *10*, 820277. [CrossRef]
2. Imanishi, K.; Ando, R.; Kuwahara, Y. Unusual shallow normal-faulting earthquake sequence in compressional northeast Japan activated after the 2011 off the Pacific coast of Tohoku earthquake. *Geophys. Res. Lett.* **2012**, *39*, L09306. [CrossRef]
3. Asano, Y.; Saito, T.; Ito, Y.; Shiomi, K.; Hirose, H.; Matsumoto, T.; Aoi, S.; Hori, S.; Sekiguchi, S. Spatial distribution and focal mechanisms of aftershocks of the 2011 off the Pacific coast of Tohoku Earthquake. *Earth Planets Space* **2011**, *63*, 669–673. [CrossRef]
4. Shebalin, P.N.; Baranov, S.V.; Dzeboev, B.A. The Law of the Repeatability of the Number of Aftershocks. *Dokl. Earth Sci.* **2018**, *481*, 963–966. [CrossRef]
5. Zaliapin, I.; Ben-Zion, Y. Earthquake clusters in southern California. I: Identification and stability. *J. Geophys. Res. Solid Earth* **2013**, *118*, 2847–2864. [CrossRef]
6. Zaliapin, I.; Ben-Zion, Y. A global classification and characterization of earthquake clusters. *Geophys. J. Int.* **2016**, *207*, 608–634. [CrossRef]
7. Gvishiani, A.D.; Vorobieva, I.A.; Shebalin, P.N.; Dzeboev, B.A.; Dzeranov, B.V.; Skorkina, A.A. Integrated Earthquake Catalog of the Eastern Sector of the Russian Arctic. *Appl. Sci.* **2022**, *12*, 5010. [CrossRef]
8. Wetmiller, R.J.; Forsyth, D.A. Seismicity of the Arctic, 1908–1975. Arctic Geophysical Review. *Publ. Earth Phys. Br.* **1978**, *45*, 15–24.
9. Burtman, V.S.; Kolodyazhnyi, S.Y. Fault systems in multiply deformed regions of Eurasia. *Russ. J. Earth Sci.* **2023**, *1*, 1–9. [CrossRef]
10. Kondorskaya, N.V.; Shebalin, N.V. (Eds.) *New Catalog of Strong Earthquakes in the USSR from Ancient Times through 1977*; Report SE-31; World Data Center A for Solid Earth Geophysics: Boulder, CO, USA, 1982; 608p.
11. Morozov, A.N.; Vaganova, N.V.; Konechnaya, Y.V. The October 14, 1908 MW 6.6 earthquake in the Barents and Kara Sea region of the Arctic: Relocation based on instrumental data. *Polar Sci.* **2019**, *20*, 160–166. [CrossRef]
12. Morozov, A.N.; Vaganova, N.V.; Asming, V.E.; Kremenetskaya, E.O. Relocation of Early Instrumental Earthquakes in the Arctic. *Seism. Instrum.* **2022**, *58*, 32–34. [CrossRef]
13. Morozov, A.N.; Vaganova, N.V.; Shakhova, E.V.; Konechnaya, Y.V.; Asming, V.E.; Antonovskaya, G.N.; Evtyugina, Z.A. Seismicity of the Arctic in the Early Twentieth Century: Relocation of the 1904–1920 Earthquakes. *Bull. Seismol. Soc. Am.* **2019**, *109*, 2000–2008. [CrossRef]
14. Morozov, A.N.; Vaganova, N.V.; Konechnaya, Y.V.; Asming, V.E.; Dulentsova, L.G.; Evtyugina, Z.A. Seismicity in the far Arctic areas: Severnaya Zemlya and the Taimyr Peninsula. *J. Seismol.* **2021**, *25*, 1171–1188. [CrossRef]
15. Rogozhin, E.A.; Kapustian, N.K.; Antonovskaya, G.N.; Konechnaya, Y.V. New seismicity map for the European sector of the Russian arctic region. *Geotectonics* **2016**, *50*, 238–243. [CrossRef]
16. Nikonov, A.A.; Shvarev, S.V. Holocene Tectonic Activity and Seismicity of the White Sea Basin. Abstract of the Paleoseismological Seminar of the IPE RAS on 16 December 2013. Available online: http://www.ifz.ru/fundamental/tektonicheskaja-aktivnost-belomorskogo-basseina/ (accessed on 1 June 2023). (In Russian).
17. Panasenko, G.D. Problems of seismic zoning of the Western sector of the Soviet Arctic. *Nat. Econ. North* **1986**, *14*, 4–6. (In Russian)
18. Avetisov, G.P. Tectonic factors of intraplate seismicity of the Western Arctic. Izvestiya. *Phys. Solid Earth* **1996**, *32*, 975–985.
19. Antonovskaya, G.N.; Kovalev, S.M.; Konechnaya, Y.V.; Smirnov, V.N.; Danilov, A.V. New information about the seismicity of the Russian Arctic based on the work of the seismic station "Severnaya Zemlya". *Arct. Antarct. Res.* **2018**, *64*, 170–181. (In Russian) [CrossRef]
20. Antonovskaya, G.N.; Kapustiana, N.K.; Konechnaya, Y.V.; Danilov, A.V. Registration capabilities of Russian island-based seismic stations: Case study of the Gakkel ridge monitoring. *Seism. Instrum.* **2020**, *56*, 33–45. [CrossRef]
21. Antonovskaya, G.; Morozov, A.; Vaganova, N.; Konechnaya, Y. Seismic monitoring of the European Arctic and adjoining regions. In *The Arctic. Current Issues and Challenges*; Nova Science Publishers, Inc.: New York, NY, USA, 2020; pp. 303–368.
22. Schweitzer, J.; Paulsen, B.; Antonovskaya, G.N.; Fedorov, A.V.; Konechnaya, Y.V.; Asming, V.E.; Pirli, M. A 24-Yr Long Seismic Bulletin for the European Arctic. *Seismol. Res. Lett.* **2021**, *92*, 2758–2767. [CrossRef]
23. Morozov, A.N.; Vaganova, N.V. The Travel Times of Regional P and S for Spreading Ridges in the European Arctic. *J. Volcanol. Seismol.* **2017**, *11*, 156–163. [CrossRef]
24. Morozov, A.N.; Vaganova, N.V.; Asming, V.E.; Peretokin, S.A.; Aleshin, I.M. Seismicity of the western sector of the Russian Arctic. *Phys. Solid Earth* **2023**, *2*, 115–148. (In Russian)
25. Shebalin, P.N.; Gvishiani, A.D.; Dzeboev, B.A.; Skorkina, A.A. Why Are New Approaches to Seismic Hazard Assessment Required? *Dokl. Earth Sci.* **2022**, *507*, 930–935. [CrossRef]
26. Giardini, D. The Global Seismic Hazard Assessment Program (GSHAP)—1992/1999. *Ann. Geofis.* **1999**, *42*, 957–974. [CrossRef]

27. Giardini, D.; Grunthal, G.; Shedlock, K.M.; Zhang, P. The GSHAP Global Seismic Hazard Map. *Ann. Geofis.* **1999**, *42*, 1225–1228. [CrossRef]
28. Ulomov, V.I. General seismic zoning of the territory of Russian Federation: GSZ-2012. *Seism. Instrum.* **2014**, *50*, 290–304. [CrossRef]
29. Ulomov, V.I. Seismic hazard of Northern Eurasia. *Ann. Geofis.* **1999**, *42*, 1023–1038. [CrossRef]
30. Ulomov, V.I.; Peretokin, S.A.; Medvedeva, N.S.; Akatova, K.N.; Danilova, T.I. Seismological aspects of general seismic zoning for the territory of the Russian Federation territory (OSR-97, OSR-2012, and OSR-2014 Maps). *Seism. Instrum.* **2015**, *51*, 311–328. [CrossRef]
31. Zavyalov, A.D.; Peretokin, S.A.; Danilova, T.I.; Medvedeva, N.S.; Akatova, K.N. General Seismic Zoning: From Maps GSZ-97 to GSZ-2016 and New-Generation Maps in the Parameters of Physical Characteristics. *Seism. Instr.* **2019**, *55*, 445–463. [CrossRef]
32. Malovichko, A.A.; Malovichko, D.A.; Golubeva, I.V.; Ivanova, Y.V. Natural and technogenic seismicity of the Ural region. *J. Min. Sci.* **2005**, *41*, 6–13. [CrossRef]
33. Adushkin, V.V. Tectonic earthquakes of anthropogenic origin. *Izv. Phys. Solid Earth* **2016**, *52*, 173–194. [CrossRef]
34. Petrov, V.A.; Minaev, V.A.; Ustinov, S.A.; Nafigin, I.O.; Lexin, A.B. Assessment of Seismogeodynamic Activity of Mining Areas on the Basis of 3D Geoinformation Modeling. *Russ. J. Earth Sci.* **2022**, *6*, 1–12. [CrossRef]
35. Manconi, A.; Picozzi, M.; Coviello, V.; De Santis, F.; Elia, L. Real-time detection, location, and characterization of rockslides using broadband regional seismic networks. *Geophys. Res. Lett.* **2016**, *43*, 6960–6967. [CrossRef]
36. Vorobieva, I.; Shebalin, P.; Narteau, C.; Beauducel, F.; Nercessian, A.; Clouard, V.; Bouin, M.-P. Multiscale mapping of completeness magnitude of earthquake catalogs. *Bull. Seism. Soc. Am.* **2013**, *103*, 2188–2202. [CrossRef]
37. Dzeboev, B.A.; Gvishiani, A.D.; Agayan, S.M.; Belov, I.O.; Karapetyan, J.K.; Dzeranov, B.V.; Barykina, Y.V. System-Analytical Method of Earthquake-Prone Areas Recognition. *Appl. Sci.* **2021**, *11*, 7972. [CrossRef]
38. Dzeboev, B.A.; Gvishiani, A.D.; Belov, I.O.; Agayan, S.M.; Tatarinov, V.N.; Barykina, Y.V. Strong Earthquake-Prone Areas Recognition Based on an Algorithm with a Single Pure Training Class: I. Altai-Sayan-Baikal Region, M $\geq$ 6.0. *Izv. Phys. Solid Earth* **2019**, *55*, 563–575. [CrossRef]
39. Dzeboev, B.A.; Soloviev, A.A.; Dzeranov, B.V.; Karapetyan, J.K.; Sergeeva, N.A. Strong earthquake-prone areas recognition based on the algorithm with a single pure training class. II. Caucasus, M $\geq$ 6.0. Variable EPA method. *Russ. J. Earth Sci.* **2019**, *19*, ES6005. [CrossRef]
40. Dzeboev, B.A.; Karapetyan, J.K.; Aronov, G.A.; Dzeranov, B.V.; Kudin, D.V.; Karapetyan, R.K.; Vavilin, E.V. FCAZ-recognition based on declustered earthquake catalogs. *Russ. J. Earth Sci.* **2020**, *20*, ES6010. [CrossRef]
41. Gorshkov, A.I.; Soloviev, A.A. Recognition of earthquake-prone areas in the Altai-Sayan-Baikal region based on the morphostructural zoning. *Russ. J. Earth Sci.* **2021**, *21*, ES1005. [CrossRef]
42. Gvishiani, A.D.; Dzeboev, B.A.; Agayan, S.M. FCAZm intelligent recognition system for locating areas prone to strong earthquakes in the Andean and Caucasian mountain belts. *Izv. Phys. Solid Earth* **2016**, *52*, 461–491. [CrossRef]
43. Gvishiani, A.D.; Soloviev, A.A.; Dzeboev, B.A. Problem of Recognition of Strong-Earthquake-Prone Areas: A State-of-the-Art Review. *Izv. Phys. Solid Earth* **2020**, *56*, 1–23. [CrossRef]
44. Kossobokov, V.G.; Soloviev, A.A. Pattern recognition in problems of seismic hazard assessment. *Chebyshevskii Sb.* **2018**, *19*, 55–90. (In Russian) [CrossRef]
45. Agayan, S.M.; Tatarinov, V.N.; Gvishiani, A.D.; Bogoutdinov, S.R.; Belov, I.O. FDPS algorithm in stability assessment of the Earth's crust structural tectonic blocks. *Russ. J. Earth Sci.* **2020**, *20*, ES6014. [CrossRef]
46. Di Giacomo, D.; Bondár, I.; Storchak, D.A.; Engdahl, E.R.; Bormann, P.; Harris, J. ISC-GEM: Global Instrumental Earthquake Catalogue (1900–2009), III. Re-computed MS and mb, proxy MW, final magnitude composition and completeness assessment. *Phys. Earth Planet. Inter.* **2015**, *239*, 33–47. [CrossRef]

applied sciences

MDPI

Article

Integrated Earthquake Catalog III: Gakkel Ridge, Knipovich Ridge, and Svalbard Archipelago

Inessa A. Vorobieva [1,2], Alexei D. Gvishiani [1,3], Peter N. Shebalin [1,2], Boris A. Dzeboev [1,*], Boris V. Dzeranov [1], Natalia A. Sergeeva [1], Ernest O. Kedrov [1] and Yuliya V. Barykina [1]

1 Geophysical Center of the Russian Academy of Sciences (GC RAS), 119296 Moscow, Russia; vorobiev@mitp.ru (I.A.V.); a.gvishiani@gcras.ru (A.D.G.); p.n.shebalin@gmail.com (P.N.S.); b.dzeranov@gcras.ru (B.V.D.); n.sergeyeva@gcras.ru (N.A.S.); e.kedrov@gcras.ru (E.O.K.); u.barykina@gcras.ru (Y.V.B.)
2 Institute of Earthquake Prediction Theory and Mathematical Geophysics of the Russian Academy of Sciences (IEPT RAS), 117997 Moscow, Russia
3 Schmidt Institute of Physics of the Earth of the Russian Academy of Sciences (IPE RAS), 119296 Moscow, Russia
* Correspondence: b.dzeboev@gcras.ru; Tel.: +7-495-930-05-46

Abstract: This paper represents the final part of a series of studies aimed at creating the most reliable and representative earthquake catalog covering the Russian and European Arctic. The earthquake catalog of the Gakkel and Knipovich ridges, as well as the Svalbard Archipelago with a unified magnitude scale, was formed based on the combination of four regional Russian catalogs and the ISC catalog. The merging of catalogs was carried out using the modification of the author's methodology, which allowed for the identification of records in different datasets related to the same seismic event. The modification was introduced due to significant changes over time in the source catalogs. The unified proxy moment magnitude scale was formed based on regression analysis of the different magnitude estimates provided by various agencies. The integrated catalog included 17,922 earthquakes that occurred during the period 1962–2022. Analysis of the integrated catalog showed that the level of registration in the studied area significantly varies over space and time. Before 1995, the catalog contained only strong and moderate earthquakes, and the magnitude of complete registration Mc was 5.0 in the Gakkel Ridge, 4.7 in the Knipovich Ridge, and 4.5 in the Svalbard Archipelago. The number of recorded events increased in the period 1995–2011, and Mc decreased to 4.0 in the Gakkel and Knipovich ridges and to 2.8 in the Svalbard Archipelago. The best level of registration in the Svalbard Archipelago and the Knipovich Ridge was achieved after 2012, when Mc reached 1.7 and 2.8, respectively. In the Gakkel Ridge, despite a noticeable increase in the number of reported events from 2012, the magnitude of complete registration did not improve and was 4.0. The presented integrated earthquake catalog is intended for a wide range of studies of the seismic regime of the Arctic.

Keywords: Arctic; Gakkel Ridge; Knipovich Ridge; Svalbard Archipelago; merging earthquake catalogs; magnitude unification; level of registration

Citation: Vorobieva, I.A.; Gvishiani, A.D.; Shebalin, P.N.; Dzeboev, B.A.; Dzeranov, B.V.; Sergeeva, N.A.; Kedrov, E.O.; Barykina, Y.V. Integrated Earthquake Catalog III: Gakkel Ridge, Knipovich Ridge, and Svalbard Archipelago. *Appl. Sci.* **2023**, *13*, 12422. https://doi.org/10.3390/app132212422

Academic Editor: Fabrizio Balsamo

Received: 22 September 2023
Revised: 7 November 2023
Accepted: 13 November 2023
Published: 16 November 2023

1. Introduction

The seismicity of the territory occupied by the Svalbard Archipelago and the Gakkel and Knipovich ridges within the Arctic Ocean is studied in the present paper. These ridges are a part of the Mid-Ocean Ridges (MOR) system. The Gakkel Ridge forms the divergent boundary between the North American and Eurasian lithosphere plates in the Arctic Ocean. It extends in a sub-parallel direction for 1800 km from the Laptev Sea continental margin, transitioning to the west through a complex combination of transform faults and rifts (Spitsbergen and Molloy fracture zones) into the Knipovich Ridge, the northernmost fragment of the Mid-Atlantic Ridge (Figure 1).

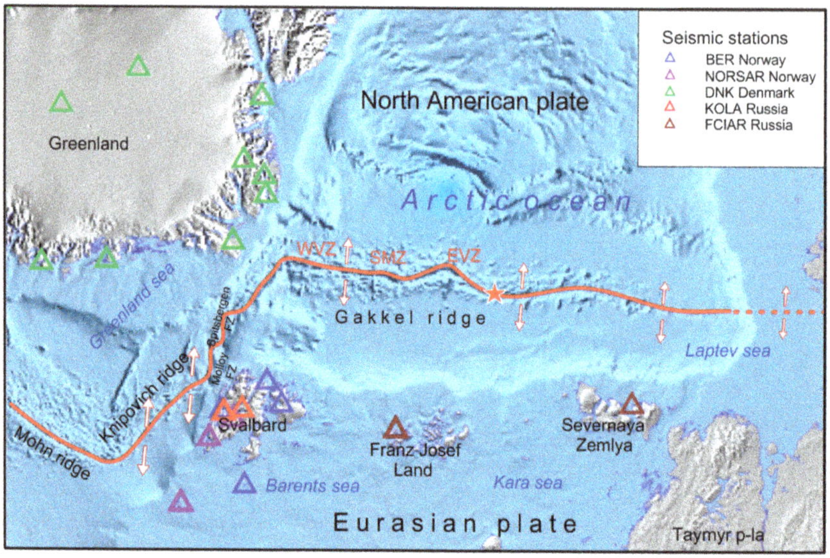

Figure 1. Scheme of the Arctic region and seismic networks. The red line is the boundary between North American and Eurasian plates. The red dotted line is the part of the boundary after the Gakkel ridge. WVZ and EVZ are the Eastern and the Western Volcanic Zones in the Gakkel ridge, and SMZ is the Sparsely Magmatic Zone [1,2]; a star marks the location of volcanic eruption and earthquake swarm in 1999 [3,4]. Arrows show the spreading direction. Triangles are seismic stations of European and Russian networks (see legend).

The Gakkel Ridge is the center of spreading for the Eurasian basin and belongs to the ultra-slow spreading rift zones. It is characterized by the lowest spreading rates among the entire MOR system. The maximum spreading rate in the orthogonal direction ranges from 6–7 mm/year in the east to 9–13 mm/year in the west [5,6]. The distance between volcanic areas along the Gakkel Ridge is 110–130 km. They are located perpendicularly to the ridge axis. It should be noted that the Gakkel Ridge is undisturbed by transform faults [7–9]. Similar characteristics of spreading rates and structure are also present in the eastern segment of the Southwest Indian Ridge. This is reflected in similar seismic activity features that distinguish these ridges from ridges with slow and medium spreading rates [10].

Based on the morphological features and deep structure, the Gakkel Ridge is divided into three provinces: the Western Volcanic Zone (WVZ) (7° W–3° E), the central Sparsely Magmatic Zone (SMZ) (3° E–30° E), and the Eastern Volcanic Zone (EVZ) (30° E–85° E) [1,2]. Despite its lower spreading rate, the WVZ is characterized by extensive magmatic processes, which are reflected in the similarity of its relief to that of the Mid-Atlantic Ridge rift zone. The maximum depth of the WVZ reaches 4.3 km. The length of volcanic ridges identified in its central part ranges from 15 to 50 km, with heights ranging from 0.4 to 1.4 km. The rift valley is composed of basalts, and several hydrothermal plumes have been identified within it [7,11,12]. In contrast, the SMZ is characterized by rare manifestations of magmatism, with a single large volcanic center located in the 19° E area. The maximum depth of the rift valley is 4.8–5.4 km [6,7,12].

Within the EVZ, extended amagmatic sections with gentle slopes and lengths of 40–120 km separated by volcanic centers can be distinguished. The maximum depth of the amagmatic sections is 4.8–5.1 km. The largest volcanic center of the Gakkel Ridge is located in the eastern part of the segment, in the 85° E area [6,7,12]. In 1999, an unprecedented surge in seismic activity was recorded there. It was caused by an effusive-explosive underwater volcanic eruption. The seismic network deployed on ice floes allowed for the registration of a swarm of 209 events with hypocenter depths up to 25 km [3,4]. Based on the data obtained in [13], a seismic tomography model was constructed. The combination of ultra-slow spreading rates with low potential mantle temperatures leads to the production of low-temperature, volatile-rich magmatic melts. Explosive underwater eruptions occur as a result of the degassing of these melts [13,14].

Maximum magnitudes are recorded in the eastern segment of the Gakkel Ridge (in the Laptev Sea). The strongest earthquake within the Gakkel Ridge, with $MS = 6.5$ (International Seismological Centre), occurred there on 25 August 1964 [15]. The epicenters of two earthquakes with $MW = 6.4$ and $MW = 6.2$ (Global Centroid-Moment-Tensor), which occurred on 21 March 1988 and 6 March 2005, respectively, were also located in this area [16].

The Svalbard Archipelago includes more than a thousand small and large islands, among which West Spitsbergen, Prince Charles Land, Barents Island, Edge, and Northeast Land can be distinguished. More than half of the archipelago's area is covered by glaciers. The first temporary seismic station was installed in 1911 in Longyearbyen. Subsequently, several analog stations were installed in Kapp Linné (1958–1963), Ny-Ålesund (1967), Hornsund (1978), Barentsburg (1979), and Pyramiden (1982–1989). In addition, several large-scale seismic surveys were conducted on Phipps Island, Southern Spitsbergen, and Northeast Land in 1976–1986 to study the tectonic structure of the region [17,18].

It should be noted that, until the 1970s, numerous researchers believed that the most seismically active zone in the Svalbard region was the area of the Mohn and Knipovich ridges. The territory of the archipelago itself was considered to be seismically quiet. However, on 18 January 1976, an earthquake with $M = 5.7$ occurred on the eastern coast of West Spitsbergen in the Storfjorden, which caused a revision of these views. On 21 February 2008, the strongest earthquake for the archipelago, with $Mw = 6.1$, was recorded in the same area [18,19].

Over the past few decades, the seismic network of the Svalbard Archipelago has continuously expanded and developed. Today, it consists of stationary stations equipped with digital seismometers and serves as a backbone network for cryoseismological research in the region. The seismic stations of NORSAR, GSN, GEOFON, and IGF PAS operate on the territory of the archipelago. In 2010, two seismic stations were installed in Barentsburg by the Kola Science Centre of the Russian Academy of Sciences, in collaboration with NORSAR [17,18].

The Knipovich Ridge stretches for 550 km along the continental margin of the Svalbard Archipelago from 73°45′ to 78°35′ N. It is a section of mid-ocean ridges with an ultra-slow spreading rate of 15–17 mm/year. Like the Gakkel Ridge, it is not disrupted by transform faults [20,21]. Within the rift valley, magmatic and amagmatic segments can be distinguished. Magmatic segments are represented by volcanic ridges oriented perpendicularly to the direction of extension. Amagmatic segments are expressed as deep depressions on the rift valley floor located parallel to its axis [9].

Earthquakes with magnitudes of $MW^{ISCGEM} = 6.54$ and $MW^{GCMT} = 6.1$ that occurred on 1 June 1915 and 9 September 1992, respectively, were the strongest events in the area of the Knipovich Ridge [15,16]. The epicenters of three earthquakes with magnitudes of $MW^{GCMT} = 5.4$, which occurred on 3 February 2000, 21 June 2009, and 22 June 2009, were also confined to the Knipovich Ridge [16].

At the same time, the strongest earthquake in the considered region was an event with a magnitude of $MW^{GCMT} = 6.7$, which occurred on 20.07.1992 in the Molloy transform fault. Also, the epicenters of four earthquakes with a magnitude of $MW^{GCMT} \geq 6.0$ were confined to the Molloy transform fault and Spitsbergen fracture zones area [16].

Until recently, earthquakes on the Gakkel and Knipovich ridges were only recorded at teleseismic distances with thresholds ranging from $M \geq 4.5$ to $M \geq 3.5$ for different segments. Information on weak seismicity is valuable for studying tectonic and volcanic processes occurring in rift zones. In this regard, it should be noted that the installation of stationary seismic stations of the Arkhangelsk seismic network on the Arctic archipelagos of Franz Josef Land and Severnaya Zemlya in 2012 allowed for the registering of low-magnitude events within the Gakkel Ridge at closer distances [22,23]. It should be also noted that stations located in Svalbard (BERGEN, NORSAR, and KOLA) and Greenland (DNK) contribute to the registration of low-magnitude events within the Knipovich Ridge.

The earthquake catalog of the Gakkel and Knipovich ridges is a fundamental basis for studying seismicity, regional segmentation, and seismotectonics of the poorly studied boundary between the Eurasian and North American plates. It should be noted that, in [24], the earthquake catalog was created, covering the area north of 72° N for the period 1955–1999. Using regional bathymetry and potential fields, a seismicity model was obtained and a segmentation of the interplate boundary was proposed.

The present paper aims to create a unified integrated earthquake catalog of the Gakkel and Knipovich ridges, as well as the Svalbard Archipelago, merging all available data from Russian and international agencies. The author's methodology of intelligent merging (with the identification of formed duplicate events and their separation from aftershocks [25]) of earthquake catalogs from different networks and seismological agencies is described in detail in [26]. The unified integrated earthquake catalogs that were created using this methodology for the eastern and western sectors of the Russian Arctic zone are available to the public at: http://www.wdcb.ru/arctic_antarctic/arctic_seism.html (accessed on 1 August 2023), and a description of the assembly process is given in [27,28].

Therefore, this article, along with the results by the authors in [26–28], completes the creation of the most reliable integrated unified earthquake catalogs that cover the whole Russian and European Arctic.

2. Materials and Methods

The region considered in the present paper (Figure 2) covers zones that were not included in the Eastern (catalog I) and Western (catalog II) sectors of the Russian Arctic. The territory is directly adjacent to the Eastern and Western sectors, but does not intersect with them. The schematic coverage map for catalogs I, II, and III is presented in (Figure S1, see Supplementary).

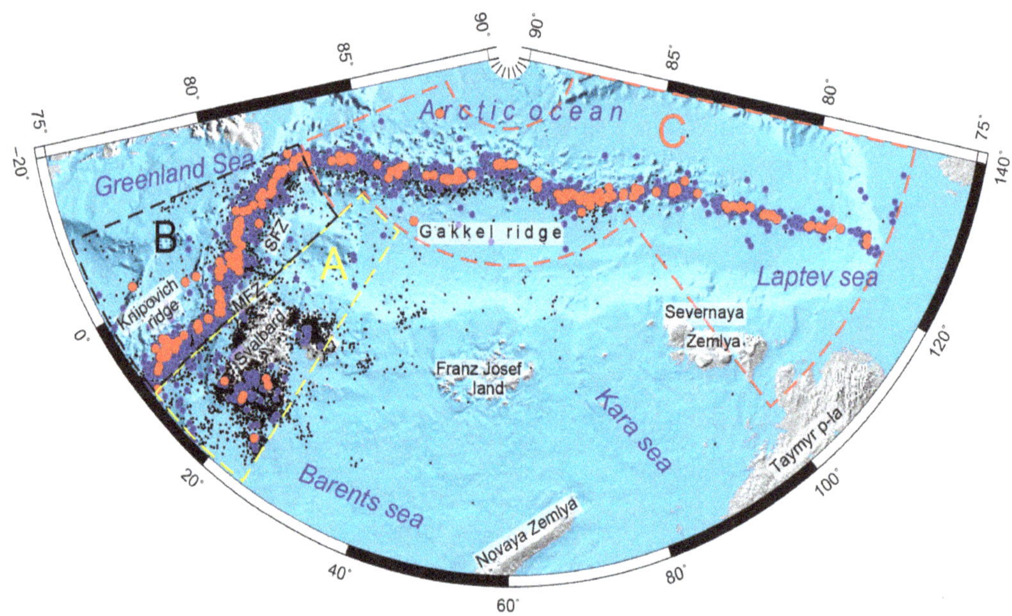

Figure 2. Map of the studied region with earthquake epicenters from the ISC catalog: events with mb^{ISC} and mb^{NEIC} magnitude determinations are marked in blue, and events with Mw^{GCMT} determinations are marked in red. MFZ and SFZ are Molloy and Spitsbergen fracture zones. Dashed lines delineate sub-regions Svalbard (**A**), Knipovich (**B**), and Gakkel (**C**).

The set of initial data was compiled from five earthquake catalogs (Table 1):

1. The Arctic catalog from the annual journals Earthquakes in the USSR 1962–1991, Earthquakes in Northern Eurasia 1992–2017, and Earthquakes in Russia 2018–2021 (hereinafter ARC);
2. The catalog of the FCIAR network (Arkhangelsk network) 2008–2017 from the annual journals of Earthquakes in Northern Eurasia (hereinafter ARKH);
3. The catalog of the Svalbard Archipelago territory for 2010–2021 from the annual journals of Earthquakes in Russia (hereinafter SHB);
4. The ISC 1962–2022 catalog, which is a composite and contains data from many world and also Russian agencies (Table S1, see Supplementary);
5. The catalog Seismicity of the western sector of the Russian Arctic for 1962–2020 [29] (hereinafter Morozov). The Morozov catalog was recently presented in [29]. In this catalog, earthquakes are relocated based on the analysis and merging of all available seismic bulletins from Russian and European seismic networks using modern velocity models. The Morozov catalog covers the shelf zone of the Western Sector of the Russian Arctic, which we included in our previous study [28], but some earthquakes were relocated [29] from the shelf to the Gakkel Ridge. We include these events in our catalog, since we consider determinations [29] to be the most reliable.

According to [30,31], a large number of small seismic events registered in the Svalbard archipelago are ice-quakes. Such events are not presented in Russian catalogs. The ISC catalog includes 16 non-earthquake events (explosions and glacial events), all of them in Svalbard. One of these events is presented in the ARC catalog and another in the ARKH catalog. These events were excluded from the source catalogs. No additional checks were performed after merging the catalogs.

According to [30,31], a large number of seismic events registered in the Svalbard Archipelago with magnitudes ranging from 1 to 2.2 are ice-quakes. This is confirmed, among other things, by a seasonal periodicity that is not typical for earthquakes. However, in the ISC catalog, only 6 events were marked as glacial events (ice-quakes). Either the ice-quakes were cleared out by the compilers, or most of the events were still earthquakes. More than 90% of the events had a significant depth, which is not typical for ice-quakes, so we considered them earthquakes and included them in the integrated catalog (Figure S1, see Supplementary).

Events with unknown magnitude/class were excluded from consideration. Earthquakes were selected within the boundaries of the studied region (Figure S2, see Supplementary).

Table 1. Input catalogs.

Catalog	Period	Number of Events	Number of Earthquakes with Energy Classes and/or Magnitudes	Number of Non-Earthquakes
ARC	1965–2021	2404	2403	1 *
ARKH	2008–2017	1493	1492	1 *
SHB	2010–2021	2634	2634	0
ISC	1962–2022	16,953	16,937	16
Morozov	1962–2020	4 **	4	0

* According to ISC data. ** Four events with known magnitudes from an additional catalog were included. They were relocated outside the studied area considered in [29], but were within the area studied in the present paper.

The methodology for identifying duplicates when merging catalogs is described in detail in [26]. It was successfully applied to create catalogs of the eastern and western sectors of the Russian Arctic [27,28]. A basic three-parameter model (1) showed efficiency for assessing the proximity between earthquakes. It takes into account the time difference DT, and the epicenter difference in the longitude DX and the latitude DY. After that, the classification of earthquakes into unique and duplicates is carried out using the threshold value of the metric *Ro*

$$Ro = \sqrt{\frac{DT^2}{\sigma_T{}^2} + \frac{DX^2}{\sigma_X{}^2} + \frac{DY^2}{\sigma_Y{}^2}}, \tag{1}$$

where σ_T, σ_X , σ_Y are the standard deviations of time, longitude, and latitude differences between the nearest events from two source catalogs.

3. Results

3.1. Merging Catalogs

As shown in (Table S1, see Supplementary), in the studied region, a significant majority of events are registered by European seismic networks, the data of which are collected in the composite ISC catalog. The number of earthquake records in the ISC catalog is approximately three times greater than the total number of events in the Russian catalogs. Moreover, the data on earthquakes presented in [29] was specified based on the analysis and combination of all available seismic bulletins of Russian and European seismic networks using modern velocity models of the environment. Thus, in the task of merging earthquake catalogs, the following priority sources of earthquake data were chosen (Table 1, Figure 3):

1. Earthquakes from the Morozov catalog (4 events);
2. Earthquakes from the ISC catalog (16,937 events);
3. Earthquakes from catalogs of ARC (2404 events), SBH (2404 events), and ARKH (1493), with preference given to data from the ARC catalog in overlapping areas.

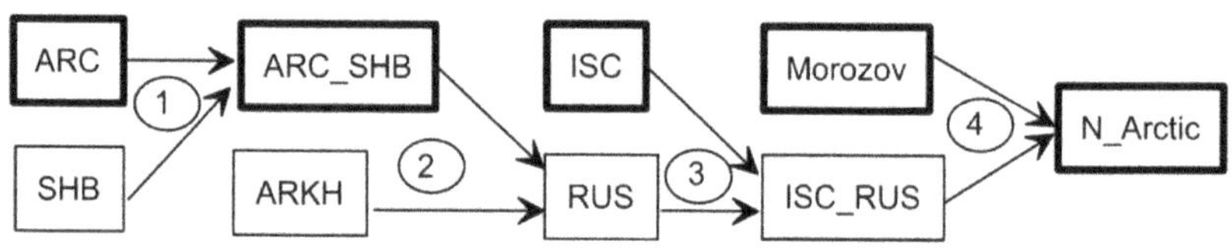

Figure 3. Assembly diagram of the integrated catalog of the studied region. At each stage, the main catalog is marked with a bold frame. The stage numbers are given in ellipses and correspond to Table 2.

Table 2. Scheme and compilation parameters of the integrated catalog.

Stage	Main Catalog	Additional Catalog	Metric Parameters σ_T min, σ_X km, σ_Y km	Threshold Value of the Metric	Estimation of the Number of Errors	Number of Duplicates	Merged Catalog
1	ARC 2403 events	SHB 2634 events	0.054; 22.5; 21.3	12	0.7%	502	ARC_SHB 4535 events
2	ARC_SHB 4535 events	ARKH 1492 events	0.048; 23.9; 22.8	14	0.3%	1136	RUS 4891 events
3	ISC 16,937 events	RUS 4891 events	0.05; 28.3; 24.7	17.5	0.9%	3906	ISC_RUS 17,922 events
4	Morozov 4 events	ISC_RUS 17,922 events				4	N_ARCTIC0 17,922

Before the merging process, each of the source catalogs (Table 1) was checked for internal duplicates. Statistical analysis did not reveal any anomalous groups of close events (Figure S3, see Supplementary).

At each stage, the merging of catalogs is performed in two steps. First, the metric parameters (1) are determined. For this, the metric (1) *Ro* between the nearest events from two source catalogs with standard parameters $\sigma_T = 0.05$ min, $\sigma_X = \sigma_Y = 15$ km is calculated. The threshold value of the metric $Ro = 10$ is used for the preliminary duplicate identification, which corresponds to the time and space difference of 0.5 min/150 km.

Standard deviations σ_T, σ_X , and, σ_Y of the variables *DT*, *DX*, and *DY* are calculated for the pre-identified duplicates (Figures 4, 6 and 8). In the next step, a threshold value of the metric is determined and the final duplicate identification takes place (Figures 5, 7 and 9).

The catalogs for the studied region consist of a mixture of data from a large number of agencies. This applies not only to the ISC catalog but also to the ARC catalog, which combines data from several regional networks KOLA, FCIAR, and the GS RAS teleseismic network. In addition, catalogs significantly change over time: the diversity of sources (contributing agencies) and the number of events significantly increase starting from the mid-1990s and then in 2010.

For this reason, metric parameters σ_T, σ_X , and, σ_Y change over time (Figure 8), and heavy tails appear in the distributions of *DT*, *DX*, and *DY*. In this situation, using the methodology applied in [27,28] to determine the threshold value of the metric leads to an increased probability of missing duplicates (Figure S4, see Supplementary). Therefore, in this paper, it was decided not to use a multivariate normal distribution model. Instead, the sampling distribution (without modeling by the normal distribution, as in the original method) of the metric for the nearest events from the two merged catalogs is used.

We assume that the maximum value of the metric for events that can be duplicates is $Ro = 30$, which corresponds to a time and space difference of about 1.5 min and 600 km. We construct the distribution of such events *F*dub. The red lines in Figures 5b, 7b and 9b represent the value of $1 - F$dub, which we consider as the probability of missing a duplicate (error of the first kind). The probability of a false duplicate (error of the second kind) is estimated in the same way as in [27,28]. For this, we calculate the values of the metric (1) *Ro* between events within an additional catalog. The blue lines in Figures 5b, 7b and 9b represent the proportion of events with a proximity less than a given value of *Ro*. The black lines show the estimate of the total probability of the first and the second kind errors. The threshold value of the metric minimizes the total number of errors. Figures 5c, 7c, and 9c show the distribution of normalized times and distances for the nearest events from the merged catalogs. The metric contour lines correspond to the chosen threshold value, which provides the optimal separation of duplicates and unique events.

The numerical parameters of the catalog merging are given in Table 2. The integrated catalog contains 17,922 events.

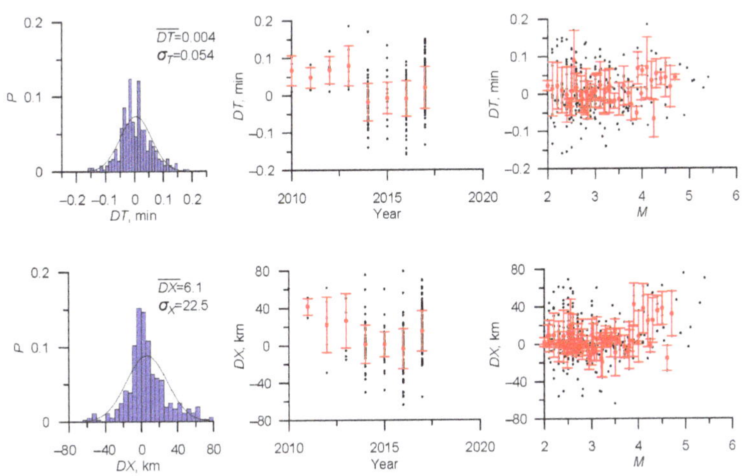

Figure 4. *Cont.*

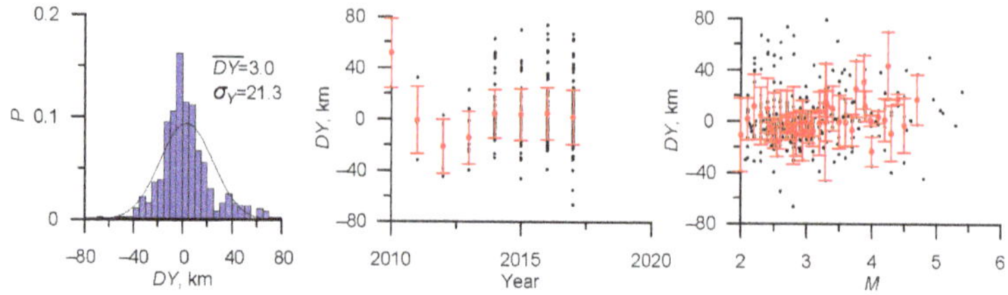

Figure 4. Stage 1. Determination of numerical parameters of the metric (1) for merging the ARC and SHB catalogs. Distributions of variables *DT*, *DX*, *DY*, and the dependence of dispersion and mean values on time and magnitude of events.

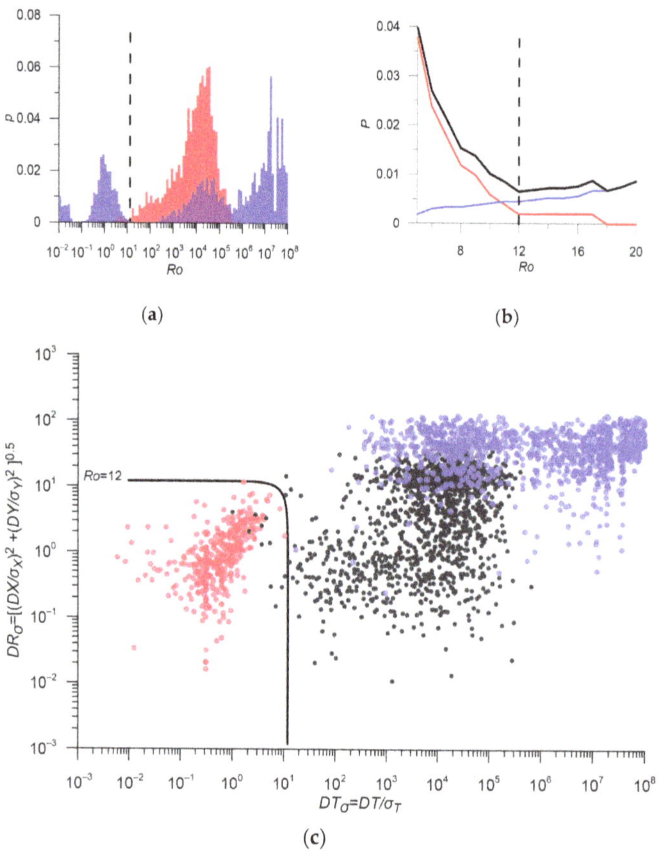

Figure 5. Stage 1: Merge of catalogs ARC and SHB. (**a**) Comparison of the metric distribution ARC/SHB pairs (blue histogram) and the same metric for SHB/SHB earthquakes (red histogram); (**b**) Threshold optimization: the red line shows the probability of missing a duplicate, the blue line shows the probability of a false duplicate, and the black line shows the total probability of the first and the second kind errors. The threshold value $Ro = 12$, which minimizes the total number of errors, approximately 0.7%, is shown by a dashed line; (**c**) Distribution of normalized *DT* and *DR*, and the contour line of metric (1). Colored dots represent ARC/SHB pairs (pink dots are duplicates, blue dots are unique events), and black dots are distances between SHB/SHB events in metric (1). The values of the metric for earthquakes in the SHB catalog are significantly larger than for ARC/SHB duplicates (pink dots). The contour line of the metric (1) $Ro = 12$ provides close to optimal separation of duplicates and naturally grouped events. Absolute duplicates are not shown.

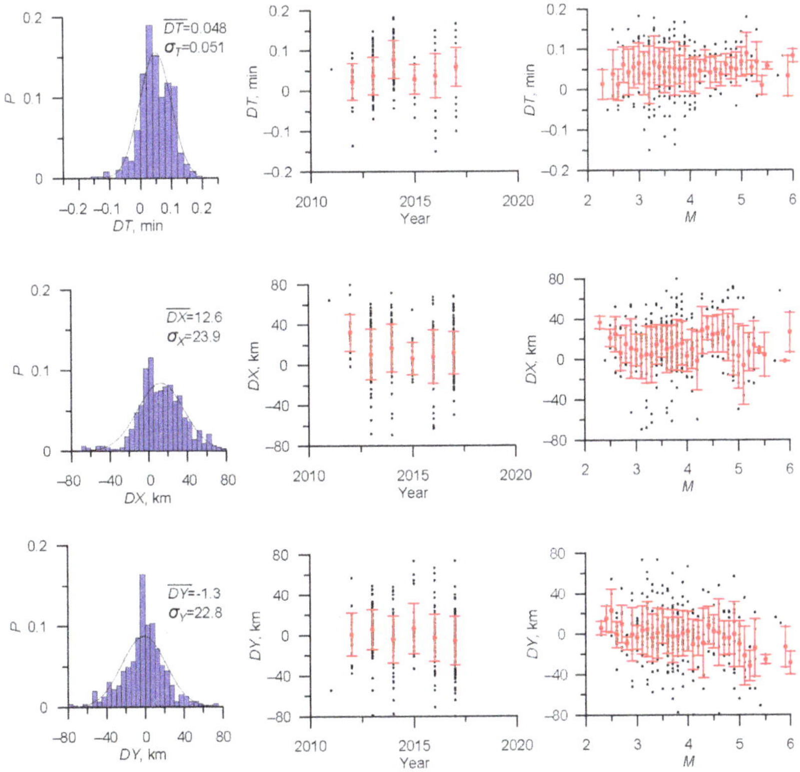

Figure 6. Determination of numerical parameters of the metric (1) for merging the ARC_SHB and ARKH catalogs. Distributions of variables *DT*, *DX*, *DY*, and the dependence of dispersion and mean values on time and magnitude of events.

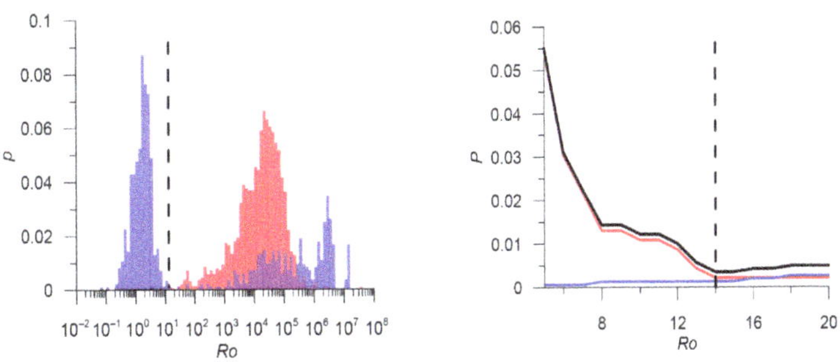

(**a**) (**b**)

Figure 7. *Cont.*

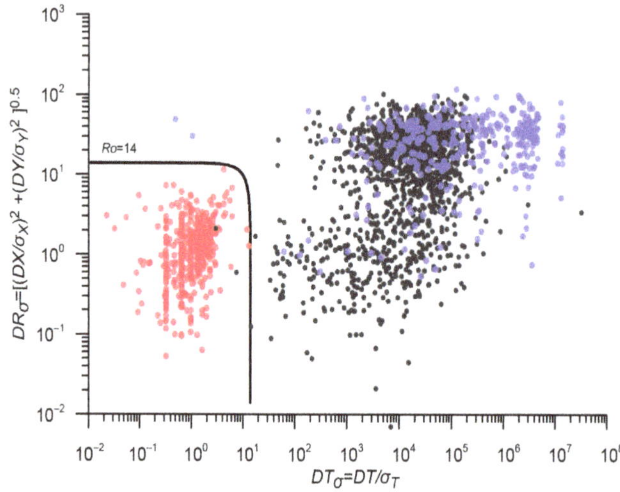

(c)

Figure 7. Stage 2: Merge of catalogs ARC_SHB and ARKH. (**a**) Comparison of the metric distribution ARC_SHB/ARKH pairs (blue histogram) and the same metric for ARKH/ARKH earthquakes (red histogram); (**b**) Threshold optimization: the red line shows the probability of missing a duplicate, the blue line shows the probability of a false duplicate, and the black line shows the total probability of the first and the second kind errors. The threshold value $Ro = 12$, which minimizes the total number of errors, approximately 0.7%, is shown by a dashed line; (**c**) Distribution of normalized DT and DR, and the contour line of metric (1). Colored dots represent ARC_SHB/ARKH pairs (pink dots are duplicates, blue dots are unique events), and black dots are distances between ARKH/ARKH events in metric (1). The values of the metric for earthquakes in the ARKH catalog are significantly larger than for ARC_SHB/ARKH duplicates (pink dots). The contour line of the metric (1) $Ro = 12$ provides close to optimal separation of duplicates and naturally grouped events. Absolute duplicates are not shown.

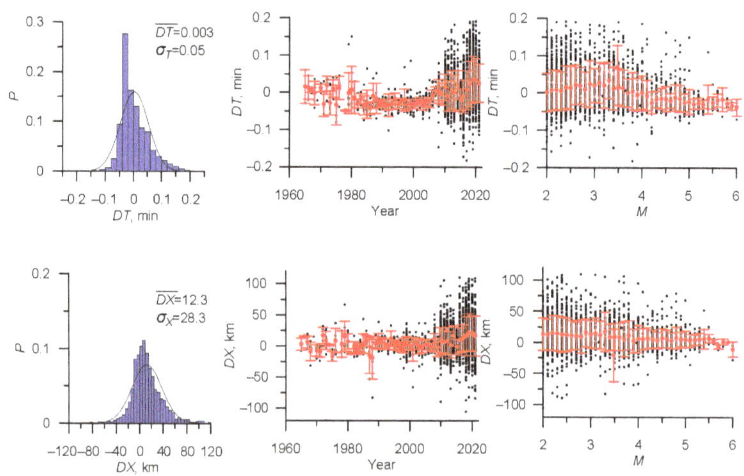

Figure 8. *Cont.*

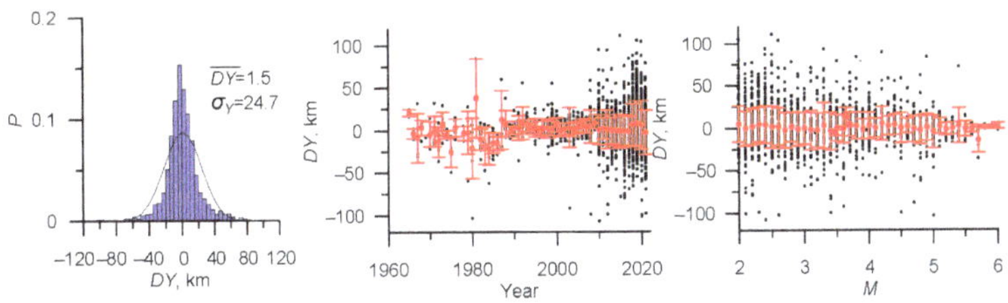

Figure 8. Stage 3. Determination of numerical parameters of the metric (1) for merging the ISC and RUS catalogs. Distributions of variables *DT, DX, DY*, and the dependence of dispersion and mean values on time and magnitude of events.

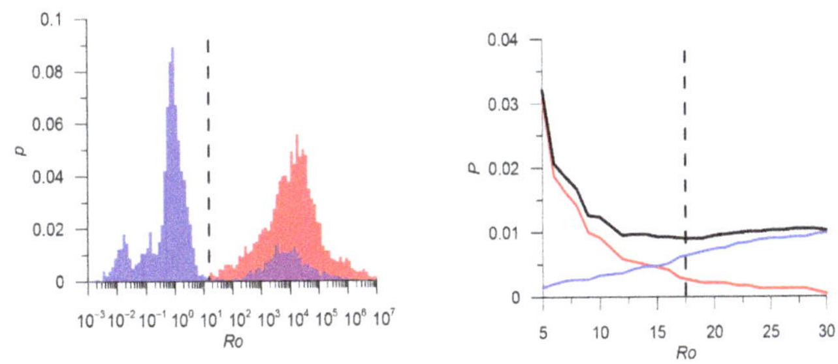

(**a**) (**b**)

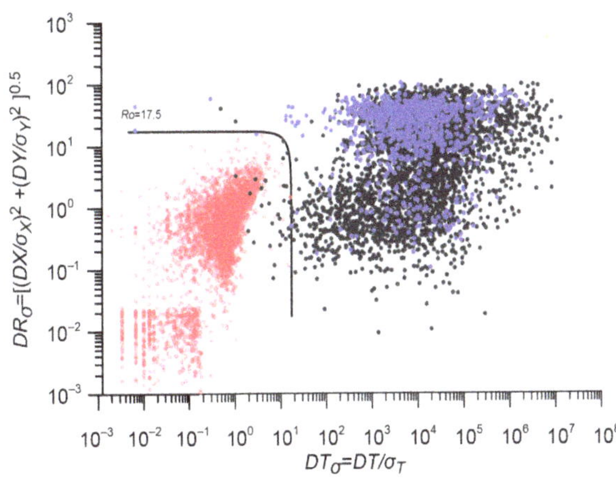

(**c**)

Figure 9. Stage 3: Merge of catalogs ISC and RUS. (**a**) Comparison of the metric distribution ISC/RUS pairs (blue histogram) and the same metric for RUS/RUS earthquakes (red histogram); (**b**) Threshold optimization: the red line shows the probability of missing a duplicate, the blue line shows the probability of a false duplicate, and the black line shows the total probability of the first and the second kind errors. The threshold value $Ro = 12$, which minimizes the total number of errors, approximately 0.7%, is shown by a dashed line; (**c**) Distribution of normalized *DT* and *DR*, and the contour line of metric (1). Colored dots represent ISC/RUS pairs (pink dots are duplicates, blue dots are unique events), and black dots are distances between RUS/RUS events in metric (1). The values of the metric for earthquakes in the RUS catalog are significantly larger than for ISC/RUS duplicates (pink dots). The contour line of the metric (1) $Ro = 12$ provides close to optimal separation of duplicates and naturally grouped events. Absolute duplicates are not shown.

3.2. Magnitude Unification in the Integrated Earthquake Catalog

During the second stage of creating the integrated earthquake catalog, the magnitude unification of events was carried out. The integrated catalog contains multiple definitions of the magnitude of different types from various agencies. Different magnitudes are presented in different periods and different parts of the considered region. A preliminary study showed that some magnitude estimates significantly changed over time: ML^{NAO} magnitude changed in 2009, MD^{BER} magnitude changed in 2010, and ML^{DNK} magnitude changed in 2015 (Figure 10).

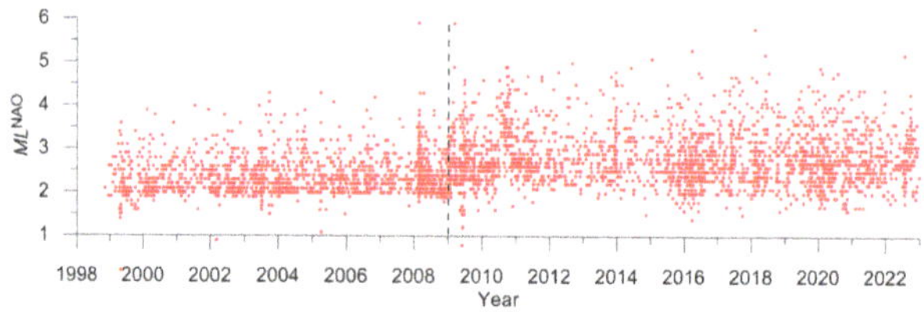

(**a**)

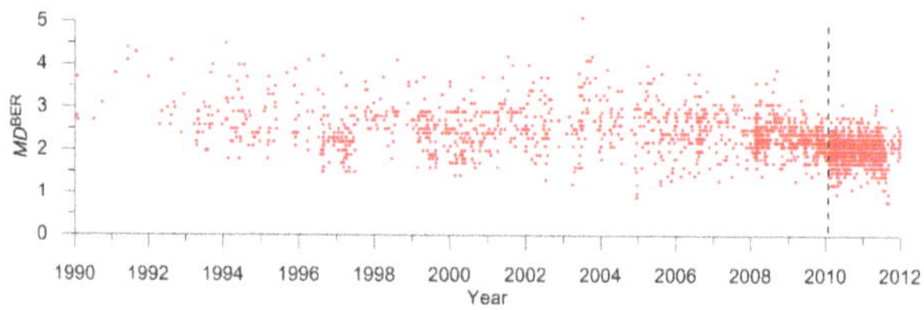

(**b**)

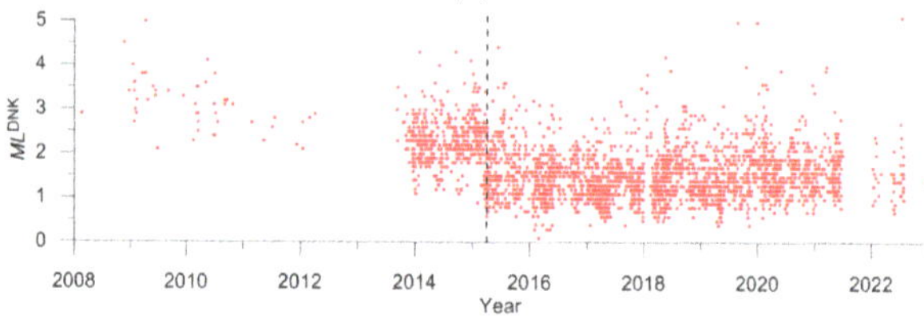

(**c**)

Figure 10. Change in magnitude estimates in time: (**a**) ML^{NAO}; (**b**) MD^{BER}; (**c**) ML^{DNK}.

Moreover, we assumed that the magnitude ratio may vary in different seismically active zones. We identified three sub-regions: Svalbard, Knipovich Ridge, and Gakkel Ridge (Figure 2). Magnitude unification was independently performed in these three sub-regions.

As the reference scale, we chose the magnitude based on the seismic moment determined by GCMT. We analyzed correlation ratios between various magnitudes and proposed conversion formulas for obtaining "proxy-*MW*" estimates. We used a simple shift-type ratio $M = m + a$ for almost all magnitudes, except for surface wave magnitudes MS^{ISC}, MS^{MOS}, and MLH^{GSR}, for which linear ratios were used.

3.2.1. Svalbard

The catalog of the Svalbard sub-region contains 6921 events. The magnitude MW^{GCMT} is determined for only 6 events, while the magnitudes mb^{ISC}, mb^{NEIC} are known for 153 events. The magnitude $mb^{ISC} \approx MW^{GCMT}$ (Figure 11a). This corresponds to the practice of ISC that uses *mb* as a proxy-*MW* for earthquakes with $M < 5.0$ [32]. This significantly increases statistics and expands the magnitude range for converting other magnitudes to proxy-*MW*. Most events (5927) have a local magnitude ML^{BER}. In total, about 88% of events have MW^{GCMT}, $mb^{ISC,\,NEIC}$, and ML^{BER} magnitudes. We consider the ratios between these magnitudes as basic (Figure 11). For the remaining magnitudes, we built ratios with

mb^{ISC} and ML^{BER}. Regressions with mb^{ISC} and ML^{BER} for most magnitudes are very close (Figures 12 and 13). This confirms the hypothesis of linearity in the ratio between different estimates over a wide range of magnitudes. The exception is the MD^{BER} magnitude in 2010–2011, for which magnitude estimates for 34 events, where other determinations are absent, were considered unreliable. For 12 events, the ratios were poorly determined due to the small number of events with both magnitude determinations (Figure 14). For 16 events, the ratios were not determined. In total, the unified magnitude was poorly determined for less than 1% of events. Statistics and formulas for converting magnitudes in the Svalbard sub-region are presented in Table 3.

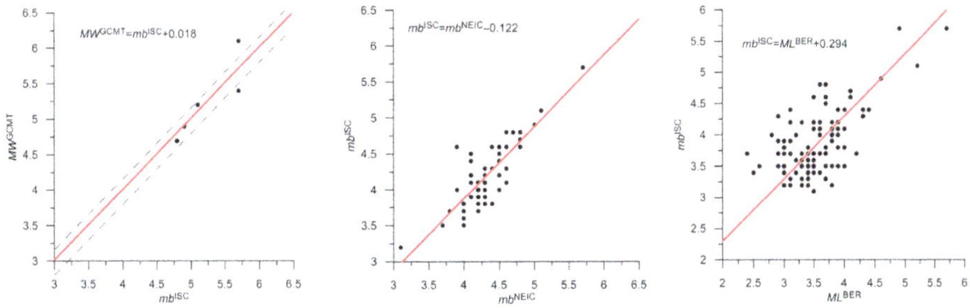

(**a**) (**b**) (**c**)

Figure 11. Basic shift-type correlation ratios for magnitudes MW^{GCMT}, $mb^{ISC,\,NEIC}$, and ML^{BER} in the sub-region of Svalbard. Dashed lines in (**a**) show 95% confidence interval, in (**b**) and (**c**), 95% confidence intervals are not shown since they are less than 0.1.

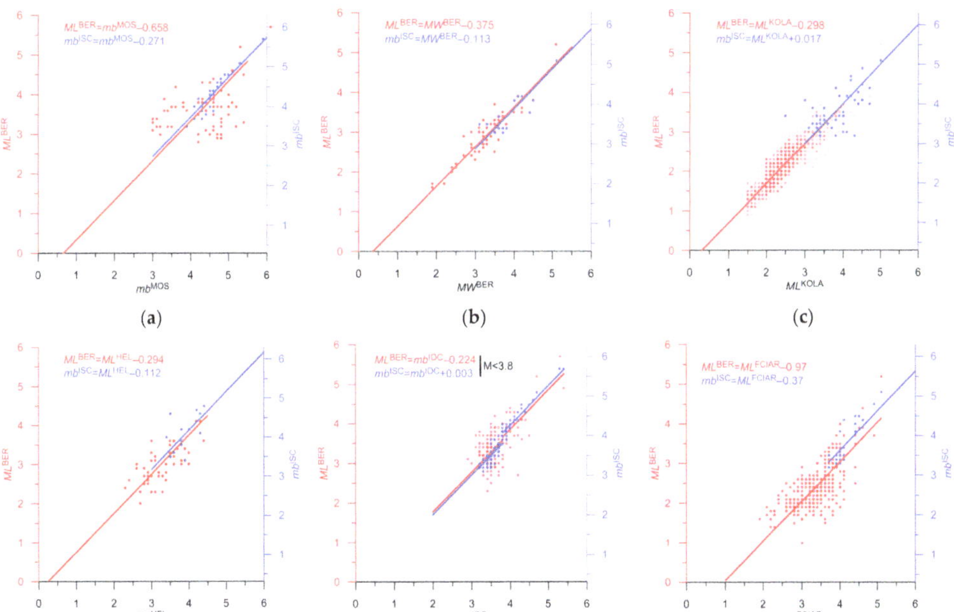

(**a**) (**b**) (**c**)

(**d**) (**e**) (**f**)

Figure 12. Shift-type correlation ratios for different magnitudes with mb^{ISC} (blue), and ML^{BER} (red) in the sub-region of Svalbard. Y-axis for mb^{ISC} is shifted to 0.3 in accordance with the ratio between mb^{ISC} and ML^{BER} in Svalbard (Table 4 and Figure 11). Dots are observations, lines are the best shift-type fits. The 95% confidence intervals are not shown since they are less than 0.1.

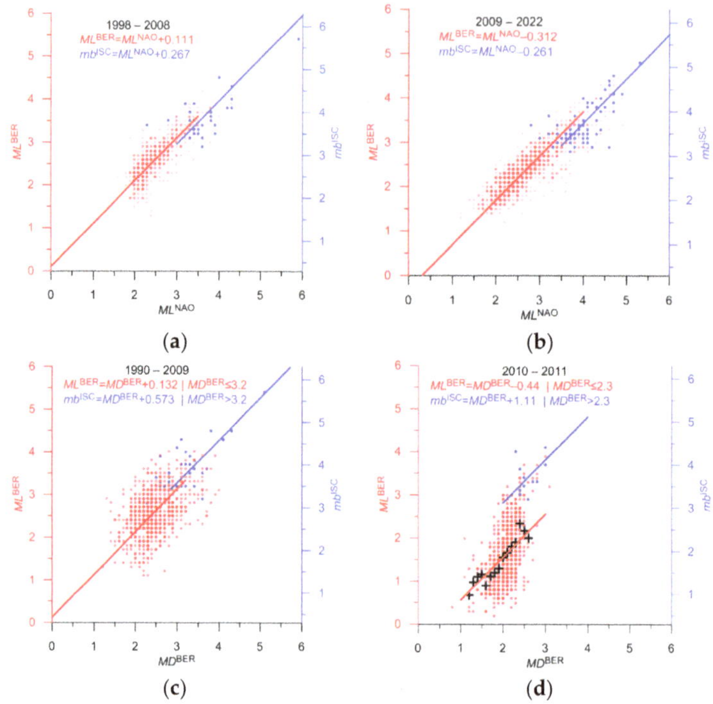

Figure 13. Shift-type correlation ratios for ML^{NAO} and MD^{BER} magnitudes with mb^{ISC} (blue), and ML^{BER} (red) in the sub-region of Svalbard in the different periods. Black crosses in (**d**) are the population mean of ML^{BER}. The Y-axis for mb^{ISC} is shifted to 0.3 in accordance with the ratio between mb^{ISC} and ML^{BER}. Dots are observations, lines are the best shift-type fits. The 95% confidence intervals are not shown since they are less than 0.1.

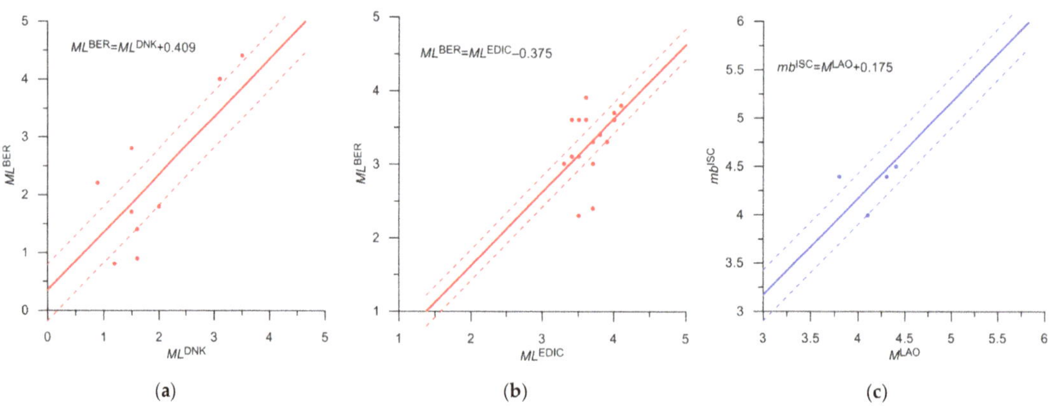

Figure 14. Shift-type correlation ratios for different magnitudes in the sub-region of Svalbard. Correlations are unreliable due to a small amount of data. Correlations with mb^{ISC} and ML^{BER} are shown in blue and red, respectively. Dots are observations, lines are the best shift-type fits. Dashed lines show a 95% confidence interval.

Table 3. Magnitude in the integrated catalog: Svalbard.

Agency	Type of Magnitude	Priority	Number of Events	Magnitude in the Integrated Catalog	Correlation	Figure	Mmin—Mmax. Initial Magnitude Scale	Note
GCMT	*MW*	1	6	$M = MW^{GCMT}$			4.7–6.1	
ISC	*mb*	2	130	$M = mb^{ISC}$	0.78	11a	3.1–5.6	
NEIC, NEIS	*mb*	2	17	$M = mb^{NEIC} - 0.1$	0.75	11b	3.3–4.9	
BER	*ML*	3	5927	$M = ML^{BER} + 0.3$	0.39	11c	0.3–3.9	
KOLA	*ML*	4	330	$M = ML^{KOLA}$	0.79	12c	1.4–3.4	
NAO	*ML*	4	250	$M = ML^{NAO} + 0.4$	0.65	13a	0.9–3.2	1998–2008
NAO	*ML*	4	47	$M = ML^{NAO}$	0.53	13b	1.2–3.5	2009–2022
FCIAR	*ML*	4	46	$M = ML^{FCIAR} - 0.7$	0.47	12f	1.8–3.8	
BER	*MD*	4	92	$M = MD^{BER} + 0.4$	0.43	13c	0.9–3.2	1990–2009
BER	*MD*	5	34	$M = MD^{BER} - 0.1$	0.25	13d	0.8–2.3	2010–2011 unreliable, Non-linear relation
MOS	*mb*	4	2	$M = mb^{MOS} - 0.3$	0.89	12a	4.9–5.8	
BER	*Mw*	4	7	$M = Mw^{BER} - 0.1$	0.85	12b	3.1–3.9	
HEL	*ML*	4	4	$M = ML^{HEL}$	0.53	12d	2.3–3.2	
IDC	*mb*	4	1	$M = mb^{IDC}$	0.78	12e	3.5	
DNK	*ML*	5	3	$M = ML^{DNK} + 0.7$	0.66	14a	1.8–2.0	Poorly determined
EDIC	*ML*	5	7	$M = ML^{EIDC} - 0.1$	0.20	14b	2.8–4.0	Poorly determined
LAO	*M*	5	2	$M = M^{LAO} + 0.2$	0.07	14c	3.4–3.8	Poorly determined
NAO	*mb*	5	1	$M = mb^{NAO}$			4	Not Determined
HEL	*MD*	5	5	$M = MD^{HEL}$			3.2–4.0	Not Determined
BER	*mb(Pn)*	5	4	$M = mb(Pn)^{BER}$			2.5–2.8	Not Determined
WAR	*M*	5	5	$M = M^{WAR}$			2.1–3.0	Not Determined
NUR	*M*	5	1	$M = M^{NUR}$			3.9	Not Determined
Total			6921					

3.2.2. Knipovich and Gakkel Ridges

The Gakkel and Knipovich ridges are structures of a similar tectonic type, known as "mid-ocean ridges". Preliminary analysis of the ratio of different magnitudes, for which there are sufficient statistics in both ridges, shows that most of the ratios are very similar. Figure 15a,b show the ratios for MW^{GCMT} and mb^{ISC}, mb^{NEIC}. In many cases, we construct ratios based on combined data from the Gakkel and Knipovich ridges. The ratios were found to be different for ML^{FCIAR}. For some magnitude types in the Gakkel Ridge, there are insufficient statistics, and the ratios from the Knipovich Ridge were used, which we considered unreliable. Detailed statistics are separately presented for the Knipovich and Gakkel ridges.

The catalog for the Knipovich Ridge sub-region contains 8912 events. The MW^{GCMT} magnitude is determined for 112 events, while magnitudes mb^{ISC} and mb^{NEIC} are known for 885 events. Magnitude $MW^{GCMT} \approx mb^{ISC} + 0.2$ (Figure 15a). A large number of events have magnitudes of ML^{BER} (2974), Mw^{BER} (1892), $mb(Pn)^{BER}$ (1521), and $M = ML^{DNK}$ (2602), which together make up over 92% of events. The ratios between MW^{GCMT}, $mb^{ISC,\,NEIC}$, and ML^{BER} magnitudes are considered basic (Figure 15). For other magnitudes, the ratios are constructed with MW^{GCMT}, mb^{ISC}, and ML^{BER} when there are sufficient statistics. As in the Svalbard region, regressions with MW^{GCMT}, mb^{ISC}, and ML^{BER} for most magnitudes are very close (Figures 16 and 17). This confirms the hypothesis of linearity in the ratio between different estimates over a wide range of magnitudes. The exception is the magnitude

MD^{BER} in 2011–2012, where estimates for 20 events with no other determinations were considered unreliable. Ratios for 16 events were poorly determined due to small statistics (Figure 18). Ratios for 11 events were not determined. In total, the unified magnitude was poorly determined for less than 1% of events. Statistics and formulas for converting magnitudes in the Knipovich Ridge sub-region are presented in Table 4.

Table 4. Magnitude in the integrated catalog: Knipovich Ridge.

Agency	Type of Magnitude	Priority	Number of Events	Magnitude in the Integrated Catalog	Correlation	Figure	Mmin—Mmax. Initial Magnitude Scale	Note
GCMT	*MW*	1	112	$M = MW^{GCMT}$		–	4.6–6.7	
ISC	*mb*	2	805	$M = mb^{ISC} + 0.2$	0.67	15a	2.9–6.3	Gakkel and Knipovich
NEIC, NEIS	*mb*	2	80	$M = mb^{NEIC} + 0.1$	0.66	15b	3.3–4.9	Gakkel and Knipovich
BER	*ML*	3	2974	$M = ML^{BER} + 1.2$	0.54	15c	0.3–3.7	Knipovich
BER	*Mw*	4	1892	$M = Mw^{BER} - 0.1$	0.72	16a	1.3–5.0	Gakkel and Knipovich
BER	*mb(Pn)*	4	1521	$M = mb(Pn)^{BER} - 0.1$	0.82	16b	1.7–4.6	Gakkel and Knipovich
DNK	*ML*	4	83	$M = ML^{DNK} + 0.6$	0.79	16c	0.9–3.1	2008–2015.2 Knipovich
DNK	*ML*	4	753	$M = ML^{DNK} + 1.3$	0.57	16d	0.1–3.1	2015.3–2022 Knipovich
KOLA	*ML*	4	82	$M = ML^{KOLA} + 0.5$	0.65	16e	1.3–3.4	Knipovich
NAO	*ML*	4	265	$M = ML^{NAO} + 1.1$	0.53	16f	0.2–3.4	1990–2008 Gakkel and Knipovich
NAO	*ML*	4	32	$M = ML^{NAO} + 0.4$	0.69	16g	1.7–4.2	2009–2022 Gakkel and Knipovich
FCIAR	*ML*	4	102	$M = ML^{FCIAR}$	0.42	16h	2.3–4.1	Knipovich
BER	*MD*	5	96	$M = MD^{BER} + 1.1$	0.62	17a	1.5–4.4	1990–2009 Knipovich
BER	*MD*	5	20	$M = MD^{BER} + 0.8 \mid MD < 2.7$	0.23	17b	1.5–2.5	2010–2011 Knipovich, unreliable
IDC	*mb*	4	21	$M = mb^{IDC} + 0.2$	0.85	17c	3.0–3.6	Gakkel and Knipovich
EIDC	*mb*	4	11	$M = mb^{EIDC} + 0.3$	0.76	17d	2.9–3.7	Gakkel and Knipovich
USCGS	*mb*	4	4	$M = mb^{USCGS} + 0.2$	0.83	18a	4.2–4.6	Gakkel and Knipovich
MOS	*MS*	4	1	$M = 0.515MS^{ISC} + 2.88$	0.71	18d	4.7	Gakkel and Knipovich
NAO	*mb*	4	3	$M = mb^{NAO} + 0.5$	0.62	18b	3.6–4.1	Knipovich
HFS	*mb*	4	1	$M = mb^{NAO} + 0.5$	0.50	18c	3.9	Knipovich
OTT	*ML*	4	27	$M = ML^{OTT} + 0.1$	0.23	18e	3.2–5.0	Gakkel and Knipovich
EIDC	*ML*	5	7	$M = ML^{EIDC} + 0.2$	0.05	18h	3.4–4.2	Poorly determined
BER	*Mc*	5	3	$M = Mc^{BER} + 0.4$	0.03	18f	1.7–3.2	Poorly determined
DNK	*Mc*	5	6	$M = Mc^{BER} - 0.5$	0	18g	2.1–3.7	Poorly determined
CGS	*M*	5	5	$M = M^{CGS}$			4.2–4.7	Not Determined
DNK	*MD*	5	1	$M = MD^{BER}$			2.4	Not Determined
PAL	*M*	5	2	$M = M^{PAL}$			4.3	Not Determined
STU	*M*	5	1	$M = M^{STU}$			5.2	Not Determined
OTT	*Mn*	5	1	$M = Mn^{OTT}$			3.1	Not Determined
WAR	*M*	5	1	$M = M^{WAR}$			2.6	Not Determined
Total			8912					

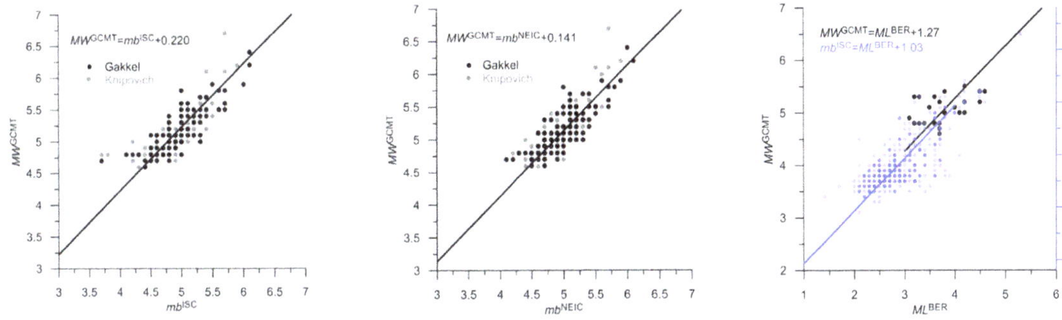

(a) (b) (c)

Figure 15. Basic shift-type correlation ratios for magnitudes MW^{GCMT}, $mb^{ISC,\ NEIC}$, and ML^{BER} in the sub-region of Knipovich Ridge. Ratios (**a**,**b**) are constructed using Knipovich (gray dots) and Gakkel (black dots) ridges data. In (**c**), correlations of ML^{BER} with mb^{ISC} (blue) and MW^{GCMT} (black) are very similar. The Y-axis for mb^{ISC} is shifted to 0.2 in accordance with the ratio between mb^{ISC} and MW^{GCMT} shown in (**a**). Dots are observations, lines are the best shift-type fits. The 95% confidence intervals are not shown since they are less than 0.1.

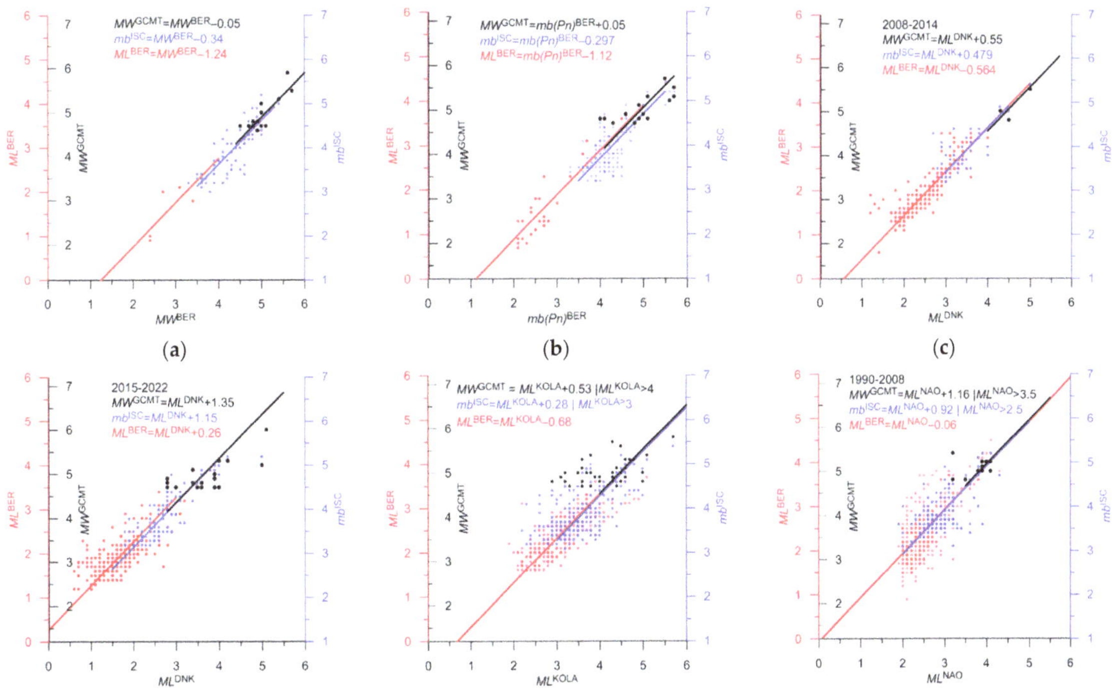

(a) (b) (c) (d) (e) (f)

Figure 16. *Cont.*

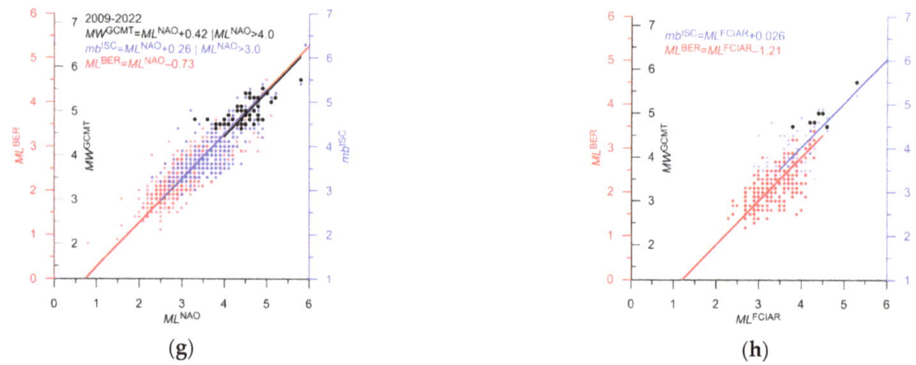

Figure 16. Correlation ratios for different magnitudes with MW^{GCMT} (black), mb^{ISC} (blue), and ML^{BER} (red) in the sub-region of Knipovich Ridge. The Y-axes for MW^{GCMT}, mb^{ISC} and ML^{BER} are shifted relative to each other in accordance with the basic ratio between the magnitudes shown in Figure 15. Dots are observations, lines are the best shift-type fits. The 95% confidence intervals are not shown since they are less than 0.1.

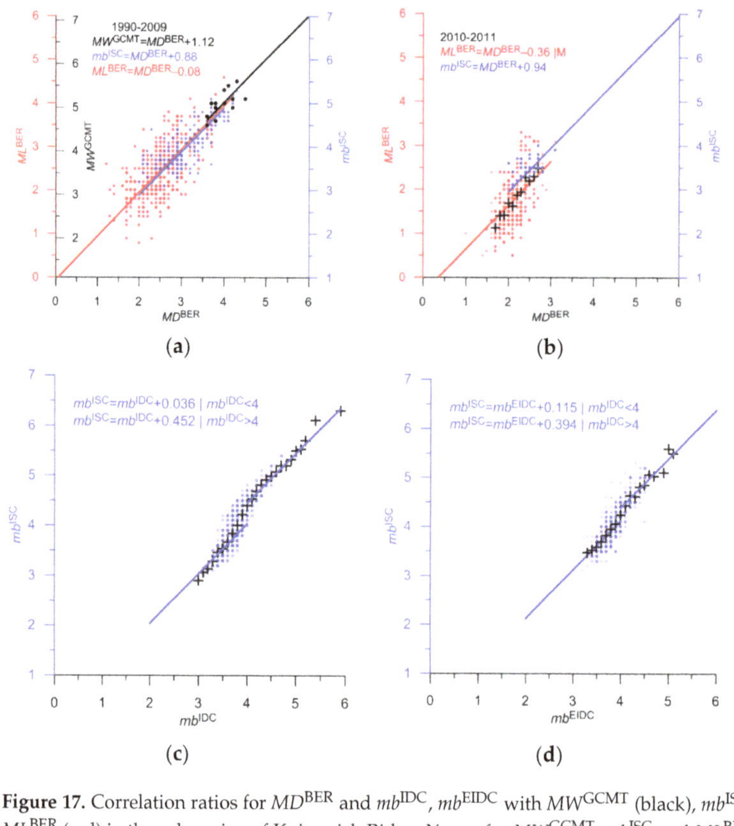

Figure 17. Correlation ratios for MD^{BER} and mb^{IDC}, mb^{EIDC} with MW^{GCMT} (black), mb^{ISC} (blue), and ML^{BER} (red) in the sub-region of Knipovich Ridge. Y-axes for MW^{GCMT}, mb^{ISC} and ML^{BER} are shifted relative to each other in accordance with the basic ratio between the magnitudes shown in Figure 15. Black crosses in (**b–d**) are the population mean of ML^{BER} and mb^{ISC}. Ratios for mb^{IDC} mb^{EIDC} are constructed using the Knipovich and Gakkel ridges data. Dots are observations, lines are the best shift-type fits. The 95% confidence intervals are not shown since they are less than 0.1.

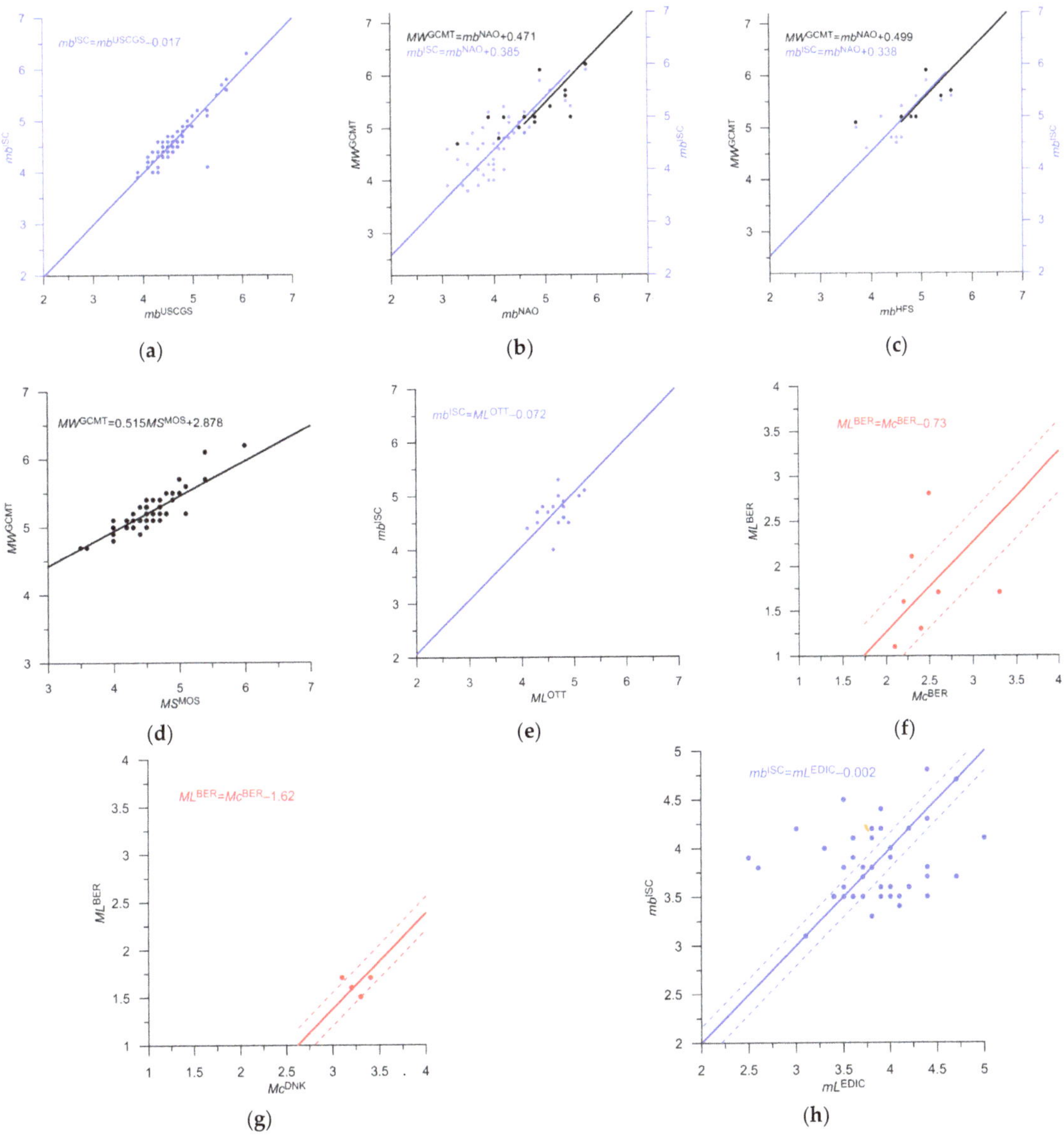

Figure 18. Correlation ratios for various magnitudes with MW^{GCMT} (black), $mb^{ISC, NEIC}$ (blue), and ML^{BER} (red) in the sub-region of Knipovich Ridge. Y-axes in (**b**,**c**) for MW^{GCMT}, mb^{ISC} are shifted relative to each other in accordance with the basic ratio between the magnitudes shown in Figure 15. All ratios are shift type, excluding MS^{MOS} (**d**), where we use linear relation. Dots are observations, lines are the best fits. Ratios are constructed using the Knipovich and Gakkel ridges data. Ratios (**f**–**h**) are poorly determined, and dashed lines show 95% confidence intervals.

The catalog of the Gakkel Ridge sub-region contains 2089 events. The magnitude MW^{GCMT} is determined for 138 events, while the magnitudes mb^{ISC} and mb^{NEIC} are known for 1053 events. A large number of earthquakes (561) have a ML^{FCIAR} magnitude, accounting, together with MW^{GCMT}, mb^{ISC}, and mb^{NEIC}, for almost 84% of the events. The magnitude $MW^{GCMT} \approx mb^{ISC} + 0.2$ (Figure 15a). The magnitude ratios, only using events from the Gakkel Ridge, are shown in Figure 19. We constructed ratios with MW^{GCMT} and mb^{ISC}. The

ML^{FCIAR} magnitude was not used due to insufficient statistics (for most events with ML^{FCIAR}, it is the only determined magnitude). For local magnitudes ML^{BER}, ML^{DNK}, and ML^{KOLA} (a total of 45 events), ratios obtained for the Knipovich Ridge events were used, but we considered them unreliable. For 21 events, the ratios were poorly determined due to the small statistics (Figures 18e and 19h). For seven events, the ratios were not determined. In total, the unified magnitude was poorly determined for 3.5% of the events. Statistics and formulas for converting magnitudes in the Gakkel Ridge sub-region are given in Table 5.

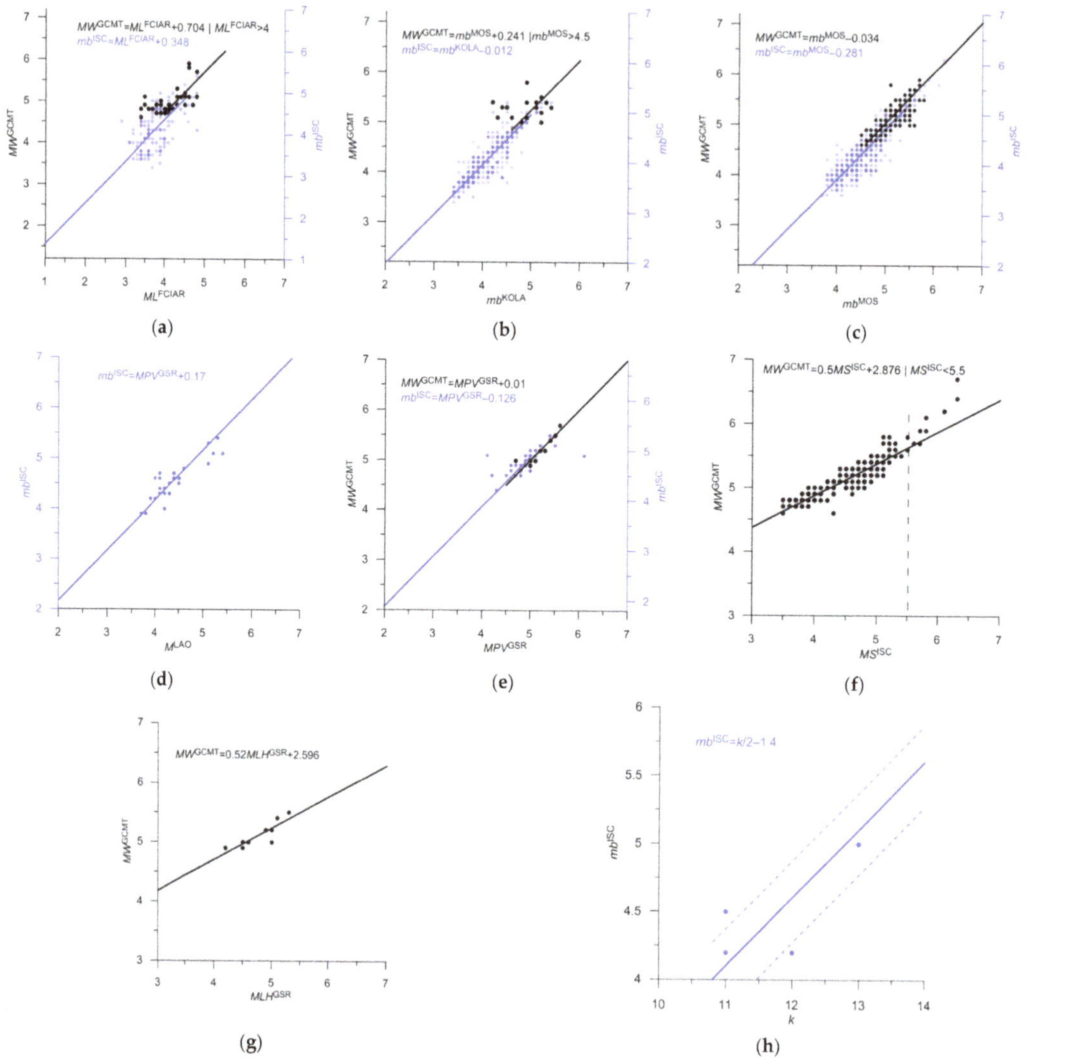

Figure 19. Correlation ratios for various magnitudes with MW^{GCMT} (black), and mb^{ISC} (blue) in the sub-region of Gakkel Ridge. Y-axes for MW^{GCMT}, mb^{ISC} in (**a–c**,**e**) are shifted relative to each other in accordance with the basic ratio between the magnitudes shown in Figure 15. All ratios are shift type, excluding MS^{ISC} (**f**) and MLH^{GSR} (**g**), where we use linear relation. Dots are observations, lines are the best fits. The ratio (**h**) is poorly determined, and dashed lines show 95% confidence intervals. Intervals are not shown in (**a–g**) since they are less than 0.1.

Table 5. Magnitude in the integrated catalog: Gakkel Ridge.

Agency	Type of Magnitude	Priority	Number of Events	Magnitude in the Integrated Catalog	Correlation	Figure	Mmin—Mmax. Initial Magnitude Scale	Note
GCMT	*MW*	1	138	$M = MW^{GCMT}$		–	4.6–6.4	
ISC	*mb*	2	979	$M = mb^{ISC} + 0.2$	0.67	15a	3.0–6.3	Gakkel and Knipovich
NEIC, NEIS	*mb*	2	74	$M = mb^{NEIC} + 0.1$	0.66	15b	3.2–4.8	Gakkel and Knipovich
FCIAR	*ML*	3	561	$M = ML^{FCIAR} + 0.6$	0.37	19a	2.0–4.3	Gakkel
KOLA	*mb*	4	32	$M = mb^{KOLA} + 0.2$	0.74	19b	3.1–4.2	Gakkel
MOS	*mb*	4	1	$M = mb^{MOS}$	0.82	19c	5.3	Gakkel
USCGS	*mb*	4	2	$M = mb^{USCGS} + 0.2$	0.83	18a	3.8–4.0	Gakkel and Knipovich
BER	*Mw*	4	23	$M = Mw^{BER} - 0.1$	0.72	16a	3.1–4.1	Gakkel and Knipovich
BER	*mb(Pn)*	4	30	$M = mb(Pn)^{BER} - 0.1$	0.82	16b	2.6–4.0	Gakkel and Knipovich
BER	*ML*	5	3	$M = ML^{BER} + 1.2$	0.54	15c	1.9–2.3	Knipovich, unreliable
DNK	*ML*	5	3	$M = ML^{DNK} + 0.6$	0.79	16c	1.7–2.3	2008–2015.2, Knipovich, unreliable
DNK	*ML*	5	15	$M = ML^{DNK} + 1.3$	0.57	16d	0.6–2.2	2015.3–2022, Knipovich, unreliable
KOLA	*ML*	5	24	$M = ML^{KOLA} + 0.5$	0.65	16e	1.3–3.5	Knipovich, unreliable
NAO	*ML*	4	3	$M = ML^{NAO} + 1.1$	0.53	16f	2.4–2.8	1990–2008, Gakkel and Knipovich
LAO	*M*	4	5	$M = M^{LAO} + 0.4$	0.79	19d	3.7–4.5	Gakkel
IDC	*mb*	4	141	$M = mb^{IDC} + 0.2$	0.85	17c	3.2–4.3	Gakkel and Knipovich
EIDC	*mb*	4	19	$M = mb^{EIDC} + 0.3$	0.76	17d	2.7–3.9	Gakkel and Knipovich
GSR	*MPV*	4	2	$M = MPV^{GSR}$	0.44	19e	4.4–5.0	Gakkel
ISC	*MS*	4	1	$M = 0.5MS^{ISC} + 2.88$	0.85	19f	5.3	Gakkel
GSR	*MLH*	4	4	$M = 0.52MLH^{GSR} + 2.6$	0.75	19g	3.0–3.4	Gakkel
OTT	*ML*	5	5	$M = ML^{OTT} + 0.1$	0.23	18e	3.2–4.2	Gakkel and Knipovich Poorly determined
GSR	*k*	5	16	$M = k/2 - 1.2$	0.51	19h	9–11	Gakkel Poorly determined
CSEM	*ML*	5	2	$M = ML^{CSEM}$		–	2.2–2.6	Not Determined
CGS	*M*	5	2	$M = M^{CGS}$		–	4.2–4.6	Not Determined
MSCGS	*M*	5	1	$M = M^{MSCGS}$		–	4.3	Not Determined
PAL	*M*	5	2	$M = M^{PAL}$		–	4.0–4.8	Not Determined
STU	*M*	5	1	$M = M^{STU}$		–	4.8	Not Determined
Total			2089					

3.3. Statistics of the Integrated Catalog for Three Sub-Regions

Figure 20 shows the distribution of earthquake epicenters from the created integrated catalog. The catalog contains 17,922 events, 989 events are from Russian catalogs, and the rest are from ISC. In the sub-regions of Svalbard and Knipovich Ridge, the addition of Russian data to ISC is insignificant. However, in the Gakkel Ridge, Russian data accounts for more than a quarter of the events, and more than half since 2012. Detailed statistics are provided in Table 6.

Table 6. Statistics of the integrated catalog.

Time Period, Catalog *	N Total	N from ISC	N from GS RAS, Morozov	Mc	N, $M \geq Mc$	Mmax
1962–2022						
N_Arctic	17,922	16,933 (94.2%)	989 (5.8%)	-		6.7
Svalbard	6921	6617 (95.6%)	304 (4.4%)	-		6.1
Knipovich Ridge	8912	8794 (98.7%)	118 (1.3%)	-		6.7
Gakkel Ridge	2089	1522 (72.9%)	567 (27.1%)	-		6.5
1962–1994						
N_Arctic	703	683 (97.2%)	20 (2.8%)	5.0	181	6.7
Svalbard	94	94 (100%)	0 (0%)	4.5	20	5.6
Knipovich Ridge	329	329 (100%)	0 (0%)	4.7	166	6.7
Gakkel Ridge	280	260 (92.9%)	20 (7.1%)	5.0	102	6.5
1995–2011						
N_Arctic	4377	4261 (97.3%)	116 (2.7%)	4.0	762	6.5
Svalbard	2209	2103 (95.2%)	105 (4.8%)	2.8	696	6.1
Knipovich Ridge	1408	1405 (99.8%)	3 (0.2%)	4.0	275	6.5
Gakkel Ridge	760	752 (98.9%)	8 (1.1%)	4.0	454	6.2
2012–2022						
N_Arctic	12,842	11,989 (95.4%)	853 (6.6%)	4.0	657	6.0
Svalbard	4618	4419 (95.7%)	199 (4.3%)	1.7	2351	5.3
Knipovich Ridge	7175	7060 (98.4%)	115 (1.6%)	2.8	3447	6.0
Gakkel Ridge	1049	510 (48.6%)	539 (51.4%)	4.0*	388	5.9

* Excluding the most eastern segment, where $Mc = 4.5$.

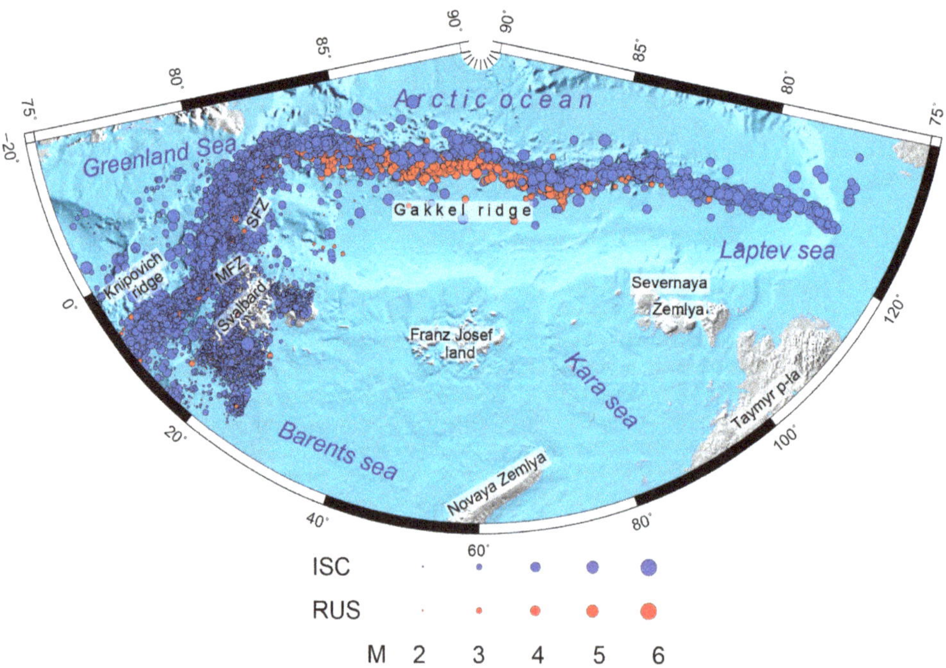

Figure 20. Map of earthquake epicenters $M \geq 2.0$ of the integrated catalog. Blue dots show events from the ISC catalog, and red dots show events from Russian catalogs.

Figures 21–23 show event distributions over time and magnitude, as well as differential magnitude-frequency graphs for various time periods for the sub-regions of Svalbard, Knipovich Ridge, and Gakkel Ridge.

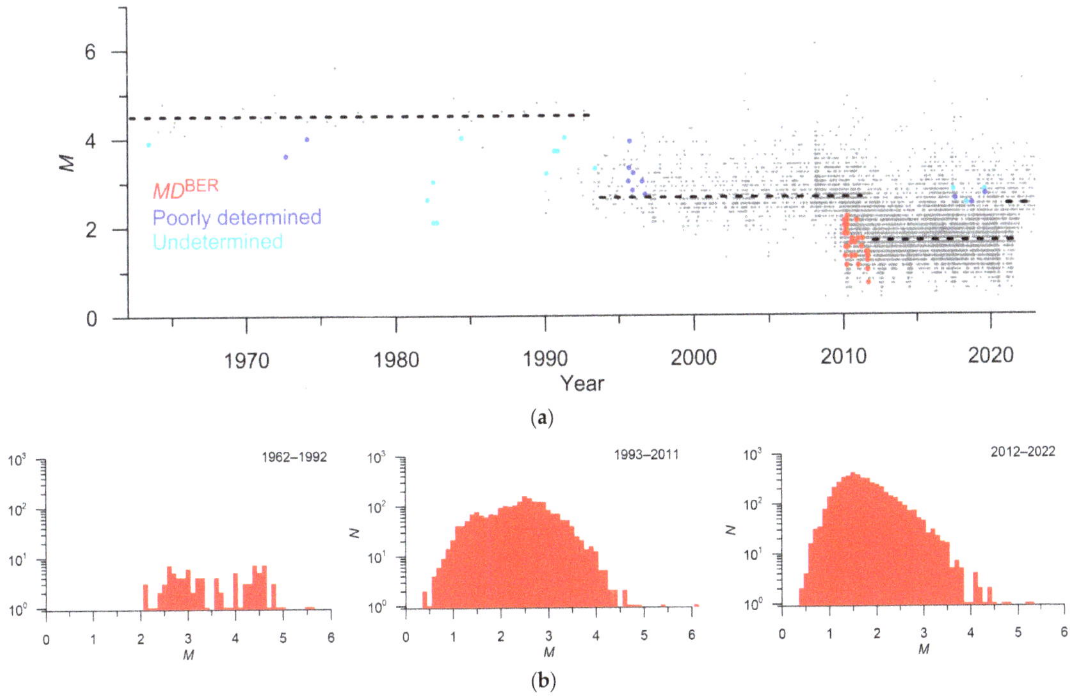

Figure 21. Svalbard. (**a**) Event distribution over time and unified magnitude; (**b**) non-cumulative frequency-magnitude distributions in different periods. Colored dots in (**a**) show events with unreliable magnitudes (Table 3), and the dashed lines show preliminary estimates of completeness magnitude *Mc*.

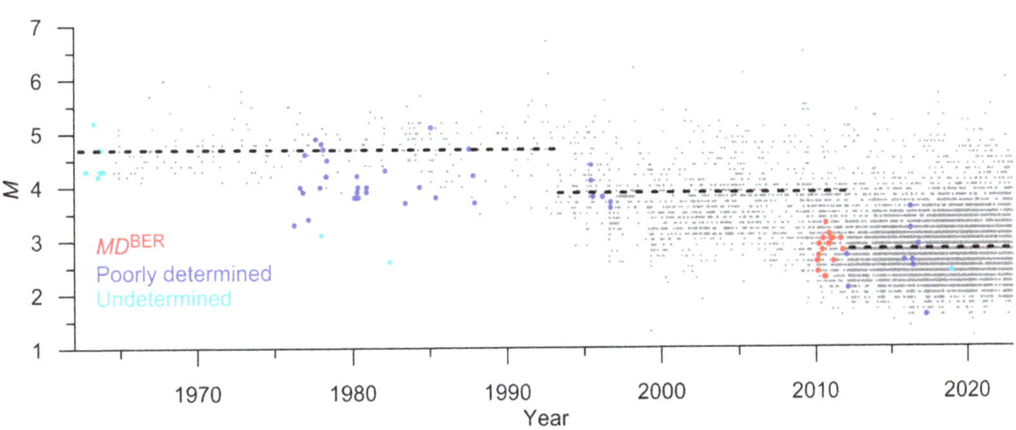

(**a**)

Figure 22. *Cont.*

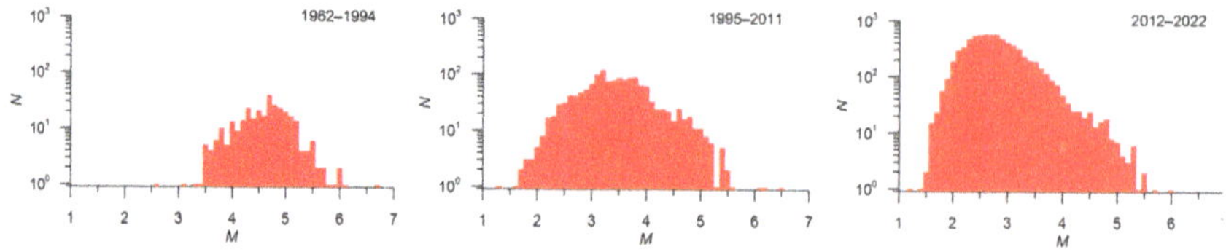

(b)

Figure 22. Knipovich Ridge. (**a**) Event distribution over time and unified magnitude; (**b**) non-cumulative frequency-magnitude distributions in different periods. Colored dots in (**a**) show events with unreliable magnitudes (Table 4), and the dashed lines show preliminary estimates of the completeness magnitude *Mc*.

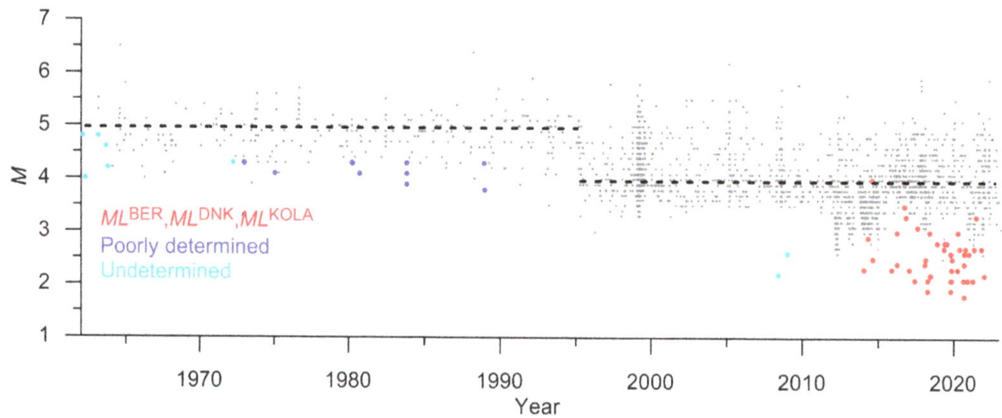

(a)

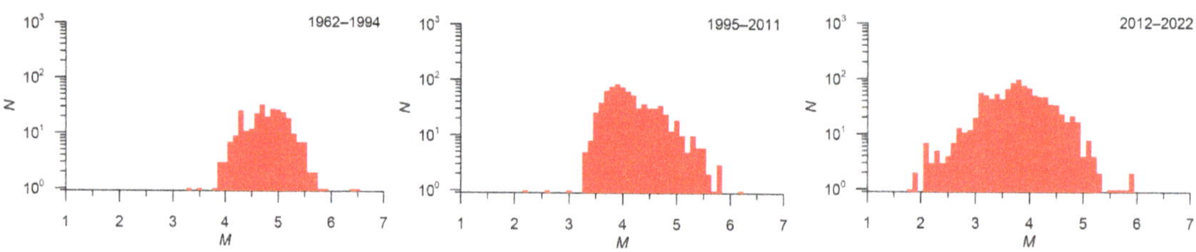

(b)

Figure 23. Gakkel Ridge. (**a**) Event distribution over time and unified magnitude; (**b**) non-cumulative frequency-magnitude distributions in different periods. Colored dots in (**a**) show events with unreliable magnitudes (Table 5), and the dashed lines show preliminary estimates of the completeness magnitude *Mc*.

Figure 24 shows spatial-temporal variations of the magnitude of complete registration *Mc*, constructed using the multi-scale method. To determine spatial-temporal variations of the magnitude of complete registration, we used the author's multi-scale method [33,34]. It was developed for the analysis of heterogeneous catalogs with significant variations in the registration level. The algorithm was based on adapting the dimension of the studied zone to the range of the event magnitude. We associated ranges of larger magnitudes with increasing areas for data selection. The high resolution of the *Mc*-value was achieved through the determination of the smallest space–magnitude scale in which the Gutenberg–Richter law is verified. The high accuracy and resolution of the method were confirmed by testing on real earthquake catalogs and on synthetic data with the prescribed Mc. The data in Table 6 are based on multi-scale analysis, the results of which are presented in Figure 24.

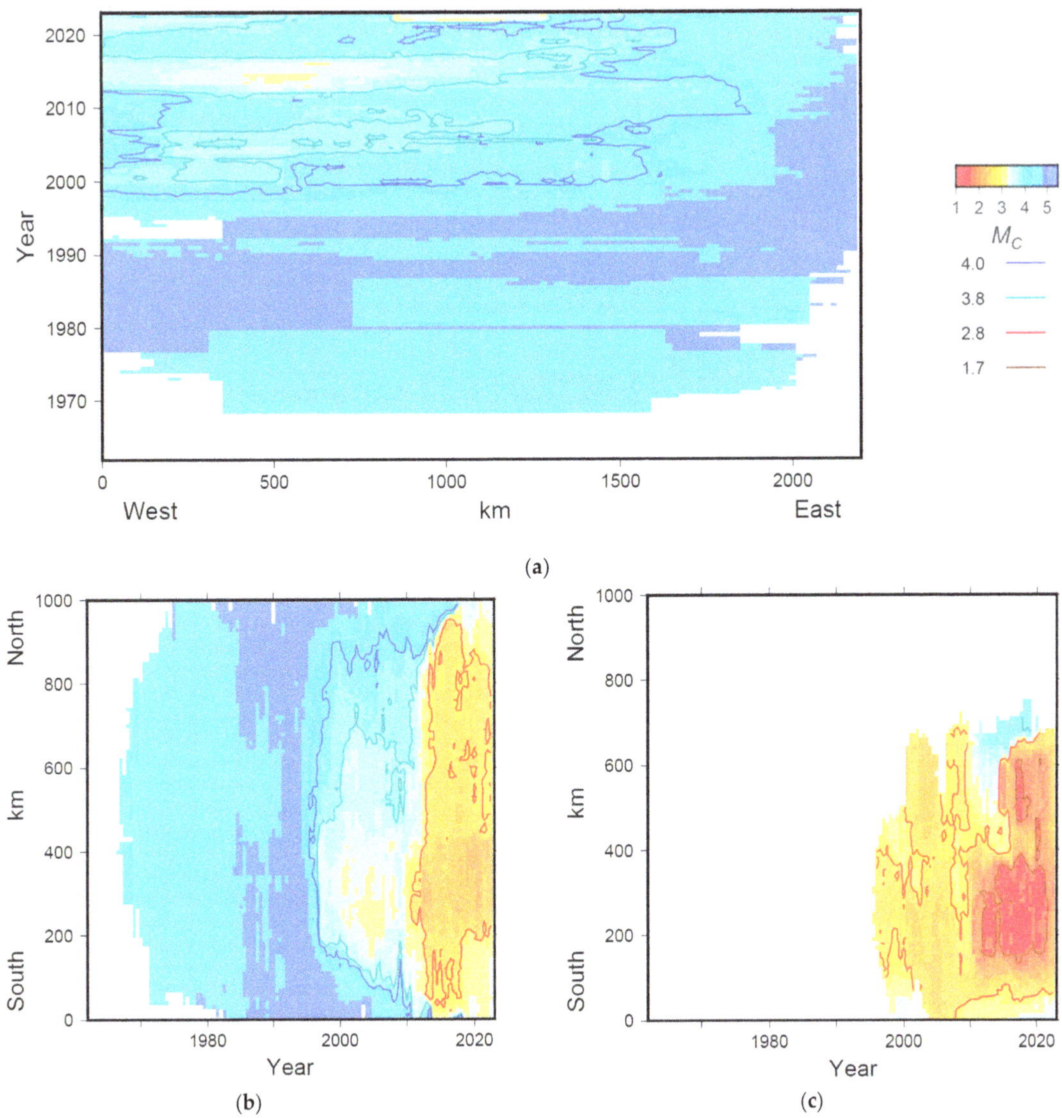

Figure 24. Spatial-temporal variations of the magnitude of complete registration Mc (**a**) Gakkel Ridge; (**b**) Knipovich Ridge; (**c**) Svalbard.

The number of recorded events in the Svalbard Archipelago and the Knipovich Ridge noticeably increased in 2010, but the best level of registration occurred in 2012. In the Gakkel Ridge, the number of weak events significantly increased after 2012 (data from the Arkhangelsk FCIAR network). However, the magnitude of complete registration Mc only decreased in certain spatial-temporal areas, with $Mc = 4.0$ for the region as a whole, except for the easternmost segment where $Mc = 4.5$. The Mc maps are in good agreement with the spatial-temporal distribution of magnitudes (Figure S5, see Supplementary).

4. Conclusions

In this paper, the data from seismic catalogs of various monitoring networks were generalized and integrated using the author's technique [26] based on statistic distribution analysis. The target was to create the most complete and representative earthquake catalog of the Gakkel and Knipovich ridges, as well as the Svalbard Archipelago. The magnitude estimates were unified in the obtained integrated catalog. The earthquake epicenter map of the integrated catalog is shown in Figure 20. The integrated catalog was made available to the public on the website of the World Data Center for Solid Earth Physics, Moscow, at http://www.wdcb.ru/arctic_antarctic/arctic_seism.html (accessed on 1 August 2023).

Based on the conducted study and obtained results, the authors consider it possible to formulate the following conclusions:

1. The earthquake catalogs of the studied region (the Gakkel and Knipovich mid-ocean ridges plus the Svalbard Archipelago) are a mixture of data from a large number of agencies. Moreover, the catalogs significantly vary over time. As a result, heavy tails appear in the *DT*, *DX*, and *DY* distributions. Therefore, determining the threshold value of the metric using the methodology applied in [27,28] leads to an increased probability of missing duplicates. For this reason, in this study, we decided not to use a multivariate normal distribution model. Instead, the actual distribution of the metric for the nearest events from two combined catalogs was used. As a result, the estimated number of errors in the integrated catalog does not exceed 1%;
2. The integrated catalog contains 17,922 events; 16,933 are from the ISC and 989 events are from Russian catalogs. The latter were not presented in the ISC, while the information regarding 578 events from Russian catalogs was used as a part of the ISC data. In the Gakkel Ridge, Russian data accounts for more than a quarter of events, and more than half after 2012. In the sub-regions of Svalbard and Knipovich, the addition of ISC data with Russian catalogs is insignificant. However, an important aspect here is the unification of magnitude;
3. The ratios between local magnitudes *ML* with the reference magnitude MW^{GCMT} and mb^{ISC} significantly differ in Svalbard and mid-ocean ridges. In Svalbard, the difference between the local and moment magnitudes is about 0.3 (Table 3). In the Knipovich and Gakkel ridges, the estimates of local magnitudes are significantly underestimated compared to the moment magnitude, with a difference exceeding 1.0 (Tables 4 and 5). Figure 25 shows the frequency-magnitude distributions constructed using original magnitudes. In Svalbard, noticeable discontinuities in the distribution are observed (Figure 25a). This can significantly affect the estimates of the b-value (the slope of the magnitude-frequency plot), which is an important parameter in seismic hazard assessment. The distribution in the Knipovich Ridge (Figure 25b) has a bimodal character, which contradicts the Gutenberg-Richter law. This is an independent confirmation of the inconsistency of magnitude estimates *ML* and MW^{GCMT}, mb^{ISC}. In the Gakkel Ridge (Figure 25c), the distribution weakly follows the Gutenberg-Richter law. After the proposed magnitude conversion, the magnitude-frequency plots acquired a common form (Figures 21b, 22b and 23b);
4. When creating the unified magnitude scale, ratios were used with three types of magnitude, MW^{GCMT}, mb^{ISC}, and ML^{BER}, which are well represented in the studied region. The shift-type ratios turned out to be very similar for most magnitudes of different types determined by various agencies. This approach allows for a significant expansion of the interval for converting magnitudes to proxy-*MW*, increases statistics, and thus increases the reliability of the conversion. The MW^{GCMT} magnitude is only determined for strong earthquakes with $M > 5.0$. The use of correlations with mb^{ISC} and ML^{BER} allows for an extension of the interval to *M* of the order of 1. Strictly mathematically, this is not proof of the linearity of ratios between different magnitudes, but it is a weighty argument in favor of such an assumption;
5. The level of registration significantly varies over time and differs in sub-regions. The best registration level is on Svalbard ($Mc = 1.7$ after 2012), where there are many

seismic stations of the BER, NORSAR, and KOLA networks. A good registration level is apparent in the Knipovich Ridge (Mc = 2.8 after 2012), which is provided by Norwegian and Russian stations on Svalbard and Danish DNK stations in Greenland. The worst registration level is in the Gakkel Ridge (Mc = 4.0), which is not surprising. The nearest seismic stations of FCIAR are located on the archipelagos of Severnaya Zemlya and Franz Josef Land, approximately 600 km from the seismic zone. The distance from Svalbard stations to the Gakkel Ridge is approximately the same;

6. The integrated earthquake catalog created and reported in this paper is intended for a wide range of researchers involved in both the study of the seismic regime of the Arctic and, in general, seismic hazard assessment [35–46]. Presented here, the integrated earthquake catalog, along with the author's Arctic catalogs [27,28], provides an important contribution to the development of an Arctic Big Data system. Its creation is one of the important requirements for starting a full-scale system analysis of geophysical dynamics in the Arctic;
7. Figure S6 (see Supplementary) presents all three created integrated earthquake catalogs for the Arctic zone of the Russian Federation. The integrated catalog of Arctic regions I, II, and III contains 45,793 events. In the Svalbard region, 12 duplicates were identified and removed. We believe that the magnitude scale is homogeneous, because all magnitudes were converted to proxi-Mw. The catalog is available to the public at: http://www.wdcb.ru/arctic_antarctic/arctic_seism.html (accessed on 1 August 2023).

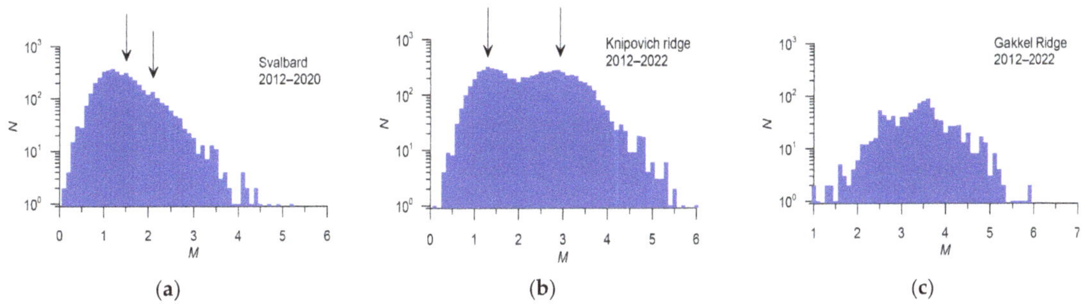

Figure 25. Non-cumulative frequency-magnitude distributions constructed using original magnitudes. (**a**) Svalbard, arrows mark discontinuities in distribution. (**b**) Knipovich Ridge, arrows mark two modes of distribution. (**c**) Gakkel Ridge, distribution weakly follows the Gutenberg-Richter law.

Supplementary Materials: The following supporting information can be downloaded at: https://www.mdpi.com/article/10.3390/app132212422/s1, Figure S1: Schematic coverage map for catalogs I (Eastern Sector), II (Western Sector) and III (75° N, 10° W; 88° N, 10° W; 88° N, 140° E; 77° N, 140° E; 77° N, 100° E; 84° N, 100° E; 84° N, 25° E; 75° N, 25° E). It includes the Svalbard Archipelago and adjacent areas (75° N–84° N, 10° E–25° E), the Knipovich Ridge, Molloy and Spitsbergen fracture zones (75° N–83° N, 10° W–10° E), and the Gakkel Ridge (the rest of the studied area); Table S1: Statistics of the ISC catalog; Figure S2: Distribution of focal depth for events with $M \leq 2.2$ in Svalbard; Figure S3: Distribution of the metric for events within the source earthquake catalogs (Table 1). The catalog name is indicated on the histogram; Figure S4: Modified Figure 9. Magenta line in (b) shows the probability of missing a duplicate in the model of multivariate normal distribution, green line shows the total probability of the first and the second kind errors. Dashed green line in (a) and (b) shows threshold value of Ro = 7, minimizing total error. (c) Green contour line of the metric (1) $Ro = 7$ shows obvious missing of duplicates; Figure S5: Distribution of earthquakes in space and time. (a) Gakkel ridge; (b) Knipovich ridge; (c) Svalbard.; Figure S6: Map of earthquake epicenters of three created integrated catalogs: I (Eastern Sector), II (Western Sector) and III (Gakkel Ridge, Knipovich Ridge and Svalbard Archipelago).

Author Contributions: Conceptualization, I.A.V., A.D.G. and P.N.S.; Data curation, P.N.S., B.A.D., B.V.D., N.A.S., E.O.K. and Y.V.B.; Formal analysis, I.A.V.; Investigation, P.N.S., B.A.D., B.V.D. and

E.O.K.; Methodology, I.A.V.; Resources, B.A.D., B.V.D., N.A.S. and E.O.K.; Software, I.A.V.; Validation, I.A.V., A.D.G., P.N.S., B.A.D. and B.V.D.; Visualization, I.A.V., B.V.D. and Y.V.B.; Writing—original draft, I.A.V., A.D.G., P.N.S., B.A.D., B.V.D., N.A.S., E.O.K. and Y.V.B.; Writing—review and editing, I.A.V., A.D.G., P.N.S., B.A.D., B.V.D., N.A.S., E.O.K. and Y.V.B. All authors have read and agreed to the published version of the manuscript.

Funding: The reported study was funded by the Russian Science Foundation, project number 21-77-30010 «System analysis of geophysical process dynamics in the Russian Arctic and their impact on the development and operation of the railway infrastructure».

Institutional Review Board Statement: Not applicable.

Informed Consent Statement: Not applicable.

Data Availability Statement: Data are contained within the article and Supplementary Materials.

Acknowledgments: This work employed data provided by the Shared Research Facility «Analytical Geomagnetic Data Center» of the Geophysical Center of RAS (http://ckp.gcras.ru/ accessed on 1 August 2023). The data used in the work were obtained with large-scale research facilities «Seismic infrasound array for monitoring Arctic cryolitozone and continuous seismic monitoring of the Russian Federation, neighboring territories and the world». This work employed data provided by the International Seismological Centre (2023), Seismological Dataset Repository, https://doi.org/10.31905/6TJZECEY (accessed on 1 August 2023).

Conflicts of Interest: The authors declare no conflict of interest.

Abbreviations

The following abbreviations for seismological agencies are used in this manuscript:

BCIS	Bureau Central International de Sismologie, France
BER	University of Bergen, Norway
CSEM	Centre Sismologique Euro-Méditerranéen (CSEM/EMSC), France
DNK	Geological Survey of Denmark and Greenland, Denmark
EIDC	Experimental (GSETT3) International Data Center, USA
FCIAR	Federal Center for Integrated Arctic Research, Russia
GFZ	Helmholtz Centre Potsdam GFZ German Research Centre For Geosciences, Germany
HEL	Institute of Seismology, University of Helsinki, Finland
HFS	Hagfors Observatory, Sweden
IDC	International Data Centre, CTBTO, Austria
IEPN	Institute of Environmental Problems of the North, Russian Academy of Sciences, Russia
INMG	Instituto Português do Mar e da Atmosfera, I.P., Portugal
ISC	International Seismological Centre, United Kingdom
ISS	International Seismological Summary, United Kingdom
MOS	Geophysical Survey of Russian Academy of Sciences (GS RAS), Russia
KOLA	Kola Regional Seismic Centre, GS RAS, Russia
MSUGS	Michigan State University, Department of Geological Sciences, USA
NAO	Stiftelsen NORSAR, Norway
NEIC	National Earthquake Information Center, USA
OTT	Canadian Hazards Information Service, Natural Resources Canada
SYKES	Sykes Catalogue of earthquakes 1950 onwards
WAR	Institute of Geophysics, Polish Academy of Sciences, Poland
ZEMSU	USSR

References

1. Thiede, J.; The Shipboard Scientific Party. Polarstern Arctis XVII/2 Cruise Report: AMORE 2001 (Arctic Mid-Ocean Ridge Expedition). *Alfred Wegener Inst.* **2002**, *421*, 297.
2. Schlindwein, V.; Muller, C.C.; Jokat, W. Microseismicity of the Ultraslow-Spreading Gakkel Ridge, Arctic Ocean: A Pilot Study. *Geophys. J. Int.* **2007**, *169*, 100–112. [CrossRef]
3. Muller, C.; Jokat, W. Seismic evidence for volcanic activity discovered in Central Arctic. *Eos Trans. AGU* **2000**, *81*, 265–269. [CrossRef]
4. Tolstoy, M.; Bohnenstiehl, D.R.; Edwards, M.H.; Kurras, G.J. Seismic character of volcanic activity at the ultraslow-spreading Gakkel Ridge. *Geology* **2001**, *29*, 1139–1142. [CrossRef]
5. Cochran, J.R.; Kurras, G.J.; Edwards, M.H.; Coakley, B.J. The Gakkel Ridge: Bathymetry, gravity anomalies and crustal accretion at extremely slow spreading rates. *J. Geophys. Res. Solid Earth* **2003**, *108*, 2116–2137. [CrossRef]
6. Cochran, J.R. Seamount volcanism along the Gakkel ridge, Arctic Ocean. *Geophys. J. Int.* **2008**, *174*, 1153–1173. [CrossRef]
7. Michael, P.J.; Langmuir, C.H.; Dick, H.J.B.; Snow, J.E.; Goldsteink, S.L.; Graham, D.W.; Lehnertk, K.; Kurras, G.; Jokat, W.; Muhe, R.; et al. Magmatic and amagmatic seafloor generation at the ultraslow-spreading Gakkel ridge, Arctic Ocean. *Nature* **2003**, *423*, 956–961. [CrossRef] [PubMed]
8. Dick, H.; Lin, J.; Schouten, H. An ultraslow-spreading class of ocean ridge. *Nature* **2003**, *426*, 405–412. [CrossRef] [PubMed]
9. Zarayskaya, Y.A. Segmentation and seismicity of the ultraslow Knipovich and Gakkel mid-ocean ridges. *Geotectonics* **2017**, *51*, 163–175. [CrossRef]
10. Dubinin, E.P.; Kokhan, A.V.; Sushchevskaya, N.M. Tectonics and magmatism of ultraslow spreading ridges. *Geotectonics* **2013**, *47*, 131–155. [CrossRef]
11. Jokat, W.; Schmidt-Aursch, M. Geophysical characteristics of the ultraslow spreading Gakkel Ridge, Arctic Ocean. *Geophys. J. Intern.* **2007**, *168*, 983–998. [CrossRef]
12. Kokhan, A.V.; Dubinin, E.P.; Grokholsky, A.L. Geodynamical peculiarities of structure-forming in Arctic and Polar Atlantic spreading ridges. *Bull. Kamchatka Reg. Assoc.* **2012**, *19*, 59–77. (In Russian)
13. Koulakov, I.; Schlindwein, V.; Liu, M.; Gerya, T.; Jakovlev, A.; Ivanov, A. Low-degree mantle melting controls the deep seismicity and explosive volcanism of the Gakkel Ridge. *Nat. Commun.* **2022**, *13*, 3122. [CrossRef] [PubMed]
14. Assinovskaya, B.A.; Panas, N.M.; Ovsov, M.K.; Antonovskaya, G.N. Preliminary seismic hazard assessment of the Arctic Gakkel ridge and surrounding. *Russ. J. Seismol.* **2019**, *1*, 35–45. [CrossRef]
15. Di Giacomo, D.; Engdahl, E.R.; Storchak, D.A. The ISC-GEM Earthquake Catalogue (1904–2014): Status after the Extension Project. *Earth Syst. Sci. Data* **2018**, *10*, 1877–1899. [CrossRef]
16. Ekström, G.; Nettles, M.; Dziewonski, A.M. The global CMT project 2004–2010: Centroid-moment tensors for 13,017 earthquakes. *Phys. Earth Planet. Inter.* **2012**, *200–201*, 1–9. [CrossRef]
17. Köhler, A.; Gajek, W.; Malinowski, M.; Schweitzer, J.; Majdanski, M.; Geissler, W.H.; Chamarczuk, M.; Wuestefeld, A. Seismological monitoring of Svalbard's cryosphere: Current status and knowledge gaps. In SESS report 2019—The State of Environmental Science in Svalbard—An annual report. *Svalbard Integr. Arct. Earth Obs. Syst.* **2020**, *1*, 136–159. [CrossRef]
18. Asming, V.E.; Fedorov, A.V.; Alenicheva, A.O.; Jevtjugina, Z.A. Usage of the NSDL Location System for the Detailed Study of the Spitsbergen Archipelago Seismicity. *Her. Kola Sci. RAS* **2018**, *3*, 120–131. (In Russian) [CrossRef]
19. Baranov, S.V. The aftershock process of the February 21, 2008 Storfjorden strait, Spitsbergen, earthquake. *J. Volcanol. Seismol.* **2013**, *7*, 230–242. [CrossRef]
20. Sokolov, S.Y. Tectonic evolution of the Knipovich Ridge based on the anomalous magnetic field. *Dokl. Earth Sci.* **2011**, *437*, 343–348. [CrossRef]
21. Crane, K.; Doss, H.; Vogt, P.; Sundvor, E.; Cherkashov, G.; Poroshina, I.; Joseph, D. The role of the Spitsbergen shear zone in determining morphology, segmentation and evolution of the Knipovich Ridge. *Mar. Geophys. Res.* **2001**, *22*, 153–205. [CrossRef]
22. Morozov, A.N.; Vaganova, N.V.; Starkov, I.V.; Mikhaylova, Y.A. Modern Low-Magnitude Earthquake Swarms of the Gakkel Mid-Oceanic Ridge, Arctic Ocean. *Russ. J. Earth. Sci.* **2023**, *23*, ES3007. [CrossRef]
23. Antonovskaya, G.; Morozov, A.; Vaganova, N.; Konechnaya, Y. Seismic monitoring of the European Arctic and Adjoining Regions. In *The Arctic: Current Issues and Challenges*; Pokrovsky, O.S., Kirpotin, S.N., Malov, A.I., Eds.; Nova Science Publishers, Inc.: New York, NY, USA, 2020; pp. 303–368.
24. Engen, Ø.; Eldholm, O.; Bungum, H. The Arctic plate boundary. *J. Geophys. Res.* **2003**, *108*, 2075. [CrossRef]
25. Shebalin, P.N.; Baranov, S.V.; Dzeboev, B.A. The Law of the Repeatability of the Number of Aftershocks. *Dokl. Earth Sci.* **2018**, *481*, 963–966. [CrossRef]
26. Vorobieva, I.; Gvishiani, A.; Dzeboev, B.; Dzeranov, B.; Barykina, Y.; Antipova, A. Nearest Neighbor Method for Discriminating Aftershocks and Duplicates When Merging Earthquake Catalogs. *Front. Earth Sci.* **2022**, *10*, 820277. [CrossRef]
27. Gvishiani, A.D.; Vorobieva, I.A.; Shebalin, P.N.; Dzeboev, B.A.; Dzeranov, B.V.; Skorkina, A.A. Integrated Earthquake Catalog of the Eastern Sector of the Russian Arctic. *Appl. Sci.* **2022**, *12*, 5010. [CrossRef]
28. Vorobieva, I.A.; Gvishiani, A.D.; Shebalin, P.N.; Dzeboev, B.A.; Dzeranov, B.V.; Skorkina, A.A.; Sergeeva, N.A.; Fomenko, N.A. Integrated Earthquake Catalog II: The Western Sector of the Russian Arctic. *Appl. Sci.* **2023**, *13*, 7084. [CrossRef]
29. Morozov, A.N.; Vaganova, N.V.; Asming, V.E.; Peretokin, S.A.; Aleshin, I.M. Seismicity of the Western Sector of the Russian Arctic. *Phys. Solid Earth* **2023**, *59*, 209–241. [CrossRef]

30. Vinogradov, Y.A.; Asming, V.E.; Baranov, S.V.; Fedorov, A.V.; Vinogradov, A.N. Seismic and infrasonic monitoring of glacier destruction: A pilot experiment on Svalbard. *Seism. Instr.* **2015**, *51*, 1–7. [CrossRef]
31. Bogorodskiy, P.V.; Demidov, N.E.; Filchuk, K.V.; Marchenko, A.V.; Morozov, E.G.; Nikulina, A.L.; Pnyushkov, A.V.; Ryzhov, I.V. Growth of land fast ice and its thermal interaction with bottom sediments in the Braganzavägen Gulf (West Spitsbergen). *Russ. J. Earth Sci.* **2020**, *20*, ES6005. [CrossRef]
32. Di Giacomo, D.; Bondár, I.; Storchak, D.A.; Engdahl, E.R.; Bormann, P.; Harris, J. ISC-GEM: Global Instrumental Earthquake Catalogue (1900–2009), III. Re-computed MS and mb, proxy MW, final magnitude composition and completeness assessment. *Phys. Earth Planet. Inter.* **2015**, *239*, 33–47. [CrossRef]
33. Vorobieva, I.; Shebalin, P.; Narteau, C.; Beauducel, F.; Nercessian, A.; Clouard, V.; Bouin, M.-P. Multiscale mapping of completeness magnitude of earthquake catalogs. *Bull. Seism. Soc. Am.* **2013**, *103*, 2188–2202. [CrossRef]
34. Vorobieva, I.; Shebalin, P.; Narteau, C. Break of slope in earthquake size distribution and creep rate along the San Andreas Fault system. *Geophys. Res. Lett.* **2016**, *43*, 6869–6875. [CrossRef]
35. Agayan, S.M.; Tatarinov, V.N.; Gvishiani, A.D.; Bogoutdinov, S.R.; Belov, I.O. FDPS algorithm in stability assessment of the Earth's crust structural tectonic blocks. *Russ. J. Earth Sci.* **2020**, *20*, ES6014. [CrossRef]
36. Dzeboev, B.A.; Gvishiani, A.D.; Agayan, S.M.; Belov, I.O.; Karapetyan, J.K.; Dzeranov, B.V.; Barykina, Y.V. System-Analytical Method of Earthquake-Prone Areas Recognition. *Appl. Sci.* **2021**, *11*, 7972. [CrossRef]
37. Dzeboev, B.A.; Karapetyan, J.K.; Aronov, G.A.; Dzeranov, B.V.; Kudin, D.V.; Karapetyan, R.K.; Vavilin, E.V. FCAZ-recognition based on declustered earthquake catalogs. *Russ. J. Earth Sci.* **2020**, *20*, ES6010. [CrossRef]
38. Gorshkov, A.I.; Soloviev, A.A. Recognition of earthquake-prone areas in the Altai-Sayan-Baikal region based on the morphostructural zoning. *Russ. J. Earth Sci.* **2021**, *21*, ES1005. [CrossRef]
39. Gvishiani, A.D.; Soloviev, A.A.; Dzeboev, B.A. Problem of Recognition of Strong-Earthquake-Prone Areas: A State-of-the-Art Review. *Izv. Phys. Solid Earth* **2020**, *56*, 1–23. [CrossRef]
40. Kossobokov, V.G.; Soloviev, A.A. Pattern recognition in problems of seismic hazard assessment. *Chebyshevskii Sb.* **2018**, *19*, 55–90. (In Russian) [CrossRef]
41. Peresan, A.; Gorshkov, A.; Soloviev, A.; Panza, G.F. The contribution of pattern recognition of seismic and morphostructural data to seismic hazard assessment. *Boll. Di Geofis. Teor. Ed Appl.* **2015**, *56*, 295–328. [CrossRef]
42. Gorshkov, A.; Kossobokov, V.; Soloviev, A. Recognition of earthquake-prone areas. In *Nonlinear Dynamics of the Lithosphere and Earthquake Prediction*; Keilis-Borok, V., Soloviev, A., Eds.; Springer: Heidelberg, Germany, 2003; pp. 239–310. [CrossRef]
43. Peresan, A.; Zuccolo, E.; Vaccari, F.; Gorshkov, A.; Panza, G.F. Neo-deterministic seismic hazard and pattern recognition techniques: Time-dependent scenarios for North-Eastern Italy. *Pure Appl. Geophys.* **2011**, *168*, 583–607. [CrossRef]
44. Gorshkov, A.I.; Panza, G.F.; Soloviev, A.A.; Aoudia, A. Morphostructural zonation and preliminary recognition of seismogenic nodes around the Adria margin in peninsular Italy and Sicily. *J. Seismol. Earthq. Eng.* **2002**, *4*, 1–24.
45. Gorshkov, A.; Novikova, O.; Parvez, I.A. Recognition of earthquake-prone areas in the Himalaya: Validity of the results. *Int. J. Geophys.* **2012**, *2012*, 419143. [CrossRef]
46. Gorshkov, A.; Novikova, O. Estimating the validity of the recognition results of earthquake-prone areas using the ArcMap. *Acta Geophys.* **2018**, *66*, 843–853. [CrossRef]

applied sciences

MDPI

Article

Strong Earthquake-Prone Areas in the Eastern Sector of the Arctic Zone of the Russian Federation

Alexei D. Gvishiani [1,2], Boris A. Dzeboev [1,3,*], Boris V. Dzeranov [1,3], Ernest O. Kedrov [1], Anna A. Skorkina [1,4] and Izabella M. Nikitina [1]

1 Geophysical Center of the Russian Academy of Sciences (GC RAS), 119296 Moscow, Russia
2 Schmidt Institute of Physics of the Earth of the Russian Academy of Sciences (IPE RAS), 119296 Moscow, Russia
3 Geophysical Institute, Vladikavkaz Scientific Center RAS (GPI VSC RAS), 362002 Vladikavkaz, Russia
4 Institute of Earthquake Prediction Theory and Mathematical Geophysics of the Russian Academy of Sciences (IEPT RAS), 117997 Moscow, Russia
* Correspondence: b.dzeboev@gcras.ru

Abstract: This paper continues the series of publications by the authors on the recognition of areas prone to the strongest, strong, and significant earthquakes using the FCAZ system-analytical method. The areas prone to earthquakes with $M \geq 5.5$ in the eastern sector of the Arctic zone of the Russian Federation were recognized. It is shown that certain potential high seismicity zones are well confined to the boundaries of the Eurasian, North American, and Okhotsk tectonic plates. In addition, according to the results of the FCAZ recognition, some areas located at a sufficient distance from the main tectonic structures of the studied region were also recognized as highly seismic. The results of the study, among other factors, justify the use of the assessment of the completeness magnitude in the catalog for choosing the set of recognition objects for the FCAZ method.

Keywords: Arctic zone of the Russian Federation; earthquake-prone areas; system-analytical method; FCAZ; pattern recognition; clustering; integrated earthquake catalogs; high seismicity zones

Citation: Gvishiani, A.D.; Dzeboev, B.A.; Dzeranov, B.V.; Kedrov, E.O.; Skorkina, A.A.; Nikitina, I.M. Strong Earthquake-Prone Areas in the Eastern Sector of the Arctic Zone of the Russian Federation. *Appl. Sci.* **2022**, *12*, 11990. https://doi.org/10.3390/app122311990

Academic Editor: Jianbo Gao

Received: 26 October 2022
Accepted: 22 November 2022
Published: 23 November 2022

1. Introduction

The development of the Arctic zone of the Russian Federation (AZRF) is an important area of economic and scientific progress for the country. This is determined, first of all, by the significant reserves of minerals in the Russian Arctic. Today, the region produces oil (80% of the total volume in the country), natural gas (93%), nickel and copper (90%), diamonds and gold (33%), platinum, palladium, cobalt, tin, manganese, coal, etc. [1]. The relevance of scientific research in the Russian Arctic is growing due to climate change caused by global warming and the accelerated melting of Arctic ice [2–5].

Assessment of the hazards of geodynamical nature plays a significant role in determining the strategy for the industrial development of promising regions, which undoubtedly includes the Russian Arctic [6]. The central role here belongs to seismic hazard, which is taken into account both in determining urban planning policy and in planning industrial and infrastructure network facilities.

In this paper, the area of seismic hazard assessment is the eastern sector of the Russian Arctic. In its oceanic part, the Eurasian basin, the Gakkel Ridge, and the Canadian basin can be identified. They are divided by the Lomonosov and the Alfa–Mendeleev Ridges, which are separated from each other by the Makarov Basin and the small Mendeleev abyssal plain.

The continental part is represented by accretion–collision structures. Among them is the Verkhoyansk Range, which frames the Siberian platform from the east. This is followed by is the Chersky Range. Further to the east, there is a series of arched ranges of Koryakia and Chukotka [7,8] (Figure 1).

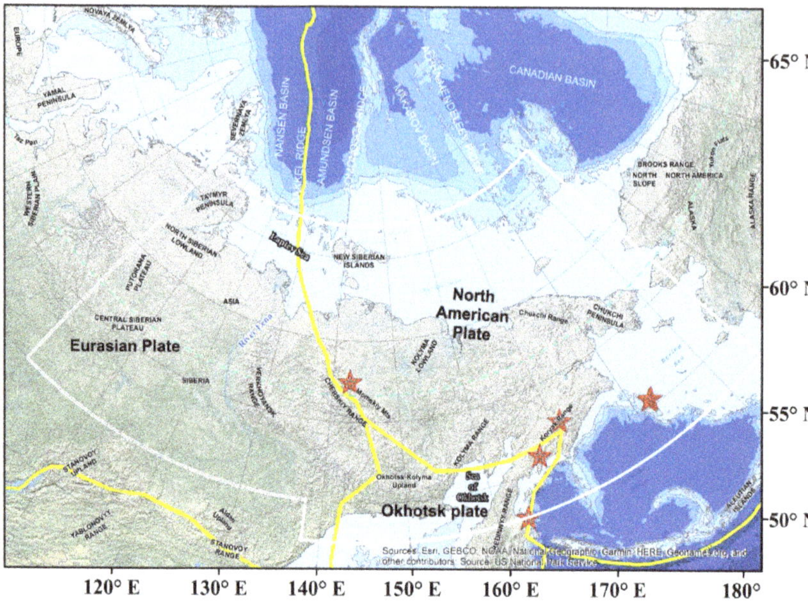

Figure 1. The main structures of the eastern sector of the Russian Arctic. The yellow line is the boundaries of tectonic plates, the light green dotted line is the polar circle, and the red stars are the epicenters of earthquakes with $M \geq 6.0$ that occurred over the past two decades. The white frame shows the boundaries of the studied region.

Seismogenic zones in the continental part are represented by the boundary between the North American and Eurasian plates (Figure 1). The latter consists of several structures extending from the Laptev Sea and the mouth of the river Lena to the coast of Okhotsk and the Isthmus of Kamchatka. The transverse extent of seismogenic structures exceeds 1000 km. Some researchers distinguish here a collage of microplates, including the Amur, Okhotsk, and Bering plates [9]. Some authors trace here several seismogenic structures, including the Laptev Sea, Kharaulakh, Cherskiy seismotectonic zones, and the Arctic–Asian seismic belt [8,10].

Over the past two decades, a number of strong earthquakes with a magnitude $M \geq 6.0$ have occurred in the studied region (Figure 2). Among them is the Olyutorsk earthquake on 20.04.2006 with $M = 7.6$; the earthquake on 24.06.2012 near the northeastern coast of the Kamchatka Peninsula with $M = 6.0$; two earthquakes with $M = 6.5$ and $M = 6.3$ that occurred consecutively on 30.04.2010 in the Bering Sea; Ilin-Tas (Abyi) earthquake on 14 February 2013 with $M = 6.6$; earthquake on 09.01.2020 with $M = 6.4$ near the border of Kamchatka and Chukotka [11–14], and others (Figures 1 and 2). Sufficiently detailed information about strong earthquakes in the Russian Arctic that occurred before the mid-1970s can be found in [15]. The foregoing emphasizes the existing high seismic hazard of the region and substantiates the urgency of the problem of determining areas in the eastern sector of the Arctic zone of the Russian Federation within which strong earthquakes can occur.

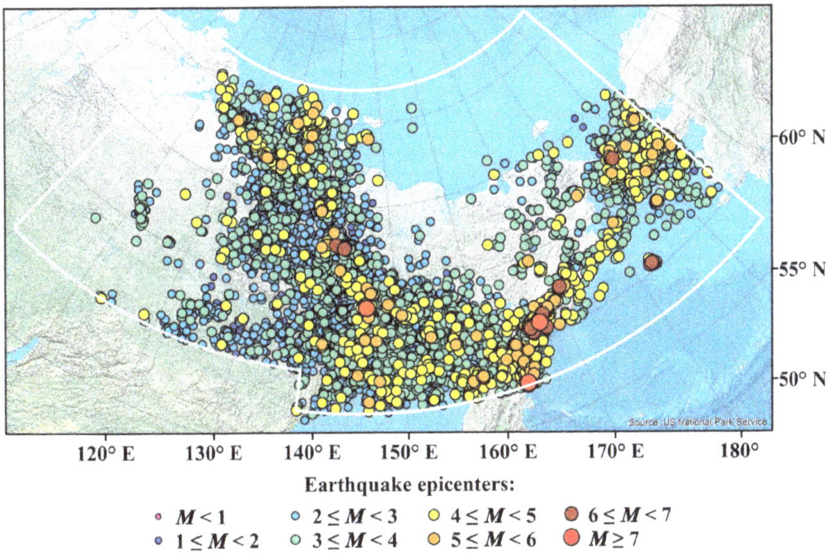

Figure 2. The integrated earthquake catalog of the Eastern Sector of the Arctic zone of the Russian Federation (1962–2020) [16]. The white frame shows the boundaries of the region (coincides with the region marked in Figure 1), for which the catalog was created and within which FCAZ recognition was performed.

A large number of scientific papers are devoted to the analysis of the seismic regime of the Arctic zone of Russia and the construction of corresponding seismic hazard maps (see, for example, [17–21], and others). Here it is necessary to note the systematically updated set of maps of the General Seismic Zoning (GSZ) [22–26] on the scale of the whole of Russia. On the GSZ maps that regulate construction in the country, the territory of the Russian Arctic belongs to zones with an estimated intensity of five to six or more points [27]. Nevertheless, the question of the boundaries of the zones prone to strong earthquakes in the Russian Arctic remains open. This article is devoted to its study.

Since the 1970s, seismic hazard assessment studies have been actively applying pattern recognition methods [28–30]. Created in the mid-2010s and currently being developed, the FCAZ (formalized clustering and zoning) system analysis method turned out to be universal and effective [29,31]. Its application is based on topological filtering of a point set of epicenters of sufficiently weak earthquakes, which serve as recognition objects. FCAZ is based on the composition of DPS (discrete perfect sets) [32,33] and E^2XT [31] algorithms. The subject of the FCAZ study is the nontrivial recognition of areas prone to strong (the strongest, significant) earthquakes in the studied region [34].

The objective of this study is to recognize strong earthquake-prone areas in the eastern sector of the Russian Arctic (Figure 1). Recognition is performed using the FCAZ method.

2. Materials and Methods

2.1. System-Analytical Method FCAZ

The system-analytical method FCAZ (formalized clustering and zoning) was created in 2012–2016 at the Geophysical Center of the Russian Academy of Sciences [31,33–41] (and others). Some ideas of the mathematical construction of the method can be observed in earlier publications [42–44]. The basic idea of the FCAZ study is the topological filtrational clustering of the seismic catalog—a finite set W of epicenters of known earthquakes in the studied region. As already noted, FCAZ makes it possible to efficiently recognize areas prone to the strongest, strong, and significant earthquakes [29].

The FCAZ system represents a sequential application of 2 algorithms: DPS (discrete perfect sets) [31–33,37,45,46] and E^2XT [31,34]. In a finite set W of the Euclidean space, DPS selects regions with a given density level α. The output is a set of points that is α-dense in each of its elements. The task of the DPS algorithm is to construct such a set $W(\alpha)$ that its density is not less than the level α at all its points. The DPS algorithm, in particular, is effective in the current task because it "attracts" recognition objects (earthquake epicenters) into dense clusters, leaving no isolated points [35].

At the next, second stage of FCAZ recognition, the E^2XT algorithm formalizes and implements the construction of a unique mapping of discrete DPS clusters into flat zones of nonzero measure, inside and at the boundaries of which earthquakes with a magnitude $M \geq M_0$ can occur [31]. Here M_0 is a given magnitude threshold, which depends on the seismic regime of the considered region [29] and determines which earthquakes we consider strong.

The algorithms that form FCAZ have a number of input parameters: DPS-$q < 0$ to calculate the localization radius and the maximum density level $\beta \in [-1, 1]$ to determine the required density level of DPS clusters $\alpha(\beta)$; E^2XT-$v < 0$ and $w < 0$ to calculate the scannability of DPS clusters and the connectivity of calculated flat high seismicity zones [33]. Another parameter for DPS is the number of iterations of applying the algorithm in one task. In 2016, the DPS and E^2XT algorithms were equipped with artificial intelligence blocks [31]. The latter allows it to automatically choose the optimal values of the free parameters. This makes the result of FCAZ recognition more objective and reproducible [34].

Previously, using the FCAZ method, earthquake-prone area recognition in the following seismically hazardous regions of the world was successfully performed:

- The mountain belt of the Andes of South America (M_0 = 7.75) [31,34];
- Pacific coast of the Kamchatka Peninsula (M_0 = 7.75) [34–36] and the Kuril Islands (M_0 = 7.75);
- California (M_0 = 6.5) [34,36,37];
- Cisbaikalia–Transbaikalia (M_0 = 5.5; M_0 = 5.75, M_0 = 6.0) [34,39];
- Altai-Sayan (M_0 = 5.5) [34,40];
- Caucasus (M_0 = 5.0) [31,33,34,37,38];
- Crimean Peninsula and northwestern Caucasus (M_0 = 4.5; M_0 = 5.0) [34,41].

In these regions, the reliability of the FCAZ results was assessed using control experiments of the "seismic history" type [29]. Naturally, the recognized high seismicity zones were also compared with the location of the actual epicenters of the strongest, strong, and significant earthquakes that occurred after the end of the catalog used for recognition [31]. Thus, in each of the 8 regions noted above, a high degree of reliability of FCAZ recognition was substantiated [34].

A detailed description of the mathematical and algorithmic construction of the FCAZ method and the results obtained earlier with its help is given in [31,33,34].

2.2. *Earthquake Catalog of the Eastern Sector of the Arctic Zone of the Russian Federation*

The basis for the recognition of potential high seismicity zones by the FCAZ method is the most complete earthquake catalog of the studied region. In this paper, epicenters from the original, integrated earthquake catalog of the eastern sector of the Arctic zone of the Russian Federation are used as recognition objects. The catalog contains information on 23,254 seismic events for the period 1962–2020 [16]. It was created by the system analysis method as a result of a formalized combination of earthquake data from the regional catalogs of the Geophysical Survey of the Russian Academy of Sciences (GS RAS, http://www.gsras.ru/new/eng/catalog/, accessed on 1 October 2022) and the International Seismological Center (ISC, http://www.isc.ac.uk/, accessed on 1 October 2022).

Different seismic agencies can both register and skip events. This happens due to network configurations, the specifics of processing methods, as well as the procedures for transferring information to data centers. When catalogs are merged, the problem of identifying and removing the resulting duplicates naturally arises. The problem is complicated by the need to distinguish duplicates from aftershocks [47] since the latter are also events close in space and time. The problem of merging these catalogs was solved using the author's technique [48]. It is based on a modification of the nearest neighbor method [49,50] for the recognition of duplicates. Namely, correspondence is built for events from two catalogs, after which earthquakes are classified into unique and duplicates using the Euclidean metric. Consistent application of the technique makes it possible to combine any number of earthquake catalogs [48].

When creating an integrated earthquake catalog of the eastern sector of the Arctic zone of the Russian Federation [16], the following were sequentially combined:

- The regional catalog of Yakutia from the annual journal "Earthquakes in Northern Eurasia"
- The regional catalog of the northeast of Russia from the annual journal "Earthquakes in Northern Eurasia"
- The earthquake catalog of Kamchatka of the Kamchatka Branch of GS RAS
- The ISC catalog, which is a composite catalog containing data from many world and Russian agencies

Different agencies define, generally speaking, different magnitude types of seismic events. At the same time, for a rigorous formulation of the problem of earthquake-prone areas recognition, which is determined by the threshold M_0, it is necessary to provide all records of the catalog with a magnitude of the same type. In this regard, the magnitude scale in the integrated catalog was unified [16]. This was undertaken for the first time with the problems of the FCAZ cycle. This was a fundamental step forward in the development of the system analysis method.

When analyzing various magnitude scales, preference is usually given to the moment magnitude: $M = M_W$ [16,51,52]. In the integrated catalog, M_W was defined only for earthquakes with $M > 5.0$. For events with magnitudes up to $M \approx 4.0$, the magnitude mb was mainly defined, and in such cases, $M = mb$. Weak earthquakes had magnitudes defined by local networks, with a large variety in their types [16]. Most of the weak earthquakes had an energy class estimate, and only for 7% of the events were other types of magnitude calculated. In this regard, the following magnitude priority system was used in the integrated catalog [16]:

1. $M = M_W{}^{GCMT}$ or
2. $M = M_S{}^{ISC}$ for strong earthquakes before 1976 or
3. $M = mb^{ISC}$ or
4. $M = b \times K_S - a$, regression coefficients a and b are given in [16]
5. $M = b \times M_L - a$, regression coefficients a and b are given in [16].

The formalized construction of the author's technique for merging earthquake catalogs is described in detail in [48]. The stages of merging the integrated catalog of the eastern sector of the Arctic zone of the Russian Federation and unifying the magnitude scales are described in [16]. The catalog is available to the public on the website of the World Data Center for Solid Earth Physics, Moscow, at http://www.wdcb.ru/arctic_antarctic/arctic_seism.html, (accessed on 24 October 2022). The map of earthquake epicenters from the integrated catalog is presented in Figure 2.

The catalog of strong earthquakes of the eastern sector of the Russian Arctic, starting from 1900, was formed based on 3 main sources: the integrated catalog [16], *New Catalog of Strong Earthquakes in the USSR from Ancient Times through 1977* (hereinafter "*New Catalog* ... ") [15] and the catalog of the International Seismological Center (ISC).

It should be noted that in the "*New Catalog*..." [15] there are many cases when coordinates of strong earthquakes correspond to low-seismicity areas, according to modern instrumental data. In the description of the "*New Catalog*..." it is said that "*the least reliable is the assessment in the presence of intensity data in one observation site. In such cases, the value obtained under the assumption that the source was located under the given observation site at the average depth for the region was taken as its lower limit. The value corresponding to the epicenter at the nearest point of a large seismically active zone was taken as the upper limit, and the average was taken as the final estimate. The error in such cases could be two units."*

Accordingly, it was decided not to include in the formed catalog of strong earthquakes the events with the accuracy of ± 2 of the indicated magnitude in the "*New Catalog*...". Thus, 53 events were taken from the "*New Catalog*...", given in the sections "Arctic Basin," "Chukotka," "Kamchatka," and "Yakutia and the North-East." At the same time, even though data from international agencies have already been used in compiling the "*New Catalog* ... ", we have identified several cases where the final magnitude estimate was lower than the available instrumental estimate. In such cases, preference was given to the highest magnitude, namely, the recalculated M_S according to the ISC data [53].

All events with $M \geq 5.0$ were selected from the integrated instrumental catalog [16]. In this case, events with $M_W \geq 5.0$ or $M_S \geq 5.0$ were included in the formed catalog of strong earthquakes automatically, and $M = M_w$ or $M = M_S$, respectively. For events with $mb \geq 5.0$, an additional check was carried out, after which such events fell into one of two categories:

- If there was an estimate of $M_S \geq 5.0$, earthquakes were included in the formed catalog of strong events, and $M = M_S$, since the saturation effect of magnitudes for $mb \geq 5.0$ is known.
- If there was an estimate of $M_S < 5.0$, earthquakes were not included in the catalog of strong earthquakes. Indeed, such a combination of M_S-mb often corresponds to deep earthquakes, for which the applicability of the FCAZ method is not clear, since devastating consequences after deep events are observed rarely.

Thus, the formed catalog of strong earthquakes includes 166 events with $M \geq 5.0$. Due to the remoteness of many seismic stations from the studied region, as well as its relatively weaker seismicity, it was decided to consider earthquakes with $M \geq 5.0$ for the studied region as strong.

Due to the a priori lack of objects for FCAZ recognition, strong events, the epicenters of which are located outside the area covered by the integrated catalog, were excluded from consideration [16] (limited by the white frame in Figures 1 and 2). Strong events in the waters of the seas and oceans at a relatively long distance from the coast were also excluded. The final catalog of strong earthquakes used in this work contains 130 events with $M \geq 5.0$ (Table 1). The epicenters of such earthquakes are shown in Figure 3.

Table 1. The catalog of earthquakes with $M \geq 5.0$ in the eastern sector of the Russian Arctic (1900–2020).

No.	Date	$\varphi,^\circ$	$\lambda,^\circ$	M
1.	18.03.1913	63.4	145.8	6.2
2.	07.06.1914	73.0	119.0	5.5
3.	30.11.1918	71.2	134.0	6.4
4.	13.03.1924	63.0	150.0	5.5
5.	27.05.1924	62.0	135.5	5.2
6.	18.02.1925	69.0	145.0	5.0
7.	07.01.1927	82.0	126.0	5.1
8.	14.11.1927	69.9	129.9	6.8
9.	14.11.1927	70.1	129.2	6.8
10.	15.11.1927	70.5	128.5	5.8
11.	03.02.1928	70.5	128.8	6.2
12.	21.02.1928	66.5	−173.0	6.9
13.	24.02.1928	67.2	−173.4	6.3
14.	26.02.1928	66.7	−172.5	6.5
15.	01.05.1928	66.8	−172.0	6.2
16.	16.08.1928	70.0	126.0	5.6
17.	15.07.1931	58.9	149.0	6.2
18.	10.10.1931	59.3	147.8	6.6
19.	14.08.1932	62.8	154.6	5.4
20.	03.11.1936	59.0	151.2	5.7

Table 1. *Cont.*

No.	Date	$\varphi,^{\circ}$	$\lambda,^{\circ}$	M
21.	21.09.1937	58.0	165.0	5.6
22.	22.01.1943	59.0	151.0	5.0
23.	20.01.1944	60.0	152.0	5.0
24.	14.02.1944	61.0	147.4	5.0
25.	25.06.1945	59.0	160.0	5.0
26.	07.01.1951	57.9	163.2	5.2
27.	09.01.1951	81.0	126.5	5.6
28.	12.02.1951	65.8	137.0	6.5
29.	04.04.1951	65.0	136.0	5.2
30.	14.04.1951	61.3	137.4	7.1
31.	29.04.1951	81.5	131.0	5.3
32.	07.1954	66.0	141.0	5.0
33.	10.12.1955	64.0	152.0	5.2
34.	27.09.1957	64.2	178.2	5.7
35.	30.10.1959	65.9019	136.8772	5.3
36.	04.03.1962	67.5	−172.9	5.0
37.	19.04.1962	69.7372	138.7082	5.7
38.	20.05.1963	72.1598	126.4704	5.0
39.	21.07.1964	71.9745	129.9732	5.4
40.	13.12.1964	64.8511	−165.719	5.5
41.	09.09.1968	66.1953	142.192	5.1
42.	22.11.1969	57.6661	163.5126	7.5
43.	27.11.1969	58.0029	163.2877	5.0
44.	23.12.1969	57.6	163.41	5.8
45.	05.06.1970	63.3638	146.1613	5.6
46.	19.06.1970	57.77	163.65	5.5
47.	18.05.1971	63.9304	145.9633	7.0
48.	30.09.1971	61.6821	140.3041	5.8
49.	05.10.1971	67.3875	−172.689	5.3
50.	13.01.1972	61.9098	147.063	5.7
51.	03.08.1972	59.4635	163.2362	5.4
52.	19.06.1974	63.1401	150.8163	5.0
53.	21.01.1976	67.6968	140.0102	5.1
54.	21.01.1976	58.8296	163.7961	6.2
55.	22.01.1976	58.8758	163.6817	5.3
56.	17.02.1977	58.9873	163.759	5.1
57.	19.08.1979	61.3151	159.334	5.3
58.	01.02.1980	73.0967	122.5227	5.3
59.	08.11.1981	61.8134	153.8102	5.4
60.	22.11.1984	68.4472	140.951	5.1
61.	10.09.1985	60.2631	169.1538	5.2
62.	01.03.1991	72.1393	126.8703	5.3
63.	08.03.1991	60.828	167.0754	6.6
64.	08.03.1991	60.8046	167.1495	5.9
65.	10.03.1991	60.9156	167.252	5.2
66.	12.03.1991	60.7718	167.1724	5.4
67.	17.04.1991	60.7056	166.9719	5.3
68.	17.04.1991	60.6803	166.9327	5.2
69.	18.04.1991	60.8177	167.0288	5.5
70.	27.04.1991	60.7949	167.1015	5.3
71.	12.06.1991	58.3428	163.2646	5.3
72.	17.07.1992	60.8576	167.3209	5.2
73.	26.03.1994	58.2648	164.1291	5.5
74.	02.10.1995	66.5262	179.2814	5.2
75.	07.07.1996	58.5323	157.8279	5.7
76.	08.08.1996	58.6572	157.7733	5.1
77.	24.10.1996	66.9183	−173.041	6.0
78.	03.01.1997	60.7922	167.4344	5.4
79.	24.03.1997	67.1386	−173.197	5.0

Table 1. *Cont.*

No.	Date	φ,°	λ,°	M
80.	15.04.1998	58.433	164.6887	5.8
81.	07.01.1999	67.6958	141.2996	5.2
82.	07.01.2001	59.4549	147.2577	5.4
83.	25.01.2005	69.6608	138.5656	5.1
84.	20.04.2006	60.8802	167.0464	7.6
85.	20.04.2006	60.8857	167.2848	5.9
86.	21.04.2006	61.1195	166.9906	5.8
87.	21.04.2006	60.5932	166.2015	5.4
88.	21.04.2006	60.4496	165.9587	6.1
89.	21.04.2006	61.3001	167.7524	6.0
90.	21.04.2006	60.6778	165.886	5.2
91.	21.04.2006	60.9465	166.8856	5.0
92.	22.04.2006	61.1619	167.3084	5.5
93.	26.04.2006	60.9509	166.6578	5.0
94.	29.04.2006	60.9127	165.799	5.2
95.	29.04.2006	60.4481	167.6232	6.6
96.	29.04.2006	60.6655	166.0764	5.2
97.	09.05.2006	60.6836	165.9512	5.7
98.	18.05.2006	60.7608	165.9845	5.0
99.	22.05.2006	60.7339	165.8081	6.6
100.	22.05.2006	60.7847	165.9961	5.3
101.	24.05.2006	60.8614	165.5075	5.0
102.	06.09.2006	61.5752	168.6293	5.3
103.	04.10.2006	60.6272	165.7793	5.0
104.	19.10.2006	64.1002	148.8337	5.2
105.	11.01.2007	60.8981	165.6173	5.0
106.	24.05.2007	62.2792	171.7404	5.3
107.	13.04.2008	67.6428	−166.86	5.2
108.	22.06.2008	67.6952	141.3933	6.1
109.	21.04.2009	64.5778	168.6858	5.0
110.	08.05.2009	58.0536	164.3764	5.4
111.	02.08.2010	61.9985	145.6676	5.3
112.	21.05.2011	65.4772	−166.857	5.0
113.	16.11.2011	65.1751	146.0932	5.1
114.	21.02.2012	67.5811	−166.561	5.1
115.	26.03.2012	66.2727	−174.545	5.1
116.	24.06.2012	57.5012	163.4145	6.0
117.	20.01.2013	64.8134	146.554	5.6
118.	14.02.2013	67.5173	142.7017	6.7
119.	05.03.2013	67.661	142.6265	5.1
120.	13.03.2013	60.1064	163.5095	5.8
121.	10.05.2013	67.5282	139.1438	5.2
122.	25.08.2014	67.6471	142.4612	5.0
123.	06.03.2017	60.8706	167.2633	5.0
124.	12.06.2017	60.9257	167.2364	5.1
125.	16.01.2018	63.1535	−172.293	5.0
126.	27.08.2018	58.7283	158.8544	5.0
127.	26.12.2019	58.8313	158.7925	5.2
128.	09.01.2020	62.358	171.0611	6.4
129.	09.01.2020	62.2438	170.9805	5.2
130.	01.09.2020	58.82	159.002	5.6

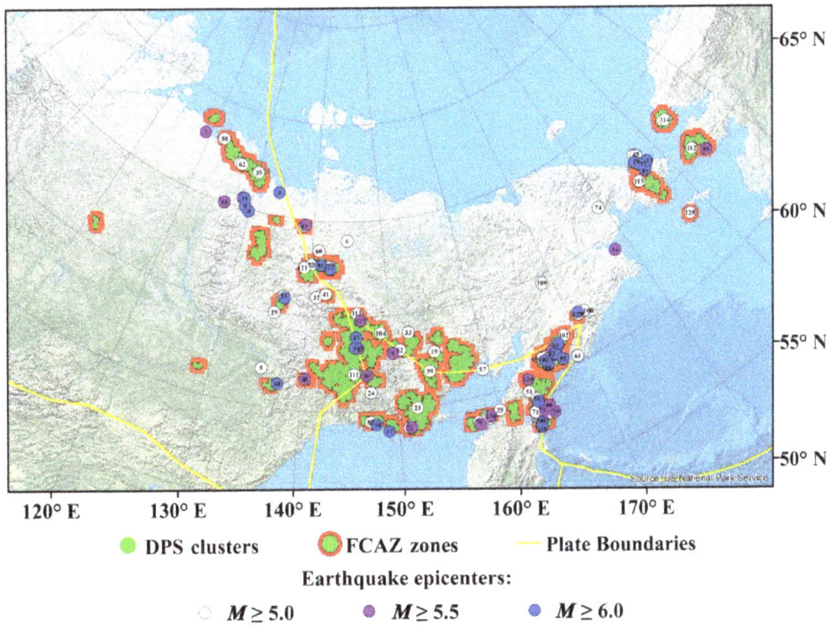

Figure 3. FCAZ zones prone to earthquakes in the eastern sector of the Russian Arctic and epicenters of earthquakes with $M \geq 5.0$.

3. Results

As the magnitude threshold M_R, starting from which the earthquake epicenters are used as objects of FCAZ recognition of strong earthquake-prone areas, was chosen, $M_R = 2.5$. The combined catalog contains 10,486 crustal events with $M \geq 2.5$, which made up the set of recognition objects W.

At the first step of the set W, which is a subset of the set of earthquake epicenters shown in Figure 2, the topological filtering algorithm DPS was applied. Three iterations of the algorithm were performed. The optimal values ($q_1 = -2.0$, $q_2 = -2.5$, $q_3 = -2.5$; $\beta_1 = 0.1$, $\beta_2 = 0.1$, $\beta_3 = 0.05$) of the input parameters of the algorithm were chosen by artificial intelligence blocks [34]. DPS clusters (Figure 3) included 67.2% of the epicenters of earthquakes with $M \geq 2.5$ used in the problem as objects of FCAZ recognition. In other words, $|DPS(W)| = 67.2\%|W|$.

In the second step, the E^2XT algorithm was applied to the recognized DPS clusters. The optimal values ($v = -4.0$ и $w = -2.25$; the geographical grid spacing is 0.25°) of its input parameters were also selected automatically using blocks [31]. In Figure 3, a combination of green and red colors shows the result—the strong earthquake-prone areas in the eastern sector of the Russian Arctic, recognized by the FCAZ method.

As can be seen from Figure 3, the recognized FCAZ zones are in good agreement with the location of the epicenters of historical and instrumental earthquakes with $M \geq 5.0$ (Table 1). In addition, certain potential high seismicity zones (Figure 3) are well confined to the boundaries of the Eurasian, North American, and Okhotsk tectonic plates. In the zone of contact of all three plates and the Okhotsk Plate with the North American Plate (Figure 3), the FCAZ system recognized very large FCAZ zones containing half of the known strong earthquakes (Table 1). This is especially true for the northeastward elongated margin of the Okhotsk Plate, where more than a third of the known events with $M \geq 6.0$ occurred. Furthermore, FCAZ zones are recognized within the Cherskiy Range, which is one of the main geomorphological structures (Figure 1) of the region. Within this range, a fairly large number of instrumentally registered strong earthquakes are known (Figure 3). The consistency of FCAZ zones with the boundaries of tectonic plates is a significant argument in favor of the reliability of the recognition [9,54].

Out of 130 events with $M \geq 5.0$, only 19 (14.6%) epicenters are located outside the FCAZ zones. Note that 14 of them occurred before 1962, i.e., long before the start of systematic instrumental observations of the seismicity of the region. Their epicenters and magnitudes were determined with a significant error [15]. Thus, for six events with $M = 5.0$–5.2, the magnitude error is ± 0.5–0.7, and for eight events, the epicenter coordinates are determined with an error of ± 1–$2°$ [15]. At the same time, exactly half of the latter is located at the very border of the recognized FCAZ zones. Thus, with a high probability, some of the earthquakes that make up "missed target" errors may not be such.

Let us now consider the epicenters of stronger earthquakes. Out of 54 considered events with $M \geq 5.5$ 8 epicenters (14.8%) are located outside the zones recognized by the FCAZ method. All such events occurred before 1957, i.e., before the beginning of the instrumental earthquake catalog, and therefore, belong to the historical part of the used catalog of strong earthquakes (Table 1). The coordinates of their epicenters were determined with an error from 0.5° to 2°. At the same time, half of the events have magnitudes $M = 5.5$–5.8, estimated with an error of ± 0.3–0.5. Thus, it is very likely that some events that did not fall into the FCAZ zones may not be missed targets of recognition.

Let us consider the case where $M_0 = 6.0$. In turn, four (14.8%) out of 27 events with $M \geq 6.0$ are errors of the "missed target" type. Here, all missed earthquakes occurred before 1930. The coordinates of their epicenters were determined with an error of $\pm 0.5°$–$1°$, and the magnitudes—± 0.5. The latter is also confirmed by the work [55], where for these events the estimates of magnitudes are given that are 0.2–0.4 less than in [15] (Table 1). It should also be noted that three of the four missed epicenters are located within a fairly small area (the maximum distance between the epicenters is $\approx$80 km). These events have the same magnitude estimates and occurred within three and a half months (the first two with a difference of less than five hours). These three events should likely be considered as one missed target of recognition. Thus, interpreting the result (Figure 3) for earthquakes with $M \geq 6.0$, we can speak of only two (8.0%) misses.

Considering only strong events that occurred after the start of systematic seismological observations in 1962 (the beginning of the used instrumental catalog), and which have reliable instrumental estimates of the magnitude and coordinates of the epicenter, we have the following positive results. Out of 95 earthquakes with $M \geq 5.0$, epicenters of five (5.26%) are located outside the FCAZ zones, and for earthquakes with $M \geq 5.5$ and $M \geq 6.0$, in this case, there are no "missed target" errors.

It should be noted that, according to the ANSS and EMSC catalogs, after the end of the integrated catalog used for FCAZ recognition, events with $M \geq 5.0$ have not yet occurred in the continental part of the studied region (the catalog ends in 2020). Therefore, the use of a direct control sample to assess the reliability of this recognition is impossible.

Figure 3 shows that the recognized FCAZ zones contain not only the epicenters of earthquakes with $M \geq 5.0$, which are directly confined to the boundaries of the Eurasian, North American, and Okhotsk tectonic plates; they also contain some events that occurred at a sufficient distance from the main tectonic structures of the studied region. This emphasizes the nontriviality of the result of the performed FCAZ recognition. Note that the FCAZ zones occupy only $\approx$ 10% of the continental part of the region within which recognition was performed (shown by the white frame in Figures 1 and 2) and contain 68.4% of events with $M \geq 3.0$, 68.9% with $M \geq 3.5$, 70.5% with $M \geq 4.0$, and 69.8% with $M \geq 4.5$ from those available in the instrumental catalog used for recognition. This once again indicates the nontriviality of the obtained result of FCAZ recognition (Figure 3).

Note that FCAZ zones for $M_0 = 5.0$, $M_0 = 5.5$, and $M_0 = 6.0$ declare the same territory as hazardous. Moreover, if the percentage of "missed target" errors is equal when considering all strong events from Table 1 their number is reduced to zero for earthquakes with $M \geq 5.5$ and $M \geq 6.0$ when the historical part of the catalog (which contains possible significant errors in determining the coordinates of epicenters and estimating magnitudes) is discarded.

The foregoing allows us to interpret the recognized FCAZ zones (Figure 3) as areas prone to earthquakes with $M \geq 5.5$ in the eastern sector of the Arctic zone of the Russian Federation.

Note that the result of FCAZ recognition shown in Figure 3 can be interpreted in another way. For example, the territory of the studied region not covered by recognized high seismicity zones can be considered as areas where only earthquakes with a magnitude $M < 5.0$ can occur. Thus, according to the results of this work, strong earthquakes cannot occur in ≈90% of the territory of the eastern sector of the Arctic zone of the Russian Federation. At the same time, the hazardous areas of ≈ 10% are outlined in Figure 3.

In the eastern sector of the Russian Arctic, it is impossible to perform control experiments of the "seismic history" type [29], which substantiate the reliability of FCAZ recognition by demonstrating the stability of the recognized zones in space and time. This is due to the lack of recognition objects (earthquake epicenters with $M \geq 2.5$) to perform control experiments on in such a large territory as the studied region of the eastern sector of the Russian Arctic. As mentioned above, there were no strong events in the region that could form material for conducting a pure experiment (checking the correspondence between the recognized FCAZ zones and the epicenters of earthquakes that occurred after the end of the catalog, which is a set of recognition objects).

To substantiate the reliability and optimality of the recognized FCAZ zones (Figure 3), we compared them with the zones recognized using other (smaller) values of the M_R threshold, which determines a set of objects. In other words, the number of recognition objects varied. Figures 4–7 show the FCAZ zones prone to strong earthquakes in the eastern sector of the Russian Arctic, mapped using thresholds: $M_R = 2.0$, $M_R = 1.5$, $M_R = 1.0$, and $M_R = 0.5$, which determine the set $\overline{W}$.

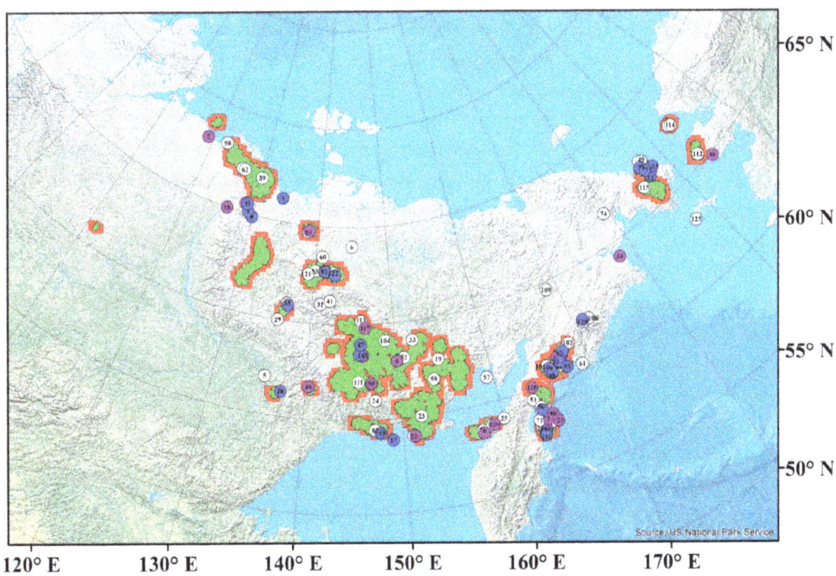

Figure 4. FCAZ zones prone to earthquakes (calculated at $M_R = 2.0$) in the eastern sector of the Russian Arctic and earthquake epicenters with $M \geq 5.0$. Designations as in Figure 3.

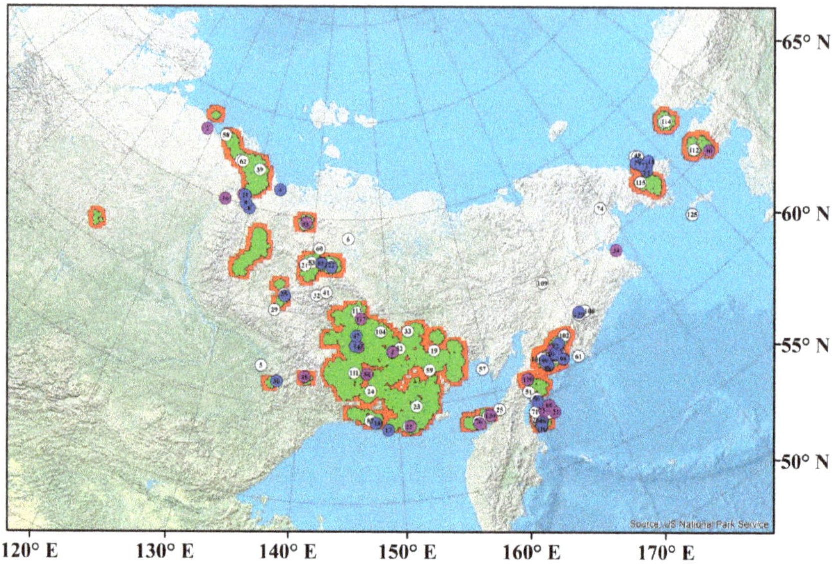

Figure 5. FCAZ zones prone to earthquakes (calculated at M_R = 1.5) in the eastern sector of the Russian Arctic and earthquake epicenters with $M \geq 5.0$. Designations as in Figure 3.

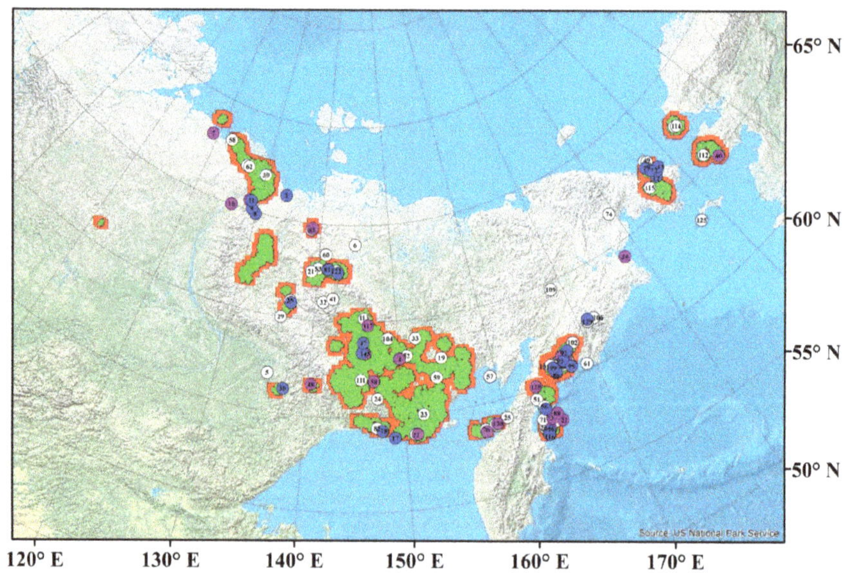

Figure 6. FCAZ zones prone to earthquakes (calculated at M_R = 1.0) in the eastern sector of the Russian Arctic and earthquake epicenters with $M \geq 5.0$. Designations as in Figure 3.

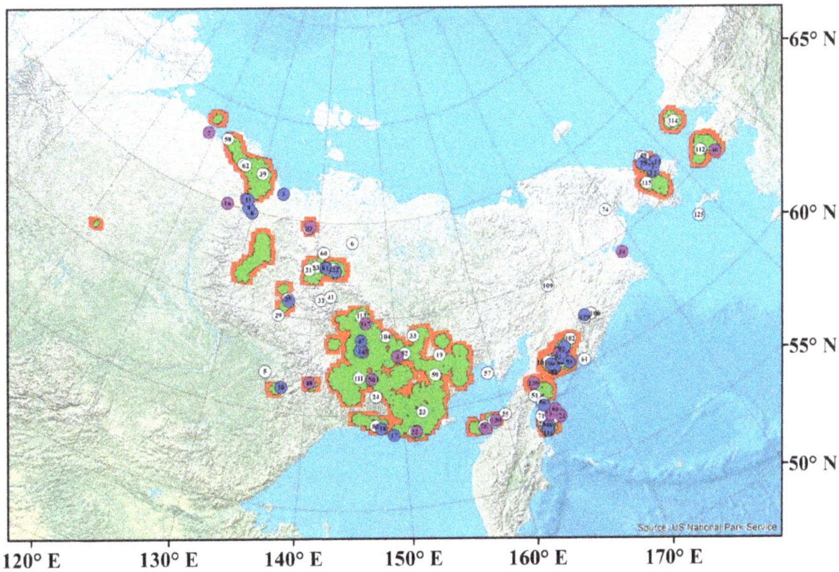

Figure 7. FCAZ zones prone to earthquakes (calculated at M_R = 0.5) in the eastern sector of the Russian Arctic and earthquake epicenters with $M \geq 5.0$. Designations as in Figure 3.

From Figures 5–7, it can be seen that FCAZ zones recognized using M_R = 1.5, M_R = 1.0, and M_R = 0.5 have almost the same number of "missed target" errors as the main variant (Figure 3). Outside the recognized zones are located 20 out of 130 earthquake epicenters with $M \geq 5.0$, 11 out of 54 with $M \geq 5.5$, and five out of 27 with $M \geq 6.0$. It should be noted that the area of these zones is 25–30% larger than the area of FCAZ zones in Figure 3. That is, in the case of low values of the M_R threshold, the number of false alarms increases, which negatively affects the recognition reliability. In this case, the values M_R = 1.5, M_R = 1.0, and M_R = 0.5 are significantly less than the magnitude M_c (completeness magnitude) in the catalog [56–58]. This means, with such values of M_R, the set of recognition objects is an incomplete sample of earthquake epicenters, with a high probability containing numerous errors in the values of seismicity parameters. The latter affects the result of FCAZ recognition.

The high seismicity zones recognized using M_R = 2.0 (Figure 4) are comparable in area to the main FCAZ zones (Figure 3). The area difference is less than 1%. At the same time, we note that these zones have the following number of "missed target" errors: 22 earthquake epicenters with $M \geq 5.0$, 14 with $M \geq 5.5$, and seven with $M \geq 6.0$, which is significantly more than the main variant.

Zone analysis in Figures 3–7, the results of comparison of their areas, and statistics of "missed target" errors show that the FCAZ zones recognized using the threshold M_R = 2.5 with the smallest area of false alarms are in the best agreement with the epicenters of known strong earthquakes in the studied region. This substantiates the reliability of the interpretation of high seismicity zones shown in Figure 3 as earthquake-prone areas with $M \geq 5.5$ in the eastern sector of the Arctic zone of the Russian Federation. It should also be noted that the obtained results substantiate the correct choice of the threshold value M_R based on the assessment of the completeness magnitude M_c in the catalog.

4. Conclusions

In this paper, using the FCAZ system-analytical method, strong earthquake-prone areas in the eastern sector of the Arctic zone of the Russian Federation were recognized. As shown in the text of the article and Figure 3, FCAZ zones are in good agreement with the epicenters of earthquakes with $M \geq 5.5$ and $M \geq 6.0$. Interpreting FCAZ zones as areas prone to earthquakes with $M \geq 5.5$, and substantiating their reliability, it should be noted that they can also be interpreted as areas prone to events with $M \geq 6.0$. This is confirmed by the statistics of "missed target" errors given in the paper.

For the first time, the high seismicity zones determined by the FCAZ method were interpreted for two thresholds M_0 when recognizing in the Crimean Peninsula and the northwestern Caucasus [41]. In the same paper [41], a mathematical theory was constructed which makes it possible to represent FCAZ zones as a fuzzy set to consider areas in which earthquakes with a magnitude of $M_0^1 \leq M < M_0^2$ can occur.

Another important result was obtained in the present paper. The classification shown in Figure 3 turns out to be stable for the variation of the threshold M_0 in the interval $\Delta M_0 \approx 1$ ($5.0 \leq M_0 \leq 6.0$). Such stability of the solution for the M_0 threshold, which was not observed in previous FCAZ regions, serves as a weighty argument in favor of the reliability of the solution presented in Figure 3.

Note that the results of this paper point to the need to use the completeness magnitude estimate M_c in the catalog as the magnitude threshold M_R (starting from which the earthquake epicenters are used as FCAZ recognition objects). This will make it possible to use in FCAZ recognition the data set that is the most complete and contains the minimum number of errors in the values of the seismicity parameters of the considered high seismicity region.

Author Contributions: Conceptualization, A.D.G.; data curation, B.A.D., B.V.D., A.A.S. and I.M.N.; formal analysis, A.D.G., B.A.D., B.V.D., E.O.K. and A.A.S.; investigation, B.A.D., E.O.K. and I.M.N.; methodology, A.D.G. and B.A.D.; software, B.A.D.; supervision, A.D.G.; validation, B.V.D., E.O.K., A.A.S. and I.M.N.; visualization, B.V.D., E.O.K. and I.M.N.; writing—original draft, A.D.G., B.A.D., B.V.D., E.O.K., A.A.S. and I.M.N. All authors have read and agreed to the published version of the manuscript.

Funding: The reported study was funded by the Russian Science Foundation, project number 21-77-30010, System analysis of geophysical process dynamics in the Russian Arctic and their impact on the development and operation of the railway infrastructure.

Institutional Review Board Statement: Not applicable.

Informed Consent Statement: Not applicable.

Data Availability Statement: Not applicable.

Acknowledgments: This work employed data provided by the Shared Research Facility, the Analytical Geomagnetic Data Center of the Geophysical Center of RAS (http://ckp.gcras.ru/ accessed on: 23 October 2022).

Conflicts of Interest: The authors declare no conflict of interest.

References

1. Shelomentsev, V.N. *The Legal Aspects of the Development and Protection of the Arctic*; Scientific Bulletin of the Moscow State Technical University: Moscow, Russia, 2015; Volume 216, pp. 113–117.
2. Bogorodskiy, P.V.; Demidov, N.E.; Filchuk, K.V.; Marchenko, A.V.; Morozov, E.G.; Nikulina, A.L.; Pnyushkov, A.V.; Ryzhov, I.V. Growth of landfast ice and its thermal interaction with bottom sediments in the Braganzavägen Gulf (West Spitsbergen). *Russ. J. Earth Sci.* **2020**, *20*, ES6005. [CrossRef]
3. Marchenko, A.V.; Morozov, E.G.; Ivanov, A.V.; Elizarova, T.G.; Frey, D.I. Ice thickening caused by freezing of tidal jet. *Russ. J. Earth Sci.* **2021**, *2*, ES2004. [CrossRef]
4. Morozov, P.A.; Berkut, A.I.; Vorovsky, P.L.; Morozov, F.P.; Pisarev, S.V. Measuring sea ice thickness with the LOZA georadar. *Russ. J. Earth Sci.* **2021**, *21*, ES4003. [CrossRef]
5. Sharmar, V.; Markina, M. Evaluation of interdecadal trends in sea ice, surface winds and ocean waves in the Arctic in 1980–2019. *Russ. J. Earth Sci.* **2021**, *21*, ES2002. [CrossRef]
6. Spyros, A.; Arnold, M.; Gert, J. Geohazards and Pipelines. State-of-the-Art Design Using Experimental, Numerical and Analytical Methodologies. In *Geohazards and Pipelines*; Springer Nature: Berlin/Heidelberg, Germany, 2021. [CrossRef]
7. Artyushkov, E.V.; Chekhovich, P.A. Geodynamics of the Lomonosov Ridge in the Central Arctic. *Russ. J. Earth Sci.* **2019**, *19*, ES1004. [CrossRef]

8. Vernikovsky, V.; Shemin, G.; Deev, E.; Metelkin, D.; Matushkin, N.; Pervukhina, N. Geodynamics and Oil and Gas Potential of the Yenisei-Khatanga Basin (Polar Siberia). *Minerals* **2018**, *8*, 510. [CrossRef]
9. Zonenshain, L.P.; Kuzmin, M.I.; Natapov, L.M. *Tectonics of Lithospheric Plates of the USSR*; Nedra Publishers: Moscow, Russia, 1990; p. 328.
10. Imaev, V.S.; Imaeva, L.P.; Kozmin, B.M.; Mackey, K.; Fujita, K. Seismotectonic processes along boundary of lithospheric plates of Northeast Asia and Alaska. *Geol. Pac. Ocean* **2000**, *15*, 211–232.
11. Chebrov, V.N. The Olyutorskii earthquake of April 20, 2006: Organizing surveys, observations, problems, and results. *J. Volcanol. Seismol.* **2010**, *4*, 75–78. [CrossRef]
12. Lander, A.V.; Levina, V.I.; Ivanova, E.I. The earthquake history of the Koryak Upland and the aftershock process of the $M_W7.6$ April 20(21), 2006 Olyutorskii earthquake. *J. Volcanol. Seismol.* **2010**, *4*, 87–100. [CrossRef]
13. Rogozhin, E.A.; Ovsyuchenko, A.N.; Marakhanov, A.V.; Novikov, S.S. A geological study of the epicentral area of the April 20(21), 2006 Olyutorskii earthquake. *J. Volcanol. Seismol.* **2010**, *4*, 79–86. [CrossRef]
14. Shibaev, S.V.; Kozmin, B.M.; Imaev, V.S.; Imaeva, L.P.; Petrov, A.F.; Starkova, N.N. The February 14, 2013 Ilin-Tas (Abyi) earthquake (Mw=6.7), Northeast Yakutia. *Russ. J. Earth Sci.* **2020**, *2*, 92–102. (In Russian) [CrossRef]
15. Kondorskaya, N.V.; Shebalin, N.V.; Khrometskaya, Y.A.; Gvishiani, A.D. *New Catalog of Strong Earthquakes in the USSR from Ancient Times through 1977*; Report SE-31; World Data Center A for Solid Earth Geophysics, NOAA: Boulder, CO, USA, 1982; 608p.
16. Gvishiani, A.D.; Vorobieva, I.A.; Shebalin, P.N.; Dzeboev, B.A.; Dzeranov, B.V.; Skorkina, A.A. Integrated Earthquake Catalog of the Eastern Sector of the Russian Arctic. *Appl. Sci.* **2022**, *12*, 5010. [CrossRef]
17. Assinovskaya, B.A. Seismic hazard assessment of the Barents sea region. *Inzhenernye Izyskaniya.* **2021**, *15*, 76–88. (In Russian) [CrossRef]
18. Assinovskaya, B.A.; Panas, N.M.; Ovsov, M.K.; Antonovskaya, G.N. Preliminary seismic hazard assessment of the Arctic Gakkel ridge and surrounding. *Russ. J. Seismol.* **2019**, *1*, 35–45. (In Russian) [CrossRef]
19. Imaev, V.S.; Imaeva, L.P.; Koz'min, B.M. *Seismotectonics of Yakutia*; GEOS: Moscow, Russia, 2000; p. 226.
20. Imaeva, L.P.; Imaev, V.S.; Koz'min, B.M. Dynamics of the Zones of Strong Earthquake Epicenters in the Arctic–Asian Seismic Belt. *Geosciences* **2019**, *9*, 168. [CrossRef]
21. Imaeva, L.P.; Kolodeznikov, I.I. *Seismotectonics of the Northeastern Sector of Russian Arctic*; Publishing House of the Siberian Branch of the Russian Academy of Sciences: Novosibirsk, Russia, 2017.
22. Giardini, D. The Global Seismic Hazard Assessment Program (GSHAP)—1992/1999. *Ann. Geofis.* **1999**, *42*, 957–974. [CrossRef]
23. Giardini, D.; Grunthal, G.; Shedlock, K.M.; Zhang, P. The GSHAP Global Seismic Hazard Map. *Ann. Geofis.* **1999**, *42*, 1225–1228. [CrossRef]
24. Ulomov, V.I. General seismic zoning of the territory of Russian Federation: GSZ-2012. *Seism. Instrum.* **2014**, *50*, 290–304. [CrossRef]
25. Ulomov, V.I. Seismic hazard of Northern Eurasia. *Ann. Geofis.* **1999**, *42*, 1023–1038. [CrossRef]
26. Ulomov, V.I.; Peretokin, S.A.; Medvedeva, N.S.; Akatova, K.N.; Danilova, T.I. Seismological aspects of general seismic zoning for the territory of the Russian Federation territory (OSR-97, OSR-2012, and OSR-2014 Maps). *Seism. Instrum.* **2015**, *51*, 311–328. [CrossRef]
27. Shebalin, P.N.; Gvishiani, A.D.; Dzeboev, B.A.; Skorkina, A.A. Why Are New Approaches to Seismic Hazard Assessment Required? *Dokl. Earth Sci.* **2022**, *507*, 930–935. [CrossRef]
28. Gorshkov, A.I.; Soloviev, A.A. Recognition of earthquake-prone areas in the Altai-Sayan-Baikal region based on the morphostructural zoning. *Russ. J. Earth Sci.* **2021**, *21*, ES1005. [CrossRef]
29. Gvishiani, A.D.; Soloviev, A.A.; Dzeboev, B.A. Problem of Recognition of Strong-Earthquake-Prone Areas: A State-of-the-Art Review. *Izv. Phys. Solid Earth* **2020**, *56*, 1–23. [CrossRef]
30. Kossobokov, V.G.; Soloviev, A.A. Pattern recognition in problems of seismic hazard assessment. *Chebyshevskii Sb.* **2018**, *19*, 55–90. (In Russian) [CrossRef]
31. Gvishiani, A.D.; Dzeboev, B.A.; Agayan, S.M. FCAZm intelligent recognition system for locating areas prone to strong earthquakes in the Andean and Caucasian mountain belts. *Izv. Phys. Solid Earth* **2016**, *52*, 461–491. [CrossRef]
32. Agayan, S.M.; Bogoutdinov, S.h.R.; Dzeboev, B.A.; Dzeranov, B.V.; Kamaev, D.A.; Osipov, M.O. DPS Clustering: New Results. *Appl. Sci.* **2022**, *12*, 9335. [CrossRef]
33. Gvishiani, A.; Dzeboev, B.; Agayan, S. A new approach to recognition of the earthquake-prone areas in the Caucasus. *Izv. Phys. Solid Earth* **2013**, *49*, 747–766. [CrossRef]
34. Dzeboev, B.A.; Gvishiani, A.D.; Agayan, S.M.; Belov, I.O.; Karapetyan, J.K.; Dzeranov, B.V.; Barykina, Y.V. System-Analytical Method of Earthquake-Prone Areas Recognition. *Appl. Sci.* **2021**, *11*, 7972. [CrossRef]
35. Dzeboev, B.A.; Agayan, S.M.; Zharkikh, Y.I.; Krasnoperov, R.I.; Barykina, Y.V. Strongest Earthquake-Prone Areas in Kamchatka. *Izv. Phys. Solid Earth* **2018**, *54*, 284–291. [CrossRef]
36. Dzeboev, B.A.; Karapetyan, J.K.; Aronov, G.A.; Dzeranov, B.V.; Kudin, D.V.; Karapetyan, R.K.; Vavilin, E.V. FCAZ-recognition based on declustered earthquake catalogs. *Russ. J. Earth Sci.* **2020**, *20*, ES6010. [CrossRef]
37. Gvishiani, A.; Dobrovolsky, M.; Agayan, S.; Dzeboev, B. Fuzzy-based clustering of epicenters and strong earthquake-prone areas. *Environ. Eng. Manag. J.* **2013**, *12*, 1–10. [CrossRef]
38. Gvishiani, A.D.; Dzeboev, B.A. Assessment of seismic hazard in choosing of a radioactive waste disposal location. *Min. J.* **2015**, *10*, 39–43. (In Russian) [CrossRef]

39. Gvishiani, A.D.; Dzeboev, B.A.; Belov, I.O.; Sergeyeva, N.A.; Vavilin, E.V. Successive Recognition of Significant and Strong Earthquake-Prone Areas: The Baikal–Transbaikal Region. *Dokl. Earth Sci.* **2017**, *477*, 1488–1493. [CrossRef]
40. Gvishiani, A.D.; Dzeboev, B.A.; Sergeyeva, N.A.; Belov, I.O.; Rybkina, A.I. Significant Earthquake-Prone Areas in the Altai–Sayan Region. *Izv. Phys. Solid Earth* **2018**, *54*, 406–414. [CrossRef]
41. Gvishiani, A.D.; Dzeboev, B.A.; Sergeyeva, N.A.; Rybkina, A.I. Formalized Clustering and the Significant Earthquake-Prone Areas in the Crimean Peninsula and Northwest Caucasus. *Izv. Phys. Solid Earth* **2017**, *53*, 353–365. [CrossRef]
42. Gvishiani, A.; Dubois, J. *Artificial Intelligence and Dynamic Systems for Geophysical Applications*; Springer: Berlin/Heidelberg, Germany, 2002; p. 350. [CrossRef]
43. Gvishiani, A.D.; Mikhailov, V.O.; Agayan, S.M.; Bogoutdinov, S.R.; Graeva, E.M.; Diament, M.; Galdeano, A. Artificial intelligence algorithms for magnetic anomaly clustering. *Izv. Phys. Solid Earth.* **2002**, *38*, 545–559.
44. Widiwijayanti, C.; Mikhailov, V.; Diament, M.; Deplus, C.; Louat, R.; Tikhotsky, S.; Gvishiani, A. Structure and evolution of the Molucca Sea area: Constraints based on interpretation of a combined sea-surface and satellite gravity dataset. *Earth Planet. Sci. Lett.* **2003**, *215*, 35–150. [CrossRef]
45. Agayan, S.; Bogoutdinov, S.; Kamaev, D.; Kaftan, V.; Osipov, M.; Tatarinov, V. Theoretical Framework for Determination of Linear Structures in Multidimensional Geodynamic Data Arrays. *Appl. Sci.* **2021**, *11*, 11606. [CrossRef]
46. Agayan, S.M.; Bogoutdinov, S.R.; Dobrovolsky, M.N. Discrete perfect sets and their application in cluster analysis. *Cybern. Syst. Anal.* **2014**, *50*, 176–190. [CrossRef]
47. Shebalin, P.N.; Baranov, S.V.; Dzeboev, B.A. The Law of the Repeatability of the Number of Aftershocks. *Dokl. Earth Sci.* **2018**, *481*, 963–966.
48. Vorobieva, I.; Gvishiani, A.; Dzeboev, B.; Dzeranov, B.; Barykina, Y.; Antipova, A. Nearest Neighbor Method for Discriminating Aftershocks and Duplicates When Merging Earthquake Catalogs. *Front. Earth Sci. SI Nat. Clust. Earthq. Process. Var. Scales Lab. Exp. Large Earthq.* **2022**, *10*, 820277. [CrossRef]
49. Zaliapin, I.; Ben-Zion, Y. Earthquake clusters in southern California. I: Identification and stability. *J. Geophys. Res. Solid Earth* **2013**, *118*, 2847–2864. [CrossRef]
50. Zaliapin, I.; Ben-Zion, Y. A global classification and characterization of earthquake clusters. *Geophys. J. Int.* **2016**, *207*, 608–634. [CrossRef]
51. Hanks, T.C.; Kanamori, H. A Moment Magnitude Scale. *J. Geophys. Res. Solid Earth* **1979**, *84*, 2348–2350. [CrossRef]
52. Kanamori, H. The Energy Release in Great Earthquakes. *J. Geophys. Res.* **1977**, *82*, 2981–2987. [CrossRef]
53. Storchak, D.A.; Di Giacomo, D.; Bondár, I.; Engdahl, E.R.; Harris, J.; Lee, W.H.; Bormann, P. Public release of the ISC–GEM global instrumental earthquake catalogue (1900–2009). *Seismol. Res. Lett.* **2013**, *84*, 810–815. [CrossRef]
54. Shebalin, N.V. Seismicity as a Tectonic Process. In *Modern Tectonic Activity of the Earth and Seismicity*; Nauka: Moscow, Russia, 1987; pp. 22–37. (In Russian)
55. Atlas of earthquakes in the USSR. In *The Results of Observations of the Network of Seismic Stations of the USSR in 1911–1957*; AS USSR: Moscow, Russia, 1962; 336p. (In Russian)
56. Mignan, A.; Werner, M.J.; Wiemer, S.; Chen, C.-C.; Wu, Y.-M. Bayesian estimation of the spatially varying completeness magnitude of earthquake catalogs. *Bull. Seismol. Soc. Am.* **2011**, *101*, 1371–1385. [CrossRef]
57. Vorobieva, I.; Narteau, C.; Shebalin, P.; Beauducel, F.; Nercessian, A.; Clouard, V.; Bouin, M.-P. Multiscale mapping of completeness magnitude of earthquake catalogs. *Bull. Seismol. Soc. Am.* **2013**, *103*, 2188–2202. [CrossRef]
58. Wiemer, S.; Wyss, M. Minimum Magnitude of Completeness in Earthquake Catalogs: Examples from Alaska, the Western United States, and Japan. *Bull. Seismol. Soc. Am.* **2000**, *90*, 859–869. [CrossRef]

applied sciences

MDPI

Article

Keyboard Model of Seismic Cycle of Great Earthquakes in Subduction Zones: Simulation Results and Further Generalization

Leopold I. Lobkovsky [1,2], Irina S. Vladimirova [1,3], Yurii V. Gabsatarov [1,3] and Dmitry A. Alekseev [1,4,*]

1 Moscow Institute of Physics and Technology (MIPT), Dolgoprudny 141701, Russia; llobkovsky@ocean.ru (L.I.L.); vladis@gsras.ru (I.S.V.); yuryg@gsras.ru (Y.V.G.)
2 P.P. Shirshov Institute of Oceanology of the Russian Academy of Sciences, Moscow 117997, Russia
3 Geophysical Survey of the Russian Academy of Sciences, Obninsk 249035, Russia
4 Schmidt Institute of Physics of the Earth, Russian Academy of Sciences, Moscow 123242, Russia
* Correspondence: alekseev.da@mipt.ru or alexeevgeo@gmail.com

Abstract: Catastrophic megaearthquakes (M > 8) occurring in the subduction zones are among the most devastating hazards on the planet. In this paper we discuss the seismic cycles of the megathrust earthquakes and propose a blockwise geomechanical model explaining certain features of the stress-deformation cycle revealed in recent decades from seismological and satellite geodesy (GNSS) observations. Starting with an overview of the so-called keyboard model of the seismic cycle by L. Lobkovsky, we outline mathematical formalism describing the motion of seismogenic block system assuming viscous rheology beneath and between the neighboring elastic blocks sitting on top of the subducting slab. By summarizing the GNSS-based evidence from our previous studies concerning the transient motions associated with the 2006–2007 Simushir earthquakes, 2010 Maule earthquake, and 2011 Tohoku earthquake, we demonstrate that those data support the keyboard model and reveal specific effect of the postseismic oceanward motion. However, since the seismogenic blocks in subduction systems are mostly located offshore, the direct analysis of GNSS-measured displacements and velocities is hardly possible in terms of the original keyboard model. Hence, the generalized two-segment keyboard model is introduced, containing both frontal offshore blocks and rear onshore blocks, which allows for direct interpretation of the onshore-collected GNSS data. We present a numerical computation scheme and a series of simulated data, which exhibits the consistency with measured motions and enables estimating the seismic cycle characteristics, important for the long-term earthquake forecasting.

Keywords: keyboard model; seismic cycle; subduction zone; fault-block structure; geomechanical model; earthquake generation

Citation: Lobkovsky, L.I.; Vladimirova, I.S.; Gabsatarov, Y.V.; Alekseev, D.A. Keyboard Model of Seismic Cycle of Great Earthquakes in Subduction Zones: Simulation Results and Further Generalization. *Appl. Sci.* **2021**, *11*, 9350. https://doi.org/10.3390/app11199350

Academic Editor: Fabrizio Balsamo

Received: 29 July 2021
Accepted: 28 September 2021
Published: 8 October 2021

1. Introduction

Large-magnitude subduction-associated earthquakes with M > 8 cause the release of enormous elastic stresses accumulated over hundreds of years or even a millennium. Forecast of such shocks, leading to substantial social, economic and ecological consequences, is one of the most crucial problems in geophysics [1,2]. To date, the most developed forecast-related approaches include deformational monitoring [3–5], identification of geophysical precursors, as well as statistical analysis of the seismicity pattern [6–8]. Machine learning (ML) techniques and big data analysis have also become a trend in recent years, and have potential to increase forecast reliability [9]. However, given the rare occurrence of catastrophic M > 8 events (around once-in-a-century), with only few such megaearthquakes recorded over a history of seismological observation, this approach is faced with difficulties due to a small size of the training sample. Thus, some physical (mechanical) model is strongly needed for appropriate description of geodynamical setting during the earthquake preparation phase, and such model might be employed in combination with ML methods to better constrain the results.

Geomechanical models used in seismology and geodynamics are mainly based on continuum mechanics, and widely employ the concept of the continuous structure being ruptured at the instant of an earthquake which results in the formation of a plane fault [10,11]. Models of seismic cycles associated with the strongest subduction earthquakes are mostly based on the concepts of the widely used "asperity" model [12], which explains the difference in characteristic sizes of sources and recurrence periods of catastrophic earthquakes in various subduction zones. Later, a set of mechanical models describing the generation of strongest earthquakes with the source length up to 1000 km was proposed assuming the idea of synchronous failure of several adjacent asperities: from a first simple model [3] to a very complex model by [4,5]. An explanation of seismic–aseismic slip patterns observed on megathrust faults proposed in [13] is also worth mentioning: the results of numerical simulations based on the quasi-dynamic asperity fault model with the slip-dependent friction and stress-dependent healing quite accurately reproduces the main features of real megathrust fault behavior.

However, this approach is still lacking accuracy when modeling the processes of the earthquake preparation, stress release, and post-quake stress relaxation. Analysis of the large-magnitude tsunamigenic earthquakes and seismotectonic features of various subduction zones suggests that these regions have essentially a fault-block structure, with certain patterns in their kinematics [14,15]. The studies based on GPS-measured displacements of the Earth's surface in the source areas of large recent earthquakes during different stages of the seismic cycle, have also confirmed the fault-block structure in the frontal part of the continental margin [16–19]. Thus, geodetic observations combined with geological and seismological studies, clearly indicate that the continental lithosphere in subduction regions has complex structure, revealing a number of separate blocks which move independently from each other.

Explanation of the observed displacements and velocities, which sometimes have opposite direction in neighboring blocks, is hardly possible in terms of the well-known asperity model [12]. At the same time, interpretation of this motion pattern might be possible assuming the so-called "keyboard" model proposed by Lobkovsky et al. [20]. This concept is essentially a geomechanical model describing seismotectonic evolution of subduction regions, taking into account the fault-block structure of the continental margin. Application of the above model implies identification of large fault-separated blocks in the considered area based on geological, geophysical, seismological and geodetic data, which allows tracing their lateral boundaries.

The model involves a numerical formalism allowing for simulation of the displacements of frontal seismically-active blocks in the subduction zone during different phases of the seismic cycle, assuming that rear portion of the continental lithosphere (which corresponds to island arc) is a single structural element that is not separated into smaller blocks and does not experience its own horizontal displacements. However, the GPS-measured velocities along with geological and seismological data clearly indicate that the structure in the rear part of the subduction region also consists of separate segments divided by large faults rooted into the contact zone of interacting lithospheric plates [18,21].

Due to the relatively weak interaction between neighboring blocks, each of them can deform nearly independently of adjacent structural segments, while the overall deformation of the entire system may exhibit a very complex pattern. During seismic events, the rear blocks, similar to frontal ones, may experience almost instantaneous offsets towards the trench. Once the stress is released, a slow straightening of the rear blocks occurs in the postseismic period, partly restrained by the viscous friction at the bottom of the block due to the underlying asthenospheric layer.

The above considerations indicate the need for modification of the original model [20] in order to take into account the discontinuities both in the frontal and rear parts of the island arc. An important motivation for model improvement is also the fact that geodetic observations of Earth's surface motions are mainly carried out onshore, that is, within the rear segment of subduction system, which troubles the direct interpretation of the available displacement data within the framework of the original single-segment model.

A very brief analysis of the generalized model has been published in [22]. In this paper, we discuss the motivation for application of keyboard model in order to explain observed motion patterns in subduction regions, with primary focus on its generalization, leading to the modified two-segment keyboard model. We also present some preliminary simulation data calculated assuming elastic spring-like interaction between the frontal and rear segments.

2. Materials and Methods

In this section we provide the description of single-segment keyboard model and then discuss the observational GNSS evidence and reasoning for the necessity of model generalization.

2.1. Keyboard Model of Seismic Cycles of Great Subduction-Associated Earthquakes

The keyboard model was originally proposed in a series of papers by L. Lobkovsky and his colleagues [20]. Here, we briefly cover this concept both in aspects of geometrical structure and numerical formalism allowing for the simulation of stresses and displacements over the system's evolution cycle.

Geometrically, the modeled region consists of a series of elastic blocks forming a keyboard-like structure aligned with the subduction trench (Figure 1a). Each block is sitting on top of subducting lithosphere, which is constantly moving underneath the continent with velocity V, and there is a thin viscous contact layer separating the block and subducting plate. Tangential stresses acting at the blocks' bottom due to subduction cause the blocks to move slowly away from the trench (continent-ward motion), leading to build up of the elastic stress within each block until the critical stress value is reached and blocks instantly jump towards the ocean. Thus, the evolution of the system follows a certain periodic pattern (Figure 1b), with repeated phases of stress accumulation, instantaneous stress release (earthquake), and postseismic relaxation.

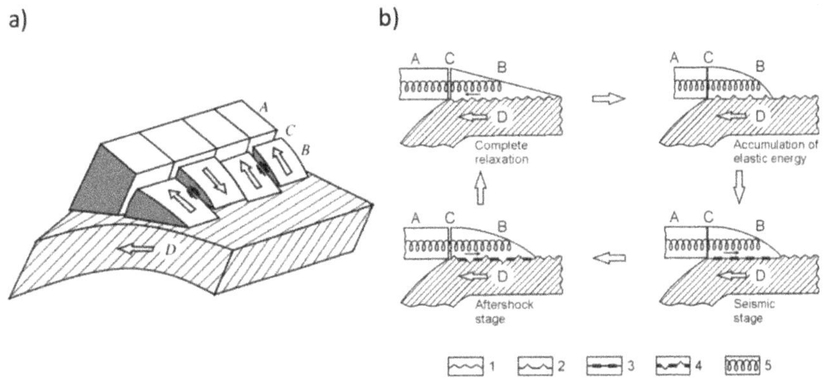

Figure 1. (**a**) Keyboard model geometry; (**b**) seismic deformation cycle phases. A is the fixed undeformable continental margin; B is the seismogenic block; C is the block-margin interface; D is the subducting plate; (1) 2-D geometry for the no deformation state; (2) stress accumulation (preseismic) phase; (3) coseismic phase (destruction of the mechanical coupling in the contact layer); (4) postseismic (aftershock) phase with continued oceanward motion of the blocks; (5) spring imitating the elastic interaction between blocks and margin.

Transition from the stress accumulation to stress release, followed by aftershock relaxation, occurs once any of the frontal blocks reaches a certain critical energy. To ensure the model's capability of predicting nonlinear kinematic behavior during the cycle, the contact layer viscosity is assumed variable, taking higher or lower values depending on the strain rate. The numerical scheme applied for simulating the cyclic dynamics, is described by a system of equilibrium equations connecting stresses in the given block and all the neighboring blocks, expressed in terms of the displacement velocities. These essentially depend on viscosities of the contact layer material and the crustal asthenosphere beneath the rear block, as well as viscosity of destructed zones separating blocks laterally.

In describing the model's mathematical formalism, we will follow notations used in the original paper [20], and present the entire calculation scheme as a flowchart in Figure 2. Model inputs (Figure 2a) include series of parameters defining both geometry and mechanical properties (rheology) of the structure to produce the spatial and temporal distributions of horizontal displacement $w_i(x,t)$ for any given (ith) block ($i = 2, 3, \ldots, \mathrm{m}-1$). The block geometry is given by its lateral dimensions l and d (along x- and y-directions, respectively), and thickness $H(x)$, which varies from H_0 at the outer side of the block to H_l at its inner side (Figure 2b). Typically, l is between 100 to 300 km, while d is 30 to 200 km, depending on a specific structure in a particular subduction zone (Table 1).

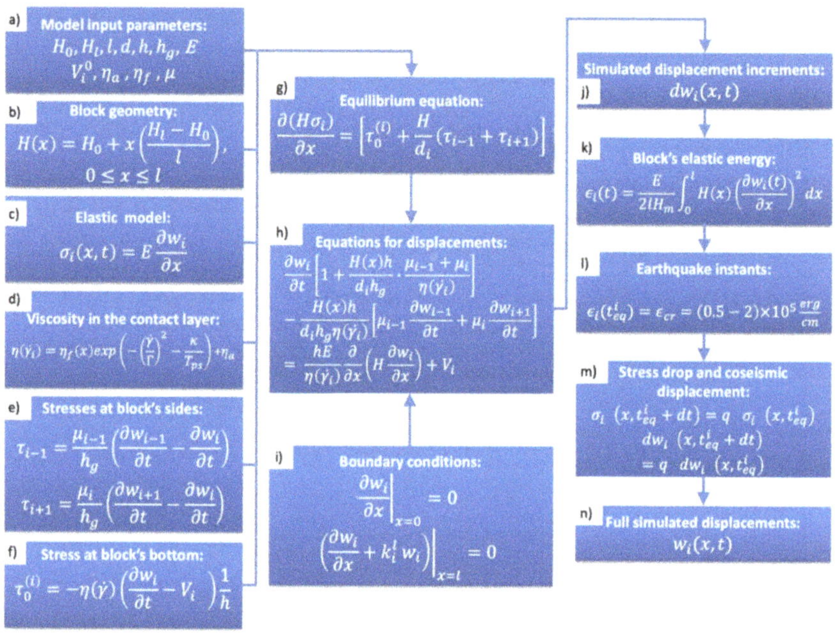

Figure 2. Flowchart showing calculation scheme of the original keyboard model. (**a**) Model input parameters; (**b**) block thickness; (**c**) 1-D elastic model; (**d**) time-dependent viscosity of the contact layer; (**e**) stresses acting on lateral sides of the block; (**f**) stress acting at the block's bottom; (**g**) stress equilibrium equation; (**h**) equation governing displacement evolution; (**i**) boundary conditions applied at the frontal ($x = 0$) and rear ($x = \mathrm{l}$) edges of the block; (**j**) calculating incremental displacements; (**k**) converting displacements into elastic energy accumulated by the block; (**l**) condition to identify the instant of the earthquake; (**m**) stress dumping conditions; (**n**) output dataset.

Table 1. Main characteristics of the structure and seismic cycle in Kuril-Kamchatka, Japan, and Chile subduction zones, according to our studies and available data.

Subduction Zone	Block Width/Length, km	Block Length (Frontal/Rear Segment), km	Cycle Period, yrs	Horizontal Coseismic/Aftershock Displacement at Rear Segment, cm
Kuril-Kamchatka	30–100 [15]	100/100 [15]	140 ± 60 [23]	50/50 [19]
Japan	100–150 [24,25]	200/300 [24,25]	100–1000 [21,26]	50–200/10–50 [18]
Chile	100–200 [27–30]	100/300 [27–30]	63–176 [31]	5–50/50–80 [17]

Mechanically, the mean compressional stress in x-direction is proportional to x-component of deformation (Figure 2c), with E being the elastic modulus of the block. The block is also experiencing tangential stresses τ_{i-1} and τ_{i+1} arising due to the friction of the blocks' lateral boundaries with neighboring blocks (Figure 2e) as well as the stress $\tau_0^{(i)}$ acting on the bottom surface due to the viscous contact layer, Figure 2f. The latter depends on the strain rate in the contact layer with thickness h, written as $\dot{\gamma} = \left(\frac{\partial w_i}{\partial t} - V_i\right)\frac{1}{h}$ (V_i is subduction rate beneath the block), and effective viscosity of the contact layer $\eta(\dot{\gamma}, x, t)$ (Figure 2d). The viscosity $\eta(\dot{\gamma}, x, t)$ is described by a monotonic function of the strain rate, reflecting the transition from the stationary viscosity η_f describing the rheology of the contact layer during the interseismic phase of the cycle to a lower postseismic phase viscosity η_a, associated with the contact layer destruction caused by the recently-occurred earthquake. Here Γ denotes the viscosity recovery parameter ($\Gamma = 10^{-4}\ \frac{1}{yr}$), κ = 0.2–0.5 yr is characteristic time required for the contact layer defects to be healed, and T_{ps} is the time passed since the last earthquake in a given block.

The lateral side stresses (Figure 2e) depend on the thickness of the viscous material filling the interblock faults h_g, as well as μ_{i-1} and μ_i, viscosities of that material between neighboring blocks number. The stress-strain state is assumed to satisfy the stress equilibrium equation (Figure 2g), where d_i is the block size along strike direction, parallel to the trench line. Table 2 shows typical values of the above parameters following [20].

Table 2. Typical values for parameters of viscous elements of the model according to [20].

Contact Layer Thickness, h	Stationary Viscosity of the Contact Layer, η_f	Postseismic Viscosity of the Contact Layer, η_a	Interblock Fault Thickness, h_g	Interblock Viscosity, μ
~0.5 km	~10^{19} Pa·s	~10^{18} Pa·s	~1 km	~4×10^{18} Pa·s

Substitution of expressions for stresses into equilibrium equation yields the series of equations in terms of displacement w (Figure 2h). Additionally, the boundary conditions are applied (Figure 2i), which have the meaning that the outer edges of the frontal blocks are free, while the rear edges undergo elastic interaction with the fixed margin controlled by the rigidity coefficient k_i^r.

The above mathematical formalism is characterized by significant nonlinearity. To solve the problem, we apply an explicit finite-difference method yielding displacement increments $\Delta W_{i,j}^k$ at nodal points x_i, y_j within each block updated at time instants t_k (Figure 2j). Evolution of the block system is controlled by the conditions for dumping the accumulated elastic energy in the form of an abrupt decrease in compressional stresses at time instants t_{eq}^i, corresponding to earthquakes, occurring when the block reaches the threshold energy ϵ_{cr} (Figure 2k–n). Note that this condition is equivalent to an instantaneous change in displacement. Specific time instants of the earthquakes, t_{eq}^i, are determined from the equality of accumulated energy to the threshold value (critical energy) ϵ_{cr} (Figure 2l) prespecified according to available estimates within range $(0.5-2)\times 10^5 \frac{erg}{cm}$ [20] ($H_m = (H_0 + H_l)/2$ is the mean thickness of the block).

Obviously, to start the iterative procedure of the solution, one needs the initial displacements to be specified at zeroth step for each point/block. Although not known in practice, since we are lacking historical geodetic data, these might be calculated from random stresses, chosen within the range of 100 to 300 Bar.

Over time, the system shows the tendency to stabilize and follow a specific cyclic pattern, no matter what specific initial stresses/displacements had been used at the starting point. The choice of initial values only affects the time instants at which the blocks reach critical energy, while other properties of the cycle (i.e., cycle period and the shape of kinematic parameters variation over time) remain constant.

The simulated motion patterns computed for a series of typical parameters mentioned above show a sufficient agreement with observational data in terms of cycle duration (around 150–250 years, Figure 3a) and contain characteristic features, with aftershock oceanward motion of the blocks being the most pronounced one (Figure 3b). Below we discuss its possible geomechanical explanations in the light of the up-to-date GNSS observational data.

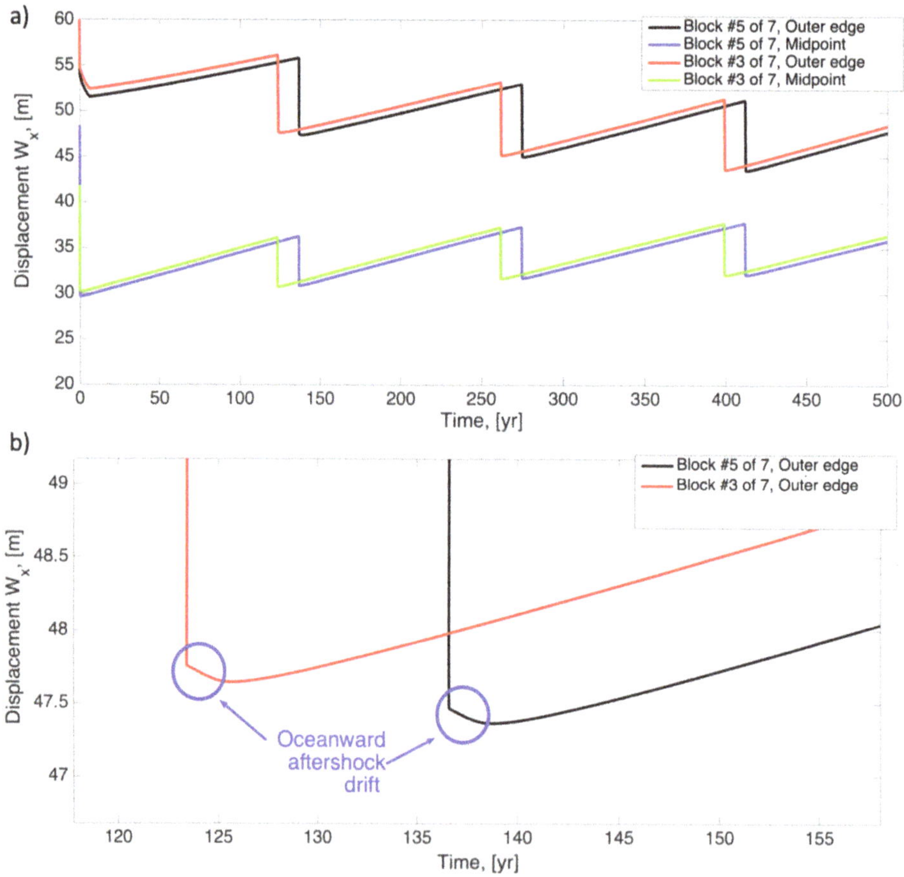

Figure 3. Displacement variations $W(x,t)$, simulated with an original single-segment model for an outer edge and midpoint of blocks. Panel (**a**) shows an overall cycle pattern for 2 different blocks (assuming 7-block structure), while panel (**b**) provides a detailed picture showing the postseismic oceanward motion, observed at the outer region of the block.

2.2. GNSS-Measured Motions and Stress-Deformation Cycle in Subduction Regions

Understanding stress-deformation cycle (SDC) [32] mechanisms in subduction zones is critically important for the long-term large earthquake prediction and seismic hazard assessment [19]. Over decades, a number of concepts have been proposed to explain the SDC features, with the so-called asperity model being the most promoted one [12]. We believe that it is crucial that those models be validated by comparing the predicted behavior with the observed motion data, which has become possible since the high-precision GNSS-based geodetic observations had been launched. GNSS stations deployed in active regions including subduction zones, had made the long-term displacement and velocity data series available, which totally changed the landscape in tectonics and seismology studies.

In a series of previous studies, we analyzed GNSS-measured motions in the Kuril subduction zone following the 2006–2007 Simushir earthquakes [16] (Figure 4), Chile subduction zone following the 2010 Maule earthquake [17] (Figure 5), and Japan subduction zone following the 2011 Tohoku earthquake [18] (Figures 6 and 7), with primary attention paid to horizontal displacements.

Main features of the observed Earth's surface motions include a slow background continent-ward motion (Figures 5a and 6a), which lasts up to 75% of the duration of the seismic cycle [32] and then is suddenly interrupted by an abrupt oceanward dislocation in source-neighboring areas at the instance of the seismic event (Figure 6b), followed by the continuous motion of the continent-side lithosphere in source-neighboring areas towards the oceanic plate. These postseismic oceanward displacements are unsteady in time and along the strike of the trench. The Figure 4a,b, Figure 5c,d and Figure 6c–f show annual variations of Earth's surface postseismic motions in the vicinity of source regions of 2006 Simushir, 2010 Maule, and 2011 Tohoku earthquakes, respectively. Despite the fact that each of the regions under consideration has a different tectonic structure and the spatial density of our data varies greatly, we observe similar patterns of postseismic motions: (a) postseismic drift occurs in the direction towards the ocean, and (b) different portions of lithosphere may experience motion in opposite directions at the same time. Such abrupt changes in the direction of motions of adjacent areas of the Earth's surface in the framework of models assuming a continuous medium will cause large deformations between them. This behavior is seen for all three events in consideration and implies the existence of geomechanical process responsible for the above motions. Obviously, the well-known asperity model lacks the ability to explain the post-quake oceanward drift, occurring in particular regions, while the neighboring territories may move in the opposite (normal) direction. At the same time, the proposed keyboard model can provide natural explanation for observed motions during preseismic, coseismic, and postseismic stages of the seismic cycle. The slow background continent-ward motion (Figures 5a and 6a) is associated with stationary elastic compression of frontal and rear blocks during the interseismic stage. These blocks at the instance of the seismic event are abruptly shifted towards the ocean, which triggers the observed continued oceanward motions of the Earth's surface. The almost independent displacements of seismogenic blocks also explain the opposite directions of Earth's surface motions along the oceanic trench during the postseismic stage of the seismic cycle (Figures 4a,b, 5c,d and 6c–f). The observed difference in postseismic motions in the direction perpendicular to the trench (Figure 5c,d) is due to the presence of viscous friction at the bottom of the rear blocks and the presence of a viscous response in the asthenospheric wedge to a coseismic stress drop [19].

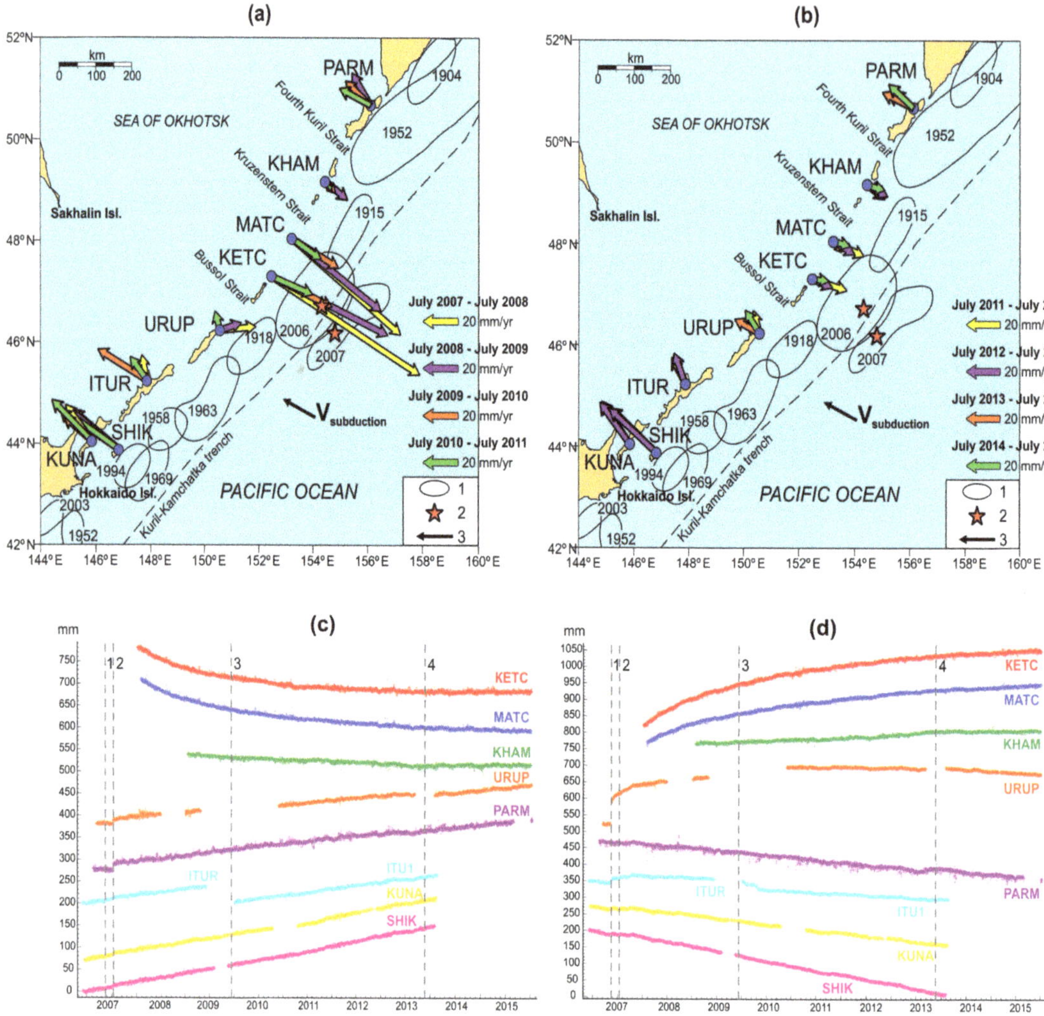

Figure 4. GNSS-observed motions in Kuril subduction zone after large Simushir earthquakes modified from [16]. Upper panel shows velocity vectors measured by the Kuril GNSS stations during time intervals from May 2007 to May 2011 (**a**) and from May 2011 to May 2015 (**b**). (1) Sources of strongest earthquakes with $M \geq 8$; (2) main shocks of earthquakes in 2006 and 2007; (3) subduction velocity equal to 80 mm/yr. The velocities indicated here are relative to the North American lithospheric plate. Lower panel shows northern (**c**) and eastern (**d**) components of the displacements of the Kuril network observation points. Dashed vertical lines denote the moments: (1) Simushir earthquake in 2006; (2) Simushir earthquake in 2007; (3) Eruption of Sarychev peak in 2009; (4) deep earthquake of the Sea of Okhotsk in 2013.

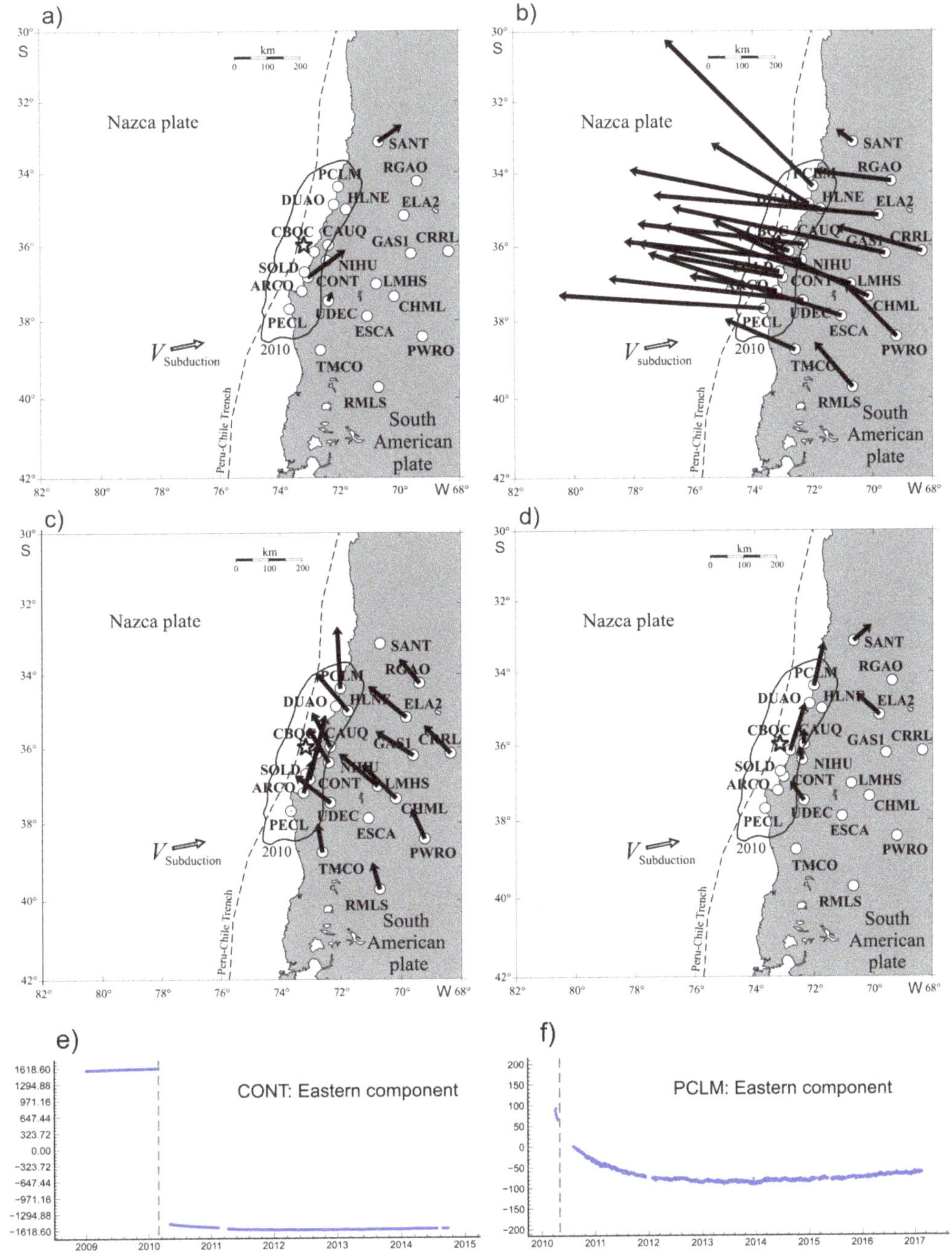

Figure 5. GNSS-observed motions in Chile subduction zone before (**a**) and after (**b**,**d**) large 2010 Maule earthquake modified from [17]. Map plots show the velocity vectors estimated from the Chilean and Argentinean regional satellite geodesy networks data for the time intervals of 27 February 2009–26 February 2010 (**a**); 28 February 2010–27 February 2011 (**b**); 28 February 2013–27 February 2014 (**c**); and 28 February 2016–27 February 2017 (**d**). Lower panel shows eastward displacement timeseries measured at CONT (**e**) and PCLM (**f**) GNSS stations. Displacements are relative to the South American lithospheric plate.

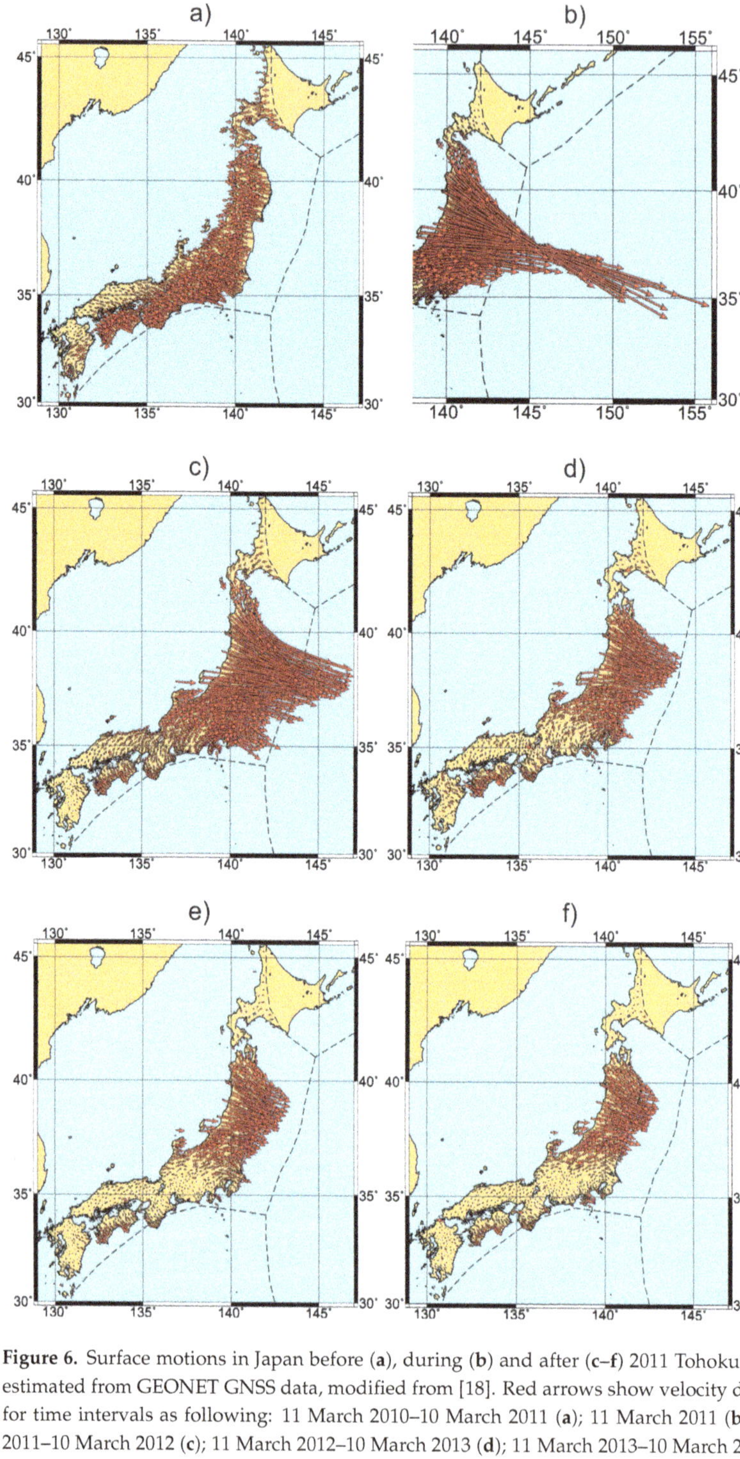

Figure 6. Surface motions in Japan before (**a**), during (**b**) and after (**c**–**f**) 2011 Tohoku earthquake estimated from GEONET GNSS data, modified from [18]. Red arrows show velocity distributions for time intervals as following: 11 March 2010–10 March 2011 (**a**); 11 March 2011 (**b**); 12 March 2011–10 March 2012 (**c**); 11 March 2012–10 March 2013 (**d**); 11 March 2013–10 March 2014 (**e**), and 11 March 2014–10 March 2015 (**f**). Velocities are relative to the South American lithospheric plate.

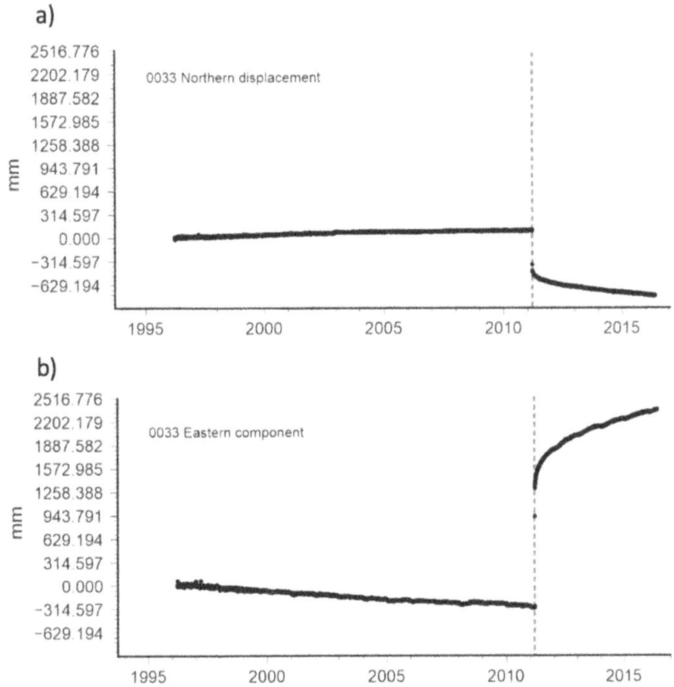

Figure 7. Timeseries for the northern (**a**) and eastern (**b**) displacement components recorded at GEONET GNSS station 0033. Dashed vertical line indicates 2011 Tohoku earthquake.

3. Results

The motion patterns discussed above provide strong GNSS data-based evidence for alternation of prevalent motion direction over the cycle. At the same time, available geological and seismological data suggest the existence of the fault interfaces separating the offshore frontal blocks from the onshore rear ones [15,33,34], which means that the complete geomechanical model should incorporate both frontal and rear structural parts, allowing for a complex interaction between them. The above considerations indicate the need for modification of the original model [20] in order to take into account the discontinuities both in the frontal and rear parts of the island arc.

In terms of geometry, the two-segment keyboard-block model is represented by a series of elastic blocks forming two chains separated by fault zones extending along the subduction axis (Figure 8). In most cases, the model allows for a relatively simple quasi one-dimensional approximation, limiting the analysis to a single strain component perpendicular to the strike (subduction axis), describing the motion of blocks relative to the subducted lithospheric plate. The submerging plate is separated from the overhanging block by a relatively thin contact layer characterized by nonlinear viscous rheology, with its viscosity essentially depending on the rate of relative motion of the block's bottom and the subducted plate surface. Nonlinear behavior implies an increase in viscosity when the strain rate is low and block is moving towards the continent (stress accumulation phase), followed by abrupt decrease during the coseismic stage due to the destruction of the material of this layer at the instant of the earthquake, which contributes to the reverse sliding of the block towards the trench in the process of partial release of accumulated elastic stress during the aftershock stage. The transition from the stress accumulation stage to coseismic stage is controlled by an abrupt change (release) of stresses at the instant of an earthquake upon reaching a certain critical stress value (Figure 8b).

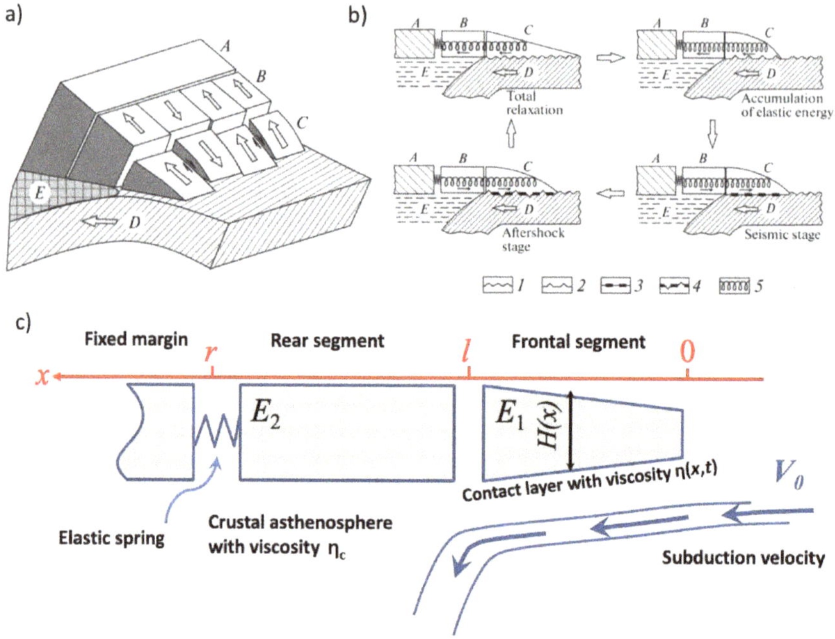

Figure 8. (**a**) Three-dimensional geometry of the two-segment keyboard-block model, (**b**) seismic deformation cycle phases, and (**c**) vertical plane section for the particular block along x-direction. A is the fixed undeformable continental margin; B is the rear-segment block; C is the frontal-segment block; D is the subducting plate; E is the crustal asthenosphere. (1) Two-dimensional geometry for the no deformation state; (2) stress accumulation (preseismic) phase; (3) coseismic phase (destruction of the crust); (4) postseismic (aftershock) phase with continued oceanward motion of the blocks.

Modification of the model's mathematical formalism involves introducing two separate model domains (corresponding to each of two blocks), and applying appropriate equations within each of them (see Figure 9). An important difference from the original model is that mechanical properties, strain rate and viscosity beneath the blocks are defined in a different way for each block.

The frontal (outer) segment falls within x-coordinate range $0 \leq x \leq l$, while the rear (inner) one lies within $l \leq x \leq r$. We assume that the gaps between the frontal and rear segments ($x = l$), as well as between the rear segment and fixed margin ($x = r$) are negligibly small compared to the block size. Similar to original single-segment model, the computational scheme applied for simulating the cyclic dynamics is described by a system of equilibrium equations connecting stresses in the given frontal-segment (seismogenic) block, rear-segment block, and all the neighboring blocks (Figure 9h).

The boundary conditions applied at the model boundaries (Figure 9j) define the behavior of not only outer edges of the frontal blocks (free moving) and elastic interaction of the rear blocks with the fixed margin, but also of the contact point separating the rear and frontal segments. This condition is variable, depending on the current state of the seismic cycle. During the stress accumulation phase (interseismic stage), the frontal block motion is fully transferred to the rear segment, while during the postseismic (aftershock) phase of the cycle, the segments are considered to be separated by a fluid-filled "gap" due to the possible different rates of backward motion of the frontal and rear blocks towards the ocean following the abrupt stress drop caused by the seismic event. These different rates are due to the viscous friction on the bottom of the rear block which may prevent its elastic straightening. Duration of the postseismic phase is defined by convergence of the segments after the frontal block changes the direction of its motion and starts moving towards the continent, which indicates the transition of the cycle to the stress accumulation phase.

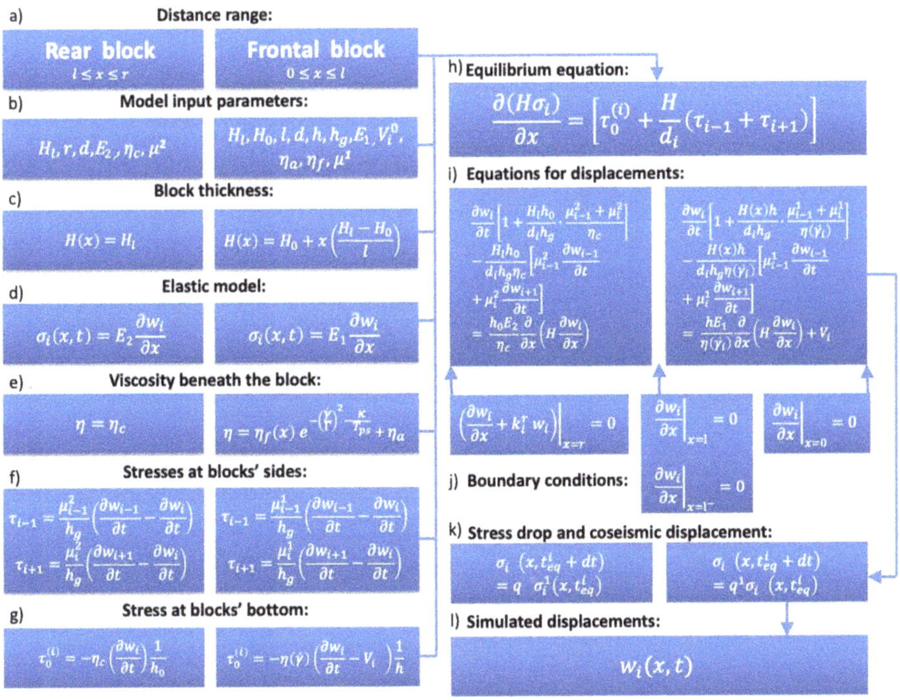

Figure 9. Flowchart showing calculation scheme of the generalized (two-element) keyboard model. (**a**) Schematic representation of rear (left) and frontal (right) modeling domains (blocks); (**b**) model input parameters for the rear (left) and frontal (right) blocks; (**c**) block thickness; (**d**) 1-D elastic model; (**e**) viscosity as function of x and t: constant viscosity beneath the rear block (left), time-dependent viscosity of the contact layer beneath the frontal block (right); (**f**) stresses acting on lateral sides of the blocks; (**g**) stress acting at the blocks' bottom; (**h**) stress equilibrium equation; (**i**) equation governing displacement evolution; (**j**) boundary conditions applied at the frontal edge of the frontal block ($x = 0$), contact between the frontal and rear blocks ($x = l$), and rear ($x = r$) edge of the rear block; (**k**) stress dumping conditions; (**l**) output dataset.

Similar to the original calculation [20], we discretize the problem both in terms of time t with step Δt ($t_k = k\cdot\Delta t$) and x with step Δx ($x_j = j\cdot\Delta x$), so that nodes x_j ($j = 1, \ldots, L$) fall within the frontal block, while those with $j = L+1, \ldots, N$ correspond to the rear block. In discrete formulation, the problem takes form.

For the frontal blocks:

$$\begin{gathered}\Delta W^k_{i,j}\left[1+\frac{H_j h}{d_i h_g \eta^k_{i,j}}\left(\mu^1_{i-1}+\mu^1_i\right)\right]-\frac{H_j h}{d_i h_g \eta^k_{i,j}}\left(\mu^1_{i-1}\Delta W^k_{i-1,j}+\mu^1_i\Delta W^k_{i+1,j}\right)\\ =\frac{hE_1}{\eta^k_{i,j}}\frac{\Delta t}{(\Delta x)^2}\left[H_j\left(W^k_{i,j+1}-2W^k_{i,j}+W^k_{i,j-1}\right)+\Delta x\left(\frac{H_l-H_0}{l}\right)\left(W^k_{i,j+1}-W^k_{i,j}\right)\right]+V^1_i\Delta t\\ j=2,\ldots,L;\ i=2,\ldots,m-1\end{gathered} \tag{1}$$

For the rear blocks:

$$\begin{gathered}\Delta W^k_{i,j}\left[1+\frac{H_l h}{d_i h_g \eta^k_{i,j}}\left(\mu^2_{i-1}+\mu^2_i\right)\right]-\frac{H_l h}{d_i h_g \eta^k_{i,j}}\left(\mu^2_{i-1}\Delta W^k_{i-1,j}+\mu^2_i\Delta W^k_{i+1,j}\right)\\ =\frac{hE_2}{\eta^k_{i,j}}\frac{\Delta t}{(\Delta x)^2}\left[H_l\left(W^k_{i,j+1}-2W^k_{i,j}+W^k_{i,j-1}\right)\right]\\ j=L+1,\ldots,N-1;\ i=2,\ldots,m-1.\end{gathered} \tag{2}$$

To ensure the scheme convergence, the time step Δt and x-spacing Δx must be related as $\Delta t < \frac{\eta_{min}(\Delta x)^2}{2HhE}$, where η_{min} is the minimum viscosity in the contact layer, and E is elastic modulus of either frontal or rear block.

Boundary conditions are written as:

$$W_{i,\,1}^{k} = W_{i,\,2}^{k}, \tag{3}$$

$$W_{i,\,L}^{k} = \frac{W_{i,\,L-1}^{k}}{1 + k_i \Delta x}, \tag{4}$$

$$W_{i,\,L+1}^{k} = W_{i,\,L}^{k},\ t_k \text{ during interseismic phase}, \tag{5}$$

$$W_{i,\,L+1}^{k} = W_{i,\,L+2}^{k},\ t_k \text{ during postseismic phase}. \tag{6}$$

Similar to the original single-segment model, the evolution of the block system described by expressions (1–2), (5–6) is controlled by the conditions for dumping the accumulated elastic energy in the form of an abrupt decrease in compressional stresses at t_{eq}^{i}, corresponding to earthquakes, occurring when the frontal block reaches the threshold energy ϵ_{cr}:

$$\sigma_i^1\left(x,\ t_{eq}^i + 1\right) = q^1 \sigma_i^1\left(x,\ t_{eq}^i\right), \tag{7}$$

$$\sigma_i^2\left(x,\ t_{eq}^i + 0\right) = q^2 \sigma_i^1\left(x,\ t_{eq}^i - 0\right), \tag{8}$$

Note that this condition is equivalent to an instantaneous change in displacement, with some coefficients that take into account the differences in the elastic modulus between the frontal and rear segments. The specific time instants of the earthquakes, t_{eq}^{i}, are determined from the expression:

$$\epsilon_i\left(t_{eq}^i\right) = \frac{E}{2lH_m} \int_0^l H(x) \left(\frac{\partial w_i\left(t_{eq}^i\right)}{\partial x} \right)^2 dx = \epsilon_{cr} \tag{9}$$

To demonstrate the main features of the model kinematics, we have performed series of simulations, showing the evolution of the model over a 500-year time interval (Figure 10).

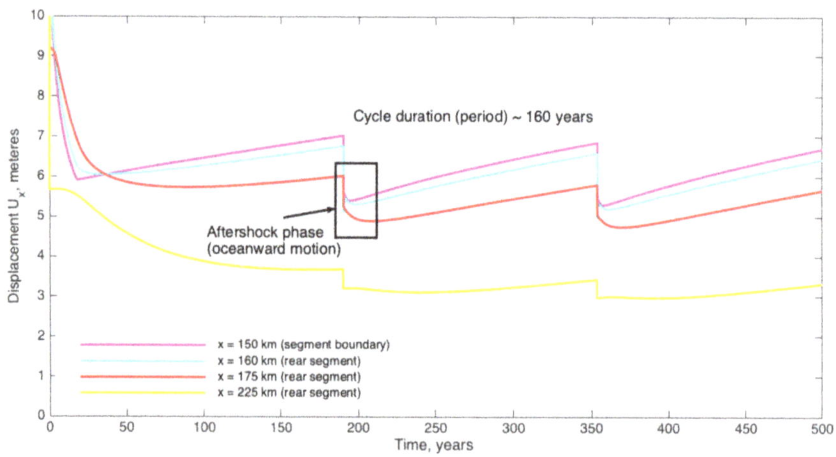

Figure 10. Displacement variations $W(x,t)$, simulated with a two-segment model for a series of points within the rear-segment blocks (x = 150, 160, 175, and 225 km), calculated for a typical set of model parameters (see Table 1).

An important feature of the model behavior is the existence of oceanward motion of the entire rear segment following every subsequent earthquake (a postseismic or aftershock stage). Modeling of this oceanward motion is of great importance for better assessment of the duration of the seismic cycle and the time moment of the transition of subduction zone to the next seismic cycle. The duration of the aftershock phase depends on the rheological parameters of the medium and can reach up to 50 years in the rear and central part of the rear block and a few years at its frontal edge. Our tests also helped us determine the limits for the viscous parameters of the model. We found that the model behaves in an expected way (accumulation of elastic stress occurs in a reasonable time range (See Table 1)) for the viscosity of the contact layer and the asthenosphere lying in the range of $0.1–1 \times 10^{19}$ Pa·s. Lower viscosity (less than 10^{18} Pa·s) causes numerical instability of the solution. In the case of higher viscosity (of the order of 10^{20} Pa·s), the slow (at around 1 cm/year) oceanward motion persists all over the cycle, and the rear block does not experience any continentward motion at all. The obtained estimates for model viscous parameters are consistent with the earlier results lying in the range $5 \times 10^{17}–5 \times 10^{19}$ Pa·s [35–40].

4. Discussion

We believe that employment of the fault-block models seems to be the most promising approach, which allows for more realistic quantitative characterization of the seismic deformation process [19], and propose the generalized two-segment keyboard model [22] as a geomechanical concept capable of explaining the SDC and kinematics in the subduction regions. It is also clear, that validation of such model is only possible through onshore GNSS observations, thus making the problem weakly constrained and unstable, given the lack of data on particular properties of the model.

To verify the above assumptions, we performed a numerical simulation of series of seismic cycles using initial and generalized formulations of keyboard model with stochastic behavior of parameters governing the critical amount of accumulated elastic energy and the amount of relaxed energy. During the simulations the model parameters were constrained to values that are typical for the Kuril subduction system (E_1 = 29 GPa, E_2 = 80 Gpa, d_i = 200 km, frontal block length = 100 km, rear block length = 200 km, effective viscosity of the contact layer and the asthenosphere = 10^{19} Pa·s). We also performed a comparison of the obtained model parameters with ones calculated earlier using the continuum medium model [19] and with geophysical data.

We assumed that the modeled coseismic slip distribution in the source zone of the 2006 Simushir earthquake [19] according to the asperity model [12] represents the preseismic mechanical coupling that was ruptured at the time of the main shock. Such suggestion helps us to construct a model of coupling distribution in the source zone of the 2006 Simushir earthquake (Figure 11). This model allows us to calculate the characteristic time of stress accumulation (i.e., seismic cycle duration) in the source zone of the 2006 Simushir earthquake (Table 3) using known coseismic slip distribution and magnitude and direction of the plate convergence vector [41].

In Table 3 and Figure 12 we present the results of modeling the dynamics of seismogenic blocks in the middle part of the Kuril island arc for 1400 years (10 times the mean duration of SDC for Kuril-Kamchatka subduction zone [32]). To simplify the model, the block sizes are assumed to be the same (100 km × 100 km). According to Global CMT Project data [42], large subduction earthquakes of the Kuril Islands are characterized by a focal length of 120 to 250 km. For example, the spatial dimensions of the source zone of the First Simushir earthquake was estimated on the basis of the aftershock distributions recorded in the first 2 months after the earthquake. This megathrust earthquake was the largest event in the central Kuril Islands since 1915 with the source zone of about 230 km × 150 km. Coseismic motions in the source zone affected three seismogenic blocks with a maximum slip along the contact zone up to 12 m [19].

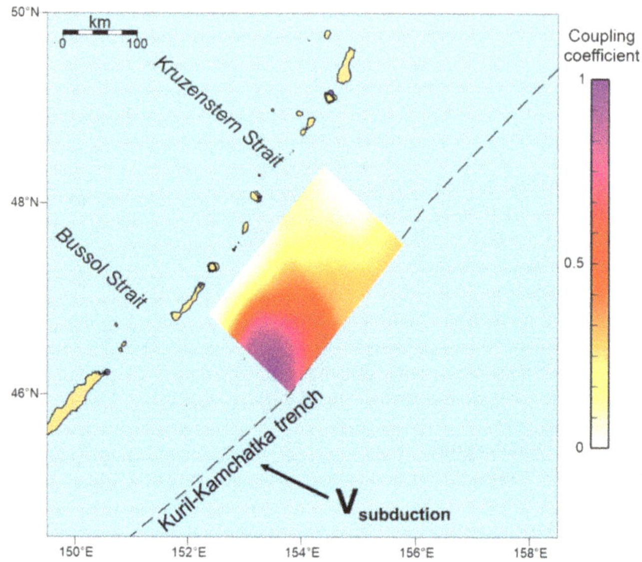

Figure 11. Model of the mechanical coupling distribution in the source zone of the 2006 Simushir earthquake. Black arrow displays magnitude and direction of the plate convergence vector [41].

Table 3. Main characteristics of the seismic deformation cycle in the Kuril-Kamchatka subduction zone.

	Geological+ Geophysical Data [32]	Continuous Medium Model [19]	Mechanical Keyboard Model	Generalized Mechanical Keyboard Model
Seismic cycle duration	140 ± 60 years	159 years	60–267 years (av. 136 years)	40–338 years (av. 162 years)
Postseismic stage duration	Up to 35–50 years	0.5 years (afterslip) up to 10 years (viscoe-lastic relaxation)	2–7 years	1–5 years
Coseismic horizontal surface displacements	-	55×10^{-2} m (GNSS observations)		10×10^{-2}–1.8 m (depends on distance from trench and magnitude)

The rheological parameters of the medium were selected during the further modeling according to the results of the previous studies of the First Simushir earthquake [19].

As a result of the simulation, we estimated the total duration of the seismic cycle, which lay between 113 and 330 years with an average value of 196 years and the duration of the postseismic stage, which corresponds to a range of 11–19 years with an average value of ~15 years. These estimates are in good agreement with the real values of the seismic cycle associated with the 2006 Simushir earthquake, which is 226 years [43] and more than 10 years [19], respectively.

For the case of the used rheological parameters of the medium during the realization of seismic cycles, episodes of simultaneous displacements of two adjacent seismogenic blocks were observed, which confirms the possibility of the formation of extended source zones, such as of the First Simushir earthquake (about 200 km).

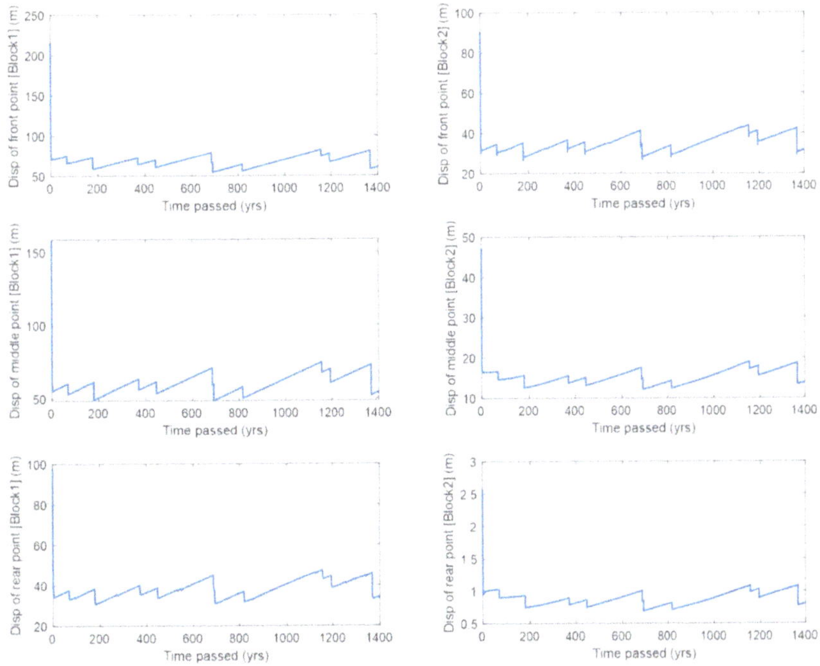

Figure 12. Simulation of several seismic deformation cycles in Kuril subduction zone for generalized keyboard model. Left column of images shows the displacements of two boundary and one middle points of frontal block, right column shows the displacements of the same points of rear block. Stochastic condition for amount of relaxed and accumulated elastic energy was applied. Initial large displacement is caused by stabilization of the numerical scheme and is ignored in further analysis.

Comparing the oceanward motion features observed in the original (Figure 3) and modified (Figure 10) models, it is important to notice that although main features may look similar, including the oceanward drift, in the original model the latter is clearly visible only within the outer (frontal) relatively narrow region of the block. At the same time, the middle and rear parts of the block show no evidence of this oceanward motion. In the modified two-segment model, the oceanward drift is observed both at the outer edges of both the frontal and rear blocks, and that is exactly what is seen in GNSS data.

Figure 13 shows the comparison of characteristic simulated motion for the rear block and GNSS-measured displacements recorded during 2007–2015 at four stations located at the Kuril Islands. It can be seen that the modeled behavior is fairly close to that recorded at KETC station, both in terms of coseismic instantaneous displacement and transient motion observed over 8 years post-earthquake.

The analysis of the two-segment model showed that for the rheological and spatial parameters typical for the Kuril subduction zone, the results of the generalized model are quite close to the results of the original keyboard model and the results of previous studies. At the same time, the generalized model allowed us to compare the model displacements with the GNSS data.

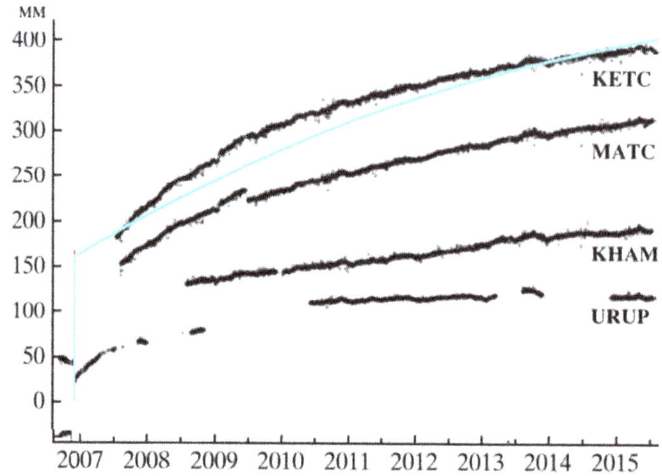

Figure 13. East-west displacement timeseries measured by Kuril GNSS network following the 2006–2007 Simushir earthquakes (black curves) and modeled displacement data (light blue line) calculated within the interior part of the rear block using the two-segment model. Positive values correspond to eastward (i.e., oceanward) motion.

5. Conclusions

Analyzing the simulated energy/stress/displacement/velocity datasets, we focus on characteristic features of the seismic-deformation cycle, presenting with repeated phases both in terms of stress accumulation/release and kinematics. Modeled displacement timeseries exhibit the pronounced cyclic behavior, with a characteristic cycle period of about 160 years, which is consistent with the known patterns of megathrust subduction earthquakes. An essential result proving the model's relevance is the manifestation of the postseismic (aftershock) phase, when the blocks' motion reverses its normal continentward direction, and a backward motion occurs towards the ocean, as observed from GPS velocity data [16–19]. In simulated data, this behavior is most clearly seen for the frontal blocks, where the duration of backward-motion phase is about 1–5 years, depending on the location (Figure 11).

Simulated data show the model's capability to predict the cyclic kinematic patterns, being in good agreement with GPS-measured velocity variations, specifically during the post-quake transient phase (Figure 12).

With the further analysis within the framework of the two-segment keyboard-block model, we expect to obtain the estimates of the relevant geomechanical parameters (material properties) through inverting GPS-measured displacement and velocity timeseries data, acquired over the last decades in subduction regions including Kuril-Kamchatka, Japan, Aleutian, and Chile. The inverse problem solution would also provide crucial data on the transition of the block system between the states of the cycle.

Author Contributions: Conceptualization, L.I.L. and I.S.V.; methodology, L.I.L. and Y.V.G.; software, Y.V.G. and D.A.A.; validation, Y.V.G. and D.A.A.; resources, Y.V.G. and I.S.V.; writing—original draft preparation, I.S.V., Y.V.G. and D.A.A.; writing—review and editing, I.S.V., Y.V.G., D.A.A. and L.I.L.; visualization Y.V.G. and D.A.A.; supervision, L.I.L.; project administration, D.A.A.; funding acquisition, L.I.L. All authors have read and agreed to the published version of the manuscript.

Funding: This research was funded by the Russian Science Foundation, grant number 20-17-00140.

Data Availability Statement: The data presented in this study are openly available in FigShare (https://figshare.com/s/ec5eb748a0fa7031cd56) at doi 10.6084/m9.figshare.16744459.

Conflicts of Interest: The authors declare no conflict of interest.

References

1. Shebalin, N.V. *Selected Publications. Great Earthquakes*; Mining Academy Publishing House: Moscow, Russia, 1997; 542p. (In Russian)
2. Keilis-Borok, V. Earthquake Prediction: State-of-the-Art and Emerging Possibilities. *Annu. Rev. Earth Planet. Sci.* **2002**, *30*, 1–33. [CrossRef]
3. Ruff, L.J. Asperity distributions and large earthquake occurrence in subduction zones. *Tectonophysics* **1992**, *211*, 61–83. [CrossRef]
4. Kaneko, Y.; Avouac, J.P.; Lapusta, N. Towards inferring earthquake patterns from 513 geodetic observations of interseismic coupling. *Nat. Geosci.* **2010**, *3*, 363–369. [CrossRef]
5. Rosenau, M.; Horenko, I.; Corbi, F.; Rudolf, M.; Kornhuber, R.; Oncken, O. Synchronization of great subduction megathrust earthquakes: Insights from scale model analysis. *J. Geophys. Res.* **2019**, *124*, 3646–3661. [CrossRef]
6. Kossobokov, V.G.; Keilis-Borok, V.I.; Smith, S.W. Localization of intermediate-term earthquake prediction. *J. Geophys. Res.* **1990**, *95*, 12763–12772. [CrossRef]
7. Soloviev, A.A.; Gvishiani, A.D.; Gorshkov, A.I.; Dobrovolsky, M.N.; Novikova, O.V. Recognition of Earthquake-Prone Areas: Methodology and Analysis of the Results. *Izv. Phys. Solid Earth* **2014**, *50*, 151–168. [CrossRef]
8. Sykes, L.R.; Menke, W. Repeat times of large earthquakes: Implications for earthquake mechanics and long-term prediction. *BSSA* **2016**, *96*, 1569–1596. [CrossRef]
9. Banna, M.H.A.; Taher, K.A.; Kaiser, M.S.; Mahmud, M.; Rahman, M.S.; Hosen, A.S.M.S.; Cho, G.H. Application of Artificial Intelligence in Predicting Earthquakes: State-of-the-Art and Future Challenges. *IEEE Access* **2020**, *8*, 192880–192923. [CrossRef]
10. Battaglia, M.; Cervelli, P.F.; Murray, J.R. Modeling crustal deformation near active faults and volcanic centers—A catalog of deformation models. In *Techniques and Methods*; U.S. Geological Survey: Menlo Park, CA, USA, 2013; Book 13, Chapter B1; 96p.
11. Madariaga, R. Seismic Source Theory. In *Treatise on Geophysics*; Elsevier: Amsterdam, The Netherlands, 2015; pp. 51–71.
12. Lay, T.; Kanamori, H. An asperity model of large earthquake sequences. In *Earthquake Prediction: An International Review*; American Geophysical Union: Washington, DC, USA, 1981; pp. 579–592.
13. Senatorski, P. Effect of slip-weakening distance on seismic–aseismic slip patterns. *Pure Appl. Geophys.* **2019**, *176*, 3975–3992.
14. Lobkovsky, L.I.; Mazova, R.K.; Garagash, I.A.; Kataeva, L.Y.; Nardin, I. To analysis of source mechanism of the 26 December 2004 Indian Ocean tsunami. *Russ. J. Earth. Sci.* **2006**, *8*, ES5001.
15. Baranov, B.V.; Ivanchenko, A.I.; Dozorova, K.A. The Great 2006 and 2007 Kuril Earthquakes, Forearc Segmentation and Seismic Activity of the Central Kuril Islands Region. *Pure Appl. Geophys.* **2015**, *172*, 3509–3535. [CrossRef]
16. Lobkovsky, L.I.; Baranov, B.V.; Vladimirova, I.S.; Gabsatarov, V.V.; Steblov, G.M.; Garagash, I.A. Post-seismic motions after the 2006–2007 Simushir earthquakes at different stages of the seismic cycle. *Dokl. Earth Sci.* **2017**, *473*, 375–379. [CrossRef]
17. Lobkovsky, L.I.; Vladimirova, I.S.; Gabsatarov, Y.V.; Baranov, B.V.; Garagash, I.A.; Steblov, G.M. Seismotectonic deformations associated with the 2010 Maule earthquake at different stages of the seismic cycle according to satellite geodetic observations. *Dokl. Earth Sci.* **2017**, *477*, 716–721.
18. Lobkovsky, L.I.; Vladimirova, I.S.; Gabsatarov, Y.V.; Steblov, G.M. Seismotectonic Deformations Related to the 2011 Tohoku Earthquake at Different Stages of the Seismic Cycle, Based on Satellite Geodetic Observations. *Dokl. Earth Sci.* **2018**, *481*, 1060–1065. [CrossRef]
19. Vladimirova, I.S.; Lobkovsky, L.I.; Gabsatarov, Y.V.; Steblov, G.M.; Vasilenko, N.F.; Frolov, D.I.; Prytkov, A.S. Patterns of the Seismic Cycle in the Kuril Island Arc from GPS Observations. *Pure Appl. Geophys.* **2020**, *177*, 3599–3617.
20. Lobkovsky, L.I.; Kerchman, V.I.; Baranov, B.V.; Pristavakina, E.I. Analysis of seismotectonic processes in subduction zones from the standpoint of a keyboard model of great earthquakes. *Tectonophysics* **1991**, *199*, 211–236. [CrossRef]
21. Ozawa, S.; Nishimura, T.; Suito, H.; Kobayashi, T.; Tobita, M.; Imakiire, T. Coseismic and postseismic slip of the 2011 magnitude-9 Tohoku-Oki earthquake. *Nature* **2011**, *475*, 373–377. [CrossRef]
22. Lobkovsky, L.I.; Vladimirova, I.S.; Alekseev, D.A.; Gabsatarov, Y.V. Two-element keyboard model of strongest subduction earthquakes generation. *Dokl. Earth Sci.* **2021**, *496*, 72–75. [CrossRef]
23. Fedotov, S.A. Regularities of distribution of large earthquakes of Kamchatka, Kuril Islands and North-Eastern Japan. *Tr. Inst. Fiz. Zemli Akad. Nauk SSSR* **1965**, *36*, 66–93. (In Russian)
24. Loveless, J.P.; Brendan, J.M. Geodetic imaging of plate motions, slip rates, and partitioning of deformation in Japan. *J. Geophys. Res.* **2010**, *115*, 1–35. [CrossRef]
25. Panayotopoulos, Y.; Hirata, N.; Sato, H.; Kato, A.; Imanishi, K.; Kuwahara, Y.; Cho, I.; Takeda, T.; Asano, Y. Investigating the role of the Itoigawa-Shizuoka tectonic line towards the evolution of the Northern Fossa Magnarift basin. *Tectonophysics* **2014**, *615–616*, 12–26. [CrossRef]
26. Minoura, K.; Imamura, F.; Sugawara, D.; Kono, Y.; Iwashita, T. The 869 Jogan tsunami deposit and recurrence interval of large-scale tsunami on the Pacific coast of northeast Japan. *J. Nat. Disaster Sci.* **2001**, *23*, 83–88.
27. Melnick, D.; Echtler, H.P. Morphotectonic and geologic digital map compilations of the south-central Andes (36°–42° S). In *The Andes—Active Subduction Orogeny*; Frontiers in Earth Science; Oncken, O., Chong, G., Franz, G., Giese, P., Götze, H.-J., Ramos, V.A., Strecker, M., Wigger, P., Eds.; Springer: Berlin/Heidelberg, Germany; New York, NY, USA, 2006; Volume 1, pp. 565–568.
28. Geersen, J.; Behrmann, J.H.; Völker, D.; Krastel, S.; Ranero, C.R.; Diaz-Naveas, J.; Weinrebe, W. Active tectonics of the South Chilean marine fore arc (35° S–40° S). *Tectonics* **2011**, *30*, 1–16. [CrossRef]

29. Moreno, M.; Melnick, D.; Rosenau, M.; Baez, J.; Klotz, J.; Oncken, O.; Tassara, A.; Chen, J.; Bataille, K.; Bevis, M.; et al. Toward understanding tectonic control on the Mw 8.8 2010 Maule Chile earthquake. *Earth Planet. Sci. Lett.* **2012**, *321–322*, 152–165. [CrossRef]
30. Jara-Muñoz, J.; Melnick, D.; Brill, D.; Strecker, M.R. Segmentation of the 2010 Maule Chile earthquake rupture from a joint analysis of uplifted marine terraces and seismic-cycle deformation patterns. *Quat. Sci. Rev.* **2015**, *113*, 171–192. [CrossRef]
31. Metois, M.; Vigny, C.; Socquet, A. Interseismic Coupling, Megathrust Earthquakes and Seismic Swarms Along the Chilean Subduction Zone (38°–18° S). *Pure Appl. Geophys.* **2016**, *173*, 1431–1449.
32. Fedotov, S.A. The seismic cycle, possibility of the quantitative seismic zoning, and long-term seismic forecasting. In *Seismic Zoning in the USSR*; Medvedev, S.V., Ed.; Nauka: Moscow, Russia, 1968. (In Russian)
33. DeMets, C. Oblique convergence and deformation along the Kuril and Japan trenches. *J. Geophys. Res.* **1992**, *97*, 17615–17625. [CrossRef]
34. Cross, R.S.; Freymueller, J.T. Evidence for and implications of a Bering plate based on geodetic measurements from the Aleutians and western Alaska. *J. Geophys. Res.* **2008**, *113*, B07405.
35. Piersanti, A. Postseismic deformation in Chile: Constraints on the asthenospheric viscosity. *Geophys. Res. Lett.* **1999**, *26*, 3157–3160. [CrossRef]
36. Hu, Y.; Wang, K.; He, J.; Klotz, J.; Khazaradze, G. Three-dimensional viscoelastic finite element model for post-seismic deformation of the great 1960 Chile earthquake. *J. Geophys. Res.* **2004**, *109*, B12403. [CrossRef]
37. Wang, K. Elastic and viscoelastic models of crustal deformation in subduction zone cycles. In *The Seismogenic Zone of Subduction Thrust Faults*; Dixon, T.H., Moore, J.C., Eds.; Columbia University Press: New York, NY, 2007; pp. 540–577.
38. Bürgmann, R.; Dresen, G. Rheology of the Lower Crust and Upper Mantle: Evidence from Rock Mechanics, Geodesy and Field Observations. *Annu. Rev. Earth Planet. Sci.* **2008**, *36*, 531–567. [CrossRef]
39. Suito, H.; Freymueller, J.T. A viscoelastic and afterslip postseismic deformation model for the 1964 Alaska earthquake. *J. Geophys. Res.* **2009**, *114*, B11404. [CrossRef]
40. Sato, T.; Larsen, C.F.; Miura, S.; Ohta, Y.; Fujimoto, H.; Sun, W.; Motyka, R.J.; Freymuller, J.T. Reevaluation of the viscosity of upper mantle beneath southeast Alaska. *Tectonophysics* **2011**, *511*, 79–88. [CrossRef]
41. Kogan, M.G.; Steblov, G.M. Current global plate kinematics from GPS (1995–2007) with the plate-consistent reference frame. *J. Geophys. Res.* **2008**, *113*, B04416. [CrossRef]
42. Ekström, G.; Nettles, M.; Dziewonski, A.M. The global CMT project 2004–2010: Centroid-moment tensors for 13,017 earthquakes. *Phys. Earth Planet. Inter.* **2012**, *200–201*, 1–9. [CrossRef]
43. Rogozhin, E.A. Application of tectonophysical approaches to the solution of seismotectonic problems by the example of the Simushir earthquakes of November 15, 2006 and January 13, 2007 in Central Kuriles. *Izv. Phys. Solid Earth* **2013**, *49*, 643–652. [CrossRef]

applied sciences

MDPI

Article

Data Management and Processing in Seismology: An Application of Big Data Analysis for the Doublet Earthquake of 2021, 03 March, Elassona, Central Greece

Alexandra Moshou [1,*], Antonios Konstantaras [1], Panagiotis Argyrakis [2], Nikolaos S. Petrakis [1], Theodoros N. Kapetanakis [1] and Ioannis O. Vardiambasis [1]

1 Department of Electronic Engineering, Hellenic Mediterranean University, 3 Romanou Str., Chalepa, 73133 Chania, Greece; akonstantaras@hmu.gr (A.K.); nik.s.petrakis@hmu.gr (N.S.P.); todokape@hmu.gr (T.N.K.); ivardia@hmu.gr (I.O.V.)

2 Department of Surveying and Geoinformatics Engineers, Faculty of Engineering, Egaleo Park Campus, University of West Attica, Ag. Spyridonos Street, Aigaleo, 12243 Athens, Greece; pargyrak@uniwa.gr

* Correspondence: amoshou@hmu.gr; Tel.: +30-694-939-9764

Abstract: On 3 March 2021 (10:16, UTC), a strong earthquake, M_W 6.3, occurred in Elassona, Central Greece. The epicenter was reported 10 km west of Tyrnavos. Another major earthquake followed this event on the same day at M_W 5.8 (3 March 2021, 11:45, UTC). The next day, 4 March 2021 (18:38, UTC), there was a second event with a similar magnitude as the first, M_W 6.2. Both events were 8.5 km apart. The following analysis shows that the previous events and the most significant aftershocks were superficial. However, historical and modern seismicity has been sparse in this area. Spatially, the region represents a transitional zone between different tectonic domains; the right-lateral slip along the western end of the North Anatolian Fault Zone (NAFZ) in the north Aegean Sea plate-boundary structure ends, and crustal extension prevails in mainland Greece. These earthquakes were followed by rich seismic activity recorded by peripheral seismographs and accelerometers. The installation of a dense, portable network from the Aristotle University of Thessaloniki team also helped this effort, installed three days after the seismic excitation, as seismological stations did not azimuthally enclose the area. In the present work, a detailed analysis was performed using seismological data. A seismological catalogue of 3.787 events was used, which was processed with modern methods to calculate 34 focal mechanisms (M_W > 4.0) and to recalculate the parameters of the largest earthquakes that occurred in the first two days.

Keywords: big data; source observations; seismological data; data management; synthetics seismograms; regional data

Citation: Moshou, A.; Konstantaras, A.; Argyrakis, P.; Petrakis, N.S.; Kapetanakis, T.N.; Vardiambasis, I.O. Data Management and Processing in Seismology: An Application of Big Data Analysis for the Doublet Earthquake of 2021, 03 March, Elassona, Central Greece. *Appl. Sci.* **2022**, *12*, 7446. https://doi.org/10.3390/app12157446

Academic Editors: Alexei Gvishiani and Boris Dzeboev

Received: 9 June 2022
Accepted: 11 July 2022
Published: 25 July 2022

1. Introduction

In-depth analyses of substantial seismological data from Central Greece are presented in this article. This extensive period of study using the latest state-of-the-art tools has never been performed before. With this analysis, we wanted to identify the faults responsible for this seismic excitation. Due to their dominance over the local seismic hazard, the faults triggered by the two main shocks must be resolved.

Whether these ruptures represent (one of) the seismic faults that generated the biggest shocks is currently debated. The hypocentral depths with the moderate-to-shallow dip angle in the InSAR images (Lekkas et al., 2021) suggest that the rupture of the primary seismic fault(s) emerged further to the south in the alpine zone, even though their direction fits the proposed nodal planes of the moment tensor solutions and the borehole findings. This study's main task was to analyze a large volume of mainly seismological data to visualize the spatiotemporal distribution of the aftershock sequence.

Regarding the tectonic setting, Thessaly is located at the western end of the North Anatolian Fault Zone (NAFZ) in the north Aegean Sea, where the right-lateral slip along this plate-boundary structure ends and crustal extension prevails in mainland Greece [1,2]. The most prominent structural and geomorphic features strike NW-SE, such as the coastal and interior mountain ranges, bounding faults, and the late Tertiary sedimentary basins [3]. The seismicity of the Thessaly Basin is detected mainly along two fault zones relative to the maximum magnitude an earthquake can provide. The first is the northern fault zone, associated with earthquakes with magnitudes up to 6.5, and the second is the southern fault zone, related to earthquakes with a magnitude of 7.0.

In early March 2021, a doublet earthquake (M_W 6.3 and 6.2) occurred in the northern Thessalian region. The spatial distance of the two earthquakes was determined at only 8.5 km, while their time difference at 32 h. The first earthquake, magnitude 6.3, occurred on 3 March 2021 (10:16:08.3, UTC). The epicenter, as calculated and presented in the next paragraph, was found at $\varphi = 39.7453°$ N, $\lambda = 22.2340°$ E at a depth of 12 km. The earthquake was felt in Larissa, Trikala, and Kozani. This fact is due to the epicenter being a short distance from large cities. The epicenter was calculated 19 km northwest of the town of Larissa, 46 km northwest of the city of Trikala, and 72 km south of Kozani. According to the above, the mainshock caused extensive damage due to its short distance in a populated urban area, according to the European Mediterranean Seismological Center [4]; https://www.emsc-csem.org/#2 (accessed on 8 June 2022). Several buildings collapsed after this earthquake. One person died from the earthquake in the worst-hit village of Mesohori, near Elassona, where the local church and some old houses were severely damaged. Of the 3.675 homes, 1575 were deemed uninhabitable in the earthquake-stricken areas of Larissa, Elassona, Tyrnavos, and Farkadona. The earthquake caused much damage in the surrounding regions, and the Damasi primary school, built in 1938, suffered significant damage, resulting in its demolition eight days after the earthquake [5].

The second major earthquake occurred one day later, on 4 March 2021 (18:38:19.1, UTC), with M_W 6.2. The epicenter of this event was calculated at 39.7916° N, 22.1274° E, and at a depth of 8 km. Eight days later (12 March 2021, 12:57:50.6, UTC), a significant event, the third chronologically, eight days after the main one, with a magnitude $M_W = 5.6$, occurred in a different hypocentral area than the other two. A rich aftershock sequence followed these earthquakes, most felt by the area's residents mainly due to their small focal depth. It is noted that in the first 24 h only, 381 events occurred, ranging from M_L 1.2 to 5.2. The intense aftershock sequence activity with $M_L > 4.0$ aftershocks continued for a few days, with a smaller magnitude but high aftershock occurrence rate. These two significant earthquakes have been the strongest in the region since 1941 when an earthquake of the magnitude M_L 6.3 occurred very close to the city of Larissa. The area of Thessaly, Central Greece, has been well known for its land subsidence phenomena since the 1980s [6–11].

Several institutions providing seismic moment tensors consistently determined an almost normal fault on the east–south and northwest plane for the first and second mainshocks. Moving in the fault, the two earthquakes had a similar focal mechanism. This means that they probably came from the same fault but different parts. From these data, we probably have the phenomenon of a seismic pair, which is extremely rare. Earthquake doublets commonly occur in subduction zones and are mainly characterized by having similar mechanisms, spatially and temporally [12–14], correlated with several tens of kilometers and hours to days, respectively [15,16]. They can mainly be explained by Coulomb stress triggering [17]. Intraplate earthquake doublets are rare and less studied.

Doublet earthquakes—and, more generally, multiple earthquakes—were initially identified as numerous earthquakes with nearly identical waveforms originating from the exact location [18]. They are now characterized as single earthquakes having two (or more) mainshocks of similar magnitude, sometimes occurring within tens of seconds but sometimes separated by years [19]. The magnitude similarity—often within four-tenths of a unit of magnitude—distinguishes multiple events from aftershocks [20], which start at approximately 1.2 magnitudes more minor than the parent shock [21,22] and decrease in magnitude and frequency according to known laws. Doublet events also have nearly identical seismic waveforms, as they come from the same rupture zone and stress field, whereas aftershocks, peripheral to the main rupture, typically reflect more diverse circumstances of origin [18]. Multiple events overlap in their focal areas (i.e., rupture zones), up to 100 km across for magnitude 7.5 earthquakes [23–25]. Doublets have been distinguished from triggered earthquakes, where the seismic waves' energy triggers a distant earthquake with a different rupture zone, although it has been suggested that such a distinction reflects "imprecise taxonomy" more than any physical reality.

Figure 1 represents a general structure map showing the prefectures of the broader region of Central Greece. For the study area, the portable seismological station sites' selection criterion was based on the existence and operation of permanent seismological and geodetic stations that were already installed. In addition, faults were collected from the international literature and digitized in QGIS.

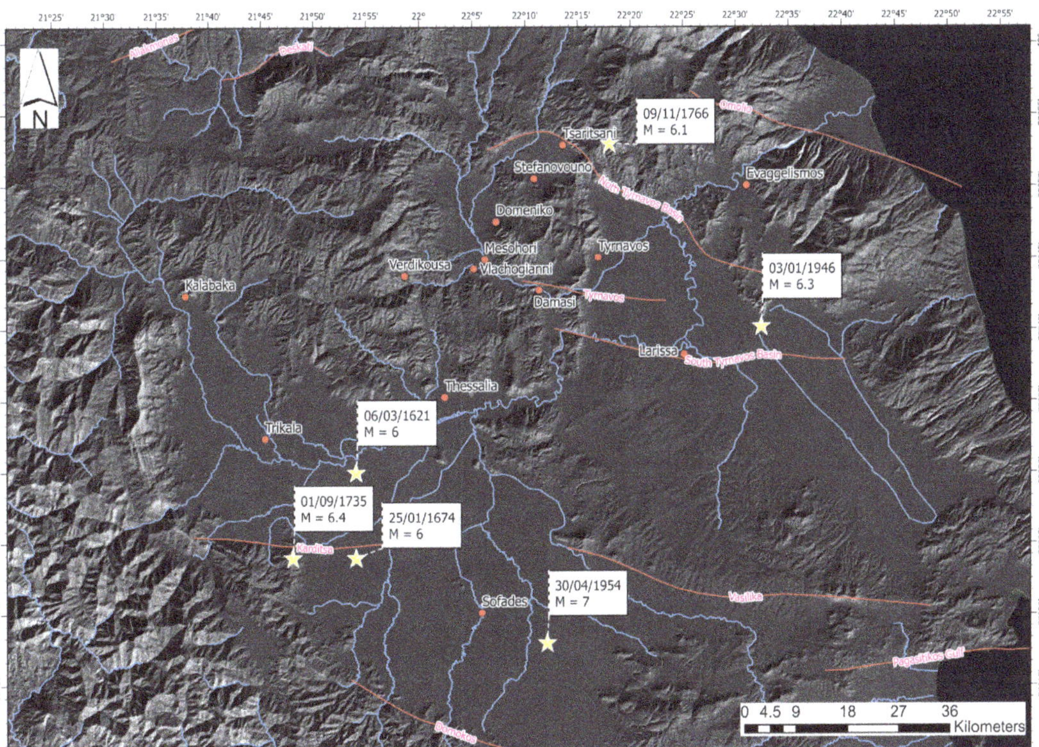

Figure 1. General structure map of Thessaly. The red circles represent the prefectures; red lines indicate the main active faults for the study area; the six yellow stars indicate the historical earthquake in the broader area of Thessaly; the blue lines represent rivers.

The contribution of the present study is the localization of the significant events, i.e., epicenter (i.e., latitude and longitude), depth (d), and local magnitude (ML), as well as the calculation of the source parameters (i.e., strike, dip, rake, depth, and moment magnitude) for the strong events and all of the significant aftershocks using moment tensor inversion.

2. Data Description

Seismological broadband data were used to calculate the epicenter, the magnitude, and the focal depth for the mainshock and the second significant event and all events with a magnitude $M_L > 5.0$, in addition to the source parameters of both main earthquakes and their significant aftershocks. The data were collected, analyzed, and processed appropriately, studied in the next section in more detail [26,27].

Seismological Data

We used waveform data from three regional component stations of the Hellenic Unified Seismological Network (HUSN); detailed information is presented in Table 1 and Figure 2. The National Observatory of Athens operates this network, composed of the Institute of Geodynamics and the Seismological Laboratories of three Universities, such as the University of Thessaloniki (A.U.Th. Seismological Network).

Table 1. Characteristics of the seismological stations in the broader area of Greece, the data from which were used in this study. Station coordinates are in decimal degrees; the elevation is in m. HL denotes the seismological network of the National Observatory of Athens; HT represents the seismological network of the University of Thessaloniki; MN designates the stations that belong to the MedNet (Mediterranean Seismological Network). Sources: Available online: https://bbnet.gein.noa.gr/HL (accessed on 15 March 2022); Available online: http://geophysics.geo.auth.gr (accessed on 15 March 2022) [28,29].

Stations	Latitude (°)	Longitude (°)	Code	Elevation (m)	Datalogger	Seismometer
THL	39.5646	22.0144	HL/MN	86	DR24-SC	STS-2
KZN	40.3033	21.7820	HL	791	DR24-SC	STS-2
NEO	39.3056	23.2218	HL	510	DR24-SC	KS2000M
JAN	39.6561	20.8487	HL	526	DR24-SC	CMG-3ESPC/60
EVR	38.9165	21.8105	HL	1037	DR24-SC	CMG-3ESPC/60
LIT	40.1003	22.4893	HT	568	Trident-B	CMG-3ESP/100
NEO	39.3056	23.2218	HL	510	DR24-SC	KS2000M
XOR	39.3660	23.1920	HT	500	TAURUS	CMG-3ESP/100
TYRN	39.7110	22.2325	HT	151	TAURUS	TRILLIUM
TYR1	39.7147	22.1684	HT	164	CENTAUR	TRILLIUM
TYR2A	39.6087	22.1291	HT	183	CENTAUR	TRILLIUM Compact
TYR3	39.7500	22.1049	HT	209	CENTAUR	TRILLIUM Compact
TYR4	39.7919	22.1202	HT	312	CENTAUR	TRILLIUM Compact
TYR5	39.8391	22.1023	HT	232	CENTAUR	TRILLIUM Compact
TYR6	39.6895	22.0781	HT	462	CENTAUR	TRILLIUM Compact
LRSO	39.6713	22.3917	HT	78	Reftek-130	CMG-3ESP/100
AGG	39.0211	22.3360	HT	622	CMG-3ESP/100	CMG-3ESP/100

In the present study, seismological research was divided into two categories. The first involved the localization of significant events, i.e., epicenter (i.e., latitude and longitude), depth (d), and local magnitude (M_L) using SeisComP4 [30]; available at: <https://www.seiscomp.de/downloader/> (accessed on 8 June 2022); and, more specifically, Scolv [31]. Scolv was the primary interactive tool for revising or reviewing origin information such as picks, location, depth, time, magnitudes, and event association. Scolv can be operated in two modes. The first is connected to messaging, where all event and station inventory parameters are received from the messaging or the database and updated in real-time. Finally, new or updated parameters can be committed to the messaging. The second used in this study was offline without messaging, where all event and station inventory parameters are read from XML files or the database and updated interactively. For the best calculation of the epicenter and the minimization of the errors in the depth and epicenter analysis, a necessary condition is selecting an appropriate velocity model. The present research used the 1D velocity model proposed by Hasslinger [32].

The next task of this paper was the calculation of the source parameters (i.e., strike, dip, rake, depth, and moment magnitude) for the strong events and all significant aftershocks. Knowledge of the seismic moment tensor effectively contributes to determining seismic parameters. For example, in the late 1960s, the calculation of the seismic moment tensor [33] began from several methodologies developed both in the field of time [34–38] and in the field of frequencies [39–42]. The seismic moment tensor describes the seismic source as a point source. Modeling is a robust method that has previously been applied successfully for earthquakes occurring in various tectonic regions of Greece, including the Hellenic subduction zone in southern Greece and strike-slip zones of shear deformation in western Greece [43], and areas of dominant extensional stress regime such as Central Greece. Determining the seismic source parameters depends on the seismic moment tensor and its reversal process.

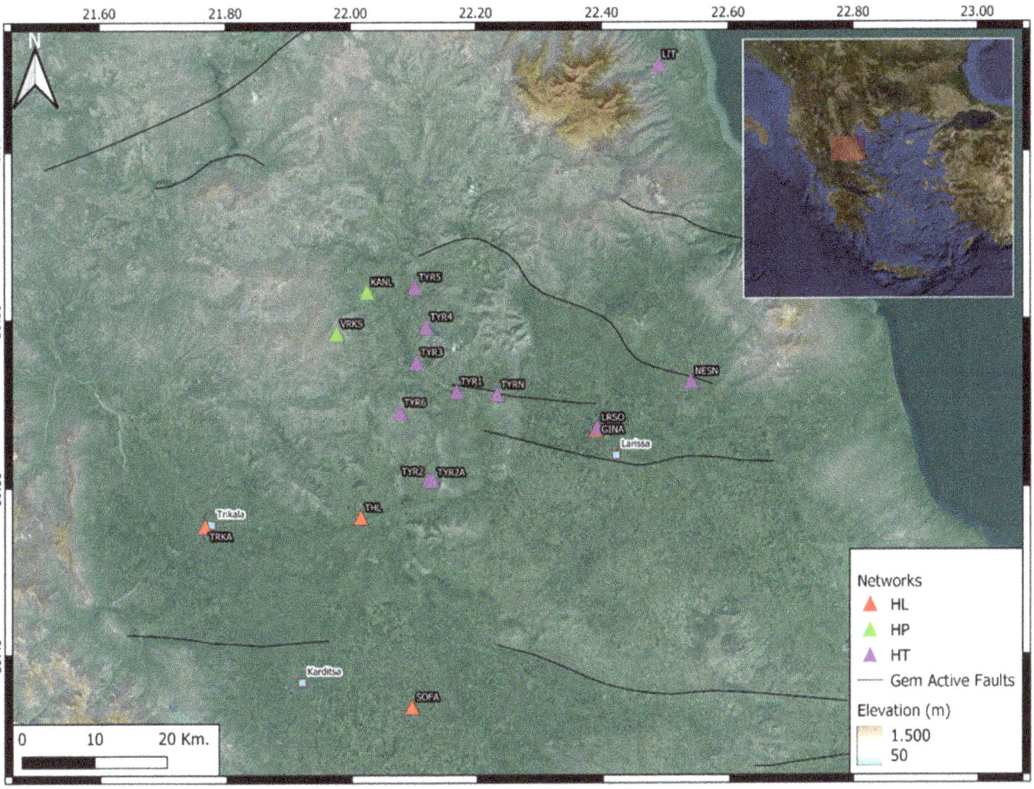

Figure 2. Geographical distribution of the Hellenic Unified Seismological Network's stations in the region of Thessaly. Green triangles represent the seismological stations belonging to the University of Patras, purple triangles denote the portable network installed by the Aristotle University of Thessaloniki immediately after the earthquake, and the red triangles are the stations of the National Observatory of Athens. The light blue color represents the cities near the epicenter.

3. Methods and Processing

Synthetics Seismograms at Regional Distances

The first deviations of P-waves are the most commonly used method to determine the focal mechanism of an earthquake. In this method [44], the information is provided by the arrival of the first P-waves and the fact that the first deviations depart as compressions at the two vertical angles formed by the plane of the fault and the auxiliary plane and as dilutions at the other two vertical angles is utilized. The above method provides reliable results for short but focused distances (<100 km) when azimuth coverage and clarity record the first arrival. This is how the focal mechanism of an earthquake is calculated. However, in most cases, this method is considered insufficient [44] for determining the focal mechanism, whether since the stations have a poor azimuth distribution or that these stations did not record the first movement. The above is evidence that the method is uncertain. Therefore, there is no reliability for the focal mechanism.

To determine the source parameters, it is required to create synthetic seismograms for each station and to compare them with corresponding observations. It is mathematically possible [45–47] to model and develops a process so that it is possible to predict the character of the seismogram, taking into account a realistic model of seismic wave propagation. This modeling is called a synthetic seismogram.

As mentioned in previous studies [26,27], the proposed methodology calculates the Green functions to determine the Earth's response. Then, the corresponding synthetic seismograms are compared with those observed to minimize their difference. For example, Ray's theory is used to create the Green functions [45,46]. The geometry of seismic rays is determined by the principle of Fermat, according to which the wave recorded at a certain point by a specific source follows the shortest path from those that can follow; that is, it follows the one that requires the least time. The above is practically equivalent to the researcher considering the focal length. This is why it is necessary to distinguish, when studying an event, at what distances the data used are located. According to the above, there are two major categories: the first includes data at distances less than 30°, and the second is for those stations with an epicentral distance of $30° < D < 90°$ [47]. Limiting the epicentral distance is considered necessary because stations with epicentral distances less than 30° have radii with departure angles dependent on the structure of the upper mantle. In this case, combining the knowledge of the Earth's response, a velocity model, and an appropriate methodology results in the calculation of Green functions.

Three different methodologies are used to calculate synthetic seismograms for earthquakes where near-regional data are used [48–51]. In the present article, Green functions were computed using the method in [48]. This method was initially presented in [52,53]. The basic idea of this method is that Green functions for an elastic layered structure can be calculated as a double integral on the frequency and horizontal waveforms [54]. Discretization was performed for each elementary point source. According to this theory, an infinite number of sources are gathered around it and distributed in equal intervals. The choice of distance depends on the time required for the point source to respond and identify the discrete number from the horizontal waveforms [55]. More specifically, the displacement of the elastic wave radiated by a pulsating point source in an infinite homogeneous medium can be expressed by the following relationship:

$$\varphi(R;\omega) = \frac{e^{i\cdot\omega(t-\frac{R}{a})}}{R} \tag{1}$$

The parameter, ω, symbolizes the angular frequency; t is the time; the parameter, R, expresses the distance between the source and the observation point; α represents the speed of the P-waves.

The above equation, in cylindrical system coordinates (i.e., r, ϑ, and z), can be written using the Sommerfeld integral as shown in Equation (2):

$$\varphi\,(r,z;\omega) = -i\cdot e^{i\cdot\omega\cdot t}\int_0^\infty \frac{k}{v} J_0(kr)e^{-i\cdot v\cdot|z|}dk \tag{2}$$

where $v = \sqrt{k_a^2 - k^2}$; Im $(v) > 0$; $k_a = \frac{\omega}{a}$; and J_0 are the Bessel functions of the first kind.

The Sommerfeld integral is a typical partial differential solution equation (wave equation):

$$\frac{\partial^2\varphi}{\partial r^2} + \frac{1}{r}\cdot\frac{\partial\varphi}{\partial r} + \frac{1}{r^2}\cdot\frac{\partial^2\varphi}{\partial\vartheta^2} + \frac{\partial^2\varphi}{\partial z^2} - \frac{1}{a^2}\cdot\frac{\partial^2\varphi}{\partial t^2} = 0 \tag{3}$$

the general solution of which is given by the equation presented below:

$$\varphi(r,\vartheta,z;\omega) = e^{i\cdot\omega\cdot t}\sum_{n=-\infty}^{\infty} e^{i\cdot n\cdot\vartheta}\int_0^\infty g_n(\omega,k)J_n(kr)e^{-iv|z|}dk \tag{4}$$

where J_n represent the Bessel functions of the first kind of order n.

Considering a different solution for the above equation results in the function:

$$\varphi_s(r,\vartheta,z;\omega) = -\frac{i}{\pi}\cdot e^{i\cdot\omega\cdot t}\sum_{n=0}^{\infty}(-1)^n\cdot\varepsilon_n\cdot e^{i\cdot 2n\cdot(\vartheta-\vartheta_s)}\int_0^{\infty}\frac{k}{\nu}J_{2n}(kr)e^{-i\nu|z|}dk \quad (5)$$

where the term ε_n expresses Neumann's factor and can take the values:

$$\varepsilon_n = \begin{Bmatrix} 2 & n\neq 0 \\ 1 & n=0 \end{Bmatrix}$$

where φ_s is the solution to Equation (3) and represents the field from a source located at a specific point. In a particular case where the source is positioned (i.e., L, ϑ_s, and 0) and the observation point is positioned (i.e., r, 0, and z), then Equation (5) is:

$$\varphi_s(r,z;L,\vartheta_s;\omega) = -\frac{i}{\pi}\cdot\sum_{n=0}^{\infty}(-1)^n\cdot\varepsilon_n e^{i\cdot 2n\cdot\psi}\int_0^{\infty}\frac{k}{\nu}J_{2n}\left(k\sqrt{r^2+L^2-2rL\cos\vartheta_s}\right)e^{-i\nu|z|}dk \quad (6)$$

where ψ represents the angular distance of the source between the source and observation point.

Using the Bessel functions:

$$e^{i\cdot 2n\cdot\psi}J_{2n}\left(k\sqrt{r^2+L^2-2rL\cos\vartheta_s}\right) = \sum_{m=-\infty}^{\infty}J_m(kr)J_{2n+m}(kL)e^{im\vartheta_s} \quad (7)$$

Equation (6) is transformed into the following:

$$\varphi_s(r,z;L,\vartheta_s;\omega) = -\frac{i}{\pi}\cdot\sum_{n=0}^{\infty}(-1)^n\cdot\varepsilon_n\int_0^{\infty}\frac{k}{\nu}\left(\sum_{0}^{\propto}J_m(kr)J_{2n+m}(kL)e^{im\vartheta_s}\right)e^{-i\nu|z|}dk \quad (8)$$

For the following, a set of elementary sources is considered as a given, and the radiation of which is symbolized by the parameters φ_s and ϑ_s. The relation provides the corresponding shift:

$$\varphi_c(r,z;L;\omega) = \int_0^{2\pi}\varphi_s(r,z;L,\vartheta_s;\omega)d\vartheta_s \quad (9)$$

or equivalents:

$$\varphi_c(r,z;L;\omega) = -2i\cdot\sum_{n=0}^{\infty}(-1)^n\cdot\varepsilon_n\int_0^{\infty}\frac{k}{\nu}J_0(kr)J_{2n}(kL)e^{-i\nu|z|}dk \quad (10)$$

Radiation from an infinite set of sources, as expressed by (10), can be represented by the following equation:

$$\sum_{m=1}^{\infty}\varphi_c(r,z;mL;\omega) \quad (11)$$

Considering that the coefficients of the Bessel functions can be replaced by trigonometric functions [41] and Fourier transforms, the elastic wave radiates through radial circular sources as circular wave sources. In the case where the source is placed in a half-space, the horizontal (u) and transverse (w) displacements are given by the equations:

$$\begin{aligned} u(r,z;\omega) &= -\tfrac{i\cdot 4\pi}{L}k_\beta^2\textstyle\sum_{n=0}^{\infty}\varepsilon_n\frac{k_n^2\cdot\gamma_n}{F(k_n)}e^{-i\nu_n|z|}\cdot J_1(kr) \\ w(r,z;\omega) &= -\tfrac{2\pi}{L}k_\beta^2\textstyle\sum_{n=0}^{\infty}\varepsilon_n\frac{k_n^2\cdot\left(2k^2-k_\beta^2\right)}{F(k_n)}e^{-i w_n|z|}\cdot J_0(kr) \end{aligned} \quad (12)$$

where parameter β symbolizes the velocity of the S-waves; γ is a scalar potential; z is the depth, and F is the Fourier transform.

A modified form of the Axitra code [50,52] was used to calculate the synthetic seismograms. Green functions were computed using the discrete wave methods in [49] and [50]. The synthetic seismogram was a linear combination of three elementary focal mechanisms according to this method.

We assumed that $u(t)$ symbolized the observed seismogram, which corresponded to any station since the effect of the instrument had already been removed. The synthetic seismogram corresponding to a double pair of forces is denoted by $s(t)$. The synthetic seismogram can then be expressed by the linear combination of three elementary mechanisms of generation of normal, reverse, oblique, and strike-slip faults according to [56], Figure 3.

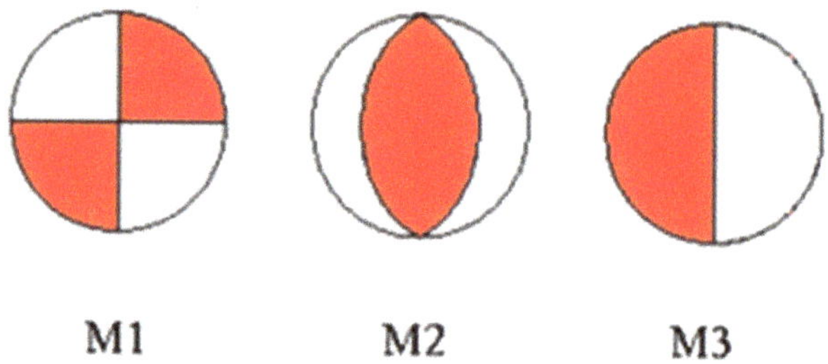

Figure 3. Elementary focal mechanisms.

The following equation defines the mathematical representation:

$$s(t) = M_0 \sum_{i=1}^{3} A_{ij}[\varphi - \varphi(\vartheta, \xi), \delta, \lambda] * G_{ij}[h, \Delta(\vartheta, \xi), t] \tag{13}$$

The index j = 1, 2, and 3 indicate the vertical, radial, and tangential components [35]; ξ is the geometric distribution of the displacement in the slipped fault; φ denotes the radiation. In addition, in (14), M_0 symbolizes the seismic moment; A_{ij} is the coefficients; G_{ij} is the Green functions, while in parameter Δ, each station's epicentral distance and the azimuth were integrated. Finally, the parameters φ, δ, λ, and h correspond to the strike, dip, slip of the rupture surface and the focus's depth, respectively.

The observed seismograms, $u(t)$, were linearly related to the seismic moment tensor M and the Green functions and are described by the mathematical equation:

$$U(w, q, v; \omega) = G \cdot M \tag{14}$$

where w, q, and v denote the vertical, radial, and tangential; r denotes the epicentral distances.

In (14), the vector U represents an array column of dimension ($n \times 1$), as observed at each station, with G denoting the matrix, which has elements computed by the Green functions, dimension ($n \times 5$); finally, M is a matrix of dimensions (5×1) with the elements of the seismic moment tensor. The data in matrix G were calculated for each epicentral distance [40]. To estimate the seismic moment tensor M data, matrix G must be inverted. The inversion of this matrix was conducted in two steps. In the first stage, it was assumed that:

$$M_{zz} = -(M_{xx} + M_{yy}) \tag{15}$$

The last equation ensures that there is no volume change. However, as already mentioned, in most cases, an earthquake is observed from more than five stations, with the result that the matrix G is non-square, and the system is called an overdetermined system. Therefore, the least squares method can solve the last equation using generalized inversion. Then, the solution is given by the formula:

$$m = \left[G^T \cdot G\right]^{-1} \cdot G^T \cdot d \tag{16}$$

where exponent T denotes the transpose operation of array G, and exponent -1 indicates the inverse of array $G^T \cdot G$. In Equation (16), the eigenvalues and eigenvectors of the matrix $G^T \cdot G$ must be calculated as well as the ratio of the minimum eigenvalue to the maximum. In this case, the coefficients, α_m, are the eigenvectors associated with corresponding eigenvalues and were calculated concerning a geographical system of coordinates (i.e., x, y, and z), where the following equations express $x, y, z > 0$:

$$\left\{ \begin{array}{c} M_{xx} = -a_4 + a_6 \\ M_{yy} = -a_5 + a_6 \\ M_{zz} = a_4 + a_5 + a_6 \\ M_{xy} = M_{yx} = a_1 \\ M_{xz} = M_{zx} = a_2 \\ M_{yz} = M_{zy} = -a_3 \end{array} \right\} \tag{17}$$

4. Results

4.1. Localization Solutions

This section presents the results of the two mainshocks, the most significant, and further light to moderate aftershocks of the sequence separately. To accurately calculate the epicenter for the mainshock and the second strong event, we used local and regional broadband data from the HUSN. Thirty-three seismological broadband stations recorded the mainshock at epicentral distances shorter than 265 km. Some of these stations record continuously; others use a trigger system. These stations are three-component broadband stations from the Hellenic Unified Seismological Network (HUSN) available in this area, mostly CMG-3ESP/100se of 100s and a smaller number from these with Trillium120P. We used these data to analyze the events reported by the HUSN for the two mainshocks on 3 and 4 March 2021. We considered a 1D velocity model, previously built by inverting more than 124 P and S arrivals from the HUSN database, using the inversion approach [48] and the earthquake location methods [57–61]. Probabilistic locations were obtained using the NonLinLoc code [62].

On 3 and 4 March 2021 (10:16:08.3 UTC and 18:38:19.1 UTC), two strong earthquakes of similar Magnitude, M_L 6.0 and M_L 5.9, occurred in Thessaly, Central Greece. These events induced extensive structural damage, and a rich aftershock sequence followed these events (Figure 4). Three thousand seven hundred eighty-seven events were recorded from the beginning of the sequence until 31 December 2021, while only in the first month, 2268. For these three events, the source parameters were calculated and compared to the observed solutions from other institutes, and for the majority of them, a good agreement was found (Table 2).

Table 2. List of moment tensor solutions published by various institutions and universities for 3 March 2021 (10:16:08.3, UTC) and 4 March 2021 (18:38:19.1, UTC). Source: CSEM—EMSC Available at: <https://www.emsc-csem.org/> (accessed on 8 June 2022) [4].

Elassona Earthquake (3 March 2021, 10:16:08.39, UTC) M_w = 6.3											
Institute	Latitude (°)	Longitude (°)	M_w	M_0 (dyn*cm)	Depth (km)	Strike (°)	Dip (°)	Rake (°)	Strike (°)	Dip (°)	Rake (°)
Our Study	39.7453	22.2340	6.3	3.173×10^{25}	11.0	147	57	−86	317	39	−113
USGS	39.5594	21.9200	6.3	3.330×10^{25}	11.5	139	55	−83	307	36	−100
NOA	39.7545	22.1992	6.3	3.173×10^{25}	10.0	146	59	−79	305	33	−108
ERD	39.8055	22.2578	6.2	2.6149×10^{25}	7.1	147	48	−94	332	43	−85
GCMT	39.6500	22.1400	6.3	3.140×10^{25}	12.0	119	45	−109	324	48	−72
INGV	39.7400	22.1900	6.2	3.600×10^{25}	10.0	116	41	−114	327	53	−70
KOERI	39.7100	22.1600	6.3	2.990×10^{25}	10.0	126	37	−103	323	53	−79
GFZ	39.7700	22.1400	6.3	3.000×10^{25}	10.0	130	45	−90	310	44	−89
OCA	39.7500	22.2100	6.2	3.600×10^{25}	7.0	135	45	−90	315	45	−90
IPGP	39.7640	22.1760	6.2	3.600×10^{25}	10.0	321	33	−78	127	57	−98
CPPT	39.7760	22.1800	6.3	3.400×10^{25}	12.0	125	55	−100	321	36	−77

Table 2. *Cont.*

Elassona Earthquake (3 March 2021, 10:16:08.39, UTC) M_w = 6.3											
Institute	Latitude (°)	Longitude (°)	M_w	M_0 (dyn*cm)	Depth (km)	Strike (°)	Dip (°)	Rake (°)	Strike (°)	Dip (°)	Rake (°)
AUTH	39.7300	22.2200	6.2	2.200×10^{25}	6.0	131	54	−92	314	36	−88
UOA	39.7505	22.1980	6.3	3.120×10^{25}	19.0	130	54	−89	309	36	−91
Elassona Earthquake (4 March 2021, 18:38:19.19, UTC) M_w = 6.2											
Institute	Latitude (°)	Longitude (°)	M_w	M_0 (dyn*cm)	Depth (km)	Strike (°)	Dip (°)	Rake (°)	Strike (°)	Dip (°)	Rake (°)
Our Study	39.7916	22.1274	6.2	2.6149×10^{25}	18.0	140	42	−86	300	40	−90
GFZ	39.8000	22.2000	6.3	3.300×10^{25}	17.0	146	48	−91	329	41	−88
NOA	39.7710	22.0958	6.0	1.364×10^{25}	11.3	112	59	−87	287	31	−95
AUTH	39.7800	22.1200	5.9	9.300×10^{24}	7.0	109	60	−89	287	30	−92
UOA	39.7708	22.0918	6.1	2.0300×10^{25}	15.0	131	40	−88	308	50	−92

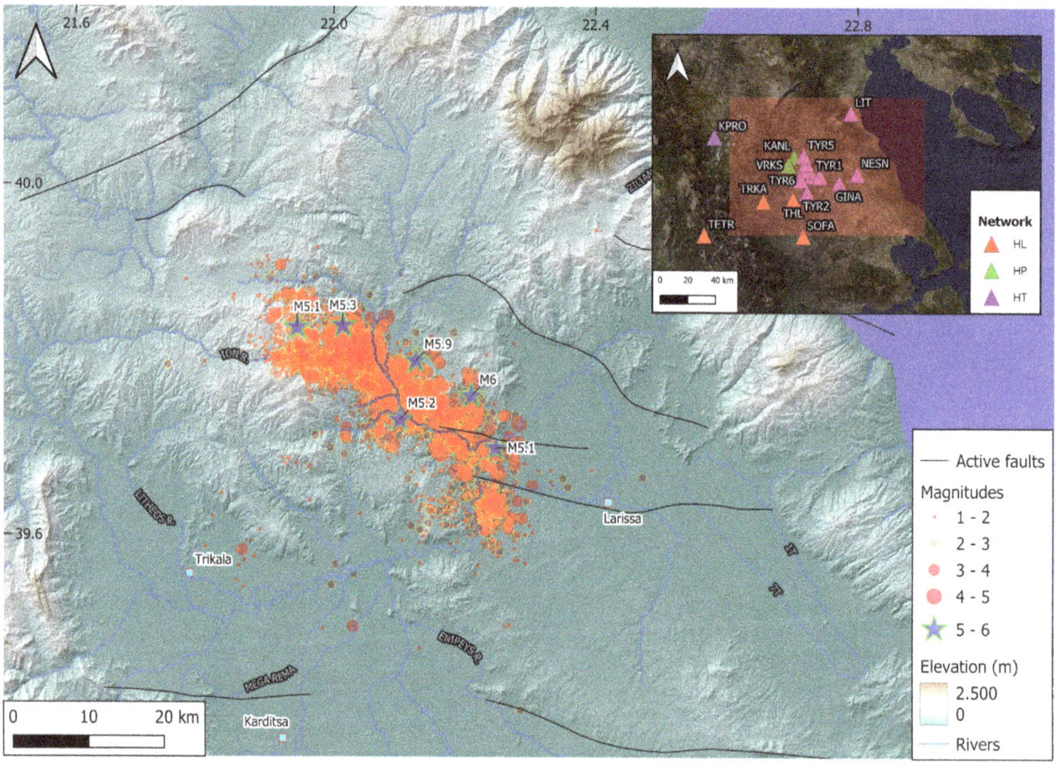

Figure 4. Black lines correspond to the main active faults for the study area—a general seismicity map of the broader area of the Thessalian, Central Greece. Light blue rectangles denote the cities of the study area. The stars represent the events in the study area for 3 March 2021–31 December 2021.

4.1.1. The M_L 6.0, Elassona, 3 March 2021 (10:16:08.3 UTC)

A strong earthquake, M_L = 6.0, occurred in the broader area of Thessaly, Central Greece, on 3 March 2021 (10:16:08.39, UTC). The geographical coordinates, as they were calculated manually for the scope of this study, were φ = 39.7453° N, λ = 22.2340° E at a depth of 12 km (Figure 5). The phases used to calculate the epicenter were 124 phases from P- and S-waves at epicentral distances between 3 and 265 km and an azimuthal gap of 24°.

More distant stations indicated a longer extensive duration for P- and S-waves because of the simultaneous arrival of refracted and reflected P- and S-waves as well as sP-waves.

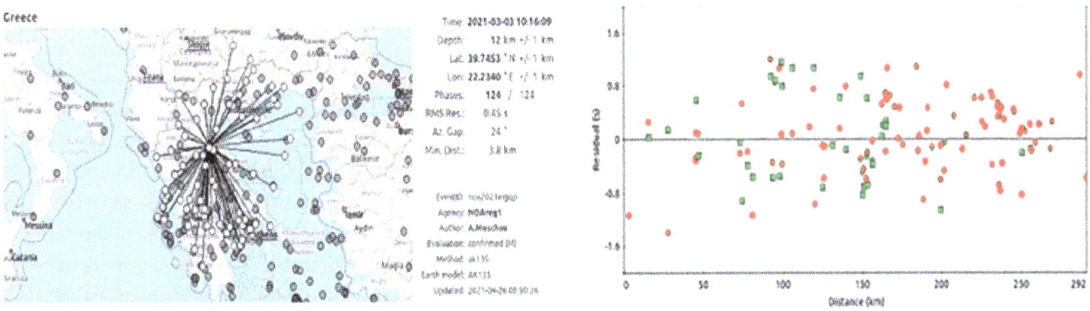

(a) (b)

Figure 5. (**a**) Azimuthal distribution of the stations used to calculate the epicenter of the 3 March 2021 (10:16:08.39, UTC) and the characteristics of the solution; (**b**) graph of the distance (km)–residual (s) stations. The green squares represent the stations where only vertical components (P-waves) were used, while the red circles represent the stations where both vertical and horizontal components (P-waves and S-waves) were used.

Figure 5a represents the geographical distribution of the stations that were used to calculate the epicenter, and Figure 5b shows a graph of the distance (km) versus residual (s) stations. Finally, the green circles denote the stations where only vertical components (P-waves) were used, while the green squares represent the stations where vertical and horizontal components (P-waves and S-waves) were used.

Figure 6 shows a section of the displacement record of the broadband seismological data from the mainshock. In local station records, both P and S direct-body waves exhibited durations shorter than 2 s.

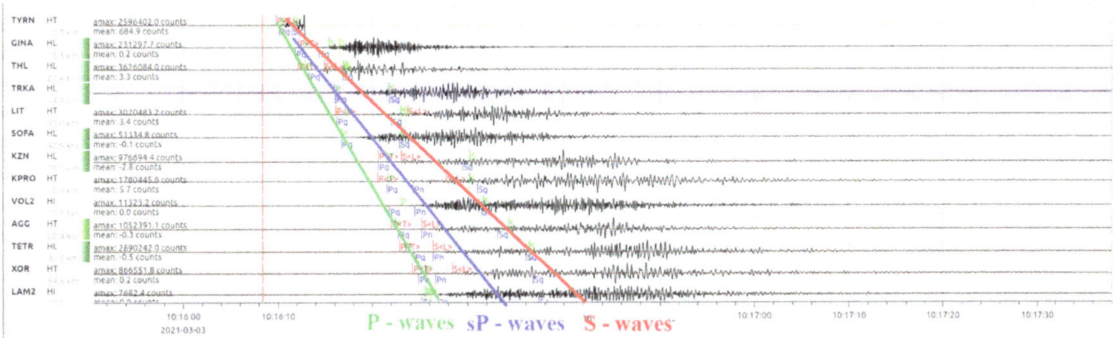

Figure 6. A section of the displacement record of the stations. The horizontal components (east–west and north–south traces) were rotated into radial and transverse components. The sP-wave was clear in the radial and vertical components, but only the S-wave was observed along with the surface waves in the transverse component.

The first strong aftershock occurred a few hours later at 11:45:45.99 (UTC) with a magnitude $M_L = 5.1$. A rich sequence followed these events, more specifically, from the main event's time, and for the next 24 h, there were 268 seismic vibrations with a magnitude greater than 1 (Figure 7a). When monitoring the seismic sequence over the first month, 2268 aftershocks occurred, 34 with a magnitude $M_L \geq 4.0$. The distribution of magnitudes concerning the number of earthquakes is presented in Figure 7a,b. As can be seen from the following diagrams, for the first day of the sequence, a significant number of earthquakes (~Nr = 268) were of a magnitude of $2.0 < M_L < 3.0$, while over the first month, the number of the respective magnitudes that fluctuated was Nr = 2062.

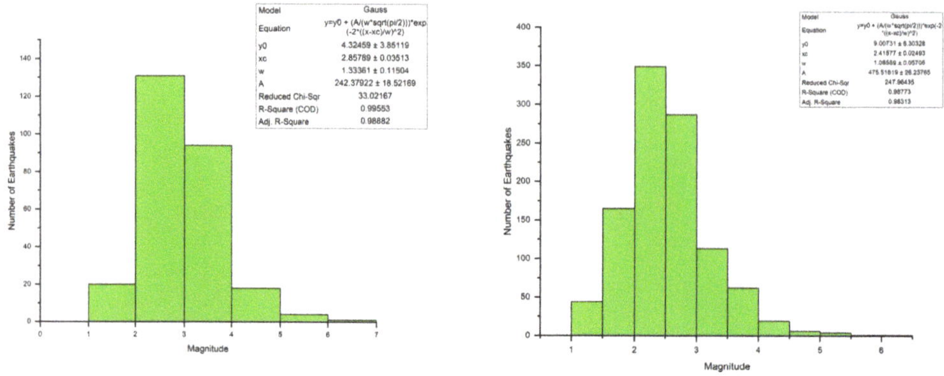

(**a**) (**b**)

Figure 7. (**a**) Magnitude distribution of the aftershocks recorded for the first 24 h after the ML 6.0, number of events (Nr) = 268; (**b**) the distribution of magnitudes concerning the number of earthquakes for the first month after the main event. Nr = 2062.

4.1.2. The M_L 5.9, Elassona, 4 March 2021 (18:38:19.1 UTC)

One day after the first event, a second (4 March, 18:38:19.1 UTC) event with a similar magnitude as the first (M_L 5.9) occurred on the NW section of the fault. The epicenter was determined to be 10 km north of the first mainshock (39.7916° N, 22.1274° E), while the depth was d = 8 km, as shown in Figure 8.

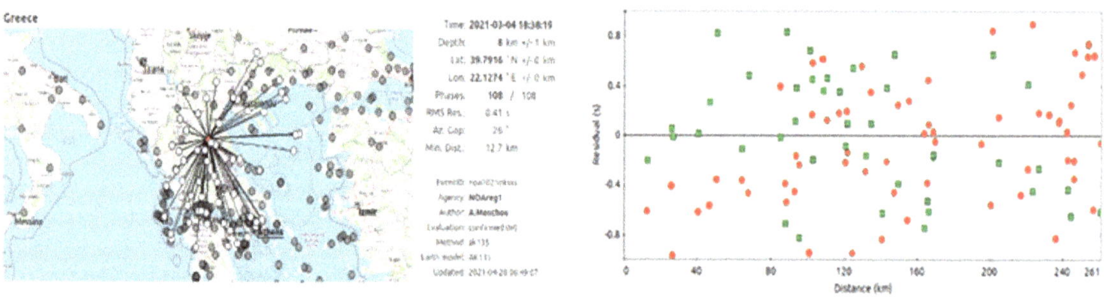

(**a**) (**b**)

Figure 8. (**a**) Azimuthal distribution of the stations used to calculate the epicenter of the 4 March 2021 (18:38:19.19, UTC) and the characteristics of the solution; (**b**) graph of the distance (km)–residual (s) stations. The green squares represent the stations where only vertical components (P-waves) were used, while the red circles represent the stations where vertical and horizontal components (P-waves and S-waves) were used.

In calculating the epicenter, 108 phases were used from P- and S-waves at epicentral distances between 5 and 261 km and an azimuthal gap of 26°. More distant stations indicated a more extensive duration for P- and S-waves because of the simultaneous arrival of refracted and reflected P- and S-waves as well as sP-waves

4.2. Moment Tensor Solutions

The second goal of this study was to calculate the source parameters for the two strong and moderate events with a local magnitude $M_L \geq 4.0$. The selected events presented a high signal-to-noise ratio, and we computed the moment tensor. Using the method of inversion for the seismic moment tensor, the Ammon code determined the seismic parameters [63]. It is noted that particular emphasis was given to stations close to the fault plane. However, due to their location, these stations contributed positively to the outcome of the inversion procedure.

Seismological broadband data (in epicentral distances, $\Delta < 6°$) were selected and analyzed to determine the focal mechanism and the source parameters. In most cases, regional broadband data were selected, processed, and applied to the proposed methodology of at least six stations at an excellent azimuthal coverage. This was based on creating synthetic seismograms and directly comparing them with the observed ones for a given velocity structure. The reflectivity method of Kennett [64], as implemented by Randall [65], was used, and a bandpass filter was applied to the data. Filter frequencies varied depending on the type of waveform: a bandpass filter of 0.1–0.2 Hz corner frequencies for traces recorded by local stations and 0.01–0.1 Hz for regional waveforms. The same bandpass filter was applied both on the observed waveforms and synthetics, having a fixed length of 70 s. The next step was the deconvolution of the instrument response from the waveforms, and the waveform conversion through integration produced pure displacement. Finally, the horizontal components with rotation were converted to radial and transverse waveforms. The three previous displacement components must be cut in the same starting time (nearly 10 s before the first arrival) with a total duration of 300 s.

The inversion results indicate that inverting waveforms longer than 70 s resulted in higher misfits. The quality of the moment tensor solutions can be evaluated by considering the average misfit and the compensated linear vector dipole (CLVD). A quality code consists of the letters A–D [65] for each solution for the minimum misfit and between the numbers 1 and 4 for the percent of CLVD [66].

The results of the inversion for the two main events as well as their comparison with those of other institutes/universities, are presented in Table 2. Only the DC part of the moment tensors is discussed because we did not consider the CLVD parts as having a physical meaning.

4.2.1. The M_W 6.3, Elassona, 3 March 2021 (10:16:08.3 UTC)

We applied the proposed methodology to examine the active fault that ruptured on the 3 March, the mainshock, the type of focal mechanism, and the source parameters. 21 seismological records were used in the final solution at seven broadband stations with epicentral distances between 70 and 280 km, having an adequate azimuthal distribution (maximum gap of 70°) and covering all quadrants of different polarity on the focal sphere to constrain the nodal planes. The waveform inversion indicated the activation of a normal type faulting with source parameters equal to $\varphi = 147°$, $\delta = 57°$, and $\lambda = -86°$; the depth was calculated at d = 11 km; the seismic moment was M = 3.173e+25 dyn*cm. In addition, this technique showed a high percentage of double pairs of forces concerning the double vector dipole with DC = 85% and CLVD = 15%.

The obtained focal mechanism was found to be in good agreement with the one proposed solution by the USGS (Table 2). The focal mechanism of an earthquake and the simulation of seismic waves for the selected stations are presented in Figure 9.

4.2.2. The M_W 6.2, Elassona, 4 March 2021 (18:38:19.1 UTC)

On 4 March 2021 (18:38:19.1, UTC), an extreme event with a similar magnitude occurred in the same region one day later. It applied the same methodology with the same criteria, as the two events had similar magnitude and waveforms. The focal mechanism was also a normal faulting plane solution with directivity on the east–northeast–west-southwest-oriented planes and closely matched the observed first-motion body-wave polarities. The seismic moment magnitude was calculated as $M_W = 6.2$, with the likely fault plane having the strike of $\varphi = 140°$, the dip of $\delta = 42°$, and the rake of $\lambda = -86°$. The following calculations of other agencies for the mainshock indicated pure NW-SE normal dip-slip faulting. The calculated depth from moment tensor inversion was 18 km (Figure 10). The focal mechanism was found to be in good agreement with the one proposed solution by the GFZ (Table 2).

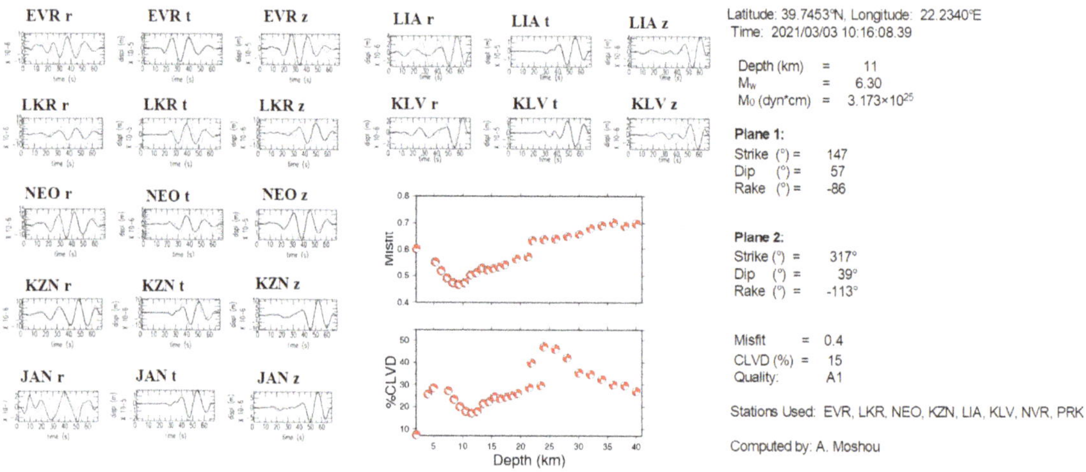

Figure 9. Inversion results for the first mainshock of 3 March 2021 (10:16:08.3 UTC). The left of the misfit/CLVD diagrams observed and synthetic displacement waveforms (continuous and dotted lines, respectively) are shown at the inverted stations for the radial, tangential, and vertical components. The solution's summary and the fault plane solution as a lower hemisphere equal-area projection are depicted in the right part of the figure.

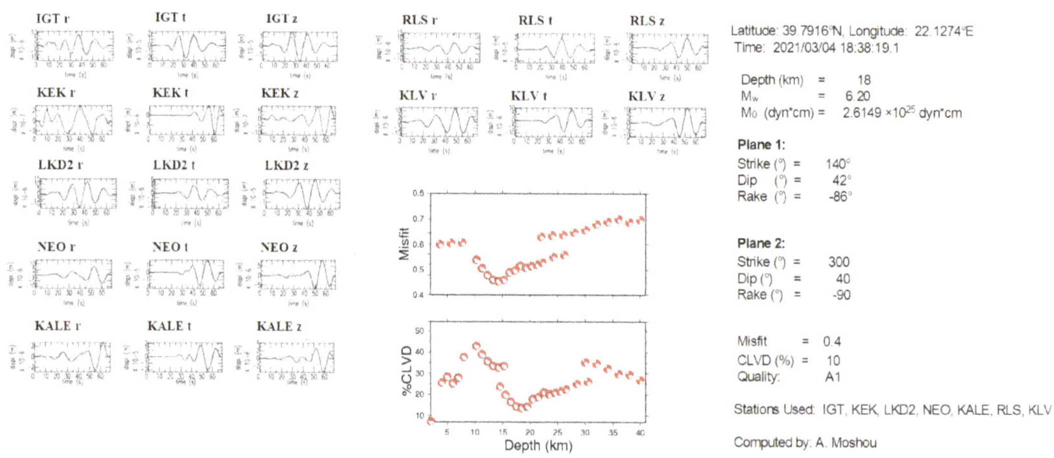

Figure 10. Inversion results for the second major mainshock of 4 March 2021 (18:38:19.19 UTC). To the left of the misfit/CLVD diagrams, observed and synthetic displacement waveforms (continuous and dotted lines, respectively) are shown at the inverted stations for the radial, tangential and vertical components. The solution's summary and the fault plane solution as a lower hemisphere equal-area projection are depicted in the right part of the figure.

Various seismological observatories and institutes routinely publish moment tensor solutions for strong events in Greece. Therefore, the German Research Centre for Geosciences (GFZ), the United States Geological Survey (USGS), the National Institute of Geophysics and Volcanology (INGV), the Kandilli Observatory and Earthquake Research Institute (KOERI), and others, as shown in Table 2, were selected in the present study to compare their solutions with those derived from this work. In addition, ordinary moment tensor solutions were available from the European–Mediterranean Seismological Centre [4], and a comparison of the moment magnitude (M_W) and focal mechanism results was performed.

5. Discussion

The historical seismicity of the region of Thessaly is related to the North Transform Fault (NTF), approximately 50 km southeast of the double earthquake. Therefore, knowledge regarding a dynamic error in this area is minimal. Thessaly, Central Greece, is a region of active crustal extension but low strain [67]. This was confirmed both empirically and by the study in [68]. Although the Thessaly region has a well-known history of large earthquakes, the studied region showed only weak seismicity over the last 15 years, which the Hellenic Unified Seismological Network covers. A subcrustal presence suggests that the upper crust deformation was probably caused by large-scale tectonics and plate dynamics.

On 3–4 March, two similar magnitudes (M_W = 6.3 and 6.2, respectively) occurred in the broader area of the Thessaly area. A few hours later, two earthquakes with similar magnitudes (M_W = 5.3 and 5.2, respectively) were observed in the same region. A rich seismic sequence followed these earthquakes. More specifically, in the first month alone, 2062 earthquakes took place. We used the Hellenic Unified Seismological Network (HUSN) (Figure 2) to study the two 2021 earthquakes in the Elassona in Thessaly, Central Greece. We accurately calculated the epicenter of these two events and computed the moment tensor of the mainshock and other more significant magnitude aftershocks. Two datasets were used for inversion: one using regional stations and the other including only stations close to the mainshock's epicenter. These analyses show that the observed inversion indicates the activation of a normal type faulting and shallow depth for both events.

The first-motion body-wave polarities provide information only regarding the onset of the rupture, whereas the inversion of surface waveforms includes information regarding the earthquake centroid (point of highest slip release). Especially for more significant earthquakes, these two points do not necessarily coincide spatially. According to the study in [67], resolving moment tensors, especially strike-slip mechanisms, is challenging when the source depth is much shallower than the wavelength considered during inversion. On the contrary, this is not always the case for moderate events due to the extensive period component's poor quality. Analysis of moderate events is crucial for seismogenic volumes, where many such earthquakes occur, allowing the calculation of their source parameters. Thus, the seismotectonic characteristics of the study area can be determined. Knowing the source parameters for moderate earthquakes is very important for seismically active regions, mainly when no significant events occur. In general, it allows analytical studies to reveal a specific area's tectonics and seismogenic characteristics. In the present work, the applied waveform inversion methodology was calculated using data at regional distances, the source parameters for the most decisive events, and 33 moderate earthquakes with magnitudes of $M_W > 4.0$. Finally, the results for the moment tensor solutions of the analyzed aftershocks are shown in Appendix A.

Almost all aftershocks of the sequence occurred north of the surface rupture. Interestingly, the aftershocks were concentrated along the edges of the presumptive rupture plane. This effect is visible on the map view as a gap of aftershocks north of the central part of the surface rupture. The seismicity gap is visible at the shallow depth of the first mainshock on the east-west depth section. Considering only the aftershocks that occurred in March 2021, they spread in the east-west direction. From the different depths for the first mainshock and the east-west extent of the aftershock distribution, we estimated the size of the rupture plane and recorded and visualized its spreading.

Figure 11 represents the relation time versus the event's magnitude in the study area. It was observed that most aftershocks occurred within the first three months in the first 75 days. It was observed that for the first two months after the occurrence of the main earthquake, both the density and magnitude of the aftershocks were much higher than the rest.

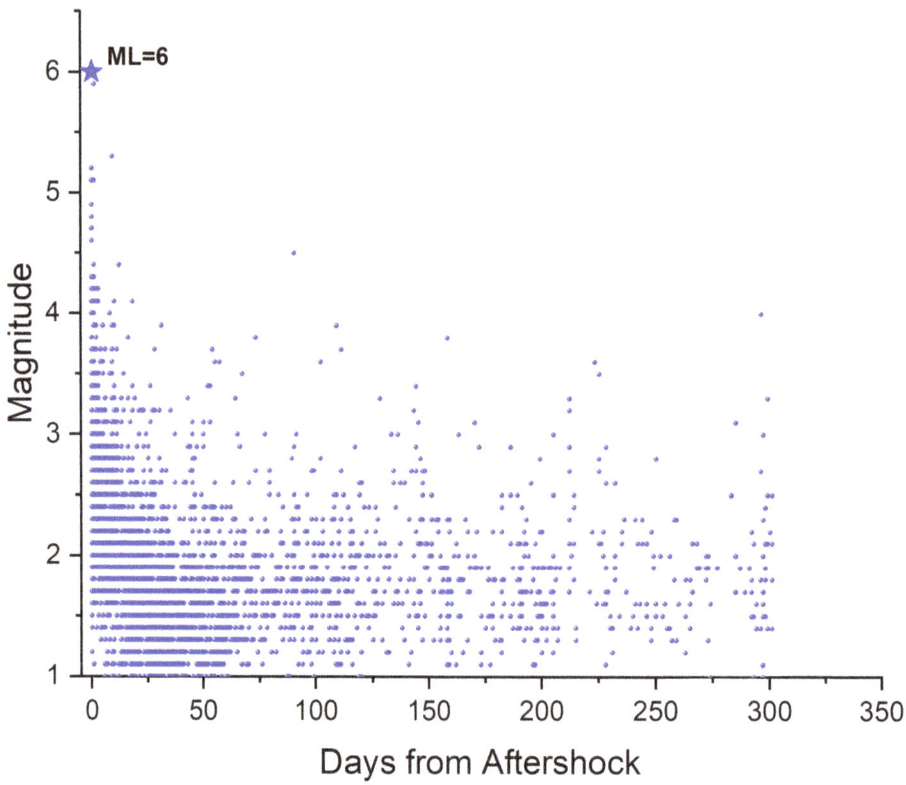

Figure 11. Distribution of magnitudes with time as a function of days from the mainshock of 3 March 2021 (10:16:08, UTC) for the 3787 records. On the x-axis is the magnitude time (days from mainshock) of each event from Elassona, Central Greece, while on the y-axis is the local magnitude (M_L).

Figure 12 represents the relation distance from the mainshock versus the event's magnitude in the Elassona, Central Greece study area. From the diagram, one can observe that a higher density of epicenters accumulated in the first 20 km from the epicenter of the main earthquake. Magnitude variations ranged from $1.0 < M_L < 4.0$, as when seismic monitoring stations are added or removed, our ability to detect small earthquakes changes reliably.

The seismic sequence's monitoring and study showed a shift in the epicenters to the northwest (Figures 13 and 14). The spatial distribution versus date indicated a displacement of earthquake density to the NW of the first three dates after the main event. Then, the density of the earthquakes decreased, and after eight days, a high-density event occurred at latitude = 39.80° and longitude = 21.90°, without any further indication of spatial displacement. This fact proves the claim that the two strong earthquakes belonged to two different fault zones.

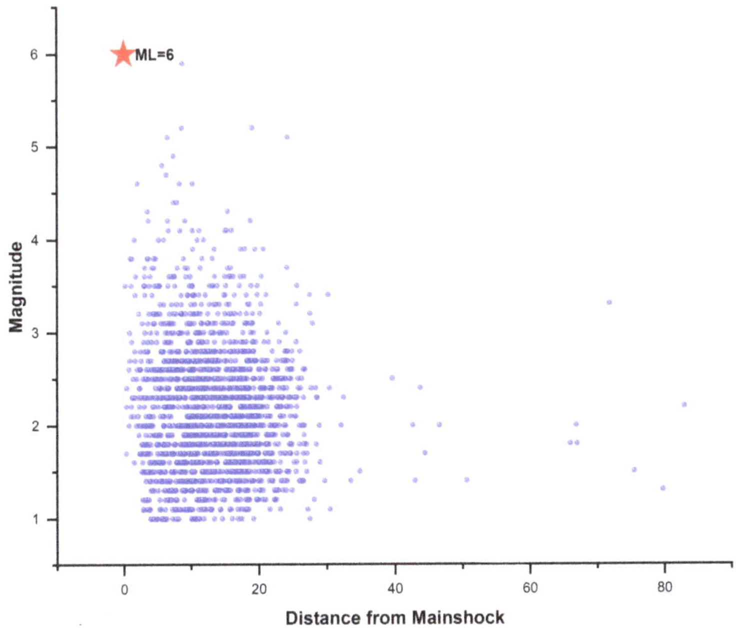

Figure 12. Location with distance (km) from mainshock as a magnitude (M_L) function for the 3.787 events recorded. On the *x*-axis is the distance from the mainshock, while on the y-axis is the magnitude of each aftershock.

We determined deviatoric moment tensor solutions for 34 earthquakes with $M_W \geq 4.0$ at depths between 6 and 14 km. The depths decreased in the eastern and especially in the western termination of the sequence. The proposed methodology's application gave rise to the activation of normal faulting with the direction N-S. The depth varied 11 km < d < 15 km for the five strongest sequences, a fact that agrees with the study in [67], while the magnitudes were $5.2 < M_W < 6.3$. Here, they were concentrated at approximately 7 km in depth, in contrast to about 14 km in the central part of the sequence. The aftershocks occurred along the entire surface rupture length but mainly north. A very high percentage of a double couple (DC) in relation emerged to the compensated linear vector dipole (CLVD) for the most significant percentage of solutions. More specifically, 80% of the solutions presented a DC > 90% solution, while the remaining 20% had a DC between 75% and 89%. The focal mechanism solutions for the strongest earthquakes determined in this study appear in Figure 15.

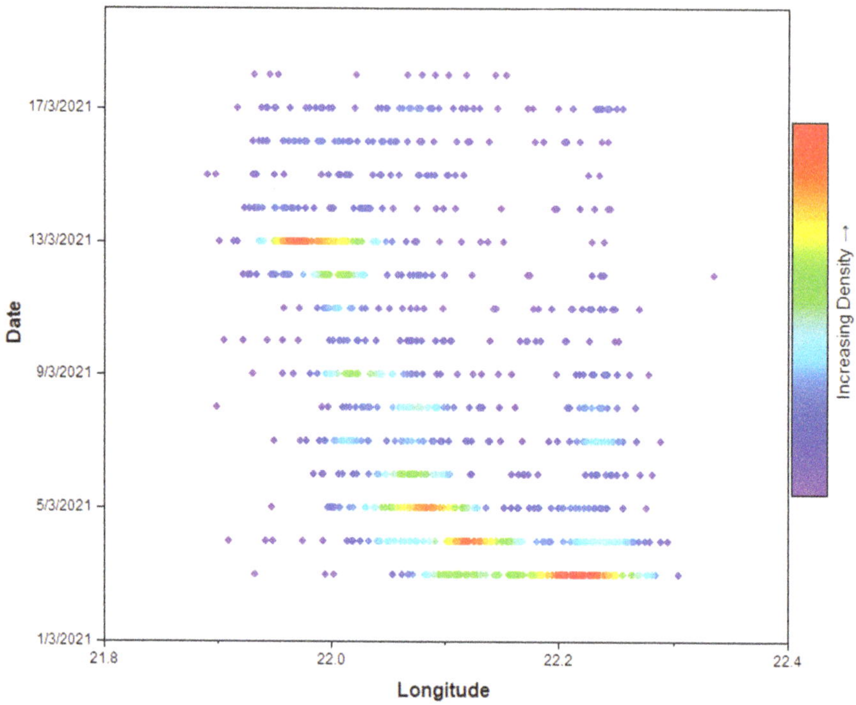

Figure 13. Spatial distribution (longitude) of earthquake density for different dates.

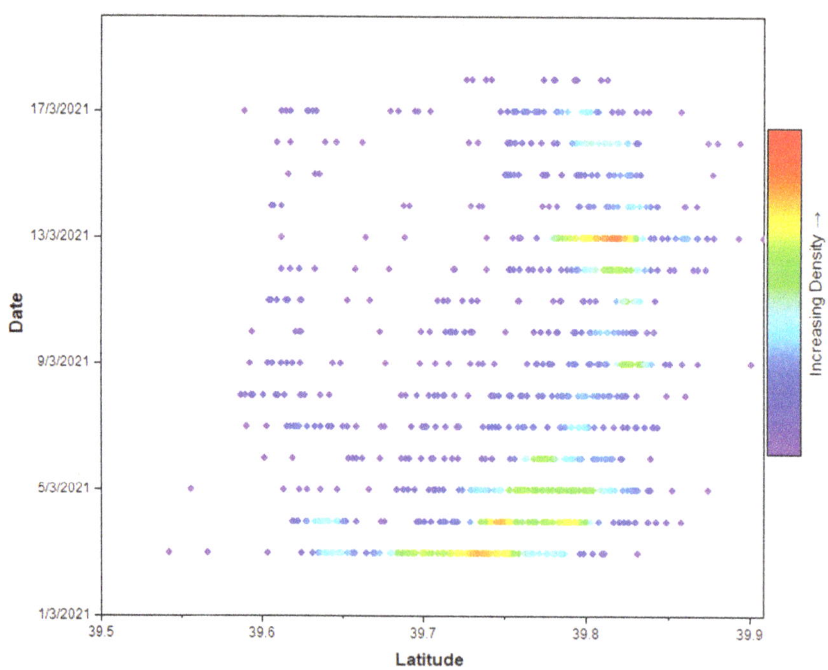

Figure 14. Spatial distribution (latitude) of earthquake density for different dates.

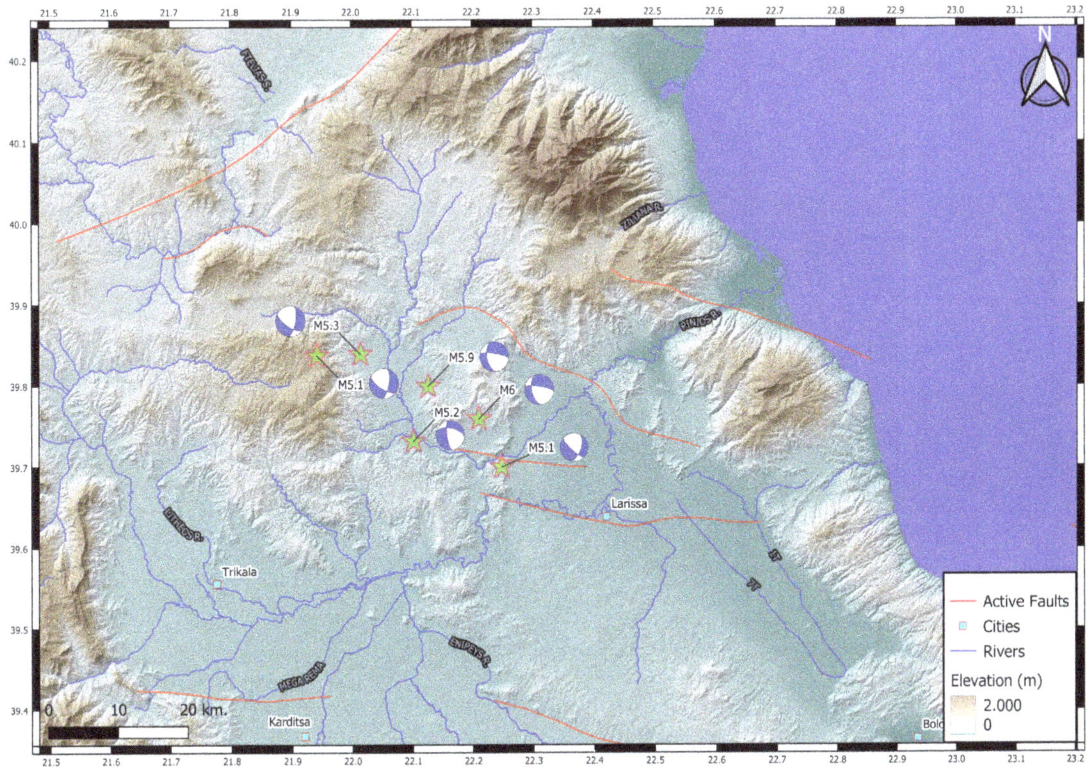

Figure 15. A seismotectonic map of the study area. On the map, the epicenters appear for the study area's more significant events, corresponding focal mechanisms, and significant faults.

6. Conclusions

Thessaly (Central Greece) experienced a rich seismic sequence in March 2021. The mainshock, measuring $M_L = 6.3$, struck on 3 March and was quickly followed by a powerful aftershock, measuring $M_L = 6.2$. Using a large dataset, high-resolution catalogs of earthquakes, and hundreds of focal mechanisms, we provided a thorough investigation of the March–April 2021 seismic sequence in Northern Thessaly (Damasi–Tyrnavos region). More than 30 focal mechanisms for earthquakes with M 4.0 were computed using local–regional data. The occurrences primarily showed normal faulting in an NW-SE direction, but there was also evidence of WNW-ESE to E-W normal faulting, especially after the last significant event on 12 March 2021. The aftershocks' spatial distribution and focal mechanisms indicated normal faulting with a strike direction of NW-SE.

Additionally, the focal mechanisms indicated an extensional stress field oriented differently than previously known, from NE to SW. Strong evidence points to the rupture of two neighboring fault segments based on the two most significant shocks' size, location, and spatiotemporal evolution. We propose that this fault system, a steeper splay of a deeper low-to-moderate angle normal fault, was partially reactivated as a secondary surface rupture throughout the episode.

For the study period, more than 3,500 events were recorded and analyzed. The spatial distribution and the timely distribution of high-density aftershocks indicate the activation of two different fault zones. The depth value varied between 10 and 15 km for the aftershocks where the source parameters were calculated. The error depth was ~1 km; the depth error depended on various factors such as the software used, the velocity model, the number of available stations, the azimuth distribution of the existing stations, and the analyst's ability.

For the first two months after the main earthquake, the density and magnitude of the aftershocks were much higher than the rest. A higher density of epicenters accumulates in the first 20 km from the epicenter of the main earthquake. Magnitude variations were $1.0 < M_L < 4.0$.

According to our findings, the 2021 doublet ruptured previously unmapped fault segments, with most of the slide occurring during the two big shocks to the west of Tyrnavos town. From the first main shock's epicenter, the activity predominantly moved northward, rupturing a crustal volume between 4 and 15 km thick and producing shallower off-fault seismicity.

Author Contributions: Conceptualization, A.M. and A.K.; methodology, A.M., P.A., A.K. and I.O.V.; software, A.M. and P.A.; validation, A.M., A.K. and I.O.V.; formal analysis, A.M.; investigation, A.M.; data curation, A.M. and A.K.; writing—original draft preparation, A.M., P.A., A.K., I.O.V., N.S.P. and T.N.K.; writing—review and editing, A.M., P.A., A.K., I.O.V., N.S.P. and T.N.K.; visualization, A.M., A.K., I.O.V. and T.N.K.; supervision, A.M.; project administration, A.M. All authors have read and agreed to the published version of the manuscript.

Funding: This research received no external funding.

Institutional Review Board Statement: Not applicable.

Informed Consent Statement: Not applicable.

Data Availability Statement: The datasets presented in this study and the contents of the tables are openly available at: https://bbnet.gein.noa.gr/HL/real-time-plotting/noa-stations-list/hl-network-and-collaborative-stations-information (accessed on 8 June 2022); http://www.geophysics.geol.uoa.gr/stations/gmaps3/leaf_stations.php?map=1&lng=el (accessed on 8 June 2022); http://geophysics.geo.auth.gr/ss/Book_LOG.htm (accessed on 8 June 2022); http://seismo.geology.upatras.gr/heliplots/StationsInfo.htm (accessed on 8 June 2022); https://www.emsc-csem.org/Earthquake/; https://bbnet.gein.noa.gr/HL/ (accessed on 8 June 2022); http://www.orfeus-eu.org/eida (accessed on 8 June 2022).

Acknowledgments: The authors would like to thank the editors for allowing us the chance to publish our work as well as the three anonymous reviewers for their helpful criticism that helped improve the initial draft of the manuscript. We thank the European permanent seismic network operators who made their data available through EIDA at http://www.orfeus-eu.org/eida (accessed on 8 June 2022). In this study, data from the following Institutes were used: (1) HL (NOA, Hellenic Seismic Network), doi:10.7914/SN/HL; (2) HT (Aristotle University of Thessaloniki Seismological Network), doi:10.7914/SN/HT; (3) HP (University of Patras, Seismological Laboratory), doi:10.7914/SN/HP; (4) HA (National and Kapodistrian University of Athens, Seismological Laboratory), doi:10.7914/SN/HA. We also thank the scientists and personnel who installed and/or maintained the permanent and temporary stations belonging to the HUSN.

Conflicts of Interest: The authors declare no conflict of interest.

Appendix A

Table A1. Source Parameters of the main events and intermediate magnitudes for the Elassona seismic sequence for 03 March 2021–31 December 2021. Nr is the event number; Lat and Lon are the geographical coordinates of each event as calculated by the National Observatory of Athens; M_0 is the seismic moment in dyn*cm; M_w is the moment magnitude; the strike, dip, and rake of the two nodal planes were the seismic parameters as calculated from the inversion; CLVD is the percentage of the compensated linear vector dipole, which describes seismic sources with no volume changes; Ns is the number of stations used in inversion; finally, Q is the quality of the solution depending on the misfit and the percentage CLVD.

N_r	Origin		Location		M_0	M_w	Depth (km)		Plane 1			Plane 2			CLVD (%)	Nr of Stations	Quality
	Date	Time	Lat (°)	Lon (°)	(dyn*cm)		Catalog	MT	Strike (°)	Dip (°)	Rake (°)	Strike (°)	Dip (°)	Rake (°)			
1	24 December 2021	11:41:47	39.8131	22.0482	0.135×10^{23}	4.1	8.2	13	178	55	−58	307	40	−85	8	5	A1
2	1 June 2021	15:56:14	39.8112	22.0601	9.602×10^{22}	4.6	10.9	12	130	50	−87	300	46	−98	12	5	A1
3	21 March 2021	17:15:53.4	39.7801	22.0903	0.296×10^{23}	4.3	9.8	10	176	54	−91	357	35	−89	10	6	A1
4	15 March 2021	15:43:37.0	39.7581	22.1159	7.173×10^{22}	4.5	6.6	11	126	42	−87	301	48	−93	5	6	A1
5	13 March 2021	15:09:13.3	39.8158	22.0372	0.135×10^{23}	4.1	7.0	12	123	43	−90	304	46	−89	12	5	A2
6	12 March 2021	12:57:50.6	39.8387	22.0134	1.187×10^{24}	5.3	7.0	13	153	31	−91	334	58	−89	7	7	A1
7	11 March 2021	14:19:40.4	39.7801	22.0811	0.240×10^{23}	4.2	6.0	11	130	50	−89	309	44	−95	10	5	A1
8	6 March 2021	19:47:40.2	39.8387	22.0738	0.240×10^{23}	4.2	5.8	10	178	56	−60	312	44	−126	13	5	A2
9	6 March 2021	16:36:17.8	39.6730	22.2505	0.135×10^{23}	4.1	7.7	15	110	45	−92	313	50	−90	11	5	A2
10	5 March 2021	10:01:14.8	39.7801	22.0775	0.296×10^{23}	4.3	8.5	13	130	46	−85	300	45	−94	7	6	A1
11	5 March 2021	09:59:59.2	39.8524	22.0308	0.296×10^{23}	4.3	9.0	12	115	40	−90	315	54	−77	7	6	A1
12	4 March 2021	20:03:08.8	39.7472	22.1581	0.296×10^{23}	4.3	6.1	10	175	48	−78	337	43	257	7	6	A1
13	4 March 2021	19:31:32.2	39.8227	22.0505	0.296×10^{23}	4.3	6.9	11	135	45	−87	320	55	−78	5	6	A1
14	4 March 2021	19:23:51.0	39.8373	21.9424	8.339×10^{23}	5.2	9.4	12	120	40	−85	295	50	−95	3	7	A1
15	4 March 2021	18:45:26.7	39.8483	22.0816	0.296×10^{23}	4.3	6.4	10	130	55	−100	290	52	−100	10	6	B1
16	4 March 2021	18:38:19.1	39.7993	22.1260	2614×10^{25}	6.2	4.8	15	149	46	−95	336	44	−85	3	7	A1
17	4 March 2021	11:10:08.2	39.7417	22.0619	0.135×10^{23}	4.1	8.0	13	140	56	−76	300	46	−87	13	7	A2
18	4 March 2021	09:36:15.8	39.7911	22.1297	7.173×10^{22}	4.5	5.6	12	156	44	−83	327	46	−96	10	6	A1
19	4 March 2021	02:43:38.1	39.7362	22.2597	0.135×10^{23}	4.1	5.2	14	130	46	−85	300	45	−95	11	7	A2
20	3 March 2021	21:00:54.9	39.7563	22.1324	0.135×10^{23}	4.1	6.2	10	183	55	−46	304	53	−136	12	5	A2
21	3 March 2021	18:49:49.1	39.7394	22.1246	0.296×10^{23}	4.3	8.8	11	115	40	−99	307	50	−81	4	5	A1
22	3 March 2021	18:24:08.0	39.7316	22.1013	1.187×10^{24}	5.3	6.4	13	100	41	−90	305	51	−84	6	6	A1
23	3 March 2021	11:49:03.9	39.7037	22.2812	9.602×10^{22}	4.6	7.6	13	160	29	−74	323	62	−99	7	7	A1
24	3 March 2021	11:45:45.9	39.6996	22.2478	1.187×10^{24}	5.3	7.1	12	171	51	−71	322	42	−112	5	7	A1
25	3 March 2021	11:35:57.0	39.7000	22.2318	0.163×10^{24}	4.8	6.4	11	112	45	−94	299	50	−87	6	7	A1
26	3 March 2021	11:19:02.0	39.7458	22.2189	0.402×10^{23}	4.0	9.4	12	145	50	−90	330	43	−85	10	5	B1
27	3 March 2021	11:12:22.3	39.7755	22.2130	0.296×10^{23}	4.3	5.8	11	110	55	−120	320	55	−87	11	5	A2
28	3 March 2021	10:34:08.1	39.7270	22.2835	0.420×10^{24}	5.0	8.0	10	171	46	−71	325	47	−109	10	7	B1
29	3 March 2021	10:26:18.7	39.6703	22.2020	0.402×10^{23}	4.0	8.1	10	109	60	−89	287	30	−92	11	4	A2
30	3 March 2021 3 March 2021	10:23:08.5	39.6927	22.1640	0.402×10^{23}	4.0	7.6	14	126	41	−90	306	48	−89	13	5	A2
31	3 March 2021	10:21:37.2	39.7343	22.2194	7.173×10^{22}	4.5	8.4	11	138	31	−76	301	60	−98	10	5	A2
32	3 March 2021	10:20:46.5	39.6991	22.1576	2.243×10^{23}	4.8	7.6	13	140	45	−98	313	62	−86	5	5	A1
33	3 March 2021	10:19:54.5	39.7966	22.1013	1.505×10^{23}	4.7	9.3	12	151	71	−84	315	20	−106	6	6	A1
34	3 March 2021	10:16:08.3	39.7712	22.1566	3.173×10^{25}	6.3	8.5	11	177	37	−56	317	59	−113	5	7	A1

References

1. Hatzfeld, D.; Ziazia, M.; Kementzetzidou, D.; Hatzidimitriou, P.; Panagiotopoulos, D.; Makropoulos, K. Microseismicity and focal mechanisms at the western termination of the North Anatolian Fault and their implications for continental tectonics. *Geophys. J. Int.* **1999**, *137*, 891–908. [CrossRef]
2. Kahle, H.-G.; Cocard, M.; Pete, Y.; Geiger, A.; Reilinger, R.P.; Barka, A.; Veis, G. GPS-derived strain rate field within the boundary zones of the Eurasian, African and Arabian Plates. *J. Geophys. Res.* **2000**, *105*, 23353–23370. [CrossRef]
3. Caputo, R.; Pavlides, S. Late Cenozoic geodynamic evolution of Thessaly and surroundings (Central-Northern Greece). *Tectonophysics* **1993**, *223*, 339–362. [CrossRef]
4. Emsc-csem.org. Earthquakes—Earthquake today—Latest Earthquakes in the World—EMSC. 2022. Available online: https://www.emsc-csem.org/#2 (accessed on 11 May 2022).
5. Lekkas, E.; Agorastos, K.; Mavroulis, S.; Kranis, C.; Skourtsos, E.; Carydis, P.; Thoma, T. The early March 2021 Thessaly earthquake sequence. *Newsl. Environ. Disaster Cris. Manag. Strateg.* **2021**, *22*, ISSN 2653–9454. Available online: https://www.researchgate.net/publication/350091329_The_early_March_2021_Thessaly_Greece_earthquake_sequence (accessed on 8 June 2022).
6. Zouros, N.; Pavlides, S.; Chatzipetros, A. Recent movement on the Larissa plain neotectonic faults (Thessaly, C. Greece). Water-level fluctuation or tectonic creep. In Proceedings of the ESC XXIV General Assembly, Athens, Greece, 19–24 September 1994; Volume 67.
7. Soulios, G. Subsidence de terrains alluviaux dans le sud-est de la plaine de Thessalie, Grece. In *Engineering Geology and the Environment*; Marinos, P.G., Koukis, G.C., Tsiambas, G.C., Stournaras, G.C., Eds.; A.A. Balkema, Brookfield: Rotterdam, The Netherlands, 1997; pp. 1067–1072.
8. Rapti-Caputo, D.; Caputo, R. Some remarks on the formation of ground fissures in Thessaly, Greece. In Proceedings of the 5th International Symposium on Eastern Mediterranean Geology, Thessaloniki, Greece, 14–20 April 2004; pp. 1016–1019.
9. Kontogianni, V.; Pytharouli, S.; Stiros, S. Ground subsidence, Quaternary faults and vulnerability of utilities and transportation networks in Thessaly, Greece. *Environ. Geol.* **2007**, *52*, 1085–1095. [CrossRef]
10. Apostolidis, E.; Koukis, G. Engineering-geological conditions of the formations in the Western Thessaly basin, Greece. *Cent. Eur. J. Geosci.* **2013**, *5*, 407–422. [CrossRef]
11. Ilia, I.; Loupasakis, C.; Tsangaratos, P. Assessing ground subsidence phenomena with persistent scatterer interferometry data in western Thessaly, Greece. *Bull. Geol. Soc. Greece* **2016**, *50*, 1693–1702. [CrossRef]
12. Georgoulas, G.; Konstantaras, A.; Katsifarakis, E.; Stylios, C.D.; Maravelakis, E.; Vachtsevanos, G.J. "Seismic-mass" density-based algorithm for spatio-temporal clustering. *Expert Syst. Appl.* **2013**, *40*, 4183–4189. [CrossRef]
13. Konstantaras, A.J.; Katsifarakis, E.; Maravelakis, E.; Skounakis, E.; Kokkinos, E.; Karapidakis, E. Intelligent Spatial-Clustering of Seismicity in the Vicinity of the Hellenic Seismic Arc. *Earth Sci. Res.* **2012**, *1*, 1–10. [CrossRef]
14. Konstantaras, A. Deep Learning and Parallel Processing Spatio-Temporal Clustering Unveil New Ionian Distinct Seismic Zone. *Informatics* **2020**, *7*, 39. [CrossRef]
15. Kagan, Y.Y. 3-D rotation of double-couple earthquake sources. *Geophys. J. Int.* **1991**, *106*, 709–716. [CrossRef]
16. Konstantaras, A.; Vallianatos, F.; Varley, M.R.; Makris, J.P. Soft-Computing Modelling of Seismicity in the Southern Hellenic Arc. *IEEE Geosci. Remote Sens. Lett.* **2008**, *5*, 323–327. [CrossRef]
17. Astiz, L.; Kanamori, H. An earthquake doublet in Ometepec, Guerro, Mexico. *Phys. Earth Planet. Inter.* **1984**, *34*, 24–45. [CrossRef]
18. Poupinet, G.; Ellsworth, W.L.; Frechet, J. Monitoring Velocity Variations in the Crust Using Earthquake Doublets: An Application to the Calaveras Fault, California. *J. Geophys. Res. Atmos.* **1984**, *89*, 5719–5731. [CrossRef]
19. Beroza, G.C.; Spudich, P. Linearized inversion for fault rupture behavior: Application to the 1984 Morgan Hill, California, earthquake. *J. Geophys. Res* **1988**, *93*, 6275–6296. [CrossRef]
20. Felzer, K.R.; Becker, T.W.; Abercrombie, R.E.; Ekström, G.; Rice, J.R. Triggering of the 1999 MW 7.1 Hector Mine earthquake by aftershocks of the 1992 MW 7.3 Landers earthquake. *J. Geophys. Res.* **2002**, *107*, 2190. [CrossRef]
21. Bath, M. Earthquake energy and magnitude. *Phys. Chem. Earth* **1966**, *7*, 117–165. [CrossRef]
22. Richter, C.F. *Elementary Seismology*; Freeman & Co.: San Francisco, CA, USA, 1958.
23. Kagan, Y.; Jackson, D. Worldwide doublets of large shallow earthquakes. *Bull. Seismol. Soc. Am.* **1999**, *89*, 1147–1155. [CrossRef]
24. Konstantaras, A.J. Classification of Distinct Seismic Regions and Regional Temporal Modelling of Seismicity in the Vicinity of the Hellenic Seismic Arc. *IEEE J. Sel. Top. Appl. Earth Obs. Remote Sens.* **2012**, *6*, 1857–1863. [CrossRef]
25. Konstantaras, A. Expert knowledge-based algorithm for the dynamic discrimination of interactive natural clusters. *Earth Sci. Inform.* **2016**, *9*, 95–100. [CrossRef]
26. Moshou, A. Strong Earthquake Sequences in Greece during 2008-2014: Moment Tensor Inversions and Fault Plane Discrimination. *Open J. Earthq. Res.* **2020**, *9*, 323–348. [CrossRef]
27. Moshou, A.; Argyrakis, P.; Konstantaras, A.; Daverona, A.-C.; Sagias, N.C. Characteristics of Recent Aftershocks Sequences (2014, 2015, 2018) Derived from New Seismological and Geodetic Data on the Ionian Islands, Greece. *Data* **2021**, *6*, 8. [CrossRef]
28. Bbnet.gein.noa.gr. NOA Seismic Network (HL)—Introduction. 2022. Available online: https://bbnet.gein.noa.gr/HL (accessed on 15 March 2022).
29. Geophysics.geo.auth.gr. Τομέας Γεωφυσικής Α.Π.Θ. 2022. Available online: http://geophysics.geo.auth.gr (accessed on 15 March 2022).

30. Seiscomp.de. Scolv—SeisComP Release Documentation. 2022. Available online: https://www.seiscomp.de/doc/apps/scolv.html (accessed on 15 March 2022).
31. Seiscomp.de. 2022. Available online: https://docs.gempa.de/seiscomp3/current/apps/scolv.html (accessed on 25 April 2022).
32. Haslinger, F.; Kissling, E.; Ansorge, J.; Hatzfeld, D.; Papadimitriou, E.; Karakostas, V.; Makropoulos, K.; Kahle, H.; Peter, Y. 3D crustal structure from local earthquake tomography around the Gulf of Arta (Ionian region, NW Greece). *Tectonophysics* **1999**, *304*, 201–218. [CrossRef]
33. Kanamori, H. Determination of effective tectonic stress associated with earthquake faulting. The Tottori earthquake of 1943. *Phys. Earth Planet. Inter.* **1972**, *5*, 426–434. [CrossRef]
34. Haskell, N. Total energy and energy spectral density of elastic wave radiation from propagating faults. *Bull. Seismol. Soc. Am.* **1964**, *54*, 1811–1841. [CrossRef]
35. Helmberger, D.V. Theory and application of synthetic seismograms. In Proceedings of the International School of Physics <<Enrico Fermi>>, Course LXXXV. pp. 174–222. 1983. Available online: https://eurekamag.com/research/020/477/020477221.php (accessed on 8 June 2022).
36. Langston, C. A body wave inversion of the Koyna, India, earthquake of December 10, 1967, and some implications for body wave focal mechanisms. *J. Geophys. Res.* **1976**, *81*, 2517–2529. [CrossRef]
37. Golub, H.; Van, L.C. *Matrix Computations*; The Johns Hopkins Univeristy Press: Baltimore, Maryland, 1983; p. 476.
38. Das, S.; Kostrov, B. Inversion for seismic slip rate history and distribution with stabilizing constraints: Application to the 1986 Andreanof Islands Earthquake. *J. Geophys. Res.* **1990**, *95*, 6899. [CrossRef]
39. Brune, J. Tectonic stress and the spectra of seismic shear waves from earthquakes. *J. Geophys. Res.* **1970**, *75*, 4997–5009. [CrossRef]
40. Langston, C. Source inversion of seismic waveforms: The Koyna, India, earthquakes of 13 September 1967. *Bull. Seismol. Soc. Am.* **1981**, *71*, 1–24. [CrossRef]
41. Langston, C.; Barker, J.; Pavlin, G. Point-source inversion techniques. *Phys. Earth Planet. Inter.* **1982**, *30*, 228–241. [CrossRef]
42. Cotton, F.; Campillo, M. Frequency domain inversion of strong motions: Application to the 1992 Landers earthquake. *J. Geophys. Res. Solid Earth* **1995**, *100*, 3961–3975. [CrossRef]
43. Kiratzi, A.; Sokos, E.; Ganas, A.; Tselentis, A.; Benetatos, C.; Roumelioti, Z.; Serpetsidaki, A.; Andriopoulos, G.; Galanis, O.; Petrou, P. The April 2007 earthquake swarm near Lake Trichonis and implications for active tectonics in western Greece. *Tectonophysics* **2008**, *452*, 51–65. [CrossRef]
44. Usgs.gov. 2022. FPFIT, FPPLOT and FPPAGE; Fortran Computer Programs for Calculating and Displaying Earthquake Fault-plane Solutions | U.S. Geological Survey. Available online: https://www.usgs.gov/publications/fpfit-fpplot-and-fppage-fortran-computer-programs-calculating-and-displaying (accessed on 8 June 2022).
45. Hardebeck, J.L.; Shearer, P.M. A New Method for Determining First-Motion Focal Mechanisms. *Bull. Seismol. Soc. Am.* **2002**, *92*, 2264–2276. [CrossRef]
46. Meju, M. *Geophysical Data Analysis: Understanding Inverse Problem Theory and Practise*; Course Notes Series; Society of Exploration Geophysicists: Houston, TX, USA, 1994; Volume 6.
47. Aki, K.; Richards, P.G. *Quantitative Seismology*, 2nd ed.; Freeman: San Francisco, CA, USA, 1980.
48. Dreger, D. *Time—Domain Moment Tensor INVerse Code (TDMT_INV) Version 1.1*; Berkley Seismological Laboratory: Berkeley, CA, USA, 2002; p. 18.
49. Ichinose, G. Source Parameters of Eastern California and Western Nevada Earthquakes from Regional Moment Tensor Inversion. *Bull. Seismol. Soc. Am.* **2003**, *93*, 61–84. [CrossRef]
50. Bouchon, M. A simple method to calculate Green's functions for elastic layered media. *Bull. Seismol. Soc. Am.* **1981**, *71*, 959–971. [CrossRef]
51. Bouchon, M. A Review of the Discrete Wavenumber Method. *Pure Appl. Geophys.* **2003**, *160*, 445–465. [CrossRef]
52. Bouchon, M.; Aki, K. Discrete wave-number representation of seismic-source wave fields. *Bull. Seismol. Soc. Am.* **1977**, *67*, 259–277. [CrossRef]
53. Bouchon, M. Discrete wave number representation of elastic wave fields in three-space dimensions. *J. Geophys. Res. Solid Earth* **1979**, *84*, 3609–3614. [CrossRef]
54. Athanasiadis, C.; Martin, P.; Spyropoulos, A.; Stratis, I. Scattering relations for point sources: Acoustic and electromagnetic waves. *J. Math. Phys.* **2002**, *43*, 5683–5697. [CrossRef]
55. Athanasiadis, C. On the acoustic scattering amplitude for a multi-layered Scatterer. *J. Aust. Math. Soc. Appl. Math.* **1998**, *39*, 431–448. [CrossRef]
56. Herrmann, R.B.; Wang, C.Y. A comparison of synthetic seismograms. *Bull. Seismol. Soc. Am.* **1985**, *75*, 41–56. [CrossRef]
57. Jost, M.; Hermann, R. A student's Guide to and Review of Moment Tensors. *Seismol. Res. Lett.* **1989**, *60*, 37–57. [CrossRef]
58. Tarantola, A.; Valette, B. Inverse Problems = Quest for Information. *J. Geophys.* **1982**, *50*, 159–170.
59. Tarantola, A.; Valette, B. Generalized Nonlinear Inverse Problems Solved Using the Least Squares Criterion. *Rev. Geophys. Space Phys.* **1982**, *20*, 219–232. [CrossRef]
60. Moser, T.J.; Van Eck, T.; Nolet, G. Hypocenter determination in strongly heterogeneous Earth models using the shortest path method. *J. Geophys. Res.* **1992**, *97*, 6563–6572. [CrossRef]
61. Wittlinger, G.; Herquel, G.; Nakache, T. Earthquake location in strongly heterogenous media. *Geophys. J. Int.* **2007**, *115*, 759–777. [CrossRef]

62. Lomax, A.; Virieux, J.; Volant, P.; Berge-Thierry, C. Probabilistic earthquake location in 3D and layered models: Introduction of a Metropolis-Gibbs method and comparison with linear locations. In *Advances in Seismic Event Location*; Thurber, C.H., Rabinowitz, N., Eds.; Kluwer: Amsterdam, The Netherlands, 2000; pp. 101–134.
63. Ammon, C.J.; Randall, G.E.; Zandt, G. On the non-uniqueness of receiver function inversions. *J. Geophys. Res.* **1990**, *95*, 15303–15318. [CrossRef]
64. Kennett, B.N.L. *Seismic Wave Propagation in Stratified Media*; Cambridge University Press: Cambridge, UK, 1983.
65. Randal, G.E. Efficient calculation of complete differential seismograms for laterally homogeneous earth models. *Geophys. J. Int.* **1994**, *118*, 245–254. [CrossRef]
66. Konstantinou, K.; Melis, N.; Boukouras, K. Routine Regional Moment Tensor Inversion for Earthquakes in the Greek Region: The National Observatory of Athens (NOA) Database (2001–2006). *Seismol. Res. Lett.* **2010**, *81*, 750–760. [CrossRef]
67. Tolomei, C.; Caputo, R.; Polcari, M.; Famiglietti, N.A.; Maggini, M.; Stramondo, S. The Use of Interferometric Synthetic Aperture Radar for Isolating the Contribution of Major Shocks: The Case of the March 2021 Thessaly, Greece, Seismic Sequence. *Geosciences* **2021**, *11*, 191. [CrossRef]
68. De Novellis, V.; Reale, D.; Adinolfi, G.; Sansosti, E.; Convertito, V. Geodetic Model of the March 2021 Thessaly Seismic Sequence Inferred from Seismological and InSAR Data. *Remote Sens.* **2021**, *13*, 3410. [CrossRef]

applied sciences

MDPI

Article

Investigation of Structural Seismic Vulnerability Using Machine Learning on Rapid Visual Screening

Ioannis Karampinis [1], Lazaros Iliadis [1] and Athanasios Karabinis [2,*]

[1] Lab of Mathematics and Informatics (ISCE), Department of Civil Engineering, Democritus University of Thrace, 67100 Xanthi, Greece; ikarampi@civil.duth.gr (I.K.); liliadis@civil.duth.gr (L.I.)

[2] Lab of Reinforced Concrete and Seismic Design, Department of Civil Engineering, Democritus University of Thrace, 67100 Xanthi, Greece

* Correspondence: akarabin@civil.duth.gr

Abstract: Seismic vulnerability assessment is one of the most impactful engineering challenges faced by modern societies. Thus, authorities require a reliable tool that has the potential to rank given structures according to their seismic vulnerability. Various countries and organizations over the past decades have developed Rapid Visual Screening (RVS) tools aiming to efficiently estimate vulnerability indices. In general, RVS tools employ a set of structural features and their associated weights to obtain a vulnerability index, which can be used for ranking. In this paper, Machine Learning (ML) models are implemented within this framework. The proposed formulation is used to train binary classifiers in conjunction with ad hoc rules, employing the features of various Codes (e.g., the Federal Emergency Management Agency, New Zealand, and Canada). The efficiency of this modeling effort is evaluated for each Code separately and it is clearly demonstrated that ML-based models are capable of outperforming currently established engineering practices. Furthermore, in the spirit of the aforementioned Codes, a linearization of the fully trained ML model is proposed. ML feature attribution techniques, namely SHapley Additive exPlanations (SHAP) are employed to introduce weights similar to engineering practices. The promising results motivate the potential applicability of this methodology towards the recalibration of the RVS procedures for various types of cases.

Keywords: rapid visual screening; machine learning; SHAP; recalibration; seismic vulnerability

Citation: Karampinis, I.; Iliadis, L.; Karabinis, A. Investigation of Structural Seismic Vulnerability Using Machine Learning on Rapid Visual Screening. *Appl. Sci.* **2024**, *14*, 5350. https://doi.org/10.3390/app14125350

Academic Editor: Nicholas Vassiliou Sarlis

Received: 15 May 2024
Revised: 10 June 2024
Accepted: 15 June 2024
Published: 20 June 2024

1. Introduction

The destructive potential of an earthquake constitutes one of the most severe and impactful engineering challenges faced by modern societies. On the one hand, the global population has increased dramatically during the last century. Coupled with a corresponding increase in urbanization, this has led to a concentration of humans and valuable infrastructure in large urban areas. On the other hand, many structures in the available building stock do not adhere to modern safety standards and, thus, seismic events over the past decades (USA, Turkey, Pakistan, India, Greece, New Zealand) have demonstrated the potential catastrophic results these earthquakes might cause.

Government bodies should be able to identify the most seismically vulnerable buildings in the population. This should enable the allocation of the available resources for preventative actions. In this context, seismic vulnerability refers to the potential of a structure to suffer a certain degree of damage under an earthquake of a predefined intensity level [1,2]. However, the seismic vulnerability of all buildings in a given population cannot be ascertained using the highest possible scrutiny (e.g., using analytical methods like step-by-step dynamic analysis) due to the limitation in resources. To this end, most countries have developed multi-step processes, wherein an initial screening step is performed to identify the potentially most vulnerable structures, which are subjected to further analysis later.

The initial step in the aforementioned processes usually consists of a so-called Rapid Visual Screening (RVS) procedure where structural characteristics that are critical to the overall seismic behavior of the building are rapidly documented by an expert. Such characteristics could include the structural type, the underlying soil, or the height of the building. These are then weighted by expert-defined coefficients to produce a vulnerability index which can be used to rank the given population of structures. The subsequent analysis focuses on the top structures that were obtained according to the respective resource constraints.

The first RVS procedure was developed by the USA Federal Emergency Management Agency (FEMA) in 1988 [3], which was subsequently revised and updated in 2002 [4]. Since the initial version of FEMA 1988, many countries and organizations that face similar challenges have developed their own adaptations of RVS. The most notable ones include the New Zealand Society for Earthquake Engineering [5], the Indian Institute of Technology [6], the Greek Earthquake Planning and Protection Organization (OASP) [7], and the Canadian Office of Critical Infrastructure Protection and Emergency Preparedness (OCIPEP) [8]. Inspired by FEMA, these adaptations share some common elements; however, as their variety indicates, they also have some differences with each other. Indeed, different countries and organizations consider different structural characteristics to be the most crucial and, thus, employ different sets of features in their respective RVS procedures. Even when the same sets of features are employed, different weights can be assigned to them according to the respective experts' opinions. The accuracy and efficiency of these methodologies can be assessed using data from near-field earthquakes [9,10] which have also been used to calibrate the above procedures [11,12].

On the other hand, the emergence and rapid development of Machine Learning (ML) techniques has highlighted the potential applicability of ML-based methodologies to tackle many complex challenges. To this end, researchers have proposed different approaches that combine the demonstrated power of ML with the established RVS procedures. Yu et al. [13] trained a Convolutional Neural Network (CNN) to ascertain the presence of a soft story, given street view images. Similarly, Ruggieri et al. [14] implemented an ML-based image recognition system to identify the structural parameters of buildings from images which were then employed within the RVS framework to compute a vulnerability score. Bektaş and Kegyes-Brassai [15] employed a Multi-Layer Perceptron (MLP) to predict the damage degree. This was achieved by developing respective classification models using data from the 2015 Nepal earthquake with the corresponding set of features. Finally, Harirchian et al. [16] used data based on RVS features from earthquakes in Ecuador, Haiti, Nepal, and South Korea to assess the predictive power of different ML algorithms, including Support Vector Machines (SVMs), k-Nearest Neighbors (k-NN) and Random Forests.

Novelty of the Research

The novelty and contributions of the present research are twofold. On the one hand, we implement Machine Learning classification models for seismic vulnerability ranking. As a matter of fact, these models perform pairwise comparisons between distinct structures. This methodology is applied to a variety of the most well-known and widely applicable RVS procedures (FEMA, Canada, India, New Zealand, Greece-OASP), to demonstrate its strong potential to outperform the currently established engineering practices. On the other hand, inspired by the linear approaches employed by the existing Codes for the computation of the respective RVS scores, this paper introduces a linearization of the fully trained ML models. Machine learning feature attribution methods are employed to compute the weights for each structural parameter. This procedure aims to provide a technique which can be used to recalibrate the existing methodologies.

The case study of this research is based on real measurements, performed on 457 reinforced concrete structures after the 1999 Athens earthquake. Given the promising results presented herein, the proposed methodologies could be implemented using datasets from different countries and organizations and even different structural types. Even though the features could be different, the data could be used to train the ML models and to obtain recalibrated weights that are more closely attuned to the individual conditions of each Seismic Code.

2. Materials and Methods

2.1. Dataset Description

The dataset employed in the present work consists of a collection of Rapid Visual Screening measurements pertaining to 457 structures. These were drawn after the 7 September 1999 Athens earthquake [17]. The samples in the dataset exhibited a varying degree of damage across the spectrum, ranging from partial or total collapse to moderate or minimal damages. To enhance the dataset diversity, the structures were drawn from various regions in the greater Athens Metropolitan area. This mitigated the influence of local effects, such as distance from the earthquake epicenter or ground conditions, on the observed damage states. The authors in [17] took further steps to mitigate these effects; when one structure was selected from a specific building block, structures across the damage spectrum were also selected from the same area [18]. The observed damage states were classified into four distinct classes, namely:

- "Black": The category suffering the most severe damages. Structures belonging to this label suffered partial or total collapse during the earthquake, crossing the so-called Collapse Limit State (CLS).
- "Red": While the structures belonging to this label did not cross the CLS, they suffered extensive damages to their structural members and crossed the so-called Ultimate Limit State (ULS).
- "Yellow": Structures of this category suffered only moderate damages to their structural members. While these structures did not cross the ULS, they did cross the so-called Serviceability Limit State (SLS).
- "Green": The structures of this class suffered only minor damages.

The distribution of structures across the damage spectrum is shown in the following Table 1. Table 2 shows which attributes have been considered by each Seismic Code. It is a fact that each code considers its own vector of structural aspects [3–8,17].

1. Structural type: In general, Seismic Codes distinguish buildings according to their structural type, e.g., framed structures or frames with shear walls [8].
2. Significant height: This is an attribute that most Seismic Codes take into consideration. It affects structures whose height exceeds a predefined threshold. However, this threshold varies between Seismic Codes. For example, FEMA sets this threshold at 7 stories, while this threshold for OASP equals 5. In addition to the above, some Seismic Codes, e.g., that of India, distinguish structures with "moderate height" as a separate category.
3. Poor condition: This attribute accounts for potential deterioration in the design seismic capacity of the building. This may be due to corroded reinforcement bars or due to poor concrete quality (e.g., aggregate segregation or erosion) [8].
4. Vertical irregularity: This attribute pertains to structures with significant variations in their height, which leads to discontinuities in the paths of the vertical loads.
5. Horizontal irregularity: This feature affects structures whose floor plans exhibit sharp, re-entrant corners, such as "L", "T", or "E", which have the potential to develop higher degrees of damage [8].
6. Soft story: This pertains to structures wherein one story has significantly less stiffness than the rest; for example, due to a discontinuity of the shear walls.
7. Torsion: Structures with high eccentricities suffer from torsional loads during an earthquake. This could lead to higher degrees of damage.
8. Pounding: This attribute is considered when two adjacent structures do not have a sufficient gap between them and especially when the buildings have different heights, which may lead to the slabs of one structure ramming into the columns of the other.

9. Heavy non-structural elements: While they are non-structural, the displacement of such elements during an earthquake can lead to eccentricities and to additional torsion.
10. Short columns: This attribute refers to columns wherein modifications have resulted in a reduction in their active length. Such modifications include the addition of spandrel beams, or wall sections that do not cover the full height of the story [8].
11. Benchmark year: Various Seismic Codes define the so-called "benchmark years" where an improved version of the Code took effect.
12. Soil type: Generally, Seismic Codes classify soil types ranging from rock and semi-rock formations to loose fine-grained silt [19]. The quality of the underlying soil is a significant factor of the overall seismic behavior of the building.
13. Lack of Seismic Code design: This attribute pertains to structures that were designed without adhering to the provisions of a dedicated Seismic Code.
14. Wall filling regularity: When the infill walls are of sufficient thickness and with few openings, they can support the surrounding frames during an earthquake, leading to improved overall performance.
15. Previous damages: This attribute pertains to structures whose previous damages have not been adequately repaired, leading to a deterioration of their overall seismic capacity.

Table 1. Distribution of structures across the damage spectrum.

Damage Label	Number of Structures
Black	93
Red	201
Yellow	69
Green	94

Table 2. Attributes employed by each Seismic Code.

Attribute	Seismic Code					
	FEMA 1988	FEMA 2002	OASP (2000–2004)	India	Canada	New Zealand
Structural type	✓	✓	✓	✓	✓	✓
Significant height	✓	✓	✓	✓	✗	✓
Poor condition	✓	✗	✓	✗	✓	✗
Vertical irregularity	✓	✓	✓	✓	✓	✓
Horizontal irregularity	✓	✓	✓	✓	✓	✓
Soft story	✓	✓	✓	✓	✓	✓
Torsion	✓	✗	✓	✗	✓	✓
Pounding	✓	✗	✓	✗	✓	✓
Heavy non-structural elements	✓	✗	✓	✗	✗	✓
Short columns	✓	✗	✓	✗	✓	✓
Benchmark year	✓	✓	✓	✓	✓	✓
Soil type	✓	✓	✓	✓	✓	✓
Moderate height	✗	✓	✗	✓	✗	✗
Lack of Seismic Code design	✗	✓	✓	✗	✗	✗
Wall filling regularity	✗	✗	✓	✗	✗	✗
Previous damages	✗	✗	✓	✗	✓	✗

As it is evident from the above, the existing Seismic Codes consider distinct feature vectors in their respective RVS procedures. The selection of these features is based on experts' knowledge, which varies from country to country and from institution to institution. Even when the same features are used, different weights might be assigned to them and the calculation of the final vulnerability score might be carried out using different procedures. This calls for the need for the use of Machine Learning data-driven models. These models can learn from the spectrum of the available data vectors, and they can lead to ranking improvements, which is the aim of this study.

2.2. Overview of the Proposed Formulation

At high levels, the employed formulation is based on the development of a binary classification model $f : \mathbb{R}^n \times \mathbb{R}^n \rightarrow \{-1, +1\}$ that can effectively consider a pair of feature vectors $\mathbf{x}_i, \mathbf{x}_j \in \mathbb{R}^n$, corresponding to a respective pair of structures s_i, s_j in order to rank them. More specifically, the formulation predicts whether s_j should rank higher than s_j ("+1") or vice versa ("−1"). Overall, the algorithm proceeds as shown in Algorithm 1 and Figure 1 presented below [20]:

Algorithm 1 Algorithmic seismic vulnerability ranking process using ML model [20].

for $i = 1, 2, \ldots, m$ **do**:

Initialize $\text{wins}(s_i) = 0, \text{sum\_proba}(s_i) = 0$

for $j = 1, 2, \ldots, m, j \neq i$ **do**:

Compute $\mathbf{x}_{\text{new}}$ and predict the class probabilities $[p_{-1}, p_{+1}]$

if $p_{+1} > p_{-1}$:

$\text{wins}(s_j) = 0 += 1, \text{sum\_proba}(s_j) += p_{+1}$

else:

$\text{wins}(s_i) = 0 += 1, \text{sum\_proba}(s_i) += p_{-1}$

end for

end for

Rank the given population of structures based on the total number of "wins", using the running sum of probabilities as a secondary metric.

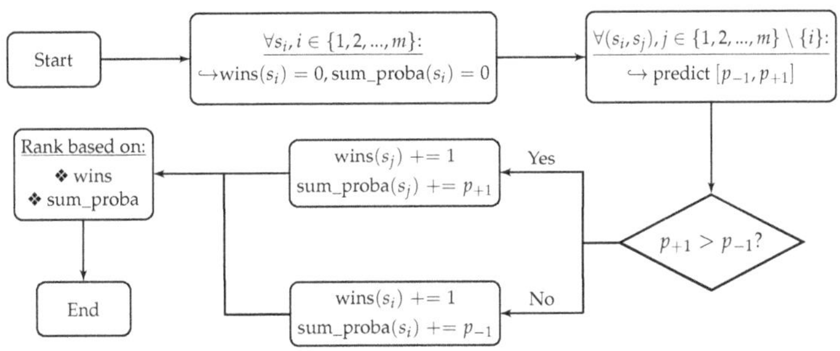

Figure 1. Flowchart of the proposed formulation.

Using the above algorithm, originally introduced in [20], the problem of seismic vulnerability ranking can be effectively reduced to that of binary classification.

2.3. Data Preprocessing

Data preprocessing is a vital step in a ML model, which not only prepares the dataset in a form that can be considered by the ML models, but it can also directly affect their performance. The following steps were undertaken in this study.

First, as can be seen in Table 1, the "Red" class heavily dominates the dataset. This leads to the well-known "class imbalance problem". This problem significantly deteriorates the performance of any ML modeling effort. The developed model is always skewed towards the majority class and it cannot effectively adapt to the feature values related to the minority class. This leads to biased models models which lose the ability to generalize [21,22]. To this end, we undersampled the majority class by a factor of 50% to produce a more balanced dataset [23,24].

Subsequently, the features were transformed using numerical values. Indeed, most of the features in Table 2 are binary, indicating whether the structure had the specific attribute. We modeled these features using the 0, 1 values. Similarly, features like soil or structural type, which had distinct values, were modeled using integers. The damage grades were also modeled with integer values, i.e., $\{\text{Green, Yellow, Red, Black}\} \rightarrow \{1,2,3,4\}$ [18,20].

Finally, as shown in the previous section, the ML models were trained to receive a pair of structures s_i, s_j with the corresponding feature vectors $\mathbf{x}_i, \mathbf{x}_j$ as the input and to predict either "−1" or "+1" as the output, i.e., which of the two given structures should rank higher. To achieve this, the pairwise transformation $T : \mathbb{R}^n \times \mathbb{R}^n \rightarrow \mathbb{R}^n$ given by $T(\mathbf{x}_i, \mathbf{x}_j) = \mathbf{x}_j - \mathbf{x}_i$ was employed [20].This resulted in a transformed input dataset, $\mathbf{X}^{new}$, with rows corresponding to the pairwise transformed vectors $T(\mathbf{x}_i, \mathbf{x}_j)$. It should be noted that for each individual seismic Code, the corresponding vectors $\mathbf{x}_i$ contain only the features that the respective Code utilizes, as shown in Table 2. Similarly, a transformed output variable y^{new} was obtained. For a pair of structures s_i, s_j with the original target variables $y_i, y_j, y_i \neq y_j$, the corresponding entry in y^{new} is defined as $sign(y_j - y_i)$, where $sign$ denotes the sign() function. Thus, $\mathbf{X}^{new}$ and y^{new} were the input and output variables of the trained ML models.

2.4. Machine Learning Algorithm

As has already been mentioned in Section 2.2, using the proposed formulation, the problem of seismic vulnerability ranking was converted to a binary classification problem. There are many available algorithms in the literature for this task [25,26]. Working on the same dataset as OASP (Greece), the authors in [18,20] have compared a variety of different ML models and they have concluded that the Gradient Boosting (GB) classifier offers the highest performance. Extending the previously established results, we will also employ GB in the sequel and to this end we are presenting a short overview of the algorithm.

Gradient Boosting [27] belongs to the group of the ensemble classifiers [28]. It combines a large number of individual "weak" learners, most often Decision Trees [29]. Specifically, the algorithm incrementally learns a function with the form

$$f(\mathbf{x}) = \sum_{i=1}^{N} \alpha_i h_i(\mathbf{x}; \theta_i). \tag{1}$$

The $h_i(\mathbf{x}; \theta_i)$ stands for the "weak" learners whose parameters θ_i are trained during the iterations, N is a hyperparameter denoting the maximum number of such models and a_i is the learned weights of the ensemble. These are learned via a process known as "boosting" [30], wherein the model iteratively computes the gradient of an arbitrary differentiable loss function $\mathcal{L}$ which is used as an input to train the next iteration. An overview of the process is shown in the following Algorithm 2 [27]. The algorithm was implemented using the Python programming language (v. 3.11.5) and the dedicated Machine Learning library scikit-learn (v. 1.3.0) [31].

Algorithm 2 Gradient Boosting learning process [27].

Initialize $F_0(\mathbf{x})$

for $i = 1, 2, \ldots, N$ **do**:

Compute $w_j(\mathbf{x}_j) = \left.\frac{\partial \mathcal{L}(y_j, F(\mathbf{x}_j))}{\partial F(\mathbf{x}_j)}\right|_{F(\mathbf{x})=F_{i-1}(\mathbf{x})}, \quad j = 1, 2, \ldots M$

Compute $\theta_i = \arg\min_{\theta,\mu} \sum_{j=1}^{M} \left[-w_j(\mathbf{x}_j) - \mu h_i(\mathbf{x}; \theta_i)\right]^2$

Compute $\alpha_i = \arg\min_{\alpha} \sum_{j=1}^{M} \mathcal{L}(y_j, f_{i-1} + \alpha h_i(\mathbf{x}; \theta_i))$

Update $f_i(\mathbf{x}) = f_{i-1}(\mathbf{x}) + \lambda \alpha_i h_i(\mathbf{x}; \theta_i))$

end for

2.5. Proposed Methodology for the Recalibration of RVS Using Machine Learning

As we will examine in the next paragraphs, a hybrid approach that employs the algorithm presented in Section 2.2, in conjunction with the Gradient Boosting classifier presented in Section 2.4, can significantly improve the performance of the currently established RVS procedures for the examined Seismic Codes. In this Section, we examine the potential extension of the above methodology towards a closer alignment with the spirit of the aforementioned Codes. Our motivation stems from the fact that, as is well known, most of the Seismic Codes examined herein and presented in Table 2 are additive. Specifically, each of the features they utilize is assigned a weight, ΔV_m, which can be either negative or positive, and the final vulnerability index is obtained as

$$V_I = V_I^b + \sum \Delta V_m, \tag{2}$$

wherein V_I^b is a basic vulnerability index which varies in different Seismic Codes. The structures in the population can then be ranked in ascending order of the above computed index. One notable exception is the Canadian Code. In this RVS approach, the employed features are first grouped into intermediate variables pertaining to seismicity, soil conditions structural type, and any potential irregularities. These intermediate variables are obtained in an additive way, while the final index is the product of the intermediate variables. Furthermore, the final ranking is obtained in descending order of the index, unlike most other Codes.

The above observations, especially in the form of (2), provide the motivation for a potential recalibration of RVS using the examined novel pairwise formulation. Indeed, for each Seismic Code, we can define a weight vector $\mathbf{w} = \{w_1, w_2, \ldots, w_n\}$, where n is the number of features of each Seismic Code. Thus, these weights are completely analogous to ΔV_m. Furthermore, these weights allow us to modify Algorithm 1 presented in Section 2.2 as shown in Algorithm 3.

This procedure is fundamentally different from the way RVS rankings are obtained in the established literature. However, it is not independent from the Machine Learning methodology examined in Section 2.2, since, as we will analyze, these weights are obtained from the fully trained ML models using the so-called SHapley Additive exPlanations (SHAP) values. It should be reiterated that the aim of this study is not to provide the values of the weights themselves, but rather to demonstrate a procedure for their estimation. Given the promising results presented in the present study, it is hoped that the proposed methodology will be applied to larger datasets, specifically tailored to each country's seismic demands, which will yield a more accurate calibration of the models for each individual Seismic Code.

Algorithm 3 Algorithmic seismic vulnerability ranking process using the proposed recalibration procedure.

for $i = 1, 2, \ldots, m$ **do**:

Initialize $\text{wins}(s_i) = 0, \theta(s_i) = 0$

for $j = 1, 2, \ldots, m, j \neq i$ **do**:

Compute $\mathbf{x}_{\text{new}}$ and compute $\theta = \mathbf{w} \cdot (\mathbf{x}_j - \mathbf{x}_i)$

if $\theta > 0$:

$\text{wins}(s_j) = 0 \mathrel{+}= 1, \theta(s_j) \mathrel{+}= \theta$

else:

$\text{wins}(s_i) = 0 \mathrel{+}= 1, \theta(s_i) \mathrel{+}= \theta$

end for

end for

Rank the given population of structures based on the total number of "wins", using the running sum θ as a secondary metric.

2.6. Explainability and SHAP Values

SHapley Additive exPlanations (SHAP) is a feature attribution methodology recently introduced by Lundberg and Lee [32]. They are based on the so-called Shapley values proposed by Lloyd Shapley in the field of cooperative game theory [33]. They are used to analyze the predictions of a trained Machine Learning model into components pertaining to each individual feature. Specifically, given a trained model f, a local approximation h is constructed as follows [32]:

$$h(\mathbf{z}) = \phi_0 + \sum_{i=1}^{m} \phi_i z_i, \tag{3}$$

where m is the number of features employed by the machine learning model, $\mathbf{z} \in \{0,1\}^m$ is a binary vector indicating whether the ith feature was used in the prediction, and ϕ_i is the SHAP values corresponding to that feature. The values of ϕ_i are computed in such a way that the so-called "local accuracy" property holds [32], i.e., that they approximate the behavior of the underlying model f. To this end, they are given by the formula [32,34]

$$\phi_i = \sum_{S \subseteq N \setminus \{i\}} \frac{|S|!(n - |S| - 1)!}{n!} [f(S \cup \{i\}) - f(S)]. \tag{4}$$

Following the notation of Lundberg et al., in the above Equation (4) $M = 1, 2, \ldots, m$ is the set of features employed by f and $S \subset M$ is a subset of M called "coalition", using the term from cooperative game theory. Intuitively, the contribution of each feature is measured as a weighted average of the model's predictions with and without the inclusion of the particular feature.

The above provides a motivation to employ these SHAP values as the weights $\mathbf{w}$ in our proposed formulation for the recalibration of RVS, which was presented in Section 2.5. However, these SHAP values are computed for each individual data point in the dataset, i.e., for each individual pair of structures. Thus, they are given in the form of a matrix $\mathbf{\Phi}^{k \times m}$, where k is the number of pairs of structures in the transformed dataset and m is the number of features each Seismic Code employs. In order to provide a single weight for the whole dataset, the computed values must be aggregated. To this end, we employed a normalized (dividing by the dimension of the vector) norm of each column in the matrix. The $\mathcal{L}_1$ norm, i.e., the sum of absolute squares, is one of the most commonly employed ones in the literature. As an alternative, we employed the $\mathcal{L}_2$ norm, i.e., the well-known Euclidean norm. Specifically, we computed [18]

$$\overline{\phi}_j = \begin{cases} \frac{1}{m}\sum_{i=1}^{m} |\phi_{ij}|, & \text{using } \mathcal{L}_1 \\ \frac{1}{m}\sqrt{\sum_{i=1}^{m} \phi_{ij}^2}, & \text{using } \mathcal{L}_2 \end{cases} . \tag{5}$$

A fundamental difference between the two is that the $\mathcal{L}_2$ norm tends to reduce the contribution of small, noisy components, while simultaneously highlighting the effect of the most prominent features [18]. This is due to the fact that it is based on squaring the terms before summation. Thus, using the $\overline{\phi}_j$ as defined in (5), we can define the weights $\mathbf{w}$ of our proposed recalibration procedure for RVS. In the present study, the formulations using $\mathcal{L}_2$ outperformed the formulation using $\mathcal{L}_1$. Thus, the results we will present are based on this approach. The computation of the SHAP values was carried out using the dedicated Python library by Lundberg et al. [35].

3. Results

In this section, we present the results of the present study. As was mentioned in the Introduction, the goal of this study is twofold: on the one hand, it aims to demonstrate that the proposed formulation, as presented in Section 2.2, outperforms the currently established engineering practices in terms of the seismic vulnerability ranking for all the studied Seismic Codes, some of which are the most applicable at an international level. On the other hand, it aims to employ the fully trained Machine Learning models in order to propose a potential recalibration procedure of the aforementioned Codes.

In order to measure the efficiency of the ranking, we compare it with the "ideal" ranking, i.e., how a human would rank the given structures, having perfect knowledge of their damage labels. Indeed, according to this "ideal" ranking, the first 93 structures should be the ones that suffered total or partial collapse (Black), the following 100 structures (post undersampling) should be the ones that were severely damaged, but did not cross the CLS (Red) and so on. Using the above, the predicted ranking can be subdivided into bins, where each bin's size is the number of structures of the corresponding damage label. In addition, given a damage index $d \in \{\text{Black, Red, Yellow, Green}\}$, let bin_d^{true}, bin_d^{pred} denote the structures that belong to this label and the structures that the algorithm ranked in the corresponding bin, respectively. The efficiency of the ranking was measured using the bin accuracy (BAC) defined as [20]

$$\text{BAC}_d = \frac{|bin_d^{true} \cap bin_d^{pred}|}{|bin_d^{true}|}. \tag{6}$$

Other alternative metrics have also been proposed in the literature. For example, in [36], the authors performed a linear regression to obtain a correlation metric between the observed damage labels and the computed RVS score. However, in the present study, we obtained the ranking directly, without computing a similar score beforehand.

Using the metric defined in (6), we computed the bin accuracy for all damage labels and for each individual Seismic Code. First, we computed the accuracy that was obtained using the RVS scores as given by the Codes in their current form. Subsequently, we obtained the accuracy using the fully trained Machine Learning model. Finally, we computed the accuracy obtained using the ML-based recalibrated RVS procedure, as described in Sections 2.5 and 2.6. The results are shown in Tables 3–5 as well as Figure 2, for comparison. It should be reiterated that after undersampling, the dataset contained exactly 100 Red structures ($|bin_d^{true}| = 100$), which is why the corresponding entries in Tables 3–5 have integer values.

Comparing the ranking obtained using the established RVS procedures, it can be readily observed that the combination of features with their established weights used in OASP 2000 and 2004 outperformed the other Seismic Codes for this dataset. This improvement in performance was evident for all damage labels. In the label corresponding to collapse (Black), the Codes of New Zealand, FEMA 2002, Canada, and India exhibited slightly lower and comparable performances. It can be observed that, in agreement with the established literature, FEMA 2002 significantly improved over FEMA 1988 in this regard, as the latter exhibited a significantly deteriorated accuracy. Furthermore, it should be noted that a successful ranking should prioritize accuracy in the labels corresponding to "collapse" and "severe damages" (Black and Red, respectively). These are the worst-case damage categories and the most impactful to society. In this regard, all the examined Seismic Codes were successful. In spite of the fact that some performed better than others, the accuracy of the Black bin was significantly higher, with the exception of the FEMA 1988 case, where it was only slightly higher.

Table 3. Bin accuracy obtained using the established RVS scores.

Seismic Code	Bin Accuracy (%)			
	Black	Red	Yellow	Green
FEMA 1988	24.73	22.00	5.80	22.58
FEMA 2002	37.63	24.00	11.59	19.35
OASP 2000	43.01	32.00	14.49	27.42
OASP 2004	48.39	35.00	24.64	32.26
India	33.33	24.00	18.84	19.35
Canada	36.56	31.00	14.49	16.13
New Zealand	38.71	23.00	30.43	12.90

Table 4. Bin accuracy obtained using the fully trained Machine Learning model.

Seismic Code	Bin Accuracy (%)			
	Black	Red	Yellow	Green
FEMA 1988	52.69	41.00	30.43	41.94
FEMA 2002	61.29	38.00	26.09	30.64
OASP (2000 and 2004)	53.76	37.00	28.99	46.77
India	56.99	41.00	24.66	29.03
Canada	55.94	42.00	28.99	35.48
New Zealand	53.76	38.00	31.88	45.16

Table 5. Bin accuracy obtained using the proposed recalibrated RVS.

Seismic Code	Bin Accuracy (%)			
	Black	Red	Yellow	Green
FEMA 1988	56.99	44.00	30.43	19.35
FEMA 2002	60.22	37.00	27.54	33.87
OASP (2000 and 2004)	53.76	32.00	21.74	30.65
India	53.76	30.00	27.54	30.65
Canada	37.63	24.00	20.29	25.81
New Zealand	46.24	39.00	30.43	19.35

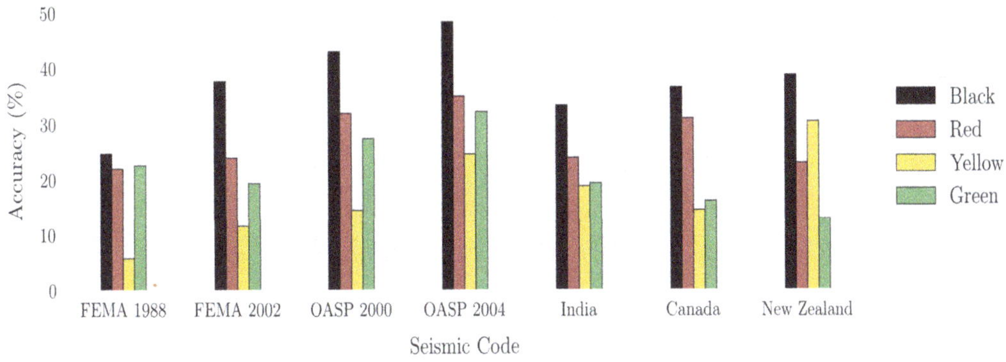

(**a**)

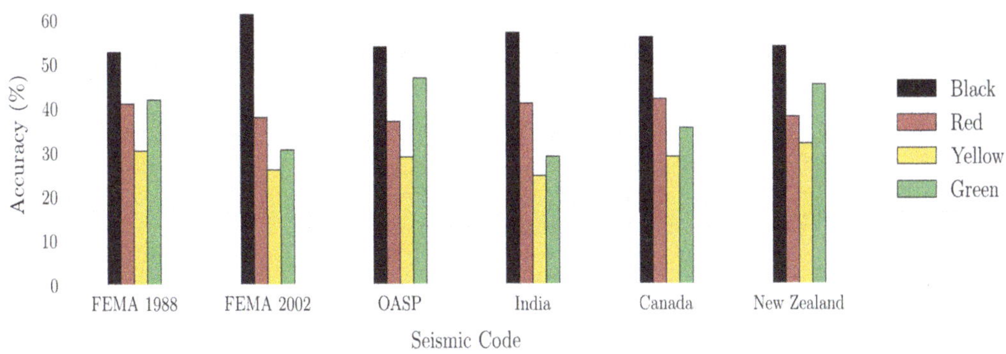

(**b**)

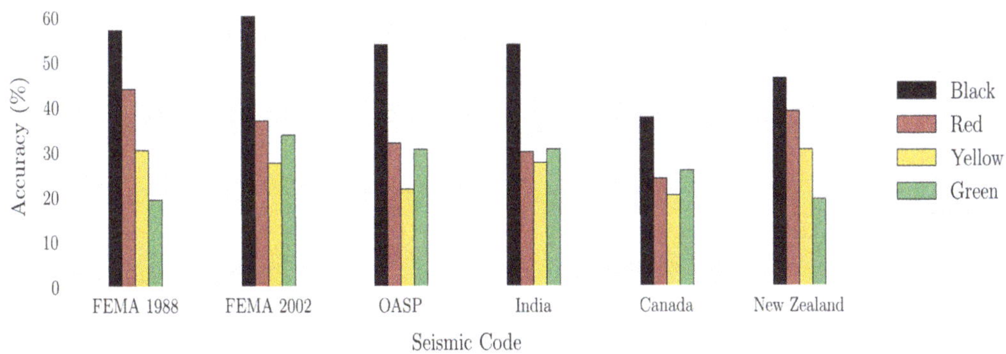

(**c**)

Figure 2. Comparison of the obtained bin accuracies for the examined Seismic Codes and methodologies. (**a**) Ranking obtained using RVS. (**b**) Ranking obtained using the fully trained Machine Learning models. (**c**) Ranking obtained using the proposed recalibrated RVS.

We have compared the rankings obtained after running the fully trained Machine Learning models for all the examined Seismic Codes and all the damage categories. We can readily observe that there is a significant increase in the accuracy in all cases. More specifically, the accuracy for the Black class (worst case) was higher than 50% for all employed models. FEMA 2002 achieved a significant accuracy as high as 61%. It should be noted that the same Seismic Code in its current form achieved an accuracy of slightly less than 40% without the use of ML classification algorithms. The models based on the Codes of India, Canada, and New Zealand also exhibited performance improvement. Their accuracy for the Black class bin increased more than 10% with the use of the ML models. OASP exhibited a slight increase of approximately 5% in the accuracy of the Black bin. The highest performance increase for this Code was achieved for the Green class bin, corresponding to buildings that suffered minimal damages. For the Red bin (severe damages), the second most impactful, almost all Seismic Codes exhibited an increase close to 15%. The OASP Code seemed to have benefited the least in this bin as well, with an increase of only 5%. However, it should be reiterated that OASP's performance was the highest to begin with. Finally, the bin accuracy of the Black label in FEMA 1988 exhibited a noteworthy increase, from approximately 25% to more than 50% when the fully trained ML model was employed.

All of the above clearly indicate the significant benefit of employing ML models for the task of seismic vulnerability ranking. As was demonstrated, using the same features, ML models are capable of learning complex, non-linear relationships that significantly outperform currently established engineering practices.

Finally, we compared the rankings obtained using the proposed ML-based recalibration procedure for RVS. In general, most Seismic Codes benefited from the proposed recalibration, since their accuracy in the most severe damage classes was improved compared to the currently established literature. Furthermore, most of the Codes seemed to have retained high levels of accuracy in the Black and Red bins when compared to the respective scores obtained from the fully trained ML models. On the other hand, most of the loss in accuracy was related to the Green and Yellow bins, which are the least severe and thus the least impactful to society.

The Canadian Code is a notable exception. It exhibited a minimal improvement compared to its currently established form and a considerable loss of accuracy for the Black and Red bins. This is attributed to the fact that the Canadian Code is fundamentally different than the rest. As was mentioned in Section 2.5, most of the examined Seismic Codes are additive, i.e., the computed RVS score is obtained by the summation of (2). On the other hand, the Canadian Code uses summation to obtain a set of intermediate features, which are then multiplied to obtain the final score. The recalibration procedure proposed in this study is linear, which is why it did not fit the Canadian Code as properly as the others.

4. Summary and Conclusions

In the framework of the present study, we have utilized ML-based methodologies to address one of the most severe and impactful engineering challenges; the pre-earthquake seismic vulnerability ranking of a large number of structures within a building stock. As was examined, different countries and organizations have proposed similar yet distinct methodologies. We have examined the approach employed by FEMA (as formulated in 1988 and revised in 2002), India, Canada, New Zealand, and the Greek OASP. As it was shown in Table 2, a different set of features is employed by each of the aforementioned Codes. Even when the same features are used, different weights can be assigned to each of them, based on experts' opinions. To mitigate this, we have employed a ML-based pairwise methodology, specifically targeting the problem of ranking. More specifically, the Gradient Boosting Machine algorithm was employed to perform comparisons between pairs of structures.

As it can be seen from Tables 3 and 4, as well as from Figure 2a,b, the proposed methodology was able to outperform currently established engineering practices and to improve on the results obtained with each separate Seismic Code. Regarding the structures that crossed CLS ("Black"), all the examined ML models exhibited an accuracy of over 50%. In some cases, like FEMA 2022 or New Zealand's Code, we have achieved an accuracy increase between 10 and 20%, compared to the ranking results obtained by the use of the respective Codes' coefficients. The accuracy improvement for the structures that crossed the ULS ('Red") was as high as 15% for most of the examined Codes. The performance improved for the other two damage labels as well (moderate—"Yellow", minimal—"Green"), although these are less impactful than the first two. The above highlight that Gradient Boosting was able to adapt better to the available features and to produce ranking results with higher accuracy and demonstrate its applicability for the task of seismic vulnerability ranking.

Subsequently, the proposed methodology was extended using ML feature attribution techniques, as examined in Sections 2.5 and 2.6. Inspired by the linearity in which the examined Codes compute their RVS scores, we proposed a linearization of the fully trained ML models using the aggregated metrics of the so-called SHAP values to produce weights similar to the ones employed in a standard RVS procedure. As these values are designed to approximate the behavior of the underlying ML model, we observed that, in most cases, employing this recalibrated and reformulated RVS methodology retained increased performance metrics, compared to the respective currently established approaches.

The benefit of the proposed formulation is that the trained ML models are capable of learning complex, non-linear patterns inherent in the data, which in turn leads to their ability to outperform currently established engineering practices of different countries and organizations. Furthermore, the proposed linearization procedure is also based on the fully trained ML models via the SHAP values feature attribution technique. Thus, even though it inspired by the currently established engineering practices, it provides a basis with which they can be recalibrated to obtain increased accuracy.

Our case study was based on a dataset consisting of real measurements from 457 reinforced concrete structures obtained after the 1999 Athens earthquake. However, given the promising results that were obtained, the proposed methodologies could be expanded. Indeed, different countries and organizations employ different features of interest for pre-earthquake seismic vulnerability ranking in an RVS context. In addition, buildings with different structural types, e.g., masonry buildings commonly found in traditional communities, employ a fundamentally different set of features. Using the respective expanded datasets, individual ML models can be trained for each Seismic Code and structural type, which will be more closely attuned to their individual conditions.

Author Contributions: Conceptualization, I.K. and A.K.; methodology, I.K., L.I. and A.K.; software, I.K., L.I. and A.K.; validation, I.K., L.I. and A.K.; formal analysis, I.K., L.I. and A.K.; investigation, I.K., L.I. and A.K.; resources, A.K.; data curation, I.K., L.I. and A.K.; writing—original draft preparation, I.K. and A.K.; writing—review and editing, L.I.; visualization, I.K.; supervision, L.I. and A.K. All authors have read and agreed to the published version of the manuscript.

Funding: This research received no external funding.

Institutional Review Board Statement: Not applicable.

Informed Consent Statement: Not applicable.

Data Availability Statement: The raw data supporting the conclusions of this article will be made available by the authors on request.

Conflicts of Interest: The authors declare no conflicts of interest.

Abbreviations

The following abbreviations are used in this manuscript:

RVS	Rapid Visual Screening
ML	Machine Learning
SHAP	SHapley Additive exPlanations
FEMA	Federal Emergency Management Agency
OASP	Greek Earthquake Planning and Protection Organization
OCIPEP	Office of Critical Infrastructure Protection and Emergency Preparedness

CNN	Convolutional Neural Network
MLP	Multi-Layer Perceptron
SVM	Support Vector Machine
k-NN	*k*-Nearest Neighbors
GB	Gradient Boosting

References

1. Vicente, R.; Parodi, S.; Lagomarsino, S.; Varum, H.; Silva, J.M. Seismic vulnerability and risk assessment: Case study of the historic city centre of Coimbra, Portugal. *Bull. Earthq. Eng.* **2011**, *9*, 1067–1096. [CrossRef]
2. Lang, K.; Bachmann, H. On the seismic vulnerability of existing buildings: A case study of the city of Basel. *Earthq. Spectra* **2004**, *20*, 43–66. [CrossRef]
3. Federal Emergency Management Agency (FEMA). *Rapid Visual Screening of Buildings for Potential Seismic Hazards: A Handbook*, 1st ed.; Technical Report; Applied Technology Council: Redwood City, CA, USA, 1988.
4. Federal Emergency Management Agency (FEMA). *Rapid Visual Screening of Buildings for Potential Seismic Hazards: A Handbook*, 3rd ed.; Technical Report; Applied Technology Council: Redwood City, CA, USA, 2002.
5. Brundson, D.; Holmes, S.; Hopkins, D.; Merz, S.; Jury, R.; Shephard, B. *The Assessment and Improvement of the Structural Performance of Earthquake Risk Buildings*; Technical Report; New Zealand Society for Earthquake Engineering: Wellington, New Zealand, 1996.
6. Sinha, R.; Goyal, A. *A National Policy for Seismic Vulnerability Assessment of Buildings and Procedure for Rapid Visual Screening of Buildings for Potential Seismic Vulnerability*; Technical Report; Indian Institute of Technology, Department of Civil Engineering: New Delhi, India, 2002.
7. Earthquake Planning and Protection Organization (OASP). *Provisions for Pre-Earthquake Vulnerability Assessment of Public Buildings (Part A)*; Technical Report; Greek Society of Civil Engineers: Athens, Greece, 2000.
8. Foo, S.; Naumoski, N.; Saatcioglu, M. *Seismic Hazard, Building Codes and Mitigation Options for Canadian Buildings*; Technical Report; Canadian Office of Critical Infrastructure Protection and Emergency Preparedness (CCIPEP): Ottawa, ON, Canada, 2001.
9. Madariaga, R.; Ruiz, S.; Rivera, E.; Leyton, F.; Baez, J.C. Near-field spectra of large earthquakes. *Pure Appl. Geophys.* **2019**, *176*, 983–1001. [CrossRef]
10. Alhan, C.; Öncü-Davas, S. Performance limits of seismically isolated buildings under near-field earthquakes. *Eng. Struct.* **2016**, *116*, 83–94. [CrossRef]
11. Rossetto, T.; Elnashai, A. Derivation of vulnerability functions for European-type RC structures based on observational data. *Eng. Struct.* **2003**, *25*, 1241–1263. [CrossRef]
12. Eleftheriadou, A.; Karabinis, A. Damage probability matrices derived from earthquake statistical data. In Proceedings of the 14th World Conference on Earthquake Engineering, Beijing, China, 12–17 October 2008; pp. 07–0201.
13. Yu, Q.; Wang, C.; McKenna, F.; Yu, S.X.; Taciroglu, E.; Cetiner, B.; Law, K.H. Rapid visual screening of soft-story buildings from street view images using deep learning classification. *Earthq. Eng. Eng. Vib.* **2020**, *19*, 827–838. [CrossRef]
14. Ruggieri, S.; Cardellicchio, A.; Leggieri, V.; Uva, G. Machine-learning based vulnerability analysis of existing buildings. *Autom. Constr.* **2021**, *132*, 103936. [CrossRef]
15. Bektaş, N.; Kegyes-Brassai, O. Development in machine learning based rapid visual screening method for masonry buildings. In Proceedings of the International Conference on Experimental Vibration Analysis for Civil Engineering Structures, Milan, Italy, 30 August–1 September 2023; Springer: Cham, Switzerland, 2023; pp. 411–421.
16. Harirchian, E.; Kumari, V.; Jadhav, K.; Rasulzade, S.; Lahmer, T.; Raj Das, R. A synthesized study based on machine learning approaches for rapid classifying earthquake damage grades to RC buildings. *Appl. Sci.* **2021**, *11*, 7540. [CrossRef]
17. Karabinis, A. Rating of the First Level of Pre-Earthquake Assessment. 2004. Available online: https://oasp.gr/sites/default/files/program_documents/261%20-%20Teliki%20ekthesi.pdf (accessed on 13 May 2024).
18. Karampinis, I.; Iliadis, L.; Karabinis, A. Rapid Visual Screening Feature Importance for Seismic Vulnerability Ranking via Machine Learning and SHAP Values. *Appl. Sci.* **2024**, *14*, 2609. [CrossRef]
19. Greek Code for Seismic Resistant Structures–EAK. 2000. Available online: https://iisee.kenken.go.jp/worldlist/23_Greece/23_Greece_Code.pdf (accessed on 13 May 2024).
20. Karampinis, I.; Iliadis, L. A Machine Learning Approach for Seismic Vulnerability Ranking. In Proceedings of the International Conference on Engineering Applications of Neural Networks, León, Spain, 14–17 June 2023; Springer: Cham, Switzerland, 2023; pp. 3–16.
21. Abd Elrahman, S.M.; Abraham, A. A review of class imbalance problem. *J. Netw. Innov. Comput.* **2013**, *1*, 9.
22. Longadge, R.; Dongre, S. Class imbalance problem in data mining review. *arXiv* **2013**, arXiv:1305.1707.
23. Liu, B.; Tsoumakas, G. Dealing with class imbalance in classifier chains via random undersampling. *Knowl.-Based Syst.* **2020**, *192*, 105292. [CrossRef]
24. Hasanin, T.; Khoshgoftaar, T. The effects of random undersampling with simulated class imbalance for big data. In Proceedings of the 2018 IEEE International Conference on Information Reuse and Integration (IRI), Salt Lake City, UT, USA, 6–9 July 2018; IEEE: New York, NY, USA, 2018; pp. 70–79.
25. Kumari, R.; Srivastava, S.K. Machine learning: A review on binary classification. *Int. J. Comput. Appl.* **2017**, *160*. [CrossRef]

26. Singhal, Y.; Jain, A.; Batra, S.; Varshney, Y.; Rathi, M. Review of bagging and boosting classification performance on unbalanced binary classification. In Proceedings of the 2018 IEEE 8th International Advance Computing Conference (IACC), Greater Noida, India, 14–15 December 2018; pp. 338–343.
27. Friedman, J.H. Greedy function approximation: A gradient boosting machine. *Ann. Stat.* **2001**, *29*, 1189–1232. [CrossRef]
28. Ganaie, M.A.; Hu, M.; Malik, A.K.; Tanveer, M.; Suganthan, P.N. Ensemble deep learning: A review. *Eng. Appl. Artif. Intell.* **2022**, *115*, 105151. [CrossRef]
29. Kingsford, C.; Salzberg, S.L. What are decision trees? *Nat. Biotechnol.* **2008**, *26*, 1011–1013. [CrossRef] [PubMed]
30. Schapire, R.E. The boosting approach to machine learning: An overview. *Nonlinear Estim. Classif.* **2003**, 171, 149–171.
31. Pedregosa, F.; Varoquaux, G.; Gramfort, A.; Michel, V.; Thirion, B.; Grisel, O.; Blondel, M.; Prettenhofer, P.; Weiss, R.; Dubourg, V.; et al. Scikit-learn: Machine learning in Python. *J. Mach. Learn. Res.* **2011**, *31*, 2825–2830.
32. Lundberg, S.M.; Lee, S.I. A unified approach to interpreting model predictions. *Adv. Neural Inf. Process. Syst.* **2017**, *30*, 4765–4774.
33. Shapley, L.S. *Notes on the n-Person Game: The Value of an n-Person Game*; RAND Corporation: Santa Monica, CA, USA, 1951.
34. Lundberg, S.M.; Erion, G.G.; Lee, S.I. Consistent individualized feature attribution for tree ensembles. *arXiv* **2018**, arXiv:1802.03888.
35. Lundberg, S.M.; Erion, G.; Chen, H.; DeGrave, A.; Prutkin, J.M.; Nair, B.; Katz, R.; Himmelfarb, J.; Bansal, N.; Lee, S.I. From local explanations to global understanding with explainable AI for trees. *Nat. Mach. Intell.* **2020**, *2*, 56–67. [CrossRef] [PubMed]
36. Kapetana, P.; Dritsos, S. Seismic assessment of buildings by rapid visual screening procedures. *Earthq. Resist. Eng. Struct. VI* **2007**, *93*, 409.

applied sciences

MDPI

Article

Prospects of Geoinformatics in Analyzing Spatial Heterogeneities of Microstructural Properties of a Tectonic Fault

Stepan Ustinov [1,2,*], Alexey Ostapchuk [2,3,*], Alexey Svecherevskiy [1], Alexey Usachev [2,3], Grigorii Gridin [2], Antonina Grigor'eva [1,2] and Igor Nafigin [1]

1 Institute of Geology of Ore Deposits, Petrography, Mineralogy and Geochemistry of the Russian Academy of Sciences, 119017 Moscow, Russia; svecherevskiy@igem.ru (A.S.); grig@igem.ru (A.G.); nafigin@igem.ru (I.N.)
2 Sadovsky Institute for Dynamics of Geospheres of the Russian Academy of Sciences, 119334 Moscow, Russia; usachev.ad@phystech.edu (A.U.); gridin.grigorii@phystech.edu (G.G.)
3 Moscow Institute of Physics and Technology, 141701 Dolgoprudny, Russia
* Correspondence: ustinov@igem.ru (S.U.); ostapchuk.aa@phystech.edu (A.O.)

Abstract: The paper proposes a special technique for microstructural analysis (STMA) of rock samples based on two provisions. The first one is an algorithm for the automatic detection and digitalization of microstructures in images of oriented thin sections. The second one utilizes geographic information system (GIS) tools for an automatized analysis of objects at the micro scale. Using STMA allows the establishment of geometric features of fissure and pore space of rock samples to determine the parameters of stress–strain fields at different stages of rock massif deformation and to establish a relationship between microstructures and macrostructures. STMA makes it possible to evaluate the spatial heterogeneity of physical and structural properties of rocks at the micro scale. Verification of STMA was carried out using 15 rock samples collected across the core of the Primorsky Fault of the Baikal Rift Zone. Petrographic data were compared to the quantitative parameters of microfracture networks. The damage zone of the Primorsky Fault includes three clusters characterized by different porosity, permeability, and deformation type. Findings point to the efficiency of STMA in revealing the spatial heterogeneity of a tectonic fault.

Keywords: microstructural analysis; GIS technology; image analysis; image filtering; microfracture mapping; porosity; permeability; Primorsky Fault; tectonite; Baikal Rift Zone

Citation: Ustinov, S.; Ostapchuk, A.; Svecherevskiy, A.; Usachev, A.; Gridin, G.; Grigor'eva, A.; Nafigin, I. Prospects of Geoinformatics in Analyzing Spatial Heterogeneities of Microstructural Properties of a Tectonic Fault. *Appl. Sci.* **2022**, *12*, 2864. https://doi.org/10.3390/app12062864

Academic Editors: Alexei Gvishiani and Boris Dzeboev

Received: 29 January 2022
Accepted: 8 March 2022
Published: 10 March 2022

1. Introduction

One of the main tasks of tectonophysics and structural geology is to establish the hierarchical rank of tectonic stress fields from local [1–3] and subregional [4,5] to regional [6–8] and global [9–12] ones. Depending on the tasks being solved and the scale of the objects under study, various methods from microstructural and petrophysical analyses to numerical methods of continuum mechanics and physical modeling are used [13–15]. A reconstruction of the stress–strain field can be successful only after detailed structural studies, mapping smallest changes in tectonic deformations and taking into account the lithological factors and the interaction of structures at different scales [16,17]. Additional information in many cases can decide the interpretation of the obtained data. For example, a comprehensive study of the mechanical, hydraulic and seismological properties of a fault allows the consideration of various components of the fault structure and permits the creation of a most accurate model of the fault [18–20]. Actually, the structure of a fault zone being determined provides information forecasting its mechanical behavior [19]. Determining orientation and distribution of microfractures along a large fault helps in verifying the tectonophysical data obtained in macrofracture measurements [21]. Additionally, studying microfractures in oriented rock samples may be used to define paleostress conditions in the rock at the time when the microfractures formed [22–24].

Microstructural analysis in geology is mainly used to establish the spatial and temporal relationship between microfracture ensembles of different genetic types, as well as to reconstruct the stress–strain fields, which act within small volumes of mineral aggregates [25–28]. The technique of traditional microstructural analysis is reduced to statistical measurements of the position of certain crystallographic or optical elements of grains of some minerals in the rock. Resting on the typification of orientations for a single mineral and different minerals in polymineral formations observed in stereograms, S-, B- and R-tectonites are identified. For S-tectonites the optical axes and other crystallographic elements of mineral grains are directed in different ways. This is expressed in one or two maxima in stereograms. B-tectonites in stereograms are characterized by a series of maxima located within a certain belt, the axis of which is perpendicular to the deformation plane. Sometimes, rocks are characterized by an orientation with a relatively uniform belt density of R-tectonite type [29]. One accumulated an extensive database on spatial orientations of various minerals, textural and structural features of rocks of various genesis and composition. However, the main problem of the traditional microstructural analysis method is the technical complexity of its implementation, which implies the use of a four-axis Fedorov stage mounted on a microscope, the lack of a computer automated software, the need to obtain a large amount of data and the extremely ambiguous interpretation of the data [10]. Measurements of crystallographic and optical elements of mineral grains in this case have to be carried out manually, which takes a lot of time.

Cracks and microcracks form mainly in the zone of brittle deformations as a response to the external stress [9,30–33]. Tectonic stresses result in the formation of permeable fractures or cause shear disturbances in pre-existing open fractures [34]. In the shallow crust the brittle deformation is predominant, and the intensity of deformation increases with depth and is relatively independent on temperature, rock type and strain rate [2]. Brittle deformation significantly depends on the coefficient of friction, the pressure of pore fluids and the orientation of the structure relative to the axes of the stress–strain field [33,34]. In contrast, at deep crust, ductile deformation is predominant and is highly dependent on temperature, strain rate and rock type [35]. The temperature increase results in a brittle–ductile transition, and rocks are characterized by brittle properties at low temperatures and acquire ductile characteristics at high temperatures (hundreds of degrees). The brittle–ductile transition usually takes place in the temperature range of 370–400 °C [36,37].

Porosity and permeability determine physical properties of the rock and characterize the fissure-pore space available for the pathway of fluids and deposition of minerals [38]. Porosity and permeability are directly related to the structural properties of the rock and can be calculated using geometric analysis of fractures [38,39]. Geometric analysis includes the determination of mean fracture orientation, length, aperture and volumetric density [40–50]. Fracture network parameters can be estimated manually by using semi-automatic and automatic measurements in rock samples [51–54] and indirectly by using electrical conductivity, Vp and Vs velocity measurements [38]. Other approaches, involving techniques such as X-ray tomography [55,56] or confocal scanning laser microscopy (CSLM) [57] are also used.

Methods of microstructural analysis always rest on the determination of geometric parameters of all microfractures, for which its number can reach hundreds of thousands in a single-oriented thin section. Obviously, without using any special technique of automatic detection, the solution of this problem is not only technically complex but is also time consuming. Many works present results of successful automatic lineament detection at microscales [27,58–60], mesoscales [61,62] and macroscales [61–63]. The development of digital technologies allows taking a fresh look at microstructural analyses.

The geospatial binding of thin sections allows the performance of microfracture detection by mapping objects at the area of thin section, which can be a conditional cartographic basis. GIS techniques can be used to solve spatial problems. In terms of territorial coverage, it is customary to distinguish between global GIS, subcontinental GIS, national GIS, regional GIS, subregional GIS and local GIS [64–66]. Geospatial binding of thin sections allows the performance of the transition from macro-GIS projects to micro-GIS.

This paper presents a new approach for the microstructural analysis of rock samples based on an algorithm of automatically mapping microfractures and their geospatial binding using GIS tools. The advantages of STMA method are shown by analyzing the microfractures of rock samples collected across the core of the Primorsky Fault. In the context of ongoing microstructural studies, given that the studied rock samples belong directly to the damage zone of the Primorsky Fault, the rocks were characterized by the type of tectonites among other parameters. In most cases, tectonites are defined as rocks that have experienced tectonic stress [67,68]. Rocks with pronounced schistosity are presented by protomylonites, blastomylonites, mylonites and ultramylonites. Rock samples without obvious schistosity are subdivided into protocataclasites, blastocataclasites, cataclasites and ultracataclasites. Mylonites are fine-grained, siliceous, very densely banded or vein-like rocks resulting from the fine crushing of coarse-grained rocks without noticeable chemical transformations. Mylonites are the end product of intensive dislocation metamorphism [67,68]. Cataclasites are rocks that formed during discontinuous deformations (brecciation and crushing) of brittle parent rocks. Cataclasites gradually transform into mylonites via a decrease in grain size and the development of banded textures.

2. Materials and Methods

The research methodology is aimed at identifying the heterogeneity of the fault zone based on the analysis of the spatial distribution of microfracture parameters. The main emphasis is conducted on a detailed description of the developed automatic identification and mapping of microfractures. Figure 1 shows a flowchart of the research protocol.

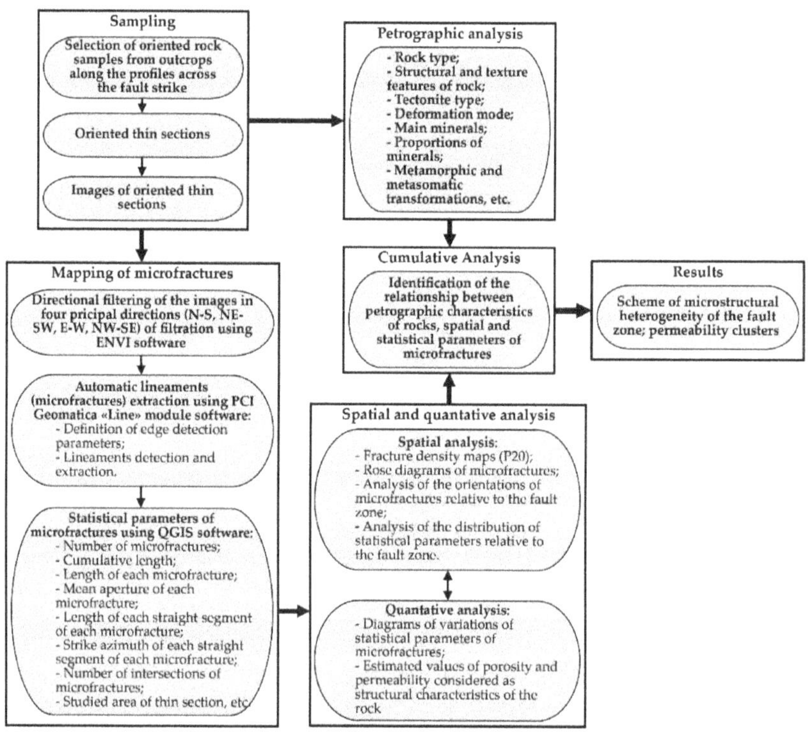

Figure 1. Flowchart of the research protocol.

2.1. Area of Study

Oriented rock samples were taken from outcrops across the strike of the Primorsky deep fault of the Baikal Rift Zone (Irkutsk region, Russia) in the area of the Sarma river

(Figure 2). The Primorsky Fault is the largest and best-documented neotectonic structure [69–75] in the region. It belongs to the Obruchev fault system [76,77].

The main thrust event occurred in the Paleozoic [78], but the compression structures became active again in later tectonic events. The age of the last activity remains poorly established and varies from Mesozoic to Miocene–Pliocene in different models [78–81]. The setting of neotectonic regional stress changed from early orogenic shear strike-slip about 30 million years ago to late orogenic extension about 3 million years ago [80]. The faults of the Obruchev system, which stretch for hundreds of kilometers in the northeasterly direction, have experienced different stages of evolution and can be studied via field structural measurements. The main trace of the Primorsky Fault, which was formed at the final phase of the collapse, is geomorphically expressed as a ledge and is available for geological surveys for over more than 200 km.

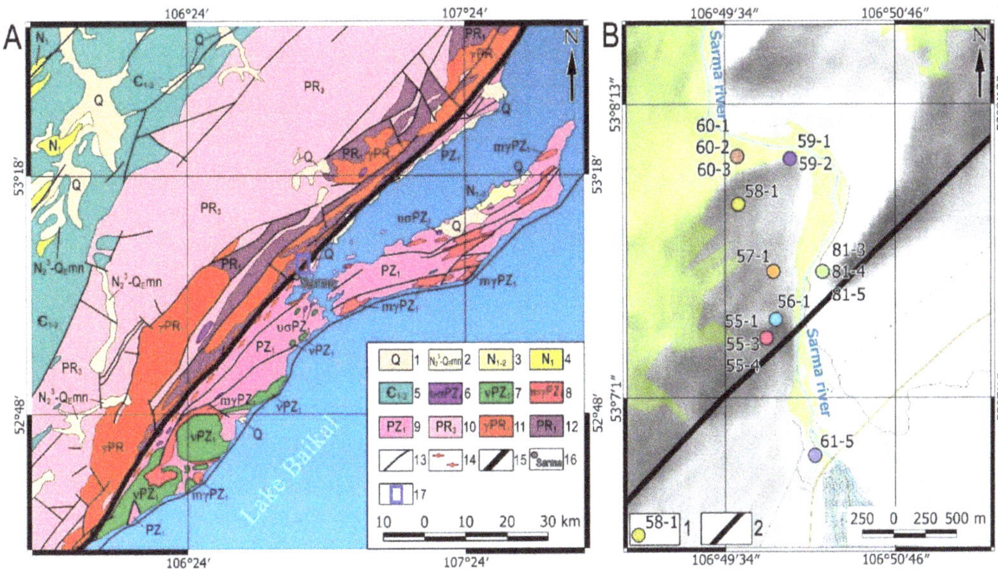

Figure 2. (**A**) Geological map of the Primorsky Fault [74,75,81–84]: 1—beach lacustrine deposits of Lake Baikal and alluvium of river valley terraces; 2—Manzurka Fm. alluvium; 3—Golumets Fm. alluvium; 4—Bayanday Fm. alluvium; 5—Lower and Middle Cambrian sediments on the Siberian craton; 6—Atarkhan peridotite–gabbro complex; 7—Ozersk gabbro–diorite complex; 8—Olkhon migmatite–granite complex; 9—Early Paleozoic metamorphic complexes of the Olkhon terrane; 10—Upper Proterozoic sediments (Baikal Group); 11—Primorsky granite complex; 12—Early Proterozoic craton basement; 13—main faults; 14—blastomylonite from the marginal suture of the Siberian craton; 15—Primorsky Fault; 16—Sarma village; 17—area of study. (**B**) Scheme of the oriented rock samples selection within the Primorsky Fault zone near Sarma village: 1—sampling points and their numbers; 2—Primorsky Fault.

The Primorsky deep fault has been well studied via geological, structural and tectonophysical methods [85]. The internal structure and dimensions of the zone of influence of this discontinuity have been studied carefully [73–75]. Due to its geomorphological severity, excellent exposure and accessibility for study, it is a reference point for paleotectonic reconstructions. Previously, it was established that, at the neotectonic stage, the main interface of the discontinuity was formed as a normal fault with an insignificant strike-slip component. It has been established that the Primorsky Fault had formed under conditions of crustal extension, as evidenced by the corresponding structural fault pattern. During Cenozoic rifting, the main displacement of the Primorsky Fault became active as a normal fault with a very insignificant sinistral strike-slip component [73]. By now, horizons of the Earth's crust within the zone of the Primorsky Fault, which were previously located at seismic focal depths, have become accessible for visual geological study at the Earth's surface [86].

2.2. Collection of Rock Samples and Petrographic Analysis

An obligatory condition for the implementation of any microstructural analysis technique is the collection of spatially oriented samples, for which spatial fixation must be strictly observed throughout the entire duration of their processing [13,22,24–31]. Samples for petrographic and STMA studies were collected mainly in the outfall parts of the Sarma river. Rocky outcrops contain dislocated series of rocks of the Neo-Proterozoic that had undergone multistage physical and chemical transformations while being exposed to stress metamorphism. It is specific for amphibolite and greenshale facies of regional metamorphism.

The collected and analyzed samples of rocks are represented by numerous varieties of predominantly metamorphic and igneous rocks, as well as their metasomatically altered varieties. Shales (55-1, 55-3, 56-1, 57-1, 60-3, 81-3, 81-4, 81-5), gneissose granite (58-1, 59-1, 59-2, 60-1), plagiogranites (55-3, 55-4), amphibolites (60-2) and gabbro (61-5), varied in mineral composition, were distinguished (Table 1). The studied rock samples are classified as mylonites (81-4), blastomylonites (55-1, 55-3, 81-3, 81-5), cataclasites (55-4, 60-2, 60-3), blastocataclasites (56-1, 57-1, 61-5) and host rocks (58-1, 59-1, 59-2, 60-1), which were not subjected to intensive deformations. Among other characteristics, the type of deformation (brittle or ductile) was assessed, and metamorphic and metasomatic changes were marked.

Table 1. Petrographic description of rock samples.

Sample Number	Petrographic Type	Tectonite Type	Deformation Mode	Main Minerals * and Proportions (%)	Metamorphic and Metasomatic Transformations
55-1	Quartz-Sericite-Chlorite shale	Blastomylonite	Brittle	Q(60), Fsp(20), Src(10), Cl(5)	Complete recrystallization of quartz; sericite is formed by plagioclase; actinolite is chloritized
55-3	Contact rock: Quartz-Actinolite-Chlorite shale–Plagiogranite	Blastomylonite	Brittle and ductile	Fsp(60), Q(30), Cl(5), Ac(3)	Partial recrystallization of quartz; sericite is formed by plagioclase; actinolite is chloritized
55-4	Plagiogranite	Mylonite	Brittle	Q(60), Fsp(30), Ac(5), Cl(3), Src(2)	Partial recrystallization of quartz; sericite is formed by plagioclase
56-1	Plagioamphibolite shale	Mylonite	Brittle	Q(50), Fsp(40), Src(5), Cl(3), Ac(2)	Partial recrystallization of quartz; sericite is formed by plagioclase; actinolite is chloritized
57-1	Chlorite-Sericite-Plagioquartzite shale	Blastocataclasite	Brittle	Q(50), Fsp(30), Src(15), Cl(5), Ac(<1)	Partial recrystallization of quartz; sericite is formed by plagioclase; actinolite is chloritized
58-1	Leucocratic gneissose granite	Host rock	Brittle	Q(55), Fsp(35), Mu(2), Bt(2) Am(1)	Quartz is slightly granular
59-1	Leucocratic gneissose granite	Host rock	Brittle	Q(50), Fsp(35), Mu(5), Bt(4) Am(1)	Quartz is slightly granular
59-2	Leucocratic gneissose granite	Host rock	Brittle	Q(50), Fsp(35), Mu(7), Bt(5) Am(3)	Quartz is slightly granulated; about 10% potassium feldspar saussuritized

Table 1. *Cont.*

Sample Number	Petrographic Type	Tectonite Type	Deformation Mode	Main Minerals * and Proportions (%)	Metamorphic and Metasomatic Transformations
60-1	Gneissose granite	Host rock	Brittle	Q(55), Fsp(30), Bt(10)	Saussurite developed according to potassium feldspar; quartz is slightly granulated
60-2	Amphibolite	Cataclasite	Ductile	Am(55), Ep(25) Q(15), Fsp(5)	Recrystallization of quartz by schistosity
60-3	Amphibolite shale	Cataclasite	Ductile	Bt(50), Am(40), Q(5), Fsp(5)	Recrystallization of quartz by schistosity; amphibole is replaced by biotite
61-5	Gabbro	Blastocataclasite	Brittle	Am(35), Fsp(30), Px(25), Bt(5), Q(5)	Amphibole is substituted with biotite; quartz is recrystallized and switched into the matrix
81-3	Epidote shale	Blastomylonite	Brittle and ductile	Ep(60), Cl(30), Q(5),	Zone of stress metamorphism; a matrix of ground material of hydromicas formed over amphiboles; quartz is granulated
81-4	Amphibolite shale	Mylonite	Brittle and ductile	Cl(40), Fsp(20), Am(10), Ep(20), Q(10)	Zone of stress metamorphism; amphibole is replaced by chlorite and stretched into ribbons; Matrix–ground materials of potassium feldspar, chlorite, iron hydroxide
81-5	Epidote shale	Blastomylonite	Brittle and ductile	Ep(35), Q(15), Cl(40), Ca(10)	Zone of stress metamorphism; a matrix of ground material of hydromicas formed over amphiboles; quartz is granulated

* Main minerals: Q—Quartz; Fsp—Potassium feldspat; Am—Amphibole; Px—Pyroxene; Ep—Epidote; Cl—Chlorite; Ac—Actinolite; Src—Sericite; Bt—Biotite; Mu—Muscovite; Ca—Calcite.

2.3. Fotos of Thin Sections

Images of oriented thin sections for microstructural mapping were obtained using an Olympus BX-51 optical microscope with an Olympus Camedia C7070zoom digital camera. To determine the orientation of microfractures in a GIS project and to calculate their geometric parameters (strike azimuth, length and aperture), all studied thin sections must be strictly oriented in space. This requirement must also be taken into account when taking photographs of individual parts of an oriented thin section. Separate parts of oriented thin sections were photographed in such a manner that a panoramic image of the entire section could be automatically constructed for them. Images of thin sections should have a resolution sufficient for analysis, which corresponds to at least 10,000 pixels along the smallest side. Photomosaics collected from separate sections should be provided with a scale ruler. Images were saved in TIFF format. Figure 3 shows thumbnail images of the studied thin sections.

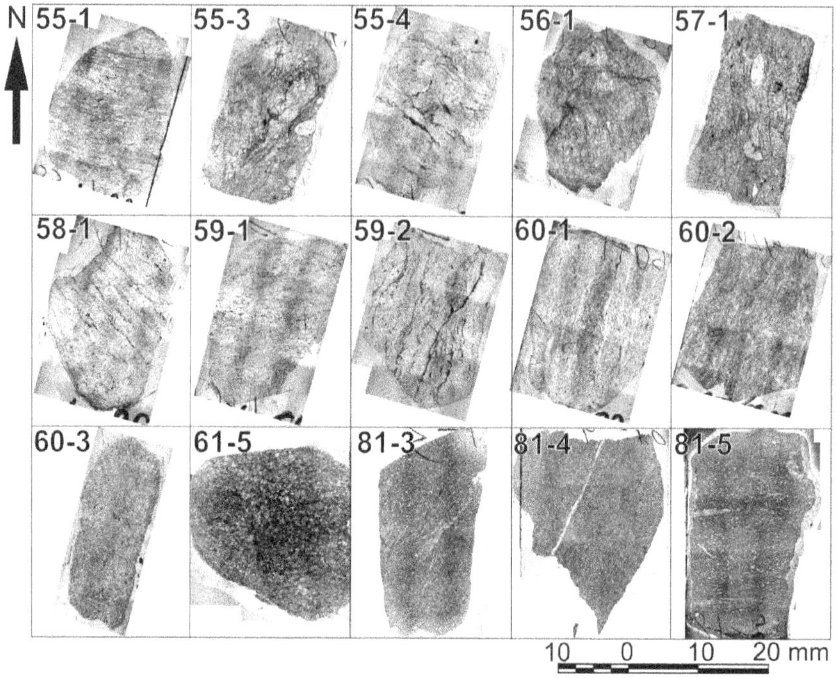

Figure 3. Images of the studied oriented thin sections obtained using the Olympus BX-51 optical microscope with the Olympus Camedia C7070zoom digital camera.

2.4. Special Technique of Microstructural Analysis (STMA)

Given that the focus of the research is studying microfractures, the software used in STMA should implement and automate the following basic operations:

- Implementation of binding the rasters (images) of individual parts of a thin section, as well as their filling and docking in a single coordinate system;
- Determination of relative coordinates X and Y of any point in a thin section;
- Marking all microstructures within the thin section;
- Automatic determination and measurement of geometrical parameters of microstructures (strike azimuth, length and aperture);
- Identification of various ensembles and microstructure generations depending on the values of azimuth and their separate marking;
- Possibility of marking objects over type (microcracks filled with secondary fluid inclusions (FIPs), open or partially mineralized microcracks and microcracks filled with ore material);
- Calculation of porosity and permeability in paleo and modern conditions at various stages of deformation of geological bodies;
- Determination of quantitative and percent ratios of different types of microstructures and presentation of the results in the form of diagrams (graphs, rose-diagrams and histograms).

2.4.1. Automatic Mapping of Microfractures

At the first stage of STMA, images of thin sections were mapped in a local rectangular coordinate system using QGIS (a free and open-source cross-platform GIS application). The global Mercator projection of Google Maps (EPSG–900913) [87] was used to display the image of the thin section.

At the second stage, a directional filtering of the image was carried out to remove the noise from sections located within a certain area of the image but without blurring its edges. For this purpose, the Sobel operator was used. It is a discrete differential operator that calculates the approximate value of the image brightness gradient. The result of applying the Sobel operator at each point of the image is the vector of brightness gradient at that point.

The directional filters induce an optical effect of shadow at the image, as if it was illuminated by an oblique light. Directional filters are applied to the image using a convolution process by means of constructing a window, as a rule a (3 × 3) pixels box (Table 2) of Sobel kernel filters (Table 3). Sobel kernels generate an effective and faster method for evaluating lineaments (linear objects) in four principal directions [88–91].

Table 2. Floating image matrix, size 3 × 3.

Z1	Z2	Z3
Z4	Z5	Z6
Z7	Z8	Z9

Table 3. Sobel operator masks for four principal directions of filtration.

N-S (0°)			**NE-SW (45°)**			**E-W (90°)**			**NW-SE (135°)**		
−1	0	1	−2	−1	0	−1	−2	−1	0	1	2
−2	0	2	−1	0	1	0	0	0	−1	0	1
−1	0	1	0	1	2	1	2	1	−2	−1	0

The Sobel operator for horizontal and vertical brightness differences is defined by the following functions.

$$Gx = (Z7 + 2 \times Z8 + Z9) - (Z1 + 2 \times Z2 + Z3) \quad (1)$$

$$Gy = (Z3 + 2 \times Z6 + Z9) - (Z1 + 2 \times Z4 + Z7) \quad (2)$$

Based on these data, it is possible to calculate the value of the brightness difference.

$$G = \sqrt{(Gx^2 + Gy^2)} \quad (3)$$

The directional filtering operation narrows the boundary that exists between the values of neighboring pixels. In this study, directional filtering is used to improve the boundaries of gradient transition between pixel values in order to highlight certain characteristics of the image resting on their spatial frequency associated with textural features [92]. The original image of the section consists of three spectral channels: R, G and B. The R channel must be fed to the filter input, since the data obtained in red and infrared spectral ranges are more sensitive to structural features. Four filtered images were produced by ENVI software related to the directions N-S, NE-SW, E-W and NW-SE (Figure 4).

At the third stage, the automatic extraction of lineaments is carried out. At this stage, the first step is the edge detection procedure, which provides information about areas of sharp changes in the intensity of neighboring pixels by applying the Canny edge detector, and the second step is the detection of lineaments [93]. These steps can be performed using the "LINE" module of the PCI Geomatica software, which is a widely used module for automatic extraction of lineaments [93]. There are two categories of parameters in this module: the first category controls the edge detection step, the second one detects and extracts curves (Table 4).

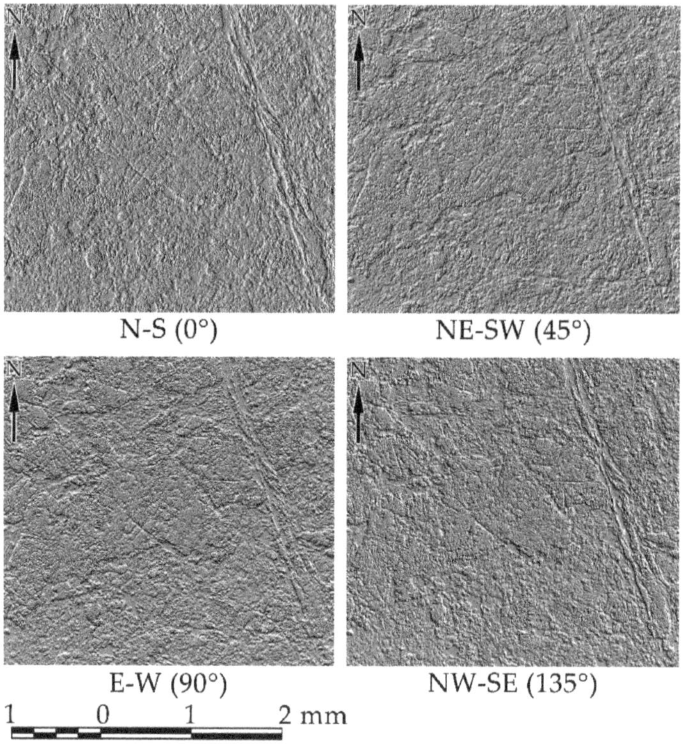

Figure 4. The result of filtering a part of the image of an oriented thin section with the Sobel operator in four principal directions.

Table 4. Parameters used for the PCI Geomatica LINE module.

Parameter	Description	Range and Unit
	Edge Detection	
RADI	Filter radius. It specifies the radius of the edge detection filter (Filter de Canny).	0–8192 (pixel)
GTHR	Gradient threshold. It specifies the threshold for the minimum gradient level for an edge pixel to obtain a binary image (Filter de Canny).	0–255
	Curve extraction	
LTHR	Length threshold: It specifies the minimum length of curve to be considered as lineament	0–8192 (pixel)
FTHR	Line fitting error threshold: It specifies the maximum error (in pixels) allowed in fitting a polyline to a pixel curve.	0–8192 (pixel)
ATHR	Angular difference threshold: It is the maximum angle between two vectors for them to be linked.	0–90 (degrees)
DTHR	Linking distance threshold: It specifies the minimum distance between the end points of two vectors for them to be linked.	0–8192 (pixel)

A number of experimental tests was performed using various combinations of values for each parameter of the LINE module to make an automatic selection of lineaments (microfractures) directly related to the structural and texture features in thin section images.

Parameters of the LINE module were verified visually. Default and calculated values of parameters of the LINE module are presented in Table 5.

Table 5. Default and verified values of the PCI Geomatica LINE module parameters for microfracture extraction.

Thresholds and Units	Default Value	Verified Value
RADI (pixels)	10	10
GTHR	100	35
LTHR (pixels)	30	60
FTHR (pixels)	3	5
ATHR (degrees)	30	45
DTHR (pixels)	20	20

Operating the program gives results that highlight linear objects, marking microfractures (Figure 5).

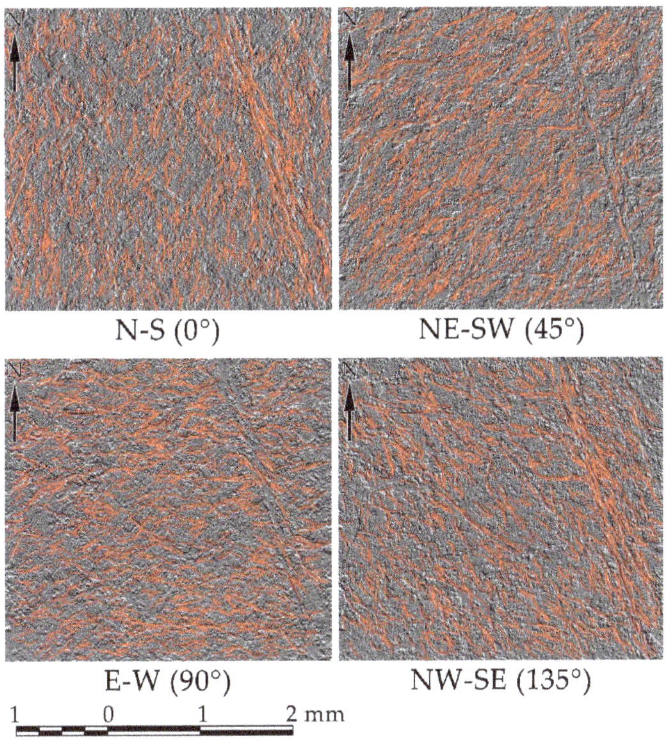

Figure 5. The result of detecting microfractures (red lines) by the LINE module in the PCI Geomatica according to the verified parameters of the part of the filtered image of the oriented thin section in four principal directions.

Directional filtering and Canny edge detecting results in constructing 4 groups of linear objects. The general ensemble of lineaments, obtained by a simple addition of the 4 groups, can be interpreted as the ensemble of detected microfractures. Figure 6 shows the image of a thin section with highlighted microfractures.

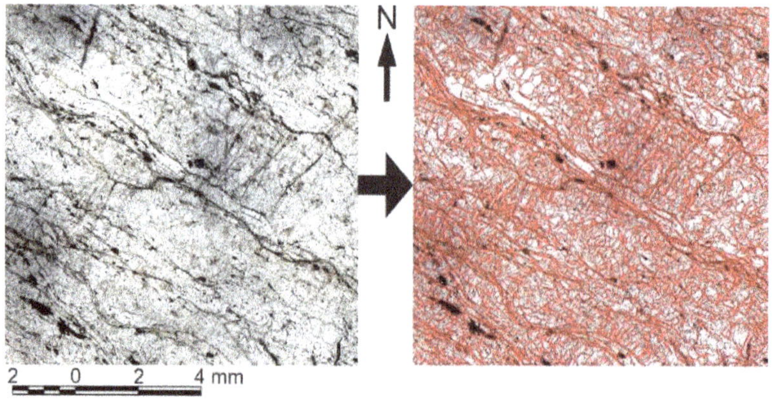

Figure 6. The result of detecting and marking microfractures in a part of the oriented thin section of the sample 58-1 using STMA.

2.4.2. Microfracture Characteristics

The main quantitative parameters of the identified microfractures were calculated in QGIS using the Field Calculator tool [94]. The Field Calculator in the attribute table performs calculations based on existing attribute values or specific functions, such as calculating the length or the area of geometric objects. The results can be written to a new attribute column or used to update an existing column. The following parameters were automatically calculated for microfractures:

- Number of objects (microfractures);
- X and Y coordinates of the ends;
- Cumulative length;
- Length of each microfracture;
- Mean aperture of each microfracture;
- Length of each straight segment of each microfracture;
- Strike azimuth of each straight segment of each microfracture;
- Number of intersections of microfractures;
- Studied area of thin section, etc.

The spatial distribution of microfractures over the area of the section was analyzed by constructing density maps (P20) [95] using the Line Density tool. Rose diagrams of microfractures were built according to the values of their spatial orientation and taking into account their lengths. This allowed the visual assessment of the nature of distribution, the significance and severity of the selected linear objects in the thin section and to compare the intensity of deformations that resulted in the origination of a particular ensemble of microfractures at different stages of tectogenesis. Constructing rose diagrams has made it possible to carry out a comparative analysis of microfracture orientations and to identify different systems and generations of microfractures.

2.4.3. Determination of Porosity and Permeability

Porosity and permeability of rocks were determined via the quantitative method proposed in [27,49]. The mapped microfractures were considered as a network of fractures. Cracks were approximated by disks. It was assumed that the maximum crack length corresponds to the disk diameter. In this study, the assumption was made that all the open microcracks that formed at a certain stage of deformation are interconnected. Thus, an equivalent crack with a total length of the disk diameter could be introduced. The effective porosity can be determined as follows:

$$\Phi = (L \cdot \pi \cdot e)/S \quad (4)$$

where L is the total length of a certain crack population, S is the area of the studied surface and e is the mean aperture of the crack ensemble [27,49].

The definition of fracture permeability is also based on the assumption that all fractures are connected and can be approximated by disks with an aperture equivalent to the fracture aperture. An assumption was introduced that cracks propagated perpendicularly to the plane of the oriented section and formed a non-uniform spatial distribution. This assumption allows to determine the microcrack permeability (K_f) [27]:

$$K_f = (\varepsilon \cdot \pi \cdot N_f \cdot c^2 \cdot e^3)/12 \quad (5)$$

where ε is the crack wall roughness coefficient, N_f is the volume density of crack population f, c is the mean value of half-length of the cracks in the ensemble and e is the mean aperture of cracks.

If the surface density and half-length of all cracks are known, it is possible to determine their volumetric density [27,94]:

$$N_f = 2/(S \cdot \pi \cdot c \cdot \sin\theta) \quad (6)$$

where θ is the angle of crack incidence. If we assume that the cracks are perpendicular to the plane of the thin section, then $\sin\theta = 1$.

Thus, Formula (5) can be simplified as follows [27]:

$$K_f = (\varepsilon \cdot c \cdot e^3)/6S \quad (7)$$

where K_f is the total or directive permeability (for a certain population of cracks), expressed in square meters (m^2) or in darcy (1 darcy = $0.987 \cdot 10^{-12}$ m^2).

It should be noted that the results of laboratory tests are in good correlation with the values of porosity and permeability calculated with this method [96–98].

3. Results

Using the STMA, quantitative parameters of microfractures in thin sections were calculated. The mean length of one object (microfracture segment) was about 7.5 μm. The investigated area varied from 150 to 370 mm^2. Microstructural parameters are presented in Table 6 and in Figure 7. Samples 57-1 and 60-2 show maximum cumulative length and maximum number of microfractures (see Table 6 and Figure 7A,B). Minimal values of these parameters are observed in samples 55-3, 55-4, 81-4 and 81-5. The remaining samples are characterized by average values. The L/S ratios (see Table 6 and Figure 7C) reached maxima in samples 57-1, 60-2, 60-3, 61-5 and minima in samples 55-4, 81-3, 81-4, 81-5. Maximum values of the fracture intensity (F = N/L) [99] were observed in samples 61-5, 81-3, 81-4, 81-5 and the minimum values–in samples 55-4, 60-2 (see Table 6 and Figure 7D). In general, the distribution of the fracture intensity is extremely nonuniform. Porosity and permeability can be most informative parameters for assessing the degree of deformations, since they are directly related to the physical properties of rocks. Porosity and permeability vary in a similar manner (see Table 6 and Figure 7E,F). Four clusters can be distinguished. The first cluster shows increased values of porosity and permeability. It includes samples 60-2, 60-3 and 61-5. The second cluster shows average values of porosity/permeability typical for samples 57-1, 58-1, 59-1 and 59-2. The third cluster includes only sample 60-1 and is associated with below-average values. Minimum values characterize the fourth cluster, which includes samples 55-1, 55-3, 55-4, 56-1, 81-3, 81-4 and 81-5. Figure 8 shows distributions of L/S and N/L versus permeability.

Table 6. Statistical parameters of microfractures.

Sample Number	Cumulative Length, mm (L)	Mean Aperture, Microns (e)	Number of Objects (N)	Studied Area, mm^2 (S)	Fracture Intensity F = N/L	L/S	Porosity, %	Permeability, m^2
55-1	2416.76	3.1	326,112	282.52	0.0367	0.1156	1.13	1.44×10^{-15}
55-3	1349.20	2.7	177,193	151.99	0.0357	0.1200	1.02	9.84×10^{-16}
55-4	1168.48	3.3	142,813	287.83	0.0332	0.0549	0.57	8.22×10^{-16}
56-1	1849.79	2.6	251,385	243.86	0.0369	0.1025	0.84	7.52×10^{-16}
57-1	3579.96	4.5	484,590	337.23	0.0368	0.1435	2.03	5.45×10^{-15}
58-1	2233.05	4.8	292,253	251.44	0.0356	0.1200	1.81	5.53×10^{-15}
59-1	2801.10	4.9	379,502	306.55	0.0369	0.1235	1.90	6.05×10^{-15}
59-2	2387.49	4.8	309,648	251.95	0.0353	0.1281	1.93	5.90×10^{-15}
60-1	2881.68	3.9	384,124	354.89	0.0363	0.1098	1.34	2.71×10^{-15}
60-2	3444.97	5.4	424,009	326.80	0.0335	0.1425	2.42	9.35×10^{-15}
60-3	2648.34	5.6	356,331	242.95	0.0366	0.1473	2.59	1.08×10^{-14}
61-5	2501.05	5.3	346,100	259.61	0.0376	0.1302	2.17	8.08×10^{-15}
81-3	1958.90	2.9	278,241	325.88	0.0386	0.0812	0.74	8.26×10^{-16}
81-4	1241.22	3.2	173,286	276.85	0.0380	0.0606	0.61	8.27×10^{-16}
81-5	1173.13	3.1	166,625	371.26	0.0386	0.0427	0.42	5.30×10^{-16}

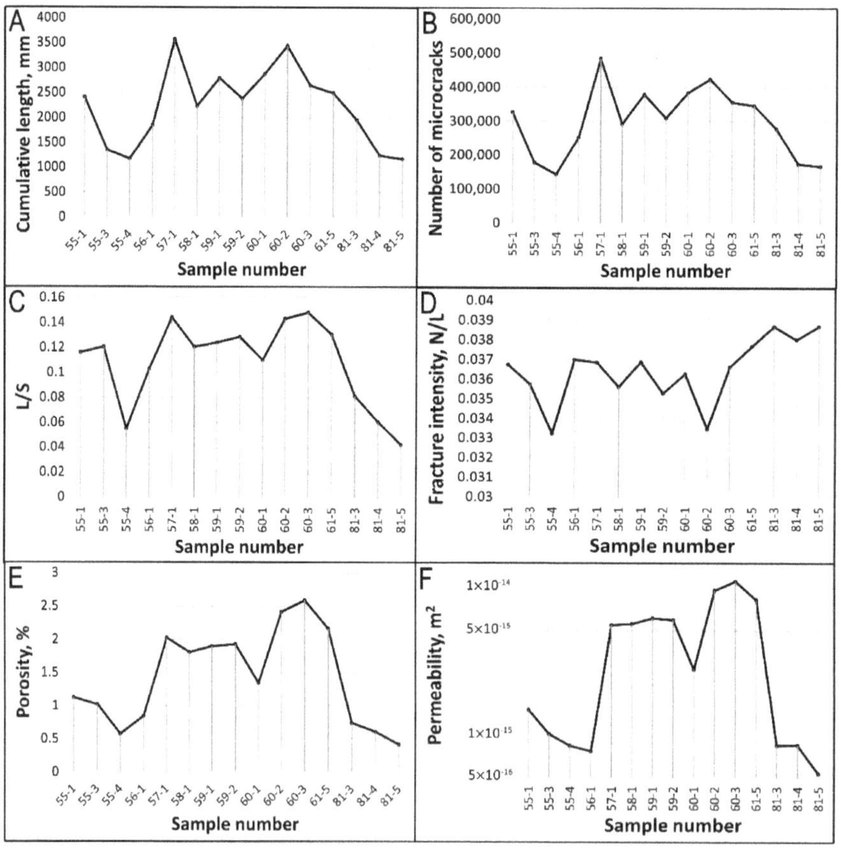

Figure 7. Graphs of the distribution of cumulative length (**A**), number of microfractures (**B**), L/S (**C**), Fracture intensity (**D**), porosity (**E**) and permeability (**F**) of microfractures in oriented thin sections.

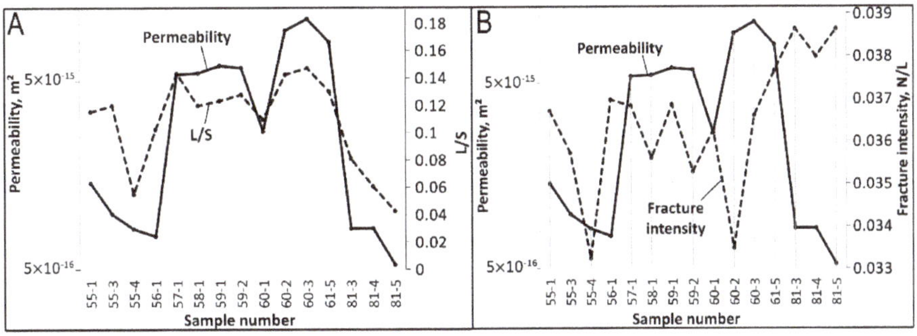

Figure 8. Graphs of the distribution of L/S (**A**) and N/L (**B**) versus permeability.

Fracture density maps (P20) [100] of the identified microfractures were built to assess the most permeable zones in the thin section. These maps show the areal distribution of microfractures (Figure 9).

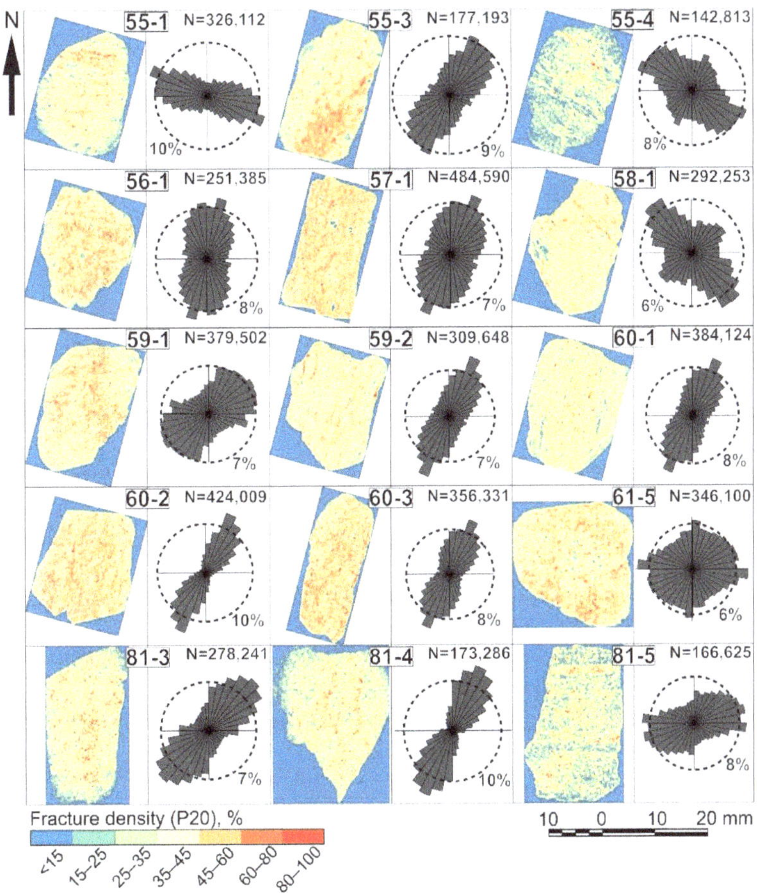

Figure 9. Fracture density (P20) maps and rose diagrams of microfractures in oriented thin sections.

Microfractures in the samples 56-1, 57-1, 59-1, 60-2, 60-3 and 61-5 have relatively uniform spatial distributions. In other thin sections, the distribution of microfractures is extremely uneven. Most often, the areas of maximum density of microfractures are confined to linearly elongated zones oriented in a certain manner. In sample 55-3, the density diagram clearly shows the contact of two rocks, which strike northeast. The rocks differ in the density of microfractures. Figure 9 also shows rose diagrams microfracture orientation. The results of orientation measurements can be used to reconstruct stress–strain fields, as well as to indirectly assess the degree of deformations that resulted in the origination of microfracture ensembles. In this paper, only the predominant directions of strikes of microfracture ensembles are considered. For a more detailed analysis, it is necessary to rank microfractures over length and orientation intervals, as well as structural and genetic types in order to identify individual ensembles associated with certain stages of deformations and parameters of the acting stress–strain field [6–9]. Samples 55-3, 59-1, 59-2, 60-1, 60-2, 60-3, 81-3 and 81-4 have predominantly northeastern orientation of microfractures. In samples 55-1 and 81-5, the microfractures are developed along the latitude with slight deviations to the northwest and northeast, respectively. Predominant northwestern orientations were identified in thin sections 55-4 and 58-1; a less pronounced northeastern orientation is also distinguished here. Samples 56-1 and 57-1 show north–northeastern orientations of microfractures. In the sample 61-5, microfractures are oriented almost uniformly in all directions.

A comparison of petrographic analysis data to the results of STMA reveals a high correlation. Thus, it was possible to identify clusters that characterize the heterogeneity of the fault zone. Four clusters were identified (Figure 10).

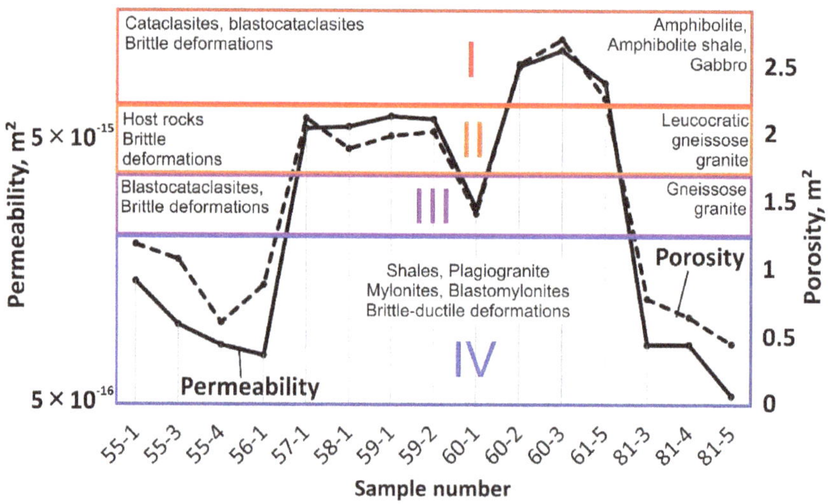

Figure 10. Integration of the data of petrographic and microstructural analyses of rock samples collected across the Primorsky Fault.

Cluster I shows the highest permeability. It includes samples of amphibolites, amphibolite schists and gabbro, which, by the type of tectonite, belong to cataclasites and blastocataclasites. Predominantly brittle deformations manifest in this area.

Cluster II shows lower permeability, which corresponds to host rocks represented by leucocratic granite gneisses. Brittle deformations also prevail here.

Cluster III was identified via only a single gneissose granite sample. It shows less pronounced brittle deformations, and it is a blastocataclasite according to the type of tectonite. The identification of this cluster is in doubt.

Cluster IV is represented by shales of different mineral composition and, to a lesser extent, by transformed plagiogranites. Cluster IV characterizes the transition from brittle to ductile deformations; it mainly includes mylonites and blastomylonites. Moreover, this cluster coincides with the zone of stress metamorphism established via petrographic analysis.

In order to characterize the fault zone, it is important to understand the change in the main quantitative characteristics of microfractures with distance from the fault trace (Figure 11). The total length and number of microstructures naturally increase with distance from the fault trace (Figure 11A). At the same time, a local maximum is fixed in the interval of 400–500 m, which was established in one sample (57-1). The distributions of L/S ratios and fracture intensity (N/L) show heterogeneous distribution (Figure 11B). In the immediate vicinity of the fault trace, the L/S values are in the range of 0.043–0.120. At a distance of more than 400 m from the fault trace, the values vary from 0.110 to 0.150. The fracture intensity near the fault trace varies in a wide range—from 0.033 to 0.039. Then, the values decrease linearly again to 0.033 with distance. It should be noted that all values of the N/L parameter are generally in a very narrow range. The porosity and permeability show a similar trend (Figure 11C). They are characterized by the lowest values near the fault trace (up to 200 m) and increase with distance from 400 to 1100 m. The maxima are reached at a distance of more than 1200 m.

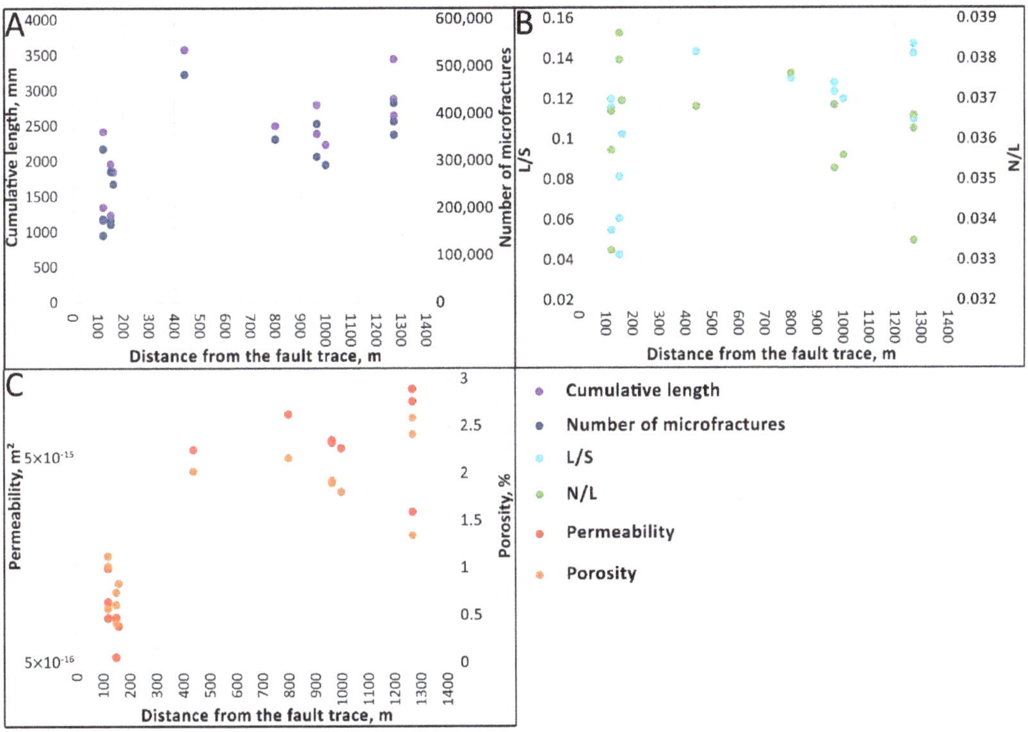

Figure 11. Graphs of the distribution of cumulative length versus number of microfractures (**A**), L/S versus N/L (**B**), permeability versus porosity (**C**) with distance from the fault trace.

4. Discussion

The damage zone of the fault is usually characterized by a spatial heterogeneity of physical properties of rocks, types of their deformations and the development of secondary structures and microstructures [21,26]. At the same time, it is assumed that there is a certain zonality within this zone, the identification of which is one of the most important tasks of structural geology [6,18,21,24,33,69]. The solution of this problem is extremely difficult, since most fault structures are characterized by a long history of evolution and a successive change of several stages of deformations with certain parameters of the stress-strain field. The developed STMA method, using the GIS tools, significantly reduces the complexity of identifying and mapping microfractures in oriented thin sections, measuring and calculating their spatial and statistical parameters. The method provides necessary opportunities for a most comprehensive description and analysis of microfractures.

The Primorsky Fault is characterized by a long geological and tectonic history difficult for reconstruction. During its evolution it passed through numerous stages, including compression, strike-slip and tension [74,75,79,80]. The main challenge for a researcher is to restore the sequence of these stages. Microstructural analysis makes it possible to obtain additional information, which allows the identification of the stages of deformations and the reconstruction of the evolution of tectonic stress fields. At the same time, the STMA technique remains quite difficult to implement, and the process of sample preparation takes a long time. The first problem is the selection of a statistically significant number of oriented rock samples from outcrops. In the presented study, the main tasks were to confirm the efficiency of the developed approach and demonstrate its advantages. Therefore, the results of the analysis of 15 rock samples are preliminary. In the future, it is necessary to increase the density of rock sampling for a more comprehensive analysis of the heterogeneity of the fault zone, as well as to carry out selection along profiles across the fault strike throughout the entire damage zone.

In this study, fluid transfer properties are used as indicators of structural features of rock samples and probable heterogeneity of the fault zone. Porosity and permeability are in agreement with the results of petrographic analysis. It allows the identification of zones that characterize the heterogeneity of the fault. It is possible to create a preliminary scheme of heterogeneity of the fault (Figure 12). Three zones, which are characterized by different values of porosity and permeability, types of deformations and types of tectonites, were identified. The closest to the fault trace zone is characterized by the minimum values of porosity and permeability. Blastomylonites and mylonites formed there. The spatial distribution of parameters indicates intensive deformations within a zone 200 m thick. This zone can be considered as the fault core. During the formation of the network of microfractures, brittle and ductile deformations presumably took place. In the transition to the middle zone, an increase in porosity and permeability is observed. Rock samples mainly consist of cataclasites, and brittle deformations predominate. The farthest zone from the fault trace is characterized by the highest porosity and permeability; cataclasites and blastocataclasites are predominant there, and brittle deformations dominate. This model is consistent with most of the previously proposed models of fault zones [11,16–21,33,69].

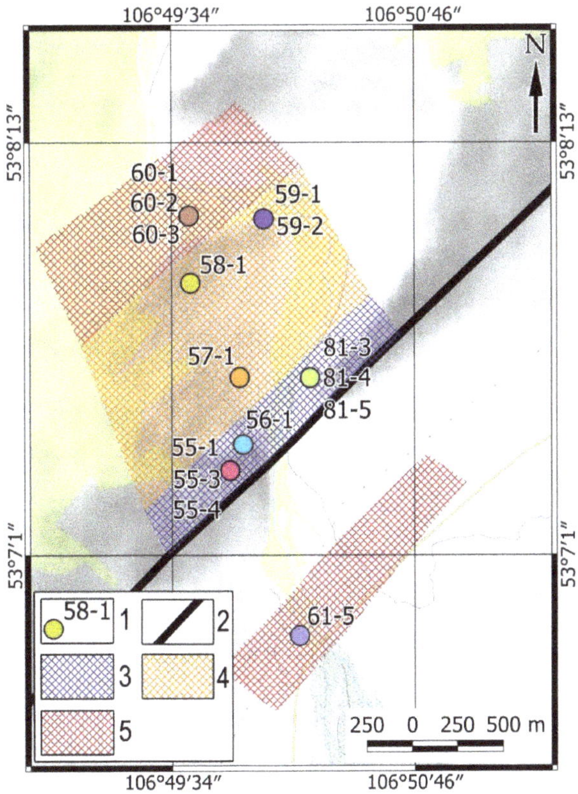

Figure 12. Scheme of heterogeneity of the Primorsky Fault zone based on microstructural analysis: 1—Sampling points and their numbers; 2—Primorsky Fault; 3—Cluster with low porosity and permeability, zone of blastomylonites and brittle and ductile deformations; 4—Cluster with medium porosity and permeability, zone of cataclasites and brittle deformations; 5—Cluster with high porosity and permeability, zone of cataclasites and blastocataclasites and brittle deformations.

Tectonophysical studies usually result in detecting variations of the parameters of the stress–strain field, which, in its turn, makes it possible to reconstruct the sequence of changes in deformations [3–12]. At the microscale, this issue can be resolved by studying fluid inclusion planes (FIPs) as tectonic indicators of changing stress–strain fields [23,27,48,49]. In studying the oriented sections of the collected samples, it was found that most of the microcracks are of mode I [34,47]. Resting on this fact, Figure 13A shows a preliminary scheme for interpreting the orientations of principal axes of stress fields. It should be noted that the identified orientations probably reflect the most pronounced stages of deformation. The preliminary interpretation of the position of stress axes for the sample 61-5 is ambiguous. Figure 13B shows the interpretation of the modern stress–strain field, identified by conducting an analysis of lineaments [75,101,102] in the area of the Sarma River. The orientation of the axes of modern stress field does not coincide with the orientation of the main axes of the stress fields identified in single samples. This suggests that main deformations observed in rock samples took place in previous geological epochs. It is worth mentioning that the presented results cannot be considered as final ones and a complete sequence of stress field changes can be reconstructed only based on a detailed study of cracks of various genetic types and an identification of their generations.

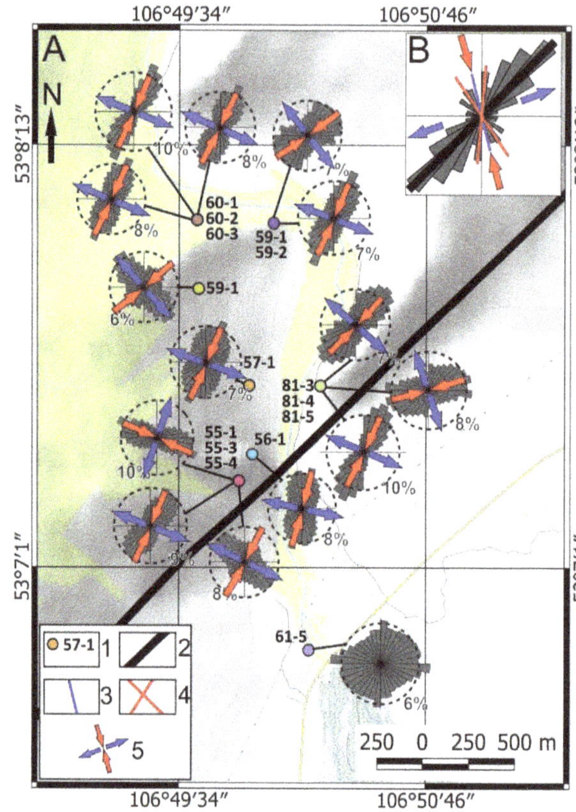

Figure 13. Scheme for interpreting the orientations of the principal axes of stress fields. (**A**) Interpretation of orientations of the main axes of the stress field performed for single rock samples based on the analysis of rose diagrams of microfractures, taking into account the fact that most studied microstructures are cracks of mode I; (**B**) interpretation of the orientations of principal axes of the modern stress field, performed via the analysis of lineaments in the area of the Sarma River [75,101,102]. 1—sampling points and their numbers; 2—Primorsky Fault; 3—cracks of mode I; 4—pair cracks of mode II; 5—orientation of the main axes of compression (red) and tension (blue).

5. Conclusions

A special technique of microstructural analysis (STMA) that utilizes advantages of GIS technologies is presented. Applying the GIS approach simplifies detection, vectorization and processing of large amounts of microstructural data (one thin section can contain up to 10^5 cracks). The GIS tools of STMA makes it possible to link all collected samples and their detected characteristics to maps of different types, as well as to simultaneously analyze microstructures and macrostructures identified via topographic and tectonic map data and satellite images. There is a unique possibility of linking and upscaling macro-, meso- and micro-characteristics of a rock massif, as well as restoring the sequence of changes in tectonic stress–strain fields at different scale levels and times of origination.

Applying the STMA algorithm, the structure of the damage zone of the Primorsky Fault was detected. Three permeable zones that are consistent with the results of petrographic studies have been identified. The thickness of the zone of most intensive deformations is estimated as 10–20 m.

Author Contributions: All authors contributed to the study conception and design. Conceptualization, S.U. and A.O.; methodology, S.U., A.S. and I.N.; software, I.N., A.U. and G.G.; validation, S.U., A.O., I.N. and A.S.; formal analysis, S.U., A.U. and A.G.; investigation, S.U., A.O. and A.G.; resources, S.U. and A.O.; data curation, G.G. and A.O.; writing—original draft preparation, S.U. and A.O.; visualization, S.U. and A.S.; supervision, S.U.; project administration, A.O.; funding acquisition, A.O. All authors have read and agreed to the published version of the manuscript.

Funding: This research was funded by the Russian Science Foundation, Grant No. 20-77-10087.

Institutional Review Board Statement: Not applicable.

Informed Consent Statement: Not applicable.

Data Availability Statement: All the data presented in this study are available from corresponding author on request.

Acknowledgments: Rock samples for analyses were delivered by IDG RAS in the frames of the Program No. FMWN-2022-0020 of the Ministry of Science and Higher Education of Russian Federation. The development of the special technique of microstructural analysis (STMA) was carried out by IGEM RAS in the frames of the Program of the Ministry of Science and Higher Education of Russian Federation.

Conflicts of Interest: The authors declare no conflict of interest.

References

1. Stephens, T.L.; Walker, R.J.; Healy, D.; Bubeck, A.; England, R.W. Mechanical models to estimate the paleostress state from igneous intrusions. *Solid Earth* **2018**, *9*, 847–858. [CrossRef]
2. Yamaji, A.; Sato, K. Clustering of fracture orientations using a mixed Bing ham distribution and its application to paleostress analysis from dike or vein orientations. *J. Struct. Geol.* **2011**, *33*, 1148–1157. [CrossRef]
3. De Guidi, G.; Caputo, R.; Scudero, S. Regional and local stress field orientation inferred from quantitative analyses of extension joints: Case study from southern Italy. *Tectonics* **2013**, *32*, 239–251. [CrossRef]
4. Saintot, A.; Angelier, J. Tectonic paleostress fields and structural evolution of the NW-Caucasus fold-and-thrust belt from Late Cretaceous to Quaternary. *Tectonophysics* **2002**, *357*, 1–31. [CrossRef]
5. Sim, L.A. Some methodological aspects of tectonic stress reconstruction based on geological indicators. *Comptes Rendus Geosci.* **2012**, *344*, 174–180. [CrossRef]
6. Bott, M.H.P. The mechanics of obique slip faulting. *Geol. Mag.* **1959**, *96*, 109–117. [CrossRef]
7. Gushchenko, O.I. The method of kinematic analysis of destruction structures in reconstruction of tectonic stress fields. In *Fields of Stress and Strain in the Lithosphere*; Grigoriev, A.S., Osokina, D.N., Eds.; Nauka: Moscow, Russia, 1979; pp. 7–25. (In Russian)
8. Rebetskii, Y.L.; Marinin, A.V. Stressed state of the Earth's crust in the western region of the Sunda subduction zone before the Sumatra-Andaman earthquake on December 26, 2004. *Dokl. Earth Sci.* **2006**, *407*, 321–325. [CrossRef]
9. Gzovsky, M.V. *Fundamentals of Tectonophysics*; Nauka: Moscow, Russia, 1975; p. 536. (In Russian)
10. Nikolaev, P.N. *Methodology of Tectonodynamic Analysis*; Nedra: Moscow, Russia, 1992; p. 295. (In Russian)
11. Angelier, J. Fault slip analysis and paleostress reconstruction. In *Continental Deformation*; Hancock, P.L., Ed.; Pergamon Press: Oxford, UK, 1994; pp. 53–100.
12. Lacombe, O. Do fault slip data inversions actually yield "paleostresses" that can be compared with contemporary stresses? A critical discussion. *Comptes Rendus Geosci.* **2012**, *344*, 159–173. [CrossRef]
13. Heap, M.J.; Kushnir, A.R.L.; Gilg, H.A. Microstructural and petrophysical properties of the Permo-Triassic sandstones (Buntsandstein) from the Soultz-sous-Forêts geothermal site (France). *Geotherm. Energy* **2017**, *5*, 26. [CrossRef]
14. Yang, G.; Leung, A.K.; Xu, N.; Zhang, K.; Gao, K. Three-dimensional physical and numerical modelling of fracturing and deformation behaviour of mining-induced rock slopes. *Appl. Sci.* **2019**, *9*, 1360. [CrossRef]
15. Unver, B.; Yasitli, N. Modelling of strata movement with a special reference to caving mechanism in thick seam coal mining. *Int. J. Coal Geol.* **2016**, *66*, 227–252. [CrossRef]
16. Petrov, V.A.; Poluektov, V.V.; Ustinov, S.A.; Minaev, V.A.; Lespinasse, M. Rescaling of fluid-conducting fault structures. *Dokl. Earth Sci.* **2017**, *472*, 130–133. [CrossRef]
17. Petrov, V.A.; Poluektov, V.V.; Ustinov, S.A.; Minaev, V.A.; Lespinasse, M. Scale effect in a fluid-conducting fault network. *Geol. Ore Depos.* **2019**, *61*, 293–305. [CrossRef]
18. Ben-Zion, Y.; Sammis, C.G. Characterization of fault zones. *Pure Appl. Geophys.* **2003**, *160*, 677–715. [CrossRef]
19. Faulkner, D.R.; Lewis, A.C.; Rutter, E.H. On the internal structure and mechanics of large strike-slip fault zones: Field observations of the Carboneras fault in southeastem Spain. *Tectonophysics* **2003**, *367*, 235–251. [CrossRef]
20. Wibberley, C.A.J.; Shimamoto, T. Internal structure and permeability of major strike-slip fault zones: The Median Tectonic Line in Mie Prefecture, southwest Japan. *J. Struct. Geol.* **2003**, *25*, 59–78. [CrossRef]

21. Mitchell, T.M.; Faulkner, D.R. The nature and origin of off-fault damage surrounding strike-slip fault zones with a wide range of displacements: A field study from the Atacama fault system, northern Chile. *J. Struct. Geol.* **2009**, *31*, 802–816. [CrossRef]
22. Lespinasse, M.; Pecher, A. Microfracturing and regional stress field: A study of the preferred orientation of fluid-inclusion planes in a granite from the Massif Central, France. *J. Struct. Geol.* **1986**, *8*, 169–180. [CrossRef]
23. Boullier, A.-M. Fluid inclusions: Tectonic indicators. *J. Struct. Geol.* **1999**, *21*, 1229–1235. [CrossRef]
24. Wilson, J.E.; Chester, J.S.; Chester, F.M. Microfracture analysis of fault growth and wear processes, Punchbowl Fault, San Andreas system, California. *J. Struct. Geol.* **2003**, *25*, 1855–1873. [CrossRef]
25. Wise, D.U. Microjointing in basement, middle Rocky Mountains of Montana and Wyoming. *Geol. Soc. Am. Bull.* **1964**, *75*, 287–306. [CrossRef]
26. Hoxha, D.; Homand, F. Microstructural approach in damage modeling. *Mech. Mater.* **2000**, *32*, 377–387. [CrossRef]
27. Lespinasse, M.; Désindes, L.; Fratczak, P.; Petrov, V. Microfissural mapping of natural cracks in rocks: Implications for fluid transfers quantification in the crust. *Chem. Geol.* **2005**, *223*, 170–178. [CrossRef]
28. Duarte, M.T.; Liu, H.Y.; Kou, S.Q.; Lindqvist, P.A.; Miskovsky, K. Microstructural modeling approach applied to rock material. *J. Mater. Eng. Perform.* **2005**, *14*, 104–111. [CrossRef]
29. Lukin, L.I.; Chernyshev, V.F.; Kushnarev, I.P. *Microstructural Analysis*; Nauka: Moscow, Russia, 1965; p. 124. (In Russian)
30. Costin, L.S. A microcrack model for the deformation and failure of brittle rock. *J. Geophys. Res.* **1983**, *88*, 9485–9492. [CrossRef]
31. Kranz, R.L. Microcracks in rocks: A review. *Tectonophysics* **1983**, *100*, 449–480. [CrossRef]
32. Lawn, B. Fracture of brittle solids. In *Cambridge Solid State Science Series*, 2nd ed.; Cambridge University Press: Cambridge, UK, 1993; p. 378.
33. Brandes, C.; Tanner, D.C. Chapter 2—Fault mechanics and earthquakes. In *Understanding Faults*; Tanner, D., Brandes, C., Eds.; Elsevier: Amsterdam, The Netherlands, 2020; pp. 11–80. [CrossRef]
34. Jaeger, J.C.; Cook, N.G.W. *Fundamentals of Rock Mechanics*, 3rd ed.; Chapman and Hall: London, UK, 1979; p. 593.
35. Buck, W.R. The Dynamics of Continental Breakup and Extensio. In *Treatise on Geophysics*, 2nd ed.; Schubert, G., Ed.; Elsevier: Amsterdam, The Netherlands, 2015; pp. 325–379. [CrossRef]
36. Tamanyu, S.; Fujimoto, K. On the models for the deep-seated geothermal system in the Kakkonda field. *Rept. Geol. Surv. Jpn.* **2000**, *284*, 133–164.
37. Fournier, R.O. Hydrothermal processes related to movement of fluid from plastic into brittle rock in the magmatic-epithermal environment. *Econ. Geol.* **1999**, *94*, 1193–1211. [CrossRef]
38. Harvey, P.K.; Brewer, T.S.; Pezard, P.A.; Petrov, V.A. *Petrophysical Propoerties of Cristalline Rocks*; The Geological Society: London, UK, 2005; p. 345.
39. Nelson, R.A. *Geological Analysis of Naturally Fractured Reservoirs*, 2nd ed.; Gulf Professional Publishing: Boston, MA, USA, 2001.
40. Snow, D.T. Anisotropie permeability of fractured media. *Water Resour. Res.* **1969**, *5*, 1273–1289. [CrossRef]
41. Moench, A.F. Double-porosity models for a fissured groundwater reservoir with fracture skin. *Water Resour. Res.* **1984**, *20*, 831–846. [CrossRef]
42. Ababou, R. *Approaches to Large Scale Unsaturated Flow in Heterogeneous Stratified and Fractured Geologic Media*; US Nuclear Regulatory Commission: Washington, DC, USA, 1991.
43. Long, J.C.S.; Gilmour, P.; Whiterspoon, P.A. A model for steady fluid flow in random three-dimensional networks of diskshaped fractures. *Water Resour. Res.* **1985**, *21*, 105–115. [CrossRef]
44. Gueguen, Y.; Dienes, J. Transport properties of rocks from statistics and percolation. *Math. Geol.* **1989**, *21*, 1–13. [CrossRef]
45. Canals, M.; Ayt Ougougdal, M. Percolation on anisotropic media, the Bethe lattice revisited. Application to fracture networks. *Nonlinear Processes Geophys.* **1997**, *4*, 11–18. [CrossRef]
46. Stauffer, D. *Introduction to Percolation Theory*; Taylor and Francis: London, UK, 1985; p. 124.
47. Homand, F.; Hoxha, D.; Belem, T.; Pons, M.N.; Hoteit, N. Geometric analysis of damaged microcracking in granites. *Mech. Mater.* **2000**, *32*, 361–376. [CrossRef]
48. Lespinasse, M.; Sausse, J. Quantification of fluid flow: Hydromechanical behaviour of different natural rough fractures. *J. Geochem. Explor.* **2000**, *69–70*, 483–486. [CrossRef]
49. Sausse, J.; Jacquot, E.; Fritz, B.; Leroy, J.; Lespinasse, M. Evolution of crack permeability during fluid-rock interaction. Example of the Brézouard granite (Vosges, France). *Tectonophysics* **2001**, *336*, 199–214. [CrossRef]
50. Sausse, J. Hydromechanical properties and alteration of natural fracture surfaces in the Soultz granite (Bas-Rhin, France). *Tectonophysics* **2002**, *348*, 169–185. [CrossRef]
51. Zhang, L.; Einstein, H.H. Estimating the intensity of rock discontinuities. *Int. J. Rock Mech. Min. Sci.* **2000**, *37*, 819–837. [CrossRef]
52. Zhang, L.; Einstein, H.H. Estimating the mean trace length of rock discontinuities. *Rock Mech. Rock Eng.* **1998**, *31*, 217–235. [CrossRef]
53. Riley, M.S. Fracture trace length and number distributions from fracture mapping. *J. Geophys. Res.* **2005**, *110*, B08414. [CrossRef]
54. Zeeb, C.; Gomez-Rivas, E.; Bons, P.D.; Blum, P. Evaluation of sampling methods for fracture network characterization using outcrops. *AAPG Bull.* **2013**, *97*, 1545–1566. [CrossRef]
55. Hirono, T.; Takahashi, M.; Nakashima, S. In situ visualization of fluid flow image within deformed rock by X-ray CT. *Eng. Geol.* **2003**, *70*, 37–46. [CrossRef]

56. Menendez, B.; David, D.; Martı'nez Nistal, A. Confocal scanning laser microscopy applied to the study of pore and crack networks in rocks. *Comput. Geosci.* **2001**, *27*, 1101–1109. [CrossRef]
57. Geraud, Y.; Caron, J.M.; Faure, P. Porosity network of a ductile shear zone. *J. Struct. Geol.* **1995**, *17*, 1757–1769. [CrossRef]
58. Passchier, C.W.; Trouw, R.A.J. *Microtectonics*, 2nd ed.; Springer: Berlin, Germany, 2005; p. 366.
59. Ustinov, S.; Petrov, V.; Poluektov, V.; Minaev, V. Calculation of filtration properties of the Antey uranium deposit rock massif at the deformation phases: Microstructural approach. In *Trigger Effects in Geosystems*; Kocharyan, G., Lyakhov, A., Eds.; Springer Nature: Cham, Switzerland, 2019; pp. 179–186.
60. Xu, J.; Zhao, X.; Liu, B. Digital image analysis of fluid inclusions. *Int. J. Rock Mech. Min. Sci.* **2007**, *44*, 942–947. [CrossRef]
61. Healy, D.; Rizzo, R.E.; Cornwell, D.G.; Farrell, N.J.C.; Watkins, H.; Timms, N.E.; Gomez-Rivas, E.; Smith, M. FracPaQ: A MATLAB™ toolbox for the quantification of fracture patterns. *J. Struct. Geol.* **2017**, *95*, 1–16. [CrossRef]
62. Zeeb, C.; Gomez-Rivas, E.; Bons, P.D.; Virgo, S.; Blum, P. Fracture network evaluation program (FraNEP): A software for analyzing 2D fracture trace-line maps. *Comput. Geosci.* **2013**, *60*, 11–22. [CrossRef]
63. Xu, C.; Dowd, P. A new computer code for discrete fracture network modelling. *Comput. Geosci.* **2010**, *36*, 292–301. [CrossRef]
64. Saha, K.; Frøyen, Y.K. *Learning GIS Using Open Source Software*, 1st ed.; Taylor and Francis Group: London, UK, 2021; p. 240. [CrossRef]
65. Longley, P.A.; Goodchild, M.F.; Maguire, D.J.; Rhind, D.W. *Geographic Information Systems and Science*, 2nd ed.; John Wiley and Sons: Chichester, UK, 2005.
66. Chang, K. *Introduction to Geographic Information System*, 4th ed.; McGraw-Hill Education: New York, NY, USA, 2016; p. 249.
67. Best, M.G. *Igneous and Metamorphic Petrology*, 2nd ed.; Blackwell Publishing Company: Oxford, UK, 2003; p. 752.
68. Letnikov, F.A.; Balyshev, S.O. *Petrophysics and Geoenergetics of Tectonites*; Nauka: Novosibirsk, Russia, 1991; p. 148. (In Russian)
69. Sherman, S.I. *Physics of Crustal Faulting*; Nauka: Novosibirsk, Russia, 1977; p. 102. (In Russian)
70. Mats, V.D. *The Cenozoic History and Structure of the Baikal Rift Basin*; GEO: Novosibirsk, Russia, 2001; p. 252. (In Russian)
71. Lunina, O.V.; Gladkov, A.S.; Cheremnykh, A.V. Fracturing in the Primorsky fault zone (Baikal Rift system). *Russ. Geol. Geophys.* **2002**, *43*, 446–455.
72. Mats, V.D.; Lobatskaya, R.M.; Khlystov, O.M. Evolution of faults in continental rifts: Morphotectonic evidence from the southwestern termination of the North Baikal basin. *Earth Sci. Front.* **2007**, *14*, 207–219. [CrossRef]
73. Cheremnykh, A.V. Stress fields in the Primorsky normal fault zone (Baikal rift). *Lithosfera* **2011**, *1*, 135–142.
74. Cheremnykh, A.V.; Cheremnykh, A.S.; Bobrov, A.A. Faults in the Baikal region: Morphostructural and structure-genetic features (case study of the Buguldeika fault junction). *Russ. Geol. Geophys.* **2018**, *59*, 1100–1108. [CrossRef]
75. Cheremnykh, A.V.; Burzunova, Y.P.; Dekabryov, I.K. Hierarchic features of stress field in the Baikal region: Case study of the Buguldeika Fault Junction. *J. Geodyn.* **2020**, *141–142*, 101797. [CrossRef]
76. Seminskii, K.Z.; Kozhevnikov, N.O.; Cheremnykh, A.V.; Pospeeva, E.V.; Bobrov, A.A.; Olenchenko, V.V.; Tugarina, M.A.; Potapov, V.V.; Burzunova, Y.P. Interblock zones of the northwestern Baikal rift: Results of geological and geophysical studies along the Bayandai Village—cape Krestovskii profile. *Russ. Geol. Geophys.* **2012**, *53*, 194–208. [CrossRef]
77. Lamakin, V.V. *Neotectonics of the Baikal Basin*; Nauka: Moscow, Russia, 1968; p. 247. (In Russian)
78. Aleksandrov, V.K. *Thrusts in the Baikal Region*; Nauka: Novosibirsk, Russia, 1990; p. 102. (In Russian)
79. Delvaux, D.; Moeys, R.; Stapel, G.; Melnikov, A.; Ermikov, V. Paleostress reconstructions and geodynamics of the Baikal region, Central Asia. Part I: Palaeozoic and Mesozoic pre-rift evolution. *Tectonophysics* **1995**, *252*, 61–101. [CrossRef]
80. Delvaux, D.; Moyes, R.; Stapel, G.; Petit, C.; Levi, K.; Miroshnitchenko, A.; Ruzhich, V.; San'kov, V. Paleostress reconstruction and geodynamics of the Baikal region, Central Asia. Part II: Cenozoic rifting. *Tectonophysics* **1997**, *282*, 1–38. [CrossRef]
81. Gladkochub, D.P.; Donskaya, T.V.; Wingate, M.T.D.; Poller, U.; Kr¨oner, A.; Fedorovsky, V.S.; Mazukabzov, A.M.; Todt, W.; Pisarevsky, S.A. Petrology, geochronology, and tectonic implications of ca. 500 Ma metamorphic and igneous rocks along the northern margin of the Central-Asian Orogen (Olkhon terrane, Lake Baikal, Siberia). *J. Geol. Soc. Lond.* **2008**, *165*, 235–246. [CrossRef]
82. Gladkochub, D.P.; Donskaya, T.V.; Mazukabzov, A.M.; Sklyarov, E.V.; Fedorovskii, V.S.; Lavrenchuk, A.V.; Lepekhina, E.N. Fragment of the Early Paleozoic (~500 Ma) island arc in the structure of the Olkhon terrane, Central Asian fold belt. *Dokl. Earth Sci.* **2014**, *457*, 905–909. [CrossRef]
83. Donskaya, T.V.; Gladkochub, D.P.; Mazukabzov, A.M.; Fedorovsky, V.S.; Cho, M.; Cheong, W.; Kim, J. Synmetamorphic granitoids (~490 Ma) as accretion indicators in the evolution of the Olkhon terrane (Western Cisbaikalia). *Russ. Geol. Geophys.* **2013**, *54*, 1205–1218. [CrossRef]
84. Donskaya, T.V.; Gladkochub, D.P.; Fedorovsky, V.S.; Sklyarov, E.V.; Cho, M.; Sergeev, S.A.; Demonterova, E.I.; Mazukabzov, A.M.; Lepekhina, E.N.; Cheong, W.; et al. Pre-collisional (& 0.5 Ga) complexes of the Olkhon terrane (southern Siberia) as an echo of events in the Central Asian Orogenic Belt. *Gondwana Res.* **2017**, *42*, 243–263. [CrossRef]
85. Mats, V.D.; Perepelova, T.I. A new perspective on evolution of the Baikal Rift. *Geosci. Front.* **2011**, *2*, 349–365. [CrossRef]
86. Ruzhich, V.V.; Kocharyan, G.G.; Savelieva, V.B.; Travin, A.V. On the structure and formation of earthquake sources in the faults located in the subsurface and deep levels of the crust. Part II. Deep level. *Geodyn. Tectonophys.* **2018**, *9*, 1039–1061. [CrossRef]
87. Lauf, G.B. The Transverse Mercator projection. *S. Afr. Geogr. J.* **2012**, *57*, 118–125. [CrossRef]
88. Suzen, M.L.; Toprak, V. Filtering of Satellite Images in Geological Lineament Analyses: An Application to a Fault Zone in Central Turkey. *Int. J. Remote Sens.* **1998**, *19*, 1101–1114. [CrossRef]

89. Gloaguen, R.; Marpu, P.R.; Niemeyer, I. Automatic extraction of faults and fractal analysis from remote sensing data. *Nonlinear Process. Geophys.* **2007**, *14*, 131–138. [CrossRef]
90. Sedrette, S.; Rebai, N. Assessment Approach for the Automatic Lineaments Extraction Results Using Multisource Data and GIS Environment: Case Study in Nefza Region in North-West of Tunisia. In *Mapping and Spatial Analysis of Socio-Economic and Environmental Indicators for Sustainable Development. Advances in Science, Technology & Innovation (IEREK Interdisciplinary Series for Sustainable Development*; Rebai, N., Mastere, M., Eds.; Springer: Berlin, Germany, 2019; pp. 63–69. [CrossRef]
91. Enoh, M.A.; Okeke, F.I.; Okeke, U.C. Automatic lineaments mapping and extraction in relationship to natural hydrocarbon seepage in Ugwueme, South-Eastern Nigeria. *Geod. Cartogr.* **2021**, *47*, 34–44. [CrossRef]
92. Paplinski, A. Directional filtering in edge detection. *IEEE Trans. Image Processing* **1998**, *7*, 611–615. [CrossRef]
93. Salui, C.L. Methodological validation for automated lineament extraction by LINE method in PCI Geomatica and MATLAB based Hough transformation. *J. Geol. Soc. India* **2018**, *92*, 321–328. [CrossRef]
94. QGIS User Guide—QGIS Documentation. Available online: https://docs.qgis.org/3.16/en/docs/user_manual/index.html (accessed on 21 January 2022).
95. Underwood, E.E. *Quantitative Microscopy*; DeHoff, R.T., Rhines, F.N., Eds.; McGraw-Hill Book Co.: New York, NY, USA, 1968; p. 149.
96. Brace, W.F. Permeability of crystalline rocks: New in situ measurements. *J. Geophys. Res.* **1984**, *89*, 4327–4330. [CrossRef]
97. Gueguen, Y.; Shubnel, Y. Elastic wave velocities and permeability of cracked rocks. *Tectonophysics* **2003**, *370*, 163–176. [CrossRef]
98. Shmonov, V.M.; Vitovtova, V.M.; Zharikov, A.V. *Fluid permeability of Rocks of the Earth's Crust*; Scientific World: Moscow, Russia, 2002; p. 216. (In Russian)
99. Ortega, O.J.; Marrett, R.A.; Laubach, S.E. A scale-independent approach to fracture intensity and average spacing measurement. *AAPG Bull.* **2006**, *90*, 193–208. [CrossRef]
100. Bisdom, K.; Gauthier, B.D.M.; Bertotti, G.; Hardebol, N.J. Calibrating discrete fracture-network models with a carbonate three-dimensional outcrop fracture network: Implications for naturally fractured reservoir modeling. *AAPG Bull.* **2014**, *98*, 1351–1376. [CrossRef]
101. Rebetsky, Y.L.; Sim, L.A.; Marinin, A.V. *From Slickensides to Tectonic Stresses. Methods and Algorithms*; GEOS Publishing House: Moscow, Russia, 2017; p. 234. (In Russian)
102. Petrov, V.A.; Sim, L.A.; Nasimov, R.M.; Shchukin, S.I. Fault tectonics, neotectonic stresses, and hidden uranium mineralization in the area adjacent to the Strel'tsovka Caldera. *Geol. Ore Depos.* **2010**, *52*, 279–288. [CrossRef]

applied sciences

MDPI

Article

Electronic Atlas of Climatic Changes in the Western Russian Arctic in 1950–2021 as Geoinformatic Support of Railway Development

Alexei D. Gvishiani [1,2], Igor N. Rozenberg [3], Anatoly A. Soloviev [1,2], Andrey G. Kostianoy [1,4,5], Sofia A. Gvozdik [1,6,*], Ilya V. Serykh [1,4], Roman I. Krasnoperov [1], Nikolay V. Sazonov [7], Irina A. Dubchak [3], Anton B. Popov [1], Evgenia A. Kostianaia [1,4] and Georgy A. Gvozdik [1,6]

1 Geophysical Center of the Russian Academy of Sciences, 119296 Moscow, Russia
2 Schmidt Institute of Physics of the Earth of the Russian Academy of Sciences, 123242 Moscow, Russia
3 Russian University of Transport, 127994 Moscow, Russia
4 Shirshov Institute of Oceanology of the Russian Academy of Sciences, 117997 Moscow, Russia
5 Sergei Witte Moscow University, 115432 Moscow, Russia
6 Geological Faculty, Lomonosov Moscow State University, 119234 Moscow, Russia
7 Research and Design Institute of Informatization, Automation and Communications in Railway Transport, 109029 Moscow, Russia
* Correspondence: s.gvozdik@gcras.ru

Abstract: The Arctic zone of the Russian Federation is one of the most intensively developing regions of the country. Amongst the major domains of economic and industrial growth and improvement is transport infrastructure and particularly the railway network. This area is being exposed to negative factors of rapid climate change that can significantly affect and compromise this activity. Thus, it is vital to take them into account during design, construction, and operation of the railway infrastructure facilities. This work details the production of a digital atlas comprising the 1950–2021 dynamics of the main hydrometeorological parameters: air and soil temperature, precipitation, wind speed, air and soil humidity, and snow cover thickness. The maps are based on climatic data derived from the MERRA-2 (Modern-Era Retrospective Analysis for Research and Applications, version 2) reanalysis. In total there are 459, which are arranged into 7 chapters. The atlas geographically covers the western part of the Russian Arctic encompassing the regions of quite intensive transport development, which includes the construction of the Northern Latitudinal Railway. Original algorithms of geospatial data processing and their further representation as well as the maps compiled in GIS environment are discussed. Comprehensive analysis of climatic changes in the region of the Russian Arctic including detailed quantitative evaluation over 40 years is given. In the Discussion, we focus on those changes of the regional climate which, from our point of view, are the most significant for consideration by railway operators. The obtained results contribute to framing the theoretical basis of design, development, and sustainable operation of the railway infrastructure in the Arctic and facilitate the decision-making process. This is the first experience of building a specialized climatic cartographic product for the needs of the Russian railways, and to our knowledge the first atlas such as that in the world. In the future, the amassed experience may be transferred to other regions of the Russian Federation as well as similar regions in Canada, Sweden and Highland China that are also subject to significant climate change.

Keywords: Russian Arctic; railway network development; Northern Latitudinal Railway; climate change; MERRA-2 reanalysis; digital atlas; geoinformatics

Citation: Gvishiani, A.D.; Rozenberg, I.N.; Soloviev, A.A.; Kostianoy, A.G.; Gvozdik, S.A.; Serykh, I.V.; Krasnoperov, R.I.; Sazonov, N.V.; Dubchak, I.A.; Popov, A.B.; et al. Electronic Atlas of Climatic Changes in the Western Russian Arctic in 1950–2021 as Geoinformatic Support of Railway Development. *Appl. Sci.* **2023**, *13*, 5278. https://doi.org/10.3390/app13095278

Academic Editors: Yosoon Choi and Stefania Pindozzi

Received: 21 February 2023
Revised: 8 April 2023
Accepted: 18 April 2023
Published: 23 April 2023

1. Introduction

The Arctic zone of the Russian Federation (AZRF) is exposed to the factors of climate change that drastically affect numerous natural (seas, lakes, rivers, forests, tundra, landscapes, soils, biodiversity, etc.) and socio-economic (population, demography, human resources, employment, health of people, industry, oil and gas production, mining of coal, metal ores, diamonds, fishery, agriculture, forestry, water management, energy production and transportation, aerial, road, railway and water transport, etc.) systems of this territory [1]. On 10 October 2022, the Russian Federal Service for Hydrometeorology and Environmental Monitoring (Roshydromet) issued the "Third Assessment Report on Climate Change and its Consequences in the Territory of the Russian Federation" [1]. The major conclusion in this report states that the entire territory of the country is warming at an average rate of 0.51 °C per decade while the AZRF is warming at an average rate of 0.71 °C per decade. According to current forecasts, the area occupied by near-surface permafrost in the territory of Russia will decrease by the middle of the 21st century according to the SSP2-4.5 scenario by 22 ± 7% and for the SSP5-8.5 scenario by 28 ± 10% as compared to the period 1995–2014. By the end of the 21st century, this reduction is expected to be at the level of 40 ± 15% and 72 ± 20%, respectively [1,2].

The subarctic zone of Russia hosts vast oil and gas, sea, railway, and pipeline transport infrastructure worth hundreds of billions of dollars. This region being covered with permafrost is also vulnerable to climate change because the thawing of frozen layers due to significant ice content may cause an average soil settlement of 10–20 cm per year [2], which is critical for pile structures and the entire transport infrastructure in general [3,4]. Serykh et al. [5] showed that in 1999–2020, the Republic of Karelia, Murmansk and Arkhangelsk Oblasts (Regions) experienced significant climate warming at a rate from +0.9 °C to +1.5 °C as compared to previous years (1977–1998). A sharp increase in air temperature at a rate from +0.4 °C to +1.0 °C per decade resulted in the displacement of the +2 °C isotherm for 550 km northwards up to the White Sea southern part and induced the total disappearance of average negative temperatures in the Republic of Karelia, Murmansk and Arkhangelsk Oblasts.

The railway infrastructure in the subarctic territories of Russia is naturally operated in extremely difficult geological and climatic conditions, being exposed to the continuous negative impact of various external factors, leading to deformation of railway tracks and damage of artificial structures [1,3]. Thawing of permafrost soils and significant increase in average temperature cause further changes in the water balance of numerous rivers and lakes in this region. These processes intensify coastal abrasion, erosion, mudflows, floods, landslides, ground creep, rockfalls, rockslides, karst sinkholes, snow avalanches, etc. [1,3,5,6].

Isolated permafrost zones may still be found on the Kola Peninsula [7]. Russian railway facilities in this region are particularly vulnerable to the negative factors of regional climate change, considering their intense development. It is planned that the Murmansk section of the Oktyabrskaya Railway will increase transportation from 28 to 44 million tons per year by 2023, and by 2035 it should grow up to 100 million tons. At the same time, some railway sections are still single-track, which limits their carrying capacity [8].

Extensive railway infrastructure is located in the Northwestern and Ural Federal Districts of Russia. The Northwestern Federal District (NFD) covers 9.8% of the country's territory; 9.5% of the Russian population lives here [9]. In this district, the railway infrastructure runs through the territory of the Republic of Karelia, Murmansk and Arkhangelsk Oblasts, the Komi Republic, and the Nenets Autonomous Okrug (District).

Let us consider the main aspects of the NFD economic development. The turnover of organizations of all types of economic activity in the first quarter of 2022 amounted to 175.7 billion dollars, or 176.2% of the level of the first quarter of 2021. Sales per capita were equal to 1110 dollars (the average parameter in Russia was 1002 dollars). The volume of investment in fixed capital in the economic and social development of the district in the first quarter of 2022 amounted to 5.4 billion dollars, or 103.4% of the level of the corresponding period in 2021. The consolidated budget of the NFD regions had a surplus of 2.3 billion dollars in the first quarter of 2022. Thus, revenues amounted to 7.5 billion dollars, and expenditures were 5.3 billion dollars.

The Ural Federal District (UFD) covers 10.6% of the territory of Russia; 8.4% of the Russian population lives here [10]. In this district, the railway infrastructure runs through the territory of Khanty-Mansi and Yamalo-Nenets Autonomous Okrugs. The turnover of organizations of all types of economic activity in the first quarter of 2022 was 147.3% compared with the same period in 2021, or 134.3 billion dollars. Sales per capita were equal to 949 dollars. The volume of investment in fixed assets of the district economy and social sphere in first quarter of 2022 amounted to 11.1 billion dollars, or 118.7% of the level of the corresponding period in 2021. The consolidated budget of the UFD regions in the first quarter of 2022 had a surplus of 1.6 billion dollars. Thus, revenues amounted to 6.6 billion dollars, and expenditures were 5.0 billion dollars.

The abovementioned economic parameters indicate the accelerated development of these territories. In this regard, an extensive network of railways and smooth operation of their infrastructure has become increasingly important. For example, the current Northern Latitudinal Railway (NLR) project is critically important for the development of the Yamalo-Nenets Autonomous Okrug (YaNAO). NLR is a 686 km long railway that is being built along the line Obskaya–Salekhard–Nadym–Novy Urengoy–Korotchaevo in Figure 1. This projected railway will connect the western and eastern parts of the YaNAO [11]. It will sustain comprehensive economic growth of the northern territories, provide infrastructure for the development of gas-condensate and oil fields, and ensure the transportation of extracted natural resources. The implementation of the NLR project will form the infrastructure that will contribute to the expansion of the tanker fleet and efficient development of the Arctic resources. NLR will facilitate the direct access to the Northern Sea Route through the port of Sabetta in the Yamal Peninsula. This project will also produce jobs for the Russian railways and servicing industries [12]. After the project completion, the estimated traffic volume will be about 24 million tons, primarily as gas condensate and oil cargo.

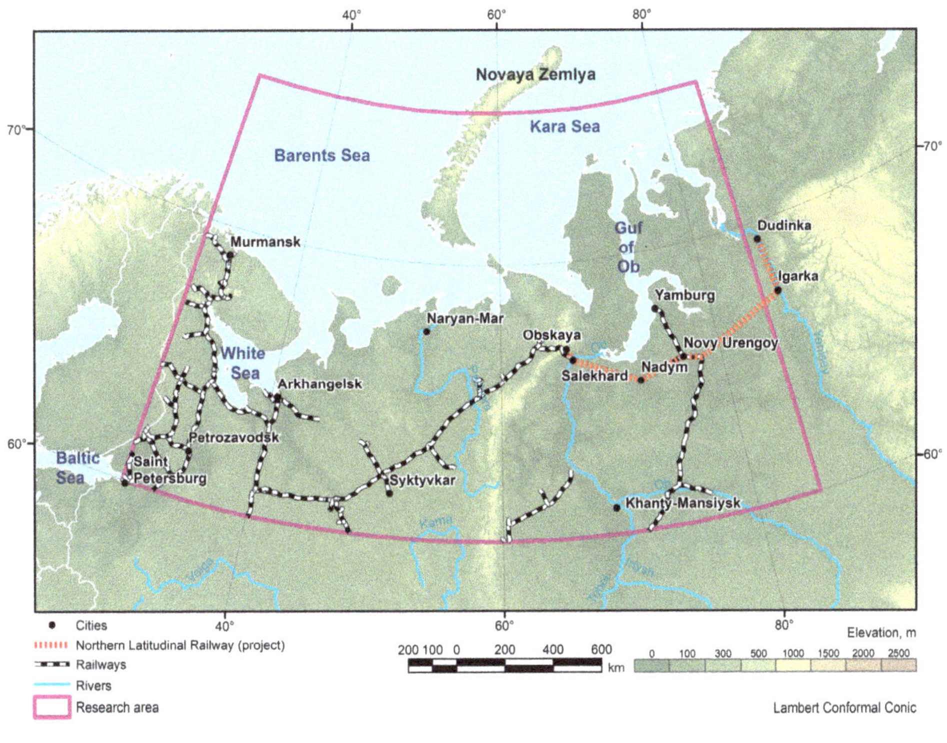

Figure 1. Studied area.

As the part of the NLR implementation, construction of the Bovanenkovo–Sabetta railway section or the so-called NLR-2 project is planned. In the future, up to 2030, it is planned to extend the railway from Korotchaevo eastwards to the Yuzhno-Russkoye oil and gas field (122 km) and from this point to the port of Igarka via Ermakovo (482 km). As a farther perspective, there are plans to build the Igarka–Dudinka railway eastwards with a subsequent connection to the Norilsk railway in Figure 1. This project is called the "eastern arm" of the NLR.

In this regard, research on climate change in the Republic of Karelia, Murmansk and Arkhangelsk Oblasts, Komi Republic, Yamalo-Nenets and Khanty-Mansi Autonomous Okrugs using modern geoinformatic tools is extremely important for the Russian railways' operation.

The core goal of the present work is the production of the digital atlas for tracking climatic variations of basic hydrometeorological parameters in the western part of the Russian Arctic (60–75° N, 30–85° E) over 1950–2021 based on the MERRA-2 atmospheric reanalysis dataset. The article describes the main hydrometeorological parameters forming the atlas: surface air temperature, total precipitation, wind speed at the Earth's surface, soil temperature, soil moisture content, air humidity, and snow cover thickness. The methodology of the map compilations is described in detail. For each of the parameters, we present its brief characteristics, methods of measurement, and provide examples of the different map types. Comprehensive assessment of regional changes in climatic parameters was previously performed for the Russian Arctic, e.g., for the Barents Sea [13]. The most significant results of studying the inter-annual variability of certain climatic parameters directly affecting the smooth railway operation are thoroughly discussed. In conclusion, we agree in favor of atlas expansion that will ensure the theoretical basis for sustainable development and operation of the Russian railways in the Arctic zone.

Thus, the novelty of the research is in building of a specialized digital climate atlas for the needs of the Russian Railways in the western part of the Russian Arctic and in the subsequent analysis of change of basic meteorological parameters which may impact operability of the existing railway network and its development in future. Building of the atlas required development of original algorithms of geospatial data processing and their further representation, as well as the maps compiled in a GIS environment based on the MERRA-2 atmospheric reanalysis for the past 40 years. Comprehensive analysis of climatic changes included 7 hydrometeorological parameters presented in different characteristics in 459 maps.

2. Data and Methods

2.1. Studied Area

The studied area includes the western part of the Russian Arctic within the geographical boundaries 60–75° N, 30–85° E in Figure 1. It includes the Republic of Karelia, Murmansk and Arkhangelsk Oblasts (Regions), Komi Republic, Yamalo-Nenets and Khanty-Mansi Autonomous Okrugs (Districts). Within this area there are railway sections between St. Petersburg and Murmansk and Arkhangelsk, railways to Salekhard and Yamburg, the 700km section of the constructed NLR from the Obskaya station (Labytnangi) to Korotchaevo (Novy Urengoy), and the next section to Igarka and further to Dudinka in Figure 1.

2.2. Initial Data for Climatic Atlas Maps

For a production of maps of hydrometeorological parameters (surface air temperature, total precipitation, wind speed at the Earth's surface, soil temperature, soil moisture content, air humidity, and snow cover thickness) presented in the atlas, the authors used the MERRA-2 reanalysis data for the period 1980–2021.

MERRA-2 dataset (Modern-Era Retrospective Analysis for Research and Applications, version 2) [14] represents the results of reprocessing of a massive catalog of information on climatic parameters, such as air temperature and humidity, atmospheric pressure, wind speed, precipitation, etc. Generally, the database is a set of global daily data on the state of the atmosphere and the Earth's surface, collected over several decades. This mission was implemented at the Global Modeling and Assimilation Office (GMAO) under the direction of the U.S. National Aeronautics and Space Administration (NASA).

MERRA was originally created for integration of satellite measurements into a unified climate catalog to facilitate the description of the global hydrological cycle from the Earth's atmosphere to the surface of the planet. This mission was launched in 1979, but by 2016 it was closed due to certain disadvantages: measurement system errors (e.g., non-physical measurement jumps when the observing system changes), imbalances in some atmospheric and terrestrial hydrological parameters, degraded representation of the upper stratosphere, and system limitations that prevented inclusion of new satellite data sources [14].

The main goal of reprocessing and refining the MERRA mission was to provide continuous near-real-time climate analysis (data are published with delays of a couple of weeks). Accordingly, MERRA-2 was designed as a reanalysis that uses the latest developments in data measurement and processing to address the known limitations and disadvantages of MERRA. The reanalysis provides long-term monitoring and comprehensive analysis of Earth conditions by means of system integration of atmospheric, ocean, land, and physical and chemical information. Other updates include information on ozone monitoring and the use of precipitation observations. On the technical side, the MERRA-2 observing system has been equipped with additional instruments that register the following parameters [15]:

- Atmospheric motion vectors from AVHRR (Advanced Very High Resolution Radiometer);
- Surface wind speed from SSMIS (Special Sensor Microwave Imager/Sounder);
- Temperature profiles and ozone characteristics from EOS (Earth Observing System) Aura MLS (Microwave Limb Sounder);
- Total amount of ozone in the column from EOS (Earth Observing System) Aura OMI (Ozone Monitoring Instrument);
- Deviation angle from GPSRO (Global Positioning System Radio Occultation);
- Microwave sounding from ATMS (Advanced Technology Microwave Sounder);
- Hyper spectral infrared radiation from IASI (Infrared Atmospheric Sounding Interferometer) on MetOp-A and MetOp-B (Meteorological Operational Satellites) and from CrIS (Cross-Track Infrared Sounder) on SNPP (Suomi National Polar-Orbiting Partnership);
- Geostationary radiation from MSG (Meteosat Second Generation) SEVIRI (Spinning Enhanced Visible Infrared Imager).

The rest of the observation system remained unchanged as in the original MERRA mission except for improvements of the sensors for measuring surface wind speed and ozone content. These sensors were changed to more advanced models during the mission development.

Today, MERRA-2 is known as one of the most accurate global representations of climatic data for land and ocean [14]. This results from the extensive volume of output data available for processing along with high spatial resolution, which is important for the subsequent modeling of regional weather and climate conditions. For example, Bosilovich [16] demonstrated the high-quality representation of precipitation and temperatures over the American continent derived from MERRA data. Tilinina et al. [17] demonstrated that MERRA in the best way reflects the activity of atmospheric cyclones in the Atlantic-Eurasian sector of the Northern Hemisphere. Bentamy et al. [18] have shown that MERRA is a good representation of the statistical characteristics of turbulent flows between air and sea over the ocean. Luo et al. [19] demonstrated that MERRA-2 displays well the values of sea surface temperature, atmospheric temperature, and humidity over the Atlantic Ocean. Sharmar et al. [20] showed that MERRA-2 uses an advanced scheme for winds interpolation which is important for the ocean surface wave climate. Schubert et al. [21] showed that MERRA-2 reflects extreme temperature events in the Northern Hemisphere very well.

The MERRA-2 output data are presented in the atlas on a regular grid of 0.5° N $\times$ 0.625° E in Figure 2. It is worth to mention that the reanalysis includes data from the updated version of the Goddard Earth Observing System Model Version 5 (GEOS-5), which has an approximate resolution of 50 × 50 km, so the final datasets were spatially interpolated onto the final grid without loss of accuracy.

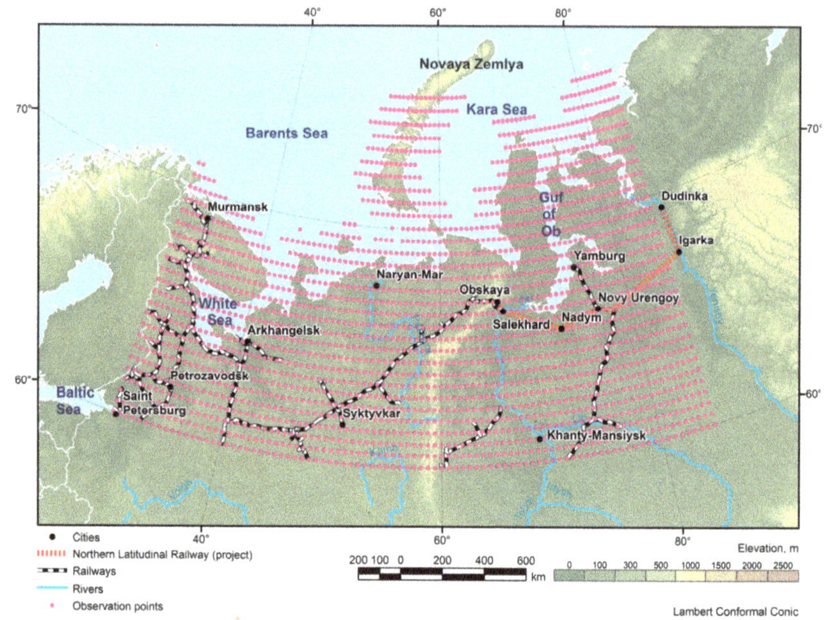

Figure 2. MERRA-2 observation points laid over the studied area.

2.3. Additional Data

This paper also presents data from the NCEP/NCAR Reanalysis 1 Project (National Centers for Environmental Prediction/National Center for Atmospheric Research Reanalysis 1 Project) [22], which is the system of analysis and prediction for air temperature and wind speed from 1948 to the present. On average, data are presented at a frequency of 4 or 8 times per day, but older data were extrapolated depending on existing databases for the selected period. Precipitation estimates were also compared to NOAA's PRECipitation REConstruction over Land (PREC/L) model, which presents data only over land on the $1^\circ \times 1^\circ$ grid. The precipitation parameter is available at 3 different spatial resolutions, and the analysis is determined by interpolating sensor data over the land and reconstructing historical observations over the ocean.

The resulting atlas maps also show the modern railway network of the studied area for the regional climate change assessment along the major railway sections and for the future transportation development planning. The layer is based on the map of the railway network of Russia. This layer was compiled using the Digital Chart of the World datasets (1993 version with 2002 updates) at a scale of 1:1,000,000. The authors of the dataset are the Russian Academy of Sciences (RAS) and International Institute for Applied Systems Analysis (IIASA), Austria [23]. For the final maps, the initial data set was clipped according to the boundaries of the studied area.

2.4. Application of GIS Technologies for Climatic Atlas Compilation

2.4.1. Main Data-Layers

The Golden Software Surfer (version 4.3) [24,25] and ESRI ArcGIS (ArcMap), (version 10.8) [26,27] software packages were chosen as the main tools for the compilation of the atlas maps. The Surfer has a considerably large set of tools for primary data processing, including a user-friendly interface and extensive interpolation capabilities. In turn, the ArcMap software package is designed to solve a wide range of tasks including the transformation of point data on a regular grid (grid-data model) and irregular grid (Triangular irregular network—TIN data models) [28,29] into raster layers. Subsequently, these data are prepared for publication as cartographic services using the geoportal approach [30] or export as formatted map.

In our work, the process of data processing and preparation is shown in Figure 3:

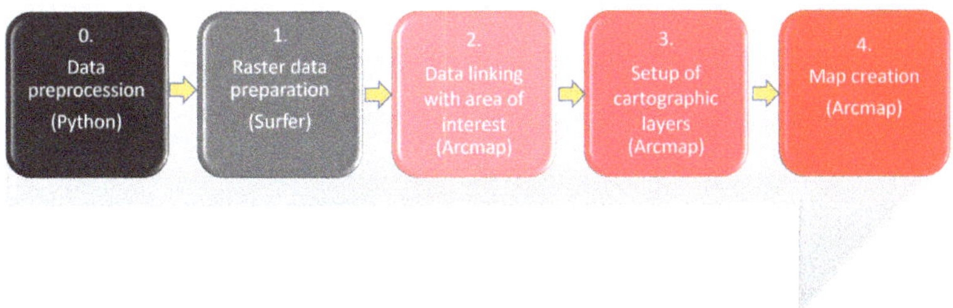

Figure 3. The data processing scheme for the atlas map compilation. The employed software is indicated in brackets.

Further, each block shown in Figure 3 will be discussed.

The preliminary data processing included the so-called "cleaning", i.e., data outside the terrestrial boundaries are removed for several categories of data, e.g., for soil temperature or soil moisture, since the latter can seriously distort the model calculations. For this purpose, the authors developed a software module in Python 2.7 using Python ArcPy and Python.os libraries. The main processing tool is the Clip tool built into the ArcGIS toolkit for raster data processing, which allows for cropping a single raster layer to the specified contour in Figure 4 [31,32].

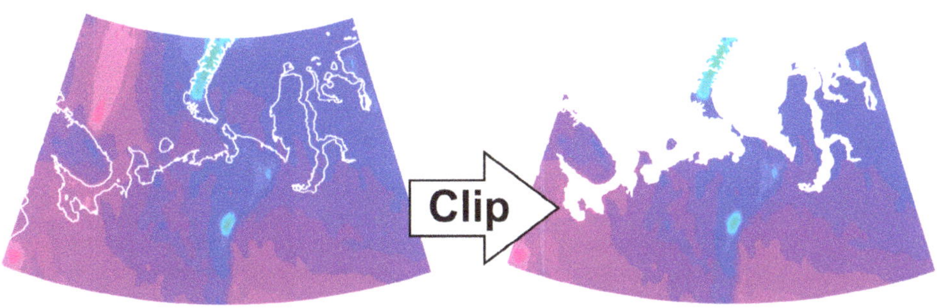

Figure 4. The Clip tool operation principle.

The module has a graphical interface and was integrated within the ArcMap package. The input parameters include selection of the directory of the processed raster data (in .tif format), specifying the vector file with contour mask in shape-file format (.shp) for cropping the data, and the directory for saving the results. The software module processes all raster files one by one in the catalog using the Python.os library. Each layer is cropped by the Clip tool using the contour mask based on the coastline within the considered territory. The results are saved to a specified directory. The source code of this tool is available online [33].

To improve the quality of the output data, the initial model data representing a regular 0.5°-grid were interpolated to reach the 0.05° resolution (1101 × 301 pixels) using kriging. This is a widely used gridding method aimed at producing maps from sparse and irregular data [34,35]. The options and parameters of the tool were selected as follows. The procedure type was the standard point kriging with a variogram estimation. The parameters of standard deviation were calculated automatically without deviation grids and external drift grids. The search options for data points for interpolating grid nodes were defined as follows: the number of sectors to search—4, maximum number of data to use—64 points, minimum number of data to use—8 points; search ellipse radius in data units—28.5, and angle 0°. The interpolated data were then exported to a raster format (.flt) for further processing. This procedure was performed using Surfer. This tool is provided by other software packages, but the choice of the Surfer software was made due to advantages in interpolation procedures operating time and data export without loss of quality. Thus, the resulting data covers the territory within the geographical boundaries 60–75° N, 30–85° E and is shown in Figure 5.

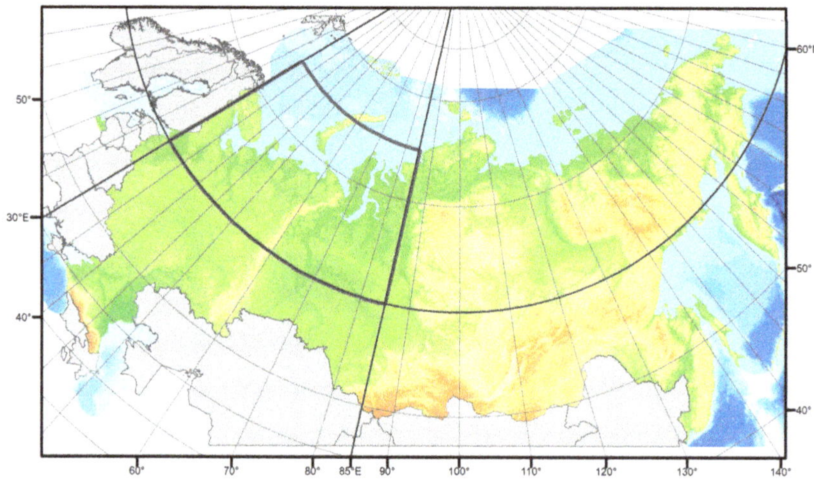

Figure 5. Geographical location of the studied area. Thickened lines show the border of the AZRF (northwards from 60° N), as well as the western and eastern borders of the studied area (30–85° E).

The georeferencing of the data in our case was determined by the choice of the map projection. At this stage, the conical Lambert projection (ESPG: 102027) was considered for the initial raster data [36,37]. The final maps of the atlas were also made in this projection.

The choice of this map projection was made based on the geographical features of the studied area, located in high latitudes. Standard map projections, such as Mercator (ESPG: 4326), distort spatial information in proportion with the increase in latitude. Accordingly, the conic Lambert projection was chosen to compile the atlas maps.

Configuration of data files included defining the color palette for groups of data, as well as selecting the color scale for their clear representation. A single scale was selected for the main data categories (temperature, wind speed, precipitation), as well as for auxiliary data (soil temperature and moisture content, air humidity, snow cover thickness) with the values of average changes in the indicators.

Numerical scales were selected for the values of average rates of changes (the first derivative), preserving a unified color palette for the entire category of data. Isolines with the same interval for the data group of average values and rates of change were constructed for the majority of data groups. For the air temperature values, isolines were selected with a 1 °C interval. For wind speed it was 2 m/s, and for precipitation data—0.2 inch.

At this stage the prepared raster file along with isolines represent a tool for the comparative analysis, both within and between the data groups. As a platform for this analysis, various geoinformation software packages can be used that have the necessary functionality (e.g., ArcGIS, QGIS, etc.). In our case, the comparative analysis was carried out using ESRI ArcGIS (ArcMap) software in Figure 6 [38,39].

(a)

(b)

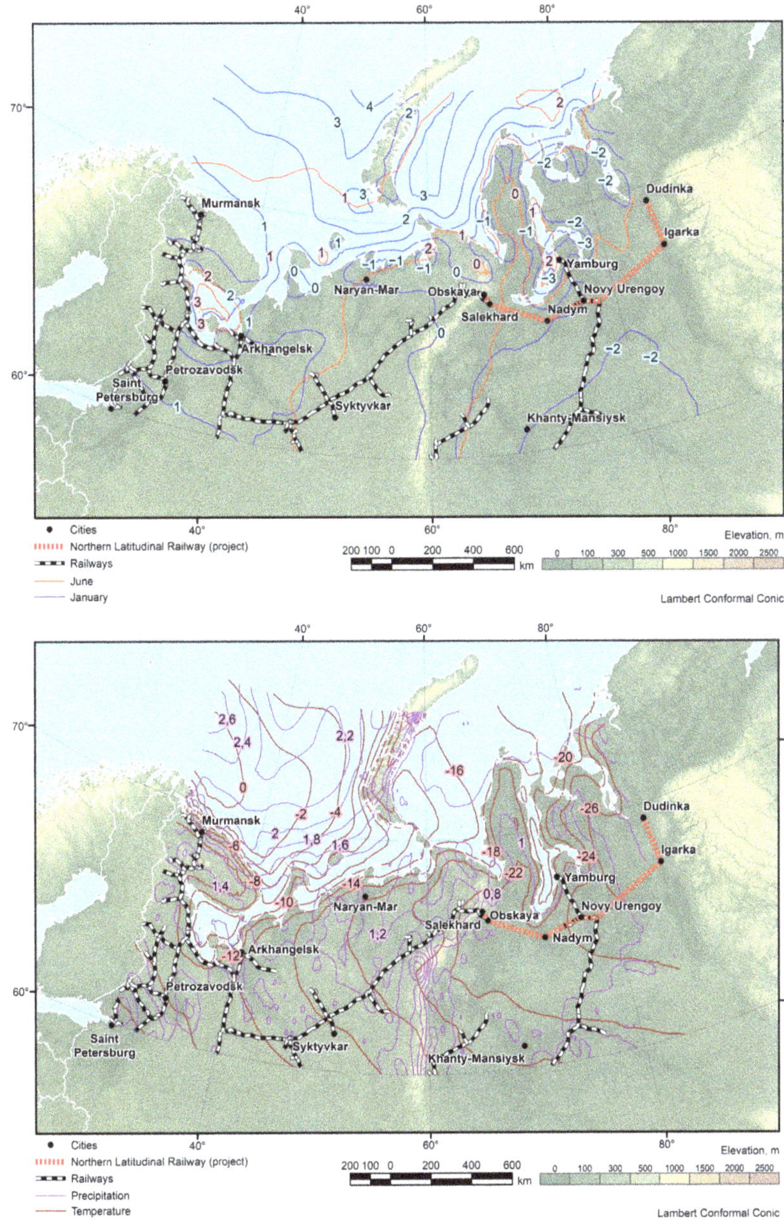

Figure 6. Comparison of spatial data: (**a**) within data groups (information on the average change in air temperature between the periods 1980–1999 and 2000–2021, comparison between the months January and July); (**b**) between data groups (comparison of average air temperature and precipitation for the period 1980–2021 for January).

The final preparation of data included their compilation into finished maps and exporting into the atlas layout. At this step, we added the coordinate grid along with the marginal information, including the legend, color scale, numerical and linear scale. For all groups of data, GIS projects in the MXD (map exchange document) format were created. Each of them contains all the cartographic materials related to a particular group of data for quick access and editing in Figure 7. Subsequently, the maps were exported into PDF for further compilation of the atlas.

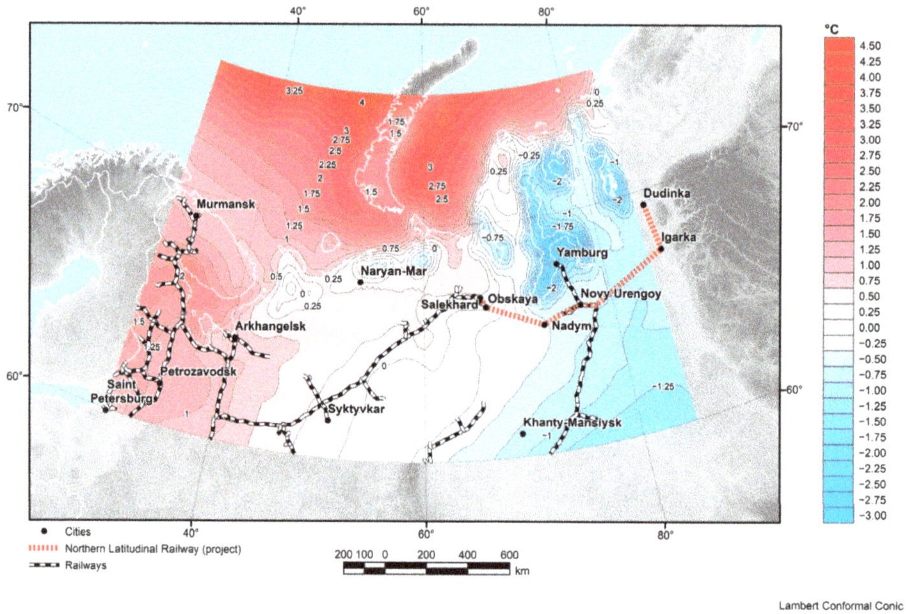

Figure 7. Example of the map design. Average air temperature changes (°C) between the periods 1980–1999 and 2000–2021 for the winter season (December–February).

2.4.2. Climatic Atlas Structure

The final version of the climatic atlas is a set of digital maps of hydrometeorological parameters, divided into categories. The scheme of the atlas compilation is presented in Figure 8.

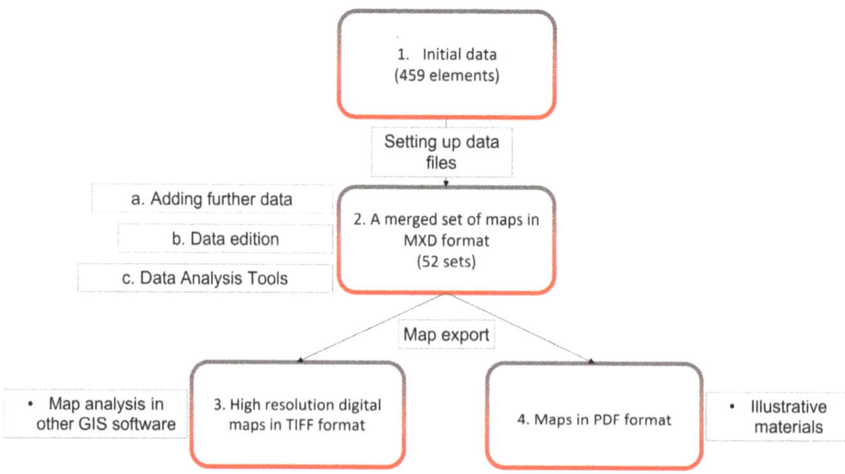

Figure 8. Scheme of the climatic atlas compilation.

For each category of data, source maps were obtained (459 files in total) and subsequently processed. At the initial stage of processing, these source cartographic materials do not represent a valuable cartographic unit. The processing and preparation of maps of hydrometeorological parameters for the subsequent atlas compilation was mentioned above.

The raster layers were combined into MXD files consisting of raster layers of basic data categories (air temperature at ground surface, total precipitation, wind speed at ground surface), as well as auxiliary data (soil temperature and moisture content, air humidity, snow cover thickness), divided into groups:

- Average values for the entire period 1950–2021 and separately for the periods 1950–1979, 1980–1999, and 2000–2021.
- Average changes between the periods 1950–1979 and 1980–1999, 1950–1979 and 2000–2021, 1980–1999 and 2000–2021 (difference in average values between these periods, with the earlier period always subtracted from the later one).
- The rate of change of mean monthly anomalies relative to the annual rate for the periods 1950–2021, 1950–1979, 1980–1999, and 2000–2021, estimated using the 1st derivative.
- Mean values for the winter (December–February) and summer (June–August) seasons and for the 12 months of the year for the entire period 1980–2021 and separately for the periods 1980–1999 and 2000–2021.
- Average changes for the winter (December–February) and summer (June–August) seasons and for the 12 months of the year between the periods 1980–1999 and 2000–2021 (the difference of average values between these periods, with the earlier period always subtracted from the later one).

In each data category group, auxiliary data are presented. Thus, each data category group represents a separate MXD project. The total amount of files of merged raster layer projects was 52, each available for processing in ESRI ArcGIS (ArcMap).

This approach is flexible, i.e., additional data (at the user's discretion) can be added to each of the prepared projects for more detailed data analysis/comparison. In this version of the climatic atlas, data editing, color palette changes, and data analysis tools are also available. As the main analysis tool, a raster calculator can be used [40,41]. The principle of the tool is to perform pixel-by-pixel arithmetic operations with two raster layers. An example of this tool operation is shown in Figure 9.

Along with this, the data isoline overlay mentioned above can be used as a tool for spatial data analysis.

The atlas maps were also presented in high resolution TIFF (Tagged Image File Format) format that can be applied for further analysis in other geographic information software. Although ArcMap is an optimal software solution with many built-in tools to address a wide range of tasks, it also has certain disadvantages. The obvious one is the commercial distribution of this software. The TIFF format is universal, ready for handling in any GIS software, including the open source one. The TIFF maps were converted into PDF (Portable Document Format) files. This format is intended for convenient visual assessment of cartographic information and can serve as illustrative material for a variety of purposes.

(a)

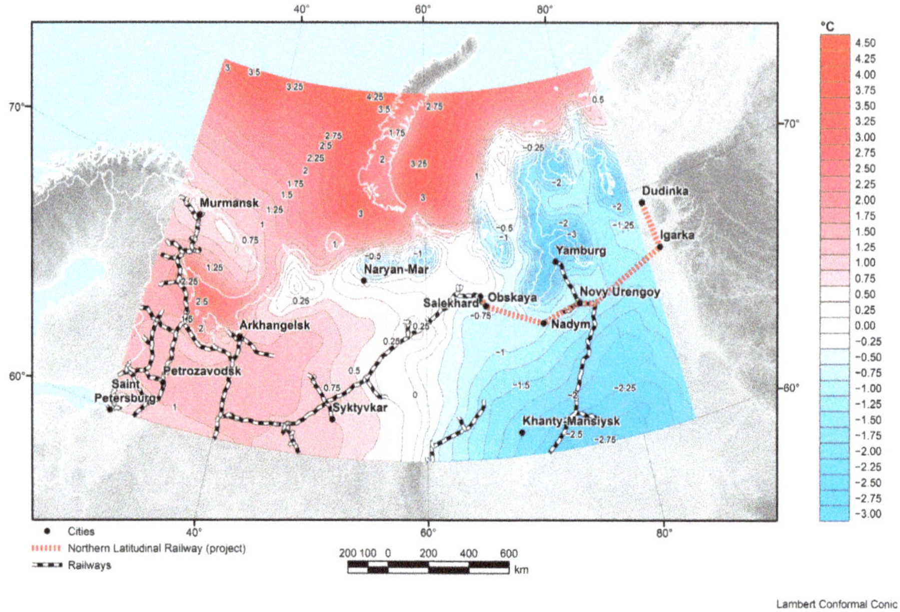

(b)

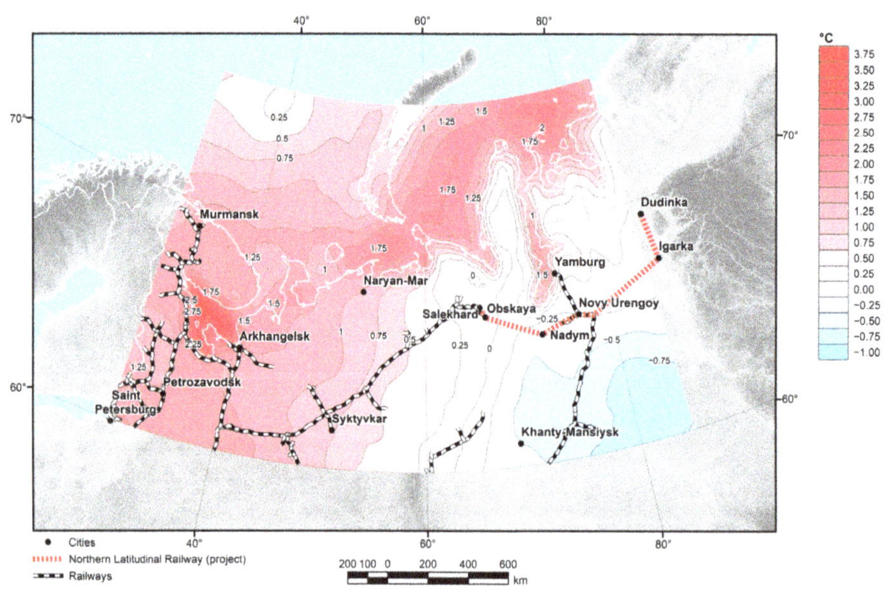

Figure 9. *Cont.*

(c)

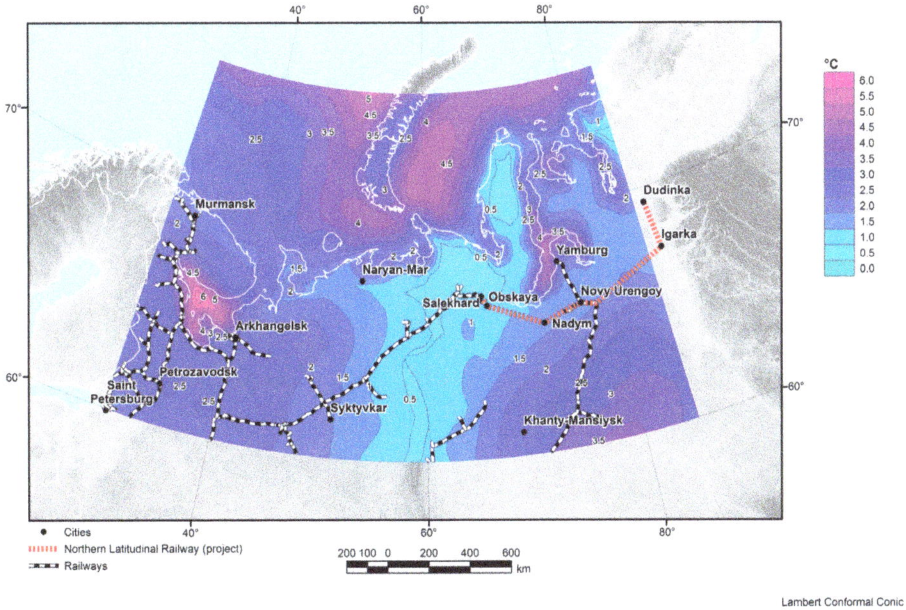

Figure 9. The result of the raster calculator tool application. The data represents the average change in air temperature (°C) between 1980–1999 and 2000–2021: (**a**) data for January; (**b**) data for July; (**c**) the resulting amplitude map of the average temperature between the two months, January and July.

3. Results

3.1. Climatic Parameters Selected for Mapping

The created atlas consists of the groups of electronic maps reflecting seven climatic parameters: air temperature, total precipitation, wind speed (main parameters); soil temperature, soil moisture content, air humidity, and snow cover thickness (auxiliary parameters). The first three parameters contain the key information necessary for the analysis and prediction of climatic changes in the region in general [42]. The auxiliary parameters facilitate the assessment of climate change in relation to soil characteristics, snow cover thickness, and air humidity [43].

Let us describe the main groups of the resulting maps and the ways of processing the information for their construction. For each of the parameters, the input data are digital arrays containing the coordinates (with a spatial resolution $0.5° \times 0.625°$) and parameter values. These arrays are further structured by time intervals: for main parameters from 1950 to 2021, and for auxiliary parameters from 1980 to 2021.

While calculating various characteristics, the used data had certain time sampling, which was as follows:

- 1950–2021—the entire time interval according to NCEP/NCAR Reanalysis 1;
- 1980–2021—the entire time interval according to MERRA-2 reanalysis data;
- 1980–1999—period when the observed values were less accurate (according to MERRA-2 reanalysis data);
- 2000–2021, the period with the best resolution and modern observation system (according to MERRA-2 reanalysis data).

By overlaying the information in each file on the grid, the characteristics were calculated in each node of the grid within the defined period [42]. Each grid node contains data averaged for its vicinity $\pm 1.25°$.

For the main parameters (air temperature, total precipitation, wind speed), the characteristics were calculated starting from 1950, since these parameters were measured in sufficient detail even before the modern satellite observation network was created. An example of such an initial digital map is shown in Figure 10.

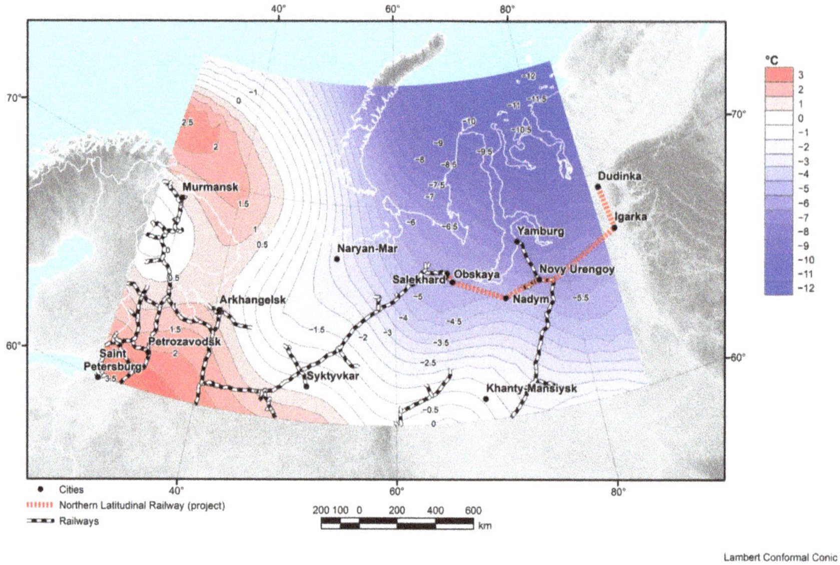

Figure 10. Map of average air temperature (°C) at 42.2 m altitude for the period 1950–2021 according to MERRA-2 data.

For each combination of climatic parameters, the following groups of characteristics were calculated when creating the initial digital map:

- Arithmetic means for individual time periods;
- Average parameter changes between the periods 1980–1999 and 2000–2021;
- Average values for each month separately (12 maps);
- Average values by season: summer (June–August), and winter (December–February);
- Average values between the periods 1980–1999 and 2000–2021 for each month;
- Average values between periods 1980–1999 and 2000–2021 by season—summer (June–August), and winter (December–February);
- Average rate of change of mean monthly characteristics.

The anomalies shown on the maps for different months allow us to distinguish local changes, as well as to correlate different parameters in time. In this group, the average values are taken as a sample for each day of a certain month for each year of the selected time interval.

Since the studied area is the northwestern part of Russia, this zone is characterized by strong temperature fluctuations and, consequently, by strong changes in other parameters. The seasonal analysis makes it possible to trace the intra-annual variability of the parameters.

Moreover, there are the maps of differences in values of individual parameters between the periods 1980–1999 and 2000–2021. As this group initially carries information about anomalies for the two certain periods and compares them, it allows for not only tracing regional changes of climatic parameters, but also giving a quantitative assessment of climate change on scale of almost half a century.

The rate of change of the specified parameters is calculated as the first derivative over time (X) for the data set in Figure 11 [5]. The rate of change (a) of any parameter (Y) was calculated using the linear regression equation:

$$Y(X) = a \cdot X + b.$$

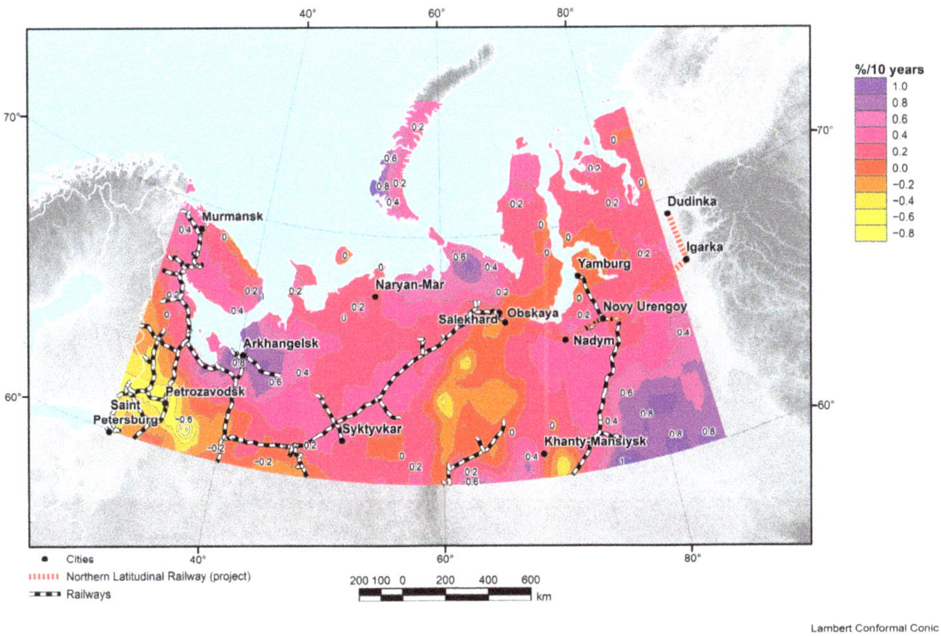

Figure 11. Average rate of change in mean monthly anomalies of the soil upper layer moisture content (% per 10 years) for the period 1980–2021.

This characteristic of temporal variability allows for defining how fast the studied parameters change.

In total, 459 maps with different characteristics of the specified climatic parameters within the selected time intervals were compiled.

3.2. Air Temperature

Air temperature is the parameter that reflects the degree to which air is heated. Historically, air temperature was one of the first climatic parameters that scientists began to measure. When satellite missions were launched, they began to measure atmospheric temperature at various altitudes, e.g., sea surface or land surface. The most common method for satellite measurements is obtaining data using radiometric sounding. For this purpose, instruments that measure radiation in different wavelength ranges, most commonly infrared radiation allowing for calculating the temperature at specified heights, have been developed. The MERRA-2 reanalysis used upgraded instruments to measure vertical profiles of parameters such as air temperature and humidity—the CrIS Cross-Track Infrared Sounder (CrIS) and the Advanced Technology Microwave Sounder (ATMS) [19].

The CrIS Cross-Track Infrared Sounder is one of the most advanced hyper spectral instruments in the National Oceanic and Atmospheric Administration (NOAA) Joint Satellite System. CrIS is a high-resolution infrared spectrometer of which the operating principle is based on the separation of infrared energy emitted by the atmosphere, resulting in high vertical resolution. The instrument provides atmospheric sounding within 2,211 spectral channels in three wavelength bands: long-wave LWIR (9.14–15.38 μm), medium-wave MWIR (5.71–8.26 μm), and short-wave SWIR (3.92–4.64 μm) [44].

The Advanced Technology Microwave Sounder (ATMS) is a 22-channel scanning microwave radiometer for atmospheric and Earth surface observations, which makes observations in the microwave part of the electromagnetic spectrum. ATMS and CrIS provide data on the water cycle, namely, water vapor, clouds, and precipitation. Since clouds are non-transparent in the infrared part of the spectrum (as measured by the CrIS instrument), the two instruments work in combination to cover a broader range of weather conditions. ATMS provides a view inside and under clouds and can be used to study storms and hurricanes from the inside [45].

The MERRA-2 catalog presents temperature values for various altitudes: 2, 10, 42, 72 m, and at "surfaces" where atmospheric pressure is 250, 500, and 850 hPa. The atlas used air temperature values at 42.2 m above the ground surface. This makes it possible to analyze the data without taking into account the influence of wind, snow, and other surface factors.

If we use the compiled maps for the purpose of climate forecasting, the general tendency of average air temperature increase since 1980 is clearly observed in Figure 12. This map shows the rate of change of the average annual air temperature. All values are positive, which means that for the specified time period, the temperature values are exclusively increasing. Additionally, we can distinguish the general trend—the more northward, the stronger the temperature changes are. In the area of the Novaya Zemlya Archipelago, the values of average annual temperature increase by 1 °C per 10 years.

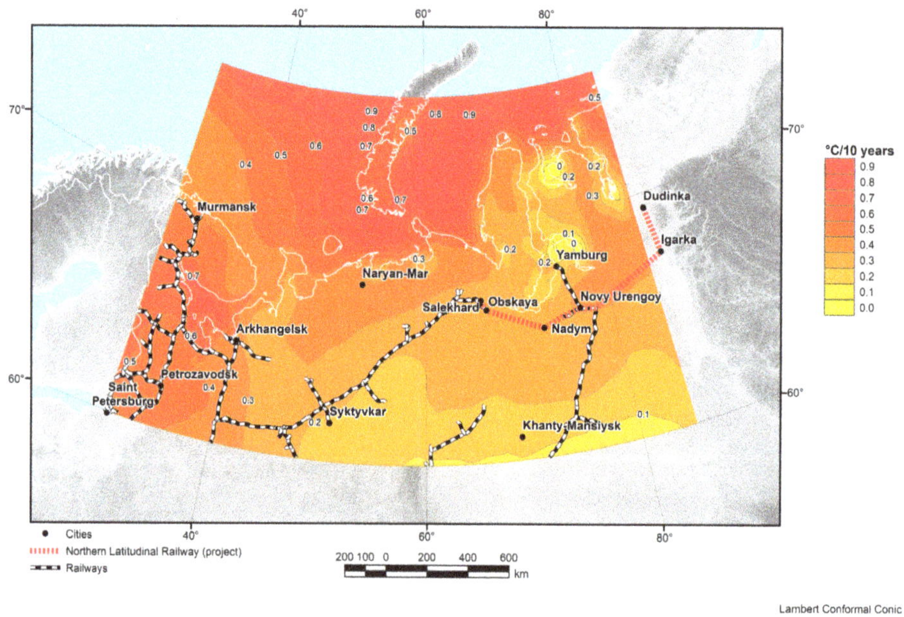

Figure 12. Average rate of change in mean monthly anomalies (relative to the annual rate) of air temperature at 42.2 m (°C per 10 years) according to NCEP/NCAR Reanalysis for the period 1980–2021.

Since the studied area is located in the north, we can note extreme temperature values in winter periods, as well as characteristic areas where the cooling is not in contrast. Figure 13 shows a map of air temperature values for the whole period for January. The data indicate that the general trend of decreasing temperatures is directed from west to east—from near-zero temperatures north of the Kola Peninsula, to −26 °C in Taimyr and from −20 °C to −24 °C in Siberia. Zero temperatures, however, are present only in the oceanic part of the territory, which is explained by the influence of warm ocean currents. On land, this influence disappears, and temperatures range from −8 °C to −10 °C.

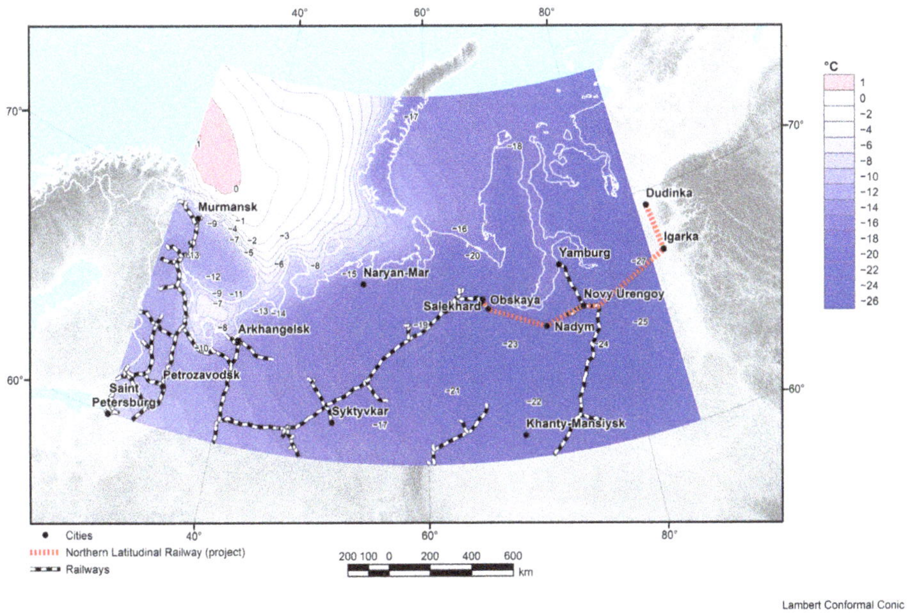

Figure 13. Mean values of air temperature (°C) at 2 m for January according to MERRA-2 data for the period 1980–2021.

3.3. Total Precipitation

Precipitation is a value that describes the height, in millimeters, of a water layer that would form on the surface of the Earth without external influence. The first measurements of this parameter were made for the purpose of storm analysis, rain distribution, and rainfall forecasting. The first satellite measurements of precipitation were made using radar to produce three-dimensional maps of storm structure and to calculate the altitudes where the phase transition from snow to rain occurs. Since one of the most important climatic processes is the water cycle on Earth and in the atmosphere, such data allowed qualitatively to complement and improve models of global atmospheric circulation. The most common methods of precipitation measuring nowadays are either microwave sensors or ground-based observatories. Microwave sensors measure the energy emitted by the atmosphere or the ground, and extract quantitative characteristics of water vapor, water in clouds, and the intensity of precipitation in the atmosphere from the signal. There are sensors that additionally allow estimating both the rate of precipitation and the geometric characteristics of water particles.

The MERRA-2 reanalysis mainly contains data from ground stations, collected in the Global Historical Climatological Network (GHCN) database, and in the Climate Anomaly Monitoring System (CAMS) database. In these catalogs, there is a basic division of information by ground and ocean stations, which is compared in parallel with similar published data sets from satellite missions.

In the reanalysis, there is a separate processing block to account for inter-annual variability in ocean evaporation [46,47].

It is also worth noting that MERRA-2 initially uses precipitation data based on ground-based observations, which are further archived as an output variable and input parametrically to values obtained from satellite data. A description of the satellite instruments obtaining atmospheric precipitation values is given above in Section 2.2. In addition, some observation areas use CMAP satellite sensor results due to limitations in available observations [48].

Consequently, the results are provided in the following variations: maximum and minimum precipitation rates for the period, total precipitation, convective precipitation, large-scale precipitation, snow, total precipitation from the atmospheric model, evaporation totals, and corrected large-scale and total precipitation. The adjusted total precipitation data were used in Figure 14. This map clearly identifies the zone with the maximum amount of precipitation, belonging to the zone of the Ural Mountains, and the minimum values are confined to the water area of the Kara Sea.

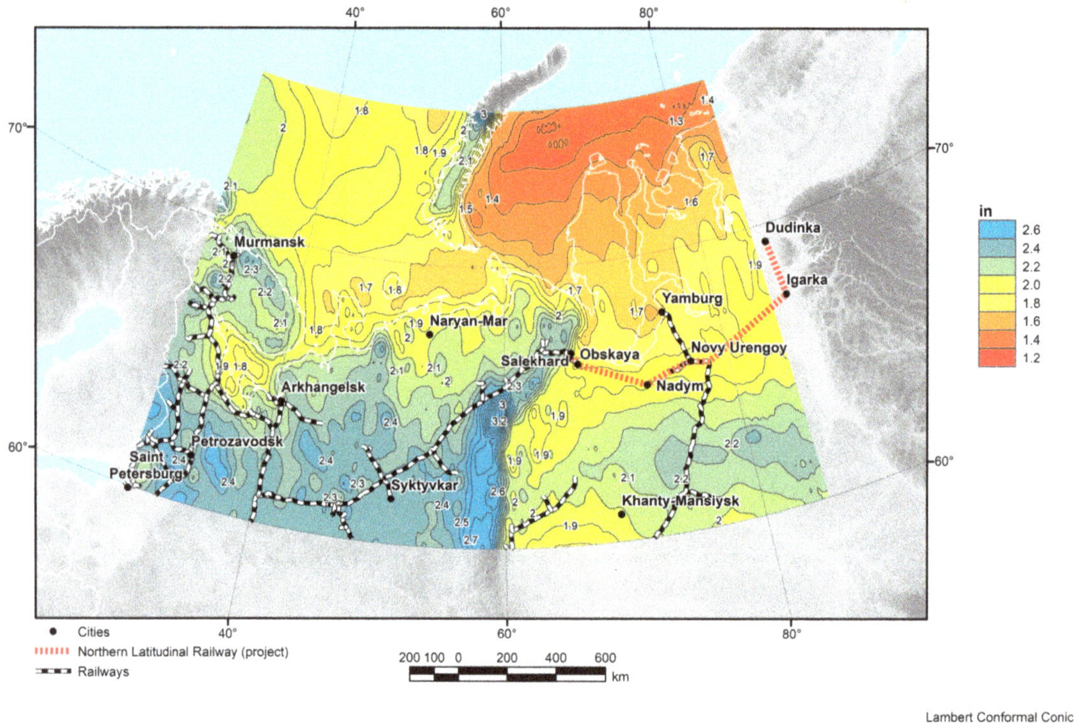

Figure 14. Average values of total precipitation per day (in) according to MERRA-2 for the period 1980–2021.

3.4. Wind Speed

Note that in satellite measurements of wind speed, this value is decomposed into orthogonal components—east and north. Using them, the total vector of wind speed is determined. The measurement methods are often similar to scattermeter devices, the principle of which is based on the reception of signals reflected from the sea surface, and further analysis of the intensity of the reflected wave [14].

Since the wind speed value is not only a climatic parameter, but also a parameter for introducing corrections to other parameters, the MERRA-2 reanalysis uses sensors from both ground-based observatories (UCAR and NCEP) and data from a large number of satellite systems [49]:

- AVHRR atmospheric motion vector, 1 October 1982–present, CIMSS;
- SSM/I surface wind speed, 9 July 1987–4 November 2009, RSS;

- ERS-1 surface wind vector, 5 August 1991–21 May 1996, ESA;
- ERS-2 surface wind vector, 19 March 1996–29 March 2011, ESA;
- QuikSCAT surface wind vector, 19 July 1999–22 November 2009, JPL;
- MODIS atmospheric motion vector, 2 July 2002–present, CIMSS and NCEP;
- SSMIS surface wind speed, 23 October 2003–29 October 2013, RSS;
- WindSat surface wind vector, 13 August 2007–4 August 2012, NCEP;
- ASCAT surface wind vector, 15 September 2008–present, NCEP.

The MERRA-2 catalogs have wind velocity values available for two vectors (meridional and zonal wind) at 2, 10, 50, and 72 m altitudes, at the daytime surface, and at "surfaces" where pressure is 250, 500, and 850 hPa [48]. There is also information on trends in inter-annual zonal and meridional wind variability. The atlas presents various wind speed characteristics at 50 m altitude. One of the maps is shown in Figure 15, and displays the average wind speed for the entire time interval (1980–2021).

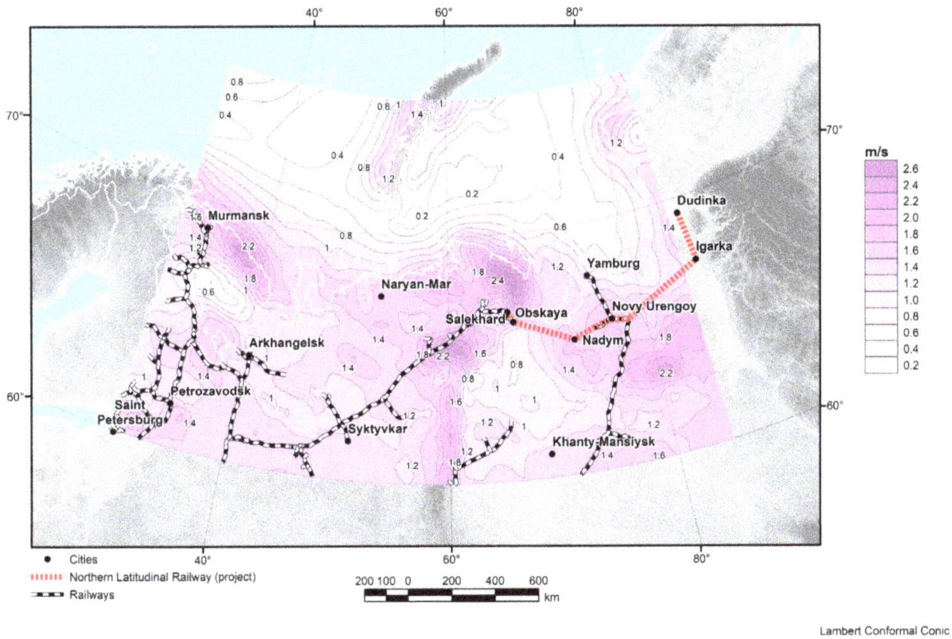

Figure 15. Average wind speed (m/s) at 50 m for the period 1980–2021.

Analysis of average wind speed by months showed a strong seasonal dependence—the highest speeds up to 5 m/s in winter (extremes in December) in Figure 16a and the lowest speeds in summer—values up to 0.4 m/s and less in the regions of the White Sea and Arkhangelsk in Figure 16b.

3.5. Soil Temperature

Soil temperature is a characteristic of the soil upper layer (up to the first meters), reflecting the temperature in a layer of selected thickness. There are special thermometers and remote sensing methods for measuring soil temperature. Physically, the devices emit a signal, which is subsequently recorded by sensors after scattering or reflection from the surface.

Soil temperature in MERRA-2 is measured remotely once per hour, which provides observation of short-scale changes in the parameter [50]. The reanalysis presents soil temperature for different layers: 0–0.1, 0–0.2, 0–0.4, 0–0.75, 0–1.5, 0–10.0 m [48]. The atlas presents soil temperature maps of the Arctic zone of the northwestern Russia, where permafrost zoning is quite apparent in Figure 17. The thickness of the measured layer was chosen as a maximum depth of 10 m, because these areas are characterized by short-term temperature changes, affecting the assessment of global trends in the parameter.

(a)

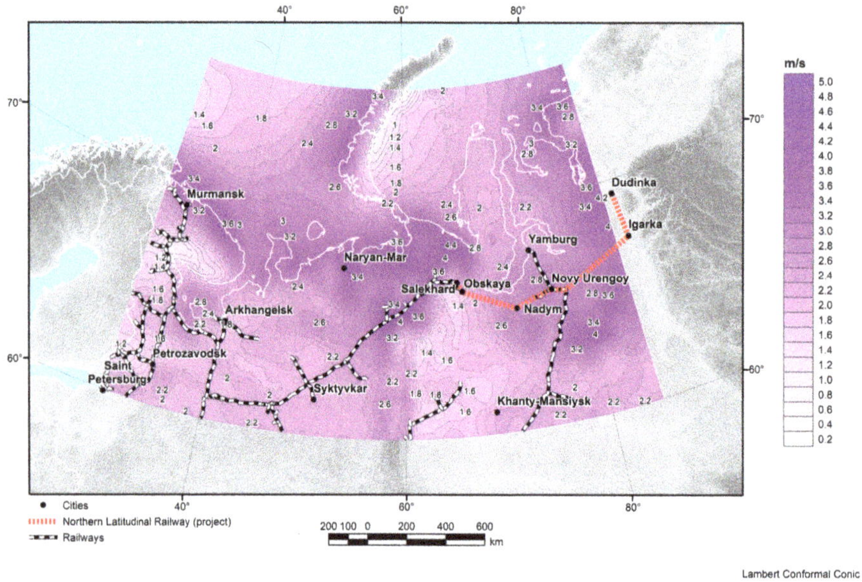

(b)

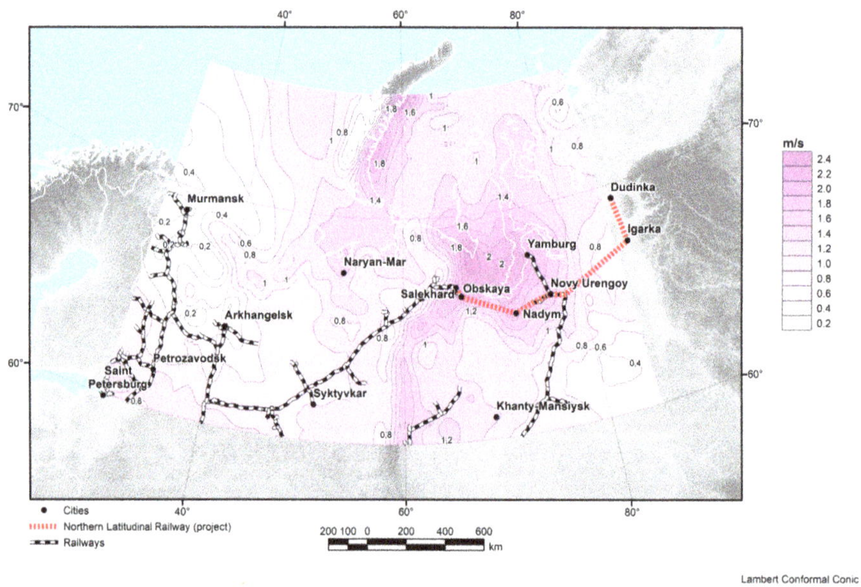

Figure 16. Average values of wind speed (m/s) at 50 m for the period 2000–2021: (**a**) for winter season (December–February); (**b**) for summer season (June–August).

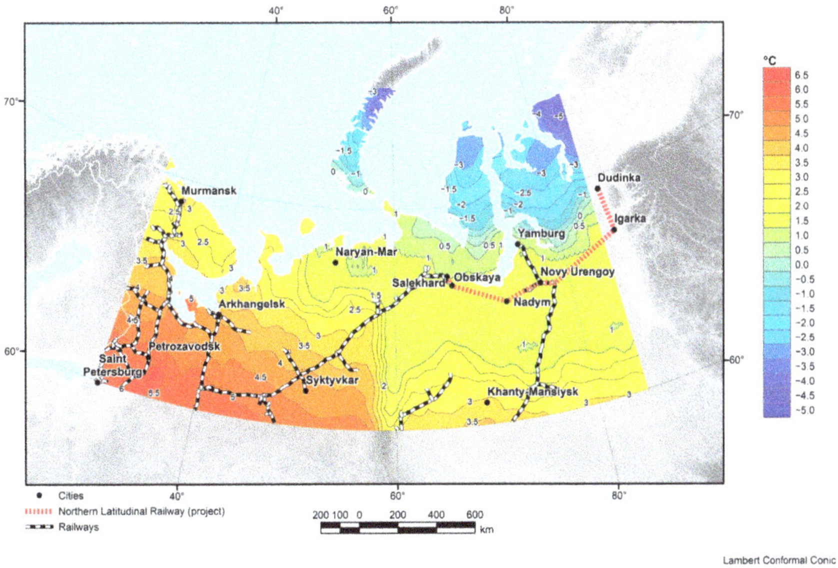

Figure 17. Mean values of temperature (°C) of the upper 10 m of soil for the period 2000–2021.

3.6. Soil Moisture Content

Soil moisture content is the percentage of water in the soil compared to dry soil.

For satellite observations, radar and a radiometer are required to obtain an adequate solution. For the MERRA-2 system, soil moisture data were provided using the Soil Moisture Active Passive Observatory (SMAP) system, which is a radiometric and radar instrument.

The approach to soil moisture measurement uses a combination of the radar spatial resolution and the radiometer accuracy with simultaneous measurements of surface radiation and backscattering. Instruments measure parameters in the upper layers of soil to provide a global assessment of the soil moisture. Since some parts of the land are heavily covered with vegetation, the calculations include an automatic algorithm for extrapolation between the values obtained at different times of the day at different positions of the satellite relative to the Earth's surface.

The final reanalysis data provide soil moisture values in different units [51]. The first is in dimensionless units of relative saturation for different layer depths. The second is soil moisture content in volumetric units of m^3/m^3, considered as the volume of water in the soil volume (including all solid material, water, and air). In both cases, soil moisture variables are provided for the top 0–100 cm layer. Summary data are also presented for different soil layers, 0–5, 10–100, and 134–853 cm [48]. An example of the soil moisture values is shown in Figure 18. The atlas includes 63 different soil moisture characteristics based on the analysis of data for the period 1980–2021.

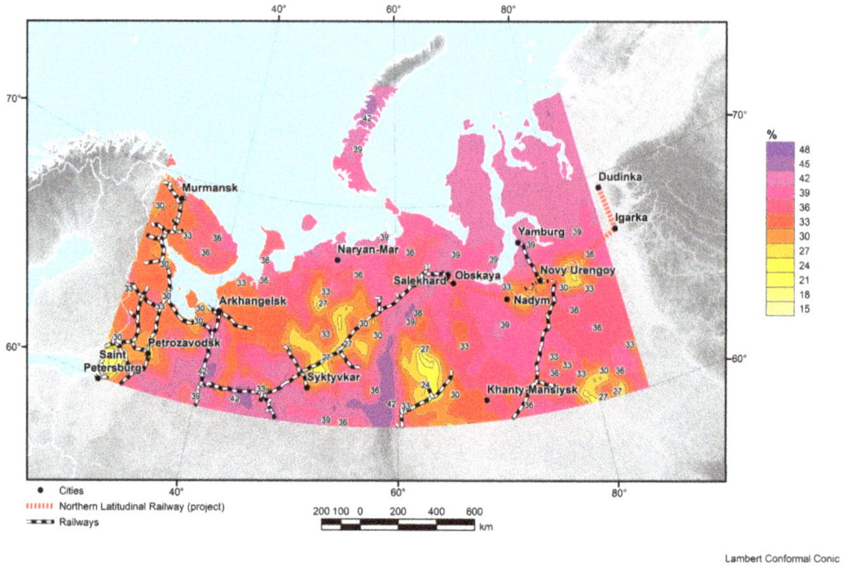

Figure 18. Average values of the soil upper layer moisture content (%) for the period 1980–2021.

3.7. *Air Humidity*

Air humidity characterizes the water vapor content in the atmosphere. The MERRA reanalysis used pseudo-relative humidity [52], which was determined through the ratio of water vapor mixing to saturation value.

However, MERRA-2 uses a new approach—the normalized pseudo-relative humidity [53], which is determined by normalizing the pseudo-relative humidity to the standard deviation of the background error, which has a near-Gaussian distribution. Physically, MERRA-2 produces consistent time series of the total amount of water in the atmospheric column and the transport of water from the ocean to the land.

It is possible to obtain types of air humidity as output parameters in the atlas, such as effective specific humidity at the surface, specific humidity using mixed estimation, relative humidity, estimates of the general trend of humidity, and humidity at different heights: 2, 10, 42, 72 m, and at "surfaces", where the pressure is 250, 500, and 850 hPa [48]. An example of the air humidity distribution is shown in Figure 19. The atlas presents an array of maps for specific air humidity at 2 m.

3.8. *Snowcover Thickness*

The thickness of the snow cover is commonly referred to the thickness of the layer of snow covering the surface of the ground. When measuring the snow cover thickness, there is a need for additional adjustments, which must take into account the characteristics of the snow. It is necessary to measure the thickness of the already compacted layer, and not at the moment of fall, when the melting process in the outer medium is possible. A solution to this problem is presented in [54], where a comprehensive dataset for the northern hemisphere on permafrost with a resolution of 81 km is evaluated. The snow cover thickness available in MERRA-2 is recorded only within the territory covered with snow [55]. The reanalysis presents characteristics such as the snow cover adjusted with displacement, total snow mass, snow mass above the ice surface, and snow thickness [48]. The atlas presents various characteristics of snow thickness in meters, e.g., the average values of snow cover thickness in Figure 20.

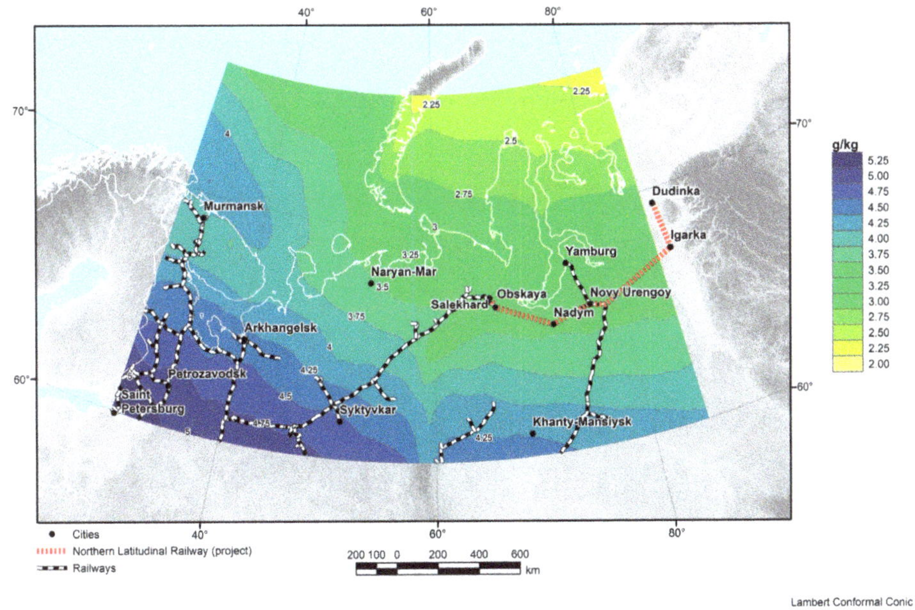

Figure 19. Average values of specific humidity (g/kg) of air for the period 1980–2021.

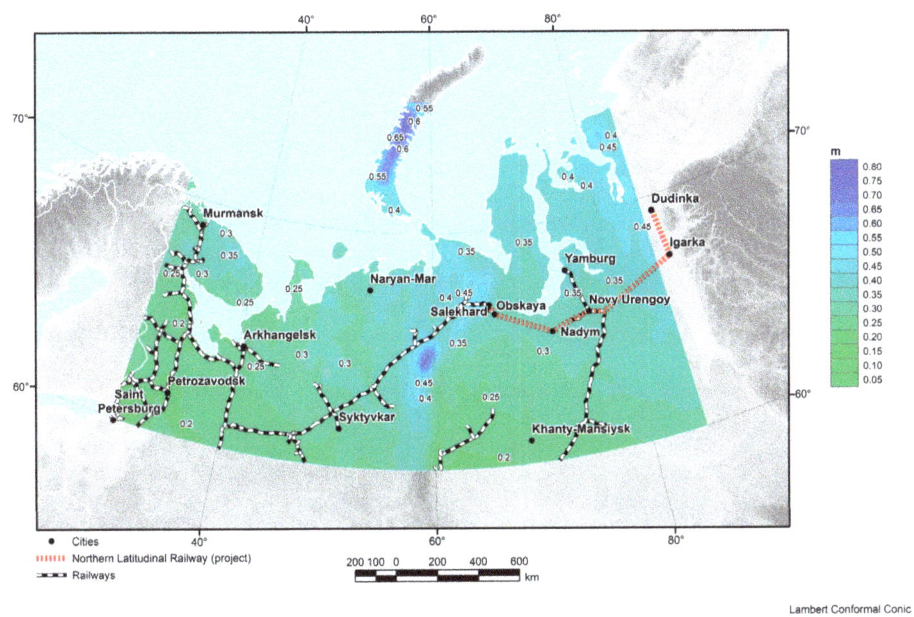

Figure 20. Average values of snow cover thickness (m) for the period 1980–2021.

Additionally, it is possible to analyze the change in the thickness of the snow cover in the Arctic zone. Figure 21 presents a map of the difference in snow cover thickness between 1980–1999 and 2000–2021. The difference in the average values of snow cover thickness between the periods allows us to conclude that in the western and eastern parts of the studied area, as well as along the coast of the Barents Sea, there is a decrease in snow cover thickness up to 10 cm, while in the central part there is a slight increase—up to 2 cm.

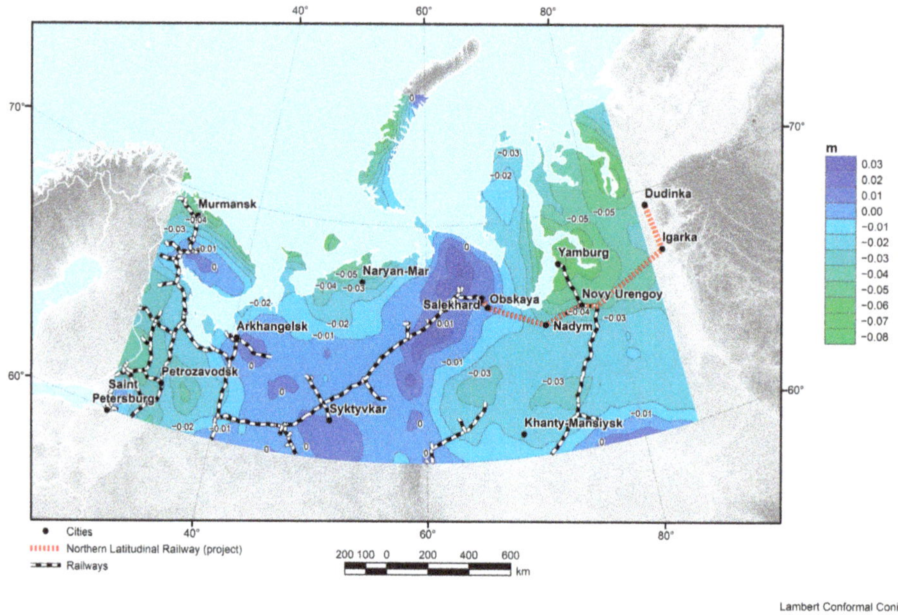

Figure 21. Average changes in snow cover thickness (m) between the periods 1980–1999 and 2000–2021 (difference of average values between these periods) according to MERRA-2 data.

3.9. Auxiliary Data-Layers of the Atlas

Auxiliary data were employed for filling the atlas maps with additional layers to facilitate the visual assessment (more accurate determination of objects' location within the map). Such information included physical and geographical parameters, administrative regions, and infrastructure objects. All operations on employing the auxiliary data were performed in ESRI ArcGIS (ArcMap) software. Let us consider this auxiliary information.

The physical map of Russia was used as the base for compiling the atlas maps. It includes topographical features of the regions of the western part of the Russian Federation: Northwestern Federal District (NFD), and Ural Federal District (UFD). The most suitable digital representation (at a scale 1:2,500,000) was provided by the Karpinsky All-Russian Research Geological Institute (VSEGEI). Digital geographic bases were prepared using ESRI ArcGIS software in conic equidistant projection [56,57].

The digital elevation model (DEM) was added to display the main terrain and relief features of the studied area. Considering the wide coverage of the territory, as well as high resolution, the GEBCO DEM with 30 m resolution was selected. The data is a global DEM for ocean and land showing elevation and depth in meters, on a grid with an interval of 1 angular second (about 30 m) [58,59]. The model at first was cropped to the territory of the Russian Federation, then to the boundaries of the studied area as part of this work.

Since there are only six administrative centers of the Russian Federation regions within the studied area, in order to increase the level of detail of the atlas maps, it was decided to also include the basic information on the settlements with a population of over ten thousand people [60]. The initial database includes: the name of the settlement, population, coordinates, administrative codes, and other information. The database was transformed into vector point geodata, and then cropped to the studied area.

4. Discussion

In this section, we discuss only those results which, from our point of view, are the most significant for consideration by railway operators.

Serykh and Tolstikov [42] analyzed climatic changes in air temperature, precipitation, and wind speed in this region. The authors showed that there were significant changes in these parameters between the periods 1980–2000 and 2001–2021. The strongest increase in temperature was observed for November and April, indicating that there was a shift in the time boundaries of the seasons—a later start and an early end of winter. It was revealed that in 2001–2021 the temperature increased most rapidly in the offshore area of the Barents and Kara Seas and this growth was accelerated. We have shown that the detected increase in the amount of precipitation is associated with a significant change in atmospheric circulation in the studied area. In the summer season and in September, there was an increase in the west wind within this territory. In the winter season of 2001–2021, there was an increase in the south wind in the Barents and Kara Seas as compared to 1980–2000.

Serykh and Tolstikov [43] found an increase in upper 1.5 m soil temperatures of about 0.5 °C in 2001–2021 as compared to 1980–2000 in the west of the studied area. This may lead to the reduction and even complete disappearance of the island permafrost on the Kola Peninsula, where average soil temperatures increased almost everywhere in 2001–2021 to +3 °C and more. In 2001–2021, an accelerating increase in soil temperature also began in the northeast of the western part of the Russian Arctic. There was a decrease in snow cover thickness in the west and east of the studied area in 2001–2021 as compared to 1980–2000. In the west of the studied area, there was also a significant reduction in the area of snow cover in November and April. An increase in specific humidity at 2 m altitude began in the west of the studied territory, and especially over the White Sea in 1980–2000. In 2001–2021, the increase in air humidity spread to the center and the east of the studied region with the highest growth rate over the waters of the seas, and this growth occurred with acceleration. These changes can be explained by the increasing influence of the North Atlantic on this territory and this process can be called "Atlantification" of the climate of the western part of the Russian Arctic. This phenomenon may lead to an increase in the number, strength, and duration of extreme weather events in this area [43].

Our research confirms that the warming of this area is significant and occurs in the direction from southwest to northeast [61]. The railway section from Syktyvkar to Salekhard and the section to Yamburg over the past two decades are in an area where average annual air temperatures remain below 0 °C in Figure 10. This means that these sections are operated under difficult weather and climatic conditions for all 12 months (as positive air temperature here is on average only from May to September), and the average monthly air temperature reaches from −20 °C to −22 °C in winter. Warming of the regional climate will occur along these sections, which will lead to thawing of permafrost, change of hydrological characteristics of numerous rivers, lakes, wetlands and may negatively affect the stability of railway tracks and bridges.

Air temperatures along these railways increased by an average of 0.4–0.6 °C between 1980–1999 and 2000–2021. Comparing these two periods, in January, the warming up to 1.0–1.5 °C was observed along the section to Murmansk and Arkhangelsk, warming up to 0.5 °C was observed along the section to Salekhard, and the cooling of 1–2 °C was observed on the section to Yamburg. In February, warming up to 0.8 °C was observed only on the section to Murmansk. In March, warming to 1 °C was observed only on the section to Yamburg. In April, warming to 1–2 °C was observed along all railway sections, with the warming being greater in more northerly sections. In May, warming to 1–2 °C was observed along all sections of the railways. In June, the greatest warming was observed along the sections to Salekhard and Yamburg, and northwards and eastwards the anomaly was greater, up to 2.4 °C. In July, August, and September, on the contrary, the temperature anomalies in the western part of the studied area were greater (up to 0.8–1.4 °C) than along the eastern sections of the railways. In October, there was a uniform warming of the entire region to 0.8 °C, and only on the northernmost railway section to Murmansk and Yamburg the anomaly reached 1.2 °C. In November, all the sections showed warming between 2 °C and 2.6 °C, except for the section to Yamburg, which reached 0.8 °C. In December, the warming increased from east to west, from 0 °C to 2.2 °C on the section to Murmansk in Figure 22. Thus, the regional climate change is very uneven within the year (by month), spatially, and even along each section of the railway separately.

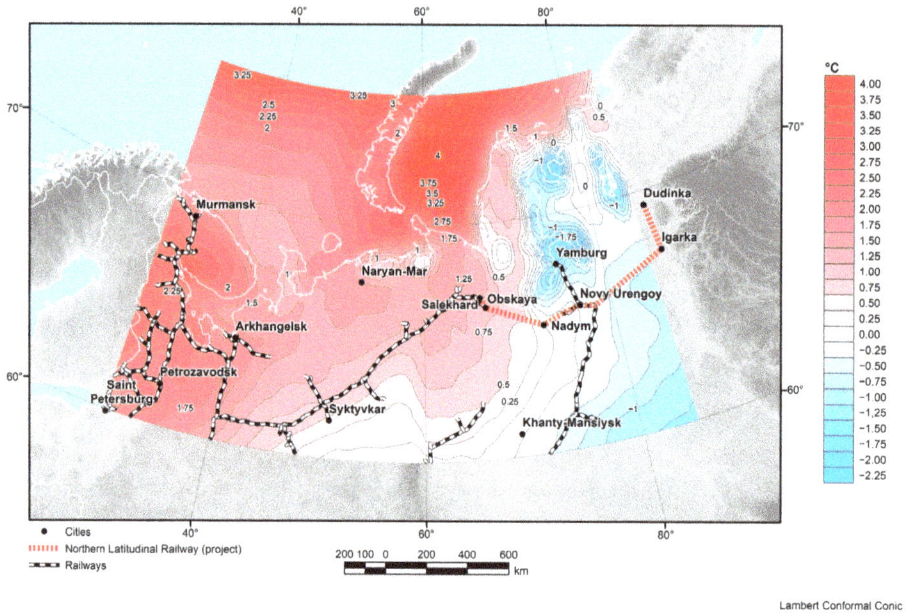

Figure 22. Average air temperature change (°C) at 2 m altitude between the periods 1980–1999 and 2000–2021 (difference of average values between these periods) for December.

Not only the rate, but also the direction of air temperature change varied radically from 1980 to 2021. If in 1980–1999, almost the whole studied area was cooling at a rate from 0 °C to 0.5 °C per 10 years. In Figure 23 (except for the Kola Peninsula and Karelia, where warming was 0.2 °C per 10 years), in 2000–2021, warming was observed everywhere from 0.1 °C to 0.5 °C per 10 years and the farther northwards, and it was faster on average in Figure 24.

Atmospheric precipitation in the studied area was on average 1.8–2.4 in/day, and the average values for 1980–2021 are distributed regularly over the territory, except for the Ob Bay area, where precipitation is less in Figure 17. Difference in mean precipitation between 1980–1999 and 2000–2021 shows that precipitation has increased by 0.06–0.12 in/day on railway sections to Murmansk, by 0.10–0.16 in/day to Arkhangelsk and Salekhard and remained virtually unchanged along the section to Yamburg in Figure 25. These changes are only 5% of the average values, so in the mean this increase is insignificant. However, in areas where average air temperatures have crossed 0 °C, this may indicate predominantly rain rather than snow, but this requires a joint analysis of air temperature and precipitation changes for individual months. Intra-annual variability shows that the greatest amount of precipitation in the studied area occurs from June to August. The most significant changes between the periods 1980–1999 and 2000–2021 occurred in March in the central part of the studied territory, where precipitation increased by 0.2–0.5 in/day. In May, near Arkhangelsk the increase was 0.5 in/day; in June, along the railway to Salekhard the increase was 0.5–1.0 in/day; in August, at some sections of the Murmansk and Arkhangelsk railways, as well as the railway to Yamburg, the increase was 0.6 in/day; in September, at some sections of the Murmansk and Arkhangelsk railways, as well as along the railway between Syktyvkar and Salekhard, the increase was 0.4–0.8 in/day. These changes in some months and in some areas are significant, increasing from 25 to 50% of the average values. In addition, the highest growth rate of precipitation was observed exactly in the last 20 years, when it reached 0.15–0.2 in/day per 10 years, i.e., approximately 10% for 10 years along almost all railway sections in Figure 26.

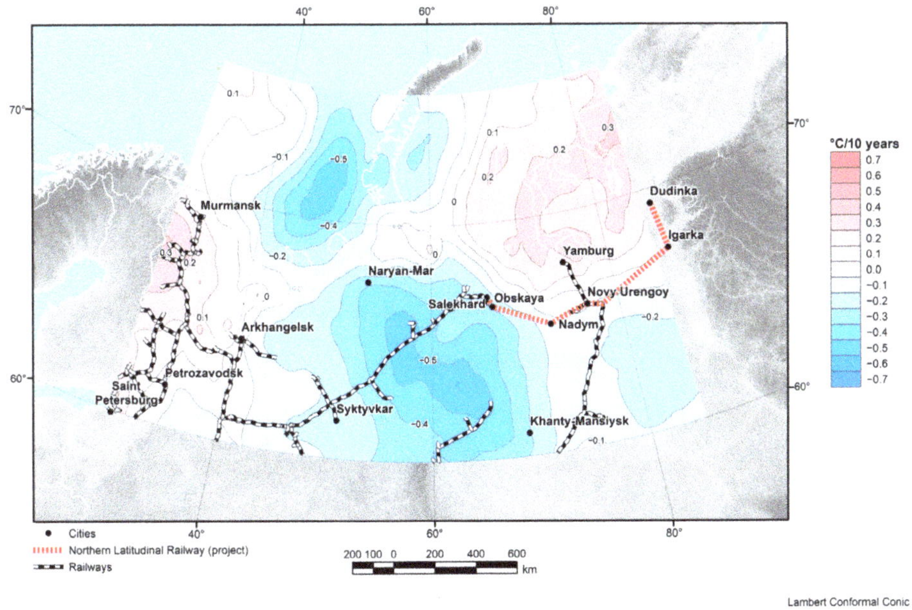

Figure 23. Average rate of change (linear trend) of mean monthly anomalies (relative to the annual rate) of air temperature at 2 m altitude (°C per 10 years) for 1980–1999.

Average wind speed in the region ranges from 1.0 to 1.5 m/s for 1980–2021. This is the southwestern wind within all the studied area. Between the 1980–1999 and 2000–2021 periods, the wind speed has increased in some areas by 0.10–0.15 m/s, which is 10% of the average values. There is a significant seasonal variability in wind speed, which changes on average from 1 m/s in summer (June–August) to 2–3 m/s in winter (December–February). In some months and in some areas, the observed changes are even more significant. For example, between the periods 1980–1999 and 2000–2021, in March, the wind speed increased by 0.5 m/s along the railway to Yamburg. In April, the same wind speed increase was observed on the same railway to Salekhard and on the Murmansk railway. In June, the same wind speed increase was observed in the vast area around Syktyvkar. In July, the wind speed increased by 0.5–1.0 m/s near Salekhard and Yamburg. In August, the wind speed increased by 0.5 m/s in Karelia, at the railway sections to Salekhard and Yamburg, and in September by 0.5 m/s everywhere in Figure 27. In the central part of the studied area, the wind speed growth rate in the last 20 years reached 0.1 m/s per 10 years, which is about 10% per decade in Figure 28.

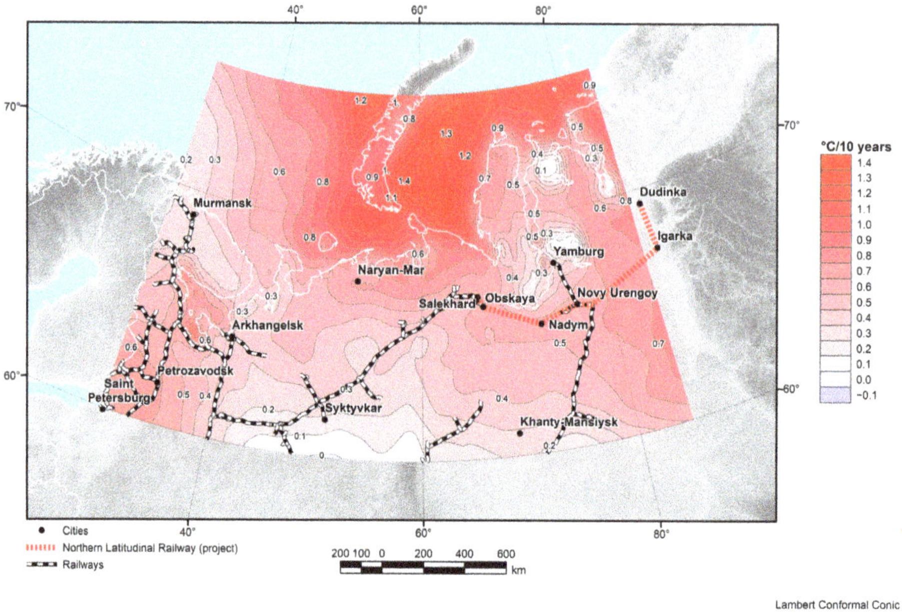

Figure 24. Average rate of change (linear trend) of mean monthly anomalies (relative to the annual rate) of air temperature at 2 m altitude (°C per 10 years) for 2000–2021.

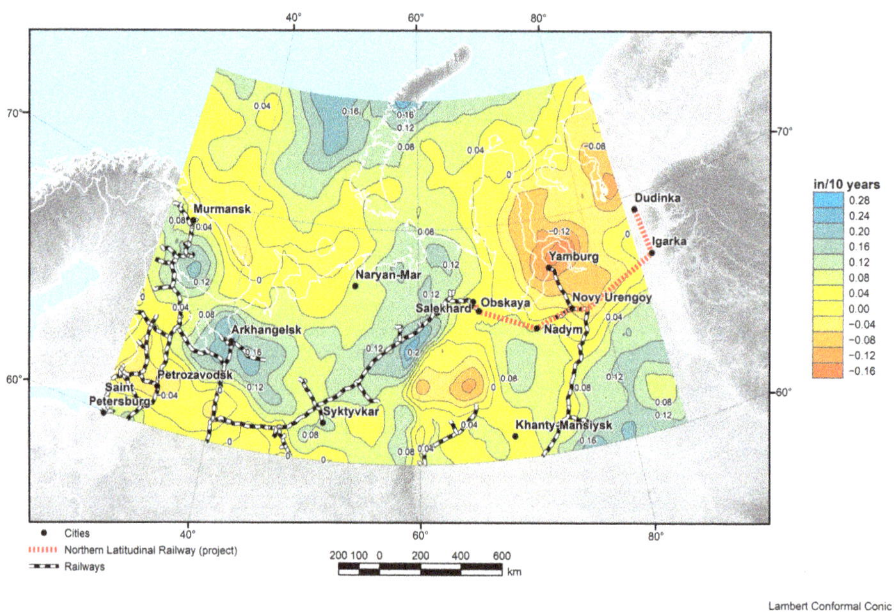

Figure 25. Average changes (difference in average values) in total precipitation per day (in) between the periods 1980–1999 and 2000–2021.

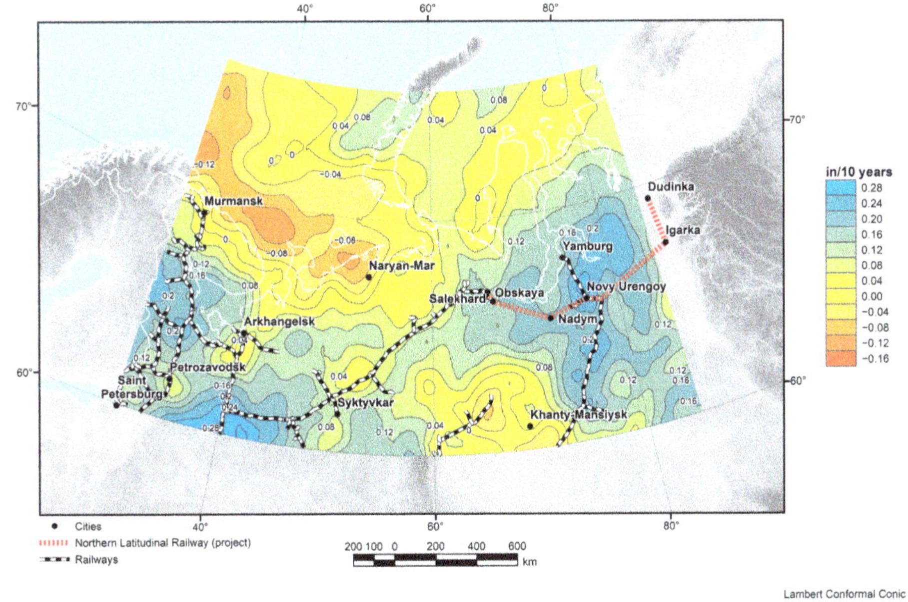

Figure 26. Average rate (linear trend) of change in mean monthly anomalies (relative to the annual rate) of total precipitation per day (in/day per 10 years) over 2000–2021.

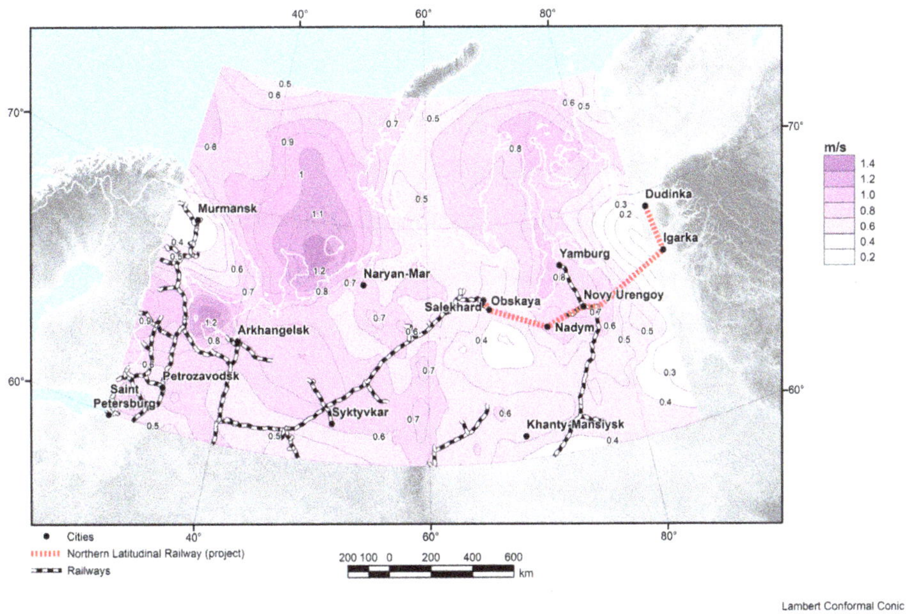

Figure 27. Average wind speed change (m/s) (difference of average values) at 50 m altitude between the periods 1980–1999 and 2000–2021 for September.

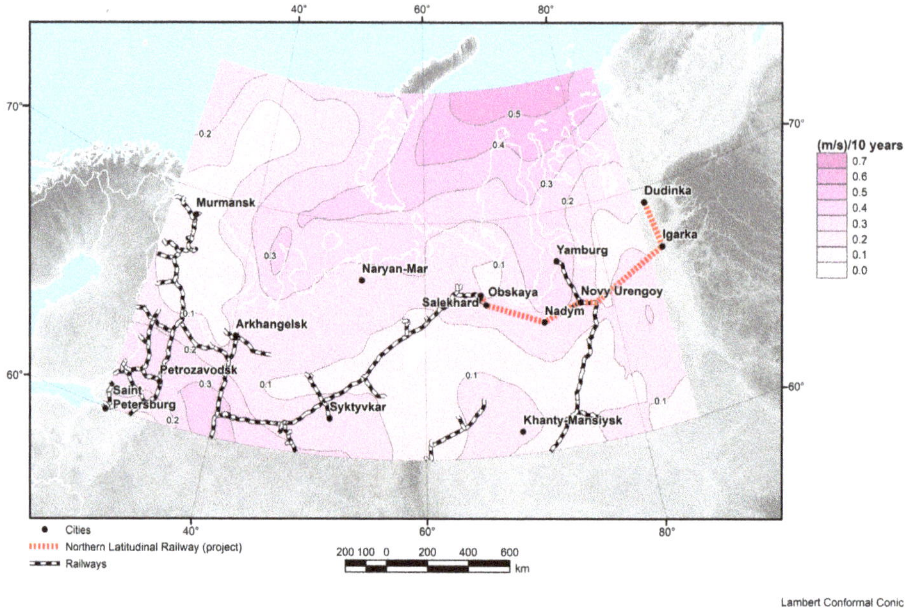

Figure 28. Average rate (linear trend) of wind speed change at 50 m altitude (m/s per 10 years) for 2000–2021.

In the last two decades, the average snow cover thickness in the studied area was 20–35 cm in Figure 29. The observed climate warming led to a decrease in snow cover thickness from 1980–1999 to 2000–2021 by 2–4 cm in the western and eastern parts of this territory, i.e., by approximately 10% in Figure 21. The highest snow cover thickness is observed in March, when it reaches 50–80 cm in different regions, and from 70 to 100 cm along the railway from Syktyvkar to Salekhard in Figure 30. From June to September, the snow cover is absent. The rate of snow cover reduction can reach 2.5–3.5 cm per 10 years in the northern part of this region, both on the Kola Peninsula and in the area of the Gulf of Ob.

Soil temperature increased significantly along with the climate warming [62]. Average changes in temperature of the upper 10 m of soil between the periods 1980–1999 and 2000–2021 varied from 0.2 °C to 0.8 °C. This warming affected the area westwards from the Ural Mountains; the area eastwards remained virtually unaffected in Figure 31. The spatial distribution of soil temperature in the last two decades is shown in Figure 17. Notably, in both average values and in values for individual months, negative values of soil temperature in the MERRA-2 database are observed only on the Novaya Zemlya Archipelago, Yamal Peninsula, and northwards from 68°N in the territory east of the Gulf of Ob. This distribution contradicts the known maps of the permafrost boundary position in northern Russia and Siberia [3,6], so this issue requires special consideration. The average rate of soil temperature increased in 2000–2021 and reached from 0.2 °C to 0.8 °C per 10 years in the central part of the studied area whereas in the far northern regions it reached from 1.2 °C to 1.6 °C per 10 years in Figure 32.

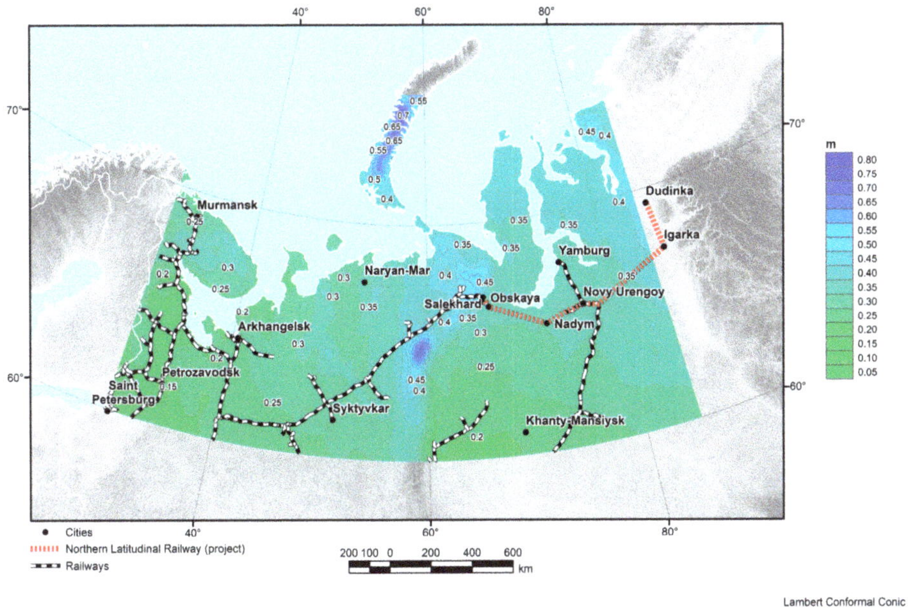

Figure 29. Average values of snow cover thickness (m) for 2000–2021.

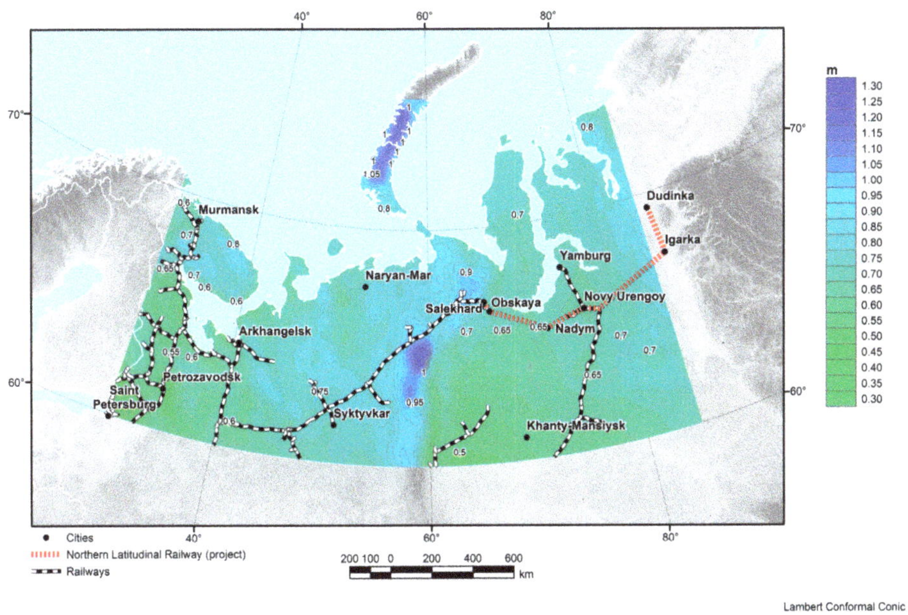

Figure 30. Average values of snow cover thickness (m) for March over 2000–2021.

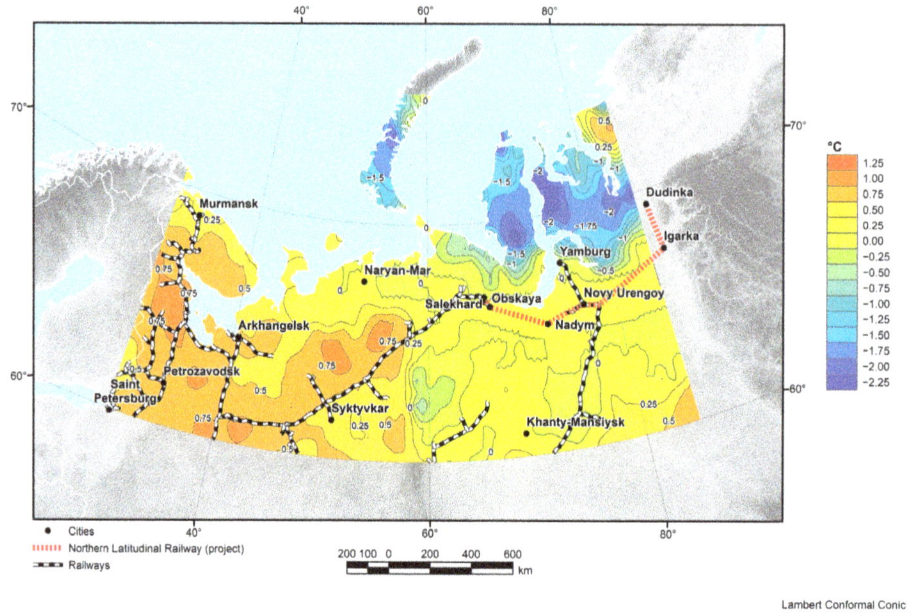

Figure 31. Average change (difference of average values) in temperature (°C) of the upper 10 m of soil between the periods 1980–1999 and 2000–2021.

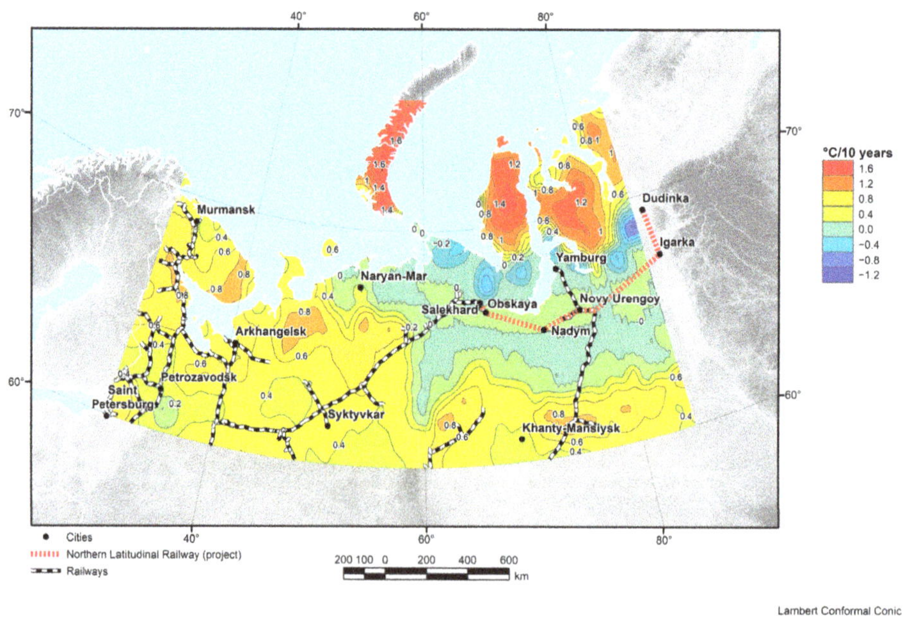

Figure 32. Average rate (linear trend) of change in mean monthly anomalies (relative to annual variation) of the temperature (°C per 10 years) of the upper 10 m of soil for 2000–2021.

In this section, we discussed only those specific features of the regional climate change in terms of seasonal and interannual variability of seven meteo-parameters which, from our point of view, are the most significant for consideration by railway operators. For the first time, such an analysis was completed specially for the western part of the Russian

Arctic where a dense railway network exists and there are plans for its development in the near future. The obtained information is unique, because we concisely discussed every parameter and presented the main key peculiarities of the regional climate change for different sections of the railway network where the greatest changes were observed. To our knowledge supported by the opinion of the experts from Research and Design Institute of Informatization, Automation and Communications in Railway Transport, which belongs to the Russian railways, this is the first digital atlas of climate change impact on operability and infrastructure of the Russian railways in the Russian Federation.

For instance, the "Third assessment report on climate change and its consequences on the territory of the Russian Federation" issued in October 2022 by the Russian Hydrometeorological Service [63], which is the most comprehensive recent analysis of climate change on the territory and aquatoria of the Russian Federation, contains only very general and very limited information about the impact on the Russian railways in the volume of 3.5 pages from 678 pages in total. Moreover, this concerns both automobile and railway transport.

There is a long list of publications and reviews devoted to climate change impact on railway transport operability and infrastructure in different countries, but all of them dis-cuss general negative issues from heat or cold waves, frosts, permafrost thawing, heavy rains, storms and high winds, extreme sea level and waves, riverine and coastal storm flooding, and show different case studies of this impact [2,64–76]. However, we could not find a detailed atlas such as this one with a comprehensive set of different maps and parameters which display the ongoing regional climate change and its impact on railway networks in other parts of the world.

5. Conclusions

The atlas depicting climatic changes of basic hydrometeorological parameters in the western part of the Russian Arctic over 1950–2021 is the first experience of building a specialized climatic cartographic product for the needs of the Russian railways [77]. It contains 459 maps of mean seasonal and decadal characteristics of surface air temperature (69 maps), total precipitation (69 maps), wind speed at ground surface (69 maps), soil temperature (63 maps), soil moisture (63 maps), air humidity (63 maps), and snow cover thickness (63 maps), along with their mean rates of change (linear trends). This article details the initial data and GIS-based methodology used for compiling the atlas; and provides its main features and examples of certain map types. In addition, it discusses the most intriguing results that come from the interannual variability of the studied parameters, which are crucial for the railway operation in the northwestern part of the Russian Arctic.

We show that the climate warming in the studied area is very irregular within a year (by months), spatially, and even along each section of the railway (from 0.5 °C to 2.6 °C between 1980–1999 and 2000–2021). The rate of air temperature increase is maximal exactly in the last 20 years and reaches 0.5 °C per 10 years. The observed climate warming led to a 2–4 cm reduction in snow cover thickness from 1980–1999 to 2000–2021 in the western and eastern parts of the studied area, i.e., by approximately 10%. As the climate is becoming warmer, soil temperatures increase significantly. Average temperature change in the upper 10 m of soil between the 1980–1999 and 2000–2021 periods varies from 0.2 °C to 0.8 °C. Between 1980–1999 and 2000–2021, there is a significant increase in precipitation, which in some months and in some areas ranges from 25% to 50% of average values. The highest rate of precipitation growth is observed precisely within the last 20 years, when it reaches 10% per 10 years along almost all railway sections.

The performed analysis has revealed significant spatial and temporal heterogeneity of the considered parameters variability. These results suggest that a thorough study of the climatic parameters along each railway section separately is needed. This will make it possible to clarify the observed changes and improve the forecast for individual railway sections. The future research in this direction will result in: (1) creation of specialized diagrams (Hovmöller diagram) of spatial and temporal variability of selected hydrometeorological parameters along the main railway sections in the northwestern part of the

Russian Arctic; (2) creation of maps and diagrams of spatial and temporal variability of extreme weather phenomena in the form of observed anomalies, occurrence frequency and duration; (3) creation of forecast maps of spatial and temporal variability of basic hydrometeorological parameters up to the end of the 21st century for the existing and planned railway infrastructure.

Particular strengths of this study concern a choice of the meteo-parameters, specification of the maps, visual representations of the maps (color scales, isolines, railway lines, rivers, coastlines, geographical projection, etc.), original algorithms of geospatial data processing and their representation in the in GIS environment, and our own recommendations derived from the research on how to improve future generations of the atlas. This is why we describe in detail technical aspects of the atlas construction that can be used by our followers to avoid mistakes and save time in building atlases for other similar regions in the world.

This is the first experience in building a specialized climatic cartographic product for the needs of the Russian railways, and to our knowledge the first atlas such as that in the world. The atlas expansion and improvement will be continued in the framework of the Russian Science Foundation Project No. 21-77-30010 (2021–2024) "System analysis of geophysical process dynamics in the Russian Arctic and their impact on the development and operation of the railway infrastructure" in close collaboration with experts from the Russian railways.

We hope that the detailed analysis of the atlas maps prepared for the Russian rail-ways, along with its future expansion, will contribute to sustainable development and adaptation of the railway infrastructure to climate change in the northwestern part of the Russian Arctic. In the future, the amassed experience may be transferred to other regions of the Russian Federation, as well as similar regions in Canada, Sweden, and Highland China that are also subject to significant climate change [72,76].

Author Contributions: Conceptualization, A.D.G., I.N.R. and A.A.S.; methodology, map validation, writing—climatic aspect, A.G.K.; software compilation and climatic data processing, I.V.S.; map design, figure preparation, writing, S.A.G., A.B.P. and G.A.G.; writing—transport aspect, N.V.S. and I.A.D.; writing—economical aspect, E.A.K.; supervision, review and editing, R.I.K. All authors have read and agreed to the published version of the manuscript.

Funding: The analysis and interpretation of climatic data and atlas maps presented in this paper were funded by the Russian Science Foundation (Project No. 21-77-30010) "System analysis of geophysical process dynamics in the Russian Arctic and their impact on the development and operation of the railway infrastructure".

Institutional Review Board Statement: Not applicable.

Informed Consent Statement: Not applicable.

Data Availability Statement: Not applicable.

Acknowledgments: The atlas of hydrometeorological parameters was compiled in the framework of the project with the Research and Design Institute of Informatization, Automation and Communications in Railway Transport. The authors wish to thank the scientific teams of the responsible agencies and institutions for the provided data.

Conflicts of Interest: The authors declare no conflict of interest.

References

1. Katzov, V.M.; Akenteva, E.M.; Anisimov, O.A.; Gruza, G.V.; Zaitsev, A.E.; Frolov, I.E.; Sirotenko, A.D.; Kattsov, V.M.; Karol, I.E.; Terziev, F.S.; et al. *Third Assessment Report on Climate Change and Its Consequences on the Territory of the Russian Federation. General Summary*; Science-Intensive Technologies (Naukoyemkie Technologii): St. Petersburg, Russia, 2022; pp. 22–30.
2. Masson-Delmotte, V.; Zhai, P.; Pirani, A.; Connors, S.L.; Péan, C.; Berger, S.; Caud, N.; Chen, Y.; Goldfarb, L.; Gomis, M.I.; et al. (Eds.) *Climate Change 2021: The Physical Science Basis. Contribution of Working Group I to the Sixth Assessment Report of the Intergovernmental Panel on Climate Change*; Cambridge University Press: Cambridge, UK; New York, NY, USA, 2021; 2391p.

3. Kostianaia, E.A.; Kostianoy, A.G.; Scheglov, M.A.; Karelov, A.I.; Vasileisky, A.S. Impact of regional climate change on the infrastructure and operability of railway transport. *Transp. Telecommun.* **2021**, *22*, 183–195. [CrossRef]
4. Yakubovich, A.N.; Yakubovich, I.A. Forecasting the impact of climatic changes on the functionality of the transport infrastructure of the permafrost zone in Russia. *Intellect. Innovation. Invest.* **2019**, *1*, 104–110. [CrossRef]
5. Serykh, I.V.; Kostianoy, A.G.; Lebedev, S.A.; Kostianaia, E.A. On the transition of temperature regime of the White Sea Region to a new phase state. *Fund. App. Hydrophys.* **2022**, *15*, 98–111.
6. Osipov, V.I.; Rumyantseva, N.A.; Eremina, O.N. Living with risk of natural disasters. *Russ. J. Earth Sci.* **2019**, *19*, 1–10. [CrossRef]
7. Romanenko, F.A.; Shilovtseva, O.A. Geomorphological processes in the mountains of the Kola Peninsula and climate change. *MSU Vestnik Ser. 5 Geog.* **2016**, *6*, 78–86. (In Russian)
8. *World Experience in the Construction, Maintenance and Repair of Railway Infrastructure Facilities in Permafrost Conditions*; Center for Scientific and Technical Information and Libraries: Moscow, Russia, 2017; 126p. (In Russian)
9. Rosstat. Socio-Economic Situation of the North-Western Federal District. Available online: https://rosstat.gov.ru/storage/mediabank/Sev-Zap-fo_1k-2022.pdf (accessed on 1 November 2022).
10. Rosstat. Socio-Economic Situation of the Ural Federal District. Available online: https://rosstat.gov.ru/storage/mediabank/Ural-fo_1k-2022.pdf (accessed on 1 November 2022).
11. Northern Latitudinal Railway. Available online: http://uralfo.gov.ru/projects/sshh/ (accessed on 1 November 2022).
12. Northern Latitudinal Railway. Available online: http://www.cupp.ru/karta-proekta/ (accessed on 1 November 2022).
13. Verezemskaya, P.; Selivanova, Y.; Tilinina, N.; Markina, M.; Krinitskiy, M.; Sharmar, V.; Razorenova, O. Projected changes in the near-surface atmosphere over the Barents Sea based on CMIP5 scenarios. *Russ. J. Earth. Sci.* **2022**, *22*. [CrossRef]
14. Gelaro, R.; McCarty, W.; Suárez, M.J.; Todling, R.; Molod, A.; Takacs, L.; Randles, C.A.; Darmenov, A.; Bosilovich, M.G.; Reichle, R.; et al. The Modern-Era Retrospective Analysis for Research and Applications, Version 2 (MERRA-2). *J. Clim.* **2017**, *30*, 5419–5454. [CrossRef]
15. McCarty, W.; Coy, L.; Gelaro, R.; Huang, A.; Merkova, D.; Smith, E.B.; Sienkiewicz, M.; Wargan, K. MERRA-2 input observations: Summary and initial assessment. *Tech. Rep. Ser. Glob. Model. Data Assim.* **2016**, *46*, 61.
16. Bosilovich, M.G. Regional Climate and Variability of NASA MERRA and Recent Reanalyses: U.S. Summertime Precipitation and Temperature. *J. Appl. Meteorol. Climatol.* **2013**, *52*, 1939–1951. [CrossRef]
17. Tilinina, N.; Gulev, S.K.; Rudeva, I.; Koltermann, K.P. Comparing cyclone life cycle characteristics and their interannual variability in different reanalyses. *J. Clim.* **2013**, *26*, 6419–6438. [CrossRef]
18. Bentamy, A.; Piollé, J.F.; Grouazel, A.; Danielson, R.; Gulev, S.; Paul, F.; Azelmat, H.; Mathieu, P.P.; von Schuckmann, K.; Sathyendranah, S.; et al. Review and assessment of latent and sensible heat flux accuracy over the global oceans. *Remote Sens. Environ.* **2017**, *201*, 196–218. [CrossRef]
19. Luo, B.; Minnett, P.; Szczodrak, M.; Nalli, M.; Morris, V. Accuracy assessment of MERRA-2 and ERA-Interim sea-surface temperature, air temperature and humidity profiles over the Atlantic Ocean using AEROSE measurements. *J. Clim.* **2020**, *33*, 6889–6909. [CrossRef]
20. Sharmar, V.; Markina, M. Evaluation of interdecadal trends in sea ice, surface winds and ocean waves in the Arctic in 1980–2019. *Russ. J. Earth. Sci.* **2021**, *21*, 3. [CrossRef]
21. Schubert, S.D.; Chang, Y.; DeAngelis, A.M.; Koster, R.D.; Lim, Y.; Wang, H. Exceptional Warmth in the Northern Hemisphere during January–March of 2020: The Roles of Unforced and Forced Modes of Atmospheric Variability. *J. Clim.* **2022**, *35*, 2565–2584. [CrossRef]
22. Kalnay, E.; Kanamitsu, M.; Kistler, R.; Collins, W.; Deaven, D.; Gandin, L.; Joseph, D. The NCEP/NCAR 40-year reanalysis project. *Bull. Am. Meteorol. Soc.* **1996**, *77*, 437–471. [CrossRef]
23. The International Institute for Applied Systems Analysis (IIASA), Russian Academy of Sciences (RAS). 2002. Land Resources of Russia. Available online: http://www.iiasa.ac.at/Research/FOR/russia_cd/download.htm (accessed on 3 November 2022).
24. Si, Z.; Li, S.; Huang, L.; Chen, Y. Visualization programming for batch processing of contour maps based on VB and Surfer software. *Adv. Eng. Softw.* **2010**, *41*, 962–965. [CrossRef]
25. Golden Software Surfer. Available online: https://www.goldensoftware.com/products/surfer (accessed on 1 November 2022).
26. ESRI ArcMap. Available online: https://www.esri.com/en-us/arcgis/products/arcgisdesktop/resources (accessed on 3 November 2022).
27. Bartuś, T. Raster images generalization in the context of research on the structure of landscape and geodiversity. *Geol. Geophys. Environ.* **2014**, *40*, 271–284. [CrossRef]
28. Watson, C.; Richardson, J.; Wood, B.; Jackson, C.; Hughes, A. Improving geological and process model integration through TIN to 3D grid conversion. *Comp. Geosci.* **2015**, *82*, 45–54. [CrossRef]
29. Theoretical Fundaments in Geoinformation Systems (GIS). Available online: https://desktop.arcgis.com/ru/arcmap/latest/extensions/3d-analyst/fundamentals-of-3dsurfaces.htm (accessed on 30 October 2022).
30. Soloviev, A.A.; Krasnoperov, R.I.; Nikolov, B.P.; Zharkikh, J.I.; Agayan, S.M. Web-oriented software system for analysis of spatial geophysical data using geoinformatics methods. *Izv. Atm. Ocean. Phys.* **2018**, *54*, 1312–1319. [CrossRef]
31. Islam, M.S. Assessing the dynamics of land cover and shoreline changes of Nijhum Dwip (Island) of Bangladesh using remote sensing and GIS techniques. *Reg. Stud. Mar. Sci.* **2021**, *41*, 101578. [CrossRef]

32. Built-In Processing Tools Technical Support, ArcMap. Available online: https://support.esri.com/en/technical-article/000013000 (accessed on 2 November 2022).
33. A Source Code of Author's Tool to Clip Unnecessary Values. The Tool for ArcGIS. Available online: https://github.com/GAGvozdik/Group_raster_clip (accessed on 1 November 2022).
34. Yang, C.S.; Kao, S.P.; Lee, F.B.; Hung, P.S. Twelve different interpolation methods: A case study of Surfer 8.0. In Proceedings of the XXth ISPRS Congress, Istanbul, Turkey, 12–23 July 2004.
35. Kriging. Available online: https://surferhelp.goldensoftware.com/griddata/idd_grid_data_kriging.htm (accessed on 27 March 2023).
36. Williams, R.T. Lambert and Mercator map projections in geology and geophysics. *Comp. Geosci.* **1995**, *21*, 353–364. [CrossRef]
37. Spatial Reference System Information for ESPG. Available online: https://products.aspose.app/gis/epsg/code-102027 (accessed on 2 November 2022).
38. Boyarshinov, G.; Popov, A.; Odintsova, A.; Gvozdik, S.; Rybkina, A.; Korolkova, A. Application of Geoportal's Web-Technologies in GIS, Case Study: Interactive Geology Atlas. *Russ. J. Earth. Sci.* **2022**, *22*, 3002. [CrossRef]
39. About Analyzing Images and Raster Data. Available online: https://desktop.arcgis.com/ru/arcmap/latest/manage-data/raster-and-images/about-analyzingimagery-raster-data.htm (accessed on 4 November 2022).
40. Habel, M.; Obodovskyi, O.; Szatten, D.; Babinski, Z.; Rozlach, Z.; Pochaievets, O. Using the raster calculator tool to appraise riverbed elevation changes nearby hydrotechnical objects on alluvial rivers. *Eur. Assoc. Geosci. Eng.* **2019**, *2019*, 1–5. [CrossRef]
41. Description of Raster Calculator Utilization. Available online: https://desktop.arcgis.com/ru/arcmap/10.5/tools/spatial-analyst-toolbox/raster-calculator.htm (accessed on 1 November 2022).
42. Serykh, I.V.; Tolstikov, A.V. Climate change in the western part of the Russian Arctic in 1980–2021. Part 1. Air temperature, precipitation, wind. *Arct. Antarct. Res.* **2022**, *68*, 258–277. [CrossRef]
43. Serykh, I.V.; Tolstikov, A.V. Climate change in the western part of the Russian Arctic in 1980–2021. Part 2. Soil temperature, snow, humidity. *Arct. Antarct. Res.* **2022**, *68*, 352–369. [CrossRef]
44. Han, Y.; Revercomb, H.; Cromp, M.; Gu, D.; Johnson, D.; Mooney, D.; Scott, D.; Strow, L.; Bingham, G.; Borg, L.; et al. Suomi NPP CrIS measurements, sensor data record algorithm, calibration and validation activities, and record data quality. *J. Geophys. Res. Atmos.* **2013**, *118*, 12734–12748. [CrossRef]
45. Kim, E.; Lyu, C.H.J.; Anderson, K.; Leslie, R.V.; Blackwell, W.J. S-NPP ATMS instrument prelaunch and on-orbit performance evaluation. *J. Geophys. Res.* **2014**, *119*, 5653–5670. [CrossRef]
46. Chen, M.; Xie, P.; Janowiak, J.E.; Arkin, P.A. Global Land Precipitation: A 50-yr Monthly Analysis Based on Gauge Observations. *J. Hydrometeorol.* **2002**, *3*, 249–266. [CrossRef]
47. Reichle, R.H.; Koster, R.D.; De Lannoy, G.J.; Forman, B.A.; Liu, Q.; Mahanama, S.P.P.; Toure, A. Assessment and enhancement of MERRA land surface hydrology estimates. *J. Clim.* **2011**, *24*, 6322–6338. [CrossRef]
48. Bosilovich, M.G.; Lucchesi, R.; Suarez, M. *MERRA-2: File Specification*; GMAO Office Note No. 9 (Version 1.1); NASA: Greenbelt, MD, USA, 2016; pp. 1–75.
49. Poli, P.; Hersbach, H.; Tan, D.; Dee, D.; Thépaut, J.; Simmons, A.; Peubey, C.; Laloyaux, P.; Komori, T.; Berrisford, P.; et al. *The Data Assimilation System and Initial Performance Evaluation of the Ecmwf Pilot Reanalysis of the 20th-Century Assimilating Surface Observations Only (ERA-20C)*; European Centre for Medium-Range Weather Forecasts: Shinfield Park, UK, 2013; pp. 1–59.
50. Ma, H.; Zeng, J.; Zhang, X.; Fu, P.; Zheng, D.; Wigneron, J.; Chen, N.; Niyogi, D. Evaluation of six satellite- and model-based surface soil temperature datasets using global ground-based observations. *Remote Sens. Environ.* **2021**, *264*, 112605. [CrossRef]
51. Reichle, R.H.; Draper, C.S.; Liu, Q.; Girotto, M.; Mahanama, P.P.; Koster, R.D.; De Lannoy, G.J. Assessment of MERRA-2 Land Surface Hydrology Estimates. *J. Clim.* **2017**, *30*, 2937–2960. [CrossRef]
52. Dee, D.P.; da Silva, A.M. The choice of variable for atmospheric moisture analysis. *Mon. Weather. Rev.* **2003**, *131*, 155–171. [CrossRef]
53. Holm, E.V. Revision of the ECMWF humidity analysis: Construction of a Gaussian control variable. In Proceedings of the ECMWF/GEWEX Workshop on Humidity Analysis, Reading, UK, 8–11 July 2003.
54. Tao, J.; Koster, R.D.; Reichle, R.H.; Forman, B.A.; Xue, Y.; Chen, R.H.; Moghaddam, M. Permafrost Variability over the Northern Hemisphere Based on the MERRA2 Reanalysis. *Cryosphere* **2019**, *13*, 2087–2110. [CrossRef]
55. Toure, A.M.; Reichle, R.H.; Forman, B.A.; Getirana, A.; De Lannoy, G.J. Assimilation of MODIS Snow Cover Fraction Observations into the NASA Catchment Land Surface Model. *Remote Sens.* **2018**, *10*, 316. [CrossRef]
56. Information Resources: Basic Geographical Backgrounds. Available online: https://vsegei.ru/ru/info/topo/ (accessed on 31 October 2022).
57. Gvishiani, A.D.; Kaftan, V.I.; Krasnoperov, R.I.; Tatarinov, V.N.; Vavilin, E.V. Geoinformatics and systems analysis in geophysics and geodynamics. *Izv. Phys. Sol. Earth* **2019**, *55*, 33–49. [CrossRef]
58. Lemenkova, P. GEBCO Gridded Bathymetric Datasets for Mapping Japan Trench Geomorphology by Means of GMT Scripting Toolset. *Geod. Cartogr.* **2020**, *46*, 98–112. [CrossRef]
59. General Bathymetric Chart of the Ocean, GEBCO. Available online: https://www.gebco.net/data_and_products/gridded_bathymetry_data/grid_production/ (accessed on 1 November 2022).
60. Russian-language Database of Names with Specific Addresses Based on VMap0 Data. Available online: https://gis-lab.info/qa/vmap0-settl-rus.html (accessed on 2 October 2022).

61. Petrunin, A.G.; Soloviev, A.A.; Sidorov, R.V.; Gvishiani, A.D. Inverse-forward method for heat flow estimation: Case study for the Arctic region. *Russ. J. Earth Sci.* **2022**, *21*, ES6004. [CrossRef]
62. Al-shateri, H.A.A. Estimation of land surface temperature and distribution across Land use/land cover in response to coal mining activity in V. D. Yelevsky coal mine area–Russia. *Russ. J. Earth Sci.* **2022**, *22*, 5. [CrossRef]
63. Katzov, V.M. (Ed.) . *The Third Assessment Report on Climate Changes and Their Consequences on the Territory of the Russian Federation*; Science-Intensive technologies (Naukoyemkie Technologii): St. Petersburg, Russia, 2022; 678p.
64. Baker, C.J.; Chapman, L.; Quinn, A.; Dobney, K. Climate Change and the Railway Industry: A Review. [Online]. 2009. Available online: http://journals.sagepub.com/doi/pdf/10.1243/09544062JMES1558 (accessed on 6 April 2023).
65. Nemry, F.; Demirel, H. Impacts of Climate Change on Transport: A Focus on Road and Rail Transport Infrastructures. [Online]. 2012. Available online: http://ftp.jrc.es/EURdoc/JRC72217.pdf (accessed on 6 April 2023).
66. Oslakovic, I.S.; ter Maat, H.W.; Hartmann, A.; Dewulf, G. Risk Assessment of Climate Change Impacts on Railway Infrastructure. [Online]. 2013. Available online: https://core.ac.uk/download/pdf/29215286.pdf (accessed on 6 April 2023).
67. Rail Safety and Standards Board. Tomorrow's Railway and Climate Change Adaptation: Executive Report. [Online]. 2016. Available online: www.rssb.co.uk (accessed on 6 April 2023).
68. Palco, K.; Lemmen, D.S. (Eds.) *Climate Risks and Adaptation Practices for the Canadian Transportation Sector*; Government of Canada: Ottawa, ON, Canada, 2017; [Online]. Available online: http://www.nrcan.gc.ca/environment/impacts-adaptation10761 (accessed on 6 April 2023).
69. Bubeck, P.; Dillenardt, L.; Alfieri, L.; Feyen, L.; Thieken, A.H.; Kellermann, P. Global warming to increase flood risk on European railways. *Clim. Change* **2019**, *155*, 19–36. [CrossRef]
70. Koks, E.E.; Rozenberg, J.; Zorn, C.; Tariverdi, M.; Vousdoukas, M.; Fraser, S.A.; Hall, J.W.; Hallegatte, S. A global multi-hazard risk analysis of road and railway infrastructure assets. *Nat. Comm.* **2019**, *10*, 2677. [CrossRef]
71. Climate Change Impacts and Adaptation for Transport Networks and Nodes. Report of United Nations Economic Commission for Europe (ECE), 2020, 216p. [Online]. Available online: https://unece.org/sites/default/files/2021-01/ECE-TRANS-283e_web.pdf (accessed on 6 April 2023).
72. Thaduri, A.; Garmabaki, A.; Kumar, U. Impact of climate change on railway operation and maintenance in Sweden: A State-of-the-art review. *Maint. Reliab. Cond. Monit.* **2021**, *1*, 52–70. [CrossRef]
73. Garmabaki, A.H.S.; Thaduri, A.; Famurewa, S.; Kumar, U. Adapting Railway Maintenance to Climate Change. *Sustainability* **2021**, *13*, 13856. [CrossRef]
74. Palin, E.J.; Stipanovic Oslakovic, I.; Gavin, K.; Quinn, A. Implications of climate change for railway infrastructure. *Wiley Interdiscip. Rev. Clim. Chang.* **2021**, *12*, e728. [CrossRef]
75. Ochsner, M.; Palmqvist, C.W. Weather and Train Disruptions in Sweden 2011–2019. In Proceedings of the 18th International Conference on Railway Engineering Design & Operation, Valencia, Spain, 21–23 September 2022.
76. Kostianaia, E.A.; Kostianoy, A.G. Railway transport adaptation strategies to climate change at high latitudes: A review of experience from Canada, Sweden and China. *Transp. Telecommun.* **2023**, *24*, 180–194. [CrossRef]
77. Rozenberg, I.N.; Gvishiani, A.D.; Soloviev, A.A.; Voronin, V.A.; Pilipenko, V.A. Influence of space weather on the reliability of railway transport in the Arctic zone of Russia. *Railw. Transp. (Zheleznodorozhny Transp.)* **2021**, *12*, 20–26. (In Russian)

applied sciences

MDPI

Article

Identification of Areas of Anomalous Tremor of the Earth's Surface on the Japanese Islands According to GPS Data

Alexey Lyubushin

Institute of Physics of the Earth, Russian Academy of Sciences, 123242 Moscow, Russia; lyubushin@yandex.ru

Abstract: Statistical properties of Earth surface tremors measured by means of GPS were investigated. This article considers measurements of the Earth's surface displacements in three orthogonal directions relayed by a network of GPS sensors with about 1200 points distributed across Japan in 2009–2021. Next, the following characteristics of the tremors were considered: the entropy of the distribution of squared orthogonal wavelet coefficients, the entropy of the distribution of power spectrum values, and the spectral index. The anomalous regions of maxima of probability densities of the distribution of extreme values of the tremor statistics were determined: entropy minima and spectral index maxima. The average density maps of the distribution of extreme value tremor statistics were found to be highly correlated with one another. This made it possible to consider a weighted average density map and identify five anomalous regions in the center and south of Japan. A trajectory of visiting anomalous regions by a sequence of points realizing local maxima of the average probability density was obtained, for which seasonal periodicity was set. Estimates of changes in the average and maximum values of the correlation coefficients of tremor properties in an auxiliary network of 16 reference points in a semi-annual time window were obtained.

Keywords: GPS tremor; entropy; spectral index; probability density; correlations

Citation: Lyubushin, A. Identification of Areas of Anomalous Tremor of the Earth's Surface on the Japanese Islands According to GPS Data. *Appl. Sci.* **2022**, *12*, 7297. https://doi.org/10.3390/app12147297

Academic Editors: Alexei Gvishiani and Boris Dzeboev

Received: 1 July 2022
Accepted: 19 July 2022
Published: 20 July 2022

Publisher's Note: MDPI stays neutral with regard to jurisdictional claims in published maps and institutional affiliations.

1. Introduction

Analysis of the spectral structure and models of the noise component of a GPS time series is traditionally of interest to researchers. In [1], spectral indexes were analyzed by fitting power laws to spectra and estimating their parameters using the maximum likelihood method. The effect of the flicker noise intensity of a GPS time series on velocity determination errors, depending on the measurement latitude, was studied in [2,3]. In [4–6], the maximum likelihood method was applied to estimate the amplitudes of GPS time series noise component models for various regions of the world. The problem of estimating the error in determining the velocity based on the spectral index was considered in [7].

The use of parametric models of a GPS time series to consider the effect of gaps in data for tectonically active regions is described in [8]. In [9,10] the high-frequency noise structure of a GPS time series for a number of areas in New Zealand and the United States was analyzed. The influence of the seasonal component of the hydrological load on the Earth's surface on the determination of the velocities of tectonic plates was studied in [11,12]. The extraction of common spatial characteristics and seasonal components of a GPS time series using the principal component method, empirical orthogonal functions, and singular spectrum analysis was considered in [13,14]. In [15], the high-frequency noise of a GPS time series was analyzed together with the records of the seismic network of accelerometers.

The relationship between the appearance of non-stationary effects in the GPS time series and the processes of slow motions at the boundaries between crustal blocks was analyzed in [16]. Estimates of the non-stationary effects of GPS time series and station positioning stability using discrete wavelet transforms were considered in [17,18]. In [19,20], the coherence of the tremors of the Earth's surface measured by GPS was studied. An assessment of the fractal characteristics of a high-frequency GPS time series to identify potentially dangerous seismically active regions in Japan and California was used in [21–23].

This article analyzes the structure of high-frequency noise in the time series of the network of GPS stations in Japan (about 1200 points) over 13 years of observations, 2009–2013, with a time step of 5 min. Four characteristics of tremors are considered: the entropy of the distribution of squared orthogonal wavelet coefficients, the entropy of the distribution of power spectrum values, and two estimates of the spectral index, which is understood as the regression coefficient between the logarithm of the power spectrum and the logarithm of the period. The first estimate of the spectral index is traditional, by calculating the power spectrum. Another estimate of the slope is based on orthogonal wavelet decompositions, where the averages of the squares of the wavelet coefficients are considered as the contribution of the frequency band of the detail level of the decomposition to the total signal energy. For successive time fragments 1 day long (288 readings with a time step of 5 min), for measurements at each station, the values of the four statistics described above are calculated after detrending by a 4th order polynomial. The removal of the trend is aimed at getting rid of the influence of external deterministic factors, such as tidal and thermal deformations, and switching to the analysis of the random high-frequency component of the Earth's surface displacements, which we will call tremors. The anomalous regions of maxima of probability densities of the distribution of extreme values of the studied tremor statistics are determined: entropy minima and spectral index maxima. This definition of tremor anomalous behavior is based on the fact that white noise is characterized by maximum entropy values and a zero spectral index. Consequently, the maximum deviation from uninformative "white-noise" behavior should be characterized by entropy minima and spectral slope maxima.

2. Materials and Methods

Ground displacement data measured by GPS are taken from the website [24]. These data represent displacements of the Earth's surface in three orthogonal directions, which we will further denote as E, N, and Up for displacements in the west-east, north-south and vertical directions, with a time step of 5 min [25].

Figure 1a shows the positions of GPS points for measuring displacements of the Earth's surface, as well as 16 numbered reference points. These points will be further needed to assess the spatial correlations between the properties of the Earth's surface tremors. Reference point positions were noted as the centers of 16 clusters of GPS stations using the K-means method [26]. The numbering was performed by decreasing the latitude of the reference points. Figure 1b presents the density of GPS stations, which were estimated by Gaussian kernel method [26] with bandwidth 0.5° ($\approx$56 km), within a regular 30 $\times$ 30 network of nodes which covered the region. Figure 1c shows a graph of the daily number of operational GPS stations for 13 years of observations, 2009–2021.

By operational stations on the current day, we mean those stations for which the total number of missing data does not exceed 10% of the duration of the daily window of 288 counts with a time step of 5 min, that is, no more than 28 values. Missing data are filled in with "plausible" artificial data, the values of which are determined by the behavior of the signals in adjacent intervals of the same length as the length of the gap. This data replenishment operation is described in detail in [19].

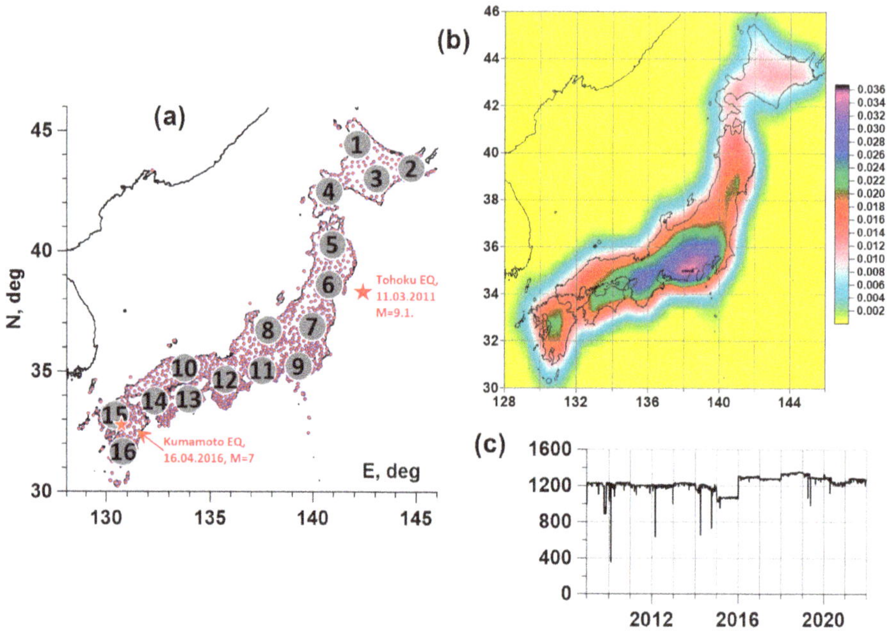

Figure 1. (**a**)—position of GPS measurement points in Japan (purple dots) and positions of 16 reference points (grey circles), red stars indicate epicenters of Tohoku (11 March 2011, M = 9.1) and Kumamoto (16 April 2016, M = 7) earthquakes; (**b**)—density of GPS stations; (**c**)—daily numbers of operational GPS stations in Japan, 2009–2021.

2.1. Minimum Normalized Wavelet-Based Entropy En_W and Wavelet-Based Spectral Index β_W

Let $c_j^{(k)}$ be wavelet coefficients of the analyzed signal $x(t), t = 1, \ldots, L$-discrete index numbering successive readings. The upper index k is the number of the detail level of the orthogonal wavelet decomposition, the lower index j numbers the sequence of time interval centers in the vicinity of which the signal convolution with finite elements of the basis is calculated. Below, 17 orthogonal Daubechies wavelets were used: 10 ordinary bases with minimal support with the number of vanishing moments from 1 to 10 and 7 so-called Daubechies symlets [27], with the number of vanishing moments from 4 to 10. For each of the bases, the normalized entropy of the distribution of squares was calculated coefficients and found a basis that provides the minimum entropy:

$$En_W = -\sum_{k=1}^{m} \sum_{j=1}^{M_k} p_j^{(k)} \cdot \ln(p_j^{(k)}) / \ln(N_r) \rightarrow \min, \; p_j^{(k)} = |c_j^{(k)}|^2 / \sum_{l,i} |c_i^{(l)}|^2 \quad (1)$$

Here, m is the number of detail levels taken into consideration, and M_k is the number of wavelet coefficients at the detail level with the number k. The number of levels m depends on the length L of the analyzed sample. For example, if $L = 2^n$, then $m = n$, and $M_k = 2^{(n-k)}$. The condition $L = 2^n$ is necessary for applying the fast wavelet transform. If the length L is not equal to a power of two, then the signal $x(t)$ is padded with zeros to the minimum length N, which is greater than or equal to L: $N = 2^n \geq L$. In this case, among the number $2^{(n-k)}$ of all wavelet coefficients at the level k, only $L \cdot 2^{-k}$ coefficients correspond to the decomposition of the real signal, while the remaining coefficients are approximately equal to zero due to the addition of zeros to the signal $x(t)$. Thus, in Formula (1) $M_k = L \cdot 2^{-k}$, and only "real" coefficients $c_j^{(k)}$ are used to calculate the entropy. The number N_r in Formula (1) is equal to the number of "real" coefficients, that is $N_r = \sum_{k=1}^{m} M_k$. By construction $0 \le En_W \le 1$.

After determining the optimal orthogonal wavelet, we can calculate the average values of the squares of the wavelet coefficients at each detail level:

$$S_k = \sum_{j=1}^{M_k} \left| c_j^{(k)} \right|^2 / M_k. \quad (2)$$

The average value (2) of the squares of the wavelet coefficients is the part of the oscillation energy corresponding to the detail level k. In other words, this value can be considered as an estimate of the signal $x(t)$ power spectrum in the frequency band corresponding to the detail level k [27]:

$$[f_{\min}^{(k)}, f_{\max}^{(k)}] = [1/(2^{(k+1)}\Delta s), 1/(2^k \Delta s)] \quad (3)$$

where Δs is the length of the discretization time interval (in our case Δs = 5 min). Consider the values of the periods corresponding to the centers of the frequency bands (3):

$$T_k = 2/(f_{\min}^{(k)} + f_{\max}^{(k)}) = 2\Delta s/(2^{-k} + 2^{-(k+1)}) \quad (4)$$

Then the quantities $S_k = S(T_k)$, $k = 1, \ldots, m$ are similar to the usual Fourier power spectra. The difference lies in the fact that the values (4) are much smoother, which is convenient when calculating the spectral index-the slope of the graph of the logarithm of the power spectrum depending on the logarithm of the period. To calculate the wavelet-based spectral index β_W, consider the following model:

$$\ln(S(T_k)) = \beta_W \cdot \ln(T_k) + c + \varepsilon_k, \; k = 1, \ldots, m \quad (5)$$

where ε_k is a sequence of independent random variables with zero mean. The parameter β_W in Formula (5) can be called the wavelet-based spectral index, the value of which can be found by the least squares method: $\sum_{k=1}^{m} \varepsilon_k^2 \to \min_{c,\beta_W}$.

2.2. *Spectral Normalized Entropy En_S and "Usual" Spectral Index β_S*

The normalized entropy (1) refers to the entropy of the energy distribution on different time scales (because of the use of finite orthogonal basis functions). At the same time, the entropy can be estimated in the frequency domain. Let $S_{xx}(\omega_j)$ be an estimate of the signal $x(t)$ power spectrum, ω_j be a sequence of discrete frequency values with a step $\Delta\omega$, $\omega_j = (j-1) \cdot \Delta\omega$, $\Delta\omega = 2\pi/(N \cdot \Delta s)$, $j = 1, \ldots, N$. Quantities

$$p_j^{(S)} = S_{xx}(\omega_j) / \sum_{k=1}^{N/2} S_{xx}(\omega_k), \; j = 1, \ldots, N/2, \quad (6)$$

can be considered as the probabilities of energy distribution in a sequence of non-overlapping frequency intervals of length $\Delta\omega$. Here, N is the minimum length, which is greater than or equal to the sample length L: $N = 2^n \ge L$, the index j in Formula (6) varies up to $N/2$ due to the symmetry of the power spectrum of real signals with respect to the Nyquist frequency $\omega_{N/2+1}$: $S_{xx}(\omega_{N/2+1-k}) = S_{xx}(\omega_{N/2+1+k})$, $k = 1, \ldots, N/2-1$. Similarly to the values (1) and (5), we introduce the spectral normalized entropy En_S and spectral index β_S according to formulae:

$$\begin{aligned} En_S &= -\sum_{j=1}^{N/2} p_j^{(S)} \ln(p_j^{(S)}) / \ln(N/2), \; 0 \le En_S \le 1; \\ \ln(S_{xx}(T_j)) &= \beta_S \cdot \ln(T_j) + d + \eta_j, \; j = 1, \ldots, N/2 \end{aligned} \quad (7)$$

where period $T_j = 2\pi/\omega_j$, the value of β_S is defined from the least squares approach: $\sum_{j=1}^{N/2} \eta_j^2 \rightarrow \min_{d,\beta_S}$, η_j is a sequence of independent random variables with zero mean.

The values of En_W, En_S, β_W, and β_S were estimated in successive time windows consisting of 288 adjacent 5-min readings, which is one day. Before their calculation, the operation of detrending by a polynomial of the fourth order was performed. Thus, for each station, time series were obtained with a time step of 1 day. The maximum entropy method [28] with autoregression order 28 (10% of the sample length in the time window) was used to calculate the power spectrum $S_{xx}(\omega_j)$ estimate. The proximity of the normalized spectral entropy En_S to the maximum value of 1 means that the energy distribution over frequencies is close to uniform, i.e., the power spectrum is close to the white noise spectrum.

Since displacements of the Earth's surface in three orthogonal directions are measured at each point of the network, then, when necessary, we will distinguish the values of the analyzed tremor statistics for different directions of surface displacements by the superscript, for example, for entropy (1) $En_W^{(E)}$, $En_W^{(N)}$, $En_W^{(U)}$, where the superscripts correspond to measurements in the E, N, and Up directions. Similar notation will be used for the quantities, En_S, β_W, and β_S.

For an auxiliary network of 16 reference points (Figure 1a) 12 time series were determined with a time step of 1 day, the values of which were calculated as medians of the values $(En_W^{(E)}, En_W^{(N)}, En_W^{(U)})$, $(En_S^{(E)}, En_S^{(N)}, En_S^{(U)})$, $(\beta_W^{(E)}, \beta_W^{(N)}, \beta_W^{(U)})$, and $(\beta_S^{(E)}, \beta_S^{(N)}, \beta_S^{(U)})$ from the ten nearest operational GPS measurement points. Figure 2 shows graphs of such time series for central reference point number 8. These graphs show a strong seasonal periodicity of tremor properties.

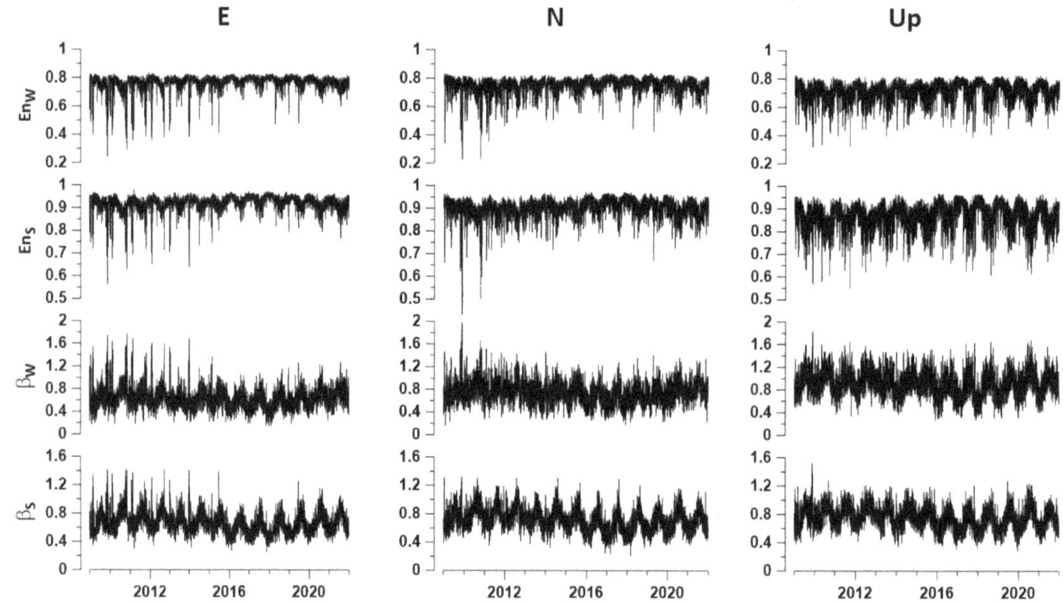

Figure 2. Plots of daily values for four GPS ground tremor properties for reference point number 8 for three ground displacement components E, N, and Up. The property values are calculated as medians from the ten nearest operable stations.

2.3. Probability Densities of Extreme Values

Further, estimates of 12 parameters of the Earth's surface tremors will be considered in the sequence of time windows 182 days long (half a year), taken with a shift of 3 days. In this case, in each window, we will consider only operable GPS stations, by which we mean those stations for which the total number of missing data in the current window does not exceed 0.1 of the window length.

Consider a regular grid of 30 × 30 nodes covering an area with a latitude from 28° N to 46° N and longitude from 128° E to 146° E. Let U be any value En_W, En_S, β_W, or β_S, whose estimates are obtained in a sliding time window. For each grid node (i, j) and for each time window labeled t at the right end of the window, find the 10 nearest seismic stations, which gives 10 values of U. Let's take their median value $U_{ij}^{(t)}$.

The values $U_{ij}^{(t)}$ form an "elementary" map corresponding to a time window of 182 days. Consider the values of the parameter as a function of two-dimensional vectors $z_{ij} = (x_i, y_j)$ of longitudes and latitudes of nodes (i, j) in an explicit form: $U_{ij}^{(t)} \equiv U^{(t)}(z_{ij})$. For each "elementary map" with a discrete time index t we find the coordinates of the nodes $z_{mn}^{(t)} = (x_m^{(t)}, y_n^{(t)})$ at which the extreme value of U is reached with respect to all other nodes of the regular grid.

If $U = En_W$ or $U = En_S$, then they will look for the minimum values. Otherwise, for $U = \beta_W$ or $U = \beta_S$, the maximum values will be searched. As noted above, the choice of a minimum for entropies and a maximum for spectral indexes is due to considerations that anomalous tremor regions should be characterized by the values of statistics that reflect the maximum deviation from the properties of white noise. A cloud of two-dimensional vectors $z_{mn}^{(t)}$ considered within a certain time interval $t \in [t_0, t_1]$ forms a random set. Let us estimate their two-dimensional probability distribution function for each node of the regular grid. For this, the Parzen–Rosenblatt estimate with the Gaussian kernel function [26] will be applied:

$$p(z_{ij}|t_0, t_1) = \frac{1}{2\pi h^2 (t_1 - t_0 + 1)} \sum_{t = t_0}^{t_1} \sum_{mn} \exp\left(-\frac{\left| z_{ij} - z_{mn}^{(t)} \right|^2}{2h^2} \right) \tag{8}$$

Here, h is the kernel averaging radius and t_0, t_1 are integer indices that enumerate the "elementary" maps of each time window. Thus, $(t_1 - t_0 + 1)$ is the number of time windows in the considered time interval. The width of the smoothing band was used $h = 0.5°$ which is approximately equal to 56 km. It should be noticed that the maximum distance from the nearest 10th stations to the reference points equals 47.5 km, the minimum equals 25.4 km, and the median maximum distance equals 37 km.

Figures 3 and 4 show the average distribution maps of the extreme values of the four tremor statistics under consideration for all three displacement components. Even a purely visual perception of these maps suggests that they have a strong similarity to each other. This visual impression is confirmed by the estimate of the matrix of pairwise correlation coefficients presented in Table 1. The minimum value of the correlation coefficient is 0.7044 for the pair, while for the pair the correlation coefficient reaches 0.9903.

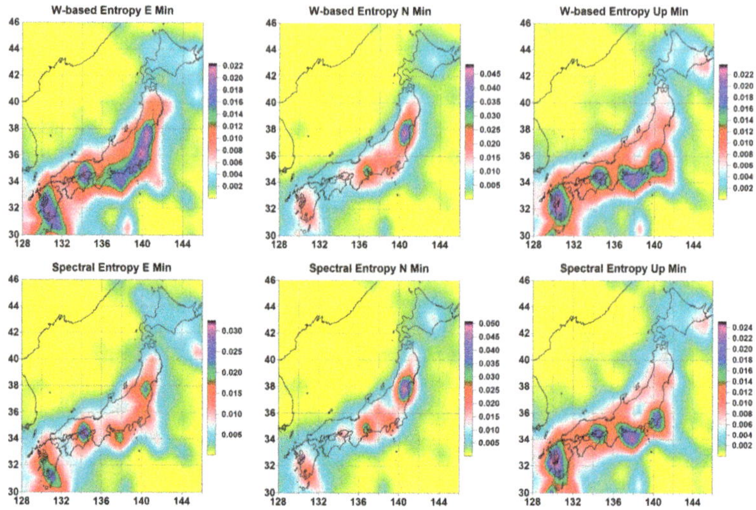

Figure 3. Averaged maps of the distribution of the minimum values of the wavelet-based and spectral entropies of the GPS tremor for the displacement components E, N, and Up on a regular grid of 30 × 30 nodes.

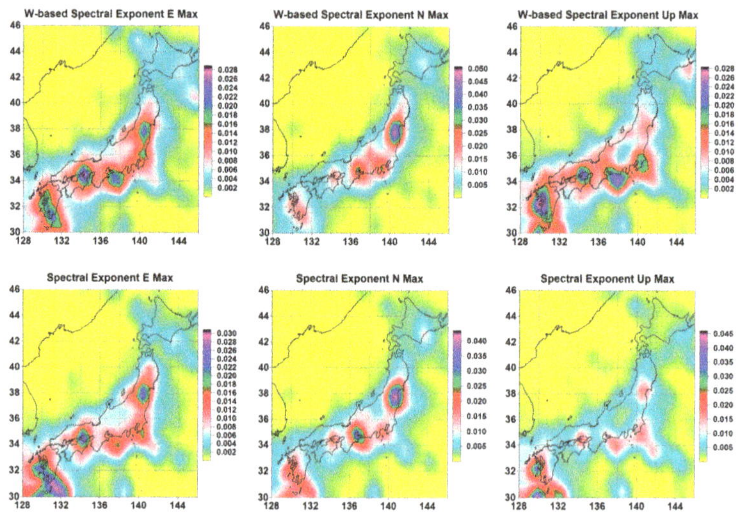

Figure 4. Averaged maps of the distribution of the maximum values of the wavelet-based and usual spectral indices of the GPS tremor for the displacement components E, N, and Up on a regular grid of 30 × 30 nodes.

The high correlation of probability density maps of the distribution of extreme values of tremor statistics suggests the use of the principal component method [29] to extract the most common distribution over the space of the most common characteristics. To do this, we find the eigenvector of the correlation matrix (Table 1), corresponding to the maximum eigenvalue of this matrix. The squares of the values of the components of this vector (by definition, their sum is equal to 1) will be taken as the weights of digital maps. The result of the weighted average of the probability densities of the distribution of extreme values in Figures 3 and 4 is shown in Figure 5. Figure 5a contains white contour lines which extract regions where the weighted probability density of minimum entropy and maximum spectral index exceeds the level of 98% quantile. These regions are considered as regions of anomalous GPS-measured Earth tremors.

Table 1. Matrix of pairwise correlation coefficients between 12 average probability density maps of extreme values of ground tremor statistics.

	$\beta_W^{(E)}$	$\beta_W^{(N)}$	$\beta_W^{(U)}$	$En_W^{(E)}$	$En_W^{(N)}$	$En_W^{(U)}$	$En_S^{(E)}$	$En_S^{(N)}$	$En_S^{(U)}$	$\beta_S^{(E)}$	$\beta_S^{(N)}$	$\beta_S^{(U)}$
$\beta_W^{(E)}$	1	0.8872	0.9315	0.9668	0.8746	0.9269	0.9884	0.8753	0.9336	0.9387	0.9049	0.8469
$\beta_W^{(N)}$	0.8872	1	0.7922	0.9052	0.9832	0.8161	0.8799	0.9903	0.8129	0.8215	0.941	0.7044
$\beta_W^{(U)}$	0.9315	0.7922	1	0.9129	0.7887	0.9764	0.9193	0.7826	0.9879	0.8994	0.8541	0.8948
$En_W^{(E)}$	0.9668	0.9052	0.9129	1	0.9141	0.9308	0.9625	0.8975	0.9253	0.9198	0.9045	0.8154
$En_W^{(N)}$	0.8746	0.9832	0.7887	0.9141	1	0.8227	0.8644	0.9851	0.815	0.8102	0.9372	0.6936
$En_W^{(U)}$	0.9269	0.8161	0.9764	0.9308	0.8227	1	0.9074	0.8041	0.9894	0.862	0.8402	0.8261
$En_S^{(E)}$	0.9884	0.8799	0.9193	0.9625	0.8644	0.9074	1	0.8701	0.9127	0.9486	0.8988	0.8565
$En_S^{(N)}$	0.8753	0.9903	0.7826	0.8975	0.9851	0.8041	0.8701	1	0.803	0.8236	0.9477	0.7055
$En_S^{(U)}$	0.9336	0.8129	0.9879	0.9253	0.815	0.9894	0.9127	0.803	1	0.8787	0.8524	0.85
$\beta_S^{(E)}$	0.9387	0.8215	0.8994	0.9198	0.8102	0.862	0.9486	0.8236	0.8787	1	0.905	0.9326
$\beta_S^{(N)}$	0.9049	0.941	0.8541	0.9045	0.9372	0.8402	0.8988	0.9477	0.8524	0.905	1	0.8453
$\beta_S^{(U)}$	0.8469	0.7044	0.8948	0.8154	0.6936	0.8261	0.8565	0.7055	0.85	0.9326	0.8453	1

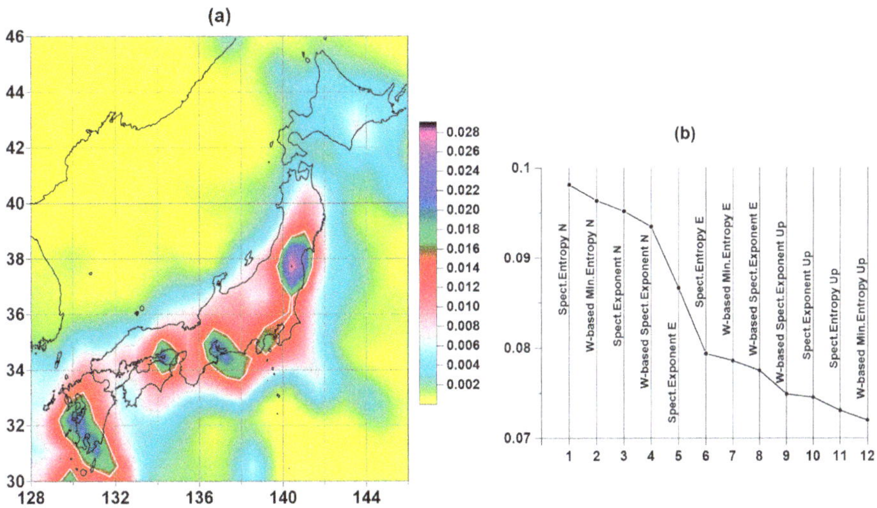

Figure 5. (**a**)—weighted average map of distribution of extreme values of four properties of the Earth's surface tremors for three displacement components E, N, and Up, white contour lines extract anomalous regions of Earth's tremor as 98% quantile of probability density values; (**b**)—values of the squares of the main eigenvector components of the correlation matrix between the distribution maps of extreme values of four daily tremor properties of the Earth's surface for three displacement components E, N, and Up on a regular 30 × 30 grid of nodes, sorted in descending order.

Figure 5b is a plot of the squares of the eigenvector components of the correlation matrix corresponding to the maximum eigenvalue after sorting in descending order. From this plot, it can be seen that the GPS noise statistics corresponding to measurements in the N direction contribute the most to the averaging weights. The smallest contribution corresponds to the vertical Up measurements. However, it should be noted that the differences in the weights are insignificant.

3. Trajectory of Extreme Mean Probability Density Maxima

According to the location of local maxima of the weighted average probability density in Figure 5a, 5 areas of anomalous tremors can be distinguished on the Japanese islands. Since the probability densities were estimated in a sliding time window 182 days long, it is of interest to obtain the trajectory of geographic coordinates corresponding to the position of the local maxima of the averaged probability density in each of the successive time windows.

To obtain such a trajectory for each time window 182 days long, we calculate the weighted average of all 12 probability densities of the distribution of extreme values of the used tremor statistics. Thus, in each time window, we calculate a 12×12 correlation matrix similar to the matrix of averaged correlations in Table 1, determine the eigenvector corresponding to the maximum eigenvalue, and use the squares of its components as weights to calculate the average weighted probability density in each window. Next, we find the coordinates of the point that realizes the maximum average density. This sequence of operations performed in each time window makes it possible to obtain the trajectory of geographical coordinates of points that realize the maximum values of the average probability density.

Figure 6(a_1,a_2) shows the graphs of changes in the geographical coordinates of the trajectory points of the maxima of the average weighted probability densities of the extreme values of the tremor statistics. Recall that, according to the construction, these points are nodes of a regular grid of size 30×30 covering the region under study. Figure 6b shows the points at which the maxima of the probability densities are realized. These points are divided into five clusters, and cluster number four consists of one point, which is repeatedly visited by the trajectory (plot in Figure 6(a_3)).

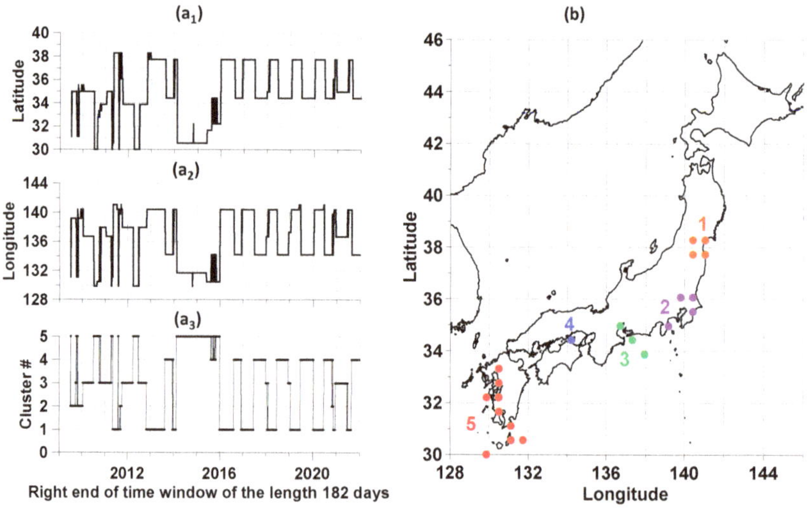

Figure 6. ($\mathbf{a_1}$,$\mathbf{a_2}$) are the latitudes and longitudes of the nodes of a regular grid of 30×30 in size, in which the maxima of the average weighted probability density of the distribution of extreme values of the properties of the tremor of the Earth's surface were realized in sliding time windows 182 days long; ($\mathbf{a_3}$) are the corresponding numbers of clusters of maximum points. (**b**)—division of points of the trajectory of maxima into five clusters.

4. Spatial Correlations of Tremor Properties

In this section of the article, spatial correlations between tremor properties in the network of 16 reference points presented in Figure 1 will be investigated. Property values are computed at 16 reference points as medians of daily values from the 10 closest operational stations. Examples of graphs of such time series for reference point #8 are shown in Figure 2.

Consider the correlation coefficients between the same tremor properties at different reference points. Denote by $\rho_{ij}^{(\tau)}(P)$ the estimates of the correlation coefficients between the values of the property P at the reference points with numbers i and j for a sequence of time windows 182 days long with time marks τ of the right ends of the windows. As P can be any of the 12 tremor properties: $En_W^{(E)}$, $En_W^{(N)}$, $En_W^{(U)}$, $En_S^{(E)}$, $En_S^{(N)}$, $En_S^{(U)}$, $\beta_W^{(E)}$, $\beta_W^{(N)}$, $\beta_W^{(U)}$, $\beta_S^{(E)}$, $\beta_S^{(N)}$, $\beta_S^{(U)}$.

For each property, we determine the average value of the correlation coefficients for all pairs of reference points:

$$\bar{\rho}^{(\tau)}(P) = \sum_{(i,j)} \rho_{ij}^{(\tau)}(P)/M \tag{9}$$

In Formula (9) M is the number of different pairs of reference points. In our case, for 16 reference points $M = 120$. Let us denote by $n^{(\tau)}(P)$ the number of pairs of reference points for which the corresponding correlation coefficients $\rho_{ij}^{(\tau)}(P)$ exceed the threshold of 0.7 in the time windows with the label of the right end τ. These quantities define some of the strong spatial correlations of tremor properties. The long-range action of strong correlations is described by the values $d^{(\tau)}(P)$ of the maximum distances between those pairs of reference points for which the correlation coefficients $\rho_{ij}^{(\tau)}(P)$ exceed the threshold of 0.7.

The obtained values $\bar{\rho}^{(\tau)}(P)$, $n^{(\tau)}(P)$, $d^{(\tau)}(P)$ average over all 12 properties, that is, consider the values:

$$\rho_\tau^* = \sum_P \bar{\rho}^{(\tau)}(P)/12,\ n_\tau^* = \sum_P n^{(\tau)}(P)/12,\ d_\tau^* = \sum_P d^{(\tau)}(P)/12 \tag{10}$$

Dependences (10) are shown in Figure 7. Quantities (10) could be called measures of the activity of the Earth's surface tremor.

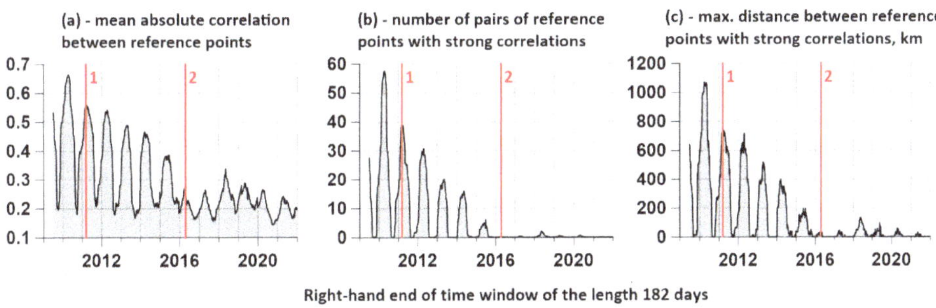

Figure 7. (**a**) is a plot of mean correlations ρ_τ^*; (**b**) is a graph of the average number n_τ^* of pairs of control points for which the correlations exceeded the threshold of 0.7; (**c**) is a graph of the average maximum distances d_τ^* between those pairs of control points for which the correlation coefficients exceed the threshold of 0.7. Vertical red lines indicate time moments of earthquakes: 1—Tohoku, 11 March 2011, M = 9.1; 2—Kumamoto, 16 April 2016, M = 7.

For each reference point with a number $j = 1, \ldots, 16$, it is possible to calculate how the average value $\rho_\tau^{(j)}$ of the correlation coefficients of the tremor statistics with the rest of the reference points changes in each time window. When calculating the values $\rho_\tau^{(j)}$, the average is made over all 12 properties of the tremor of the Earth's surface. Graphs of values $\rho_\tau^{(j)}$ are shown in Figure 8.

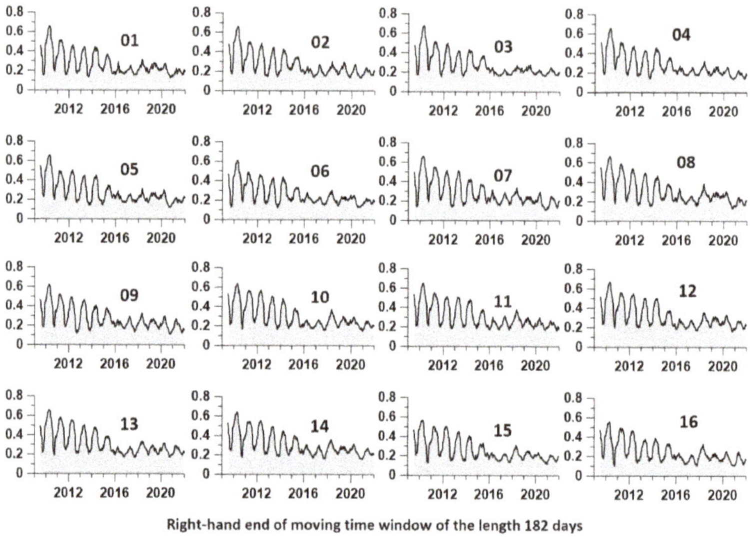

Figure 8. Graphs of average correlation coefficients $\rho_{\tau}^{(j)}$ for each reference point with all other reference points.

The presence of values in 16 reference points distributed throughout Japan allows the construction of maps of spatial values. The construction of such maps provides additional information on the features of the distribution of the correlation of tremor properties. For this, as before when constructing maps in Figures 3 and 4, consider a regular grid of 30 knots in latitude and 30 knots in longitude covering Japan. Then the values at the nodes of the regular grid are calculated by the nearest reference point neighbor method [26]. These values, calculated for each position of the time window at all nodes, make it possible to obtain "elementary" maps of the distribution over the space of average values of the correlation of the tremor property. The result of averaging all "elementary" maps for time windows lying between some 2 dates gives the corresponding averaged map. Figure 9 shows averaged correlation distribution maps for time windows before and after 2016. The distribution of values is given only for the territory of the Japanese Islands covered by the GPS network of stations and some of their surroundings.

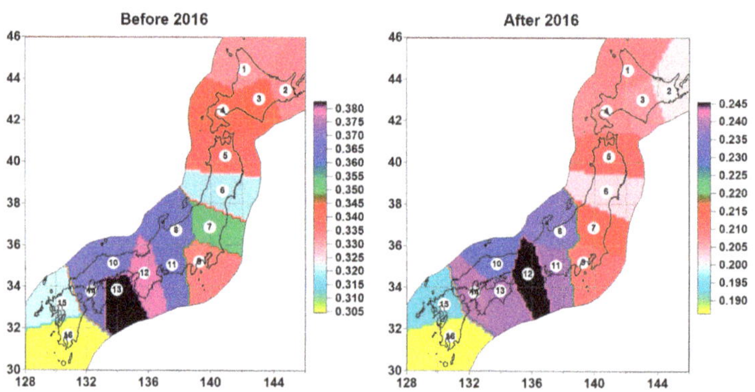

Figure 9. Average maps of interpolation of the distribution of average correlations $\rho_{\tau}^{(j)}$ from reference points for time windows before and after 2016, obtained by the nearest neighbor method. The numbered white circles represent the positions of the 16 reference points.

It follows from the maps in Figure 9 that the region of the highest spatial correlation of tremor properties is practically unchanged and occupies the southern part of Honshu Island, although the degree of correlation drops noticeably after 2016.

5. Discussion

A new method for estimating the intensity (activity measures) of the Earth's surface tremor based on measurements by a network of GPS sensors is proposed. The method is based on estimating the trajectory of points corresponding to the maxima of the average weighted probability densities of the distribution of extreme values of tremor statistics in a sliding time window. As tremor statistics, we chose the minimum values of the entropy of the distribution of squared orthogonal wavelet coefficients and the entropy of the power spectrum, as well as the maximum values of the spectral index calculated for the power spectra and the average values of the squared wavelet coefficients. A method has also been developed for estimating the variability of the spatial correlation of tremor properties based on estimates of the correlation values of statistics in an auxiliary network of reference points in a sliding time window.

Methods which used values of spectral index for GPS time series investigations were elaborated in [1–6] for identifying models of GPS noise. Similar methods were applied in [21–23] for detecting regions which are suspected for increased seismic danger although these methods are based on detrended fluctuations analysis (DFA). However, these methods do not investigate evolution of points providing maximum to probability density of extreme values of spectral index and entropy of GPS noise within moving time window what is the key tool in the current article.

The developed methods were applied to the data of the GPS network in Japan in 2009–2021. Five areas of tremor activity were identified in the center and south of Japan, and a time point (2016) for a significant restructuring of the tremor activation mode was determined. Anomalies of tremor activity were obtained, presumably associated with the preparation of the Tohoku, 11 March 2011, $M = 9.1$, $Lat = 38.32°$, $Lon = 142.37°$ and Kumamoto, 16 April 2016, $M = 7$, $Lat = 32.77°$, $Lon = 130.72°$ earthquakes.

The trajectory of the points of maximum average probability densities of the distribution of extreme values of tremor statistics and their belonging to the selected 5 activity areas in Figure 6 demonstrates a clear division into two large time fragments with significantly different behaviors; before and after 2016. The interval after 2016 is characterized by stable seasonal periodicity of switching the trajectory of the maxima of the average probability density between the first and fourth clusters. For the first time interval before 2016, we cannot notice such regularity. Thus, it could be called "chaotic," although this name simply underlines the existence of the change point in the behavior of maxima trajectory. During the two years 2014–2016, tremor activity is concentrated in the south of Japan, in the fifth cluster (Figure 6(a_3)). It is possible that this 2-year concentration of the tremor activation maximum in the fifth cluster reflects the preparation of the Kumamoto earthquake. We also note that the north of Japan in 2009–2021 does not apply to abnormal areas of GPS tremor.

It follows from the graphs in Figure 7 that all average tremor properties are subject to a strong annual periodicity. Note that the maxima of the average correlation ρ^*_τ, the average number n^*_τ of pairs of reference points with strong correlations, and the average maximum distances d^*_τ of strong correlations correspond to the time windows falling on the winter period and the beginning of spring. Up until 2016, there was a gradual decrease in the annual peaks of the mean values of these measures of tremor activity intensity. After 2016, the average correlations reached a background level of random fluctuations in the vicinity of 0.2, while the values of the measures n^*_τ and d^*_τ were almost equal to zero. The first half of 2010 displays the maximum value of annual bursts of tremor activity. It is natural to assume that these bursts reflect the process of preparing for the Tohoku mega-earthquake on 11 March 2011.

Funding: The work was carried out within the framework of the state assignment of the Institute of Physics of the Earth of the Russian Academy of Sciences.

Data Availability Statement: Publicly available datasets were analyzed in this study. This data can be found here: Nevada Geodetic Laboratory, http://geodesy.unr.edu/NGLStationPages/RapidStationList (accessed on 18 July 2022).

Conflicts of Interest: The authors declare no conflict of interest.

References

1. Langbein, J.; Johnson, H. Correlated errors in geodetic time series, implications for time-dependent deformation. *J. Geophys. Res.* **1997**, *102*, 591–603. [CrossRef]
2. Zhang, J.; Bock, Y.; Johnson, H.; Fang, P.; Williams, S.; Genrich, J.; Wdowinski, S.; Behr, J. Southern California permanent GPS geodetic array: Error analysis of daily position estimates and site velocities. *J. Geophys. Res.* **1997**, *102*, 18035–18055. [CrossRef]
3. Mao, A.; Harrison, C.G.A.; Dixon, T.H. Noise in GPS coordinate time series. *J. Geophys. Res.* **1999**, *104*, 2797–2816. [CrossRef]
4. Williams, S.D.P.; Bock, Y.; Fang, P.; Jamason, P.; Nikolaidis, R.M.; Prawirodirdjo, L.; Miller, M.; Johnson, D.J. Error analysis of continuous GPS time series. *J. Geophys. Res.* **2004**, *109*, B03412. [CrossRef]
5. Bos, M.S.; Bastos, L.; Fernandes, R.M.S. The influence of seasonal signals on the estimation of the tectonic motion in short continuous GPS time-series. *J. Geodyn.* **2010**, *49*, 205–209. [CrossRef]
6. Wang, W.; Zhao, B.; Wang, Q.; Yang, S. Noise analysis of continuous GPS coordinate time series for CMONOC. *Adv. Space Res.* **2012**, *49*, 943–956. [CrossRef]
7. Caporali, A. Average strain rate in the Italian crust inferred from a permanent GPS network—I. Statistical analysis of the time-series of permanent GPS stations. *Geophys. J. Int.* **2003**, *155*, 241–253. [CrossRef]
8. Li, J.; Miyashita, K.; Kato, T.; Miyazaki, S. GPS time series modeling by autoregressive moving average method, Application to the crustal deformation in central Japan. *Earth Planets Space* **2000**, *52*, 155–162. [CrossRef]
9. Beavan, J. Noise properties of continuous GPS data from concrete pillar geodetic monuments in New Zealand and comparison with data from U.S. deep drilled braced monuments. *J. Geophys. Res.* **2005**, *110*, B08410. [CrossRef]
10. Langbein, J. Noise in GPS displacement measurements from Southern California and Southern Nevada. *J. Geophys. Res.* **2008**, *113*, B05405. [CrossRef]
11. Blewitt, G.; Lavallee, D. Effects of annual signal on geodetic velocity. *J. Geophys. Res.* **2002**, *107*, 2145. [CrossRef]
12. Bos, M.S.; Fernandes, R.M.S.; Williams, S.D.P.; Bastos, L. Fast error analysis of continuous GPS observations. *J. Geod.* **2008**, *82*, 157–166. [CrossRef]
13. Teferle, F.N.; Williams, S.D.P.; Kierulf, H.P.; Bingley, R.M.; Plag, H.P. A continuous GPS coordinate time series analysis strategy for high-accuracy vertical land movements. *Phys. Chem. Earth Parts A/B/C* **2008**, *33*, 205–216. [CrossRef]
14. Chen, Q.; van Dam, T.; Sneeuw, N.; Collilieux, X.; Weigelt, M.; Rebischung, P. Singular spectrum analysis for modeling seasonal signals from GPS time series. *J. Geodyn.* **2013**, *72*, 25–35. [CrossRef]
15. Bock, Y.; Melgar, D.; Crowell, B.W. Real-Time Strong-Motion Broadband Displacements from Collocated GPS and Accelerometers. *Bull. Seismol. Soc. Am.* **2011**, *101*, 2904–2925. [CrossRef]
16. Hackl, M.; Malservisi, R.; Hugentobler, U.; Jiang, Y. Velocity covariance in the presence of anisotropic time correlated noise and transient events in GPS time series. *J. Geodyn.* **2013**, *72*, 36–45. [CrossRef]
17. Goudarzi, M.A.; Cocard, M.; Santerre, R.; Woldai, T. GPS interactive time series analysis software. *GPS Solut.* **2013**, *17*, 595–603. [CrossRef]
18. Khelif, S.; Kahlouche, S.; Belbachir, M.F. Analysis of position time series of GPS-DORIS co-located stations. *Int. J. Appl. Earth Observ. Geoinf.* **2013**, *20*, 67–76. [CrossRef]
19. Lyubushin, A. Global coherence of GPS-measured high-frequency surface tremor motions. *GPS Solut.* **2018**, *22*, 116. [CrossRef]
20. Lyubushin, A. Field of coherence of GPS-measured earth tremors. *GPS Solut.* **2019**, *23*, 120. [CrossRef]
21. Filatov, D.M.; Lyubushin, A.A. Fractal analysis of GPS time series for early detection of disastrous seismic events. *Phys. A* **2017**, *469*, 718–730. [CrossRef]
22. Filatov, D.M.; Lyubushin, A.A. Precursory Analysis of GPS Time Series for Seismic Hazard Assessment. *Pure Appl. Geophys.* **2019**, *177*, 509–530. [CrossRef]
23. Filatov, D.M.; Lyubushin, A.A. Stochastic dynamical systems always undergo trending mechanisms of transition to criticality. *Phys. A* **2019**, *527*, 121309. [CrossRef]
24. Nevada Geodetic Laboratory. Available online: http://geodesy.unr.edu/NGLStationPages/RapidStationList (accessed on 1 July 2022).
25. Blewitt, G.; Hammond, W.C.; Kreemer, C. Harnessing the GPS data explosion for interdisciplinary science. *Eos* **2018**, *99*, 485. [CrossRef]
26. Duda, R.O.; Hart, P.E.; Stork, D.G. *Pattern Classification*; Wiley: Hoboken, NJ, USA, 2000.
27. Mallat, S. *A Wavelet Tour of Signal Processing*, 2nd ed.; Academic Press: Cambridge, MA, USA, 1999.
28. Marple, S.L., Jr. *Digital Spectral Analysis with Applications*; Prentice-Hall, Inc.: Englewood Cliffs, NJ, USA, 1987.
29. Jolliffe, I.T. *Principal Component Analysis*; Springer: Berlin/Heidelberg, Germany, 1986.

applied sciences

MDPI

Article

Singular Points of the Tremor of the Earth's Surface

Alexey Lyubushin

Institute of Physics of the Earth, Russian Academy of Sciences, 123242 Moscow, Russia; lyubushin@yandex.ru

Abstract: A method for studying properties of the Earth's surface tremor, measured by means of GPS, is proposed. The following tremor characteristics are considered: the entropy of wavelet coefficients, the Donoho–Johnston wavelet index, and two estimates of the spectral slope. The anomalous areas of tremor are determined by estimating the probability densities of extreme values of the studied properties. The criteria for abnormal tremor behavior are based on the proximity to, or the difference between, tremor properties and white noise. The greatest deviation from the properties of white noise is characterized by entropy minima and spectral slope and DJ index maxima. This behavior of the tremor is called "active". The "passive" tremor behavior is characterized by the maximum proximity to the properties of white noise. The principal components approach provides weighted averaged density maps of these two variants of extreme distributions of parameters in a moving time window of 3 years. Singular points are the points of maximum average densities. The method is applied to the analysis of daily time series from a GPS network in California during the period 2009–2022. Singular points of tremor form well-defined clusters were found. The passive tremor could be caused by the activation of movement in fragments of the San Andreas fault.

Keywords: GPS tremor singular points; entropy; spectral slope; probability density; principal components

Citation: Lyubushin, A. Singular Points of the Tremor of the Earth's Surface. *Appl. Sci.* **2023**, *13*, 10060. https://doi.org/10.3390/app131810060

Academic Editor: Alexei Gvishiani

Received: 14 August 2023
Revised: 2 September 2023
Accepted: 4 September 2023
Published: 6 September 2023

1. Introduction

The thin structure of the time series of displacements of the Earth's surface, measured by means of GPS, is the subject of research by a large number of specialists. The maximum likelihood approach for estimating parameters of GPS time series models was used in [1–4]. The structure of the noise and estimates of errors were considered in [5–7]. Parametric models of GPS time series for the analysis of phase correlations were used in [8]. Seasonal effects in GPS time series connected with hydrological loading and tectonic movements of the crust were studied in [9–16] by different methods including the multidimensional statistic approach. The influence of ionospheric and tropospheric delays on the sensitivity of the GPS solutions were considered in [17]. Hypotheses about origin of "abnormal" harmonics in GPS time series were regarded in [18].

The coherence of the Earth's surface tremor was analyzed in [19,20]. This article presents a further development of the methods for analyzing the Earth's surface tremor, proposed in [21], and analyzes the high-frequency noise structure in the time series of the GPS station network in the western United States during the time interval 2009–2022. The US West is one of the most seismically active regions. A dense network of GPS stations makes it possible to study in detail the features of the Earth's surface oscillations and try to compare them with global seismic activity.

The term "tremor" in geophysics has many different meanings. In seismology, tremor is understood as slow movements of the Earth's crust as a result of so-called silent earthquakes. In volcanology, tremor is the shaking of the Earth's surface in the vicinity of volcanoes, usually associated with eruptions. In this article, the tremor is the residual in the values of the GPS time series after the removal of trends in polynomials of the fourth order in time windows 365 days long. It should be emphasized that time series with a step of

1 day are considered below (that is, after the removal of trends) the range of periods under consideration changes from 2 days up to about 100 days.

2. Materials

GPS ground displacement data are taken from the Nevada Geodetic Laboratory website at: http://geodesy.unr.edu/gps_timeseries/tenv3/IGS14/ (accessed on 1 May 2023).

These data represent displacements of the Earth's surface in three orthogonal directions, which we will further denote as E, N, and U (Up) for displacements in the west–east, north–south, and vertical directions, with a time step of 1 day [22].

In what follows, we will consider a regular grid of 100 × 150 nodes, covering an area with latitude from 32.5° N to 51° N and longitude from 115° W to 128.5° W. Figure 1a shows the positions of GPS points for measuring displacements of the Earth's surface. Figure 1b shows the distribution density of GPS stations obtained using the Gaussian kernel estimate [23] at the nodes of this regular grid with a smoothing radius of 0.25° (about 28 km).

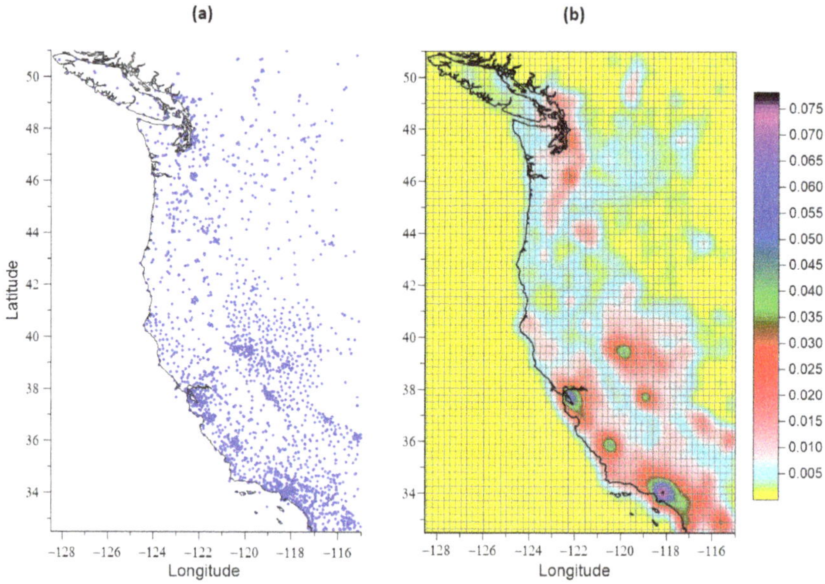

Figure 1. (**a**) Network of GPS stations (blue dots) in the US West, 2009–2022; (**b**) the density of GPS stations, estimated using a Gaussian kernel with a smoothing radius of 0.25 degrees on a regular grid of nodes with a size of 100 × 150. Plot (**b**) presents every second grid line by coordinates, because all the grid lines form too dense a network.

We will consider estimates of the properties of GPS time series in a moving time window 365 days long with a shift of 7 days. For each time window, operable stations are determined from the condition that the number of gaps in the considered time window does not exceed 5% of the time window length, that is, that this number is not more than 18 values. Missing values are filled in with artificial data, which are found from the value to the left and right of the missing time interval of the same length as the length of the gap [19]. Figure 2 shows a graph of the number of working stations within a sliding time window of 365 days for 14 years of observations, 2009–2022.

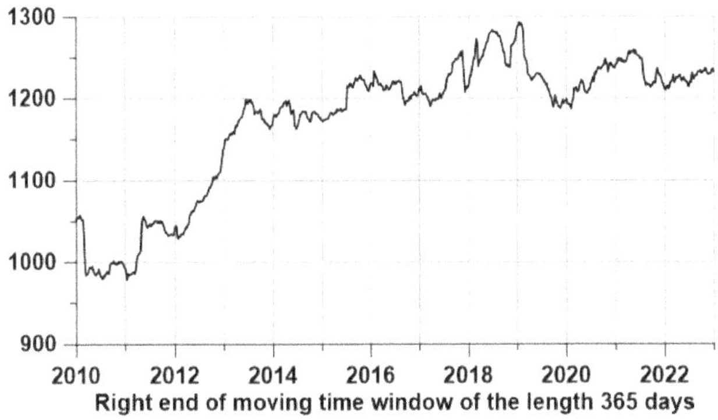

Figure 2. The number of operational GPS stations in a window of the length 365 days with mutual shift of 7 days.

3. Description of the Time Series Statistics Used

For each station in each time window, four surface tremor statistics are computed for the three ground displacement components.

Spectral exponent. Let $x(t)$, $t = 1, \ldots, N$ be a sample of the GPS time series within a window of sample lengths N. Denote by $S_{xx}(\omega_j)$ the estimate of the signal $x(t)$ power spectrum ω_j—a sequence of discrete values of frequencies with a step $\Delta\omega$, $\Delta\omega = 2\pi/(N_f \cdot \Delta s)$, $j = 1, \ldots, N_f$, Δs—a time step. Here, N_f is the minimum length, which is greater than or equal to the sample length N: $N_f = 2^n \geq N$, the index j in Formula (6) varies up to $N_f/2$ due to the symmetry of the power spectrum of real signals with respect to the Nyquist frequency $\omega_{N_f/2+1}$: $S_{xx}(\omega_{N_f/2+1-k}) = S_{xx}(\omega_{N_f/2+1+k})$, $k = 1, \ldots, N_f/2 - 1$. We introduce the spectral exponent α by the following formula:

$$\log(S_{xx}(T_j)) = \alpha \cdot \log(T_j)) + d + \eta_j,\ j = 1, \ldots, N_f/2, \tag{1}$$

where period $T_j = 2\pi/\omega_j$, the value α is determined by the least squares method: $\sum_{j=1}^{N_f/2} \eta_j^2 \rightarrow \min_{d,\alpha}$, η_j is a sequence of random variables with zero mean. In each time window, the power spectrum was computed after transition to increments using an autoregressive model (maximum entropy estimate) [24] with an autoregressive order of $N/10$, i.e., 36 for a window of length 365 samples. The transition to increments before calculating the power spectrum aims to free from the dominant effect of low frequencies in the daily time series of GPS.

We then address wavelet-based entropy and wavelet-based spectral slope. The minimum normalized wavelet information entropy for the sample $x(t)$, $t = 1, .., N$ is determined by the following formula:

$$En = -\sum_{k=1}^{N} p_k \cdot \log(p_k)/\log(N), \quad p_k = c_k^2 / \sum_{j=1}^{N} c_j^2 \tag{2}$$

Here, c_k denotes the coefficients of the orthogonal wavelet decomposition. Daubechies bases were used with the number of vanishing moments from 1 to 10 [25]. We chose such a basis from this set for which the value (2) is minimal for a given time window and for the considered component of the displacement of the Earth's surface. By Formula (2), $0 \leq En \leq 1$.

For the orthogonal wavelet, which is found from a minimum of entropy, the wavelet power spectrum can be calculated as the average values of the squares of the wavelet coefficients at each level of detail of the decomposition:

$$S_k = \sum_{j=1}^{M_k} \left| c_j^{(k)} \right|^2 / M_k. \tag{3}$$

In Formula (3), $c_j^{(k)}$ are the wavelet coefficients at the detail level with number k, $k = 1, \ldots k_{\max}$, index $j = 1, \ldots M_k$, where M_k is the number of wavelet coefficients at the detail level with number k, which corresponds to the frequency band $[\frac{1}{2^{(k+1)}\Delta s}, \frac{1}{2^k \Delta s}]$ with the central period $T_k = \frac{2\Delta s}{2^{-k} + 2^{-(k+1)}}$ [25]. The maximum detail level number $k_{\max}$ is chosen such that it contains at least a given number $L_{\min}$ of wavelet coefficients. Recall that the number of wavelet coefficients at the level of detail decreases by a factor of 2 when the level number increases by one. The value $L_{\min} = 4$ was used in the calculations, which gives the value $k_{\max} = 6$ for the window length $N = 365$. Values $S_k = S(T_k)$, are similar to conventional power spectra. The regression model is used to calculate the wavelet-based spectral slope:

$$\log(S(T_k)) = \beta \cdot \log(T_k)) + c + \varepsilon_k,\ k = 1, \ldots, k_{\max}, \tag{4}$$

where ε_k is a sequence of random variables with zero mean. The parameter β in Formula (4) is the wavelet-based spectral exponent, the value of which can be found by the least squares method: $\sum_{k=1}^{k_{\max}} \varepsilon_k^2 \to \min\limits_{\beta,c}$.

We then address the Donoho–Johnstone index (DJ-index). The threshold T_{DJ} is defined by the following formula [25,26]:

$$T_{DJ} = \sigma\sqrt{2 \cdot \ln N} \tag{5}$$

The threshold (5) separates rather large (informative), in their absolute values, wavelet coefficients from other coefficients which are considered to be noisy. Thus, we can consider the dimensionless signal characteristics γ, $0 \leq \gamma \leq 1$, as the ratio of the number of the largest wavelet coefficients, for which the inequality $|c_k| > T_{DJ}$ is fulfilled, to the number N of all wavelet coefficients.

The value σ in the Formula (5) is the noise standard deviation estimate under the assumption that the noise is most concentrated in the first-detail level of orthogonal wavelet decomposition. The estimate of standard deviation σ should be robust with respect to outliers in the values of the coefficients at the first level. To provide this, a median estimate of the standard deviation for a normal random variable is used:

$$\sigma = med\left\{ \left| c_k^{(1)} \right|,\ k = 1, \ldots, N/2 \right\} / 0.6745. \tag{6}$$

For each station in each time window, four surface tremor statistics are computed for the three ground displacement components where $c_k{}^{(1)}$ are wavelet coefficients at the first level of detail. The estimate of the standard deviation σ from Formula (6) determines the value (5) as a threshold for extracting noise wavelet coefficients. The quantity (5) is known in wavelet analysis as the Donoho–Johnstone threshold, and the expression for this quantity is based on the formula for the asymptotic probability of the maximum deviations of Gaussian white noise [25,26]. The DJ index γ values can be interpreted as a measure of the non-stationarity of seismic noise. For stationary Gaussian white noise, the index γ is zero. In [27,28], the DJ index was used as one of the main statistics for investigating properties of global seismic noise.

In summary, we will briefly describe in the list the characteristics of the values of the tremor statistics for stationary white Gaussian noise, which is considered as a standard of a completely random oscillation:

1. The "usual" spectral exponent α is the coefficient of regression between the logarithm of the power spectrum values and the logarithm of the period ("spectral slope"). For white noise, it is equal to zero (white noise has a "flat spectrum").
2. Wavelet-based spectral exponent β—the regression coefficient between the logarithm of the mean values of the squared wavelet coefficients for each level of detail and the logarithm of the period of the central frequency corresponding to the level of detail of the wavelet decomposition ("spectral slope"). For white noise, it is zero.
3. Donoho–Johnstone threshold (DJ index) γ—the ratio of the number of "large" in absolute magnitude orthogonal wavelet coefficients to the total number of coefficients. For white noise, it is zero.
4. Minimum entropy En of the distribution of squared orthogonal wavelet coefficients in the enumeration of wavelet basis functions in the class of Daubechies wavelets with the number of vanishing moments from 1 to 10. For white noise, the entropy is maximum.

Since we use four statistics for three components of the ground displacement vector, we obtain 12 statistics for each type of tremor. Let us put them in one matrix:

$$\left[\begin{pmatrix} \alpha_{E,\max} & \alpha_{N,\max} & \alpha_{U,\max} \\ \beta_{E,\max} & \beta_{N,\max} & \beta_{U,\max} \\ \gamma_{E,\max} & \gamma_{N,\max} & \gamma_{U,\max} \\ En_{E,\min} & En_{N,\min} & En_{U,\min} \end{pmatrix} \begin{pmatrix} \alpha_{E,\min} & \alpha_{N,\min} & \alpha_{U,\min} \\ \beta_{E,\min} & \beta_{N,\min} & \beta_{U,\min} \\ \gamma_{E,\min} & \gamma_{N,\min} & \gamma_{U,\min} \\ En_{E,\max} & En_{N,\max} & En_{U,\max} \end{pmatrix}\right] \tag{7}$$

On the left side of the matrix (7), there are the designations of the characteristics of active tremor, on the right side of the matrix (7) there are the designations of the characteristics of passive tremor. Recall that we are considering a regular grid of nodes with a size of 100 nodes in longitude and 150 nodes in latitude. For each time window with a length of 365 days with a shift of 7 days, the values of all 12 sequences of the distribution of tremor properties at the nodes of the regular grid are determined.

4. Extreme Value Probability Density Maps

Let U be any value from the matrix (7), the estimates of which were obtained in a moving time window. For each grid node (i, j) and for each time window labeled t at the right end of the window, we find the 10 closest seismic working stations, which gives 10 values of U. Let us take their median value $U_{ij}^{(t)}$. The values $U_{ij}^{(t)}$ form an "elementary" map corresponding to a time window of 365 days. Consider the values of the parameter as a function of two-dimensional vectors $z_{ij} = (x_i, y_j)$ of longitudes and latitudes of nodes (i, j) in an explicit form: $U_{ij}^{(t)} \equiv U^{(t)}(z_{ij})$. For each "elementary map" with a discrete time index t, we find the coordinates of the nodes $z_{mn}^{(t)} = (x_m^{(t)}, y_n^{(t)})$ at which the extreme value U is reached with respect to all other nodes of the regular grid.

As noted above, the choice of the minimum for entropies and the maximum for spectral slopes is due to the considerations that the regions of active tremor should be characterized by the values of statistics that reflect the maximum deviation from the properties of white noise. For passive tremor, the opposite is true: maxima for entropies and minima for spectral slopes. A cloud of 2D vectors $z_{mn}^{(t)}$ considered within a certain time interval $t \in [t_0, t_1]$ composes some random set. Let us estimate two-dimensional probability density function for each node of the regular grid. For this, the Parzen–Rosenblatt estimate with the Gaussian kernel function [23] will be applied:

$$p(z_{ij}|t_0, t_1) = \frac{1}{2\pi h^2(t_1 - t_0 + 1)} \sum_{t=t_0}^{t_1} \sum_{mn} \exp\left(-\frac{\left|z_{ij} - z_{mn}^{(t)}\right|^2}{2h^2}\right) \tag{8}$$

Here, h is the averaging radius, integer indices t_0, t_1 enumerate the maps of each time window. The $(t_1 - t_0 + 1)$ equals the number of time windows in the considered time interval. The width of the smoothing band $h = 0.25°$ was used, which is approximately equal to 28 km.

In our analysis, time windows of two lengths are used: a "short" window of 365 days in length, which is used to obtain estimates of tremor parameters, and a "long" window of 1095 days (3 years) in length, in which a sequence of estimates of the distribution densities of extreme values of statistics is the obtained tremor. Estimates of characteristics from matrix (7) are calculated using "short" time windows 365 days long. Next, the probability densities of extreme values are calculated from the values that fall within the current "long" time window of 1095 days, which are also shifted by 7 days. Each "long" window includes 105 "short" time windows.

As a result of such assessments, 24 maps of the distribution density of extreme values of tremor parameters were constructed, presented in Figure 3, on which 12 maps for active tremor are in the upper half, and 12 maps for passive tremor are in the lower half.

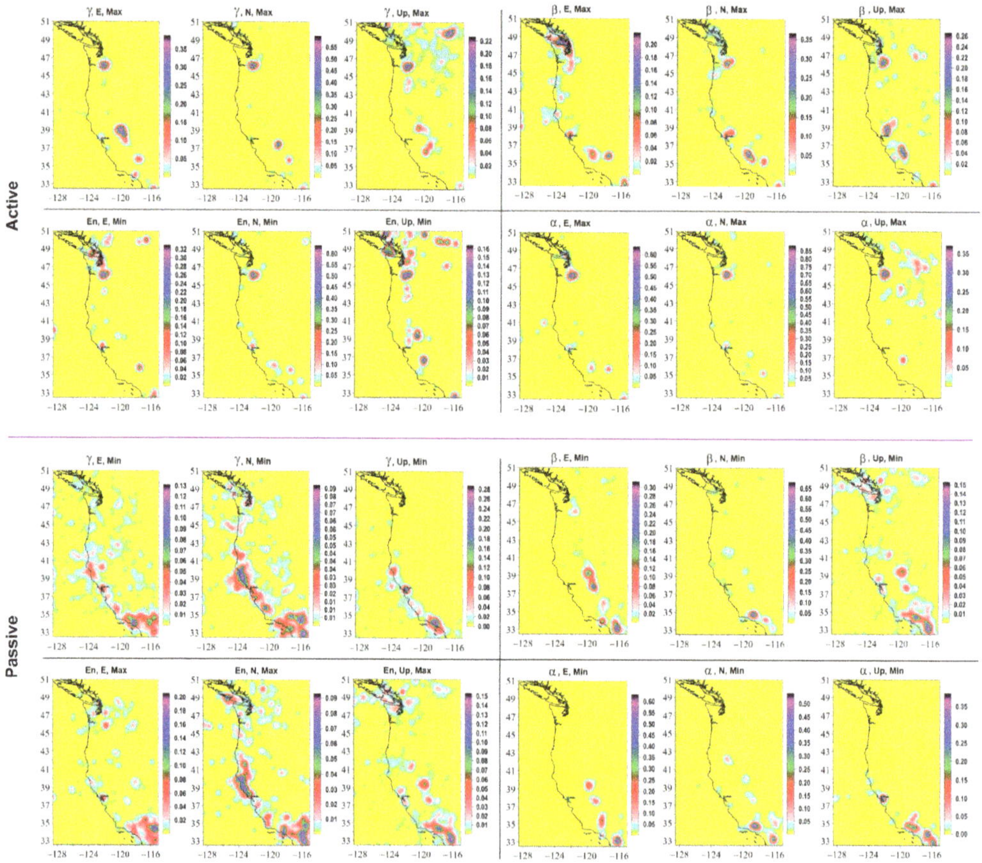

Figure 3. Average probability densities of distribution of extreme values of 4 statistics of "active" and "passive" tremor of the Earth's surface for 3 components of the displacement vector. Due to the fact that each property value is estimated over a 365-day window offset by 7 days, the average probability density estimate is obtained by averaging 548 probability density maps, each derived from 105 tremor property value maps.

Purely visually, a correlation of probability density maps of the distribution of extreme values of tremor statistics is noticeable. The principal component method [29] is used to extract the most common spatial characteristics. In each "long" 3-year time window, a correlation matrix of the size 12×12 between the probability densities of the distribution of extreme values of tremor statistics is calculated. The eigenvector corresponding to the maximum eigenvalue of the correlation matrix is found and the squares of its components were used as weights to calculate the average weighted probability density in each "long" window. The coordinates of the point that realizes the maxima of average density were found. These operations make it possible to obtain the trajectory of geographic coordinates of points that realize the maximum values of the average probability density. These points are called singular tremor points.

The results of weighting the probability density maps of the distribution of extreme properties of tremor by the method of principal components are shown in Figure 4. On Figure 4a, red circles indicate the positions of singular points of active tremor, which are combined into five clusters. The clusters are numbered in descending order of the latitude of their centers of mass: the cluster numbers are also shown. On Figure 4b, blue circles indicate the positions of singular points of passive tremor, which form three clusters. On Figure 4c,d, there are plots of histograms of the distribution of the number of singular points in each cluster for each type of tremor.

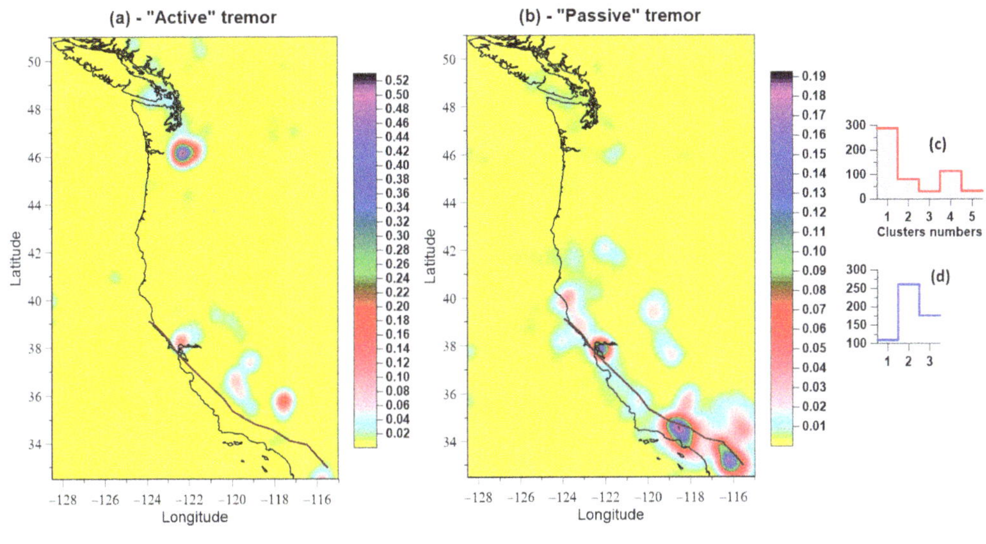

Figure 4. (**a**,**b**) are the results of averaging the weighted average probability densities of extreme values of 12 surface tremor parameters over a moving time window of 3 years for active and passive tremor. The red circles in (**a**) represent the positions of active tremor singular points forming 6 numbered clusters, the blue circles in (**b**) represent the positions of passive tremor singular points forming 3 numbered clusters. Bold brown line presents San Andreas fault. (**c**,**d**) are histograms of the numbers of singular points for "active" and "passive" tremors belonging to different clusters.

Figure 5 shows graphs of changes in the squares of the components of sequences of correlation matrices for probability density maps of the distribution of extreme values, respectively, for active (12 upper graphs) and passive (12 lower graphs). As noted above, these values are used as weight functions to obtain weighted average probability density maps. The result of such a weighing operation using the principal component method is shown in Figure 4. The graphs in Figure 5 show the average values of the weights calculated for all "long" time windows. For each tremor variant, four maximum average values of weights are highlighted in red—they select those components from matrix (7) that make the greatest contribution to obtaining the weighted average sum of probability densities.

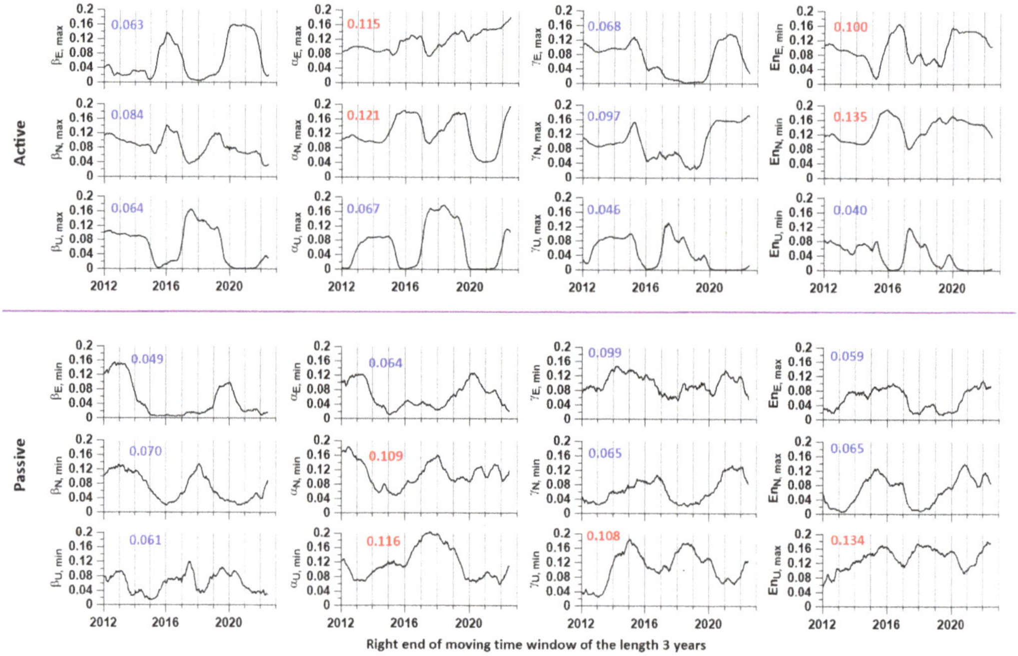

Figure 5. Plots of changes in the weights of each of the 12 probability densities in "long" time windows of 1095 days with a shift of 7 days for "active" and "passive" tremor. The numbers indicate the average values of the weights, the 4 largest averages out of 12 variants for each type of tremor are highlighted in red, other values are highlighted by blue.

The trajectories of changes in geographic latitudes and longitudes of singular points over time are shown in Figure 6a,b. Along the trajectories of change of coordinates, there are marks of belonging to clusters of stations. Figure 6c shows how the distances between the singular points of active and passive tremor change as the time window of the length of 3 years moves from left to right. From this graph, it can be seen that the concentration of singular points of both types of tremor in those regions where their positions are very close to each other, namely cluster #2 of active singular points and cluster #1 of passive singular points mainly occur in different time windows. The exception is the time intervals with labels of the right ends of the "long" 3-year windows 2012–June 2012 and June 2015–2016, when the distances from singular points of different types are less than 100 km, but do not reach zero values.

As can be seen from Figure 4b, clusters of singular points of passive tremor are concentrated in the vicinity of three fragments of the San Andreas fault. At the same time, Figure 6 shows that there is a switching mode from one cluster to another with long time intervals of belonging to one or another cluster. A hypothesis can be proposed that this behavior reflects the modes of activation of chaotic movements of crustal blocks in these three fragments of the fault, which causes the properties of high-frequency surface movements to approach white noise.

As for the singular points of active tremor, their occurrence may be associated with the processes of consolidation of small blocks of the Earth's crust and the formation of temporary larger structures that respond to external influences (such as atmospheric cyclones) with a greater amplitude, which, on the contrary, leads to an increase in the deviation from the properties of white noise.

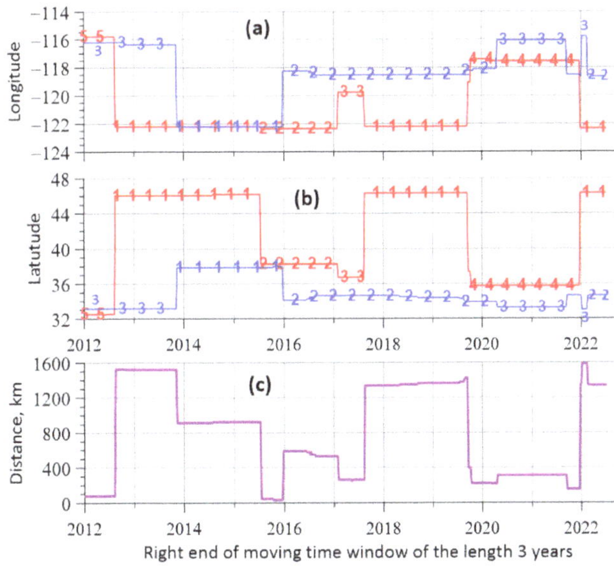

Figure 6. (**a**,**b**) represent time sequences of longitudes and latitudes of singular points of tremor with a shift in the time window of 3 years for "active" (red) and "passive" (blue) tremor with cluster membership marks. (**c**) is the distance between the singular points of active and passive tremor in each time window of 3 years. Numbers along red and blue lines indicate the numbers of the active and passive tremor clusters to which belong the current point of the trajectory. There are time intervals when longitude and latitude trajectories of singular points for different types of tremor are very close to each other.

Of particular interest are short time intervals of small distances between the singular points of active and passive tremor, which are visible in Figure 6c. They can highlight areas of the Earth's crust that are in an unstable state.

Let $\widetilde{\lambda}_1$ be the normalized maximum eigenvalue of the correlation matrix:

$$\widetilde{\lambda}_1 = \lambda_1 / \sum_{k=1}^{q} \lambda_k \tag{9}$$

In Formula (9), $\lambda_1, \ldots, \lambda_q$ are the eigenvalues of the correlation matrix, sorted in descending order: $\lambda_k \geq \lambda_{k+1}$, $\sum_{k=1}^{q} \lambda_k = q$. According to the principal component method [29], the value $\widetilde{\lambda}_1$ is equal to the share of the total variance attributable to the first principal component and can be interpreted as a measure of synchronization of changes in the scalar components of the multivariate time series of the probability density values of the tremor statistics from the matrix in Formula (7) in regular grid nodes covering the region under study.

Figure 7 shows graphs of changes in the eigenvalues of the correlation matrix for active and passive tremor. It can be seen from these plots that the active tremor (Figure 7a) is much more synchronized than the passive tremor (Figure 7b). Moreover, for active tremor, there is a very strong synchronization for the positions of the right end of the 3-year time window from 2013 to 2015. Given the length of the window, this synchronization takes a period of time from 2010 to 2015.

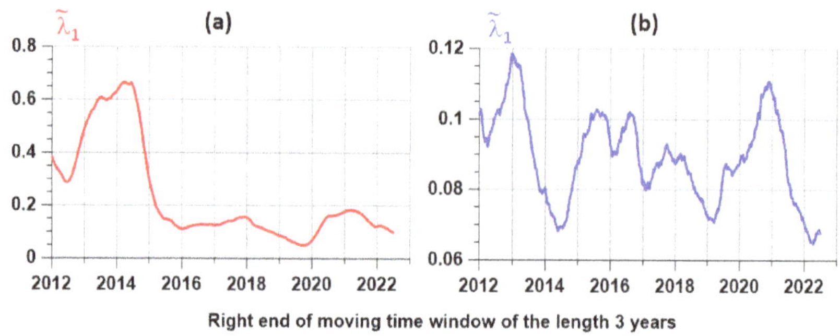

Figure 7. (**a**,**b**) represent time sequences of values of the normalized maximum eigenvalue of the correlation matrix with a shift in the time window of 3 years for "active" (**a**) and "passive" (**b**) tremor.

As was shown in [28], after a couple of the strongest earthquakes in the world, Maule in Chile on 27 February 2010 and Tohoku in Japan on 3 November 2011, there was an explosive increase in the radius of spatial correlations of global seismic noise properties. It is possible that the increase in synchronization of active tremor in the western United States in 2010–2015 is also a response to this pair of mega earthquakes. Note that a sharp jump in the magnitude of global correlations of the Earth's surface tremor in 2010–2011 was shown in [19].

5. Conclusions

A method has been proposed for studying the properties of the Earth's surface tremor, measured by means of GPS, based on the use of oscillation statistics that describe the deviation or similarity to the properties of Gaussian white noise, considered as a standard of a stationary random signal. Based on the application of the principal component method to the estimation of the probability densities of the distribution of extreme properties of statistics in a moving time window, concepts of active and passive tremor and their singular points are introduced, and the trajectories of singular points are determined. The concept of tremor synchronization measure is introduced as the maximum eigenvalue of the correlation density matrix of distribution of extreme values of tremor statistics. The method is applied to the data of daily GPS time series on a network of stations located in California during the time interval 2009–2022. It was shown that singular points of both types of tremor form well-defined clusters. A hypothesis is proposed that the difference in the types of singular points of tremor is associated with different mechanisms of consolidation of crustal blocks. In particular, passive tremor may be due to the activation of chaotic block movements in the vicinity of three fragments of the San Andreas fault. Estimates of measures of synchronization of active and passive tremor are given. A hypothesis has been put forward that the strong synchronization of active tremor in 2010–2015 caused a jump in the global synchronization of the tremor of the Earth's surface and an explosive increase in the correlations of global seismic noise in 2010–2011.

Funding: The work was carried out within the framework of the state assignment of the Institute of Physics of the Earth of the Russian Academy of Sciences (topic FMWU-2022-0018).

Institutional Review Board Statement: Not applicable.

Informed Consent Statement: Not applicable.

Data Availability Statement: The open access data from the source: http://geodesy.unr.edu/gps_timeseries/tenv3/IGS14/ (accessed on 1 May 2023) were used.

Conflicts of Interest: The author declare no conflict of interest.

References

1. Roncagliolo, P.A.; Garcнa, J.G.; Mercader, P.I.; Fuhrmann, D.R.; Muravchik, C.H. Maximum-likelihood attitude estimation using GPS signals. *Digit. Signal Process.* **2007**, *17*, 1089–1100. [CrossRef]
2. Wang, F.; Li, H.; Lu, M. GNSS Spoofing Detection and Mitigation Based on Maximum Likelihood Estimation. *Sensors* **2017**, *17*, 1532. [CrossRef]
3. Parkinson, B.W. *Global Positioning System: Theory and Applications*; AIAA: Reston, VA, USA, 1996.
4. Xu, J.; He, J.; Zhang, Y.; Xu, F.; Cai, F. A Distance-Based Maximum Likelihood Estimation Method for Sensor Localization in Wireless Sensor Networks. *Int. J. Distrib. Sens. Netw.* **2016**, *12*, 2080536. [CrossRef]
5. Agnew, D. The time domain behavior of power law noises. *Geophys. Res. Lett.* **1992**, *19*, 333–336. [CrossRef]
6. Amiri-Simkooei, A.R.; CTiberius, C.J.M.; Teunissen, P.J.G. Assessment of noise in GPS coordinate time series: Methodology and results. *J. Geophys. Res.* **2007**, *112*, B07413. [CrossRef]
7. Bos, M.S.; Fernandes, R.M.S.; Williams, S.D.P.; Bastos, L. Fast error analysis of continuous GPS observations. *J. Geod.* **2008**, *82*, 157–166. [CrossRef]
8. Kermarrec, G.; Schon, S. On modelling GPS phase correlations: A parametric model. *Acta Geod. Geophys.* **2018**, *53*, 139–156. [CrossRef]
9. Liu, B.; Xing, X.; Tan, J.; Xia, Q. Modeling Seasonal Variations in Vertical GPS Coordinate Time Series Using Independent Component Analysis and Varying Coefficient Regression. *Sensors* **2020**, *20*, 5627. [CrossRef]
10. Tesmer, V.; Steigenberger, P.; van Dam, T.; Mayer-Gurr, T. Vertical deformations from homogeneously processed GRACE and global GPS long-term series. *J. Geod.* **2011**, *85*, 291–310. [CrossRef]
11. Yan, J.; Dong, D.; Burgmann, R.; Materna, K.; Tan, W.; Peng, Y.; Chen, J. Separation of Sources of Seasonal Uplift in China Using Independent Component Analysis of GNSS Time Series. *J. Geophys. Res. Solid Earth* **2019**, *124*, 11951–11971. [CrossRef]
12. Pan, Y.; Shen, W.-B.; Ding, H.; Hwang, C.; Li, J.; Zhang, T. The Quasi-Biennial Vertical Oscillations at Global GPS Stations: Identification by Ensemble Empirical Mode Decomposition. *Sensors* **2015**, *15*, 26096–26114. [CrossRef] [PubMed]
13. Liu, B.; Dai, W.; Liu, N. Extracting seasonal deformations of the Nepal Himalaya region from vertical GPS position time series using Independent Component Analysis. *Adv. Space Res.* **2017**, *60*, 2910–2917. [CrossRef]
14. Santamarнa-Gymez, A.; Memin, A. Geodetic secular velocity errors due to interannual surface loading deformation. *Geophys. J. Int.* **2015**, *202*, 763–767. [CrossRef]
15. Fu, Y.; Argus, D.F.; Freymueller, J.T.; Heflin, M.B. Horizontal motion in elastic response to seasonal loading of rain water in the Amazon Basin and monsoon water in Southeast Asia observed by GPS and inferred from GRACE. *Geophys. Res. Lett.* **2013**, *40*, 6048–6053. [CrossRef]
16. Chanard, K.; Avouac, J.P.; Ramillien, G.; Genrich, J. Modeling deformation induced by seasonal variations of continental water in the Himalaya region: Sensitivity to Earth elastic structure. *J. Geophys. Res. Solid Earth* **2014**, *119*, 5097–5113. [CrossRef]
17. He, X.; Montillet, J.P.; Fernandes, R.; Bos, M.; Yu, K.; Hua, X.; Jiang, W. Review of current GPS methodologies for producing accurate time series and their error sources. *J. Geodyn.* **2017**, *106*, 12–29. [CrossRef]
18. Ray, J.; Altamimi, Z.; Collilieux, X.; van Dam, T. Anomalous harmonics in the spectra of GPS position estimates. *GPS Solut.* **2008**, *12*, 55–64. [CrossRef]
19. Lyubushin, A. Global coherence of GPS-measured high-frequency surface tremor motions. *GPS Solut.* **2018**, *22*, 116. [CrossRef]
20. Lyubushin, A. Field of coherence of GPS-measured earth tremors. *GPS Solut.* **2019**, *23*, 120. [CrossRef]
21. Lyubushin, A. Identification of Areas of Anomalous Tremor of the Earth's Surface on the Japanese Islands According to GPS Data. *Appl. Sci.* **2022**, *12*, 7297. [CrossRef]
22. Blewitt, G.; Hammond, W.C.; Kreemer, C. Harnessing the GPS data explosion for interdisciplinary science. *Eos* **2018**, *99*, 485. [CrossRef]
23. Duda, R.O.; Hart, P.E.; Stork, D.G. *Pattern Classification*; Wiley: Hoboken, NJ, USA, 2000.
24. Marple, S.L., Jr. *Digital Spectral Analysis with Applications*; Prentice-Hall, Inc.: Englewood Cliffs, NJ, USA, 1987.
25. Mallat, S. *A Wavelet Tour of Signal Processing*, 2nd ed.; Academic Press: Cambridge, MA, USA, 1999.
26. Donoho, D.L.; Johnstone, I.M. Adapting to unknown smoothness via wavelet shrinkage. *J. Am. Stat. Assoc.* **1995**, *90*, 1200–1224. [CrossRef]
27. Lyubushin, A. Low-Frequency Seismic Noise Properties in the Japanese Islands. *Entropy* **2021**, *23*, 474. [CrossRef]
28. Lyubushin, A. Spatial Correlations of Global Seismic Noise Properties. *Appl. Sci.* **2023**, *13*, 6958. [CrossRef]
29. Jolliffe, I.T. *Principal Component Analysis*; Springer: Berlin/Heidelberg, Germany, 1986.

applied sciences

MDPI

Article

Eco-Geophysical and Geoecological Factors in Assessing the State of the Geological Environment Based on the Analysis of Spatial Databases of the Territory of the Republic of North Ossetia–Alania

Vladislav Zaalishvili *, Olga Burdzieva, Aleksandr Kanukov and Tamaz Zaks

Geophysical Institute of Vladikavkaz Scientific Centre, Russian Academy of Sciences, 362002 Vladikavkaz, Russia; olgaburdzieva@mail.ru (O.B.); akanukov@list.ru (A.K.); tzaal@mail.ru (T.Z.)
* Correspondence: vzaal@mail.ru; Tel.: +7-867-276-4084

Abstract: The article considers the main sources of pollutionin the territory of the Republic of North Ossetia–Alania. A study of environmental geophysical factors in the city of Vladikavkaz was carried out at 126 points; indicators of noise pollution, electric fields and the level of gamma radiation were measured. A geoaccumulation index of heavy metals in soils and indices of carcinogenic and non-carcinogenic risks were calculated and corresponding maps were constructed. The obtained data supporting a high level of carcinogenic risk are consistent with a high level of cancer morbidity in the city, which indicates a close relationship between morbidity and the carcinogenic risk index. It has been determined that emissions from road transport are greater by an order of magnitude than stationary sources emissions, while there is a steady trend towards an increase in air pollution as a result of the increasing negative impact of motor vehicle emissions. It has been established that the most hazardous way for heavy metals to enter the human body from the soil is by inhalation. It has been determined that in areas where environmental pollution with heavy metals is higher, cancer morbidity is also higher.

Keywords: ecology; environmental pollution; mining; spatial databases; GIS technologies

Citation: Zaalishvili, V.; Burdzieva, O.; Kanukov, A.; Zaks, T. Eco-Geophysical and Geoecological Factors in Assessing the State of the Geological Environment Based on the Analysis of Spatial Databases of the Territory of the Republic of North Ossetia–Alania. *Appl. Sci.* **2022**, *12*, 2644. https://doi.org/10.3390/app12052644

Academic Editor: Alexei Gvishiani

Received: 3 February 2022
Accepted: 2 March 2022
Published: 3 March 2022

1. Introduction

Environmental pollution has become a global problem due to the growing anthropogenic impact. It should be noted that almost all types of natural phenomena and processes of a geological, hydrogeological, and meteorological nature are sources of risk [1]. We have developed a scheme of the main sources of environmental pollution for the territory of the Republic of North Ossetia–Alania (Figure 1). The main sources of pollution are discussed in detail below.

Noise pollution is one of the three most significant environmental disturbances in the world. With the growth of urbanization, noise has become a permanent part of human life and one of the essential parametric pollutants of the urban environment. The problem of protection of the population against increased noise is, first of all, a problem of health maintenance [2]. Acoustic pollution worldwide constitutes about 70–75% of all environmental pollution. Noise pollution leads to increased fatigability in humans and animals, a decrease in labor productivity, as well as physical and nervous diseases [3–7]. One of the main environmental problems in cities is noise pollution from road transport, which largely determines the quality of life of the population [8]. Noise pollution is assessed by the following parameters: sound level, loudness, vibration, and sound pressure [9,10].

Every person, regardless of location, is constantly exposed to various doses of radiation [11]. Radiation pollution of the environment comes from sources of both terrestrial and cosmic origin. Radiation of terrestrial origin is caused by the decomposition of uranium

235 and 238, thorium 232, and potassium 40 [11]. The extraterrene component is caused by cosmic rays, and the magnitude of this radiation depends on altitude, latitude, and the solar cycle [12]. The terrestrial component is associated with the geological structure and features of the environment [13]. In addition to natural sources of background radiation, anthropogenic activities, such as the survey and exploitation of natural resources (leading in some cases to a significant increase in the radiation impact on the population [14]), also make a significant contribution [15–19].

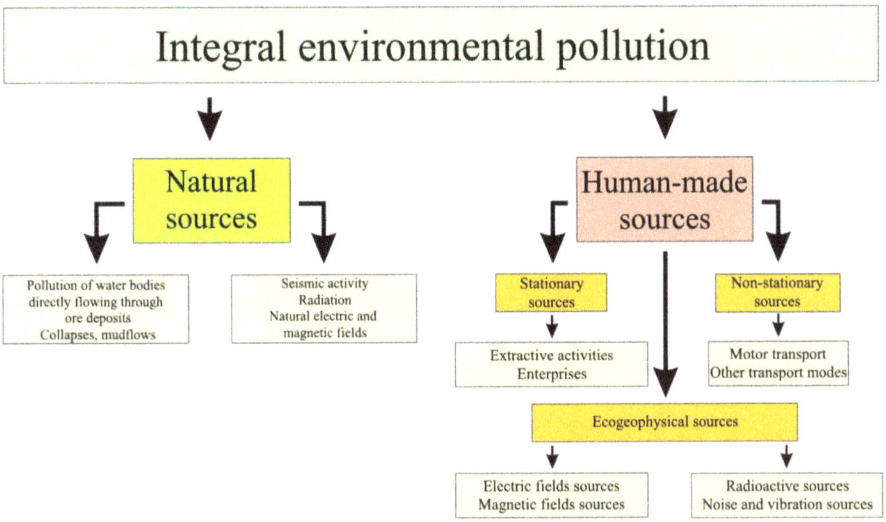

Figure 1. Scheme of the main sources of environmental pollution for the territory of the Republic of North Ossetia–Alania.

According to a number of modern studies, there are concerns that magnetic fields can affect the health of the population. The possibility of such an effect has provoked a serious scientific dispute and drawn attention among the population exposed to radiation [20–32]. Since 1998, studies have provided evidence that power-frequency electromagnetic fields have adverse biological effects on children's health, in particular, in carcinogenesis due to electric and magnetic fields exposure in residential premises [21,22] and cancer among children consequent onparental exposure [23],as well asoccupational cancer diseases, including leukemia, brain cancer, and breast cancer [24]. Some research data on the effects of electromagnetic fields on the human body also indicate an increased risk of miscarriage [25], neurodegenerative diseases, such as amyotrophic lateral sclerosis and Alzheimer's disease [26], heart disease [27], and suicide [28].

Information on outdoor background field strength in urban areas may be of interest because there is no data on the magnitude of magnetic radiation when the greater part of the population is exposed to significant levels over long periods. It was the lack of information about electromagnetic fields in the urbanized territories of the Republic of North Ossetia–Alania that prompted the present study, the purpose of which is to construct maps of the magnetic and electric fields.

In the areas where mining and processing enterprises are located, one of the main sources of disturbance and pollution of the components of the natural environment are solid mineral waste storage facilities (the total mass of waste materials in the Russian Federation reaches more than 20 billion tons). This causes the deterioration of sanitary and hygienic conditions, the emergence of areas of increased soil pollution, and surface and groundwater pollution, which leads to the continuous transformation of natural landscapes into man-made landscapes and hence to the loss of natural resources [33].

The greatest harm is caused by dusty manmade deposits—dumps of products of the imperfect technology used for the processing of non-ferrous metal ores, rock dumps and dumps of substandard ores containing hazardous concentrations of free silica (SiO_2), and toxic metals in the airborne dust (Cu, Zn, Pb, Ni), causing air pollution and the destruction of organisms. This situation has developed in the mountainous provinces of the North Caucasus, which has a high concentration of mining and metallurgical industry facilities and social infrastructures [34].

More than 30 enterprises are located in the industrial hub of the city of Vladikavkaz. Their activities are in one way or another connected with air pollution. The largest enterprise in the Republic was OJSC "Electrozinc", which has been actively polluting the environment throughout the years of its existence [35–37]. At the end of 2016, by agreement of the leadership of the Republic and OJSC "Electrozinc", the environmentally harmful lead production was terminated. At the time of writing, OJSC "Electrozinc" is in a stand-by condition [38].

A large number of air pollutants with various toxicities enter the environment during the movement or operation of road transport. Pollutants enter the environment from vehicles equipped with internal combustion engines as well as their elements (brakes, clutch discs, tires, and fuel tanks), and, in addition, from wear to road surfaces.

The level of air pollution varies over time and space depending on various features, such as proximity to roads, vehicle fleet composition, traffic flow, and the presence of other sources of pollution. Modern works investigate the spatial distribution of pollution based on geoinformation technologies [39,40]. Pollution by particulate matter contained in waste gases may be the largest potential threat to public health [41]. Numerous studies indicate that some diseases are directly related to the levels of air pollution caused by large amountsof road transport [42–46].

2. Methods

2.1. Methods for Map Construction Based on Spatial Databases

Pollutant concentrations can vary greatly across the surface of the earth, so it is very difficult to obtain an accurate spatial distribution of heavy metals. The presence of a certain proportion of samples that exceed the established regulatory threshold has been a classic method of characterizing the degree of soil contamination [47,48]. However, the classical estimation method has many limitations whichprobably cause errors or uncertainties in pollution assessments.

The classical statistical method usually requires that data be subject to a number of assumptions: independence of observations, exact or approximate normality of observations, and large and repeated sampling. Mapping the spatial distribution of pollutants requires the use of spatial interpolation methods. The most commonly used interpolation methods are inverse distance weighting (IDW), kriging, and spline.

Interpolation accuracy is related to the precise definition of the polluted area and its boundaries. Consequently, this directly affects the accuracy of pollution assessment. There have been many studies of the performance of the spatial interpolation methods mentioned above, but the results are not clear-cut [49]. Some of them found that the kriging method performed better than IDW [50], while others showed that kriging is no better than alternative methods [51].

The first considered interpolation method of kriging is based on the assumption that the interpolated parameter can be considered as a regionalized variable. As with IDW, the kriging estimation is given by a linear combination of the observed values with weights. Depending on the stochastic properties of random fields, different types of kriging are applied. The type of kriging defines the linear constraint on weights implied by the unbiased condition. There are several types of kriging, including simple kriging, ordinary kriging, universal kriging, etc., and ordinary kriging is the most commonly used method. The ordinary kriging weights are derived from the kriging equations using the semivariance function. The parameters of the semivariance function and the nugget effect can be estimated using the empirical semivariance function [52]. The unbiased assessment of the semivariance function is half the average squared difference between paired data values:

$$\gamma(p) = \frac{1}{2N(h)} \sum_{i=1}^{N(h)} [z(x_i) - z(x_i + h)]^2 \tag{1}$$

where $\gamma(h)$ is the semivariance value at distance interval h,N(h) is the number of sample pairs within the distance interval h, and $z(x_i + h)$ and $z(x_i)$ are sample values at two points separated by the distance interval h.

The next kind of interpolatorsareradial basis functions (RBF). This is the name given to a large family of exact interpolators that use a basic equation that depends on the distance between the interpolated point and the sampling points. The RBF prediction value can be expressed as the sum of two components [53]:

$$Z(x) = \sum_{i=1}^{m} a_i f_i(x) + \sum_{j=1}^{n} b_j \psi(d_j) \tag{2}$$

where (d_j) shows the radial basis functions and d_j is the distance from sample site to prediction point x. $f_i(x)$ is a trend function, a member of a basis for the space of polynomials of degree < m. The coefficients a_i and b_j are calculated by means of the resolution of the following system of n + m linear equations; n is the total number of known points used in the interpolation.

Polynomial interpolation is a process of finding a formula (often a polynomial) whose graph will pass through a given set of points. Global polynomial interpolation corresponds to the polynomial of the entire surface, while local polynomial interpolation can be seen as a combination of global polynomial methods and a moving average procedure. Instead of fitting the polynomial to the entire dataset, it is fitted to a local subset defined by the window, as in the moving average model. The size of this window should be large enough to include a reasonable number of data points in the process.

The next method is an inverse distance weighting (IDW), which is based on the assumption that the forecasts are presented by a linear combination of the available data. The following expression is an interpolating function:

$$\begin{aligned} Z(x) &= \sum_{i=1}^{n} w_i z_i / \sum_{i=1}^{n} w_i, \\ w_i &= d_i^{-u} \end{aligned} \tag{3}$$

where Z(x) is the predicted value at an interpolated point, Z_i is at a known point, n is the total number of known points used in interpolation, d_i is the distance between point i and the prediction point, and w_i is the weight assigned to point i. Greater weighting values are assigned to values closer to the interpolated point. As the distance increases, the weight decreases [54], and u is the weighting power that determines how the weight decreases as the distance increases.

It should be clear that all interpolation results contain errors. The identification of a region as polluted should not be based only on interpolation results. It is recommended to consider the natural background and human activities before making a decision. To obtaina more reliable estimate of pollution, more samples from the uncertainty region of the pollution assessment can be added. The uncertainty of the pollution assessment is mainly in the area of high local variations, so additional sampling in the area of uncertainty is recommended.

2.2. Methods for Calculating Indices of Soil Pollution and Risks to Public Health

Muller, G. [55] developed the geoaccumulation index (I_{geo}), which is widely used in European studies of trace metals [56]. This method estimates environmental pollution by comparing the difference between current and pre-industrial concentrations. It was first developed for the assessment of the quality of bottom sediments of rivers and was used to assess soil pollution [57]. To calculate I_{geo} for the soils of the surveyed plots, the following equation is proposed:

$$I_{geo} = \log_2\left(\frac{C_n}{1.5B_n}\right) \tag{4}$$

where C_n is the measured concentration of each heavy metal found in the studied soil (mg/kg) and B_n is the geochemical background value of heavy metals in soil for different provinces (mg/kg). The constant 1.5 is used due to possible variations in the initial data [57]. The geoaccumulation index consists of 7 classes or points, among which the highest 6 classes reflects 100-fold enrichment compared to background values [58].

The USEPA health risk assessment model is used to assess the non-carcinogenic and carcinogenic impact on humans exposed to heavy metals [59].

Average daily intake (ADI) is calculated using the following expression [59]:

$$ADI = \frac{C \times IR \times EF \times ED}{BW \times AT} \tag{5}$$

where ADI is an average daily intake (mg/kg per day), C is the concentration of heavy metals in a particular exposure medium (mg/kg), IR is the ingestion rate (kg/day), EF is the exposure frequency (day/year), ED is the exposure duration (year), BW is body weight (kg), and AT is the time period over which the dose is averaged (day). For non-carcinogenic effects AT = ED × 365 (days) and for carcinogenic effects AT = 25,550 (days) (70 years × 365 days/year).

Considering the characteristics of soil pollution with heavy metals, two pathways of exposure are often considered [59]:

1. Ingestion:

$$ADI_{ing} = \frac{C_S \times IR_S \times EF \times ED}{BW \times AT} \tag{6}$$

where ADI_{ing} is the average daily intake from ingestion (mg/kg per day), C_S is the heavy metal concentration in the soil (mg/kg), and IR_s is the ingestion rate of soil (kg/day).

2. Dermal absorption:

$$ADI_{dermal} = \frac{C_S \times SA \times AF \times ABS \times EF \times ED}{BW \times AT} \tag{7}$$

where ADI_{dermal} is the average daily intake from dermal absorption (mg/kg per day), C_S is the heavy metal concentration in the soil (mg/kg), SA is the exposed skin surface area (cm^2), AF is the adherence factor (mg/cm^2 per day), and ABS is the dermal absorption index (unitless).

Non-carcinogenic risk is assessed by calculating hazard quotient (HQ) values. For composite heavy metal pollution, a hazard index (HI) is used to assess the overall non-carcinogenic risk, which summarizes the HQ for each heavy metal [59]. It is assumed that the exposed population has no potential risk when the HI is below one.

$$HQ = \frac{ADI}{RfD} \tag{8}$$

$$HI = \sum HQ \tag{9}$$

where HI is the estimated non-carcinogenic health index and RfD is the reference dose of certain heavy metal (mg/kg per day).

Due to the lack of reference doses for the assessment of dermal absorption, USEPA has developed a method for extrapolating oral toxicity values for use in dermal risk assessment [59].

$$RfD_{ABS} = RfD_0 \times ABS_{GI} \tag{10}$$

where RfD_{ABS} is skin contact reference dose (mg/kg per day), RfD_0 is the oral reference dose (mg/kg per day), and ABS_{GI} is the gastrointestinal absorption factor (unitless). Carcinogenic risk is estimated by calculating lifetime carcinogenic risk (LCR) values using the equation given below.

$$LCR = ADI \times SF \quad (11)$$

where LCR is a dimensionless lifetime cancer risk and SF is a carcinogenic potential factor (per mg/kg per day). Generally, an LCR greater than 1×10^{-4} is considered to have a significant impact on health, while an LCR of 1×10^{-6}–1×10^{-4} is considered widely acceptable, and an LCR below 1×10^{-6} is considered insignificant [31].

Similar to the dermal absorption reference dose, the dermal absorption carcinogenic potential factor is extrapolated below [59]:

$$SF_{ABS} = SF_0 / ABS_{GI} \quad (12)$$

where SF_{ABS} is the dermal absorption carcinogenic potential factor (per mg/kg per day), SF_0 is the oral carcinogenic potential factor (per mg/kg per day), and ABS_{GI} is the gastrointestinal absorption factor (unitless).

3. Results

3.1. Creation of Spatial Databases

To assess the level of environmental pollution, we have developed several spatial databases.

One of the databases is the "Database of Soil Pollution in the Republic of North Ossetia–Alania". The spatial boundaries and the research area extended over aterritory of 810 km^2, with a level of soil pollution (according to the results of previous work) from medium to crisis, within the administrative boundaries of the Republic of North Ossetia–Alania.

The main geological tasks of the work carried out were the following:

- Assessment of the current state of soil pollution, including pesticides and radioactive substances, and prediction of its change under natural and man-made conditions on the territory of North Ossetia–Alania;
- Development of proposals and recommendations for the prevention of soil pollution;
- Preparation of environmental maps marking the geopathogenic zones (zones of environmental risk) in North Ossetia–Alania.

A total of 1104 sampling points were included in the specified spatial database.

To monitor ecogeophysical factors of the environment in Vladikavkaz, we measured the indicators of noise pollution and the levels of gamma and electromagnetic radiation at 126 points of the city. The noise level was measured using an Ecogeofizika-110A, a noise and vibration meter, manufactured by the Russian company Oktava (Tula, Russia). Gamma radiation levels were measured using a DBGA-OCHA gamma radiation dosimeter. To study the electric and magnetic fields, measurements were carried out with an IEP-05 electric field meter and with IMP 05/1 and 05/2 magnetic field meters. The results of the work were also formalized in the form of a spatial database for subsequent analysis and the creation of maps.

There are close correlations between the value of the received quantitiesof metals and their concentrations in the body of a person working in production. In particular, chemical production workers have accelerated development of pathologiesrelating tothe bioelectrical and contractile function of the heart, atherogenic changes in blood serum, neurocirculatory dystonia, myocardial dystrophy, atherosclerosis, and chronic heart failure [60]. When considering the situation of population that does not work in such industries but lives in the halo of their distribution, it should be noted that there is a direct relationship between chemical pollution of the environment and an increase in the frequency of allergies, bronchopulmonary pathology, thyroid hyperplasia, caries, and neuropsychiatric and physical developmental disorders. Theecological conditionality of congenital malformations and malignant growths, which are markers of chronic exposure to xenobiotics, has also been noted.

A special database was developed for processing and subsequent analysis of the obtained data. The parameters of the database included the location, gender, age of the patient, localization of the lesion inthe patient's body (brain, larynx, stomach, female genital organs, thyroid gland, intestines, skin, bones, blood, lungs, lymph nodes, face, mammary gland, male genital organs, liver, kidneys, prostate); a total of 17 items. (Full names of patients are not included in the primary datafor ethical reasons.)

Patients were divided into several groups according to age up to 20 years, 20–29 years, 30–39 years, 40–49 years, 50–59 years, 60–69 years, and 70 years and older. The obtained data were superimposed on a digital map-scheme of the development of the city. In other words, in GIS technologies, the places of residence (streets, buildings, and their numbers) of patients were plotted on the map-scheme of the building stock, thus creating a distribution of the incidence of malignant growths over the area of the city.

3.2. Development of Maps of Territory Pollution and Morbidity of the Population

Based on the conducted research and the obtained data within the North Ossetia–Alania region, we have identified areas of the territory with varying degrees of environmental stress (catastrophic, critical, crisis, intense, and relatively satisfactory) (Figure 2).

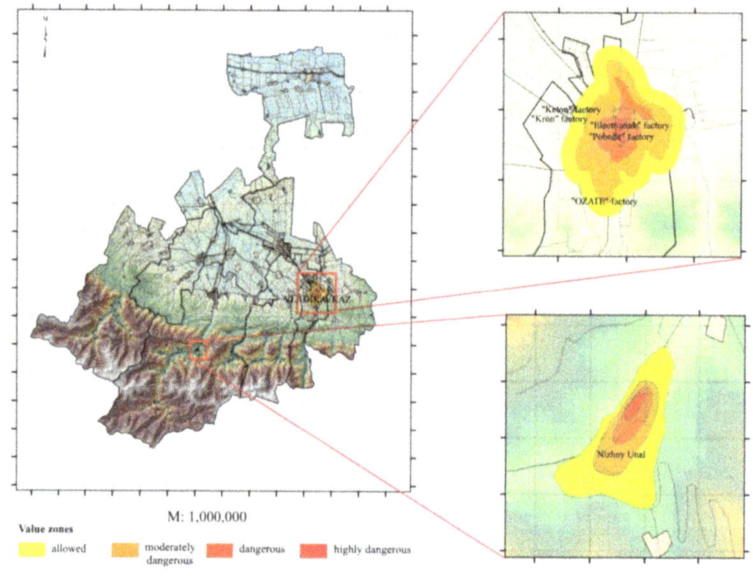

Figure 2. Map-scheme of areas of ecological situations in the mining regions of North Ossetia–Alania.

The most polluted sites in the territory of the Republic of North Ossetia–Alania are the Sadon ore district and the city of Vladikavkaz.

Until 1984, the tailings of the Sadon ores were stored in a temporary tailing dump located in a narrow side valley (left tributary of the Ardon River) above the settlement of Mizur. From 1929 to 1984 the factory worked in the mode of winter storage of tailings and dumped them into the Ardon River during the flood period.

Heavy metals (HM) enter the territory of this region as part of two anthropogenic flows: with the waters of the river Unaldon, through the feed ditches system, and due to deflation of the dry part of the tailing.

According to research data collected in the area of the Unal tailing dump, a complex anomaly of technogenic origin was defined in the soils. The spatial structure of the anomaly is determined by the morphology of the river valleys of the Ardon and Unaldon rivers and has a length of up to 1.5 km with a width of 0.2–0.8 km. The identified halos are characterized by a complex composition, the absence of longitudinal–transverse zoning, the coincidence of centers with the maximum contents of ore elements, and spatial proximity to pollution sources, which indicates their technogenic origin.

Investigations of the territory of North Ossetia–Alania pollution were carried out in two stages, which made it possible to study the dynamics of the increase in pollution of the territory over ten years. Based on the results of the studies carried out in stages 1 and 2, the soils in the vicinity of the main areas of pollution were ranked according to the value of the total index Z_c. The results of the two-stage studies form the basis of the map of soil pollution dynamics over time (Figure 3).

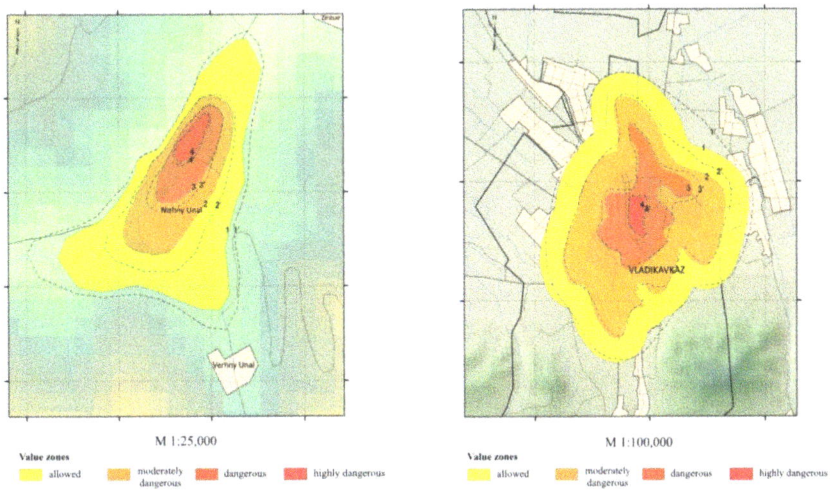

Figure 3. Maps of soil pollution dynamics over time.

By comparing the results of studies at both stages, it was found that, with a slight decrease in the concentration of pollutants at stage 2 of the studies, the area of the zones of permissible and moderately hazardous pollution zones increased by a significant amount.

Based on the developed database of Vladikavkaz territory pollution and the methods described in Section 2.2, we constructed maps of the geoaccumulation index for various metals (Figure 4).

Analyzing the constructed maps of the geoaccumulation index, the territory of Vladikavkaz was subjected to significant pollution by some heavy metals. The level of cadmium pollution for the entire territory has increased by two orders of magnitude compared to the pre-industrial era. For lead and copper, the situation is little better: in most of the territory, pollution does not go beyond the third hazard class. Pollution with mercury, zinc, and nickel, according to the geoaccumulation index, is insignificant and does not exceed the first hazard class.

In addition, maps of carcinogenic risk index (Figure 5) and non-carcinogenic risk index (Figure 6) were constructed; the indices were calculated as the ratio of the average daily dose from exposure to chemicals entering the human body from the soil to the reference concentrations for chronic exposure. Values less than one are considered as safe. For the inhalation intake of heavy metals from the soil, for all the studied indicators except mercury, a significant excess of the maximum permissible values was found.

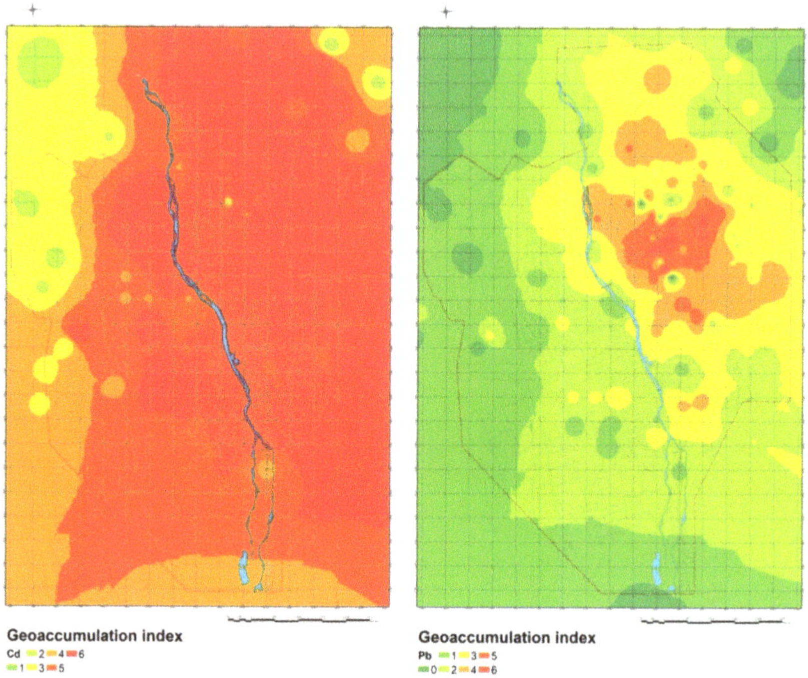

Figure 4. Geoaccumulation index of cadmium and lead for the territory of Vladikavkaz.

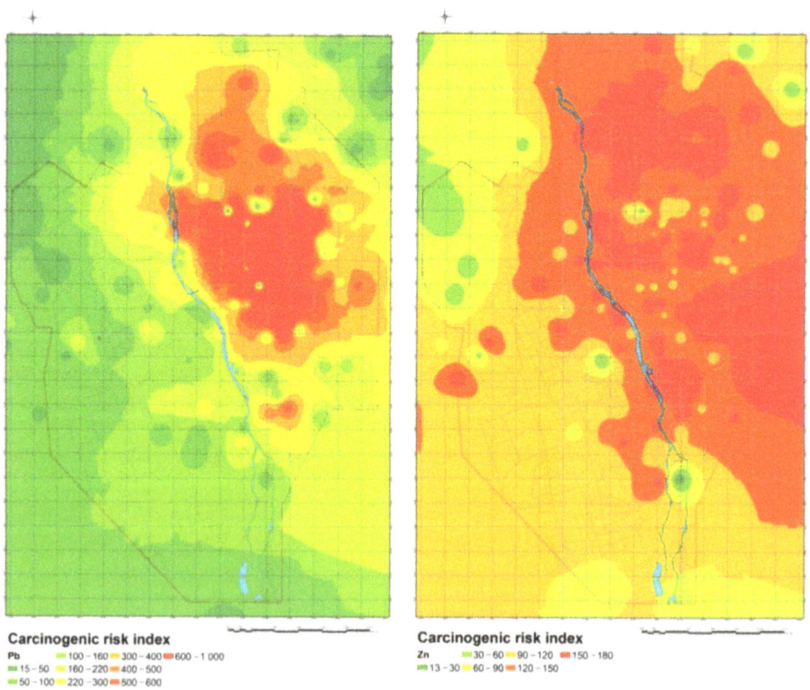

Figure 5. Carcinogenic risk index maps for lead and zinc for the territory of Vladikavkaz.

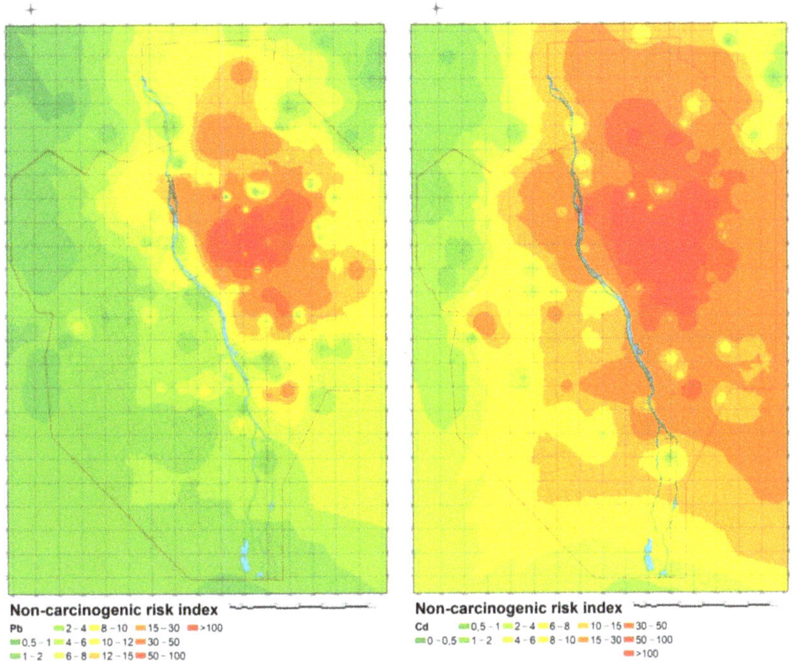

Figure 6. Non-carcinogenic risk index maps for lead and cadmium for the territory of Vladikavkaz.

For the oral intake of heavy metals from the soil into the human body, the values of the carcinogenic risk index do not exceed one for all the studied elements, except lead and 0.4 for cadmium. At the same time, for non-carcinogenic risk, the index values are significantly higher than one in the areas of the greatest contamination for cadmium and lead.

It has been determined that the most hazardous method of heavy metals intake from the soil into the human body is by inhalation, for which there is a significant excess of the permissible risk level. The oral way poses a significantly lower level of risk. The third option for heavy metals intake from the soil into the human body is dermal exposure, which gives values forthe carcinogenic risk index close to zero.

Thus, the overall level of carcinogenic risk for the territory of Vladikavkaz due to the intake of heavy metals from the soil into the human body significantly exceeds safe values. In this case, the overall risk is an integral indicator of individual risk values for various methods of heavy metals intake. However, given that the level of risk for inhalation intake significantly exceeds the risk levels for oral and dermal exposure of the soil, the maps of the carcinogenic risk index for inhalation intake of heavy metals can be considered as the final maps. At the same time, the oral way also significantly exceeds dermal exposure.

According to the results of the study of ecogeophysical pollution factors in the territory of Vladikavkaz, corresponding maps were also constructed (Figure 7). Based on the results of observations, certain orders and patterns of changes in the structure of the movement of passenger cars and trucks along the main traffic routes at different times of the day were defined; they provide a basis for the analysis and development of appropriate mathematical models as well as the prediction of environmental well-being in the zone of increased noise risk. The population of the city of Vladikavkaz is exposed to noise for a significant period of time, exceeding the permissible level by 1 to 38 dBA.

According to the obtained data for levels of gamma radiation, it can be concluded that the population of the city of Vladikavkaz is exposed to gamma radiationin the range of 0.05–0.18 μSv/h and at an average value of 0.11 μSv/h, which is below the maximum permissible level. The electric field strength reached values up to 1400 V/m at individual measurement points, with an allowable level of 800 V/m, significantly exceeding the maximum permissible values. At the same time, it should be noted that on the outskirts of the city, at points adjacent to power lines, as well as on the city streets with tram traffic, there is a significant increase in these indicators compared to other points.

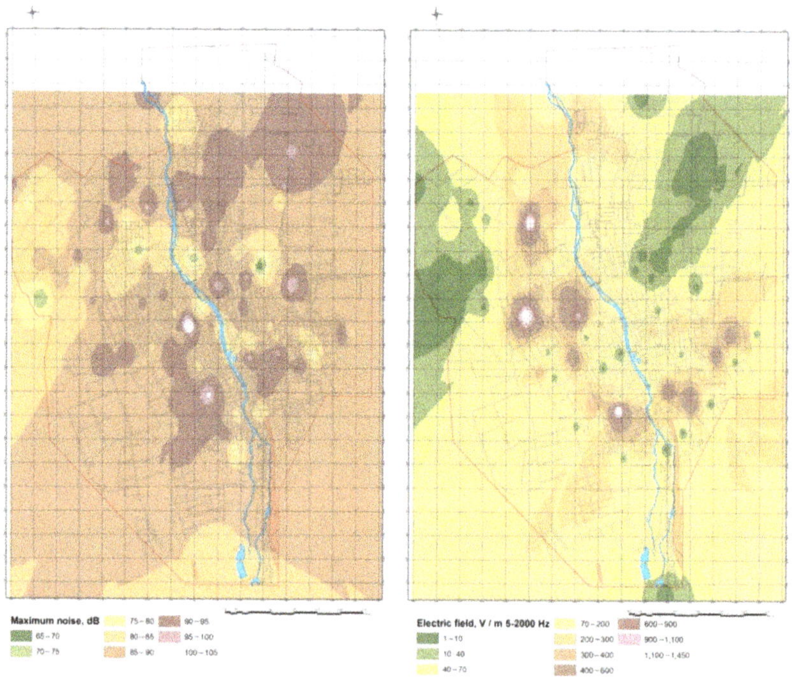

Figure 7. Maps of noise pollution and distribution of the electric field level (V/m) registered in the territory of Vladikavkaz.

In the course of this study, all available data on the dynamics of emissions from road transport in North Ossetia–Alania over the past 20 years were collected and analyzed.

The results of the analysis made it possible to establish that there is a direct relationship between traffic flow density and air pollution. At the same time, the most intense environmental pollution is observed in the area of the cinema "Druzhba" and the Arkhonsky crossroad, which are the places of the greatest traffic congestion. The studies for establishing the relationship between the number of moving vehicles and the level of environmental pollution were carried out using correlation and regression analysis (Figure 8).

It has been determined that the increase of pollutant emissions into the atmosphere from road transport is growing linearly, similar to the growth in the number of vehicles. In this case, the value of the approximation reliability R^2 is equal to 0.97. A high concentration of sulfur dioxide (up to four maximum permissible concentrations) was detected on the highways at the entrance to the city, which, in particular, may serve as an indirect sign of the use of low-quality fuel by heavy-duty vehicles. The federal highway, which leads to the Georgian Military Highway, passes right through the city. Its extremely high congestion is caused by the fact that it is the only existing land road connecting Russia with Armenia, Georgia, and Turkey.

Pollution of the city territory fromvarious sources does not occurwithout leaving a trace and there are a number of signs or factors of a changing level of pollution. Analysis of the data showed that, in our opinion, the most reliable indicator of the presence of a correlation relationship between the pollution of the territory with heavy metals can be considered malignant growths, which are markers of chronic exposure to xenobiotics.

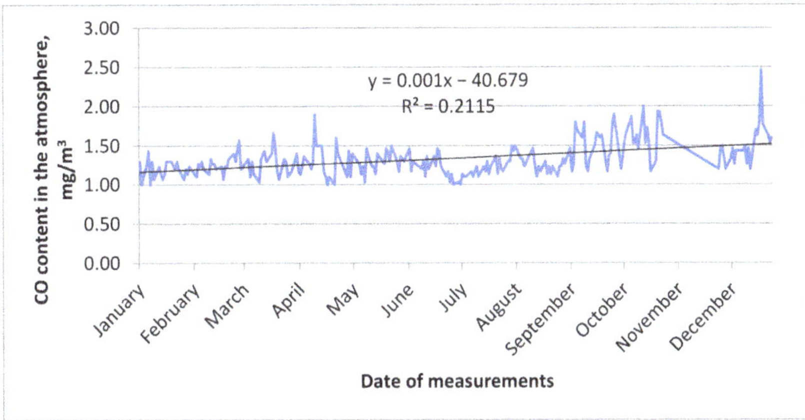

Figure 8. Average daily carbon monoxide concentrations in atmospheric air during the year, mg/m^3.

According to the generally accepted method of polyclinic territorial zoning [61–66] along the service boundaries of polyclinics, cancer incidence values were obtained for different years, as well as the average value for several years, which was put on the map (Figure 9).

Previously, we found that inhalation is the most hazardous way for heavy metals to enter the human body from soil. To assess the influence of prevailing winds on the incidence of the population, the wind rose for the territory of Vladikavkaz was considered. Figure 7 shows that the prevailing direction of airflow is south; thus, it can be concluded that, according to the constructed maps of carcinogenic risk, the highest value of cancer morbidity should be observed for the southern part of the city, which was obtained using the polyclinic zoning method. The lower morbidity rate for Polyclinic No. 1 compared to Polyclinic No. 4 can be explained by the fact that Polyclinic No. 1 serves nearby villages which are located at a greater distance from the source of pollution [67–74].

To study the possible correlation between the number of malignant growth cases and age groups, we compiled a table of the number of cases of cancer morbidity in the adult population of Vladikavkaz, arrangedaccording to age group, for the period from 2005 to 2010, along the service boundaries of polyclinics. The total number of cancer morbidity cases was calculated along the service boundaries of polyclinics (Table 1).

Table 1. Dependence of the number of the cancer morbidity cases on age groups.

Polyclinic No.	The Number of Cancer Morbidity Cases in Different Age Groups						
	Total Number	18–29	30–39	40–49	50–59	60–69	70 and Older
1	1843	37	57	217	335	439	758
3	483	8	10	34	102	109	220
4	1661	23	41	121	294	328	851
5	737	11	26	65	120	197	316
7	1045	25	37	113	236	311	323

Using the average value of the number of cancer morbidity cases in polyclinics and, having plotted the dependence of percentage of cases on age group, the shape of the approximating curve was obtained using the least squares method (Figure 10). This type of curve has a clearly expressed quadratic form, which was considered when constructing it. Taking such a high value of the approximation reliability R^2 = 0.989 into account, we can conclude that the resulting equation describes this dependence very accurately.

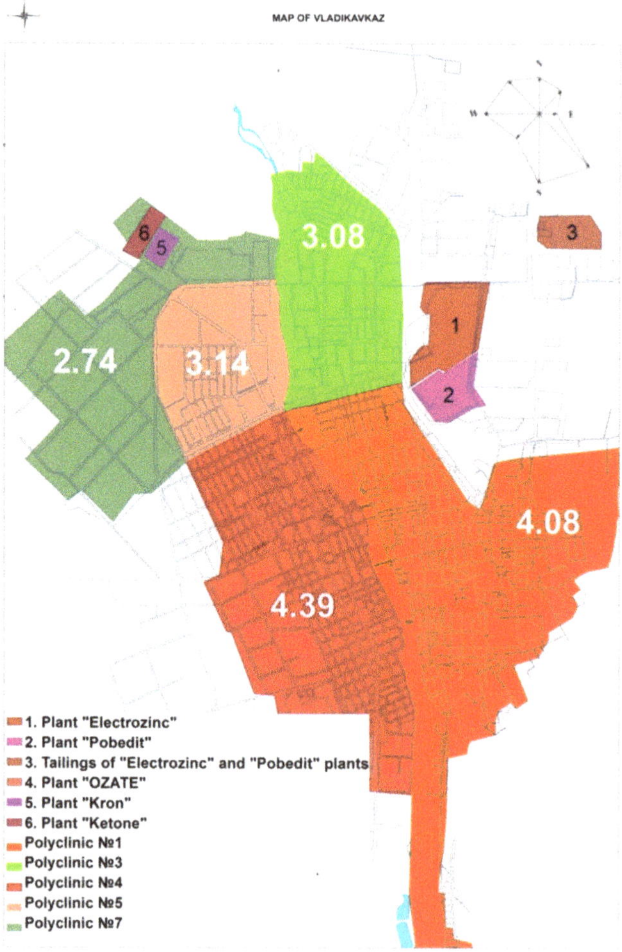

Figure 9. The incidence of malignant growths within the territory of urban polyclinics.

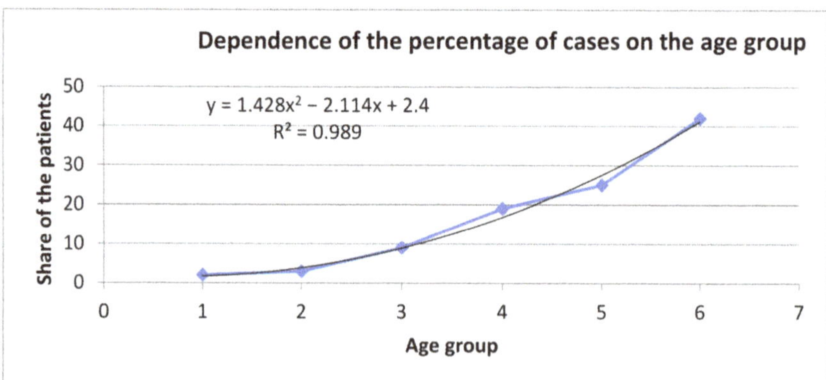

Figure 10. Dependence of the percentage of cases on age group.

Calculating the predictive value of the number of patients using the obtained regression expression, we can conclude that the established dependence of the percentage of patients on age group with a high degree of accuracy allows us to determine the number of manifestations of cancer morbidity in different age groups depending on the total number of patients. As an accuracy test of the obtained regression expression, the expected values of the number of diseases for different age groups were calculated for the territory of the Alagir district of the Republic of North Ossetia–Alania, where we also studied morbidity. On the basis ofour calculations, it was found that the predicted values of the number of cancer diseases differed from the real valuesby no more than 3 percent. The obtained results also indicate that a high degree of contamination of the territory with heavy metals and their accumulation in the human body, i.e., cumulative effect, significantly increases morbidity inthe population with age.

4. Discussion of the Results

Based on the conducted research and the obtained data within the North Ossetia–Alania region, we have identified areas of the territory with varying degrees of environmental stress: catastrophic, critical, crisis, intense, and relatively satisfactory. Investigations of the territory of North Ossetia–Alania pollution were carried out in two stages, which made it possible to study the dynamics of the increase in pollution of the territory over a ten-year period. According to the results of the studies of both stages, it was found that, with a slight decrease in the concentration of pollutants at stage 2, the area of the zones of permissible and moderately hazardous pollution zones increased by a significant amount.

Monitoring of ecogeophysical factors of the environment in Vladikavkaz was carried out: indicators of noise pollution, electric fields, and the level of gamma radiation were measured. According to the results of observations, certain orders and patterns of changes in the structure of the movement of passenger cars and trucks along the main traffic routes in Vladikavkaz, the Republic of North Ossetia–Alania at different times of the day were defined; they provided a basis for the analysis and for the development of appropriate mathematical models, as well as the prediction of environmental well-being in the zone of increased noise risk.

The population of the city of Vladikavkaz is exposed to noise for a significant period of time, exceeding the permissible level by 1 to 38 dBA.

According to the obtained data, it can be concluded that the population of the city of Vladikavkaz is exposed to gamma radiation in the range of 0.05–0.18 μSv/h and at an average value of 0.11 μSv/h, which is below the maximum permissible level. The electric field strength reached values up to 1400 V/m at individual measurement points, with an allowable level of 800 V/m, significantly exceeding the maximum permissible values. At the same time, it should be noted that on the outskirts of the city, at points adjacent to power lines, as well as on the city streets with tram traffic, there is a significant increase in these indicators compared to other points.

It has been determined that the most hazardous method of heavy metals intake from the soil into the human body is by inhalation. According to the obtained data, all the elements, except mercury, form a higher level of risk significantly exceeding one. For the oral intake of heavy metals from the soil into the human body, the values of the risk index do not exceed one for all the studied elements, except lead and 0.4 for cadmium. At the same time, the risk values for lead are exceeded for the limited area of the former plant "Electrozinc". For dermal exposure tothe soil, the risk values for all elements are close to zero. Considering that the level of risk for inhalation intake significantly exceeds the risk levels for oral and dermal exposure of the soil, the maps of the carcinogenic risk index for inhalation intake of heavy metals can be considered as final maps.

A significant contribution to environmental pollution and, in particular, atmospheric air is made by road transport. The results of the analysis made it possible to establish that there is a direct relationship between traffic flow density and air pollution. The obtained data indicate a steady trend towards an increase in air pollution as a result ofthe increasing negative impact of motor vehicle emissions, which also negatively affects the health of the population.

It has been determined that withcloser proximity to the industrial enterprises of non-ferrous metallurgy and the corresponding halos of soil contamination with heavy metals, cancer morbidity is higher amongthe population of Vladikavkaz in the localizations most susceptible to the influence of the considered heavy metals.

Using the least squares method, the functional dependence of the percentage of cases on age group was determined with the approximation reliability value $R^2 = 0.989$, which confirms that a high degree of territory contamination with heavy metals and their accumulation in the human body significantly increase the morbidity of the population with age.

5. Conclusions

This article considers various methods of interpolation of data with spatial reference. Given that the interpolation methods considered in this study lead to almost identical results, we chose the inverse distance weighted (IDW) method as the interpolation method.

Databases have been developed for the content of heavy metals in soils in the territory of North Ossetia–Alania, databases of noise, radiation, electrical and magnetic pollution of the territory of Vladikavkaz, and a database for cancer morbidity of the population of North Ossetia–Alania.

For the territory of the Republic of North Ossetia–Alania, map-schemes of areas of ecological situations in mining areas have been developed. Two highlypolluted areas were identified. It has been determinedthat the active development of polymetallic ores has led to a significant level of contamination of the territory. At the same time, even though mining activity and enterprise operations have been stopped, pollution halos continue to increase. Whence it follows that the main sources of environmental pollution, at the moment, are significant dumps from the extraction and processing of polymetallic ores, which are located to the north of the OJSC "Electrozinc" in an unpreserved state and are exposed to wind currents and leaching by precipitation.

A geoaccumulation index of heavy metals in soils and indices of carcinogenic and non-carcinogenic risks were calculated and corresponding maps were constructed. It was established that the territory of Vladikavkaz was subjected to significant pollution by a number of heavy metals. Thus, the level of cadmium pollution for the entire territory has increased by two orders of magnitude compared to the pre-industrial era. For lead and copper, the situation is a little better: in most of the territory, pollution does not go beyond the third hazard class. The levels of carcinogenic and non-carcinogenic risks significantly exceed safe values.

Maps of noise, radiation, and electrical and magnetic pollution of the territory of Vladikavkaz were developed. It has been determined that, except for radiation, a significant excess of safe values is observed for all the studied indicators.

Air pollution by road transport was studied. It has been defined that emissions from road transport exceed stationary sources by an order of magnitude.

It has been established that in areas where environmental pollution with heavy metals is higher, cancer morbidity is also higher. The functional dependence of the percentage of disease on age group was confirmed. The morbidity of the population is an integral indicator of environmental pollution, which allows us to say that the obtained high values for population morbidity are associated with an extremely unfavorable environmental situation in the territory of North Ossetia–Alania.

Author Contributions: Conceptualization, V.Z., O.B., A.K. and T.Z.; methodology V.Z., O.B., A.K. and T.Z.; software, A.K. and T.Z.; validation, O.B. and A.K.; formal analysis, V.Z., O.B., A.K. and T.Z.; investigation, V.Z., O.B., A.K. and T.Z.; resources, O.B. and A.K.; data creation, V.Z., O.B., A.K. and T.Z.; writing—original draft preparation, V.Z., O.B. and A.K.; writing—review and editing, V.Z.; visualization, A.K. and T.Z.; supervision, V.Z. All authors have read and agreed to the published version of the manuscript.

Funding: This research received no external funding.

Institutional Review Board Statement: Not applicable.

Informed Consent Statement: Not applicable.

Data Availability Statement: Not applicable.

Conflicts of Interest: The authors declare no conflict of interest.

References

1. Zaalishvili, V.B.; Beriev, O.G.; Kanukov, A.S.; Maysuradze, M.V. Mining impact on environment on the North Ossetian territory. *Int. J. Geomate* **2016**, *10*, 1693–1697.
2. Muzet, A. Environmental noise, sleep and health. *Sleep Med. Rev.* **2007**, *11*, 135–142. [CrossRef] [PubMed]
3. Alborov, I.D.; Burdzieva, O.G.; Zaalishvili, V.B. Technology for the maintenance of acoustic comfort on the transcaucasian highway in the zone of the residential areas adjusting the motor road. *Akustika* **2019**, *32*, 211–215. [CrossRef]
4. Messingher, G.; Ryherd, E.E.; Ackerman, J. Hospital noise and staff performance. *J. Acoust. Soc. Am.* **2012**, *132*, 203. [CrossRef]
5. Sørensen, M.; Andersen, Z.J.; Nordsborg, R.B.; Jensen, S.S.; Lillelund, K.G.; Beelen, R.; Schmidt, E.B.; Tjønneland, A.; Overvad, K.; Raaschou, N. Road traffic noise and incident myocardial infarction: A prospective cohort study. *PLoS ONE* **2012**, *7*, e39283.
6. Basner, M.; Babisch, W.; Davis, A.; Brink, M.; Clark, C.; Janssen, S.; Stansfeld, S. Auditory and non-auditory effects of noise on health. *Lancet* **2014**, *383*, 1325–1332. [CrossRef]
7. Verbeek, J.H.; Kateman, E.; Morata, T.C.; Dreschler, W.A.; Mischke, C. Interventions to prevent occupational noise-induced hearing loss. *Cochrane Database Syst. Rev.* **2012**, *10*, CD006396.
8. Miedema, H.M.E.; Oudshoorn, C.G.M. Annoyance from transportation noise: Relationships with exposure metrics DNL and DENL and their confidence intervals. *Environ. Health Perspect.* **2001**, *109*, 409–416. [CrossRef]
9. Ohrstrom, E.; Skanberg, A.; Svensson, H.; Gidlof-Gunnarsson, A. Effects of road traffic noise and the benefit of access to quietness. *J. Sound Vibrat.* **2006**, *295*, 40–59. [CrossRef]
10. Liang, M.J.; Zhao, F.; French, D.; Zheng, Y.Q. Characteristics of noise-canceling headphones to reduce the hearing hazard for MP3 users. *J. Acoust. Soc. Am.* **2012**, *131*, 4526–4534. [CrossRef]
11. Ugbede, F.O.; Echeweozo, E.O. Estimation of annual effective dose and excess lifetime cancer risk from background ionizing radiation levels within and around quarry site in Okpoto-Ezillo, Ebonyi State, Nigeria. *J. Environ. Earth Sci.* **2017**, *7*, 74–79.
12. UNSCEAR. *United Nations Scientific Committee on the Effect of Atomic Radiation: Exposures from Natural Radiation Sources*; Report to General Assembly, with Scientific Annexes; United Nations: New York, NY, USA, 2000.
13. Gholami, M.; Mirzaei, S.; Jomehzadeh, A. Gamma background radiation measurement in Lorestan province, Iran. *Int. J. Radiat. Res.* **2011**, *9*, 89–93.
14. Zaks, T.V.; Burdzieva, O.G.; Zaalishvili, V.B. Impact of noise, gamma radiation and other geophysical factors on population health. *Akustika* **2019**, *32*, 206–210. [CrossRef]
15. Agbalagba, E.O.; Avwiri, G.O.; Ononugbo, C.P. GIS mapping of impact of industrial activities on the terrestrial background ionizing radiation levels of Ughelli metropolis and its Environs, Nigeria. *Environ. Earth Sci.* **2016**, *75*, 1425. [CrossRef]
16. Emelue, H.U.; Jibiri, N.N.; Eke, B.C. Excess lifetime cancer risk due to gamma radiation in and around Warri refining and petrochemical company in Niger Delta, Nigeria. *Br. J. Med. Med. Res.* **2014**, *4*, 2590–2598. [CrossRef]
17. Ononugbo, C.P.; Avwiri, G.O.; Tutumeni, G. Measurement of natural radioactivity and evaluation of radiation hazards in soil of Abua/Odual districts using multivariable statistical approach. *Br. J. Environ. Sci.* **2016**, *4*, 35–48.
18. Saito, K.; Sakamoto, R.; Nagaoka, T.; Tsutsumi, M.; Moriuchi, S. Measurements of gamma dose rates in dwellings in the Tokyo metropolitan area. *Radiat. Prot. Dosim.* **1997**, *69*, 61–67. [CrossRef]
19. Sivakumar, R.; Selvasekarapandian, S.; Mugunthamanikandan, N.; Raghunath, V.M. Indoor gamma dose measurements in Gudalore (India) using TLD. *Appl. Radiat. Isot.* **2002**, *56*, 883–889. [CrossRef]
20. D'Amore, G.; Anglesio, L.; Tasso, M.; Benedetto, A.; Roletti, S. Outdoor background elf magnetic fields in an urban environment. *Radiat. Prot. Dosim.* **2001**, *94*, 375–380. [CrossRef]
21. Portier, C.J.; Wolfe, M.S. (Eds.) *Assessment of Health Effects from Exposure to Power-Line Frequency Electric and Magnetic Fields*; NIH Publication. No. 98-3981; National Institute of Environmental Health Sciences (NIEHS) Working Group Report of the National Instates of Health: Research Triangle Park, NC, USA, 1998.
22. Wartenburg, D. Residential EMF exposure and childhood leukemia: Metal-analysis and population attributable risk. *Bioelectromagn. Suppl.* **2001**, *5*, 86–104. [CrossRef]
23. Feychting, M.; Floderus, B.; Ahlbom, A. Parental occupational exposure to magnetic fields and childhood cancer (Sweden). *Cancer Causes Control* **2000**, *11*, 151–156. [CrossRef] [PubMed]
24. Savitz, D.A.; Cai, J.; van Wijngaarden, E.; Loomis, D.; Mihlan, G.; Dufort, V.; Kleckner, R.C.; Nylander-French, L.; Kromhout, H.; Zhou, H. Case-cohort analysis of brain cancer and leukemia in electric utility workers using a refined magnetic field jobexposurematrix. *Am. J. Ind. Med.* **2000**, *38*, 417–425. [CrossRef]

25. Li, D.-K.; Odoull, R.; Wi, S.; Janevic, T.; Golditch, I.; Bracken, T.D.; Senior, R.; Rankin, R.; Iriye, R. A population-based perspective cohort study of personal exposure to magnetic fields during pregnancy and the risk of miscarriage. *Epidemiology* **2002**, *13*, 9–20. [CrossRef]
26. Ahlbom, A. Neurodegenerative diseases, suicide and depressive symptoms in relation to EMF. *Bioelectromagn. Suppl.* **2001**, *5*, 132–143. [CrossRef]
27. Savitz, D.A.; Liao, D.; Sastre, A.; Fleckner, R.C.; Kavet, R. Magnetic field exposure and cardiovascular disease mortality among electric utility workers. *Am. J. Epidemiol.* **1999**, *149*, 135–142. [CrossRef]
28. Van Wijngaarden, E.; Savitz, D.A.; Kleckner, R.C.; Cai, J.; Loomis, D. Expsoure to electromagnetic fields and suicide among electric utility workers: A nested case-control study. *Occup. Environ. Med.* **2000**, *57*, 258–263. [CrossRef] [PubMed]
29. Miller, A.B.; To, T.; Agnew, D.A.; Wall, C.; Green, L.M. Leukemia following occupational exposure to 60 Hz electric and magnetic fields among Ontario electricity utility workers. *Am. J. Epidemiol.* **1996**, *144*, 150–160. [CrossRef]
30. Washburn, E.P.; Orza, M.J.; Berlin, J.A.; Nicholson, W.J.; Todd, A.C.; Frumkin, H.; Chalmers, T.C. Residential Proximity to Electricity Transmission and Distribution Equipment and Risk of Childhood Leukemia, Childhood Lymphoma, and Childhood Nervous System Tumors: Systematic Review, Evaluation, and Meta-analysis. *Cancer Causes Control* **1994**, *5*, 299–309. [CrossRef]
31. Merchant, C.J.; Renew, D.C.; Swanson, J. Exposures to Power-frequency Magnetic Fields in the Home. *J. Radiol. Prot.* **1994**, *14*, 77–87. [CrossRef]
32. Preece, A.W.; Grainger, P.; Golding, J.; Kaune, W. Domestic Magnetic Field Exposures in Avon. *Phys. Med. Biol.* **1996**, *41*, 71–81. [CrossRef]
33. Alborov, I.D.; Burdzieva, O.G.; Tedeeva, F.G.; Alborov, S.T.; Alborova, D.I. The impact of the mining and processing industry on the ecosphere. In Proceedings of the III All-Russian (National) Scientific and Practical Conference with International Participation "Modern Scientific, Technical and Social and Humanitarian Research: Current Issues, Achievements and Innovations", Vladikavkaz, Russia, 24–26 May 2021; pp. 75–81.
34. Alborov, I.D.; Tedeeva, F.G.; Glazov, A.P. Deformation of the natural landscape by the activities of mining and metallurgical production in the Republic of North Ossetia-Alania. *Life Saf.* **2018**, *2*, 9–11. (In Russian)
35. Zaalishvili, V.B.; Kanukov, A.S.; Korbesova, K.V. Pollution of urbanized territories by motor transport using the example of the republic of North Ossetia—Alania. In Proceedings of the E3S Web of Conferences, Moscow, Russia, 19–22 November 2019; Volume 164, p. 07024. [CrossRef]
36. Burdzieva, O.G.; Zaalishvili, V.B.; Beriev, O.G.; Zaks, T.V.; Kanukov, A.S. On the environmental aspects of modern urbanized territory during strong earthquakes. *Earthq. Eng. Constr. Saf.* **2012**, *3*, 62–67.
37. Burdzieva, O.G.; Zaalishvili, V.B.; Beriev, O.G.; Kanukov, A.S.; Maisuradze, M.V. Influence of mining activities on environmental pollution. *Geol. Geophys. Russ. South* **2014**, *4*, 8–13.
38. Zaalishvili, V.B.; Burdzieva, O.G.; Zaks, T.V.; Kanukov, A.S. Investigation of ecogeophysical factors of the environment of Vladikavkaz. In Proceedings of the E3S Web of Conferences, Moscow, Russia, 19–22 November 2019; Volume 164, p. 07026. [CrossRef]
39. Zhang, C.S. Using multivariate analyses and GIS to identify pollutants and their spatial patterns in urban soils in Galway, Ireland. *Environ. Pollut.* **2006**, *142*, 501–511. [CrossRef]
40. Wang, M.E.; Bai, Y.Y.; Chen, W.P.; Markert, B.; Peng, C.; Ouyang, Z.Y. A GIS technology based potential eco-risk assessment of metals in urban soils in Beijing, China. *Environ. Pollut.* **2012**, *161*, 235–242. [CrossRef]
41. Hoek, G.; Brunekreef, B.; Goldbohm, S.; Fischer, P.; van den Brandt, P.A. Association between mortality and indicators of traffic-related air pollution in the Netherlands: A cohort study. *Lancet* **2002**, *360*, 1203–1209. [CrossRef]
42. Künzli, N.; Kaiser, R.; Medina, S.; Studnicka, M.; Chanel, O.; Filliger, P.; Herry, M.; Horak, F., Jr.; Puybonnieux-Texier, V.; Quenel, P.; et al. Public-health impact of outdoor and traffic-related air pollution: A European assessment. *Lancet* **2000**, *356*, 795–801. [CrossRef]
43. Wjst, M.; Reitmeir, P.; Dold, S.; Wulff, A.; Nicolai, T.; Vonloeffelholzcolberg, E.F.; Vonmutius, E. Road traffic and adverse-effects on respiratory health in children. *Br. Med. J.* **1993**, *307*, 596–600. [CrossRef]
44. Janssen, N.A.H.; Brunekreef, B.; van Vliet, P.; Aarts, F.; Meliefste, K.; Harssema, H.; Fischer, P. The relationship between air pollution from heavy traffic and allergic sensitization, bronchial hyperresponsiveness, and respiratory symptoms in Dutch schoolchildren. *Environ. Health Perspect.* **2003**, *111*, 1512–1518. [CrossRef] [PubMed]
45. Lienesch, L.A.; Dumont, J.N.; Bantle, J.A. The effect of cadmium on oogenesis in xenopuslaevis. *Chemosphere* **2000**, *41*, 1651–1658. [CrossRef]
46. Dehghani, M.; Keshtgar, L.; Javaheri, M.R.; Derakhshan, Z.; Conti, G.O.; Zuccarello, P.; Ferrante, M. The effects of air pollutants on the mortality rate of lung cancer and leukemia. *Mol. Med. Rep.* **2017**, *15*, 3390–3397. [CrossRef] [PubMed]
47. Chen, T.B.; Wong, J.W.C.; Zhou, H.Y.; Wong, M.H. Assessment of trace metal distribution and contamination in surface soils of Hong Kong. *Environ. Pollut.* **1997**, *96*, 61–68. [CrossRef]
48. Cheng, J.-L.; Shi, Z.; Zhu, Y.-W. Assessment and mapping of environmental quality in agricultural soils of Zhejiang Province, China. *J. Environ. Sci. China* **2007**, *19*, 50–54. [CrossRef]
49. Shi, W.; Liu, J.; Du, Z.; Song, Y.; Chen, C.; Yue, T. Surface modelling of soil pH. *Geoderma* **2009**, *150*, 113–119. [CrossRef]

50. Yasrebi, J.; Saffari, M.; Fathi, H.; Karimian, N.; Moazallahi, M.; Gazni, R. Evaluation and comparison of ordinary kriging and inverse distance weighting methods for prediction of spatial variability of some soil chemical parameters. *Res. J. Biol. Sci.* **2009**, *4*, 93–102.
51. Gotway, C.A.; Ferguson, R.B.; Hergert, G.W.; Peterson, T.A. Comparison of kriging and inverse-distance methods for mapping soil parameters. *Soil Sci. Soc. Am.* **1996**, *60*, 1237–1247. [CrossRef]
52. Webster, R.; Oliver, M.A. (Eds.) *Geostatistics for Environmental Scientists*, 2nd ed.; Wiley: Hoboken, NJ, USA, 2007.
53. Aguilar, F.J.; Aguera, F.; Aguilar, M.A.; Carvajal, F. Effects of terrain morphology, sampling density, and interpolation methods on grid DEM accuracy. *Photogramm. Eng. Remote Sens.* **2005**, *71*, 805–816. [CrossRef]
54. Shepard, D. *A Two-Dimensional Interpolation Function for Irregularly-Spaced Data*; ACM: New York, NY, USA, 1968.
55. Muller, G. Index of geoaccumulation in sediments of the Rhine River. *GeoJournal* **1969**, *2*, 108–118.
56. Ji, Y.Q.; Feng, Y.C.; Wu, J.H.; Zhu, T.; Bai, Z.P.; Duan, C.Q. Using geoaccumulation index to study source profiles of soil dust in China. *J. Environ. Sci.* **2008**, *20*, 571–578. [CrossRef]
57. Loska, K.; Wiechuła, D.; Korus, I. Metal contamination of farming soils affected by industry. *Environ. Int.* **2004**, *30*, 159–165. [CrossRef]
58. Förstner, U.; Ahlf, W.; Calmano, W.; Kersten, M. *Sediment Criteria Development. Contributions from Environmental Geochemistry to Water Quality Management*; Technische Universität Hamburg: Hamburg, Germany, 1990.
59. USEPA. *Guidelines for the Health Risk Assessment of Chemical Mixtures*; USEPA: Washington, DC, USA, 1986.
60. Zaalishvili, V.B.; Burdzieva, O.G.; Kanukov, A.S.; Arkhireeva, I.G.; Dzobelova, L.V.; Korbesova, K.V.; Margoshvili, M.T. Morbidity of the population of an urbanized area in an unfavorable environmental situation. In *Modern Problems of Geology, Geophysics and Geoecology of the North Caucasus*; Collective Monograph Based on the Materials of the IX All-Russian Scientific and Technical Conference; CGI RAS: Vladikavkaz, Russia, 2019; pp. 598–603. (In Russian)
61. Burdzieva, O.G.; Beriev, O.G.; Zaks, T.V. *Influence of Environmental Factors on the Ecological Well-Being and Health of the City of Vladikavkaz Monograph*; Publishing House of the Geophysical Institute-Branch of the Vladikavkaz: Vladikavkaz, Russia, 2019; 247p, ISBN 978-5-904868-24-6. (In Russian)
62. Burdzieva, O.G.; Beriev, O.G.; Zaks, T.V.; Kanukov, A.S. The relationship between the distribution of heavy metals in an urbanized area and cancer morbidity. *Proc. Kabard.-Balkar. Sci. Cent. Russ. Acad. Sci.* **2016**, *74*, 102–108. (In Russian)
63. Zaalishvili, V.B.; Beriev, O.G.; Burdzieva, O.G.; Kanukov, A.S.; Zaks, T.V. Relationship between technogenic pollution of the urbanized area with mining waste and morbidity of the population. In *Dangerous Natural and Technogenic Geological Processes in the Mountainous and Foothill Areas of the North Caucasus*; Zaalishvili, V.B., Ed.; Publishing House of the Geophysical Institute-Branch of the Vladikavkaz: Vladikavkaz, Russia, 2010; pp. 433–438. (In Russian)
64. Zaalishvili, V.B.; Burdzieva, O.G.; Zaks, T.V.; Kanukov, A.S.; Beriev, O.G. The study of the development of neoplasms among residents of an urbanized mountainous territory. In *Young Scientists in Solving Urgent Problems of Science*; VSC RAS: Vladikavkaz, Russia, 2011; pp. 148–152. (In Russian)
65. Zaalishvili, V.B.; Beriev, O.G.; Burdzieva, O.G.; Kanukov, A.S. Pollution of the urbanized territory with industrial waste and morbidity of the population. In *Geodynamics, Volcanism, Seismicity and Exogenous Geological Processes of Natural and Technogenic Nature in the Caucasus*; Abstracts of the All-Russian Conference; CGI RAS: Vladikavkaz, Russia, 2014; p. 18. (In Russian)
66. Zaalishvili, V.B.; Chotchaev, K.O.; Shempelev, A.G.; Melkov, D.A.; Burdzieva, O.G.; Parada, S.G.; Dzeranov, B.V.; Dzhgamadze, A.K. Geodynamic situation in Central Caucasus and structural complexes on depth section of Genaldon profile. In Proceedings of the VIII Science and Technology Conference "Contemporary Issues of Geology, Geophysics and Geo-Ecology of the North Caucasus" (CIGGG 2018) "Advances in Engineering Research", Essentuki, Russia, 10–13 October 2018; pp. 325–331. (In Russian).
67. Bekuzarova, S.A.; Tebieva, D.I.; Bekmurzov, A.D.; Kebalova, L.A.; Korbesova, K.V. Phytoremediation of air toxicity of roads. *Geol. Geophys. Russ. South* **2020**, *10*, 127–135. (In Russian) [CrossRef]
68. Chotchaev, H.O.; Burdzieva, O.G.; Zaalishvili, V.B. Influence of geodynamic processes on the geoecological state of high mountain areas. *Geol. Geophys. Russ. South* **2020**, *10*, 70–100. (In Russian) [CrossRef]
69. Chotchaev, H.O.; Burdzieva, O.G.; Zaalishvili, V.B. Zoning of high mountainous areas by geoecological loads caused by geodynamic and climatic influences. *Geol. Geophys. Russ. South* **2021**, *11*, 81–94. (In Russian) [CrossRef]
70. Alborov, I.D.; Zaalishvili, V.B. Influence of wind conditions and pressure the influence of atmospheric air on the reliability of mine ventilation and the human body. *Geol. Geophys. Russ. South* **2021**, *11*, 148–159. (In Russian) [CrossRef]
71. Bosikov, I.I.; Klyuev, R.V.; Tekiev, M.V. Analysis of spatial distribution of chemical elements of apatite-stafelite ore. *Geol. Geophys. Russ. South* **2021**, *11*, 137–157. (In Russian) [CrossRef]
72. Zaalishvili, V.B.; Kanukov, A.S.; Korbesova, K.V. Assessment of the Impact of Road Traffic on the Pollution of Urbanized Territories. *Geol. Geophys. Russ. South* **2021**, *11*, 135–146. (In Russian) [CrossRef]
73. Zaalishvili, V.B.; Burdzieva, O.G.; Kanukov, A.S.; Dzobelova, L.V. Development of cancerogenic risk index maps for the territory of Vladikavkaz. *Geol. Geophys. Russ. South* **2021**, *11*, 147–160. (In Russian) [CrossRef]
74. Zalikhanov, M.C.; Kagermazov, A.K.; Sozaeva, L.T.; Bekkiev, K.M. About the possibility of using the output data of the global atmosphere model GFS NCEP in ecological research. *Geol. Geophys. Russ. South* **2021**, *11*, 161–169. (In Russian) [CrossRef]

applied sciences

MDPI

Article

Temporal and Spatial Geophysical Data Analysis in the Issues of Natural Hazards and Risk Assessment (in Example of North Ossetia, Russia)

Dmitry Melkov *, Vladislav Zaalishvili, Olga Burdzieva and Aleksandr Kanukov

Geophysical Institute, Vladikavkaz Scientific Centre, Russian Academy of Sciences, Vladikavkaz 362002, Russia; vzaal@mail.ru (V.Z.); olgaburdzieva@mail.ru (O.B.); akanukov@list.ru (A.K.)
* Correspondence: melkovd@mail.ru; Tel.: +7-8672-764084

Abstract: The paper considers the aspects of hazard assessment within the framework of a generalized approach. The aim of the study is to improve the methodology for more accurate and detailed probabilistic assessments of risks of various nature. A complex hazard map is constructed in an example of the territory of the Republic of North Ossetia-Alania and the construction site of the Mamison resort. Based on the analysis of data on Quaternary formations and quantitative estimates, it was concluded that the natural average static environmental evolution proceeds in the mode of the dynamic balance of two factors: mountain building and the equivalent increase in denudation, of which about 90% is transported and deposited by river waters and winds outside the territory. The remaining 10% is deposited in intermountain depressions and river valleys in situ. Geodynamic and climatic factors of influence on the geoenvironment create the danger of excessive environmental impact and disruption of its equilibrium development under anthropogenic impacts, which must be taken into account in designing.

Keywords: natural hazards; landslides; mudflows; avalanches; zonation; geophysical data; geomorphological conditions

Citation: Melkov, D.; Zaalishvili, V.; Burdzieva, O.; Kanukov, A. Temporal and Spatial Geophysical Data Analysis in the Issues of Natural Hazards and Risk Assessment (in Example of North Ossetia, Russia). *Appl. Sci.* **2022**, *12*, 2790. https://doi.org/10.3390/app12062790

Academic Editors: Alexei Gvishiani and Boris Dzeboev

Received: 1 February 2022
Accepted: 4 March 2022
Published: 9 March 2022

1. Introduction

The analysis of approaches to the assessment of natural hazards of various nature, used in the world, showed the following. There are no generally multi-hazard approaches: the spatial representation of natural disasters must consider all types of hazards with the help of a multi-risk approach at all spatial levels (regional and local). With few exceptions (France, Italy), the multi-risk approach is not used due to the differences in the responsibilities of sectoral planning departments for different disasters.

The intensity of the attention given to natural disasters usually depends on the experience of recent catastrophic events rather than the occurrence of catastrophic events in the more distant past (the Kolka glacier collapse in 2002, which repeated the 1902 collapse scenario). As a result, risk assessment and management are focused on more frequent hazards (river floods, avalanches, forest fires) than on less frequent events. This leads to a tendency to underestimate the hazard and risk posed by extreme events.

At present, spatial planning plays only a minor role in risk management: at the regional level, various sectoral planning departments are responsible for natural risk management. Regional planning is often just one of the subsidiary entities charged with the duty of ensuring the implementation of measures that are carried out by sectoral planning units.

Geohazards can be formed through continuous natural processes, such as weathering or erosion [1], though they are predominately created and accelerated by anthropogenic activity [2]. Geohazards that have occurred, and been thoroughly investigated to determine their source, can be used to predict vulnerability in similar formations and situations elsewhere [3] and to extend this knowledge to areas not yet developed. There are a number of works where geohazard mapping has been developed using multi-form, digital image data, permitting a regional assessment of risk concerning slope instability. Slope angle and slope aspect data were derived from a digital elevation model (DEM), produced from stereo air photographs. The geohazard map was then compiled by merging digital slope data with geotechnical characteristics, using map algebra within a geographic information system (GIS) [4–7]. An important aspect of the geohazard study is identifying the temporal and spatial distribution of zones liable (prone) to movement, including the location of potential slip surfaces.

The methodology adopted for the landslide hazard zonation of the study area includes the various geoinformatical tools comprising geographical information system (GIS) technology and the satellite remote sensing (RS) techniques [8]. Geomorphological parameters (elevation, slope, aspect, curvature) and drainage are widely extracted from a DEM based on the Shuttle Radar Topography Mission (SRTM) [5]. Furthermore, the use of GIS facilitated the extraction of geomorphic and hydrological parameters required for susceptibility assessment. More detailed DEM may be obtained using low altitude UAV photogrammetry [9]. All of the mentioned techniques are itself of data collection and should be supplemented by geophysical data, expressed in the manifestation of processes in physical fields, and on the other hand, the results of their interpretation in the form of tectonic schemes, fault schemes, geological structures, and geotechnical parameters.

A methodology for the comprehensive assessment of geological hazards of various nature on the basis of the method of expert assessments is proposed in the present paper. The relevance of the topic is caused by the fact that the territory of North Ossetia is prone to the destructive effects of almost all known types of hazardous geological processes that can lead to the death of people and enormous material losses, having a significant impact on the socio-economic situation and safety of the territory of the republic. This is due to the great variety of geological, morphological, climatic, and overall landscape conditions of the mainly mountainous territory of North Ossetia. Vivid examples are the collapse of the Kolka glacier in 2002, the collapse in the area of the Devdorak glacier in 2014, and the activation of the Matsuta landslide in 2019.

Various natural hazards and the risks of loss caused by them in different spheres are divided according to the occurrence area into geological, hydrological, meteorological, biological, and mixed. Many of them are closely interrelated, being either the main condition or the cause (impetus) of the origin and development of secondary natural hazards, often leading to greater losses than the primary hazardous natural and man-made processes that caused them. Despite significant genetic differences, all these hazards have a number of important common features (from the standpoint of forecasting and subsequent risk assessment), which consist in the uniformity of their sources in terms of the occurrence area and the nature of impacts in time and space (autochthonous or allochthonous, simultaneous or permanent).

The object of study is dangerous natural processes in the North Ossetia-Alania and adjacent regions of the North Caucasus. The aim of the study is to improve the methodology and obtain on this basis more accurate and detailed probabilistic assessments of risks of various nature.

Scientifically substantiated forecast of environmental state, based on the current endogenous geodynamic activity and the development of various genetic types of hazardous exogenous geological processes determines the relevance of ensuring the safety of the functioning and development of the tourist and recreational complex "Mamison". The territory was well investigated and provided detailed data for analysis. So different levels of complex natural hazard had been considered, from the Republic of the North Ossetia-Alania with major natural factors to Mamison resort construction region with the number of factors increased up to 27.

2. Geographical and Geological Setting

The Republic of North Ossetia-Alania is located in the south of the Russian Federation, on the northern slope of the Greater Caucasus (the eastern part of the Central Caucasus) and on the sloping plains adjacent to it between 43°50′–42°50′ north latitude and 43°25′–44°57′ east longitude (Figure 1).

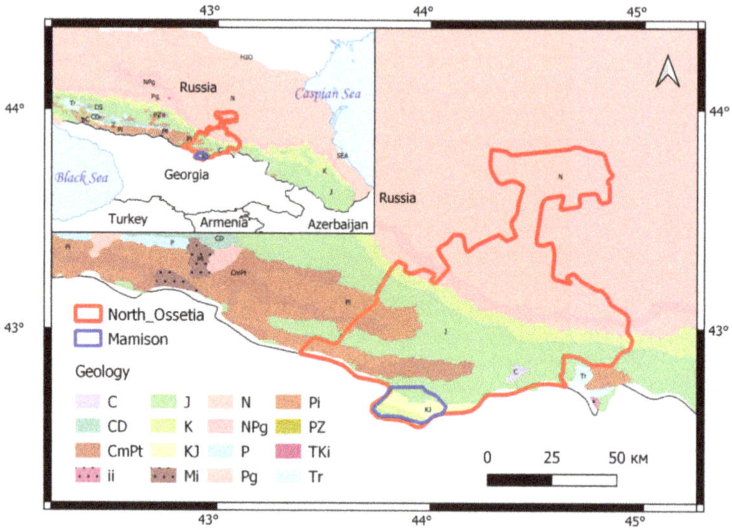

Figure 1. Geographical and geological setting of the study area: geologic scheme of the North Caucasus (modified from the work of [10]) and location of the Republic of the North Ossetia-Alania.

The Republic of North Ossetia-Alania occupies a special geopolitical and transport-geographical position in the south of Russia. It is due to the border position and the central place in the system of ciscaucasian and transcaucasian transport corridors [11]. A variety of sedimentary, igneous, and metamorphic rocks, from Precambrian to modern, take part in the geological structure of the territory of North Ossetia. There are three structural and formational levels: Caledonian-Hercynian, which includes geological formations from Precambrian to Lower Jurassic, Lower Alpine (Lower Middle Jurassic), and Upper Alpine from Upper Jurassic to modern rocks.

The Precambrian within the republic is unknown. Crystalline schists and gneisses of the river basin. The Ardon (Buron Formation) is defined as Precambrian only on the basis of analogy with the crystalline formations of the Western Caucasus. The Middle Paleozoic includes the Kasar Formation of the outcrop of metamorphic shales, which form a strip 2–3 km wide in the upper reaches of the Tsei, Ardon, and Bad rivers.

A very large and increasing in time and space for the socio-economic development of the republic, in contrast to world and Russian features, are natural emergencies, such as hazardous atmospheric phenomena, floods and avalanches, and some other hazardous natural and man-made natural processes. This is due to the great variety of geological, morphological, climatic, and overall landscape conditions of the predominantly mountainous territory of North Ossetia. The situation is aggravated by powerful anthropogenic pressure, which provokes the emergence, qualitative and quantitative growth of natural and natural-technogenic emergencies.

The territory of the proposed construction of a recreational complex "Mamison" is part of the Alagir district of North Ossetia-Alania, unites 13 settlements with a total population of 110 people for 2020 as part of the local government of the Zaramagsky rural settlement (village N. Zaramag), located in the upper reaches of the river. Ardon on the northern slope of the main range of the Greater Caucasus (Figure 1). The area is sparsely populated. There are no resources for industrial and agricultural development, and the mining industry is unprofitable. At the same time, the region has prospects for economic development due to the maximum use of the natural and climatic resources of the territory in an environmentally safe mode. This resource consists of: unique natural and climatic conditions for the organization of an all-season ski complex and tourist routes of various categories of difficulty along the objects of the geopark; numerous mineral springs (waters carbonic hydrocarbonate, sodium-calcium), the largest of which are: Kartysuar, Kalak, Kamskho, Dvugolovy, Lisri and Tib-1, Tib-2.

3. Materials and Methods

The methodological substantiation of the research is based on modern theoretical and empirical ideas about geodynamic processes, structural-tectonic and lithological features of the geological environment, a complex of physical-mechanical, geophysical, and geochemical characteristics of rocks and their physical fields [12]. The determining factor of the methodological basis of the research is also geomorphology, the influence of which on the intensity of exogenous processes formation depends on the ability of relief elements to collapse and on the conditions for the accumulation of erosive material by landforms [6].

Multi-criteria evaluation (MCE) involves the assessment, weighting, and aggregation of the attributes in a collective rather than sequential way. Certain factors (or attributes) are suspected as having a greater effect on slope stability or other exogenous processes than others. The assessment of their relative importance involves the derivation of a series of weighting coefficients. These factor weights control the effect that each factor has on the outcome. The relative significance of each factor and its influence on the system concerning the other factors need to be evaluated. The factors are ordered into a hierarchy of significance with respect to their relative influence on each process. This has been done using a comparison matrix whereby each factor is rated according to its significance relative to every other factor in the matrix [4].

Many factors (geological type, topography, presence of hazardous processes) were taken into account to form the rating of the identified taxa. According to the developed approach, the set of conditions is divided into several levels of vulnerability. Each level corresponds to the values of hazards that form vulnerability. In other words, we used the expert assessment. Further, each factor value was assigned its weight rating, also established from past experience. The following ratio was used to calculate the vulnerability rating [13]:

$$W = W_i \times D_i / S, \quad (1)$$

where W_i is the weight of i-th factor, Di represents subclass weight of ith factor, and S is the spatial unit of the final map.

The weighted overlay method (WOM) is a simple and direct tool of Arc GIS to produce susceptibility maps [5,13,14]. Many researchers used WOM to produce a landslide susceptibility map [15–17]. All layers were combined by using the weighted overlay tool based on Equation (1). In this paper, the number of the considered factors, based on the relevant databases was 27 different parameters (of 4 classes: endogenous, lithological, geomorphological factors, and factors of exogenous and natural-climatic nature). The algorithms used to recognize hazardous geodynamic and atmospheric-climatic processes affecting the environmental state of high mountain landscapes are shown in Figure 2 [17]. In the work of [18], performed for the territory of the Krasnodar Territory, the number of the considered factors was 9 indicators, for landslide hazard assessment number of parameters varies from 5 for Mussoorie Township case study [19] to8 factors in the work of [20] for Indian Himalaya.

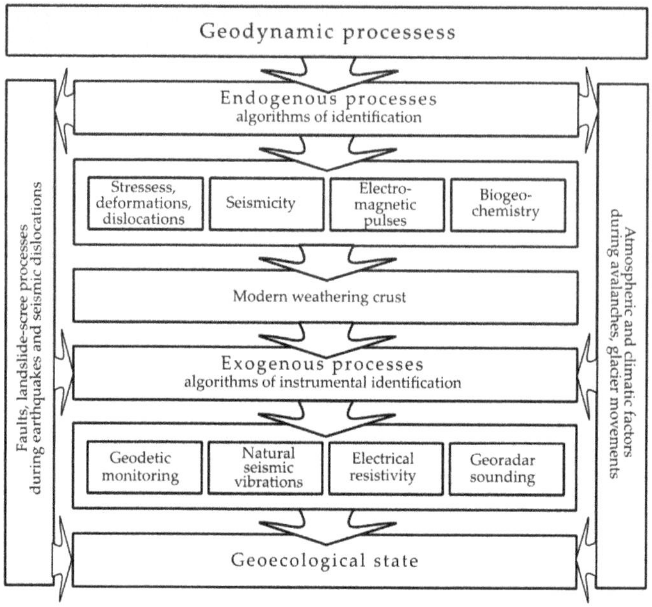

Figure 2. Algorithms for recognizing hazardous geodynamic processes affecting the environmental state of high mountain landscapes.

The weight of each factor is set based on the expert rule approach [21–29]. Or weights may be calculated on the basis of cases study (In [30], statistical procedure consisted in calculating a landslide frequency per class area expressed in percentage, i.e., number of landslide pixels.km^{-2} $\times$ 100). Logistic function may be used to weight factors is a suitable example of weights calculations based on landslides distribution data [31].

The geological environment is heterogeneous, consisting of voids, fluids, layers, blocks with a different stress state, so rhythms and catastrophes are propagated in the environment in different ways. To study all these processes, geophysical monitoring is required, and as a special case, seismic monitoring [17,32–37], gravity monitoring [38], combined geoelectric and geoelectromagnetic monitoring [39].

Promising directions of methodological solution of spatial-temporal intervals of endogenous events, their dynamic and kinematic characteristics, scales, and features of their course are considered as the basis for creating a verified database of geological, hydrogeological, seismotectonic, and technogenic processes for a comprehensive generalization of the degree of their impact on construction objects and infrastructure elements. The methodology for providing geological information stipulates the collection and processing of engineering survey materials of past years. At the same time, an important role is given to the interpretation of remote materials, including materials from traditional surveys [40].

The formation and interrelationship of erosion hazardous geological processes and their impact on the environment of high mountain areas are shown in Figure 3. The location of the investigated area in the zone of alpine tectonomagmatic activation of the Greater Caucasus determines the complex engineering and geological conditions inherent in the areas of alpine folding, caused by the diversity of the lithological composition of rocks, their intense tectonic disturbance, active neotectonics and seismicity, and the wide development of modern exogenous geological processes (Figure 3) [41,42].

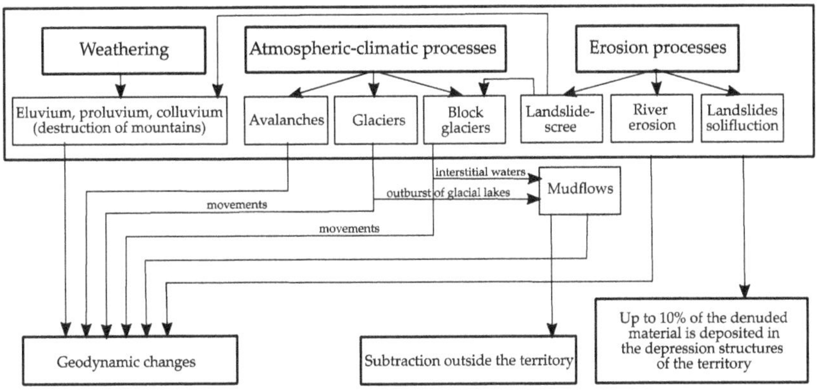

Figure 3. Typical hazardous natural exogenous geological processes of impact on the environmenta lstate of high mountain areas.

Information about genetic types, spatial and quantitative characteristics, the state of objects of hazardous exogenous geological processes are the main initial data for the pre-project solution of engineering surveys [43–45] and can be used to develop security measures and planning by security management structures [46–49]. The following exogenous geological processes are common for the territory: weathering, erosion, mudflow and gravitational (landslide, landfall), avalanche, solifluction, suffusion [43].

Weathering is the most common geological process. Liasshales and flysch rocks are especially intensively weathered. Sandstones, siltstones, and limestones break up into separate blocks and pieces. The process is widely developed over the entire area, both on its ridge sections and the slopes; it is especially manifested along rupture tectonic faults and zones of their influence. This is the main source of clastic material for the formation of mudflows, landslides, and the formation of slope deposits of various genetic types and eluvial on the ridge and peri-ridge parts of the territory [43].

Erosion processes are widely developed throughout the territory. They include sheet erosion, gully, lateral and bottom erosion of rivers. The activation of erosion processes on a regional scale is associated with the modern tectonic uplift of the Greater Caucasus orogen, estimated at rates from 2 to 14 mm/year [43]. Ascending movements increase the overall energy of the landform, which is already quite high. In high mountain areas, the length of the erosional pattern is 2.5–3.4 km/km^2, and the depth of local erosion bases reaches 1000–1500 m. Under these conditions, water flows are characterized by high speed (up to 3 m/s and more) and perform significant erosion activity [45].

The history of North Ossetia is replete with a sufficient number of facts testifying to the susceptibility of the territory of the republic to the destructive effects of almost all known types of dangerous geological, hydrometeorological, and biological processes that lead to the death of people and huge material losses, have a significant impact on the socio-economic situation and security of the territory of the republic.

The total average long-term economic damage to the republic from these dangers currently reaches 1% of the republic's GRP [11].

The cyclic nature of natural phenomena and processes creates conditions for the occurrence of emergencies that are typical for the territory of the republic. The following most significant hazardous natural processes, which are a manifestation of a number of factors (meteorological, hydrological, geomorphological), can be distinguished in Table 1, Figure 4 [11].

Table 1. Recurrence of the hazardous natural processes on the territory of North Ossetia-Alania (modified from the work of [11]).

Type of Emergency Situation	Return Period	Notes
Earthquakes	1/100 years	The catastrophic manifestation of the natural process is considered
Landslide processes	1/10–20 years	Annually, individual landslides depending on the structure
Mudflows	1/10–15 years	Individual mudflows at 5–7 years
Avalanche process	1/10 years	A number of avalanches occurs every year

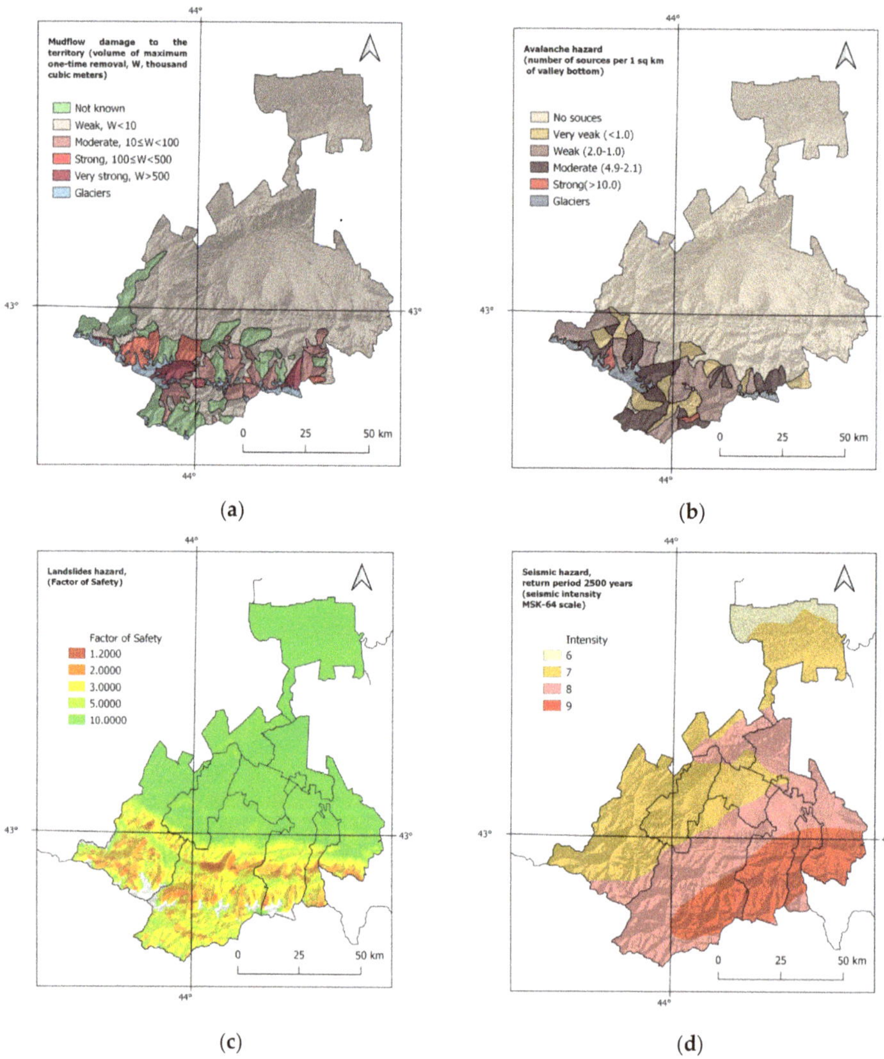

Figure 4. (**a**)Scheme of mudflow hazardous zones of North Ossetia-Alania; (**b**) scheme of avalanche hazard on the territory of North Ossetia-Alania [33]; (**c**) assessment of the potential landslide hazard of the territory of North Ossetia-Alania in the form of a safety factor F; (**d**) seismic hazard map of the territory of North Ossetia-Alania, intensity with 5% exceedance probability.

3.1. Mudflows

Mountainous Ossetia is one of the most mudflow-prone regions of the Central Caucasus. The strongly dissected topography, modern glaciation, high seismicity, huge reserves of loose-clastic material, and high moisture content create favorable conditions for mudflow formation (Figure 4a) [50]. A total of 60% of the territory of North Ossetia-Alania is subjected or may be subjected to the passage of mudflows of varying hazard levels. At the same time, huge damage is caused to settlements, agricultural lands and facilities, industrial enterprises, communication and power lines, sports and recreation facilities, and highways.

The following genetic groups of mudflow centers are developed: associated with the accumulation of loose-clastic material in the channels of temporary and small flows; associated with the damming of rivers and with the activity of modern glaciers.

Observations of mudflows in the main basins of the Mamisondon and Zrug rivers for the period of the last regular observations from 2004 to 2014 show a steady increase in the number of mudflows and their total volume (Figure 5).

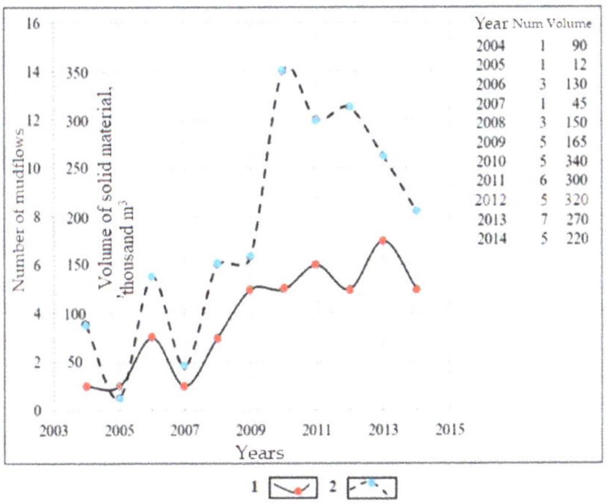

Figure 5. The frequency of mudflows in the basins of the Mamisondon and Zrug rivers by years for the period 2004–2014: 1—number of mudflows per year; 2—volume (thousand m^3) of solid material removal.

3.2. Avalanches

The altitudinal extent of climatic regions (plain, foothill, and mountainous) on the territory of the republic has a great influence on the nature of the occurrence and formation of snow cover. An analysis of the conditions of snow accumulation on the mountain slopes in the republic shows that there is a snow cover with a thickness of 40 cm or more in most of its mountainous territory. In practice, it has been determined that the existence of such a snow cover on slopes exceeding 15° causes avalanches [43].

An active zone of avalanches on the territory of the republic covers a wide range of altitudes from 1200 to 3800 m above sea level, with an average steepness of slopes from 15° to 76°. The highest relative height of the avalanching varies in a wide range from 1060 to 2060 m above sea level. The total number of avalanche sources on the territory of the republic is 688. The volume of snow carried by avalanches ranges from 500 to 1,500,000 m^3, and the maximum possible impact force of an avalanche on a fixed obstacle can reach 131 t/m^2. Each avalanche source on the territory of the republic produces an average of two avalanches per year, which indicates a high avalanche activity in the mountains of North Ossetia-Alania [43].

The duration of the avalanche period in the zone of the most active avalanche formation at altitudes above 2000 m above sea level can be up to 210 days (from November to

May). At the same time, the maximum development of avalanche processes is observed in January–February, and the minimum number of avalanches and the total volume of snow being removed falls on April–May since in most avalanche areas, the snow has already melted by this time. In the zone of average avalanche activity at lower absolute heights, the duration of the avalanche period is reduced to 90 days or less. At the same time, the maximum avalanches in this zone are usually observed in December–January. At the lower boundary of the zone of systematic avalanches, the period is usually 15–20 days [33].

3.3. Landslides

The landslide process is one of the many forms of geological matter evolution. The causes of landslides are divided into two groups: random (scholastic) and expected. The first group includes such fast-moving factors as precipitation, abrasion, erosion, earthquakes, and man-made factors. The regularity of the landslide process reflects the features and mode of rocks occurrence, tectonic disturbances that gradually develop in these rocks, and the appearance of stress fields in them. The ratio of both factors leads to the accumulation of anomalous stresses in rocks, which causes landslides.

Among 350 landslides explored in the mountains of North Ossetia, the following are distinguished by size: small (up to 1 thousand cubic meters), medium (up to 100 thousand cubic meters), large (up to 1 million cubic meters), and very large (with a volume of more than 1 million cubic meters). Examples of very large landslides are the Luar, Donifars, and N. Nar landslides.

The main factor of landslide processes activation is the excessive watering of the slopes, which occurs during snowmelt and heavy rains (Figure 6). A sharp increase in river runoff contributes significantly to the activity of landslides; as a result, the erosion and destruction of the frontal parts of landslides occur, which often leads to the displacements of the overlying masses, as was observed on the Luar landslide in 1984 and on the Matsuta landslide in 1991.

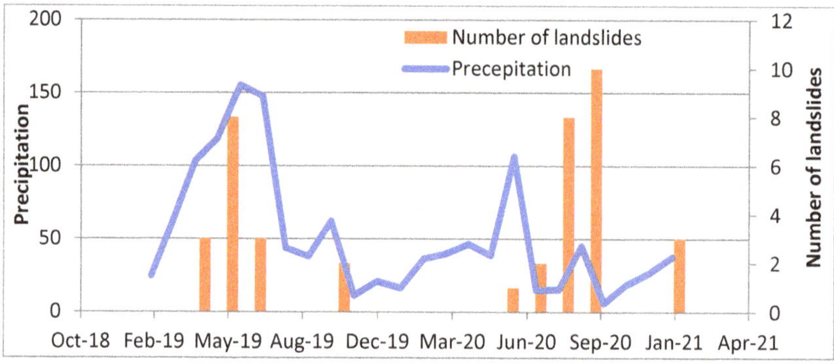

Figure 6. Monthly precipitation and the number of landslides in the Alagir region of North Ossetia-Alania in 2019–2020.

A noticeable increase in landslide activity is caused by seismic impacts associated with modern tectonic movements. In particular, the movements of the Matsuta and Donifars landslides in 1990 and the Ursdon landslide in 1993 are associated with tectonic movements. A retrospective analysis of the formation and large activation of landslides in the 20th century indicates that some of them are directly related to earthquakes (Dallagkau landslide, 1905; Korinskiy landslide, 1915; Turmonskiy landslide, 1981).

Slope stability was assessed for the territory of North Ossetia on a scale of 1:200,000. The average values of rock properties were used to identify the most hazardous zones for further detailed analysis. There was used the Scoops3D program, which makes it possible to assess the stability of slopes in a three-dimensional setting based on digital elevation models. The elevation model was built on the basis of SRTM satellite data. To perform calculations throughout the entire territory, a digital elevation model was compiled, the resolution was reduced to 200 m.

Scoops3D is a computer program developed by the United States Geological Survey (USGS) for analyzing slope stability across the entire digital landscape, represented by a digital elevation model (DEM). It detects large numbers and calculates the stability of 3D potential landslides covering a wide range of depths and volumes. Scoops3D uses a three-dimensional "column method" of limited equilibrium analysis to calculate the resilience of potential slope failures (landslides) with a spherical potential slip area. The approach was originally described in the work of [51]. The results of Scoops3D analysis show the minimum safety factor (stability index) for potential slip areas affecting each DEM cell throughout the landscape, as well as for volumes or areas associated with these potential damages of slopes.

Scoops3D allows the user to choose between two well-known geotechnical methods of the moment equilibrium for calculating the stability of a rotating surface. Bishop's method was used in our calculations. When performing a slope stability analysis, Scoops3D calculates the shear strength s on the test surface using the Coulomb-Terzaghi linear fracture rule [52].

Scoops3D calculates the safety factor F for a given test surface using moment equilibrium [52]. In general, all limited equilibrium methods (including moment equilibrium methods) define F as the ratio of the average shear strength s to the shear stress τ required for maintaining limit equilibrium along a predetermined test surface [53]:

$$F = s/\tau. \tag{2}$$

F values less than one indicate instability. A fixed 1/F ratio of the available shear strength is opposed to the shear stress at equilibrium (fundamental assumption of the method).

As a result, the assessment of the value of the safety factor F is obtained (Figure 4c) for each element of the terrain model (excluding boundary elements). For altitudes above 3500 m, the calculations were not performed (white zones in the figure). The sites characterized by the greatest danger (minimum F values) are located in the well-known areas-Tsey, Unal-Sadon, and Buron sites. In addition, separate point areas have been identified, which have not been previously studied due to their remoteness from infrastructure facilities. A detailed study of the properties of rocks and topographic survey (including the use of unmanned photogrammetry) will allow assessing their stability more accurately.

Rockslide-talus processes are widely developed in the mountainous part of North Ossetia-Alania. These processes cover almost all large outcrops in road cuts and on the slopes of valleys. In these areas, frequent falls of trees, large boulders, rubble, and whole blocks of loose formations are noted. With insufficient development of engineering protection of roads and other objects, rockslides lead to significant material losses [45].

Causes of rockslides under the conditions of sharply dissected topography and the presence of intensely dislocated rocks are the following: active geodynamic processes that are very typical for territories located at the junctions of megaplates; phenomena of accumulated seismic impact (natural and man-made); seismic impact; the mechanism of the influence of cracks on the side impact.

On the territory, any steep-sided (up to 50°–90°) section composed of rocks can be classified as rockslide-hazardous, but the most probable are the areas of powerful tectonic faults development (Tsey thrust and a series of tectonic faults of the southern slope-the Severny and Nar faults, Tib, Khalatsa, Zgil, and Saukhokh faults) and the zones of development of the newest discontinuous seismic dislocations as a way to relieve endogenous tectonomagmatic stresses [43]. Intensive manifestations of rockslide-talus processes occur in zones of concentration of tectonic disturbances, which are clearly manifested in seismic and electric fields.

3.4. Earthquakes

The high seismicity of North Ossetia-Alania is caused by a number of active regional faults of deep origin, among which one should mention the transcontinental fault of the meridional direction, which goes into the Earth's crust to a depth of 100–200 m and captures North Ossetia in the west, where there were earthquakes up to 9 points; further on, the North Caucasian subcrustal fault with a depth of 75 km, crossing a section of the territory of North Ossetia-Alania in the latitudinal direction, as well as the Ardon, Terek faults and the main thrust-crustal faults with a depth of up to 35 km, as well as a zone of deep shifts northeast-oriented, crossing the southeastern sector of the region with a depth of 30–40 km. Almost all seismic shocks of the last 50 years from 5 to 9 points were recorded within this zone [11].

The probabilistic seismic hazard maps (the maps of detailed seismic zoning) have been constructed for the total area of North Ossetia on a scale of 1:200,000 with exceedance probability for a period of 50 years (standard time of building or construction durability!) with 1%, 2%, 5%, 10% in GIS technologies, which corresponds to the reoccurrence of maximum probable earthquake for a period of 5000, 2500, 1000 and 500 years. The longer the period of time, the higher the level of possible intensity. For a period of 500 years, only a small part will be occupied by the zone of 7 intensity earthquake, for a period of 1000 years–8 intensity and at 2500 years 9 intensity earthquake appearance, correspondingly. Cornell's approach, namely the computer program SEISRisk-3, developed in 1987 by Bender and Perkins (Bender and Perkins, 1987), was used for the calculations [53,54]. The maps of 5% probability are likely to be used for the large scale building, i.e., the major type of constructions, whereas the maps of 2% probability should be used for high responsibility construction (map is shown in Figure 4d). One can see great hazards in the south of North Ossetia on the map, where exists the increased level of seismic hazard (due to the powerful Vladikavkaz fault lying nearby).

4. Results

For a quantitative assessment of the environmental impact of geodynamic factors, a selection of all manifested geodynamic and climatic factors was made. The intensity of the impact of each of these factors on the environment is given according to a five-point system. At the same time, the rank is based not on the intensity of the shaking, but on the ability of one or another geodynamic factor to accelerate the denudation of rocks, to weaken their engineering properties, to develop Quaternary deposits, dangerous geological processes and, ultimately, accelerated changes in the environmental situation.

The maximum rank of the impact of an individual factor was determined based on the methodology of expert assessments [21,22] in the form of taking into account mass transfer (landslides-screes, mudflows, landslides, etc.), landscape and geomorphological changes (glacier movements, avalanches), long-term or short-term expected consequences (tectonic, volcano-plutonic manifestations, earthquakes, seismic dislocations). It is accepted that the impact on the environmental situation of N factors, each with an intensity of one rank or a single rank, corresponds to the least influence of the factor on the environmental state, and the level of the fifth rank corresponds to the greatest one, by analogy with the works of Zaalishvili V.B. [55,56].

The ratio (1) was used to calculate the risk R of impact realization. According to the developed approach, the set of conditions is divided into several levels of vulnerability. Each level corresponds to the values of hazards that form vulnerability. This classification is based on the experience of past earthquakes. In other words, as was noted above, we used the expert assessment. Further, each factor value was assigned its weight rating, also determined based on past experience. The following ratio was used to calculate the vulnerability rating (Table 2). Intermediate data and calculation results are shown in Figure 7.

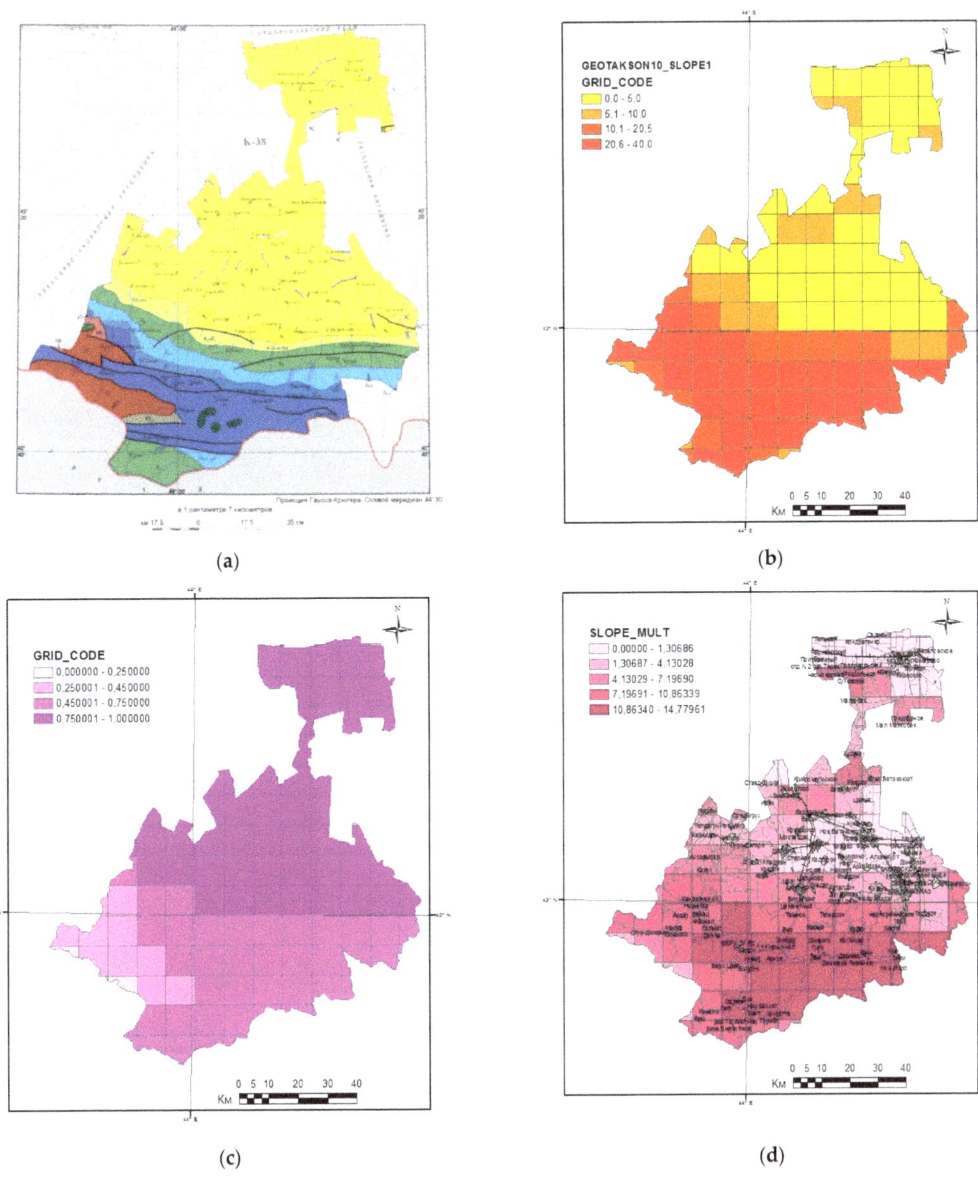

Figure 7. (**a**) Geological map of the North Ossetia-Alania territory (according to VSEGINGEI data); (**b**) rating index of relief slope angle; (**c**) rating index of geological structure; (**d**) integral rating index and location of settlements and infrastructure facilities.

Table 2. Rating indicators of factors of various nature.

Factor	Units	Vulnerability, D			WeightRating, W
		1	2	3	
Seismicity of the territory	Intensity, MSK-64	7	7–8	8	2.0
Geology	Category	I	II	III	0.5
Topography	Slope degree	<5°	5°–15°	>15°	1.0

According to the results of the rating assessment, two urbanized zones can be distinguished (Figure 7d):

(1) The zone of minimal impact is the flat northeastern part, built up with cities and industrial facilities. Beslan can be considered as the center of this zone because it is the optimal area for the development of the republic. The largest settlement on the territory of North Ossetia-Alania, Vladikavkaz, to a greater extent, is potentially exposed to natural impacts;
(2) The mountain part with the presence of historical settlements and objects associated with mining. This part is prone to the highest level of natural impact. The zone of the "Mamison" complex is also located in the zone of high impact.

For a quantitative assessment of the environmental impact of geodynamic factors, a selection of all manifested geodynamic and climatic factors was made; the intensity of the impact of each of these factors on the environment is given according to a five-point system. At the same time, the basis of the intensity degree is not the intensity of shaking, but the ability of the geodynamic factor to accelerate the denudation of rocks, weaken their engineering properties, develop Quaternary deposits, hazardous geological processes, and, finally, accelerate the change in the environmental situation.

The maximum point of the impact of a single factor is determined concerning the mass transfer (landslides, talus, mudflows), landscape and geomorphological changes (surge of glaciers, avalanches), expected long-term or short-term consequences (tectonic, volcano-plutonic manifestations, earthquakes, seismic dislocations). It is accepted that the impact on the environmental situation of N factors (each with an intensity of 1 point (single point)) is equal to the change corresponding to the stabilized platform development condition. It is necessary to note that when assessing the intensity degree of the geodynamic factor, an element of subjectivity is presented; however, increasing of factors leads to minimization of the total error for subjective approaches.

According to the complex of geodynamic factors observed on the territory of the projected complex, two tables have been developed. Table 3 shows the levels of the impact of geodynamic processes on the environmental state of high mountain territories depending on the geological and geomorphological conditions, and Table 4 represents ranking in points of the potential impact of geodynamic processes on the environmental state of high mountain territories according to the levels of their manifestation in various geological and geomorphological conditions.

Table 3. Levels of the impact of geodynamic processes on the environmental state of high-mountain territories, depending on the geological and geomorphological conditions.

No.	Types of Impact	Level of Impact			
		Weak	Middle	Hight	Regressive
1	Earthquakes in the near-field (R = 50 km), seismic intensity MSK	<6	6–7	7–8	8–9
2	Active fault zone (density in km/km^2)	0.01	0.02	0.03	0.04
3	Zone of intersection of sublatitudinal and submeridional faults from one to four and higher (number of crossings)	1	2	3	4
4	Fissure tectonics development zone, 0.03 to 0.10 km/km^2 and higher	0.03	0.05	0.07	>0.1
5	Zones affected by static geophysical, geochemical fields within an event radius of up to 50 km, distance in km: 10–20; 20–30; 30–40; 40–50	3	4	5	6
6	Manifestation zones of one to three or more seismic dislocation events within a structural block	1	2	3	3

Table 3. *Cont.*

No.	Types of Impact	Level of Impact			
		Weak	Middle	Hight	Regressive
7	Distribution of Quaternary formations at altitude intervals: 1700–2000 m; 2000–2300 m; 2300–2600 m; 2600 m and above	1700–2000	2000–2300	2300–2600	>2600
8	Soft bedrock (up to 50 MPa), exposure 40%; 50%; 60%; 70%; or more	40%	50%	60%	>70%
9	Erosion basis 1000 m or more, at altitudes of 2000 and above, taking into account glaciation	2000–2500	2500–3000	3300–3500	3500–4000
10	Erosion basis 500 m at altitudes of 1500 and above	1500–1700	1700–1800	1800–1900	1900–2000
11	Erosion basis up to 500 m at altitudes from 1500 m and below	1300–1500	1100–1300	1000–1100	>1000
12	Terrain surface inclination angles up to 15° according to genetic soil types:	landslide-scree	moraine deposits	debris flows	landslide
13	Terrain surface inclination angles 15°–20° according to genetic soil types:	landslide-scree	landslide-scree	debris flows	landslide
14	Terrain surface inclination angles >20° according to genetic soil types:	landslide-scree	landslide-scree	debris flows	landslide
15	Thickness of Quaternary deposits 0–5 m, at slope angles of 12°–25° and more	12–15	15–20	20–25	>25
16	Thickness of Quaternary deposits 5–12 m, at slope angles of 12°–25° and more	12–150	15–20	20–25	>25
17	Thickness of the Quaternary deposits more than 12 m, at slope angles of 12°–25° and more	12–15	15–20	20–25	>25
18	Number of mudflows, from one to 5 and more per season, volumes from 10 thousand m^3 and more	2	3	4	5
19	Landslides in the total volume for the year from 10 thousand to 2 million m^3	50	75	100	150
20	Landslides-scree up to 5000 m^3	75	100	150	200
21	Flat wash, area from 0.05 to 0.10 km^2 and more, per km^2	0.05	0.06	0.07	>0.1
22	Erosive activity of rivers from 1 to 5 m/hour of lateral erosion	1.0	2.5	4.0	5.0
23	Solifluction with vertical capture power from 1 to 4 m	1.0	2.0	3.0	4.0
24	Stream erosion on slopes of 200 or more	20	25	30	>35
25	Karst formation, from one to three or more manifestations per km^2 of area	1	2	3	>3
26	Avalanches, up to 4 or more descents per season	1	2	3	4
27	Glaciers and glaciers up to three or more flood-type events per area	1	2	3	>3

Table 4. Ranking in points of the potential impact of geodynamic processes on the environmental state of high-mountain territories and the levels of their manifestation in various geological and geomorphological conditions.

No.	Types of Impact	Weight Rating	Level of Impact			
			Weak	Middle	Hight	Regressive
	I Endogenous impact factors					
1	Earthquakes in the near-field (R = 50 km), seismic intensity MSK	5	1	2	3.5	5
2	Active fault zone (density in km/km^2)	3	1	1.5	1	3

Table 4. *Cont.*

No.	Types of Impact	Weight Rating	Level of Impact			
			Weak	Middle	Hight	Regressive
3	Zone of intersection of sublatitudinal and submeridional faults from one to four and higher (number of crossings)	4	1	2	3	4
4	Fissure tectonics development zone, 0.03 to 0.10 km/km^2 and higher	5	1	2	3.5	5
5	Zones affected by static geophysical, geochemical fields within an event radius of up to 50 km, distance in km: 10–20; 20–30; 30–40; 40–50	2	1	1	2	2
6	Manifestation zones of one to three or more seismic dislocation events within a structural block	3	1	2	3	3
	II Lithological factors					
7	Distribution of Quaternary formations at altitude intervals: 1700–2000 m; 2000–2300 m; 2300–2600 m; 2600 m and above	3	1	1.5	2	3
8	Soft bedrock (up to 50 MPa), exposure 40%; 50%; 60%; 70%; or more	3	1	1.5	2	3
	III Geomorphological factors					
9	Erosion basis 1000 m or more, at altitudes of 2000 and above, taking into account glaciation	4	4	3	2	1,5
10	Erosion basis 500 m at altitudes of 1500 and above	3	1	1.5	2	3
11	Erosion basis up to 500 m at altitudes from 1500 m and below	2	1	1.5	2	2
12	Terrain surface inclination angles up to 15° according to genetic soil types:	2	1	1	2	2
13	Terrain surface inclination angles 15°–20° according to genetic soil types:	3	1	1.5	2.5	3
14	Terrain surface inclination angles >20° according to genetic soil types:	4	2	2.5	3	4
	IV Factors of exogenous and natural-climatic nature					
15	Thickness of Quaternary deposits 0–5 m, at slope angles of 12°–25° and more	2	1	1	1.5	2
16	Thickness of Quaternary deposits 5–12 m, at slope angles of 12°–25° and more	3	1	1	2	3
17	Thickness of the Quaternary deposits more than 12 m, at slope angles of 12°–25° and more	3	1	1.5	1.5	3
18	Number of mudflows, from one to 5 and more per season, volumes from 10 thousand m^3 and more	5	2	3	4	5
19	Landslides in the total volume for the year from 10 thousand to 2 million m^3	3	1	1.5	2	3
20	Landslides-scree up to 5000 m^3	4	1.5	2	3	4
21	Flat wash, area from 0.05 to 0.10 km^2 and more, per km^2	3	1	1	2	3
22	Erosive activity of rivers from 1 to 5 m/hour of lateral erosion	4	1.5	2.5	3	4
23	Solifluction with vertical capture power from 1 to 4 m	3	1	1.5	2.5	3
24	Stream erosion on slopes of 200 or more	3	1	1.5	2.5	3
25	Karst formation, from one to three or more manifestations per km^2 of area	3	1	1.5	2	3
26	Avalanches, up to 4 or more descents per season	5	2	3	4	5
27	Glaciers and glaciers up to three or more flood-type events per area	4	1.5	2.5	3	4
	Sum of scores of all impacts/Percent of contribution	**95** **100%**	**34.5** **36.3%**	**48.0** **50.5%**	**64.0** **67.4%**	**91.5** **96.35%**

The ranking of the territory and its adjacent areas (about 400 sq. km) according to 27 geodynamic factors of endogenous, exogenous, and climatic nature is performed by squares of 4 sq. km (about 100 squares). In these tables, the maximum total level of influence of all factors (the sum of the weight ratings of the event) with simultaneous participation is 95 points. This value can be less by the weight level of the geodynamic factor that does not participate in the process. The environmental impact on a square is estimated by the ratio of the sum of points for this square to the potentially possible sum of the maximum determined points of the impact of the geodynamic factors involved in the process of environmental transformation.

Zoning of the environmental impact of the territory was carried out based on the obtained values of these relations, which are classified into four groups: 30–40%—weak impact; 40–55%—average impact; 55–70%—high impact, above 70%—regressive impact. When zoning the territory according to the level of environmental impact, the data of block differentiation of the territory and the development of active tectonic dislocations [57] of the Kazbek segment of the Greater Caucasus mega-arch were used. The scheme of the territory zoning according to the intensity of the environmental impact is shown in Figure 8. The site plots where the greatest observed environmental impacts are marked with a special dash.

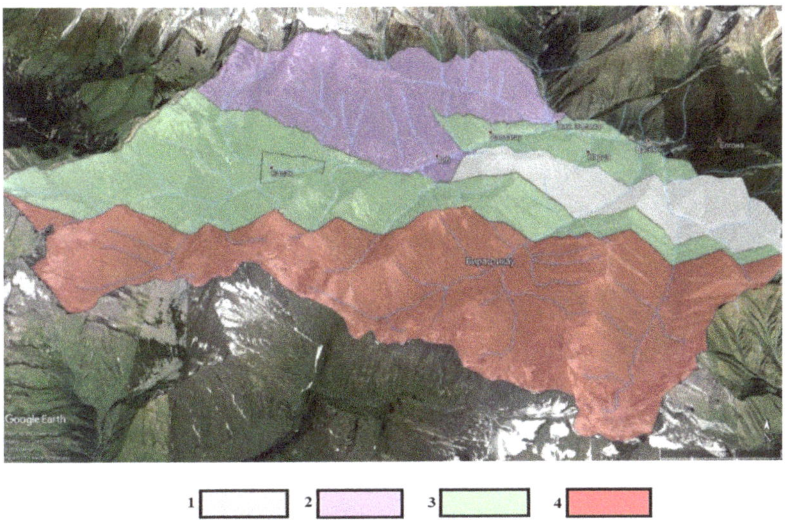

Figure 8. The scheme of the zoning of the environmental impact of the territory of the tourist and recreational complex "Mamison". Levels of environmental impact from geodynamic impacts: 1—weak; 2—average; 3—high; 4—regressive.

5. Discussion

Indicator structures that record the characteristic features of the impact of deep geodynamic energy on the environment are manifestations of the geodynamic situation as a cyclic stage of geological transformation, mountain building, volcanism, tectonic discontinuities, earthquakes in the form of local zones of origin of sources, seismic dislocations, dynamic impacts in the form of dislocations, the influence of physical fields and geochemical halos, and, finally, exogenous consequences for the environment of endogenous geodynamic activity (almost a complete set of genetic types of dangerous exogenous geological processes accompanying the tectonic structure). Exogenous dangerous geological processes typical for the mountainous territory are the following: mudflows, avalanches, landslides, glaciers, rockfalls, landslides, breakthroughs of open and closed water reservoirs, etc.).

It should be noted that during the weight contribution assessment of the geodynamic factor, an element of subjectivity is present to a certain extent. However, the more factors are used in combination, the more the total error for subjectivity approaches a minimum.

Based on the analysis of the results of studying the complex of geodynamic factors observed on the territory of the projected complex, two tables were developed.

Table 3 shows the quantitative characteristics of the impact of geodynamic processes on the environmental state of the site, caused by the lithological, geomorphological, endogenous, exogenous geological, and climatic factors. Table 4 shows the results of calculating the risk Ri of the impact of an individual factor with the potential Wi on the site and its environmental vulnerability Di on the corresponding factor.

Thus, the ranking of the risks of the potential impacts on the environmental state of high-mountain territories is realized according to the levels of their manifestation from lithological, geomorphological, endogenous, exogenous geological, and natural-climatic factors or conditions [58,59]. Given the interdisciplinary nature of the content of environmental and environmental problems, it is no wonder that both directions are so intertwined in scientific works that it is sometimes difficult to determine the solution of the problem of which discipline they are devoted. Traditionally, there is an idea that geoecology is an interdisciplinary direction, a kind of metascience, which summarizes all existing knowledge about the ecological state of the Earth, including environmental biology, although it is known that the biosphere partially determines the state of geoecology, forming the soil layer, participating in the transformation of the weathering crust; likewise, the geoecology affects the development and modification of living organisms.

Claims to the geographical orientation of geoecology can also be recognized as substantive when considering the consequences of environmental changes due to exogenous geological processes when landscapes and ecosystems change, but geography often does not have answers to changes in the environmental state of the environment caused by endogenous geological processes.

According to V.I. Osipov [60], the object of geoecology is the geospheric shells of the Earth, i.e., not only the lithosphere or geological environment, but also the hydrosphere, atmosphere, and biosphere. In this case, the subject of environmental science is the totality of all knowledge about the geospheres, including changes occurring under the influence of natural and technogenic factors.

The influence of geodynamic processes is considered only in the context of processes occurring in the geospheres of the Earth and affecting the environmental state at the boundary of the lithosphere and atmosphere.

Geodynamic hazard (endogenous and exogenous processes) is a threat caused by dangerous geological processes of an endogenous and exogenous nature, defined as the probability of a certain event or set of events manifestation in a given territory with a predictable intensity and within a given time interval.

The level of geodynamic hazard of the territory depends on the geodynamic activity of the lithosphere (endogenous geodynamics), the lithological composition of the sedimentary cover, the conditions for the interaction of the geological environment with the atmosphere, and climatic changes (exogenous geodynamics). Geodynamic and climatic processes determine the environmental state of the environment and the evolution of the Earth as a whole. Seismicity (rarely tectonic movements) and the exposure of the territory to exogenous processes [61–63], being the factors available to a person for research, are considered as the criteria for such an assessment. However, the relationship of even these two manifestations of the Earth's evolution at a subtle level (of a complex of physical fields-deformation, dislocation, electromagnetic, geochemical, and thermal fields) remains unclear.

An analysis of any exogenous natural geological process, regardless of the impact agent, shows that the object is being prepared for an event for a long time by endogenous geodynamic impact. Active tectogenesis, dislocation, tension, fracturing, fluid inflow, seismicity, magmatism, heat transfer, geochemical and geophysical fields are all agents that disrupt the structural bonds of the massif, increase porosity, contribute to intensive weathering and activation of erosion processes, affect the evolution of biota limited to a certain space [64].

Endogenous activity is a causative factor in the activation of exogenous processes, and one of the signs of this is the stable confinement of the latter to tectonic faults and nodes of discontinuous structures of different ranks, focal zones of seismic and volcano-plutonic activity [65,66], observed throughout the development of the alpine folding.

The state of endogenous geodynamics activity is manifested by lineaments of deep tectonic faults of the Caucasian strike (controlling outcrops of mineral springs), manifestations of earthquakes, folded dislocation of the massif, fracturing, geochemical and geophysical fields, seismicity, and a special indicator characteristic of the active state, i.e., seismic dislocations.

The level of impact on the environment of endogenous and exogenous geodynamic processes characterizes various genetic types of Quaternary formations widely developed over the area, forming numerous sources of dangerous exogenous geological processes. The surface topography, being a structure of endogenous manifestation, reflects many elements of tectonic activity expressed in geomorphology. With the introduction of technologies for remote sensing of the Earth's surface with different resolutions, the possibilities of mapping geological structures of different orders, including tectonic ones, have opened up.

One of the geomorphological indices characterizing the tectonics of a territory is the asymmetry factor [67], which is determined by the ratio of the area to the right of the mainstream to the entire catchment area. If the water system develops on terrain with homogeneous lithology and insignificant development of structural disturbances, the value of the asymmetry coefficient will be equal to 50. However, for the territory associated with active tectonic structures, the asymmetry coefficient will significantly differ from this value. This is due to the emergence of the Coriolis force in moving water systems, acting on the right banks in the northern hemisphere and, conversely, on the left banks in the southern hemisphere. With the general tectonic fragmentation of the basin part of the river, the displacement of the channel to the right will be the more noticeable, the more intense the tectonic processes are. According to the level of spatial asymmetry, it is possible to judge the degree of development of tectonic processes in the territory.

The main energy-conducting channels are interblock crustal tectonic faults, along which during the intrusion, a magma-fluid jet can manifest itself in different ways: it can penetrate by an intrusion, displace blocks, create conditions for earthquakes, break out, and erupt into a volcano.

The tectonic fault connects the lithosphere with the atmosphere with all ensuing consequences of the transit of gases and dissolved elements in upward and downward modes. In the downward stream, oxygen enters the lithosphere at the expense of the organic substances of the biosphere. In the upward stream, dissolved elements and volatile components, together with juvenile and mantle waters, move upward due to the energy of the convective flow of mantle jets.

Soluble and volatile elements of geochemical aureoles of concealed deposits crossed by a fault rise to the day surface along tectonic faults; they may contain vapors of heavy and radioactive elements.

Experimental studies of modern movements of the Earth's crust carried out on geodynamic polygons of different purposes defined two types of geodynamic movements: cyclic with different duration periods and trends, maintaining a constant direction for long periods. Both types of geodynamic movements influence the safety of surface structures and underground space objects. Intense local anomalies of vertical and horizontal movements are associated with fault zones of various types and ranks [68,69]. These anomalous movements occur cyclically and sometimes have pulse character, against the background of trend tectonic movements; they are characterized by high-amplitude displacements (up to 50–70 mm/year), are spatially localized in intervals of 100–1000 m, have a periodicity of manifestation from 0.1 to one year and have a pulsating and alternating directivity.

Manifestations of alternating geodynamic movements in active tectonic structures lead to the liquefaction of the rock mass in the tectonic zone [70].

The whole range of observed cyclic alternating and trend movements indicates that the main property of the geological environment, especially in fault zones, is its continuous movement. Movement acts as a form of existence of the geological environment. In this regard, the role and place of modern geodynamics in the fundamental sphere of Earth sciences is difficult to overestimate, but its role in solving practical problems of ensuring the safety of objects associated with the development of territories with active neotectonics is even more important.

6. Conclusions

The stable confinement of manifestations of hazardous exogenous geological processes to active deep tectonic disturbances and associated fissure tectonics of the post-glacial manifestation of neotectonic activity confirms the concept of the decisive role of endogenous processes in the activation of exogenous processes and in the intensification of changes in the environment.

Impacts of endogenous factors (ruptures, fracturing, earthquakes, static stresses, seismic dislocations, deformations) on rocks are enhanced due to the influence of changes at the material, structural-textural and molecular levels due to physicochemical and geophysical agents accompanying endogenous factors, which leads to a weakening of the structural bonds of the rock and destruction under the influence of atmospheric agents.

Signs-indicators of a mechanical, geochemical, geophysical, hydrogeological nature, accompanying geodynamic factors, are used as a methodological basis for determining and localizing geodynamic factors by solving direct or inverse problems.

The cause-and-effect dependence of the development of exogenous geological processes and environmental changes on the activity of endogenous processes in the work is the author's concept, and it is confirmed in a stable spatial relationship between endogenous and exogenous processes.

Geophysical methods have shown high efficiency in identifying regional faults, manifestations of neotectonic activity (expressed by the development of fissure tectonics), and seismic dislocation zones.

Typical geodynamic processes were determined, and there was assessed their level of impact, considering geological, lithological, endogenous, exogenous, geomorphological conditions and factors of the natural and climatic nature of development, ranked according to their maximum calculated impacts on the environmental situation, the sum of which represents the potential of the expected environmental impact on area unit.

For the Mamison node, the algorithms for the interconnections of geodynamic processes are compiled on the assumption of a simultaneous multi-factor or single-factor impact on the environmental situation.

On the basis of the obtained data, a schematic map of zoning of the environmental impact of the territory of the tourist and recreational complex "Mamison" (which includes four rather extensive zones with different values of environmental impacts) was compiled.

The environmental state of the territory before the design decisions for the organization of the tourist and recreational complex "Mamison" is determined by the activity of natural processes, practically without the influence of any technogenic and man-made factors.

Geodynamic and climatic factors of influence on the geoenvironment create the danger of excessive environmental impact and disruption of its equilibrium state under anthropogenic impacts, which must be taken into account in designing.

Author Contributions: Conceptualization, D.M. and V.Z.; methodology, D.M. and V.Z.; software, D.M., A.K. and O.B.; validation, V.Z. and O.B.; formal analysis, D.M., V.Z. and O.B.; investigation, D.M., V.Z., O.B. and A.K.; resources, D.M.; data creation, D.M. and A.K.; writing—original draft preparation, D.M. and V.Z.; writing—review and editing, V.Z.; visualization, D.M. and A.K.; supervision, V.Z. All authors have read and agreed to the published version of the manuscript.

Funding: This research received no external funding.

Institutional Review Board Statement: Not applicable.

Informed Consent Statement: Not applicable.

Data Availability Statement: Not applicable.

Conflicts of Interest: The authors declare no conflict of interest.

References

1. Ahmed, A.; Nawaz, R.; Woulds, C.; Drake, F. Influence of hydro-climatic factors on future coastal land susceptibility to erosion in Bangladesh: A geospatial modelling approach. *J. Geovisualization Spat. Anal.* **2020**, *4*, 6. [CrossRef]
2. Hird, R.; Di Matteo, N.; Gulerce, U.; Babu, V.L.S.; Rafiq, A. Geohazards of Saudi Arabia. *J. Maps* **2019**, *15*, 626–634. [CrossRef]
3. Griffiths, J.S. *Engineering Geological Mapping*; Geological Society; Engineering Geology Special Publications: London, UK, 2001; Volume 18, pp. 39–42. [CrossRef]
4. Mason, P.J.; Rosenbaum, M.S. Geohazard mapping for predicting landslides: An example from the Langhe Hills in Piemonte, NW Italy. *Q. J. Eng. Geol. Hydrogeol.* **2002**, *35*, 317–326. [CrossRef]
5. Ali, S.; Biermanns, P.; Haider, R.; Reicherter, K. Landslide susceptibility mapping by using a geographic information system (GIS) along the China–Pakistan Economic Corridor (Karakoram Highway), Pakistan. *Nat. Hazards Earth Syst. Sci.* **2019**, *19*, 999–1022. [CrossRef]
6. Kern, K.; Lieb, G.K.; Seier, G.; Kellerer-Pirklbauer, A. Modelling geomorphological hazards to assess the vulnerability of alpine infrastructure: The example of the grossglockner-pasterze area, Austria. *Austrian J. Earth Sci.* **2012**, *105*, 113–127.
7. Ayele, S.; Raghuvanshi, T.K.; Kala, P.M. Application of Remote Sensing and GIS for Landslide Disaster Management: A Case from Abay Gorge, Gohatsion–Dejen Section, Ethiopia. In *Landscape Ecology and Water Management*; Springer: Tokyo, Japan, 2014; pp. 15–32. [CrossRef]
8. Chandak, P.G.; Sayyed, S.S.; Kulkarni, Y.U.; Devtale, M.K. Application of geoinformatics for landslide susceptibility analysis in Konkan region, Maharashtra. *J. NX—Multidiscip. Peer Rev. J.* **2021**, 3, 25–29.
9. Kumar, N.S.; Ismail, M.A.M.; Sukor, N.S.A.; Cheang, W. Geohazard reconnaissance mapping for potential rock boulder fall using low altitude UAV photogrammetry. *IOP Conf. Ser. Mater. Sci. Eng.* **2017**, *352*, 012033. [CrossRef]
10. Persits, F.M.; Ulmishek, G.F.; Steinshouer, D.W. *Maps Showing Geology, Oil and Gas Fields and Geologic Provinces of the Former Soviet Union*; Open-File Report 97-470E; U.S. Department of the Interior: Washington, DC, USA, 1997. [CrossRef]
11. *Annual State Report*; Ministry of Emergency Situations of the Republic of North Ossetia-Alania: Vladikavkaz, Russia, 2000; p. 118. (In Russian)
12. Zaalishvili, V.B.; Chotchaev, K.O. Classification of natural electromagnetic fields. Researches by the method of telluric currents. *Geol. Geophys. Russ. South* **2016**, *2*, 27–35. (In Russian) [CrossRef]
13. Intarawichian, N.; Dasananda, S. Analytical hierarchy process for landslide susceptibility mapping in lower Mae Chaem watershed, Northern Thailand, Suranaree. *J. Sci. Technol.* **2010**, *17*, 277–292.
14. Basharat, M.; Shah, H.R.; Hameed, N. Landslide susceptibility mapping using GIS and weighted overlay method: A case study from NW Himalayas, Pakistan. *Arab. J. Geosci.* **2016**, *9*, 292. [CrossRef]
15. Roslee, R.; Mickey, A.C.; Simon, N.; Norhisham, M.N. Landslide susceptibility analysis lsa using weighted overlay method wom along the gentingsempah to bentong highway pahang. *Malays. J. Geosci.* **2017**, *1*, 13–19. [CrossRef]
16. Shit, P.K.; Bhunia, G.S.; Maiti, R. Potential landslide susceptibility mapping using weighted overlay model (WOM). *Model. Earth Syst. Environ.* **2016**, *2*, 21. [CrossRef]
17. Chotchaev, K.O.; Burdzieva, O.G.; Zaalishvili, V.B. Influence of geodynamic processes on the geoecological state of high mountain areas. *Geol. Geophys. Russ. South* **2020**, *10*, 70–100. (In Russian) [CrossRef]
18. Stogny, V.V.; Stogny, G.A.; Lyubimova, T.V. Geoecological risks of the Krasnodar region: The problem of integrated assessment of the geological hazards degree. *Geol. Geophys. Russ. South* **2021**, *11*, 121–133. (In Russian) [CrossRef]
19. Peethambaran, B.; Anbalagan, R.; Shihabudheen, K.V. Landslide susceptibility mapping in and around Mussoorie Township using fuzzy set procedure, MamL and improved fuzzy expert system-A comparative study. *Nat. Hazards* **2018**, *96*, 121–147. [CrossRef]
20. Peethambaran, B.; Anbalagan, R.; Shihabudheen, K.V.; Goswami, A. Robustness evaluation of fuzzy expert system and extreme learning machine for geographic information system-based landslide susceptibility zonation: A case study from Indian Himalaya. *Environ. Earth Sci.* **2019**, *78*, 231. [CrossRef]
21. Goltsman, F.M. Integration of observations in the recognition of geophysical objects. *FizikaZemli* **1976**, *7*, 40–54. (In Russian) [CrossRef]
22. Goltsman, F.M. Information richness of the experiment and accounting for the correlation of non-numeric features in the recognition of geophysical objects. *FizikaZemli* **1980**, *10*, 71–85.
23. Daniel, M.T.; Ng, T.F.; Kadir, M.F.A.; Pereira, J.J. Landslide Susceptibility Modeling Using a Hybrid Bivariate Statistical and Expert Consultation Approach in Canada Hill, Sarawak, Malaysia. *Front. Earth Sci.* **2021**, *9*, 616225. [CrossRef]

24. García, C.L.; Teich, I.; Gonzalez-Roglich, M.; Kindgard, A.F.; Ravelo, A.C.; Liniger, H. Land degradation assessment in the Argentinean Puna: Comparing expert knowledge with satellite-derived information. *Environ. Sci. Policy* **2018**, *91*, 70–80. [CrossRef]
25. Vanwindekens, F.M.; Gobin, A.; Curnel, Y.; Planchon, V. New Approach for Mapping the Vulnerability of Agroecosystems Based on Expert Knowledge. *Math. Geol.* **2018**, *50*, 679–696. [CrossRef]
26. Di Febbraro, M.; Sallustio, L.; Vizzarri, M.; De Rosa, D.; De Lisio, L.; Loy, A.; Eichelberger, B.; Marchetti, M. Expert-based and correlative models to map habitat quality: Which gives better support to conservation planning? *Glob. Ecol. Conserv.* **2018**, *16*, e00513. [CrossRef]
27. Kaziukonytė, K.; Lesutienė, J.; Gasiūnaitė, Z.R.; Morkūnė, R.; Elyaagoubi, S.; Razinkovas-Baziukas, A. Expert-Based Assessment and Mapping of Ecosystem Services Potential in the Nemunas Delta and Curonian Lagoon Region, Lithuania. *Water* **2021**, *13*, 2728. [CrossRef]
28. Rohrbach, B.; Weibel, R.; Laube, P. Parameter-free aggregation of value functions from multiple experts and uncertainty assessment in multi-criteria evaluation. *J. Spat. Inf. Sci.* **2018**, *16*, 27–51. [CrossRef]
29. Senouci, R.; Taibi, N.-E.; Teodoro, A.; Duarte, L.; Mansour, H.; Meddah, R. GIS-Based Expert Knowledge for Landslide Susceptibility Mapping (LSM): Case of Mostaganem Coast District, West of Algeria. *Sustainability* **2021**, *13*, 630. [CrossRef]
30. Bălteanu, D.; Micu, M.; Jurchescu, M.; Malet, J.-P.; Sima, M.; Kucsicsa, G.; Dumitrică, C.; Petrea, D.; Mărgărint, M.C.; Bilaşco, I.; et al. National-scale landslide susceptibility map of Romania in a European methodological framework. *Geomorphology* **2020**, *371*, 107432. [CrossRef]
31. Ghiasi, V.; Ghasemi, S.A.R.; Yousefi, M. Landslide susceptibility mapping through continuous fuzzification and geometric average multi-criteria decision-making approaches. *Nat. Hazards* **2021**, *107*, 795–808. [CrossRef]
32. Popov, M.G.; Popova, O.V. The possibility of complex studying for the geological environment during seismic-ecological monitoring in areas of increased environmental danger. *Geol. Geophys. Russ. South* **2021**, *11*, 152–164. (In Russian)
33. Popov, M.G.; Popova, O.V. Study of Changes in Time and in Space of Stressed State of Geological Environment. *Geol. Geophys. Russ. South* **2020**, *10*, 79–93. (In Russian) [CrossRef]
34. Chotchaev, K.O.; Burdzieva, O.G.; Zaalishvili, V.B. Zoning of high mountainous areas by geoecological loads caused by geodynamic and climatic influences. *Geol. Geophys. Russ. South* **2021**, *11*, 81–94. (In Russian)
35. Zaalishvili, V.B.; Kanukov, A.S.; Fidarova, M.I. GIS technologies in geophysical information databases processing. *IOP Conference Series Mater. Sci. Eng.* **2020**, *913*, 052050. [CrossRef]
36. Zaalishvili, V.B.; Pinar, A.; Erdik, M.; Burdzieva, O.G.; Melkov, D.A. Issues of seismic risk assessment of Vladikavkaz city. *Geol. Geophys. Russ. South* **2020**, *10*, 94–113. [CrossRef]
37. Idarmachev, S.G.; Cherkashin, V.I.; Idarmachev, I.S. Seismometric Monitoring of an Unstable Rock Mass in the Gunib district of Dagestan (North Caucasus). *Geol. Geophys. Russ. South* **2021**, *11*, 74–86. (In Russian)
38. Kerimov, I.A. Gravity Anomalies, Fault Tectonics and Seismicity of the Terek-Caspian Trough. *Geol. Geophys. Russ. South* **2021**, *11*, 30–42. (In Russian) [CrossRef]
39. Papadopoulos, T.D.; Alexopoulos, J.D.; Dilalos, S. Combined geoelectrical and geoelectromagnetic survey for contributing to local hydrogeological regime—The case study of Delfini basin (Chios isl.—Greece). *Geol. Geophys. Russ. South* **2020**, *10*, 68–80. [CrossRef]
40. Yanvarev, G.S. Latest structure and geodynamics of Western Caucasus (based on decoding of satellite images). *Geol. Geophys. Russ. South* **2020**, *10*, 30–40. (In Russian) [CrossRef]
41. Chotchaev, K.O.; Oganesyan, S.M.; Goncharenko, O.A. Hazardous geological processes in the territories of economic development in the mountainous part of the Republic of North Ossetia-Alania. *Sustain. Dev. Mt. Territ.* **2017**, *9*. [CrossRef]
42. Burdzieva, O.; Chotchaev, K.; Mammadli, T. Model of the North Caucasus Mountain Territories Supported Development. *Sustain. Dev. Mt. Territ.* **2016**, *8*, 348–358. [CrossRef]
43. Chotchaev, K.; Zaalishvili, V.; Dzeranov, B. Natural endogenous factors of geoecological transformation of the mountain part of North Ossetia. *E3S Web Conf.* **2020**, *164*, 07025. [CrossRef]
44. Espiner, S.; Becken, S. Tourist towns on the edge: Conceptualising vulnerability and resilience in a protected area tourism system. *J. Sustain. Tour.* **2013**, *22*, 646–665. [CrossRef]
45. Kääb, A.; Reynolds, J.M.; Haeberli, W. Glacier and permafrost hazards in high mountains. In *Global Change in Mountain Regions: An Overview of Current Knowledge*; Huber, U.M., Bugmann, H.K.M., Reasoner, M.A., Eds.; Springer: Dordrecht, The Netherlands, 2005; pp. 225–234.
46. Minaev, V.A.; Faddeev, A.O. Assessment of geoecological risks. In *Modeling the Security of Tourist and Recreational Areas*; M.: "Infra-M": Moscow, Russia, 2009; pp. 2–16. (In Russian)
47. Makhov, S.Y. Security in the field of recreation and tourism. In *Organization of the Safety of Active Tourism: Teaching Aid for Higher Professional Education*; FGBOU VPO "State University-UNPK": Orel, Russia, 2012; pp. 4–5. (In Russian)
48. Batugin, A.S.; Bolotny, R.A. Geodynamic risk management as a way to improve environmental safety in the metropolis. *Min. Inf. Anal. Bull.* **2009**, *4*, 132–134. (In Russian)
49. Gardner, J.S.; Saczuk, E. Systems for hazards identification in high mountain areas: An example from the Kullu District, Western Himalaya. *J. Mt. Sci.* **2004**, *1*, 115–127. [CrossRef]

50. *Cadastre of Avalanche-Mudflow Danger of the Northern Caucasus*; Razumov, V.V. (Ed.) Gidrometeoizdat: St. Petersburg, Russia, 2001; p. 112. (In Russian)
51. Reid, M.E.; Christian, S.B.; Brien, D.L. Gravitational stability of three-dimensional stratovolcano edifices. *J. Geophys. Res.* **2000**, *105*, 6043–6056. [CrossRef]
52. Reid, M.E.; Christian, S.B.; Brien, D.L.; Henderson, S. *Scoops3D-Software to Analyze Three-Dimensional Slope Stability throughout a Digital Landscape*; U.S. VolcanoScienceCenter, USGS: Menlo Park, CA, USA, 2015.
53. Zaalishvili, V.B.; Dzeranov, B.V.; Gabaraev, A.F. Seismic hazard assessment of the territory and construction of probability maps. *Geol. Geophys. Russ. South* **2011**, *1*, 48–58. (In Russian)
54. Zaalishvili, V. Assessment of Seismic Hazard of Territory on Basis of Modern Methods of Detailed Zoning and Seismic Microzonation. *Open Constr. Build. Technol. J.* **2011**, *5*, 30–40. [CrossRef]
55. Zaalishvili, V.B.; Gogmachadze, S.A.; Otinashvili, M.G.; Zaalishvili, Z.V. Territory assessment rating method for insurance purposes. In Proceedings of the IV Russian National Conference on Seismic Construction and Seismic Zoning with International Participation, Sochi, Russia, 9–13 October 2001; p. 166; Abstracts of the report.
56. Zaalishvili, V.B. Rating estimation of the urbanized territory for purposes of the insurance. Natural and technogenic risks. *Build. Saf.* **2014**, *4*, 29–36.
57. Nesmeyanov, S.A.; Bogolyubova, N.P.; Voeikova, O.A.; Mindel, I.G. Active ruptures and rupture creep on the mountainous part of the route of the Blue Stream main pipeline. *Geoecology* **2003**, *5*, 387–404. (In Russian)
58. Bondarenko, N.; Lyubimova, T.; Stogny, V. Integrated assessment of potential geological risks in the Krasnodar region. *InterCarto InterGIS* **2020**, *26*, 189–200. [CrossRef]
59. Stogny, V.V.; Stogny, G.A.; Volkova, T.A.; Lyubimova, T.V. Geoecological risks of the North-Western Caucasus and their assessment. In *Collective Monograph: Hazardous Natural and Technogenic Processes in Mountainous Regions: Models, Systems, Technologies*; Nikolaev, A.V., Zaalishvili, V.B., Eds.; GPI VSC RAS: Vladikavkaz, Russia, 2019; pp. 768–772.
60. Osipov, V.I. Geoecology is an interdisciplinary science about the environmental problems of the geospheres. *Geoecology* **1993**, *1*, 14–18. (In Russian)
61. Trofimov, V.T.; Ziling, D.G. *Ecological Geology*; CJSC "Geoinformmark": Moscow, Russia, 2002; p. 376.
62. Abalakov, A.D. *Ecological Geology*; Publishing House of the Irkutsk State University: Irkutsk, Russua, 2007; pp. 28–32.
63. Beniston, M. The risks associated with climatic change in mountain regions. In *Global Change in Mountain Regions: An Overview of Current Knowledge*; Huber, U.M., Bugmann, H.K.M., Reasoner, M.A., Eds.; Springer: Dordrecht, The Netherlands, 2005; pp. 511–519.
64. Trewick, S.A. Plate Tectonics in Biogeography. In *The International Encyclopedia of Geography*; Richardson, D., Castree, N., Goodchild, F.M., Kobayashi, A., Liu, W., Richard, A.M., Eds.; John Wiley & Sons, Ltd.: Hoboken, NJ, USA, 2017. [CrossRef]
65. Yarotsky, G.P.; Chotchaev, K.O. Tectonic deterministic nature of Khailinsky earthquake on the southwest of the Koryak upland. *Geol. Geophys. Russ. South* **2016**, *4*, 129–140.
66. Yarotsky, G.P.; Chotchaev, K.O. Volcanogens of the marginal orogenic belts of the Northwestern sector of Northeast Asia. *Geol. Geophys. Russ. South* **2019**, *9*, 18–35. [CrossRef]
67. Keller, E.A.; Pinter, N. *Active Tectonics: Earthquakes, Uplift, and Landscape*; Prentice Hall: New York, NY, USA, 2002; p. 432.
68. Makeev, V.M.; Makarova, N.V.; Sukhanova, T.V. Mismatch of deep crustal deformations of the East European Platform. Deep structure of geodynamics, thermal field of the Earth, interpretation of geophysical fields. In *Tenth Scientific Readings in Memory of Yu.P. Bulashevich*; Perm State National Research University: Yekaterinburg, Russia, 2019; pp. 171–176.
69. Nikolaev, N.I. *Recent tectonics and geodynamics of the lithosphere*; Nedra: Moscow, Russia, 1988; p. 491. (In Russian)
70. Sashurin, A.D. Modern geodynamics and safety of objects in underground space. In *Geotechnological Problems of Complex Development of Mineral Resources: Collection of Scientific Works*; Mining Institute UB RAS (Ural Branch of the Russian Academy of Sciences): Yekaterinburg, Russia, 2010; pp. 332–336.

applied sciences

MDPI

Article

Dynamic Activity Index for Feature Engineering of Geodynamic Data for Safe Underground Isolation of High-Level Radioactive Waste

Sergei M. Agayan [1], Ilya V. Losev [1,2,*], Ivan O. Belov [1], Victor N. Tatarinov [1,3], Alexander I. Manevich [1,2] and Maksim A. Pasishnichenko [1]

1 Geophysical Center RAS, Molodezhnaya Str., 3, 119296 Moscow, Russia; s.agayan@gcras.ru (S.M.A.); i.belov@gcras.ru (I.O.B.); v.tatarinov@gcras.ru (V.N.T.); a.manevich@gcras.ru (A.I.M.); m.pasishnichenko@gcras.ru (M.A.P.)

2 Mining Institute of NUST «MISiS», Leninsky Prospekt 4, Building 1, 119049 Moscow, Russia

3 Russia Institute of Physics of the Earth, O.Yu. Schmidt RAS, Moscow, B. Gruzinskaya Str., 10, Building 1, 123242 Moscow, Russia

* Correspondence: i.losev@gcras.ru

Abstract: In this study, we developed a new approach for feature engineering in geosciences. The main focus of this study was feature engineering based on the implementation of the dynamic activity index (MDAI) as a function of the anomaly of the spatial distribution of data, using systems and discrete mathematical analysis. The methodology for calculating MDAI by groups, geomorphological variability, the density of tectonic faults, stress-strain state, and magnetic field anomalies, is presented herein for a specific area. A detailed analysis of the correlation matrix of MDAI revealed weak correlations between the development features. This showed that the considered properties of the geological environment are independent sets and can be used in the analysis of its geodynamic stability. As a result, it was found that most of the territory where high-level radioactive waste (HLRW) disposal is currently planned is in a relatively stable zone.

Keywords: system analysis; DMA algorithms; dynamic activity index; structural tectonic block; geodynamic data; safety

Citation: Agayan, S.M.; Losev, I.V.; Belov, I.O.; Tatarinov, V.N.; Manevich, A.I.; Pasishnichenko, M.A. Dynamic Activity Index for Feature Engineering of Geodynamic Data for Safe Underground Isolation of High-Level Radioactive Waste. *Appl. Sci.* **2022**, *12*, 2010. https://doi.org/10.3390/app12042010

Academic Editors: Alexei Gvishiani and Boris Dzeboev

Received: 30 January 2022
Accepted: 12 February 2022
Published: 15 February 2022

1. Introduction

Methods of discrete mathematical analysis, machine learning, and big data analysis use "feature" in their terminology. The feature of a study object is the result of measuring or modeling some individual property of the object, and their aggregate reflects the model of the object. The solution to urgent problems in assessing natural and man-made risks, such as searching for anomalies in geophysical fields [1–3], recognizing strong earthquake-prone areas [4–8], geodynamic zoning [9,10], etc., requires the creation of effective methods for the formalized analysis of a complex of geological and geophysical features. When analyzing the data, the features are synthesized using mathematical modeling methods [11–13] and may contain complex mathematical constructions. It is difficult to interpret them physically. In this case, the informativeness of geological and geophysical features is assessed [14–16]. Therefore, the feature model of the object under study should be adequately analyzed, and the results of this analysis should correctly reflect the real natural-technical system.

A unique project is being implemented in Russia to create an underground research laboratory (URL). The main goal of this project is to confirm the possibility of final isolation of high-level radioactive waste (HLRW) in geological formations. Considering international experience, the concept of deep geological repository (DGR) was chosen, the main parameters of which are given in [17]. DGR is planned to be created in granite gneiss rocks of the Nizhnekansky massif 4–5 km from the Yenisei River at a depth of 500–600 m.

HLRW will be placed in large-diameter wells drilled from horizontal workings on horizons with proposed dimensions of 1.0 × 1.5 km. The multibarrier protection system consists of classical elements used in various DGR projects (Sweden, Finland, Japan, USA, etc.): geological environment, bentonite, borosilicate glass, and container body. Currently, the geodynamic and seismic regime is being studied in this area [18–20].

The geological and tectonic features of the DGR location area are described in [20–24]. The time of the radiobiological hazard of HLRW exceeds ten thousand years, so the main focus is to assess and predict the geodynamic stability of a structural tectonic block containing HLRW. A structural tectonic block is a dynamically active system in which internal and external energy sources determine the spatiotemporal evolution of its structural forms and stress-strain state. The activity degree of the system is reflected in the features (morphology) of the distribution of geological and geophysical characteristics, including the relief of the Earth's surface, the scheme of tectonic faults and stress-strain state, etc. When such information does not provide an obvious sign of instability [21], system analysis is the most constructive assessment. This allows long-term forecasts in conditions of information uncertainty, uses fundamental geological patterns as the only criteria, and guarantees reliability [25].

For this purpose, and based on the methods and algorithms of discrete mathematical analysis (DMA) and fuzzy logic, a method of formalized analysis was developed [26,27]. It considers the relationship of geodynamics and the morphological features of the distribution of geological and geophysical parameters (including a digital relief model, stress-strain state, geophysical fields, and other characteristics of the environment) [1,26,28,29].

The calculation of dynamic indicators' activity measurements is presented in Section 2. The results of correlation analysis and the final measure of geodynamic safety are presented in Section 3. Assessment of the stability of the geological environment is detailed in Section 4. Finally, in Section 5, conclusions are summarized.

2. Methodology for Assessing the Integral Index of Dynamic Stability of a Structural Tectonic Block

Information about the methodology that was developed by the Geophysical Center RAS is detailed below. Its algorithms are based on the language of fuzzy sets and fuzzy logic [26,30]. DMA algorithms have successfully proven themselves in solving a wide range of geological and geophysical problems in the field of Earth sciences (strong recognition of earthquake-prone areas, monitoring of volcanoes and geomagnetic activity, etc.) [26,31–34].

In [26], a finite system of functions of geological and geophysical parameters was studied on a two-dimensional grid. The challenge of calculating the anomaly measurements for a group of geological and geophysical features was formalized. The mathematical component of the sustainability assessment methodology for the structural-tectonic block is described in detail in [1,26] and tested at the HLRW disposal site. Its main provisions are as follows.

The area of interest $\Pi = \{o \leq x_1 \leq t; u \leq x_2 \leq v\}$ exists on the coordinate plane $\mathbb{R}^2(x_1, x_2)$. A set of geological and geophysical fields F from m datasets (in the form of digital maps of various parameters: geographical, geological, geophysical, geodynamic, economic, and others) is selected for this area for use in evaluating the stability of the region Π.

$$\begin{gathered} F = \{f_1, \ldots, f_m\}, \\ f_i : \Pi \to \mathbb{R};\ i = 1, \ldots, m \end{gathered} \tag{1}$$

The goal is to divide the region Π into relatively unstable (conditionally dangerous) and stable (conditionally safe) elements. This ranking is also called geodynamic zoning [10].

In the region Π a regular grid $W = W(h_1, h_2)$ with nodes w is given:

$$W = \left\{ w = (o + ih_1; u + jh_2) \middle| \begin{array}{l} i = 0, \ldots, N;\ h_1 = \frac{t-o}{N} \\ j = 0, \ldots, M;\ h_2 = \frac{v-u}{M} \end{array} \right\} \tag{2}$$

On the grid W it is necessary to analyze the spatial distribution of systems of functions F in the neighborhoods of the node w. To do this, we defined a fuzzy measure of activity $\mu(w)$ in the range from 0 to 1, according to the rules described below.

Step 1. Calculation of the dynamic indicator.

Each parameter f from the set F is a distribution function on the grid W. For each parameter f, it is possible to determine the dynamic indicator D_f, which is a functional characteristic of the measurement of f. The value of $D_f(w)$ is interpreted as a quantitative assessment of the behavior of the function f in the node $w \in W$, calculated according to the specified rules. In terms of data analysis, the dynamic indicator $D_f(w)$ is a feature.

Step 2. Calculation of measure of dynamic activity index (MDAI).

For each dynamic indicator D_f, a measure of activity (anomaly) μD_f is determined in the range from 0 to 1. It shows the degree of expression of the property f in the node w, defined by the indicator D_f. The measure of dynamic activity index μD_f is calculated from the dynamic indicator D_f in the methodology of discrete mathematical analysis. The transformation $D_f \to \mu D_f$ translates the analysis of the measurement of f into the language of fuzzy logic: measures of dynamic activity index μD_f for different dynamic indicators of D_f are fuzzy structures on the grid W, and they can be combined in any compositions and any quantities using fuzzy logic operations and averaging.

Step 3. Calculation the integral measure of activity μ_F.

As the last step of the algorithm, all the measures of dynamic activity index μD_f are combined into a single integral indicator of μ_F. The formula of the combination is the arithmetic mean of all measures of dynamic activity index μD_f. Depending on the research task, weighting coefficients for activity measures or other compound formulas can be used [35]. To show the measures of geodynamic safety, the inverse of the integral measure of activity is used:

$$S_F = 1 - \mu_F \quad (3)$$

The transformation $F \to \mu_F$ translates vector analysis concerning the system of functions F into scalar analysis for the final measure of anomaly μ_F, which, in terms of decision theory, is the reduction of a multicriteria problem to a scalar choice of a utility function. In terms of geodynamic zoning [36], the criteria for estimating the value of S_F according to the system of features F are determined according to Table 1. This approach to the ranking is adequate; if an integral measure is expressed in the range 0–1, then the conditions $S_F \le 0.25$ и $S_F \ge 0.75$ mean the absence or the presence of safety (anomalies), respectively. The interval (0.25; 0.75) indicates uncertainty and the need for additional research.

Table 1. Ranking of the integral measure of geodynamic safety S_F.

Node (Cell, Structural Block) w	Measures of Geodynamic Safety S_F
hazardous	≤ 0.25
neutral	$\in (0.25;\ 0.75)$
safe	≥ 0.75

3. Feature Engineering Based on Measurement of Dynamic Stability Index

3.1. Needed Data

According to the methodology described in [26], when selecting parameters and related dynamic indexes, we used a set of parameters necessary for assessing geodynamic stability. All the initial data were collected in a GIS project [37]. We used a set of the following data:

1. Digital terrain model based on radar interferometric survey of the Earth's surface, Shuttle radar topographical mission (SRTM-4);
2. Schemes of tectonic faults [38,39];
3. Scheme of neotectonic structures of the joint zone of the Siberian platform and the West Siberian plate [40];

4. Kinematic model of modern horizontal movements and rates of deformation of the Earth's crust according to GNSS monitoring data [19,41];
5. Data on the stress-strain state of the Nizhnekansky massif obtained as a result of finite element modeling [42,43];
6. Map of the anomalous magnetic field [38].

Dynamic indexes were obtained from the above datasets based on the study of the following characteristics: relief, faults, stress-strain state, and anomalous magnetic field.

3.2. Feature Calculation

For the realization of the methodology, software modules based on the calculation of morphometric indicators and normalization of geological and geophysical parameters were developed. These parameters reflect patterns of relief $L^1_{Re}(w)$, $L^2_{Re}(w)$, and $|\nabla_{Re}|(w)$; fault densities $\rho(w,\mathcal{P})$ and $\rho(w,\rho)$; stress-strain states $\sigma_{xx}(w)$, $\sigma_{yy}(w)$, $\sigma_{int}(w)$, $E_{xx}(w)$ $E_{yy}(w)$, and $dil(w)$; and magnetic field anomalies $L^2_{Mag}(w)$, $|\nabla_{Mag}|(w)$.

The first two indicators, L^1_{Re} and L^2_{Re}, characterize the geomorphological variability landform sections at node w, and the third, $|\nabla_{Re}|$, is the gradient of relief:

$$L^1_{Re}(w) = \frac{\sum_{\overline{w}\in C(w)}|Re(\overline{w}) - Re(w)|}{4} \tag{4}$$

$$L^2_{Re}(w) = \frac{2 + cos\theta_\Gamma + cos\theta_B}{2} \tag{5}$$

$$cos\theta_\Gamma = \frac{-1 + (Re(w_4) - Re(w)) \times (Re(w_6) - Re(w))}{\sqrt{1 + (Re(w_4) - Re(w))^2} \times \sqrt{1 + (Re(w_6) - Re(w))^2}} \tag{6}$$

$$cos\theta_B = \frac{-1 + (Re(w_2) - Re(w)) \times (Re(w_8) - Re(w))}{\sqrt{1 + (Re(w_2) - Re(w))^2} \times \sqrt{1 + (Re(w_8) - Re(w))^2}} \tag{7}$$

The third relief gradient index is the gradient modulus $|\nabla_{Re}(w)|$; it is responsible for the maximum change in relief at a node w and calculated by the Sobel operator:

$$|\nabla_{Re}(w)| = \left|\nabla^\Gamma_{Re}(w)\right| + \left|\nabla^B_{Re}(w)\right| \tag{8}$$

$$\nabla^\Gamma_{Re}(w) = (Re(w_7) + 2Re(w_8) + Re(w_9)) - (Re(w_1) + 2Re(w_2) + Re(w_3)) \tag{9}$$

$$\nabla^B_{Re}(w) = (Re(w_3) + 2Re(w_6) + Re(w_9)) - (Re(w_1) + 2Re(w_4) + Re(w_7)) \tag{10}$$

The MDAI for L^1_{Re}, L^2_{Re}, and $|\nabla_{Re}|$ are calculated as:

$$\begin{aligned} \mu L^1_{Re}(w) &= \frac{L^1_{Re}(w)}{L^1_{Re}(w) + \overline{L^1_{Re}}} \\ \mu L^2_{Re}(w) &= \frac{L^2_{Re}(w)}{L^2_{Re}(w) + \overline{L^2_{Re}}} \\ \mu \nabla_{Re}(w) &= \frac{\nabla_{Re}(w)}{\nabla_{Re}(w) + \overline{\nabla_{Re}}} \end{aligned} \tag{11}$$

In studying the stability of a structural-tectonic block, the key issue is the macroscopic manifestation of the geodynamic process. In the Earth's crust, this macroscopic manifestation is most often represented by the relative displacement of parts of a massif concerning each other. Analysis of the processes occurring at the boundaries between structural-tectonic blocks is a necessary step for the creation of modern analytical and numerical geodynamic models [44]. For this, we added the indicators $\rho(w)$, which characterizes fault density, and $\rho(\pi,\mathcal{P})$, which characterizes proximity to tectonic faults.

The MDAI of fault density was determined using the linear density, which was obtained in a circular vicinity within each cell of the grid. The length of the segment of each line crossed by the circular neighborhood was multiplied by the line weight factor. Then all the length values were summed up and divided by the area of the circle. This process was repeated for all cells in the grid. The calculation of the values $\rho(\text{п},\mathcal{P})$ from a cell п to the fault $\mathcal{P}$ was obtained using the Kolmogorov mean criterion with a negative index:

$$\rho(\text{п},\ \mathcal{P}) = \begin{cases} \text{0,</mo>} \in \mathcal{P} \\ M_q\left(\rho(\text{п},\ P_k)|_1^N\right) \end{cases}, \tag{12}$$

where

$$q < 0$$
$$M_q\left(\rho(\text{п},\ P_k)|_1^N\right) = \left(\frac{\sum_{k=1}^{N} \rho(\text{п},\ P_k)^q}{N}\right)^{\frac{1}{q}} \tag{13}$$

The MDAIs of $\rho(\text{п},\mathcal{P})$ are calculated using the formula:

$$\mu\rho(,\ \mathcal{P}) = \rho(,\ \mathcal{P})\rho(,\ \mathcal{P}) + \rho(,\ \mathcal{P}) \tag{14}$$

Figure 1 shows the original digital elevation model of the HLRW disposal area. Figures 2–4 show the calculated maps of elevation variability for $L^1_{Re}(w)$, $L^2_{Re}(w)$, and $|\nabla_{Re}|(w)$, respectively. An example of the calculation of the proximity to tectonic faults $\rho(w,\mathcal{P})$ as the shortest distance to a tectonic fault is given in Figure 5; an example of the measure of fault density $\rho(w)$ is given in Figure 6.

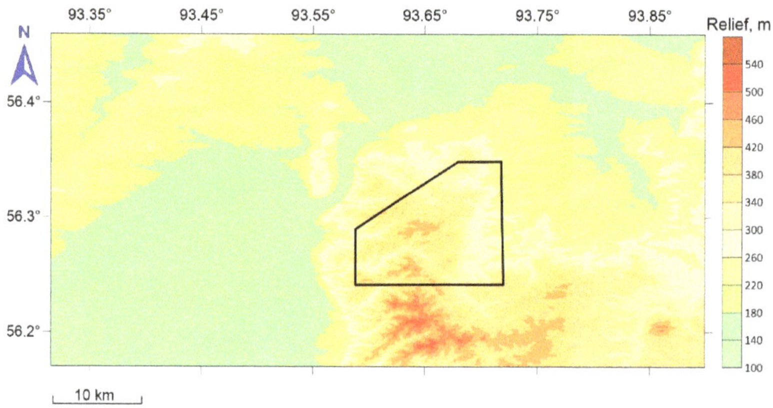

Figure 1. Original digital elevation model.

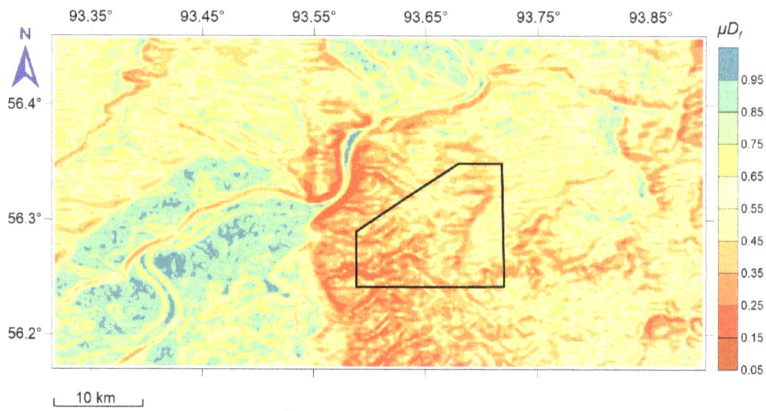

Figure 2. Measure of dynamic activity index of relief, $\mu L^1_{Re}(w)$.

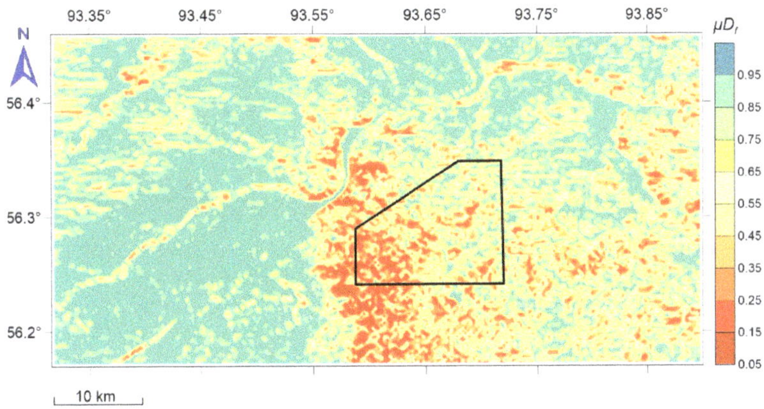

Figure 3. Measure of dynamic activity index of relief, $\mu L^2_{Re}(w)$.

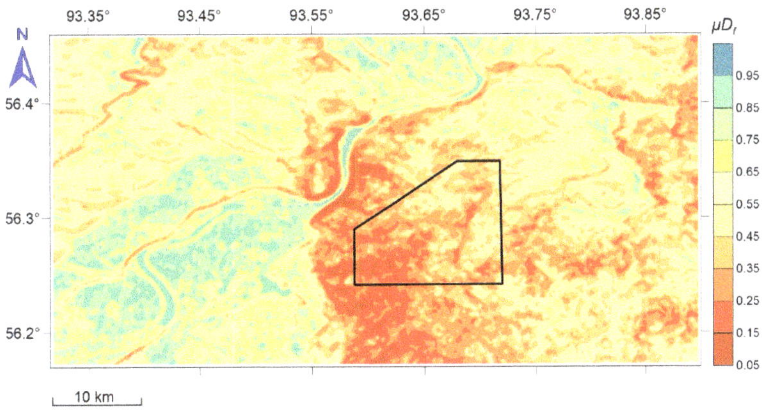

Figure 4. Measure of dynamic activity index of relief, $\mu \nabla_{Re}(w)$.

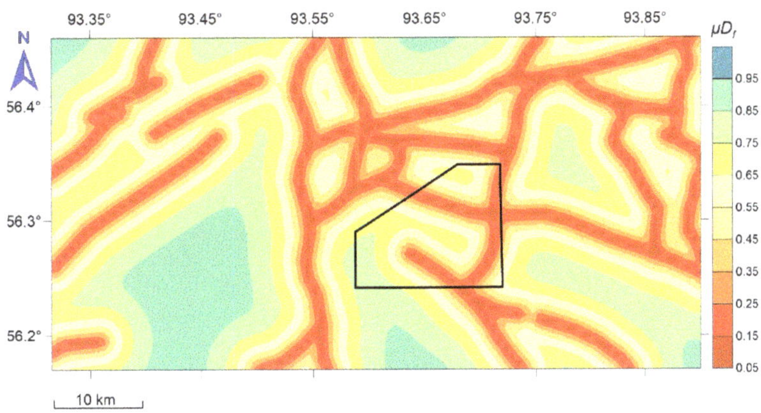

Figure 5. Measure of dynamic activity index of proximity to tectonic faults, $\mu^{\rho}(w,\mathcal{P})$.

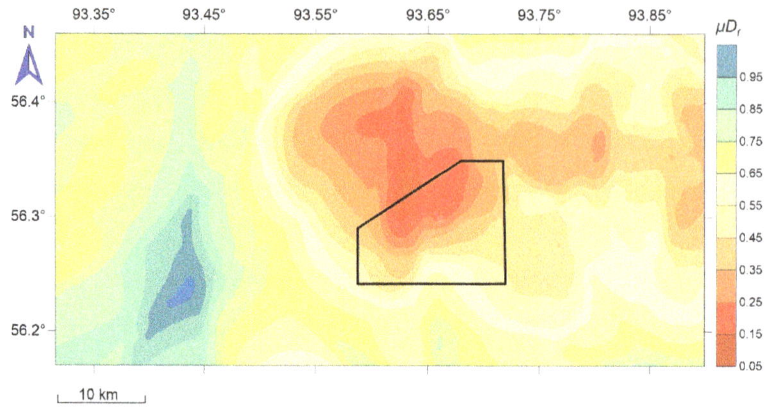

Figure 6. Measure of dynamic activity index of fault density, $\mu^{\rho}(w)$.

The next MDAI referred to the stress-strain state. The indexes for stress tensor components σ_{xx}, σ_{yy} and stress intensity σ_{int}, which were obtained from the results of the numerical modeling in [30], were calculated. The indexes for deformation parameters of the Earth's crust, derived from GNSS measurements, were also calculated [19,41]. The components of the strain tensor e_{xx}, e_{yy} and the dilatation strain E_{Dil} were used as input data. The transformation of the stress-strain state parameters was carried out based on the conversion of these parameters to a gradient scale in the range 0 to 1. For this purpose, indicators a, b, c, and d were determined. They correspond to the following measurements of the dynamic activity in Table 2: a, maximum value of stresses/strains; b, boundary value of stresses or strains on the positive part of the scale identified by expert judgment; c, boundary value of stresses or strains on the negative part of the scale identified by expert judgment; d, minimum value of stresses/strains. For each x_i interval, a transformation was performed according to Equation (14). Thus, each interval of stress and strain values x_i corresponds to a certain calculation formula and interval of the dynamic activity measure.

$$x_i \in \begin{matrix} [a;b] \\ \left[b;\frac{b+c}{2}\right] \\ \left[\frac{b+c}{2};c\right] \\ [c;d] \end{matrix} \rightarrow \left\{ \begin{matrix} 0.5 \times \frac{x-a}{b-a} \\ 0.5 + \frac{x-b}{c-b} \\ 1.5 - \frac{x-b}{c-b} \\ 0.5 \times \frac{d-x}{d-c} \end{matrix} \right. \quad (15)$$

Table 2. Dynamic activity measure intervals based on stress and strain valuation.

Intervals	Value
$[a;b]$	0–0.5
$\left[b;\frac{b+c}{2}\right]$	0.5–1
$\left[\frac{b+c}{2};c\right]$	1–0.5
$[c;d]$	0.5–0

Below are shown MDAIs of stress-strain state. Stress values σ_{xx}, σ_{yy}, and σ_{int} are given in Figures 7–9 and strain values e_{xx}, e_{yy}, and E_{Dil} in Figures 10–12.

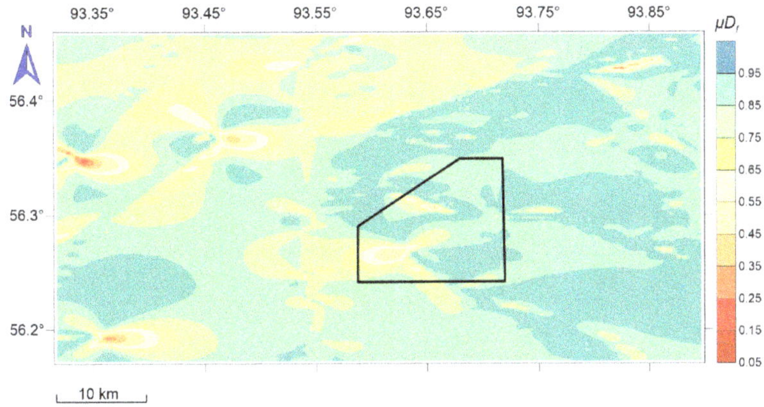

Figure 7. Measure of dynamic activity index of stress tensor component, $\mu\sigma_{xx}(w)$.

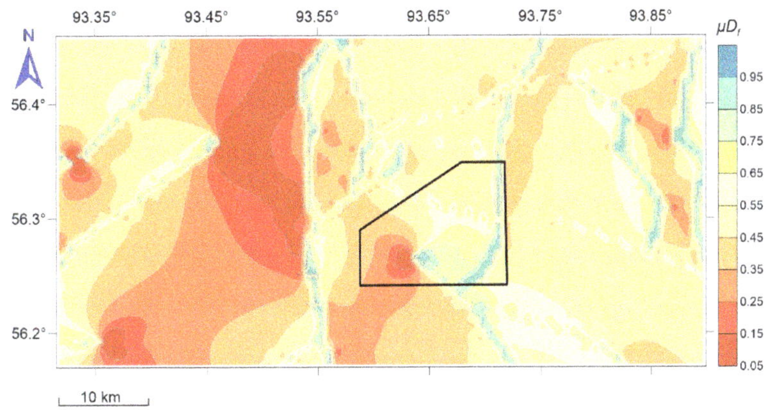

Figure 8. Measure of dynamic activity index of stress tensor component, $\mu\sigma_{yy}(w)$.

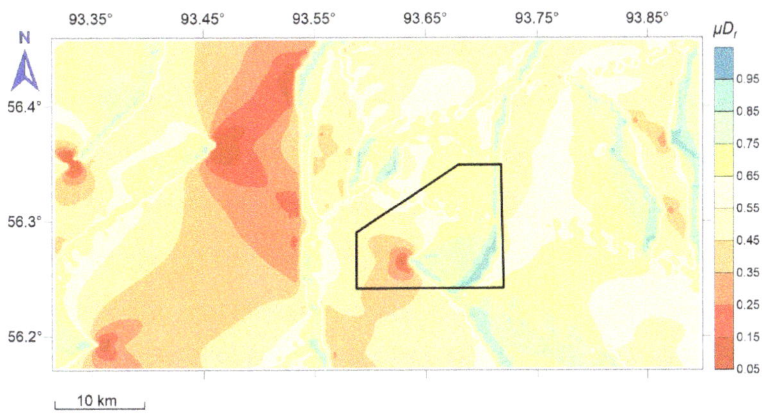

Figure 9. Measure of dynamic activity index of stress intensity, $\mu\sigma_{int}(w)$.

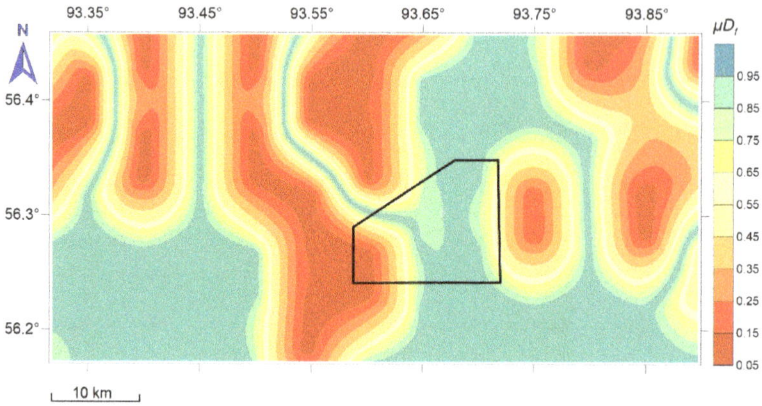

Figure 10. Measure of dynamic activity index of strain tensor component, $\mu e_{xx}(w)$.

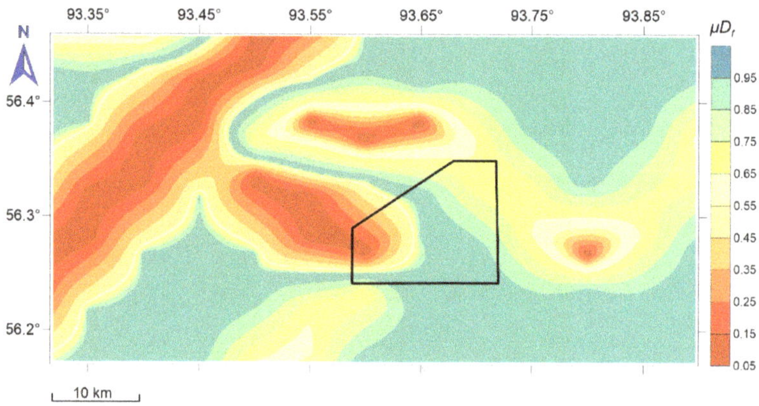

Figure 11. Measure of dynamic activity index of strain tensor component, $\mu e_{yy}(w)$.

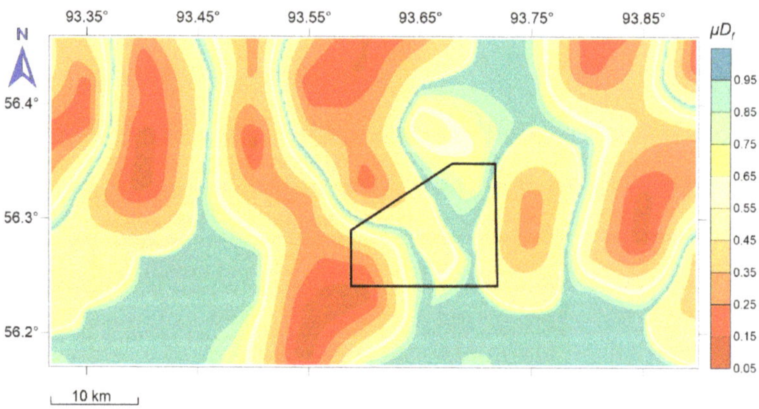

Figure 12. Measure of dynamic activity index of dilatation strain, $\mu E_{Dil}(w)$.

To determine the MDAI of the magnetic field, we used Equation (5) for magnetic field variability index $L^2_{Mag}(w)$ and Equation (8) for magnetic field variation gradient $\left|\nabla_{Mag}\right|(w)$.

Figure 13 shows distributions of integral measures of geodynamic safety and graphs of the distribution of their values by intervals within selected groups: geomorphological variability $S_{Re}(w)$, tectonic faults density $S_{\varrho}(w)$, stress-strain state $S_{sss}(w)$, and magnetic field anomalies $S_{Mag}(w)$:

$$S_{Re}(w) = 1 - \frac{\mu L^1_{Re}(w) + \mu L^2_{Re}(w) + \mu \nabla_{Re}(w)}{3} \tag{16}$$

$$S_{\varrho}(w) = 1 - \frac{\mu \varrho_{\rho}(w) + \mu \varrho(\Pi, \mathcal{P})}{2} \tag{17}$$

$$S_{sss}(w) = \frac{1}{2} \times \left(\frac{\mu\sigma_{xx}(w) + \mu\sigma_{yy}(w) + \mu\sigma_{int}(w)}{3} + \frac{\mu e_{xx}(w) + \mu e_{yy}(w) + \mu E_{Dil}(w)}{3} \right) \tag{18}$$

$$S_{Mag}(w) = 1 - \frac{\mu L^2_{Mag}(w) + \mu \nabla_{Mag}(w)}{2} \tag{19}$$

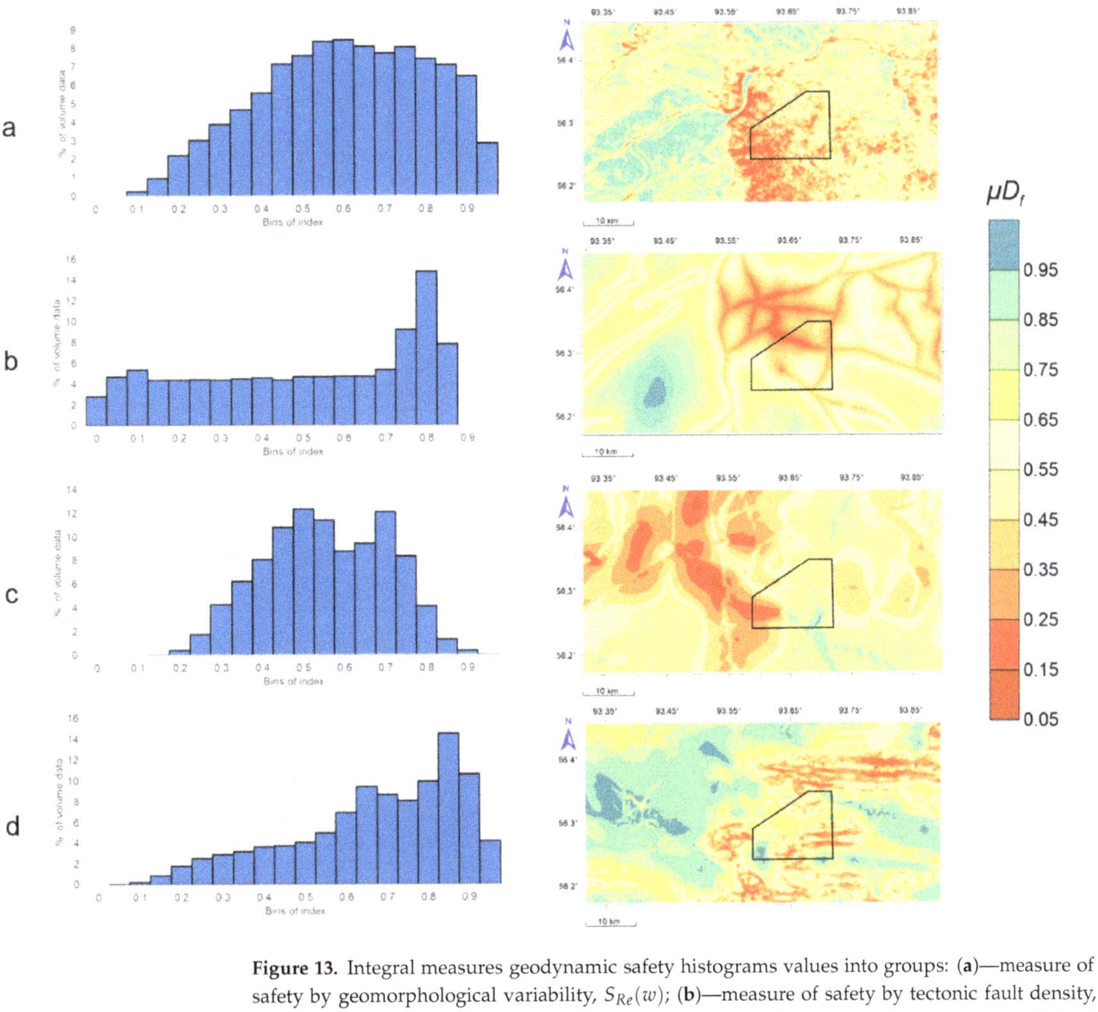

Figure 13. Integral measures geodynamic safety histograms values into groups: (**a**)—measure of safety by geomorphological variability, $S_{Re}(w)$; (**b**)—measure of safety by tectonic fault density, $S_{\varrho}(w)$; (**c**)—measure of safety by stress-strain state, $S_{sss}(w)$; (**d**)—measure of safety by magnetic field anomalies, $S_{Mag}(w)$.

The distributions of the integral measures within the groups differ significantly. The integral measure of geomorphological variability $S_{Re}(w)$ is characterized by a uniform distribution of data in the measurement interval from 0.45 to 0.9 (Figure 13a). The geomorphological variability in the study area is not high. A large part of the area is flat with low elevation differences (Figure 1), but small areas of high-degree geomorphological variability can also be observed with a relative geodynamic instability measurement in the range of 0.1–0.25. The integral measure of tectonic fault density $S_{\wp}(w)$ has a uniform distribution of values between 0 and 0.75. About 31% of all data fall within the relative geodynamic stability interval of 0.75–0.9 (Figure 13b). The integral measure of the stress-strain state $S_{sss}(w)$ has a distribution close to normal, and about 85% of all data fall into the 0.25–0.75 transition range (Figure 13c). Up to 14% of all nodes are in the area of relative geodynamic stability. The integral measure of magnetic field anomaly variability $S_{Mag}(w)$ is not uniformly distributed over the intervals. The 0.75–1 interval of relative geodynamic stability accounts for up to 47% of all data, the 0.25–0.75 interval of transient values for about 50% of all data, and the 0–0.25 interval of relative geodynamic instability for about 3%.

4. Discussion

Correlation matrices and correlation strength thresholds were calculated for dynamic indexes D_f and their MDAI μD_f. Pearson's correlation coefficient is presented in the matrix:

$$r = \frac{\Sigma((x_i - \overline{x}) \times (y_i - \overline{y}))}{\sqrt{\Sigma\left((x_i - \overline{x})^2 \times \Sigma(y_i - \overline{y})^2\right)}} \tag{20}$$

The lower threshold for the presence of correlation was determined using Student's t-test Equation (21) and the intervals for the strength of the correlation according to Equation (22):

$$r_0 = \frac{t}{\sqrt{t^2 + n - 2}} \tag{21}$$

$$r_{int} = \frac{1 - r_0}{3} \tag{22}$$

Correlation strength intervals were determined for the dataset used (27,495 rows in each of the indicators and a significance level of 0.95): weak correlation in the 0.012–0.341 interval, medium correlation in the 0.341–0.671 interval, and strong correlation in the 0.671–1 interval. The strength of the correlation was represented as a discrete color scale.

Figure 14 shows the correlation matrix of MDAI μD_f. As shown in the figure, 65 out of 78 correlation values of MDAI μD_f have a weak correlation relationship. This is a good indicator in terms of data analysis as the features must be noncollinear, otherwise, the aggregation ability of the integral measure of geodynamic safety is reduced due to the high dispersion of the data $S_F(w)$. Medium and strong correlations are found within index group μD_f. In the relief group, indexes of geomorphological variability $\mu L^1_{Re}(w)$ and $\mu L^2_{Re}(w)$ have a strong correlation with the index of relief gradient $\mu \nabla_{Re}(w)$. MDAIs $\mu L^1_{Re}(w)$ and $\mu L^2_{Re}(w)$ have a medium correlation. The measures of proximity to tectonic faults index $\mu \wp(w,\mathcal{P})$ and fault density $\mu \wp(w)$ have a strong correlation, as the indices show high spatial correspondence [42,45] (Figures 5 and 6). The fault group measurements have an average correlation strength with the stress measurements $\mu\sigma_{yy}(w)$ and $\mu\sigma_{int}(w)$. This is primarily due to the mathematical model for calculating the stress state [43], which considers the fault tectonics of the area. In the stress group, $\mu\sigma_{yy}(w)$ and $\mu\sigma_{int}(w)$ have a strong correlation relationship, as the stress σ_{yy} has the greatest contribution to the stress intensity σ_{int} [44]. The MDAIs of the strain tensor components $\mu e_{xx}(w)$ and $\mu e_{yy}(w)$, determined from GNSS observations [19], have a weak correlation. The MDAI of dilatation strain $\mu E_{Dil}(w)$ has a strong correlation with the measure $\mu e_{xx}(w)$ and a medium correlation with the measure $\mu e_{yy}(w)$; therefore, their spatial relationship can be seen (Figures 7–9). This is explained by the formula for calculating strain dilatation, which includes the components of the strain tensor e_{xx} and e_{yy} [41]. The average correlation strength was found between the measure of magnetic field anomaly variability $\mu L^2_{Mag}(w)$ and the gradient of magnetic field anomaly variability $\mu \nabla_{Mag}(w)$. Internal correlation of feature groups was due to either interdependent formulas for deriving the initial properties of the dynamic indicators D_f, or the same set of initial feature data F (as in the case of topography or magnetic field anomalies). Table 3 shows the correlation of the integral geodynamic safety groups: safety measure of geomorphological variability, $S_{Re}(w)$; safety measure of tectonic faults density, $S_\rho(w)$; safety measure of a stress-strain state, $S_{sss}(w)$; and safety measure of magnetic field anomalies, $S_{Mag}(w)$. Their correlation everywhere indicates a weak correlation or no correlation at all. This characterizes these indexes as reflecting different properties of the geological environment and as independent datasets.

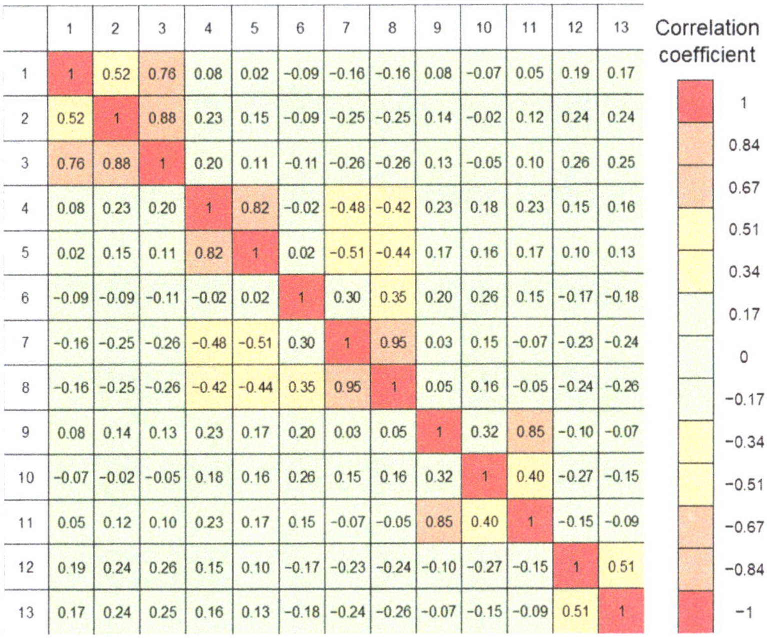

Figure 14. Correlation matrix the measures of dynamic activity indexes μD_f: 1—relief index, $\mu L^1_{Re}(w)$; 2—relief index, $\mu L^2_{Re}(w)$; 3—relief index, $\mu \nabla_{Re}(w)$; 4—proximity to tectonic faults index, $\mu^{\rho}(w, \mathcal{P})$; 5—fault density index, $\mu^{\rho}(w)$; 6—index of stress tensor component $\mu\sigma_{xx}(w)$; 7—index of stress tensor component, $\mu\sigma_{yy}(w)$; 8—index of stress intensity, $\mu\sigma_{int}(w)$; 9—index of strain tensor component, $\mu e_{xx}(w)$; 10—index of strain tensor component, $\mu e_{yy}(w)$; 11—index of dilatation strain $\mu E_{Dil}(w)$; 12—magnetic field variability index, $\mu L^2_{Mag}(w)$; 13—magnetic field variation gradient, $\mu \nabla_{Mag}(w)$.

Table 3. Correlation matrix of integral measures of geodynamic safety.

	$S_{Re}(w)$	$S_\rho(w)$	$S_{sss}(w)$	$S_{Mag}(w)$
$S_{Re}(w)$	1	0.13	−0.03	0.28
$S_\rho(w)$	0.13	1	0.04	0.16
$S_{sss}(w)$	−0.03	0.04	1	−0.28
$S_{Mag}(w)$	0.28	0.16	−0.28	1

The integral measure of geodynamic safety $S_F(w)$ was calculated based on a combination of four integral measures for relief, faults, stress-strain state, and magnetic field anomalies:

$$S_F(w) = \frac{S_{Re}(w) + S_{\rho}(w) + S_{sss}(w) + S_{Mag}(w)}{4} \tag{23}$$

The spatial distribution of the integral measure of geodynamic safety $S_F(w)$ is shown in Figure 15, and the distribution of measurements by intervals is shown in Figure 16. Green corresponds to the most stable state, red to the least stable, and yellow to the intermediate zones.

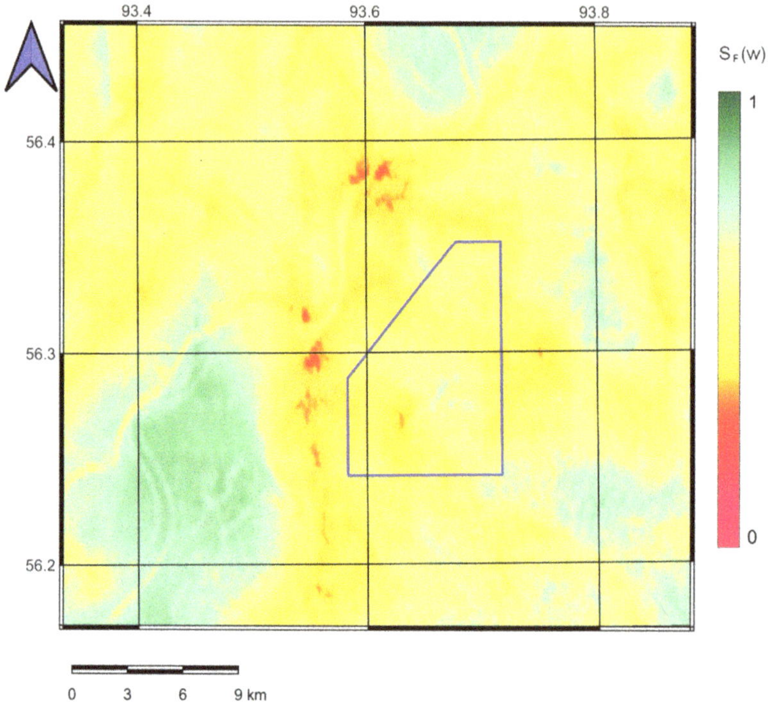

Figure 15. Map of the integral measure of geodynamic safety, $S_F(w)$; blue contour is boundary of HLRW site isolation.

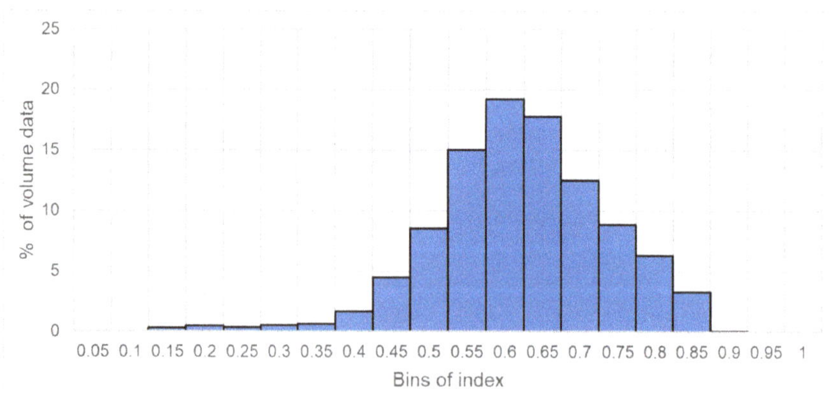

Figure 16. Histogram of the integral measure of geodynamic safety, $S_F(w)$.

According to the results of the integral measure ranking (Table 1), nodes of relative geodynamic instability $S_F(w) \leq 0.25$ account for an area of about 14 km^2 (305 nodes).

Most of these nodes belong to the area of interaction between major tectonic structures and sharp variations in height gradient, which are actively expressed in abnormal values of all measures of activity of dynamic indexes μD_f. The range of hazardous nodes is very small and occupies about 1.1% of the total volume. Nodes of geodynamic stability $S_F(w) \geq 0.75$ correspond to flat areas of relief located in the centers of main structural-tectonic blocks, equidistant from tectonic faults and outside zones of anomaly stress-strain state. Nodes of relative geodynamic stability occupy an area of about 9.6%. Transition zones $S_F(w) \in (0.25; 0.75)$ are not uniformly distributed. Most of the transition interval nodes are associated with areas of local tectonic interaction, areas of moderate terrain variability, and stress state within background values. The transition interval nodes occupy the largest study area, about 89.3%.

5. Conclusions

Assessment of the stability of the geological environment during disposal HLRW for the period of their radiobiological hazard (more than ten thousand years) is a very difficult task in the field of Earth sciences. Currently, this complex problem is being solved in several countries. For example, in Sweden («Forsmark») and Finland («Aspo»), similar facilities are already under construction. The proposed methodology for assessing sustainability complements the methods used in these and other countries to assess the suitability of the geological environment for the disposal of HLRW. The degree of stability of the environment as a geodynamic active system is related to the distribution features of the geospatial data complex, including the characteristics of the relief, tectonic faults, geophysical fields, stresses, etc. The search for anomalies and morphological patterns in the distribution of data allows us to identify possible sites of the destruction of structural blocks.

The presented methodology for assessing the stability of the geological environment based on the methods of DMA and fuzzy logic is informed by the analysis of dynamic indicators as functions of geodynamic activity of the environment by groups: geological and geophysical data, geomorphological variability of relief, the density of tectonic faults, stress-strain state, and magnetic and gravity field anomalies for the area [46–49]. A detailed analysis of the correlation matrix of the MDAI for four groups' characteristics of the environment μD_f demonstrated that the absolute majority have a weak correlation. This is a positive point, as the features should not be collinear. Otherwise, due to the high variance of the data, the generalizing ability of the integral measure of geodynamic safety $S_F(w)$ would decrease. Medium and strong correlations were found within the groups of indicators μD_f.

The integral measure of geodynamic safety $S_F(w)$, which combines different MDAIs, allows geodynamic zoning of the HLRW disposal area and specifying places for field instrumental observations.

The effectiveness of the presented methodology depends on the quality and volume of the source data. In our study, the methodology only demonstrates a possible approach to the systemic assessment of stability, and it could be expanded in the future both in terms of increasing features and by creating new algorithms. It could also be used for the identification of linear anomalies associated with hidden tectonic faults and dangerous zones of geodynamic instability.

It was found that most of the studied territory, where HLRW disposal is currently planned, is located in an intermediate zone in terms of the degree of geodynamic stability $S_F \geq 0.75$. The number of dangerous clusters $S_F \leq 0.25$ is very small and occupies about 1.1% of the entire territory.

Author Contributions: Conceptualization, S.M.A. and V.N.T.; methodology, S.M.A. and I.V.L.; software, I.O.B.; validation, I.V.L., I.O.B. and M.A.P.; formal analysis, A.I.M.; investigation, M.A.P.; resources, I.V.L. and A.I.M.; data curation, I.V.L.; writing—original draft preparation, S.M.A., I.V.L. and A.I.M.; writing—review and editing, V.N.T., M.A.P., I.O.B. and I.V.L.; visualization, I.V.L. and A.I.M.; supervision, V.N.T.; project administration, S.M.A.; funding acquisition, V.N.T. All authors have read and agreed to the published version of the manuscript.

Funding: This research was funded by the Russian Science Foundation (project No. 18-17-00241).

Institutional Review Board Statement: Not applicable.

Informed Consent Statement: Informed consent was obtained from all subjects involved in the study.

Data Availability Statement: The data presented in this study are available upon request from the corresponding author. The data are temporarily not publicly available due to research policies and implementation of research programs.

Acknowledgments: This work employed data provided by the Shared Research Facility «Analytical Geomagnetic Data Center» of the Geophysical Center of RAS (http://ckp.gcras.ru/ (accessed on 27 January 2022)).

Conflicts of Interest: The authors declare no conflict of interest.

References

1. Gvishiani, A.D.; Kaftan, V.I.; Krasnoperov, R.I.; Tatarinov, V.N.; Vavilin, E.V. Geoinformatics and systems analysis in geophysics and geodynamics. *Izvestiya Phys. Solid Earth* **2019**, *55*, 42–60. [CrossRef]
2. Glikson, A.Y.; Meixner, A.J.; Radke, B.; Uysal, I.T.; Saygin, E.; Vickers, J.; Mernagh, T.P. Geophysical anomalies and quartz deformation of the Warburton West structure, central Australia. *Tectonophysics* **2015**, *643*, 55–72. [CrossRef]
3. Adetunji, A.Q.; Ferguson, I.J.; Vayavur, R.; Cheraghi, S.; Naghizadeh, M.; Whymark, W.; Craven, J.A. Evidence of magmatism and rifting in the southern superior craton from the Temagami geophysical anomaly. *Precambrian Res.* **2021**, *362*, 106310. [CrossRef]
4. Dzeboev, B.A.; Gvishiani, A.D.; Belov, I.O.; Agayan, S.M.; Tatarinov, V.N.; Barykina, Y.V. Strong-earthquake-prone areas recognition based on an algorithm with a single pure training class I Altai-Sayan-Baikal region, M $\geq$ 6.0. *Izvestiya. Phys. Solid Earth* **2019**, *55*, 563–575. [CrossRef]
5. Gvishiani, A.D.; Dzeboev, B.A.; Agayan, S.M. FCAZm intelligent recognition system for locating areas prone to strong earthquakes in the Andean and Caucasian mountain belts. *Izv. Phys. Solid Earth* **2016**, *52*, 461–491. [CrossRef]
6. Fonsêca, J.A.S.; Ferreira, J.M.; do Nascimento, A.F.; Bezerra, F.H.R.; Lima Neto, H.C.; de Menezes, E.A.S. Intraplate earthquakes in the Potiguar Basin, Brazil: Evidence for superposition of local and regional stresses and implications for moderate-size earthquake occurrence. *J. S. Am. Earth Sci.* **2021**, *110*, 103370. [CrossRef]
7. Dzeboev, B.A.; Karapetyan, J.K.; Aronov, G.A.; Dzeranov, B.V.; Kudin, D.V.; Karapetyan, R.K.; Vavilin, E.V. FCAZ-recognition based on declustered earthquake catalogs. *Russ. J. Earth Sci.* **2020**, *20*, 1–9. [CrossRef]
8. Karapetyan, J.K.; Sargsyan, R.S.; Kazaryan, K.S.; Dzeranov, B.V.; Dzeboev, B.A.; Karapetyan, R.K. Current state of exploration and actual problems of tectonics, seismology and seismotectonics of Armenia. *Russ. J. Earth Sci.* **2020**, *20*, 3. [CrossRef]
9. Chesalova, E. The use of GIS technologies for the classification of underwater topography and estimation of their ore production (on the example of the Magellan seamounts, the Pacific Ocean). *Russ. J. Earth Sci.* **2021**, *21*, 1–10. [CrossRef]
10. Batugina, I.M.; Petukhov, I.M. *Geodynamic Zoning of Deposits in the Design and Operation of Mines*; Nedra: Moscow, Russia, 1988; 166p.
11. Jing, L.; Tsang, C.-F.; Stephansson, O. DECOVALEX—An international co-operative research project on mathematical models of coupled THM processes for safety analysis of radioactive waste repositories. *Int. J. Rock Mech. Min. Sci. Geomech. Abstr.* **1995**, *32*, 389–398. [CrossRef]
12. Pérez-Aguirre, X.J.; Chávez-Cabello, G.; Ramírez-Peña, C.F.; Méndez-Delgado, S.; Romero-de la Cruz, O.M. Geophysical modeling of the crustal boundary between the Central and Oaxaquia terranes in northern Mexico. *J. S. Am. Earth Sci.* **2021**, *110*, 103288. [CrossRef]
13. Petrov, V.A.; Minaev, V.A.; Ustinov, S.A.; Nafigin, I.O.; Lexin, A.B. Assessment of Seismogeodynamic Activity of Mining Areas on the Basis of 3D Geoinformation Modeling. *Russ. J. Earth Sci.* **2021**, *21*, 1–12. [CrossRef]
14. Jia, R.; Lv, Y.; Wang, G.; Carranza, E.; Chen, Y.; Wei, C.; Zhang, Z. A stacking methodology of machine learning for 3D geological modeling with geological-geophysical datasets, Laochang Sn camp, Gejiu (China). *Comput. Geosci.* **2021**, *151*, 104754. [CrossRef]
15. Jiang, L.; Zhao, Y.; Golsanami, N.; Chen, L.; Yan, W. A novel type of neural networks for feature engineering of geological data: Case studies of coal and gas hydrate-bearing sediments. *Geosci. Front.* **2020**, *11*, 1511–1531. [CrossRef]
16. Niño-Adan, I.; Manjarres, D.; Landa-Torres, I.; Portillo, E. Feature weighting methods: A review. *Expert Syst. Appl.* **2021**, *184*, 115424. [CrossRef]
17. Abramov, A.A.; Bolshov, L.A.; Dorofeeev, A.N.; Igin, I.M.; Kazakov, K.S.; Krasilnikov, V.Y.; Linge, I.I.; Trokhov, N.N.; Utkin, S.S. Underground Research Laboratory in the Nizhnekanskiy Massif: Evolutionary Design Study. *Radioact. Waste* **2020**, *1*, 9–21. [CrossRef]
18. Tatarinov, V.N.; Morozov, V.N.; Kamnev, E.N.; Manevich, A.I. Geodynamic aspects of high-level radioactive waste disposal: A case-study of Nizhnekansky massif. *Gorn. Zhurnal* **2021**, *3*, 108–112. [CrossRef]
19. Gvishiani, A.D.; Tatarinov, V.N.; Kaftan, V.I.; Manevich, A.I.; Dzeboev, B.A.; Losev, I.V. The velocities of modern horizontal movements of Earth crust in the South sector of Yenisei ridge according to GNSS observations. *Dokl. Earth Sci.* **2020**, *493*, 544–547. [CrossRef]

20. Kishkina, C.B.; Tatarinov, V.T.; Bugaev, E.G.; Gupalo, V.S.; Zabrodin, S.M. Underground Research Laboratory: Overcoming Uncertainties in the Assessment of Seismic Conditions for the Yeniseiskiy Site. *Radioact. Waste* **2021**, *3*, 80–93. [CrossRef]
21. Morozov, V.N.; Kolesnikov, I.Y.; Tatarinov, V.N. Modeling the hazard levels of stress-strain state in structural blocks in Nizhnekanskii granitoid massif for selecting nuclear waste disposal sites. *Water Resour.* **2012**, *39*, 756–769. [CrossRef]
22. Nguyen-Van, H.; Van Phong, T.; Trinh, P.T.; Van Liem, N.; Thanh, B.N.; Pham, B.T.; Bui, D.T.; Bieu, N.; Vinh, H.Q.; Xuyen, N.Q.; et al. Recent tectonics, geodynamics and seismotectonics in the Ninh Thuan Nuclear Power plants and surrounding regions, South Vietnam. *J. Asian Earth Sci.* **2020**, *187*, 104080. [CrossRef]
23. Young, R.P.; Nasseri, M.H.B.; Sehizadeh, M. Mechanical and seismic anisotropy of rocks from the ONKALO underground rock characterization facility. *Int. J. Rock Mech. Min. Sci.* **2020**, *126*, 104190. [CrossRef]
24. Tatarinov, V.N.; Morozov, V.N.; Manevith, A.I.; Tatarinova, T.A. Underground Research Laboratory: To the Program of Geomechanical Research. *Radioact. Waste* **2019**, *2*, 101–118.
25. Federal Safety Regulations in the Field of the Use of Atomic Energy. Disposal of Radioactive Waste. Principles, Criteria and Main Safety Requirements. NP-055-14: Approved by Rostechnadzor Order № 379 of August 22, 2014. Available online: http://en.gosnadzor.gov.ru/framework/nuclear/NP-055-04%20(to%20be%20updated%20to%20NP-055-14)%20pdf (accessed on 10 February 2022).
26. Gvishiani, A.D.; Agayan, S.M.; Bogoutdinov, S.R. Investigation of systems of real functions on two-dimensional grids using fuzzy sets. *Chebyshevskii Sbornik* **2019**, *20*, 94–111.
27. Agayan, S.M.; Tatarinov, V.N.; Gvishiani, A.D.; Bogoutdinov, S.R.; Belov, I.O. FDPS algorithm in stability assessment of the Earth's crust structural tectonic blocks. *Russ. J. Earth Sci.* **2020**, *20*, 2. [CrossRef]
28. Agayan, S.M.; Bogoutdinov, S.R.; Dobrovolsky, M.N. Discrete perfect sets and their application in cluster analysis. *Cybern. Syst. Anal.* **2014**, *50*, 176–190. [CrossRef]
29. Skrzypkowski, K. Case Studies of Rock Bolt Support Loads and Rock Mass Monitoring for the Room and Pillar Method in the Legnica-Głogów Copper District in Poland. *Energies* **2020**, *13*, 2998. [CrossRef]
30. Agayan, S.M.; Bogoutdinov, S.R.; Ivanchenko, O.V.; Kamaev, D.A. DMA algorithms for registration of a tsunami arrival in sea-level records. *J. Volcanol. Seismol.* **2021**, *15*, 126–132. [CrossRef]
31. Dzeboev, B.A.; Dzeranov, B.V.; Pasishnichenko, M.A. Recognition of strong earthquake–prone areas with a single learning class. Caucasus, M $\geq$ 6.0. In *Heat-Mass Transfer and Geodynamics of the Lithosphere. Innovation and Discovery in Russian Science and Engineering*; Springer: Cham, Germany, 2021; pp. 441–453.
32. Gvishiani, A.D.; Dzeboev, B.A.; Nekhoroshev, S.A. Recognition of earthquake-prone areas for seismic hazard evaluation. In *Disaster Risk Reduction and Resilience. Disaster and Risk Research: GADRI Book Series*; Springer: Berlin/Heidelberg, Germany, 2020; pp. 9–24.
33. Gvishiani, A.; Soloviev, A.; Krasnoperov, R.; Lukianova, R. Automated Hardware and Software System for Monitoring the Earth's Magnetic Environment. *Data Sci. J.* **2016**, *15*, 18. [CrossRef]
34. Gvishiani, A.D.; Mikhailov, V.O.; Agayan, S.M.; Bogoutdinov, S.R.; Graeva, E.M.; Diament, M.; Galdeano, A. Artificial intelligence algorithms for magnetic anomaly clustering. *Izv. Phys. Solid Earth* **2002**, *38*, 545–559.
35. Oshchenko, A.A.; Sidorov, R.V.; Soloviev, A.A.; Solovieva, E.N. Overview of anomality measure application for estimating geomagnetic activity. *Geophys. Res.* **2020**, *21*, 51–69.
36. Gvishiani, A.D.; Agayan, S.M.; Losev, I.V.; Tatarinov, V.N. Geodynamic hazard assessment of a structural block holding an underground radioactive waste disposal facility. *Min. Inf. Anal. Bull.* **2021**, *12*, 5–18.
37. Gvishiani, A.D.; Tatarinov, V.N.; Kaftan, V.I.; Losev, I.V.; Manevich, A.I. GIS-oriented database as basis for system analysis of geodynamic stability the Nizne-Kansky massif. *Issled. Zemli Iz Kosm.* **2021**, *1*, 53–66.
38. Nikulov, L.P.; Babkin, A.N.; Kolyamkin, V.M. *State Geological Map of the Russian Federation. Scale 1: 200,000*, 2nd ed.; The Yenisei series; Sheet O-46-XXXIV (Sosnovoborsk); Moscow Branch Office VSEGEI: Moscow, Russia, 2015.
39. Belov, S.V.; Morozov, V.N.; Tatarinov, V.N.; Kamnev, E.N.; Hammer, J. Study of the structure and geodynamic evolution of the Nizhnekansky Massif in connection with the disposal of highly radioactive waste. *Geoecology* **2007**, *2*, 248–266.
40. Lobatskaya, R.M. Neotectonic fault-block structure of junction of Siberian Platform and West Siberian Plate. *Geol. I Geofiz.* **2005**, *46*, 141–150.
41. Tatarinov, V.N.; Morozov, V.N.; Kaftan, V.I.; Manevich, A.I. Modern geodynamics of the southern of the Yenisei ridge derived from the results of satellite observations. *Geophys. Res.* **2018**, *19*, 64–79.
42. Anderson, E.B.; Belov, S.V.; Kamnev, E.N.; Kolesnikov, I.Y.; Lobanov, N.F.; Morozov, V.N.; Tatarinov, V.N. *Underground Isolation of Radioactive Waste*; Izdatel'stvo Gornaya Kniga: Moscow, Russia, 2011; p. 592.
43. Morozov, V.N.; Tatarinov, V.N.; Manevich, A.I.; Losev, I.V. Analogy method to determine the stress-strain state of structural-tectonic blocks of the Earth's crust for the disposal of radioactive waste. *Russ. J. Earth Sci.* **2019**, *19*, ES6001. [CrossRef]
44. Faber, R.G.; Domej, G. 3D computer-assisted geological mapping: Testing WinGeol's FaultTrace for semi-automatic structural geological assessment. *Russ. J. Earth Sci.* **2021**, *21*, 1. [CrossRef]
45. Kocharyan, G.G. *Geomechanics of Faults*; GEOS: Moscow, Russia, 2016; 424p.
46. Gvishiani, A.D.; Lukianova, R. Geoinformatics and observations of the Earth's magnetic field: The Russian segment. *Izvestya Phys. Solid Earth* **2015**, *51*, 157–175. [CrossRef]

47. Soloviev, A.A.; Bogoutdinov, S.R.; Agayan, S.M.; Gvishiani, A.D.; Kihn, E. Detection of hardware failures at INTERMAGNET observatories: Application of artificial intelligence techniques to geomagnetic records study. *Russ. J. Earth Sci.* **2009**, *11*, ES2006. [CrossRef]
48. Soloviev, A.A.; Agayan, S.M.; Gvishiani, A.D.; Bogoutdinov, S.R.; Chulliat, A. Recognition of disturbances with specified morphology in time series: Part 2. Spikes on 1-s magnetograms. *Izv. Phys. Solid Earth* **2012**, *48*, 395–409. [CrossRef]
49. Widiwijayanti, C.; Mikhailov, V.; Diament, M.; Deplus, C.; Louat, R.; Tikhotsky, S.; Gvishiani, A. Structure and evolution of the Molucca Sea area: Constraints based on interpretation of a combined sea-surface and satellite gravity dataset. *Earth Planet. Sci. Lett.* **2003**, *215*, 35–150. [CrossRef]

applied sciences

MDPI

Article

Theoretical Framework for Determination of Linear Structures in Multidimensional Geodynamic Data Arrays

Sergey Agayan [1], Shamil Bogoutdinov [1,2], Dmitriy Kamaev [3], Vladimir Kaftan [1], Maxim Osipov [4] and Victor Tatarinov [1,2,*]

1 Geophysical Center of the Russian Academy of Sciences, 119296 Moscow, Russia; s.agayan@gcras.ru (S.A.); shm@gcras.ru (S.B.); v.kaftan@gcras.ru (V.K.)
2 Schmidt Institute of Physics of the Earth of the Russian Academy of Sciences, 123242 Moscow, Russia
3 Federal State Budgetary Institution Research and Production Association "Typhoon", 249038 Obninsk, Russia; post@typhoon.obninsk.ru
4 Obninsk Institute for Nuclear Power Engineering, Branch of Federal State Autonomous Educational Institution of Higher Education "National Research Nuclear University MEPhI", 249039 Obninsk, Russia; osipovmax282@gmail.com
* Correspondence: victat@wdcb.ru

Abstract: The article addresses the issue of clustering of multidimensional data arrays with a noise using the methods of discrete mathematical analysis (DMA clustering). The theory of DMA clustering through the logical densities calculus is detailed, and the new algorithm Linear Discrete Perfect Sets (LDPS) is described. The main objective of the LDPS algorithm is to identify linearly stretched anomalies in a multidimensional array of geo-spatial data (geophysical fields, geochemistry, satellite images, local topography, maps of recent crustal movements, seismic monitoring data, etc.). These types of anomalies are associated with tectonic structures in the upper part of the Earth's crust and pose the biggest threat for integrity of the isolation properties of the geological environment, including in regions of high-level radioactive waste disposal. The main advantage of the LDPS algorithm as compared to other cluster analysis algorithms that may be used in arrays with a noise is that it is more focused on searching for clusters that are linear. The LDPS algorithm can apply not only in the analysis of spatial natural objects and fields but also to elongated lineament structures.

Keywords: finite metric space; density; solidity; clusters; discrete perfect sets; linear structures

Citation: Agayan, S.; Bogoutdinov, S.; Kamaev, D.; Kaftan, V.; Osipov, M.; Tatarinov, V. Theoretical Framework for Determination of Linear Structures in Multidimensional Geodynamic Data Arrays. *Appl. Sci.* **2021**, *11*, 11606. https://doi.org/10.3390/app112411606

Academic Editor: Jianbo Gao

Received: 25 October 2021
Accepted: 30 November 2021
Published: 7 December 2021

1. Introduction

In 2019, the construction of an underground research laboratory (URL) was started in granitic gneiss rocks of the Nizhne–Kansky rock mass (Russia, Krasnoyarsk Territory) to justify the safety of disposal of high-level radioactive waste (HLRW). The safety of HLRW underground insulation for a period of ten thousand years or more is guaranteed due to a geological barrier. The main threat of disturbance of the isolation properties of the geological environment where HLRW are disposed is associated with large-scale geodynamic processes and phenomena.

Therefore, a priority task in the field of geo-sciences includes the analysis of multidimensional geological and geophysical data as well as the creation of a geodynamic model based on such data, which provides a forecast of the safety of the rock isolation properties over the whole period of time when the radiobiological danger of a radioactive nucleus persists [1].

In order to resolve this task, we must determine linearly stretched anomalies in a multidimensional array of geo-spatial data (geophysical fields, geochemistry, satellite images, local topography, maps of recent movements, seismic monitoring data, etc.). As is known, these types of anomalies are associated with tectonic structures in the upper part of the earth crust—faults, the boundaries of large blocks, linear structures, potential zones of possible earthquakes, etc.

These are geodynamic zones that pose the biggest threat to the isolation properties of the greatest [2]. Their search is mandatorily regulated by the statutory documents applicable in the field of HLRW disposal.

It should be emphasized that the developed methodology is very versatile and can be applied to a wide range of practical tasks of the earth sciences—in geology, geodynamics, mineral exploration, etc. Thus, this methodology is used where there is a problem in identifying linear extended anomalies from spatially referenced data of field observations. A specific link to the problem of HLRW burial in geological formations is due to the fact that these algorithms were developed in the framework of the project on this problem.

The available geospatial data arrays are almost always insufficient, uncertain and distorted due to the noise, which dictates the need for developing effective analysis and interpretation algorithms [3–5]. This issue is resolved in the article within the framework of discrete mathematical analysis (DMA), an original data analysis approach developed at the Geophysical Center of the Russian Academy of Sciences.

One of the development areas of discrete data analysis and discrete math is substantially related to modeling the researcher's data analyzing skills. An experienced researcher will—better than any formal technique—distinguish any anomalies within physical fields with a small number of dimensions, move from their local level to the global one for holistic interpretation, find signals of the required form (morphology) on records of small length and many other things.

However, the researcher is helpless if faced with a large number of dimensions and volumes; therefore, a task teaching the computer in date analysis to act like a human being becomes ever more topical. When solving this task, it was considered that when the researcher thinks and operates not with numbers, but with fuzzy concepts; therefore, a technical framework for modeling includes fuzzy math and fuzzy logic along with classical math [6,7].

The advantage that researcher has in the analysis of discrete data over formal techniques is due to his or her more flexible, adaptive and stable attitude to real discrete-stochastic manifestations of fundamental mathematical properties (proximity, limitation, continuity, connectivity, trend, etc.) as compared with formal techniques, since the data analysis algorithms are built precisely on this basis as from a constructor. Hence, the plan for computer learning in the researcher's skills is as follows: building fuzzy models for discrete counterparts of fundamental math properties and then using them according to classical math scenarios to create data analysis algorithms.

The said tasks were implemented as a researcher-oriented data analysis approach, which takes an intermediate position between hard math methods and soft combinatoric methods. This is called discrete mathematical analysis (DMA) [8–11].

This paper addresses the study of stationary data arrays representing the sets in multidimensional spaces, using the DMA methods by means of clustering. The initial concept in DMA clustering is a fuzzy model of fundamental mathematical properties, such as "limitations". This is called density in DMA and represents a non-negative relationship between an arbitrary subset and any point in the initial finite space where the clustering is assumed to be carried out.

The value of density should be understood as a binding force between the subset and the point and interpreted as the degree of effect from the subset on the point, or, ambiguously, as the degree of limitation of the point for the subset. This view of density automatically requires that is shall be monotone over a subset: the larger the subset, the stronger its effect on the point, and it is more limiting for such subset.

Recording the density level α and understanding it as an ideality level, we can define any topological concept in the initial space, in particular, discrete perfection with level α: a subset is called discrete-perfect with the level of limitation (density) α if it is comprised precisely of all points of the initial space α that are of limiting kind for such a subset.

Taking into account what was said about density above, we give an equivalent formulation of the concept of discrete-perfectness: a subset is discrete-perfect with level α, if the strength of its connection with each of its points is not less than α, and with any point outside the subset is strictly less than α. Precisely this understanding of clustering formed the basis of the article and became the subject of this study. DMA has a strict theory of discrete perfect (DPS-) sets [12,13]. This serves as a methodological framework for DMA clustering and is summarized in Section 3. DMA clustering algorithms and their operation examples are shown in Section 4.

2. Review

Although there is no unified understanding of cluster, Everitt's empirical definition of cluster is one of the largest known and most convincing definitions in cluster analysis with the following wording: "Clusters are 'continuous' areas of a (certain) space with a relative higher density of points, separated from other similar areas by the areas with a relatively low density of points" [14]. Subsequently, this interpretation of the cluster is referred to as empirical. It has an advantage as it does not reduce the concept of cluster to a simple form.

One possible approach to formalizing an empirical cluster is as follows: first, we introduce the idea of a dense subset against the background of the entire source space, and then a maximum subset is distinguished inside of such space, which, in turn, is broken down into connected components. The latter will be dense, isolated regions in the original space, i.e., clusters.

It is precisely this scenario that underlies the SDPS DMA-clustering algorithm, which came into the spotlight due to its effective applications in seismology [15]. The space where it operates is assumed to be a finite metric space (FMS), and the density of a subset at a point is equal to the number of points at its intersection with its spherical neighborhood. We will call this "sets" and designate its value by S. The concept of density relative to a set is explained in Figure 1.

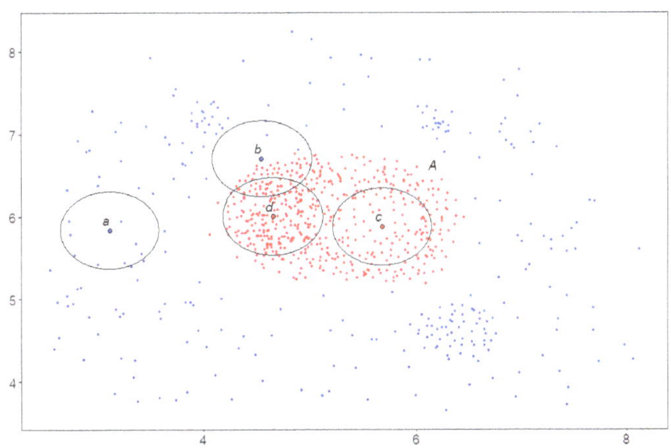

Figure 1. The concept of density relative to a set. A—a set of red dots, in which four points a, b, c and d are highlighted. The density S of the subset A at the selected points is equal to the number of red points included in the balls described around them. The densest point relative to A will be point c, followed by points b, d and a.

With the density S and setting its level to α, the most natural answer to the question about the density against the general background seems to be an answer in the form of a level solution, i.e., the subset of all points in the original space where the density level S is not lower than α. However, this is only a first approximation, and it may not be sufficient: a dense point may be isolated because space is dense here and not in its neighbors. Therefore, density against the general background requires more, i.e., that each dense point has a sufficient number of dense neighbors capable of ensuring the desired level of density α on its own here (using its own resources).

The discrete α-perfection of the density of "Sets" is exactly a formal expression of the above. The DMA theory of α-perfect sets guarantees that there is such a maximum subset available within the original space and that α-perfection will retain its property when passing to its connected components. The SDPS algorithm finds this subset and splits it into connected components. The fuzzy comparisons developed within DMA [8] allow us to effectively choose the level of limitation α so that the SDPS results are indeed internally indiscrete and externally dense, thus, embodying an empirical understanding of the cluster.

The application of the SDPS algorithm is illustrated in Figure 2. With the viewing radius for the density S and its level α, the SDPS algorithm begins its work on the array X (Figure 2a). SDPS acts on X iteratively, sequentially in four steps (Figure 2b–e) carving out of it the desired result—a local α-perfect subset $X(\alpha)$ in X (Figure 2e). The green points in Figure 2b–d show the points that did not pass the next iteration in SDPS. SDPS further splits $X(\alpha)$ into connected components (Figure 2f, yellow and black subsets).

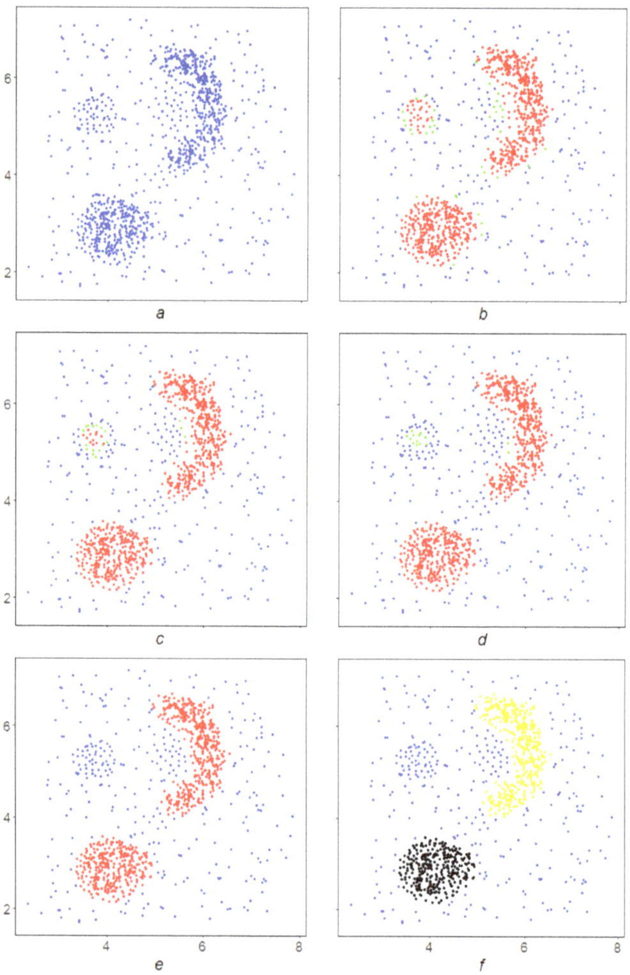

Figure 2. Application of the SDPS algorithm to the array X (**a**). Four iterations are shown in figures (**b–e**). The result is a local α-perfect subset of $X(\alpha)$ in X (**e**). The green points in figures (**b–d**) show the points that did not pass the next iteration in SDPS. SDPS further splits $X(\alpha)$ into connected components ((**f**), yellow and black subsets).

Let us detail the SDPS algorithm within the framework of classical cluster analysis: SDPS-density algorithm of direct, free, parameter-dependent classification that does not require human involvement and does not depend on the order of space scanning [16,17]. The SDPS algorithm, like the well-known modern algorithms DBSCAN, OPTICS, and RSC [18–20], represents a new stage in cluster analysis, since it not only breaks the original space into homogeneous parts but also pre-clears it of noise (filters), passing to the maximum α-perfect subset.

The use of the construction of α-perfect sets is an essential difference of the SDPS algorithm. For example, the SDPS and DBSCAN algorithms have the same initial parameters: the radius r of the view and the density level α (the minimum number of points that must be in a ball of radius r). Then, they act in different ways. As mentioned above, SDPS cuts out the maximum α-perfect set from the original space, parses it into connected components and considers them to be clusters. The DBSCAN algorithm uses an asymmetric reachability ratio, searches for regions of such reachability with centers at dense points, combines under the condition of r-proximity and considers them to be clusters.

Figure 3 shows examples of clustering a complex array (Figure 3a) using algorithms: MDPS (Figure 3b), DBSCAN (Figure 3c) and OPTICS (Figure 3d). Removing noise from space and then partitioning the balance part into connected components is possible based on a large and most important class of local monotone densities. This scenario is called a DPS-scheme, and its specific implementations at a particular density are called DPS-series algorithms (DPS-algorithms). They represent the state-of-the-art DMA clustering.

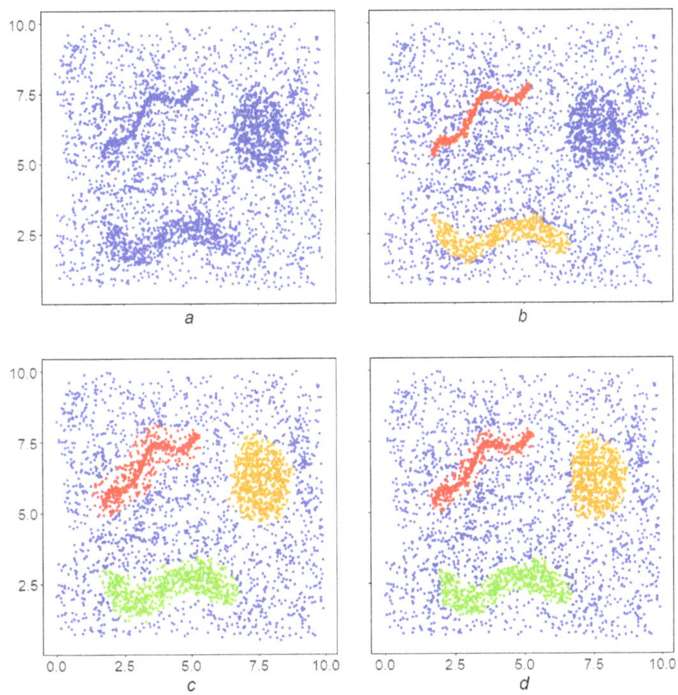

Figure 3. Examples of clustering a complex array (**a**) using algorithms: MDPS (**b**), DBSCAN (**c**) and OPTICS (**d**).

Previous DMA clustering algorithms Rodin, Crystal and Monolith [21–23] were based on non-monotone densities, and, despite successful applications, theoretically, they had drawbacks that are characteristic of density-based cluster analysis algorithms: dependence on the order of space scanning and, as a consequence, ambiguity of results, issues with convergence, etc.

In conclusion of the general description of DMA clustering, we note that the DPS-series algorithms are extremely versatile: they are capable of working with any kind of similarity in cluster analysis (distance, correlation and associativity factors). The point is that DMA has effective procedures in place that construct monotone densities.

Let us go back to the SDPS algorithm. The resulting clusters do not have any special geometry: they are simply "continuous" and contain "much space" locally. It seems that the natural extension of research should be the search for clusters with a particular geometry. Doubtless, the first step in that process is deemed to be linear structures.

3. Materials and Methods

As mentioned in the Introduction, by clustering in the initial finite space, we mean discrete perfection with respect to the density set on it. Density is an expression in the language of fuzzy mathematics of the "limit" property. By fixing the level of the selected density in the original study, it is possible to define the reference through the normal topology. Thus, in the original dimension, there is an indexed by non-negative numbers and an ascending family of topologies. It starts at zero with the minimal inseparable topology of concatenated points and ending with the separable maximal topology of all subsets.

On an arbitrary finite metric space, one can define densities that reflect various fuzzy interpretations of the "limit" property, and some of them are used in present paper. Thus, each density on a finite metric space sets its own view of it and the corresponding research program. Density, which expresses a fuzzy interpretation of the "limit" property in a finite space, is a new concept that is not reducible to the concepts of classical mathematics, for which finite metric spaces are topologically arranged in the same way–zero-dimensional, separable.

Within the framework of this work, in the family of topologies generated on the basis of density, the property of "perfection" is of interest. We will provide a brief summary of the theory of discrete perfect sets. Its complete proof can be found in [12].

3.1. Discrete Perfect Sets

Suppose X is a finite set, and $A, B, \ldots$ and $x, y, \ldots$ are its subsets and points, respectively.

Definition 1. *Let us call a mapping of the product of $2^X \times X$ into non-negative numbers $\mathbb{R}^+$ in the set X, increasing by the first argument and trivial-on-zero inputs as the density P:*

$$\begin{gathered} P(A,x) = P_A(x) \\ \forall x \in X, A \subset B \Rightarrow P_A(x) \leq P_B(x),\ P_\varnothing(x) = 0 \end{gathered} \tag{1}$$

For a density P set on X and a level $\alpha \in \mathbb{R}^+$, we make a sequence of α-n-hulls of A in X by P:

$$\begin{gathered} A^1 = \{x \in X : P_A(x) \geq \alpha\} \\ \ldots\ldots\ldots \\ A^n = \{x \in X : P_{A \cup A^{n-1}}(x) \geq \alpha\} \\ \ldots\ldots\ldots \end{gathered}$$

Induction on n, using the increasing monotonicity of P, establishes the following

Statement 1. $A^1 \subseteq \ldots \subseteq A^n \ldots$

Due to the finiteness of the set X, in the non-decreasing and bounded sequence of α-n-hulls, starting from some number n^*, stabilization occurs:

$$A^1 \subset \ldots \subset A^{n^*} = A^{n^*+1} = \ldots \tag{2}$$

Definition 2. *Let us call the set A^{n^*} α-∞-hull of the set A and represent it by A^∞.*

The A^∞ set demonstrates semi-invariance: its first density hull $(A^\infty)^1$ does not fall beyond the A^∞ set.

Statement 2. *A^∞ contains its first α-hull by the density:*

$$(A^\infty)^1 \subseteq A^\infty$$

Hence, it immediately follows that for a set A^∞ a series (2) of its α-n-hulls is constant.

Consequence 1. $(A^\infty)^n = (A^\infty)^1 \; \forall n \geq 2$

Let us designate α-∞-hull for A^∞ through $A^{2\infty}$. Therefore:

$$A^{2\infty} = (A^\infty)^\infty = (A^\infty)^1 \subseteq A^\infty$$

Sequentially plotting the α-∞-hulls based on the density P, we obtain the following scheme:

$$\begin{array}{ccccccc} A & \rightarrow & A^1 & \subseteq & \dots & = & A^\infty \\ A^\infty & \supseteq & (A^\infty)^1 & = & \dots & = & A^{2\infty} \\ \dots & \dots & \dots & \dots & \dots & \dots & \dots \\ A^{m\infty} & \supseteq & (A^{m\infty})^1 & = & \dots & = & A^{(m+1)\infty} \\ \dots & \dots & \dots & \dots & \dots & \dots & \dots \end{array}$$

Due to the X finiteness in a non-increasing sequence

$$A^\infty \supseteq \dots \supseteq A^{m\infty} \supseteq \dots$$

starting with some number m^*, stabilization occurs:

$$A^\infty \supset \dots \supset A^{m^*\infty} = A^{(m^*+1)\infty} = \dots$$

Let us designate the set $A^{m^*\infty}$ by $A(\alpha)$. The process of constructing $A(\alpha)$ has a stage of increasing from A^1 to A^∞ and a stage of decreasing from A^∞ to $A(\alpha)$:

$$A \rightarrow A^1 \subset \dots \subset A^{n^*} = A^\infty \supset \dots \supset A^{m^*\infty} = A(\alpha) \quad (3)$$

Statement 3. *$A(\alpha)$ matches its first α-hull.*

Remark 1. *Statement 3 means that the set $A(\alpha)$ is comprised exactly of those points x of space X where its density is more than α or equal to it:*

$$A(\alpha) = \left\{ x \in X : P_{A(\alpha)}(x) \geq \alpha \right\}.$$

Let us treat the density $P_A(x)$ as a limiting measure of the point x for the set A. The point x with sufficiently large density $P_A(x) \geq \alpha$ is considered to be the limiting one for A. Thus, the first hull of A^1 represents the set of all limiting points for A in X in that sense. The points from the second hull A^2 will, in general, be limiting points for A in X through A^1—that is, of the second order, etc. Statement 2 means that A^∞ is a closed set as it contains all its α-limiting points in X. Moving inside A^∞ leads to the already α-perfect set $A(\alpha)$, for restated Statement 3 means that $A(\alpha)$ consists exactly of all finite points to it in X, i.e., it is a perfect one.

Definition 3. *The set A consisting of exactly all α-limiting with regard to this set points of the initial space X is called an α-discrete perfect (simply perfect, DPS-) set in X:*

$$A \text{ – DPS-set in } X \leftrightarrow A = \{x \in X : P_A(x) \geq \alpha\}$$

Numerous studies and the examples below show that DPS-sets are condensations in X and are closely related to clustering therein. We have a way of generating them in X, i.e., $A \to A(\alpha)$ construction. It depends on four parameters: the initial space X, set A, density P and level α:

$$A(\alpha) = A_P(\alpha|X).$$

Statement 4. *Dependencies for A, P and X are increasing dependencies, and dependence for α are decreasing dependence:*

1. *If $A \subseteq B$, then $A(\alpha) \subseteq B(\alpha)$.*
2. *If P, Q densities on X and $P_A(x) \leq Q_A(x)$ $\forall x \in X, A \subseteq B$, then $A_P(\alpha) \leq A_Q(\alpha)$.*
3. *If $\alpha < \beta$, then $A(\beta) \subseteq A(\alpha)$.*
4. *If $A \subseteq X \subset Y$ and measure P are set to Y, then $A(\alpha|X) \subseteq A(\alpha|Y)$.*

3.2. Complete $\mathbb{DPS}$: Scheme and Algorithms

Definition 4. *The construction process for the set A in the universe X based on the density P of its hull $A(\alpha) = A_P(\alpha|X)$ is called the complete Discrete Perfect Sets algorithm and is designated through $\mathbb{DPS}$*

$$\mathbb{DPS}(\cdot) = \mathbb{DPS}(\cdot\,|X, P, \alpha) : 2^X \to 2^X$$
$$A \to \mathbb{DPS}(A|X, P, \alpha) \to A_P(\alpha|X) = A(\alpha)$$

Remark 2. *On a fixed space X, the $\mathbb{DPS}$ algorithm depends on two parameters, the major one being the density P. In order to emphasize this fact, we will write $\mathbb{DPS}(P)$, omitting the level α, though keeping it in mind. Furthermore, we will need a broader understanding of $\mathbb{DPS}$ as a correspondence $P \to \mathbb{DPS}(P)$ between densities and algorithms on X. In this case, we will speak of $\mathbb{DPS}$ as a scheme on X.*

Generally, the DPS algorithm has two stages (3):
increasing

$$A^n \uparrow A^\infty \leftrightarrow A \subset A^1 \subset \ldots = A^\infty$$

and decreasing

$$A^\infty \downarrow A(\alpha) \leftrightarrow A^\infty \supset A^{2\infty} \supset \ldots = A(\alpha)$$

There are situations when it works "faster" and has no more than one stage. The trivial "zero stages" case takes place for α-perfect A.

The $\mathbb{DPS}$ algorithm constructs for each $A \subseteq X$ its perfect hull $A(\alpha)$. Given that $A(\alpha)$ is of a non-trivial kind, we consider it as a promising set in X, playing a reference role and most naturally related to A. Hull $A(\alpha)$ answers the question of the role and effect of A in X. By substituting A, through the set $\{A(\alpha); A \subseteq X\}$, we obtain the information about the structure of X at the selected level of limitation α.

Therefore, the $\mathbb{DPS}$ algorithm is required for a thorough study of the space X through perfect hulls of its subsets, and for cluster analysis in X is too redundant and unnecessarily clumsy. Further research will show that clusters should be considered "connected pieces" of the X-maximal perfect subset $X(\alpha)$. They will be searched using a simplified version of the $\mathbb{DPS}$ algorithm DPS.

3.3. Simple DPS: Scheme and Algorithms

Throughout the entire space X, the $\mathbb{DPS}(P, \alpha)$ algorithm has only a decreasing stage, iteratively carving from X its maximum α-perfect subset $X_P(\alpha)$, playing a major role, and therefore it has a separate name "simple Discrete Perfect Sets" algorithm and designation DPS = DPS(P, α).

The DPS(P, α) algorithm is antagonistic by its nature to the DPS(P, α) algorithm to a certain extent: a simple DPS(P, α) algorithm is of global kind, intercepts the maximum perfect subset $X_P(\alpha)$ from X, whereas full DPS(P, α) is largely of local king, passing from A to $A_P(\alpha)$ by "auto-critical crystallization".

Remark 3. *As in Remark 2, the correspondence $P \to DPS(P)$ is called a DPS scheme. $DPS(P)$ algorithms are resulted from matching the densities P and a DPS scheme, and therefore they are of the same nature (DPS scheme), independent of P. There will be five such matching instances, e.g., $SDPS = DPS(S)$, $MDPS = DPS(M)$, $FDPS = DPS(F)$, $GDPS = DPS(G)$ and their complex LDPS-combination.*

In this context, we are talking about them as DPS-algorithms, DPS-set algorithms.

If there is a d-metric on X space and the density P (5) is consistent with it, the α-perfection property is inherited by "connected" components $X(\alpha)$. They are those that most accurately correspond to the idea of empirical clusters.

Subsequently, a metric d is set on X; therefore, (X, d) is a FMS. For $D_A(x,r)$ we designate a full-sphere in A with the center in x radius r:

$$D_A(x,r) = \{a \in A : d(x,a) \leq r\}$$

Definition 5. *Given that P is the density on X (1), $r > 0$ is the proximity radius. We assume that P has r-local influence (r-local) if*

$$\forall x \in X. A \subseteq X \to P_A(x) = P_{D_A(x,r)}(x)$$

Based on equivalence $d(x, A) > r \leftrightarrow D_A(x,r) = \varnothing$ and normalization on P, it follows that

Statement 5. *If the density P is r-local, then the implication is valid*

$$d(x, A) > r \to P_A(x) = 0.$$

3.3.1. Topological Retreat

Two points x and y in A are called r-connected if there is a chain of r-close points in A– $x_0, \ldots, x_n$ with the starting point $x_0 = x$ and terminus $x_n = y$ $(d(x_i, x_{i+1}) \leq r, i = 0, \ldots, n-1)$. The ratio of r-connectivity is an equivalence splitting the set A into components of r-connectivity $C_rA(1), \ldots, C_rA(k^*)$, $k^* = k^*(A,r)$:

$$A = C_rA(1) \vee \ldots \vee C_rA(k^*). \tag{4}$$

Algorithmically, the split (4) is achieved as follows: let a point be in A and $C_rA(a)$ component of r-connectivity that contains it. Then,

$$C_rA(a) = \cup_{i=1}^{\infty} C_r^i A(a)$$

where

$$\begin{aligned} &C_r^0 A(a) = a \\ &C_r^1 A(a) = D_A(a,r) \\ &\ldots\ldots\ldots\ldots\ldots \\ &C_r^{i+1} A(a) = \cup_{\bar{a} \in C_r^i A(a)} D_A(\bar{a}, r) \\ &\ldots\ldots\ldots\ldots\ldots \end{aligned}$$

By virtue of finiteness A everything is balanced and makes sense. Let us consider $C_rA(a)$ as the first component $C_rA(1)$ in (4). If it is not the last one, the same reasoning applies to $A_1 = A \setminus C_rA(a)$. As a result, we have the second component $C_rA(2)$ and so on.

Statement 6. *If the density P is r-local, then every r-link component of the set $X(\alpha)$ is α-perfect.*

In the event of r-local density, we will understand the DPS(P, α, r) algorithm through a broader lens.

Definition 6. *The process of construction for the finite metric space (X, d) based on the r-local density P of the α-hull $X(\alpha)$ with its subsequent splitting into r-connected components is called a "simple DPS algorithm":*

$$DPS = DPS(P, \alpha, r) \to 2^{2^X}$$
$$DPS(X) = \{C_r X(\alpha)(1), \ldots, C_r X(\alpha)(k^*)\}$$

Let us summarize our conversation about DPS(P, α, r) with its flow charts and comments to it (Figure 4).

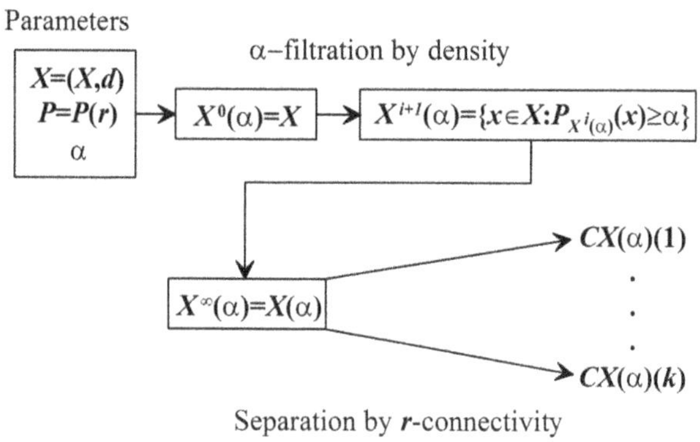

Figure 4. Block diagram of the DPS.

The first stage of DPS intercepts the maximal subset $X(\alpha)$, dense against the general background, from the initial space X. The second DPS stage splits $X(\alpha)$ into components $C_r X(\alpha)(k)$. Each component combines density against the background and connectivity, that is, it formally expresses empirical clustering.

3.3.2. Parameter Selection: Localization Radius r

Suppose that dX be the set of all non-trivial distances in X:

$$dX = \{d = d(x, y) : x \neq y \in X\}$$

The localization radius r is defined as a power mean with a negative exponent d of all distances from $d(X)$:

$$r = r_q(X) = \left(\frac{\sum_{d \in d(X)} d^q}{|dX|}\right)^{1/q} \quad (5)$$

3.3.3. Parameter Selection: Density Level α

The selection of level α greatly affects the result of the DPS algorithm. A convenient means for selecting the level α is fuzzy comparisons [8]. They allow us to effectively construct the limitation level so that the DPS results are really dense against the general background, that is, they are empirical clusters.

The fuzzy comparison $n(a, b)$ of two non-negative numbers a and b is a measure of the superiority of number b over number a, expressed as a scale of segment $[-1, 1]$:

$$n(a, b) = \text{mes}(a < b) \in [-1, 1]$$

A fuzzy comparison of a number a and a finite set B ($a \in \mathbb{R}^+, B \subset \mathbb{R}^+$) can be defined as the mean of fuzzy comparisons a with all numbers from B:

$$n(a, B) = \frac{\sum_{b \in B} n(a, b)}{|B|}, \quad n(B, a) = \frac{\sum_{b \in B} n(b, a)}{|B|}$$

and understood as a measure of minimality $\text{mes}\min_B a$ and a measure of maximality $\text{mes}\max_B a$ of the number a against the background B:

$$\text{mes}\min_B a = n(a, B), \quad \text{mes}\max_B a = n(B, a).$$

The measure of maximality $\text{mes}\max_B a$ enables formulating the necessary requirement for the DPS algorithm results: its density at each of its points must be significant (maximum enough) against the background X.

To do this, it is necessary first to calculate the density of the entire space X at all its points

$$P_X(X) = \{P_X(x) : x \in X\}.$$

This is a background of X. If $\beta \in [-1, 1]$ is the required level of density extremeness P against the background of X, then the immediate level $\alpha = \alpha(\beta)$ for P is uniquely determined by β from equation

$$n(P_X(X), \alpha) = \beta, \tag{6}$$

since the relation $\alpha \to n(P_X(X), \alpha)$ is of continuous and monotone kind. Equation (6) can be solved by dividing the segment at halves.

Therefore, the DPS algorithm must find a subset $X(\beta)$ in X, that is β-extremely P-dense against the general background X at each of its points:

$$x \in X(\beta) \leftrightarrow n(P_X(X), P_{X(\beta)}(x)) \geq \beta \leftrightarrow P_{X(\beta)}(x) \geq \alpha(\beta)$$

and split it by the components of the r-connectivity for r from (5).

3.3.4. Quality Criterion

The DMA methods allow us to evaluate the quality $\tau(P, \alpha, r)$ of the DPS(P, α, r) algorithm in a different way, as an advantage of the result $X_P(\alpha, r)$ over complement $\overline{X_P(\alpha, r)}$.

One of the options for the quality criterion will be discussed in Example 6.

3.4. *Density*

If, in comparison (6) α is replaced by $P_A(x)$ with arbitrary $x \in X$ and $A \subseteq X$, then we obtain a variable density alternating in sign on X with values on the scale $[-1, 1]$.

Definition 7. *Density*

$$\text{mes}\max P_A(x) = n(P_X(X), P_A(x)) \tag{7}$$

is called the extreme density generated by P (extreme P-density).

The value of $\text{mes}\max P_A(x)$ does not clearly answer the following question: "To what extent is the subset of A dense at the point x against the general background of space X?"

It is convenient for us to consider the segment $[-1, 1]$, rather than the segment $[0, 1]$, as the base scale in fuzzy mathematics and fuzzy logic, and given (7), all densities are normalized to the scale $[-1, 1]$. Following these assumptions, the density $P_A(x)$ at fixed A is a fuzzy structure on X. Therefore, with the help of fuzzy logic operations, as well as some others, it is possible to obtain new densities on the basis of the existing ones. This extends the capabilities of the $\mathbb{DPS}$ and DPS algorithms in space X.

Statement 7. 1. *If P and Q are densities on X and $R = R(y_1, y_2) : [-1, 1] \times [-1, 1] \to [-1, 1]$ nondecreasing mapping, then superposition*

$$R(P, Q)_A(x) = R(P_A(x), Q_A(x))$$

will be the density on X.

2. *If* $\neg$ *fuzzy negation on* $[-1,1]$, *then the superposition*

$$\neg P_A(x) = \neg(P_{\bar{A}}(x))$$

will be the density on X.

3. *If n is a fuzzy comparison on* $\mathbb{R}^+$, *then the superposition*

$$C(P,Q)_A(x) = n(Q_{\bar{A}}(x), P_A(x))$$

will be the density on X.

Consequence 2. 1. *R-connection P and* $\neg P$ *will be the density on X*

$$R(P,\neg P)_A(x) = R(P_A(x), \neg P_A(x))$$

2. *If* $\top$ $(\bot, M_p)$ *is t-norm (t-co-norm, generalized averaging operator) [7], then superpositions* $\top(P_A(x), Q_A(x))$, $\bot(P_A(x), Q_A(x))$, $M_p(P_A(x), Q_A(x))$ *will be densities on X.*
3. *If* $\lambda \in [0,1]$, *then* λ*-connection* $\lambda P_A(x) + (1-\lambda)Q_A(x)$ *will be density on X.*
4. *A "fuzzy comparison" CP will be the density on X:*

$$CP_A(x) = C(P,P)_A(x) = n(P_{\bar{A}}(x), P_A(x))$$

3.4.1. The Logical Densities Calculus

Suppose that $\mathfrak{P}_1, \ldots, \mathfrak{P}_K$ properties of elements of space X, which clusters can be obtained by the DPS algorithm with respect to densities $P_1, \ldots, P_K$.

If $\mathfrak{P}$ is a complex property obtained from properties $\mathfrak{P}_1, \ldots, \mathfrak{P}_K$ using the fuzzy logic formula Φ containing only monotone operations: $\mathfrak{P} = \Phi(\mathfrak{P}_1, \ldots, \mathfrak{P}_K)$ then clusters for $\mathfrak{P}$ in X can be obtained using DPS algorithm with density $P = \Phi(\text{mes}\max P_1, \ldots, \text{mes}\max P_K)$.

Remark 4. *The schemes* $\mathbb{DPS}$ *and DPS depend on parameters, the main of which is the density P. Connecting with it, they become algorithms* $\mathbb{DPS}(P,\alpha)$ *and* $DPS(P,\alpha)$ *with a subordinate parameter* α.

Therefore, $\mathbb{DPS}$ and DPS induce relations $P \to \mathbb{DPS}(P)$ and $P \to \text{DPS}(P)$, that to a certain extent resemble "functors" from the "category of densities" to the "category of algorithms". This enables correct understanding ("through functors") of the results of Statement 7–Section 3.4.1: the operations described therein can be considered "functors" on densities. Their superpositions with $\mathbb{DPS}$ and DPS provide new mappings of densities into algorithms, that is, new algorithmic schemes that depend on density.

Example 1. 1. *Scheme* $\neg DPS : P \to DPS(\neg P)$
2. *Scheme* $CDPS : P \to DPS(CP)$
3. *Scheme* $(\lambda, 1-\lambda)DPS : P \to DPS(\lambda P + (1-\lambda)\neg P)$

Algorithms representing implementations of these schemes on specific densities play an important role in the FMS analysis and will be discussed below.

Remark 5. *Combination of fuzzy logic with densities gives great expressive power at the local level in studying of FMS X. On the other hand, the DPS scheme is very effective in connecting local data. These two circumstances make the DPS algorithms a powerful tool in studying of FMS X at the global scale.*

The final part of the article will address the empirical evidences of this scheme by giving examples of DPS with different densities thus describing versions of DPS.

4. Results

4.1. SDPS Algorithm

Historically, the set-theoretical SDPS was the first in a series of DPS-algorithms. It is based on the density S with the name "Number of points" ("Number of space") [12,13] and conveying the degree of concentration of space X round each of its points x (the most natural understanding of density X in x).

The density $S_A(x)$ depends on the localization radius $r = r_q(X)$ (5) and the non-negative parameter p, considering the distance to x in the full-sphere $D_A(x, r)$:

$$S_A(x) = S_A(x|q, p) = \sum_{y \in D_A(x,r)} \left(1 - \frac{d(x,y)}{r}\right)^p$$

When $p = 0$, we have the usual number of points, explaining the name S:

$$S_A(x|q, 0) = |D_A(x, r_q)|$$

The S density is r-local, and the SDPS algorithm is the implementation of the DPS scheme based on S, described in Definition 6–Section 3.3.3: SDPS = DPS(S, r, β). The result of SDPS is condensations in $X \equiv$ sets locally containing "many X". They correspond to empirical clusters in terms of the most formal criteria. By varying the SDPS parameters, it is possible to obtain a fairly complete picture of the hierarchy of clusters in X.

Example 2. *Figure 5b shows the result of selection by level $\beta = -0.3$ for density S on the initial array X (Figure 5a), that is, the first iteration $X^1(-0.3)$ of the SDPS algorithm. It contains isolated points that needs to be removed, and in this sense is inferior to the final result $X(-0.3)$ of the SDPS algorithm on X (Figure 5c.)*

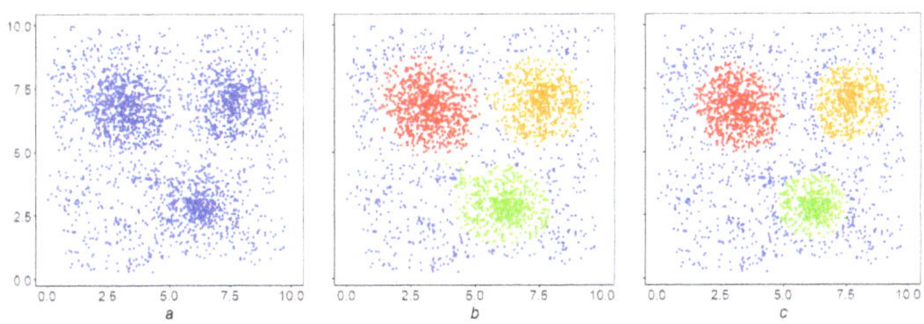

Figure 5. Application of the SDPS algorithm ($q = -2$, $\beta = -0.3$, $p = 0$): (**a**) the original array; (**b**) the result of the first iteration $X_1(-0.3)$ containing isolated points; and (**c**) the final result of applying SDPS.

Example 3. *In the conditions of the Example 2, the inverse correlation of the SDPS algorithm performance with the parameter β is shown. By increasing it, we go inside the condensation, finding dense nuclei inside them (Figure 6a–c).*

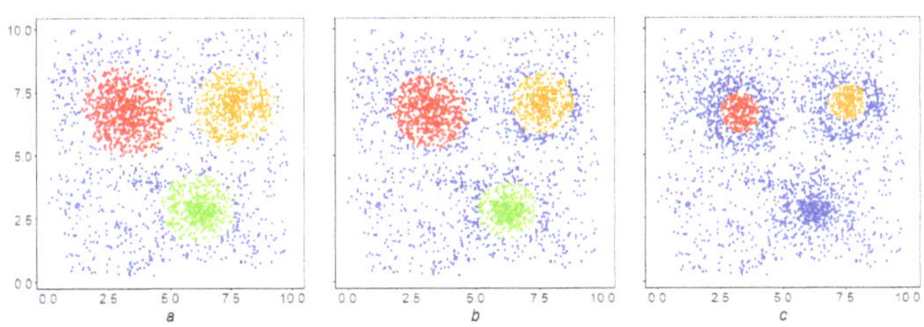

Figure 6. Inverse dependence of the result of the SDPS algorithm on the parameter β: (**a**) the result of the SDPS algorithm at $\beta = -0.35$; (**b**) the result of work at $\beta = -0.15$; and (**c**) the result of work at $\beta = 0.05$. (In all cases $q = -2$, $p = 0$).

Example 4. *In the conditions of Example 2 the direct correlation of the SDPS algorithm performance with the parameter q is shown. By lowering it, we make the SDPS algorithm more local, focusses on finding smaller condensations (Figure 7a–c). All small condensations in Figure 7c are shown in black.*

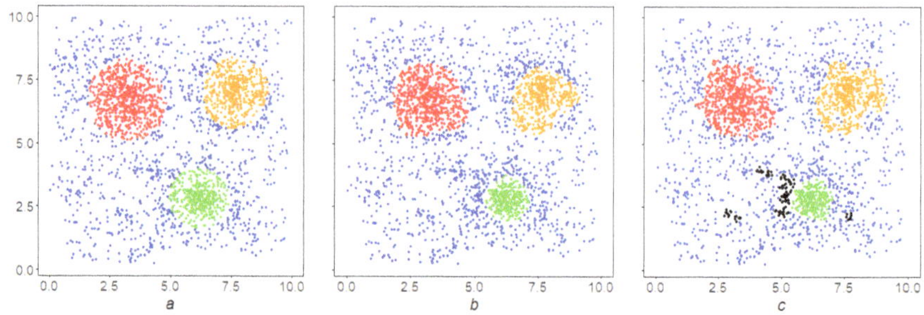

Figure 7. Dependence of the result of the SDPS algorithm on the parameter q: (**a**) the result of the SDPS algorithm at $q = -2$; (**b**) the result of work at $q = -2.8$; and (**c**) the result of work for $q = -3.5$ (in all cases $\beta = -0.2$, $p = 0$).

Example 5. *In the conditions of the Example 2, the inverse correlation of the SDPS algorithm performance with the parameter p is shown. By increasing it, we make the SDPS algorithm more stringent (Figure 8a–c).*

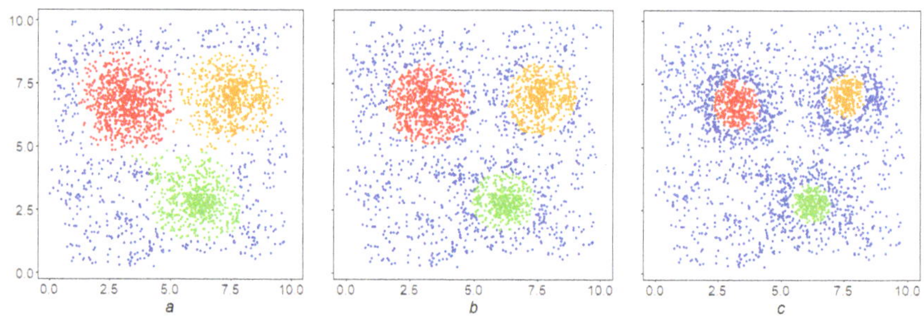

Figure 8. The inverse nature of the dependence of the SDPS algorithm on the parameter p: (**a**) the result of the SDPS algorithm at $p = 0$; (**b**) the result of work at $p = 0.5$; and (**c**) the result of work for $p = 1$ (in all cases $q = -2$, $\beta = -0.5$).

The above examples illustrate the general property of SDPS algorithm dependence on parameters: the stronger the localization (p, q) and the density level β is, the stricter the SDPS algorithm is, and its results are denser and finer.

Example 6. *Let us illustrate the clustering quality* $\tau(S, r, \alpha)$ *introduced in Section 3.3.4 on the SDPS work in the array shown in Figure 9. Let us designate by* $M(\beta, r)$ *and* $\overline{M}(\beta, r)$ *the mean densities of the sets* $X_S(\beta, r)$ *and* $\overline{X_S(\beta, r)}$ *at their points, then the result of their fuzzy comparison* $\tau(\beta, r) = n(\overline{M}(\beta, r), M(\beta, r))$ *can be considered a version of the quality of the SDPS*(β, r) *algorithm on the space X. From left to right it is equal to* 0.858, 0.595, 0.510 *respectively. This is true: the clustering in Figure 9a is clearly better, and Figure 9b,c are fairly the same.*

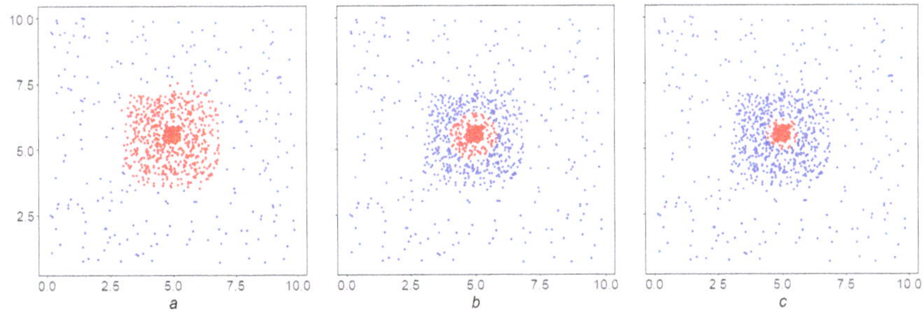

Figure 9. Illustration from left to right of the quality of the SDPS algorithm at $\beta = -0.3; 0.1; 0.3$. Clustering in figure (**a**) is clearly better, and in figures (**b**,**c**) is approximately the same. (As a fuzzy comparison, we used $n(a,b) = (b-a)/(b+1)$.

4.2. MDPS Algorithm

The SDPS algorithm is especially effective in heterogeneous, irregular spaces, where the property "density against the background" is strongly pronounced. If it is weakly expressed, there may be disadvantages in the work of SDPS caused by the density S.

Example 7. *1. Suppose that X is a uniform finite grid. The nodes at the edge X have a lower density S than the central nodes, although space X looks equally homogenous in both cases.*

2. If in a full-sphere $D_A(x,r)$ all points other than x, are concentrated on the circle $C_A(x,r)$ and there are many of them, then the density $S_A(x)$ is significant, regardless ther-isolation x.

Another construct of the density M, which is also expressing the concentration of space X at the point x does not have such disadvantages, It is called solidity and is part of the main DMA-clustering algorithm with the correspondent name [21]. Let us talk about it.

Fix natural number $m \in \mathbb{N}$ and construct a uniform grid of nodes $r_i = \frac{ir}{m}, i = 0, \ldots, m$ in the interval $[0, r]$. Then, we define a concentric in x semi-open ring $T_i(x, A)$ for each $i \neq 0$:

$$T_i(x,A) = \{y \in A : r_{i-1} < d(x,y) \leq r_i\}$$

For each $T_i(x, A)$, we assign the relevant weight $\psi_i : 1 \geq \psi_1 \geq \ldots \geq \psi_m > 0$. Solidity $M_A(x)$ is defined as the ratio of the sum of the weights of non-empty rings to the sum of the weights of all rings:

$$M_A(x) = \frac{\sum_{T_i(x) \neq 0} \psi_i}{\sum_{i=1}^{m} \psi_i}$$

The solidity M is r-local, and the MDPS algorithm is the implementation of the M-based DPS scheme described in Definition 6–Section 3.3.3: MDPS = DPS(M, r, β).

Remark 6. *Constructs S and M express the density of X in x in a different way, and this difference is shown in their names: construct S is focused on the "quantity" $D_A(x,r)$, is concentrated around x, while construct M is focused on "uniformity" $D_A(x,r)$, around x, expressed through the presence in rings $T_i(x)$.*

Example 8. *The dumb-bell shaped in Figure 10a has a monolithic but weak handle, so MDPS highlights it cleanly (Figure 10c), while SDPS cannot do so (Figure 10b). This example illustrates the independence of the MDPS and SDPS algorithms.*

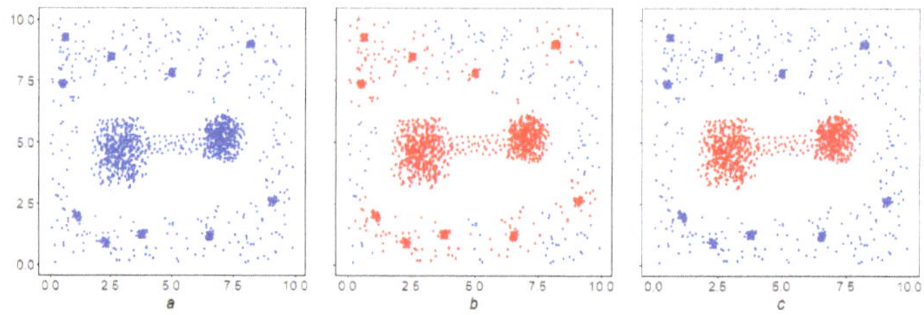

Figure 10. An illustration of the independence of the MDPS and SDPS algorithms: the "dumbbell" in the original array (**a**) has a sparse handle, which MDPS highlights cleanly (**c**); SDPS cannot do this (**b**).

4.3. FDPS Algorithm

The functional version of the DPS algorithm is related to a special r-local density $F = F(\nu)$, based on function weighting $\nu : X \to \mathbb{R}^+$:

$$F_A(x) = \frac{\sum \nu(y) : y \in D_A(x,r)}{|D_X(x,r)|}$$

The FDPS algorithm is the operation of the DPS circuit on F, as described in Definition 6–Section 3.3.3: FDPS $=$ DPS(F, r, ν, β) [24]. It aims at finding subsets in X with r-local high weights ν, and is capable to work on regular spaces and successfully complements the SDPS and MDPS algorithms.

Remark 7. *Weight ν can be thought of as a non-negative relief on X. The FDPS algorithm efficiently searches for the bases of these elevations, which is fundamental in data analysis, in particular in time series analysis (DRAS, FC ARS algorithms etc.) [25].*

Example 9. *Figure 11a shows how the FDPS algorithm works: space X in this case is a regular grid on the horizontal axis, where the weight ν of each point $x \in X$ is plotted vertically. The result of the FDPS algorithm will be two red bars on the horizontal axis, serving as the bases of the two most significant stochastic ν-elevations on X.*

As can be seen from this figure, the FDPS algorithm is stable: it disregards to insignificant drops of contour ν below the set level, as well as to insignificant rises of ν above it. This property of the FDPS explains the solidity of its highlighted elevations and is essential in decision-making issues: the selected areas must be massive and resistant to minor disturbances within them.

To compare, Figure 11b shows a classical selection on grid X with respect to a given level for relief ν. As we can see from the figure, it is unstable, it gives a lot of weak elevations.

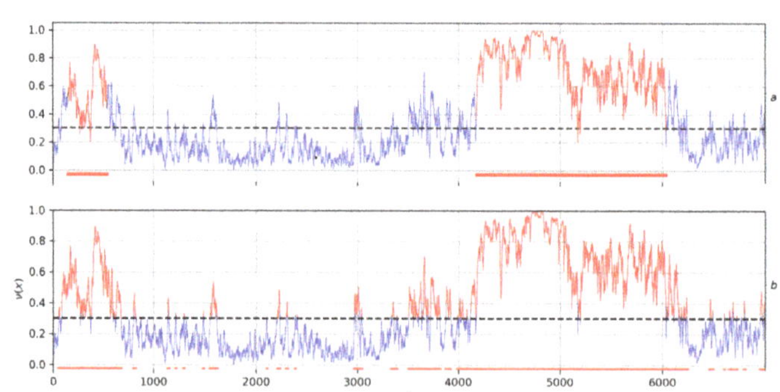

Figure 11. Operation of the FDPS algorithm on a regular grid (**a**). The FDPS algorithm results in two red lines on the horizontal axis, which serve as the bases of the two most significant stochastic heights. Figure (**b**) shows a classic choice with respect to a given level, highlighting many weak heights.

4.4. GDPS Gluing: Scheme and Algorithms

Suppose $\mathfrak{P}$ the local property of space X at each of its points x, $P(x) \geq 0$ its quantification, a $U(x) = U(x|P)$ is a subset in the full-sphere $D(x,r)$, where it is reached. In other words, $U(x)$ is the subset in $D(x,r)$ where the property is most clearly pronounced. It does not necessarily coincide with $D(x,r)$.

Task 1. *For a fixed level of α property $\mathfrak{P}$, find the subset $A = X(\mathfrak{P}, \alpha)$ in X, whose each full-sphere $D_A(a,r)$ would have the property $\mathfrak{P}$ at each point $a \in A$ to the power $\geq \alpha$.*

If the quantification of property $\mathfrak{P}$ function P is the density on X Definition 1, then the result of the DPS(P,r,α) algorithm can be taken as A. Otherwise, we consider a set of local data in X

$$U_\alpha = U_\alpha(P) = \{U(x|P) : P(x) \geq \alpha\}$$

and try to comprehensively fit into U_α the global subset $A \subset X$. Let us formulate the requirements for A:

$$(P(a) \geq \alpha) \wedge (D_A(a,r) \subseteq U(a)) \wedge (\text{difference } U(a) \setminus D_A(a,r) \text{ is minimal}) \; \forall a \in A \quad (8)$$

Under the natural assumption of "continuity" of the property $\mathfrak{P}$ we can expect its level of occurrence on $D_A(a,r)$ to be close to α.

The mismatch $D_A(a,r)$ with $U(a)$ in (8) can be understood differently. Some variants of it make it possible to find a solution of (8) using the DPS scheme (Definition 6–Section 3.3.3) with respect to densities specifically constructed by covering U_α. Let us focus on one of them.

The initial space Y will carry the covering U_α:

$$Y = \text{Supp}\, U_\alpha = \{x \in X : P(x) \geq \alpha\}$$

The difference in (8) is expressed through the intersection and induces a density G on Y:

$$G_B(y) = \frac{|D_B(y,r) \cap U(y)|}{|U(y)|} \quad y \in Y, B \subseteq Y \quad (9)$$

The density G is normalized: $G_B(y) \in [0,1]$, and its level γ is the proximity index in (9).

The dependence $P \to G(P)$ (9), connecting with the DPS scheme (Definition 6–Section 3.3.3), leads to another dependence $P \to \text{DPS}(G(P))$, which we designate as GDPS and will be understood like $\mathbb{DPS}$ Remark 2 and DPS Remark 3 ambiguously:

- as a schema if we are talking about a dependency given above;
- as the GDPS algorithm when it comes to the operation of the DPS scheme (Definition 6–Section 3.3.3) on density $G(P)$ with parameter γ: GDPS $=$ DPS(G,γ). Its GDPS(Y) result solves problem Task 1 by "gluing" local data U_α in a certain way. This explains the name of the algorithm (gluing).

If the quantification of P property $\mathfrak{P}$ is the density on X, then the space Y is the first α-hull $X^1(\alpha)$ of space X. The first solution Task 1 (let us call it "a strong one") is the operation of SDPS on X with respect to P with level α. It may differ from the second solution Task 1 (let us call it "a weak one"), which represents the operation of GDPS.

The point is that the SDPS algorithm in its interception on $X^1(\alpha)$ is guided primarily by preserving the density level α, while the GDPS algorithm seeks to preserve the proximity level γ on $X^1(\alpha)$. The weak version is more versatile, since the quantification of P property $\mathfrak{P}$ shall not be necessarily the density on X and it is also more "save" when carving. Therefore, it is the GDPS algorithm that will be part of the DPS-scheme Definition 6–Section 3.3.3 operation when the property $\mathfrak{P}$ represents a local linearity in X.

Let us explain the above on two examples relating to the density $S = S(q,p)$. In this case, the property $\mathfrak{P}$ will be the r-local "space count", and S will be its formal expression.

Since S is a density Definition 1, this Task 1 has two possible solutions: a strong and weak one. The basis for these is the space $Y = X^1(S, \beta)$—the first iteration of the initial space X of relative density S for the level of extremeness β.

The strong solution is the result of the SDPS algorithm on X with parameter β (a subset of SDPS(X, β)) in Y. The weak solution is the result of the GDPS algorithm in this setting, i.e., the result of a DPS-scheme with density $G = G(S, \beta)$ constructed on the basis of local data $U(S, \beta) = \{D_X(y, r) : y \in Y = X^1(S, \beta)\}$:

$$G_B(y) = \frac{|D_B(y.r) \cap D_X(y, r)|}{|D_X(y, r)|}$$

and a given level of proximity γ (a subset of DPS$(G, \gamma) \subseteq Y$).

4.5. LDPS Algorithm

We believe that the initial FMS lies in the Euclidean plane. It is convenient to designate it by Q rather than X for reasons that will be clear below. In this paragraph we implement in detail the previous scenario for the local linearity property in Q. The result of this work done will be the LDPS algorithm from the series of DPS-algorithms, aimed at finding global linear structures in Q.

4.5.1. Initial Data and Designations

Π is the universe plane

xOy—fixed orthogonal coordinate system on Π,

$x_\varphi O^* y_\varphi$—loose orthogonal coordinate system on Π, obtained by moving coordinate origin O to point $O^* = (x^*, y^*)$ and turning the axes x_φ, y_φ by the angle $\varphi \in [0, \pi]$,

Relation of coordinates

$$\begin{aligned} x_\varphi &= \cos\varphi(x - x^*) + \sin\varphi(y - y^*) \\ y_\varphi &= -\sin\varphi(x - x^*) + \cos\varphi(y - y^*) \end{aligned},$$

Q is a finite-state array in Π: $Q = \{q\} = \{q_i|_{i=1}^N\}$,

$r = r_s(Q), s < 0$ (5),

$\mathfrak{L}$—the property of local linearity in Q.

4.5.2. Quantification $\mathfrak{L}$

Let q^* be an arbitrary fixed point in Q, φ arbitrary angle in $[0, \pi]$. Let us move to the coordinates $x_\varphi q^* y_\varphi$ and we denote the "square" neighborhood Q in $q^*$of radius r $K_Q(q^*|\varphi, r)$:

$$K_Q(q^*|\varphi, r) = \{q \in Q : |x_\varphi(q)| \le r, |y_\varphi(q)| \le r\}.$$

The additional parameter "height" $h \in (0, r]$ enables defining the corridor $K_Q(q^*|\varphi, r, h)$ in $K_Q(q^*|\varphi, r)$

$$K_Q(q^*|\varphi, r, h) = \{q \in Q : |x_\varphi(q)| \le r, |y_\varphi(q)| \le h\}.$$

Using $K_Q(q^*|\varphi, r)$ and $K_Q(q^*|\varphi, r, h)$ we define a measure of local linearity $L_Q(q^*|\varphi, r, h)$ of space Q at point q^* to the direction φ as the density $K_Q(q^*|\varphi, r, h)$ against the background of $K_Q(q^*|\varphi, r)$ by serial relation:

$$L_Q(q^*|\varphi, r, h) = \frac{|K_Q(q^*|\varphi, r, h)|}{K_Q(q^*|\varphi, r)|}.$$

Maximum value

$$L_Q(q^*|r, h) = \max_\varphi L_Q(q^*|\varphi, r, h)$$

will be considered a quantitative expression of the property $\mathfrak{L}$ for Q in q^*, and the neighborhood of $U(q^*, \mathfrak{L})$ is the best corridor $K_Q(q^*|r, h)$ where this maximum is reached.

For geometrical reasons, the relation $hr^{-1} \in (0, 1/2]$ must be considered as satisfied.

4.5.3. Search for Global Linear Structures

A quantification L of the property $\mathfrak{L}$ is made. Hence, it is possible to involve a GDPS scheme based on L for the weak solution of Task 1 in this case, which leads to the GDPS(P, α, γ) algorithm, where α is the expression level of property L and γ is its representativity degree.

Research shows that in the generic case its result Z on space $Y = Q^1(L, \alpha)$ needs additional filtering, which is done by the MDPS algorithm with a solidity level ε. Its result MDPS(Z, ε) is considered final in the search for linear structures within the space.

The LDPS algorithm is the described superposition of GDPS and MDPS:

$$Q \to Q^1(L, \alpha) = Y \to \text{GDPS}(Y, \gamma) = Z \to \text{MDPS}(Z, \varepsilon) = \text{LDPS}(Q).$$

Its parameters will be (parameters L) + (parameters M)+$(\alpha, \gamma, \varepsilon)$, and the result is the global linear structures in Q relative to them.

The LDPS algorithm has four stages:

- the first of them with the selected parameters of local linearity r and h constructs its quantification $L_Q(q|r, h)$ at each point q of space Q, and the best corridor $K_Q(q|r, h)$, where the estimate $L_Q(q|r, h)$ is reached;
- the second stage includes constructing the basis for application of the GDPS scheme, namely coverage $U_\alpha(L) = \{K_Q(q|r, h) : L_Q(q|r, h) \geq \alpha\}$ for a given level of local linearity α;
- the third stage is GDPS scheme working on $U_\alpha(L)$ data. Its result will represent linear structures in Q. On space

$$Y = \text{Supp}\, U_\alpha(L) = \{q \in Q : L_Q(q|r, h) \geq \alpha\}$$

a measure $G = G(U_\alpha(L))$ is constructed (9)

$$G_B(y) = \frac{|D_B(y, r) \cap K_Q(q|r, h)|}{|K_Q(q|r, h)|}.$$

The result will represent the raw linear structures on Q;
- the fourth stage is their filtering by solidity using the MDPS algorithm with a level of ε.

In conclusion, we will address the operation of the LDPS algorithm on two arrays, while the operation in the first instance will be explained in detail, and in the second instance only the result is shown.

Example 10. *The initial array Q is shown in Figure 12a*

- *At the first stage with the chosen parameters at each blue point q a linear corridor $K_Q(q|r, h)$ is constructed with parameters $r = 1.09$ and $h = 0.44$, then its separability $L_Q(q|r, h)$ is calculated. Figure 12b,c show corridors in green with centers at black points. Their separability equals to 0.64 and 0.5, respectively. In the second instance, it proves to be insufficient to overcome the second stage.*
- *Second stage. The separability level α is assumed to be 0.6. Figure 13a shows the points y in red, that passed this selection and formed the basis Y—the first half for application of the GDPS scheme. The second half GDPS scheme: corridors $K_Q(q|r, h)$ are shown for the already familiar point on the left (Figure 13b) and at some point on the right (Figure 13c). It can be seen from the figures that the relative density of red dots in the left corridor is higher than in the right one. This circumstance will help the left point overcome the third stage and move into the lower linear structure, while the right point will not stand the test with GDPS operation and will not be included in the final result.*

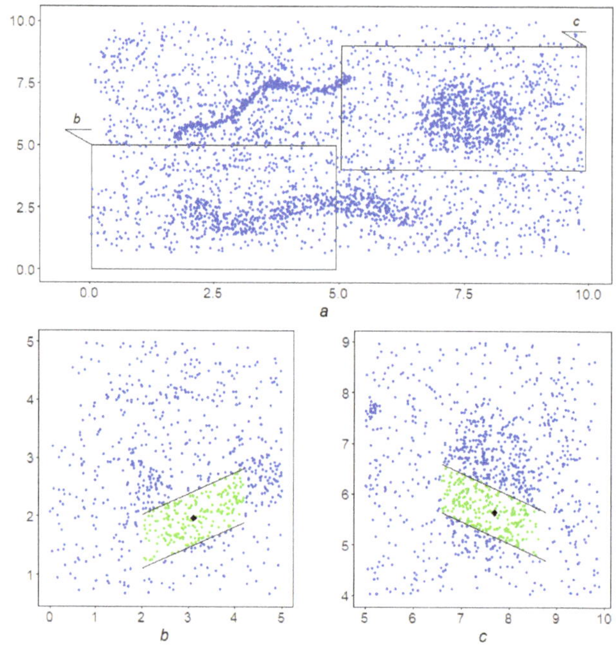

Figure 12. The first stage of the LDPS algorithm. (**a**) original array; (**b**) a corridor with a separability equal to 0.64; and (**c**) a corridor with a separation of 0.5. The position of the fragments of the original array shown in figures (**b**,**c**) is indicated by the leaders in figure (**a**).

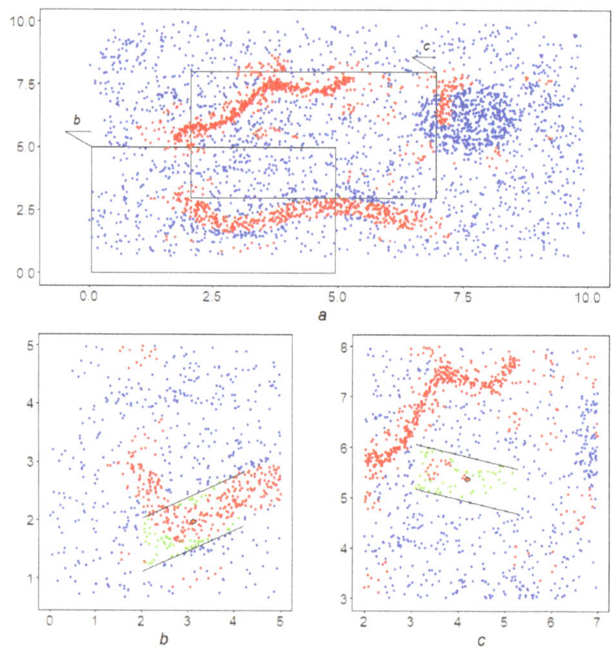

Figure 13. The second stage of the algorithm. (**a**) The original array with the selected base (red dots); (**b**) points from in the corridor from Figure 12b; and (**c**) points from in the corridor from Figure 12c. The position of the fragments of the original array shown in figures (**b**), (**c**) is indicated by the leaders in figure (**a**).

- *Third scheme. The GDPS scheme operation on the red points from the Y. Its result Z is shown in Figure 14a. It needs to be filtered.*

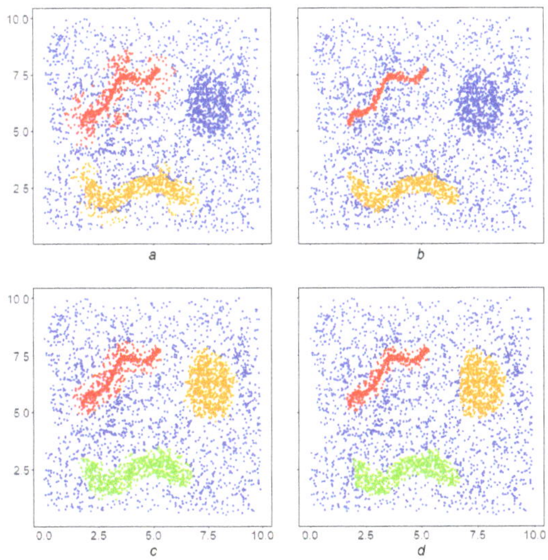

Figure 14. Result of the third and fourth stages of the LDPS algorithm. (**a**) The result of the third stage–the work of the GDPS algorithm; (**b**) the result of the fourth stage–the work of the MDPS algorithm; (**c**) the result of the DBSCAN algorithm; and (**d**) the result of the OPTICS algorithm.

- *Fourth stage. This is implemented by the MDPS algorithm on Z. The result is shown in Figure 14b. Figure 14c,d show how the well-known DBSCAN [18] and OPTICS [19] algorithms operates in this instance.*

Example 11. *The initial array is shown in Figure 15a, and the LDPS result is shown in Figure 15b.*

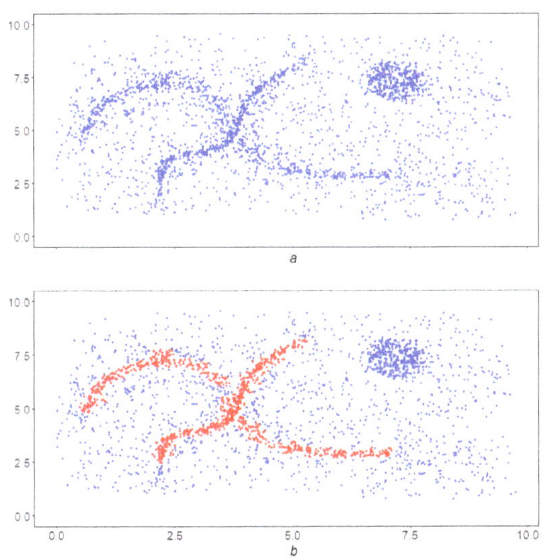

Figure 15. Operation of the LDPS algorithm. (**a**) The original array. (**b**) The linear structure indicated by red dots, built by the LDPS algorithm.

5. Discussion

This paper addressees the study of stationary data arrays, which are finite sets in multidimensional spaces, using the DMA methods by means of clustering.

A complex local condition is a conjunction of the conditions of local linearity and local representativeness. Local linearity: each point L has a linear corridor containing it that is dense against the background of Q, i.e., each point $q \in L$ on the plane has a rectangle $K(q)$ centered at q, which is a local corridor for L, and the intersection $K(q) \cap Q$ is dense against the background Q. Local representativeness: the intersection $K(q) \cap \mathrm{L}$ is dense in $K(q) \cap Q$.

The global condition for L consists of the requirement that L has no isolated points, i.e., in its discrete perfection. The detection of local linearity in the implementation of LDPS consists of a direct check for all points of the original space Q of this property. Analytical procedures are available to reduce routine calculations. In addition, in the future, we propose to modify the algorithm so that the dimensions of the corridors generally change from point to point.

The points in Q that pass the local linearity test form a subset of Y—the first approximation in Q to linear structures. We checked the linear representativity on Y in the LDPS algorithm as an implementation of the DPS scheme with respect to the special density G (clause 4.5.3). In the future, we plan to implement other variants of the LDPS algorithm based on a change in the interpretation of linear representativeness.

Studies have shown that, among the subset of points $Z \subset Q$ that have passed the local test for representativeness, there may be isolated points. The global condition in LDPS eliminates this drawback: its check for Z is also organized as an implementation of the DPS scheme with respect to the density "monolithicity" (MDPS algorithm). The points that pass the global test will be the result of applying the LDPS—that is, the union of all linear structures in Q.

In the general case, the set of linear structures L is divided into connected components. For example, in the examples given in Figures 14b and 15b, two connectivity components are obtained. In the first example, the components of connectivity should be considered independent linear structures, and in the second example, as part of a single whole. In the future, we plan to introduce an additional procedure for joining the results of applying the LDPS algorithm in order to obtain global linear structures in the original space Q.

Comparison of the LDPS algorithm with the well-known new generation cluster analysis algorithms DBSCAN and OPTICS, as well as with the previously created DMA clustering algorithms, shows that the LDPS algorithm is not inferior to them in detecting clumps; however, at the same time, it is more focused on recognizing clumps with a linear structure.

The implementation of the formal approach as shown in the article can be very effective in various fields of Earth sciences where linear structures play a special role in the investigation of spatial patterns of geographical location and geometric configuration of natural objects. This is vital when solving the issue of predicting the isolation properties of the geological environment for the preparation of a rationale for the geodynamic stability over long periods of time (ten to one hundred thousand years), arising in the selection of HLRW disposal sites. Further to this task, the algorithm can apply to the analysis of elongated artificial structures, such as road networks, etc.

Author Contributions: All authors contributed to the study conception and design. Conceptualization, original draft preparation: S.A., S.B. and V.T.; conceptualization, methodology, review and editing and validation: D.K. and V.K.; material preparation, formal analysis, data curation, algorithm development: S.B. and M.O. All authors have read and agreed to the published version of the manuscript.

Funding: The research was supported by Russian Science Foundation No 18-17-00241 "Study of the rock massifs stability by system analysis of geodynamic processes for geoecologically safe underground of radioactive waste isolation".

Informed Consent Statement: Informed consent was obtained from all subjects involved in the study.

Conflicts of Interest: The authors declare no conflict of interest.

Abbreviations

The following abbreviations are used in this manuscript:

URL	Underground research laboratory
HLRW	High-level radioactive waste
DMA	Discrete mathematical analysis
DPS	Discrete perfect sets
FMS	Finite metric space

References

1. Gvishiani, A.D.; Kaftan, V.I.; Krasnoperov, R.I.; Tatarinov, V.N.; Vavilin, E.V. Geoinformatics and systems analysis in geophysics and geodynamics. *Phys. Earth* **2019**, *1*, 42–60. [CrossRef]
2. Laverov, N.P.; Omelyanenko, B.I.; Velichkin, V.I. Geological aspects of the problem of radioactive waste disposal. *Geoecology* **1994**, *6*, 3–20. (In Russian)
3. Dzeboev, B.A.; Karapetyan, J.K.; Aronov, G.A.; Dzeranov, B.V.; Kudin, D.V.; Karapetyan, R.K.; Vavilin, E.V. FCAZ-recognition based on declustered earthquake catalogs. *Russ. J. Earth. Sci.* **2020**, *20*, ES6010. [CrossRef]
4. Gorshkov, A.I.; Soloviev, A.A. Recognition of earthquake-prone areas in the Altai-Sayan-Baikal region based on the morphostructural zoning. *Russ. J. Earth. Sci.* **2021**, *21*, ES1005. [CrossRef]
5. Belov, S.V.; Gvishiani, A.D.; Kamnev, E.N.; Morozov, V.N.; Tatarinov, V.N. Development of complex model of evolution of structural-tectonic blocks of the Earth's crust for choosing storage sites of high level radioactive waste. *Russ. J. Earth. Sci.* **2008**, *10*, ES4004. [CrossRef]
6. Zadeh, L. The concept of a linguistic variable and its application to approximate reasoning. *Inf. Sci.* **1975**, *8*, 199–249. [CrossRef]
7. Averkin, A.N.; Batyrshin, I.Z.; Blishun, A.F.; Silov, V.B.; Tarasov, V.B. *Fuzzy Sets in Models of Control and Artificial Intelligence*; Publ. Nauka: Moscow, Russia, 1986; 312p. (In Russian)
8. Agayan, S.M.; Bogoutdinov, S.R.; Krasnoperov, R.I. Short introduction into DMA. *Russ. J. Earth Sci.* **2018**, *18*, ES2001. [CrossRef]
9. Gvishiani, A.D.; Dzeboev, B.A.; Agayan, S.M. FCAZm intelligent recognition system for locating areas prone to strong earthquakes in the Andean and Caucasian mountain belts. *Izv. Phys. Solid Earth* **2016**, *52*, 461–491. [CrossRef]
10. Widiwijayanti, C.; Mikhailov, V.; Diament, M.; Deplus, C.; Louat, R.; Tikhotsky, S.; Gvishiani, A. Structure and evolution of the Molucca Sea area: Constraints based on interpretation of a combined sea-surface and satellite gravity dataset. *Earth Planet. Sci. Lett.* **2003**, *215*, 135–150. [CrossRef]
11. Gvishiani, A.; Soloviev, A.; Krasnoperov, R.; Lukianova, R. Automated Hardware and Software System for Monitoring the Earth's Magnetic Environment. *Data Sci. J.* **2016**, *15*, 18. [CrossRef]
12. Agayan, S.M.; Bogoutdinov, S.R.; Dobrovolsky, M.N. On one algorithm for searching the dense areas and its geophysical applications. In Proceedings of the Materials of 15th Russian National Workshop "Mathematical Methods of Pattern Recognition, MMRO-15", Petrozavodsk, Russia, 11–17 September 2011; Maks Press: Moscow, Russia, 2011; pp. 543–546. (In Russian)
13. Agayan, S.M.; Bogoutdinov, S.R.; Dobrovolsky, M.N. Discrete Perfect Sets and Their Application in Cluster Analysis. *Cybern. Syst. Anal.* **2014**, *50*, 176–190. [CrossRef]
14. Everitt, B.S. *Cluster Analysis*; Halsted-Heinemann: London, UK, 1980; 170p.
15. Dzeboev, B.A.; Gvishiani, A.D.; Agayan, S.M.; Belov, I.O.; Karapetyan, J.K.; Dzeranov, B.V.; Barykina, Y.V. System-Analytical Method of Earthquake-Prone Areas Recognition. *Appl. Sci.* **2021**, *11*, 7972. [CrossRef]
16. Mandel, I.D. *Cluster Analysis*; Publ. Finansy i Statistika: Moscow, Russia, 1988; 176p. (In Russian)
17. Mark, S.A.; Roger, K.B. *Cluster Analysis (Quantitative Applications in the Social Sciences)*; SAGE Publications, Inc.: Newbury Park, CA, USA, 1984; 88p.
18. Ester, M.; Kriegel, H.-P.; Sander, J.; Xu, X. A density-based algorithm for discovering clusters in large spatial databases with noise. In Proceedings of the Second International Conference on Knowledge Discovery and Data Mining (KDD-96), Portland, OR, USA, 2–4 August 1996; Simoudis, E., Han, J., Fayyad, U.M., Eds.; AAAI Press: Palo Alto, CA, USA, 1996; pp. 226–231.
19. Ankerst, M.; Breunig, M.; Kriegel, H.-P.; Sander, J. OPTICS: Ordering Points To Identify the Clustering Structure. In Proceedings of the ACM SIGMOD International Conference on Management of Data, Philadelphia, PA, USA, 31 May–3 June 1999; ACM Press: New York, NY, USA, 1999; pp. 49–60. [CrossRef]
20. Bojchevski, A.; Matkov, Y.; Günnemann, S. Robust Spectral Clustering for Noisy Data: Modeling Sparse Corruptions Improves Latent Embeddings. In Proceedings of the 23rd ACM SIGKDD International Conference on Knowledge Discovery and Data Mining (KDD-17), Halifax, NS, Canada, 13–17 August 2017; Association for Computing Machinery: New York, NY, USA, 2017; pp. 737–746. [CrossRef]
21. Gvishiani, A.D.; Agayan, S.M.; Bogoutdinov, S.R.; Soloviev, A.A. Discrete mathematical analysis and geological and geophysical applications. *Bull. Earth Sci.* **2010**, *2*, 109–125. (In Russian)
22. Mikhailov, V.; Galdeano, A.; Diament, M.; Gvishiani, A.; Agayan, S.; Bogoutdinov, S.; Graeva, E.; Sailhac, P. Application of artificial intelligence for Euler solutions clustering. *Geophysics* **2003**, *68*, 168–180. [CrossRef]

23. Agayan, S.M.; Soloviev, A.A. Highlight dense areas in metric spaces based on crystallization. *Syst. Res. Inf. Technol.* **2004**, *2*, 7–23. (In Russian)
24. Agayan, S.M.; Tatarinov, V.N.; Gvishiani, A.D.; Bogoutdinov, S.R.; Belov, I.O. Strong-Earthquake-Prone Areas FDPS algorithm in stability assessment of the Earth's crust structural tectonic blocks. *Russ. J. Earth Sci.* **2020**, *20*, ES1005. [CrossRef]
25. Agayan, S.; Bogoutdinov, S.; Soloviev, A.; Sidorov, R. The Study of Time Series Using the DMA Methods and Geophysical Applications. *Data Sci. J.* **2016**, *15*, 16. [CrossRef]

applied sciences

MDPI

Article

Real-Time Adjustment and Spatial Data Integration Algorithms Combining Total Station and GNSS Surveys with an Earth Gravity Model

Krzysztof Karsznia [1,*], Edward Osada [2] and Zbigniew Muszyński [3]

[1] Faculty of Geodesy and Cartography, Warsaw University of Technology, Plac Politechniki 1, 00-661 Warszawa, Poland
[2] Faculty of Technical Sciences, University of Lower Silesia, Wagonowa St. 9, 53-609 Wrocław, Poland; edward.osada@dsw.edu.pl
[3] Faculty of Geoengineering, Mining and Geology, Wrocław University of Science and Technology, Wyb. Wyspiańskiego 27, 50-370 Wrocław, Poland; zbigniew.muszynski@pwr.edu.pl
* Correspondence: krzysztof.karsznia@pw.edu.pl

Featured Application: The authors have developed a spatial data-fusion algorithm to adjust and effectively integrate data sets from total station and GNSS surveys with a universal EGM geoid model. The solution's high effectiveness can be applied to structural monitoring, risk management, architectural designing, and building information modelling (BIM).

Abstract: During the dynamic development of modern technologies based on advanced algorithmic and instrumental solutions, it is essential to integrate geospatial data efficiently. Such an approach is applied in all geo-information services, especially mobile ones, and is helpful in, for example, precise navigation or effective risk management. One leading application is deformation monitoring (structural monitoring) and displacement control surveying. In addition, spatial data integration methods are used in modern accessibility analysis, Smart City ideas, tracing utility networks, and building information modelling (BIM). The last aforementioned technology plays a crucial role in architectural design and construction. In this context, it is crucial to develop efficient and accurate algorithms supporting data fusion, which do not strain the computing resources and operate efficiently online. This paper proposes an algorithm for real-time adjustment of integrated satellite GNSS (global navigation satellite system), total station, and Earth Gravitational Model (EGM) vertical direction data in a geocentric coordinate system based on a statistical general linear mixed model. A numerical example shows that the proposed algorithm of the online adjustment works correctly. The results of the online adjustment are the same as those of the offline adjustment. It is also shown that the GNSS measurements are necessary only at the total station points in the spatial total station traverse. There is no need to add additional merging points of the total station positions because the differences between the results of the online adjustment, including and excluding the merging points, are very small (around 1–2 mm in standard deviation).

Keywords: algorithms; data adjustment; data integration; integrated geodesy; structural monitoring

Citation: Karsznia, K.; Osada, E.; Muszyński, Z. Real-Time Adjustment and Spatial Data Integration Algorithms Combining Total Station and GNSS Surveys with an Earth Gravity Model. *Appl. Sci.* **2023**, *13*, 9380. https://doi.org/10.3390/app13169380

Academic Editors: Alexei Gvishiani and Boris Dzeboev

Received: 20 July 2023
Revised: 12 August 2023
Accepted: 15 August 2023
Published: 18 August 2023

1. Introduction

Integrating geospatial measurements is essential in modern information services and geoportals using precise navigation, inventory registries of technical infrastructure, and studying the ongoing deformations of engineering objects. Such geo-information platforms are now fundamental when conducting construction works on, e.g., subway lines, highway tunnels, or other technical facilities. They enable effective risk management and structure the work in progress. Nowadays, almost every civil investment uses geoportals and advanced algorithms, allowing for real-time data fusion. Such effective management of spatial data has increasingly been involving mobile technologies. In such cases, the activities of system analysts focus on the operation of smartphone or tablet devices, which requires the optimization of algorithmic solutions, thereby ensuring efficient switching between functions.

In the case of structural monitoring systems, notifications of potential threats in the form of predefined alarm thresholds are generated based on previously captured and properly integrated data. The user can receive such alerts in an ongoing form, except for simple visualization on screens, also employing signals, text messages, or sound calls. Reliable generation of the warnings mentioned above is required because property and often human lives depend on the speed of response to appearing circumstances. Moreover, such systems often use artificial intelligence and machine learning to initiate and aid appropriate reactions. The topic currently belongs to the mature systems and is reflected in many publications. For example, articles [1–4] present an overview of existing solutions in this field, explaining the theoretical basis in detail. The cited works also include an in-depth analysis of the available literature, considering new solutions and presenting them as complementary. In the context of spatial data acquisition, photogrammetric techniques—mainly images taken using UAVs (unmanned aerial vehicles) and close-range photogrammetry [5]—are used to a remarkable extent. Data-driven applications are used in mobile solutions, open visions, or complex services, e.g., Smart City concepts [6]. In this context, it is worth mentioning the so-called Internet of Things (IoT), through which multiple sources, especially mobile data, can be acquired and processed online [7,8].

However, this article focuses on analytical solutions to present a mechanism for effectively integrating spatial data obtained from different sensors, particularly geodetic observables. Such an approach makes it possible to visualize and interpret the survey results with millimeter accuracy. That, in turn, makes it possible to conduct assessments of geometric changes in engineering structures while examining their displacements and defining related risk management [9]. Numerous publications have also described research on the development of computational algorithms and data adjustment in integrated geodesy. Such works date even back to the late 1980s and early 1990s, for example [10,11]. In one author's solution, which was the topic of a doctoral dissertation and the results of which are presented in the article [12], a methodology was developed for the adjustment and integration of spatial traverses surveyed using total stationtechniques with reference to GNSS satellite measurements and the geoid. The measurement experiment was conducted both in urban areas and in the mountains. The total stationand GNSS results were further enriched by modelling the mutual plumb line arrangement in successive observation stations. Various polynomial functions were tested for this purpose, and numerical algorithms were assessed, including Levenberg–Marquardt's [13]. The use of numerical methods in landslide monitoring is also described in the work [14]. Depending on the type of landslide under study, an appropriate interpolation grid was defined in the test area, distinguishing sectors of the phenomenon's low, medium, and high activity. The level of activity then determines the selection of appropriate modelling parameters. The method was developed utilizing the total station data; however, it can be modified depending on other geodetic measurement technologies. For example, such integration can be performed by employing photogrammetric methods of correlating digital images [15].

Research on integrating total station and GNSS satellite measurements is also a vibrant subject amongst researchers. For example, the publication [16] presents and discusses study results using the methods above for deformation surveys of a large dam, obtaining promising results in the long-term evaluation of its technical condition. Article [17], on the other hand, demonstrates the analysis of a mathematical model for the adjustment and integration of geodetic measurements in the local spatial system, without the need for prior projection of GNSS vectors onto a reference ellipsoid. The author proposed a computational model that integrates data from two sources—total station and static GNSS measurements—and their joint adjustment in a single, coherent coordinate system. However, this approach considers any local geoid model. In addition, the implementation of GNSS measurements requires a maximally open horizon, and in the case of other than static positioning methods (e.g., Real-Time), also the fulfillment other conditions such as an adequate number of available satellites, reducing interfering elements (multipath mitigation), and many more [18].

As previously mentioned, photogrammetric methods based on appropriate modelling of the relative and absolute positions of image projection centers while ensuring adequate coverage of the pictured object [19] are currently the subject of exciting research. Based on many available studies in this area, the authors conducted research employing photogrammetry in precise object dimensioning and structural monitoring [20]. In this case, however, one can talk about pure local network-specific micro-grids with reference points deployed directly on the examined object. Similar studies can also be found in [21]. Here, for assessing the condition of historical objects, the authors proposed the integration of multi-source photogrammetric data, including historical photographs. In this context, the publication [22] is also noteworthy, in which the authors proposed using geomatics technologies for the digital reconstruction of historical objects currently undergoing renewal.

The above considerations motivated the authors to conduct studies on developing computational methods for integrating multi-source data processing, significantly increasing the accuracy and reliability of the results obtained. Due to the integration of GNSS measurements given directly in the geocentric reference frame and the vertical directions from the EGM model (also computed in the geocentric reference frame), this frame is used in the joint adjustment of the total station and GNSS measurements.

The effects of the relevant research are summarized and organized as follows: the literature review is provided in the Introduction, the theoretical basis of the conducted research is presented in Materials and Methods and its subsequent sections, and the results of experimental works are further discussed (Discussion) and concluded (Conclusions), showing some propositions for further research.

2. Materials and Methods

Based on theoretical assumptions, in this section, we present the workflow of the developed method.

2.1. General Assumptions

Three-dimensional models of terrestrial objects are built based on a set of points with coordinates (X, Y, Z) obtained using direct (total station measurements, 3D laser scanning) and indirect methods (close-range photogrammetry, remote sensing). The accuracy of the constructed model depends on the quality of the coordinates obtained. In the case of direct methods, the accuracy of a measured point Q (Figure 1) depends on sound knowledge of six total station or laser scanner external orientation parameters:

- X_s, Y_s, Z_s, the geocentric GRS80 coordinates of the origin P of the total stationtotal station or a terrestrial laser scanner (TLS) measuring frame (x, y, z);
- Σ, ξ, η, the orientation angles of the (x, y, z) measuring frame with respect to the external reference frame (X, Y, Z).

The ξ and η orientation angles are the components of the total station or TLS vertical axis deflection from the normal to the GRS80 ellipsoid. The directional horizontal angle S is called the instrument orientation constant.

The coordinates x, y, z of the point Q measured by the total station can be expressed using the known form [23,24]:

$$x = d \cdot \cos\alpha \cdot \sin\beta \tag{1}$$

$$y = d \cdot \sin\alpha \cdot \sin\beta \tag{2}$$

$$z = d \cdot \cos\beta + i - j \tag{3}$$

where d–spatial distance, α–horizontal direction, β–vertical angle (corrected due to refraction), i–the height of the total station above ground point P; and j–the reflector (surveying prism)'s height above ground point Q.

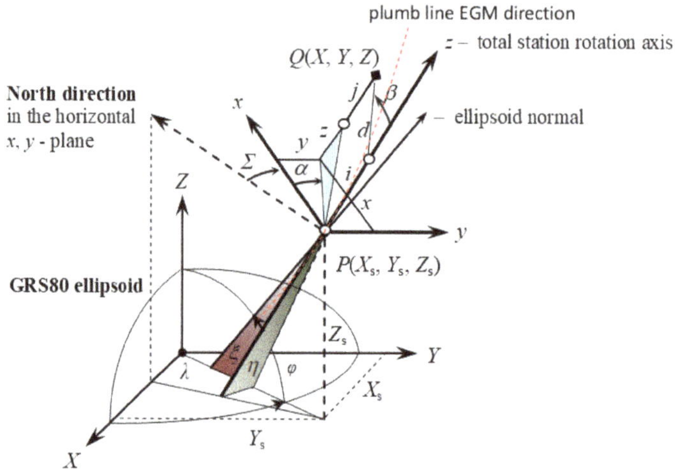

Figure 1. Geocentric vs. topocentric (instrumental) reference system: $X_s, Y_s, Z_s, \Sigma, \xi, \eta$ are six parameters of the total station external orientation; φ, λ are geodetic latitude and longitude; α, β are measured horizontal and vertical angles to the point $Q(X, Y, Z)$; d is the measured distance to the surveyed point Q; and i, j are the instrument and reflector heights.

The coordinates (x, y, z) are converted to the (X, Y, Z) external reference system of the point Q, according to the well-known formula [25]:

$$\begin{pmatrix} X \\ Y \\ Z \end{pmatrix} = \begin{pmatrix} X_S \\ Y_S \\ Z_S \end{pmatrix} + (\mathbf{R}(\Sigma)\mathbf{Q}(\xi, \eta, \varphi)\mathbf{P}(\varphi, \lambda))^T \cdot \begin{pmatrix} x \\ y \\ z \end{pmatrix} \quad (4)$$

where

$$\mathbf{P}(\varphi, \lambda) = \begin{pmatrix} -\sin\varphi\cos\lambda & -\sin\varphi\sin\lambda & \cos\varphi \\ -\sin\lambda & \cos\lambda & 0 \\ \cos\varphi\cos\lambda & \cos\varphi\sin\lambda & \sin\varphi \end{pmatrix} \quad (5)$$

$$\mathbf{Q}(\xi, \eta, \phi) = \begin{pmatrix} 1 & -\eta\tan\phi & -\xi \\ \eta\tan\phi & 1 & -\eta \\ \xi & \eta & 1 \end{pmatrix} \quad (6)$$

$$\mathbf{R}(\Sigma) = \begin{pmatrix} \cos\Sigma & \sin\Sigma & 0 \\ -\sin\Sigma & \cos\Sigma & 0 \\ 0 & 0 & 1 \end{pmatrix} \quad (7)$$

The five external total station orientation parameters $(X_s, Y_s, Z_s, \xi, \eta)$ can be obtained from GNSS measurements (X_s, Y_s, Z_s) and computed from the EGM gravity model [26]. The sixth parameter S can be determined by the solution of Equation (4) for the given total station measurements (d, α, β, i, j) and the coordinates from GNSS measurements at the total station point $P(X_s, Y_s, Z_s)$ and the target point $Q(X, Y, Z)$ using, e.g., the Levenberg–Marquardt method of conjugate gradients [27].

The integrated total station/GNSS/EGM geocentric points' positioning method can be applied to measurements on a single total station position, as well as on a few merged total station positions along a spatial traverse (Figure 2).

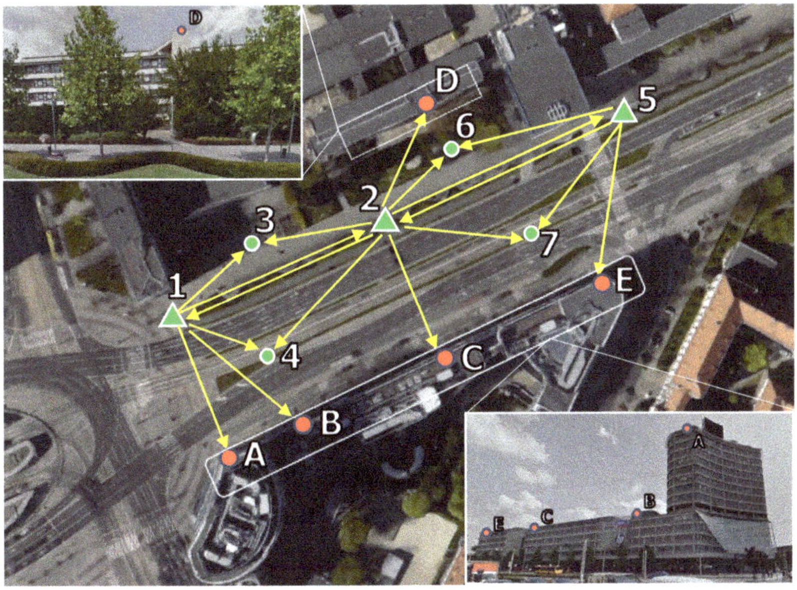

Figure 2. The total station spatial traverse points: the station points 1, 2, 5, the station merging points 3, 4, 6, 7, and the measured control points A, B, C, D, E, deployed on office buildings in Wroclaw (Poland): arrow pointers indicate the measured spatial directions (d, α, β) (picture by E. Osada, background: Google Maps). The station and merging points 1–7 are also measured using the GNSS-RTK technique.

The offline adjustment algorithm and the significance of the deflection of the vertical components (ξ, η) for the ultimate adjustment results are widely discussed in prior papers [28–30]. In the current work, we focus on the online procedures of adjusting the total station spatial traverse in real time. The GNSS and EGM data are used for spatial orientation of the total station traverse with respect to the geocentric reference frame. The online adjustment methods presented in this paper are based on the general linear–mixed statistical model, e.g., [31]. It is defined based on linearized observational equations of the integrated data, as explained in the following Section 2.2. Alternative formulations of the real-time sequential adjustment algorithm can be found in papers showing applications of the Kalman filter and sequential regularization methods, e.g., [9,32], and in the case of the Moore–Penrose pseudo-inverse in [33].

2.2. The Observational Equations of the Integrated Data

The linear observational equation of the coordinates vector determined using total station measurements (formulas (1)–(3)) is given by

$$\begin{bmatrix} x+\varepsilon_x \\ y+\varepsilon_y \\ z+\varepsilon_z \end{bmatrix} = \begin{bmatrix} x_0+dx \\ y_0+dy \\ z_0+dz \end{bmatrix} \tag{8}$$

Hence,

$$\begin{bmatrix} \varepsilon_x \\ \varepsilon_y \\ \varepsilon_z \end{bmatrix} = \begin{bmatrix} dx \\ dy \\ dz \end{bmatrix} - \begin{bmatrix} x-x_s \\ y-y_s \\ z-z_s \end{bmatrix} \tag{9}$$

where

- dx, dy, dz are differentials of the coordinates x, y, z;

- $\varepsilon_x, \varepsilon_y, \varepsilon_z$ are the measurement random errors with zero expected value and known covariance matrix:

$$\begin{bmatrix} \sigma_x^2 & \sigma_{xy} & \sigma_{xz} \\ \sigma_{xy} & \sigma_y^2 & \sigma_{yz} \\ \sigma_{xz} & \sigma_{yz} & \sigma_z^2 \end{bmatrix} \tag{10}$$

- x_s, y_s, z_s are approximate values of the measured coordinates:

$$\begin{bmatrix} x_s \\ y_s \\ z_s \end{bmatrix} = \mathbf{R}(\Sigma)\mathbf{Q}(\xi,\eta,\varphi)\mathbf{P}(\varphi,\lambda)\cdot \begin{bmatrix} X_c - X \\ Y_c - Y \\ Z_c - Z \end{bmatrix} \tag{11}$$

The elements of the covariance matrix (10), which stand for the functions of the direct observables (1)–(3), are computed using the covariance propagation law [34]:

$$\sigma_x = \sqrt{\sin^2\beta\cos^2\alpha\cdot\sigma_d^2 + d^2\cos^2\beta\cos^2\alpha\cdot\sigma_\beta^2 + d^2\sin^2\beta\sin^2\alpha\cdot\sigma_\alpha^2} \tag{12}$$

$$\sigma_y = \sqrt{\sin^2\beta\sin^2\alpha\cdot\sigma_d^2 + d^2\cos^2\beta\sin^2\alpha\cdot\sigma_\beta^2 + d^2\sin^2\beta\cos^2\alpha\cdot\sigma_\alpha^2} \tag{13}$$

$$\sigma_z = \sqrt{\cos^2\beta\cdot\sigma_d^2 + d^2\cdot\sin^2\beta\cdot\sigma_\beta^2 + \sigma_i^2 + \sigma_j^2} \tag{14}$$

$$\sigma_{xy} = \frac{1}{2}\sin 2\alpha\cdot\left(\sin^2\beta\cdot\sigma_d^2 + d^2\cos^2\beta\cdot\sigma_\beta^2 - d^2\cdot\sin^2\beta\cdot\sigma_\alpha^2\right) \tag{15}$$

$$\sigma_{xz} = \sin\beta\cos\beta\cos\alpha\cdot\left(\sigma_d^2 - d^2\cdot\sigma_\beta^2\right) \tag{16}$$

$$\sigma_{yz} = \sin\beta\cos\beta\sin\alpha\cdot\left(\sigma_d^2 - d^2\cdot\sigma_\beta^2\right) \tag{17}$$

where $\sigma_d, \sigma_\alpha, \sigma_\beta, \sigma_i, \sigma_j$ are the standard deviations of the directly measured values d, α, β, i, j.

The differential change of the total station coordinates equal the following:

$$\begin{bmatrix} x \\ y \\ z \end{bmatrix} = \mathbf{R}(\Sigma)\mathbf{Q}(\xi,\eta,\varphi)\mathbf{P}(\varphi,\lambda)\cdot \begin{bmatrix} X_c - X \\ Y_c - Y \\ Z_c - Z \end{bmatrix} \tag{18}$$

due to differential changes of the total station standpoint coordinates X, Y, Z, the target point coordinates X_c, Y_c, Z_c and total station spatial orientation angles Σ, ξ, η are given by

$$\begin{aligned} \begin{pmatrix} dx \\ dy \\ dz \end{pmatrix} &= \mathbf{R}(\Sigma)\mathbf{Q}(\xi,\eta,\phi)\mathbf{P}(\phi,\lambda)\cdot \begin{bmatrix} 1 & & \\ & 1 & \\ & & 1 \end{bmatrix} \begin{pmatrix} dX_c - dX \\ dY_c - dY \\ dZ_c - dZ \end{pmatrix} + \mathbf{R}(\Sigma)\frac{\partial\mathbf{Q}(\xi,\eta,\phi)}{d\xi}\mathbf{P}(\phi,\lambda)\begin{pmatrix} X_c - X \\ Y_c - Y \\ Z_c - Z \end{pmatrix} d\xi \\ &+ \mathbf{R}(\Sigma)\frac{\partial\mathbf{Q}(\xi,\eta,\phi)}{d\eta}\mathbf{P}(\phi,\lambda)\begin{pmatrix} X_c - X \\ Y_c - Y \\ Z_c - Z \end{pmatrix} d\eta + \frac{\partial\mathbf{R}(\Sigma)}{\partial\Sigma}\mathbf{Q}(\xi,\eta,\phi)\mathbf{P}(\phi,\lambda)\begin{pmatrix} X_c - X \\ Y_c - Y \\ Z_c - Z \end{pmatrix} d\Sigma \\ &+ \mathbf{R}(\Sigma)\frac{\partial[\mathbf{Q}(\xi,\eta,\phi)\mathbf{P}(\phi,\lambda)]}{d\phi}\begin{pmatrix} X_c - X \\ Y_c - Y \\ Z_c - Z \end{pmatrix} d\phi + \mathbf{R}(\Sigma)\frac{\partial[\mathbf{Q}(\xi,\eta,\phi)\mathbf{P}(\phi,\lambda)]}{d\lambda}\begin{pmatrix} X_c - X \\ Y_c - Y \\ Z_c - Z \end{pmatrix} d\lambda \end{aligned} \tag{19}$$

where the differentials $d\varphi, d\lambda$ are functions of the differentials dX, dY, dZ [35]:

$$\begin{pmatrix} d\phi \\ d\lambda \\ dh \end{pmatrix} = \begin{bmatrix} \frac{1}{M+h} & 0 & 0 \\ 0 & \frac{1}{(N+h)\cos\phi} & 0 \\ 0 & 0 & 1 \end{bmatrix} \mathbf{P}(\phi,\lambda)\begin{pmatrix} dX \\ dY \\ dZ \end{pmatrix} \tag{20}$$

and M, N are the radii of curvature of the ellipsoid in the meridian and in the prime vertical, respectively.

The rotation (Σ, ξ, η) of the total station reference frame (x, y, z) due to differential displacement $d\varphi, d\lambda$ of the total station point is very small and can be neglected. Thus, the terms containing differentials $d\varphi, d\lambda$ can be omitted, and Equation (19) is reduced to

$$\begin{pmatrix} dx \\ dy \\ dz \end{pmatrix} = \mathbf{R}(\Sigma)\mathbf{Q}(\xi,\eta,\phi)\mathbf{P}(\phi,\lambda)\begin{bmatrix} 1 & & \\ & 1 & \\ & & 1 \end{bmatrix}\begin{pmatrix} dX_c - dX \\ dY_c - dY \\ dZ_c - dZ \end{pmatrix} + \mathbf{R}(\Sigma)\frac{\partial \mathbf{Q}(\xi,\eta,\phi)}{d\xi}\begin{pmatrix} x_g \\ y_g \\ z_g \end{pmatrix} d\xi$$
$$+\mathbf{R}(\Sigma)\frac{\partial \mathbf{Q}(\xi,\eta,\phi)}{d\eta}\begin{pmatrix} x_g \\ y_g \\ z_g \end{pmatrix} d\eta + \frac{\partial \mathbf{R}(\Sigma)}{\partial \Sigma}\mathbf{Q}(\xi,\eta,\phi)\begin{pmatrix} x_g \\ y_g \\ z_g \end{pmatrix} d\Sigma \tag{21}$$

where

$$\begin{pmatrix} x_g \\ y_g \\ z_g \end{pmatrix} = \mathbf{P}(\varphi,\lambda)\begin{pmatrix} X_c - X \\ Y_c - Y \\ Z_c - Z \end{pmatrix} \tag{22}$$

and

$$\frac{\partial \mathbf{Q}(\xi,\eta,\varphi)}{d\xi} = \begin{bmatrix} 0 & 0 & -1 \\ 0 & 0 & 0 \\ 1 & 0 & 0 \end{bmatrix} \tag{23}$$

$$\frac{\partial \mathbf{Q}(\xi,\eta,\varphi)}{d\eta} = \begin{bmatrix} 0 & -\tan\varphi & 0 \\ \tan\varphi & 0 & -1 \\ 0 & 1 & 0 \end{bmatrix} \tag{24}$$

$$\frac{\partial \mathbf{R}(\Sigma)}{\partial \Sigma} = \begin{bmatrix} -\sin\Sigma & \cos\Sigma & 0 \\ -\cos\Sigma & -\sin\Sigma & 0 \\ 0 & 0 & 0 \end{bmatrix} \tag{25}$$

Finally, the reduced linearized observational equation of the total station-measured coordinates x, y, z (8) and (9) is given by the following formula:

$$\begin{pmatrix} \varepsilon_x \\ \varepsilon_y \\ \varepsilon_z \end{pmatrix} = \begin{bmatrix} -\sin\lambda\sin\Sigma - \sin\phi\cos\lambda\cos\Sigma & \cos\lambda\sin\Sigma - \sin\phi\sin\lambda\cos\Sigma & \cos\phi\cos\Sigma \\ -\sin\lambda\cos\Sigma + \sin\phi\cos\lambda\sin\Sigma & \cos\lambda\cos\Sigma + \sin\phi\sin\lambda\sin\Sigma & -\cos\phi\sin\Sigma \\ \cos\phi\cos\lambda & \cos\phi\sin\lambda & \sin\phi \end{bmatrix}\begin{pmatrix} dX_c - dX \\ dY_c - dY \\ dZ_c - dZ \end{pmatrix}$$
$$+\begin{bmatrix} -z_g\cos\Sigma & x_g\sin\Sigma\tan\phi - y_g\cos\Sigma\tan\phi - z_g\sin\Sigma & -x_g\sin\Sigma + y_g\cos\Sigma \\ z_g\sin\Sigma & x_g\cos\Sigma\tan\phi + y_g\sin\Sigma\tan\phi - z_g\cos\Sigma & -x_g\cos\Sigma - y_g\sin\Sigma \\ x_g & y_g & 0 \end{bmatrix}\begin{bmatrix} d\xi \\ d\eta \\ d\Sigma \end{bmatrix} - \begin{pmatrix} x - x_0 \\ y - y_0 \\ z - z_0 \end{pmatrix} \tag{26}$$

The linear observational equation of the vector of geocentric coordinates X, Y, Z (obtained from GNSS measurements) is given by

$$\begin{pmatrix} X + \varepsilon_X \\ Y + \varepsilon_Y \\ Z + \varepsilon_Z \end{pmatrix} = \begin{pmatrix} X_s + dX \\ Y_s + dY \\ Z_s + dZ \end{pmatrix} \Rightarrow \begin{pmatrix} \varepsilon_X \\ \varepsilon_Y \\ \varepsilon_Z \end{pmatrix} = \begin{pmatrix} dX \\ dY \\ dZ \end{pmatrix} - \begin{pmatrix} X - X_0 \\ Y - Y_0 \\ Z - Z_0 \end{pmatrix} \tag{27}$$

where dX, dY, dZ are corrections of the coordinates X, Y, Z; X_s, Y_s, Z_s are their approximate values, and $\varepsilon_x, \varepsilon_y, \varepsilon_z$ are measurements random errors with zero expected value and a known covariance matrix.

$$\begin{bmatrix} \sigma_X^2 & \sigma_{XY} & \sigma_{XZ} \\ \sigma_{XY} & \sigma_Y^2 & \sigma_{YZ} \\ \sigma_{XZ} & \sigma_{YZ} & \sigma_Z^2 \end{bmatrix} \tag{28}$$

The linear observational equations of the vertical deflection components ξ, η are given by

$$\begin{pmatrix} \xi + \varepsilon_\xi \\ \eta + \varepsilon_\eta \end{pmatrix} = \begin{pmatrix} \xi_0 + d\xi \\ \eta_0 + d\eta \end{pmatrix} \Rightarrow \begin{pmatrix} \varepsilon_\xi \\ \varepsilon_\eta \end{pmatrix} = \begin{pmatrix} d\xi \\ d\eta \end{pmatrix} - \begin{pmatrix} \xi - \xi_0 \\ \eta - \eta_0 \end{pmatrix} \quad (29)$$

where $d\xi, d\eta$ are corrections of the vertical component deflection ξ, η; ξ_0, η_0 are their approximate values, and $\varepsilon_\xi, \varepsilon_\eta$ are measurement random errors with zero expected value and a covariance matrix:

$$\begin{bmatrix} \sigma_\xi^2 & \sigma_{\xi\eta} \\ \sigma_{\xi\eta} & \sigma_\eta^2 \end{bmatrix} \quad (30)$$

2.3. *Adjustment of the Integrated Data at the Current Position of the Total Station Using the Statistical General Linear Mixed Model*

The statistical general linear mixed model (e.g., [31]) for all possible data measured at the current total station position $(d, \alpha, \beta, X, Y, Z, \xi, \eta)$, represented by linear observational Equations (8)–(17), including corrections to all the parameters adjusted at the previous total station position, is defined as

$$\boldsymbol{\varepsilon} = \mathbf{X}\boldsymbol{\beta} + \mathbf{U}\boldsymbol{\gamma} - \mathbf{y} \quad (31)$$

where

$$\begin{bmatrix} \gamma \\ \varepsilon \end{bmatrix} \sim \left(\begin{bmatrix} 0 \\ 0 \end{bmatrix}, \begin{bmatrix} \boldsymbol{\Sigma}_\gamma & 0 \\ 0 & \boldsymbol{\Sigma}_\varepsilon \end{bmatrix} \right) \quad (32)$$

In this model, X and U are known design matrices, y is a known data vector, β is a vector of unknown corrections to all new parameters included at the current position of the total station, $\gamma \sim (\mathbf{0}, \boldsymbol{\Sigma}_\gamma)$ is a vector of unknown corrections to all parameters adjusted at previous total-station positions with zero expected value, and the known covariance matrix $\boldsymbol{\Sigma}_\gamma$, $\varepsilon \sim (\mathbf{0}, \boldsymbol{\Sigma}_\varepsilon)$ is a vector of unknown observational residuals with zero-expected value and a known covariance matrix $\boldsymbol{\Sigma}_\varepsilon$.

The least squares solution $\varepsilon\boldsymbol{\Sigma}_\varepsilon^{-1}\varepsilon^{\mathrm{T}} + \gamma\boldsymbol{\Sigma}_\gamma^{-1}\gamma^{\mathrm{T}} = \min$ of the mixed model $\varepsilon = \mathbf{X}\boldsymbol{\beta} + \mathbf{U}\boldsymbol{\gamma} - \mathbf{y}$ (31) is given by [31]

$$\hat{\boldsymbol{\beta}} = \left(\mathbf{X}^T\boldsymbol{\Sigma}_{\mathbf{y}}{}^{-1}\mathbf{X}\right)^{-1}\mathbf{X}^T\boldsymbol{\Sigma}_{\mathbf{y}}{}^{-1}\mathbf{y} \quad (33)$$

$$\hat{\boldsymbol{\gamma}} = \boldsymbol{\Sigma}_\gamma\mathbf{U}^T\boldsymbol{\Sigma}_{\mathbf{y}}{}^{-1}(\mathbf{y} - \mathbf{X}\hat{\boldsymbol{\beta}}) \quad (34)$$

where

$$\boldsymbol{\Sigma}_{\mathbf{y}} = \boldsymbol{\Sigma}_\varepsilon + \mathbf{U}\boldsymbol{\Sigma}_\gamma\mathbf{U}^T \quad (35)$$

Or, in a block matrix notation:

$$\begin{bmatrix} \hat{\boldsymbol{\beta}} \\ \hat{\boldsymbol{\gamma}} \end{bmatrix} = \begin{bmatrix} \boldsymbol{\Sigma}_{\hat{\beta}} & \boldsymbol{\Sigma}_{\hat{\beta}\hat{\gamma}} \\ \boldsymbol{\Sigma}_{\hat{\beta}\hat{\gamma}}^T & \boldsymbol{\Sigma}_{\hat{\gamma}} \end{bmatrix} \begin{bmatrix} \mathbf{X}^T\boldsymbol{\Sigma}_\varepsilon^{-1}\mathbf{y} \\ \mathbf{U}^T\boldsymbol{\Sigma}_\varepsilon^{-1}\mathbf{y} \end{bmatrix} \quad (36)$$

where

$$\begin{bmatrix} \boldsymbol{\Sigma}_{\hat{\beta}} & \boldsymbol{\Sigma}_{\hat{\beta}\hat{\gamma}} \\ \boldsymbol{\Sigma}_{\hat{\beta}\hat{\gamma}}^T & \boldsymbol{\Sigma}_{\hat{\gamma}} \end{bmatrix} = \begin{bmatrix} \mathbf{X}^T\boldsymbol{\Sigma}_\varepsilon^{-1}\mathbf{X} & \mathbf{X}^T\boldsymbol{\Sigma}_\varepsilon^{-1}\mathbf{U} \\ \mathbf{U}^T\boldsymbol{\Sigma}_\varepsilon^{-1}\mathbf{X} & \mathbf{U}^T\boldsymbol{\Sigma}_\varepsilon^{-1}\mathbf{U} + \boldsymbol{\Sigma}_\gamma^{-1} \end{bmatrix}^{-1} \quad (37)$$

$$\boldsymbol{\Sigma}_{\hat{\beta}} = \left(\mathbf{X}^T\boldsymbol{\Sigma}_{\mathbf{y}}^{-1}\mathbf{X}\right)^{-1} \quad (38)$$

$$\boldsymbol{\Sigma}_{\hat{\beta}\hat{\gamma}} = -\boldsymbol{\Sigma}_{\hat{\beta}}\mathbf{X}^T\boldsymbol{\Sigma}_{\mathbf{y}}^{-1}\mathbf{U}\boldsymbol{\Sigma}_\gamma \quad (39)$$

$$\boldsymbol{\Sigma}_{\hat{\gamma}} = \left(\mathbf{U}^T\boldsymbol{\Sigma}_\varepsilon^{-1}\mathbf{U} + \boldsymbol{\Sigma}_\gamma^{-1}\right)^{-1} - \boldsymbol{\Sigma}_{\hat{\beta}\hat{\gamma}}^T\mathbf{X}^T\boldsymbol{\Sigma}_{\mathbf{y}}^{-1}\mathbf{U}\boldsymbol{\Sigma}_\gamma \quad (40)$$

is the covariance matrix of the adjusted parameters $\hat{\boldsymbol{\beta}}$ and $\hat{\boldsymbol{\gamma}}$.

2.4. Experimental Works

The experimental works were performed in Wrocław, Poland. The proposed algorithms were tested on data obtained during measurements at the points of the spatial traverse (Figure 2). At the total station traverse positions 1, 2, 5, there were spatial distances s, horizontal and vertical angles α, β (Figure 1), surveyed with standard deviations σ_d = 0.006 m, $\sigma_\alpha = \sigma_\beta$ = 0.0010 grad, and the instrument height i and reflector (prism) height j, with $\sigma_i = \sigma_j$ = 0.002 m. Measurements were performed with the use of the Leica FlexLine TS02 total station [36]. The X, Y, Z coordinates of the station points 1, 2, 5 and merging points 3, 4, 6, 7 were measured with a Leica GS10 GNSS receiver [37] with standard deviations $\sigma_X = \sigma_Y = \sigma_Z$ = 0.008 m. At each point, we performed real-time kinematic (RTK) GNSS surveys with direct differential corrections broadcast by the GNSS permanent station WROC, located at the Wrocław University of Environmental and Life Sciences. The observation data can be downloaded from http://www.asgeupos.pl/index.php?wpg_type=syst_descr&sub=ref_st&st_id=wroc (accessed on 18 July 2023). The distance from the mobile receiver to the reference station WROC was less than 250 m.

The northern ξ and eastern η components of the deflection of the total station vertical axis from normal to the GRS80 ellipsoid were computed at the points 1, 2, 5 from the EGM2008 geoid model data, available from https://www.usna.edu/Users/oceano/pguth/md_help/html/egm96.htm (accessed on 18 July 2023) and presented in Table 1.

Table 1. Northern and eastern components of the plumb line deviations in the total station stand points (EGM2008), see Figure 1.

Point No.	ξ-Component [arc sec ″]	η-Component [arc sec ″]	Σ [grad]
1	5.9926	6.2033	209.0750
2	5.9852	6.1967	296.7144
5	5.9775	6.1896	50.2572

The approximate value Σ_1 was computed by solving Equation (4) for GNSS-surveyed point coordinates X, Y, Z of the points 1 and 2, and x, y, z coordinates of the point 2 measured using total station set up in point 1. Analogically, we computed the values Σ_2 and Σ_5.

According to information from total station manufacturers, the direction of the vertical axis of an instrument is consistent with the real plumb line direction to the order of 0.5″–2.0″ [arc seconds]. So, in the adjustment process, the standard deviations of the components of deflection of the vertical are assumed to be 1″ ($\sigma_\xi = \sigma_\eta$ = 0.0003 [grad]). The covariance $\sigma_{\xi\eta}$ is unknown; it is assumed as 0 in the matrix (31). Research has shown [38] that in mountainous areas, in the Alps, the actual direction of the vertical can differ up to 3.5″ from the direction of the vertical computed in the EGM2008 model. This error is similar to the accuracy of the vertical angles used in the experiment's total-station instrument (3.0″). In the case of unique high-accuracy measurements using total stations in mountain areas, the accuracy of the vertical deflection computed from EGM2008 model can be improved up to 0.8″ using a correction computed from the numerical terrain model [38].

An alternative method of joint total station and GPS positioning with the use of digital terrain and gravity models is given in the paper [39]. In this case, the spherical model of the local vertical distribution is extended using local digital models of the terrain and gravity. Then, the residual field of the vertical deflection, based on spherical harmonic polynomials' expansion, is determined via real-time adjustment of the total station and GPS data.

A method of adjustment of the total station data in real-time, assuming the spherical distribution of the local vertical, is given in [33]. In this case, the relative position vectors and height differences measured with total station along a traverse are adjusted in real-time in three independent modes: spatial, planar, and height adjustment. If there are some relative position GPS vectors, GPS ellipsoidal height differences, and leveling height differences, they are included in the adjusting processes at the current position of the total station. In the case of the height adjustment, it is assumed that the equipotential surfaces are spherical in the local area. The adjustments are made using a recursive formulation of the least squares method featuring the Moore–Penrose pseudoinverse.

In the method considered in this article, the accuracy and spatial orientation of the total station traverse can also be improved, including additional measured height differences between the traverse points using a leveling instrument. The relationship between ellipsoidal and leveling orthometric or normal heights, which has been studied and developed in many publications, e.g., [40–47], should be considered.

2.4.1. Classical Offline Adjustment

The classical offline adjustment of the total station traverse is performed, having completed all measurements. In our numerical experiment, it has been performed in two variants: by "including" and "excluding" the merging points 3, 4, 6, 7 (Figure 2). In the first case, the traverse consists of three stations (1, 2, and 5) and four merging points (3, 4, 6, and 7), with coordinates determined using a GNSS receiver, and five measured control points on the high buildings A, B, C, D, and E. In the second case, the traverse consists of only three station points (1, 2, and 5), as determined by the GNSS receiver, and five measured control points (A, B, C, D, and E).

The differences between the geocentric coordinates of all points obtained from the "including" and "excluding" computation modes are not greater than 4.2 mm, with a mean value of 1.0 mm and a standard deviation of ±1.4 mm. Comparing these with the accuracy of the total station and GNSS measurements, we can conclude that the merging points are not necessary for the spatial orientation of the total station traverse. As we can see, the spatial orientation is very well provided by only the GNSS-measured coordinates of the total station points 1, 2, and 5, and the deflections of the vertical obtained from the EGM model.

The algorithms and results of the classical offline adjustment obtained via the measured data in the same experiment shown in Figure 2 are presented in the article [28].

An alternative to the offline adjustment of all observations after making measurements is to adjust the observations online directly during measurements, in real-time, and sequentially.

In Section 2.4.2, we present the online algorithm for sequential adjustment of the sets of observations performed at successive total station positions.

In Section 2.4.3, we present the online algorithm for sequential adjustment of the direct observations performed at successive total station positions.

Online adjustment of observations provides real-time information about the accuracy of the currently measured point, and about improving the accuracy of all previously surveyed points. Thus, it is possible to conduct additional observations in real-time, and select new total station positions to increase the accuracy of the measured points. These are the main practical advantages of the real-time online sequential adjustment method, in comparison with the offline method.

2.4.2. Sequential Online Adjustments of Total Station Positions

At the first total station position over the point 1 (see Figures 2 and 3), the observational–mixed model is given by the equation

$$\boldsymbol{\varepsilon} = \mathbf{X}\boldsymbol{\beta} - \mathbf{y} \tag{41}$$

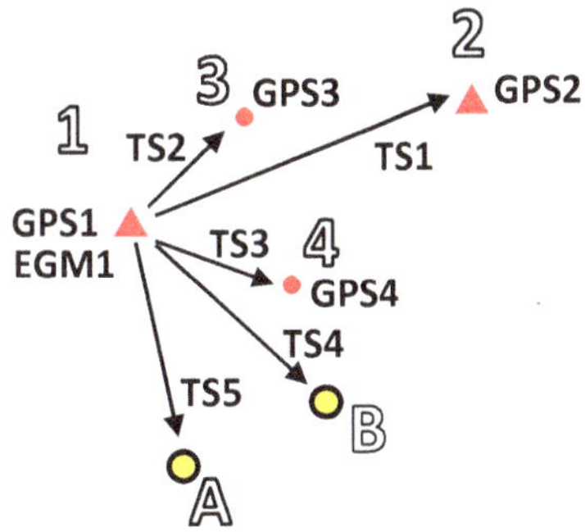

Figure 3. Surveying geometry at the total station position 1. The numbered points marked TS relate to total station surveys; GPS are measured using GNSS (GPS) technology, whereas EGM stands for the Earth Gravitational Model.

$$\begin{bmatrix} \boldsymbol{\varepsilon}_{TS1} \\ \boldsymbol{\varepsilon}_{TS2} \\ \boldsymbol{\varepsilon}_{TS3} \\ \boldsymbol{\varepsilon}_{TS4} \\ \boldsymbol{\varepsilon}_{TS5} \\ \hline \boldsymbol{\varepsilon}_{GPS1} \\ \boldsymbol{\varepsilon}_{GPS2} \\ \boldsymbol{\varepsilon}_{GPS3} \\ \boldsymbol{\varepsilon}_{GPS4} \\ \hline \boldsymbol{\varepsilon}_{EGM1} \end{bmatrix} = \mathbf{X} \begin{bmatrix} \mathbf{d}_1 \\ \mathbf{d}_2 \\ \mathbf{d}_3 \\ \mathbf{d}_4 \\ \mathbf{d}_A \\ \mathbf{d}_B \\ \hline d\xi_1 \\ d\eta_1 \\ d\Sigma_1 \end{bmatrix} - \mathbf{y} \tag{42}$$

where

- $\boldsymbol{\beta} = \begin{bmatrix} \mathbf{d}_1^T & \mathbf{d}_2^T & \mathbf{d}_3^T & \mathbf{d}_4^T & \mathbf{d}_A^T & \mathbf{d}_B^T & | & d\xi_1 & d\eta_1 & d\Sigma_1 \end{bmatrix}^T$ is the vector of corrections to the approximate geocentric coordinates X, Y, Z of the points 1, 2, 3, 4, A, B (e.g., $\mathbf{d}_1^T = \begin{bmatrix} dX_1 & dY_1 & dZ_1 \end{bmatrix}$) and corrections $d\xi_1, d\eta_1, d\Sigma_1$ to the approximate orientation angles of the total station ξ_1, η_1, Σ_1,
- $\boldsymbol{\varepsilon} = \begin{bmatrix} \boldsymbol{\varepsilon}_{TS1}^T & \boldsymbol{\varepsilon}_{TS2}^T & \boldsymbol{\varepsilon}_{TS3}^T & \boldsymbol{\varepsilon}_{TS4}^T & \boldsymbol{\varepsilon}_{TS5}^T & | & \boldsymbol{\varepsilon}_{GPS1}^T & \boldsymbol{\varepsilon}_{GPS2}^T & \boldsymbol{\varepsilon}_{GPS3}^T & \boldsymbol{\varepsilon}_{GPS4}^T & | & \boldsymbol{\varepsilon}_{EGM1}^T \end{bmatrix}^T$ is the vector of total station measurement random errors $\varepsilon_x, \varepsilon_y, \varepsilon_z$ of the x, y, z coordinates of the points 2, 3, 4, B, A, respectively (e.g., $\boldsymbol{\varepsilon}_{TS1}^T = \begin{bmatrix} \varepsilon_{x_1} & \varepsilon_{y_1} & \varepsilon_{z_1} \end{bmatrix}$); the GNSS measurements' random errors $\varepsilon_X, \varepsilon_Y, \varepsilon_Z$ of the X, Y, Z coordinates of the points 1, 2, 3, 4, respectively (e.g., $\boldsymbol{\varepsilon}_{GPS1}^T = \begin{bmatrix} \varepsilon_{X_1} & \varepsilon_{Y_1} & \varepsilon_{Z_1} \end{bmatrix}$); and the random errors $\varepsilon_\xi, \varepsilon_\eta$ of the deflection of the vertical components ξ_1, η_1 at the first total station position (at point 1): $\boldsymbol{\varepsilon}_{EGM1}^T = \begin{bmatrix} \varepsilon_{\xi_1} & \varepsilon_{\eta_1} \end{bmatrix}$.
- **X, y** is the known design matrix and free terms' vector.

The solution $\hat{\boldsymbol{\beta}}$, $\boldsymbol{\Sigma}_{\hat{\boldsymbol{\beta}}}$ of the observational model $\boldsymbol{\varepsilon} = \mathbf{X}\boldsymbol{\beta} - \mathbf{y}$ is given by Equations (33)–(40).

In the next total station standpoint 2 (see Figures 2 and 4), the observational–mixed model $\boldsymbol{\varepsilon} = \mathbf{X}\boldsymbol{\beta} + \mathbf{U}\boldsymbol{\gamma} - \mathbf{y}$ is given by

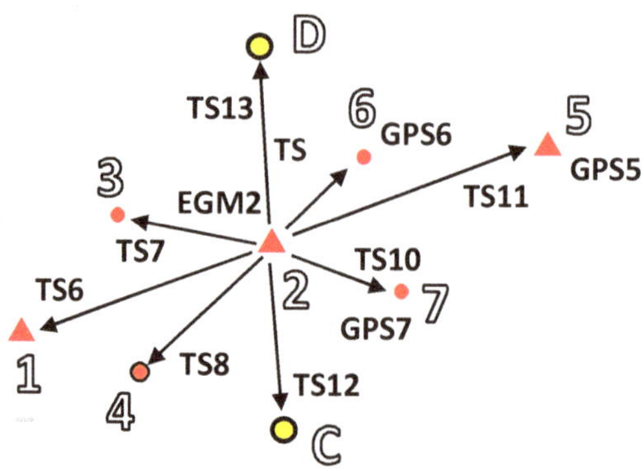

Figure 4. Surveying geometry at the total station position 2. The numbered points marked as TS relate to total station surveys; GPS are measured using GNSS (GPS) technology, whereas EGM stands for the Earth Gravitational Model.

$$\begin{bmatrix} \varepsilon_{TS6} \\ \varepsilon_{TS7} \\ \varepsilon_{TS8} \\ \varepsilon_{TS9} \\ \varepsilon_{TS10} \\ \varepsilon_{TS11} \\ \varepsilon_{TS12} \\ \varepsilon_{TS13} \\ \hline \varepsilon_{GPS5} \\ \varepsilon_{GPS6} \\ \varepsilon_{GPS7} \\ \hline \varepsilon_{EGM2} \end{bmatrix} = \mathbf{X} \begin{bmatrix} \mathbf{d}_5 \\ \mathbf{d}_6 \\ \mathbf{d}_7 \\ \mathbf{d}_C \\ \mathbf{d}_D \\ \hline d\xi_2 \\ d\eta_2 \\ d\Sigma_2 \end{bmatrix} + \mathbf{U} \begin{bmatrix} \mathbf{d}_1 \\ \mathbf{d}_2 \\ \mathbf{d}_3 \\ \mathbf{d}_4 \\ \mathbf{d}_A \\ \mathbf{d}_B \\ \hline d\xi_1 \\ d\eta_1 \\ d\Sigma_1 \end{bmatrix} - \mathbf{y} \tag{43}$$

where

- $\gamma = \begin{bmatrix} \mathbf{d}_1^T & \mathbf{d}_2^T & \mathbf{d}_3^T & \mathbf{d}_4^T & \mathbf{d}_A^T & \mathbf{d}_B^T & | & d\xi_1 & d\eta_1 & d\Sigma_1 \end{bmatrix}^T$ is the vector of unknown corrections to all parameters adjusted on the first total station position, with zero expected value and a known covariance matrix $\boldsymbol{\Sigma}_\gamma = \boldsymbol{\Sigma}_{\hat{\beta}}$.
- $\beta = \begin{bmatrix} \mathbf{d}_5^T & \mathbf{d}_6^T & \mathbf{d}_7^T & \mathbf{d}_C^T & \mathbf{d}_D^T & | & d\xi_2 & d\eta_2 & d\Sigma_2 \end{bmatrix}^T$ is the vector of corrections dX, dY, dZ to the approximate geocentric coordinates X, Y, Z of the new points 5, 6, 7, C, D included in the model at the second total station position, and corrections $d\xi_2, d\eta_2, d\Sigma_2$ to the total station orientation angles;
- $\varepsilon = \begin{bmatrix} \varepsilon_{TS6}^T & \varepsilon_{TS7}^T & \cdots & \varepsilon_{TS13}^T & | & \varepsilon_{GPS5}^T & \varepsilon_{GPS6}^T & \varepsilon_{GPS7}^T & | & \varepsilon_{EGM3}^T \end{bmatrix}^T$ is the vector of total station measurement random errors $\varepsilon_x, \varepsilon_y,\ \varepsilon_z$ of the x, y, z coordinates of the points 1, 3, 4, 6, 7, 5, C, D, respectively; the GNSS measurement's random errors $\varepsilon_X, \varepsilon_Y,\ \varepsilon_Z$ of the X, Y, Z coordinates of the points 5, 6, 7, respectively; and the random errors $\varepsilon_\xi, \varepsilon_\eta$ of the deflection of the vertical components ξ_2, η_2 at the second total station position (at point 2): $\varepsilon_{EGM2}^T = \begin{bmatrix} \varepsilon_{\xi_2} & \varepsilon_{\eta_2} \end{bmatrix}$;
- **X**, **y** is the known design matrix and the free terms vector.

The solution of the observational–mixed model at the second total station position $\varepsilon = \mathbf{X}\beta + \mathbf{U}\gamma - \mathbf{y}$ is given by Equations (33)–(40).

At the third total station position over the point 5 (see Figures 2 and 5), the observational-mixed model $\varepsilon = \mathbf{X}\beta + \mathbf{U}\gamma - \mathbf{y}$ is given by

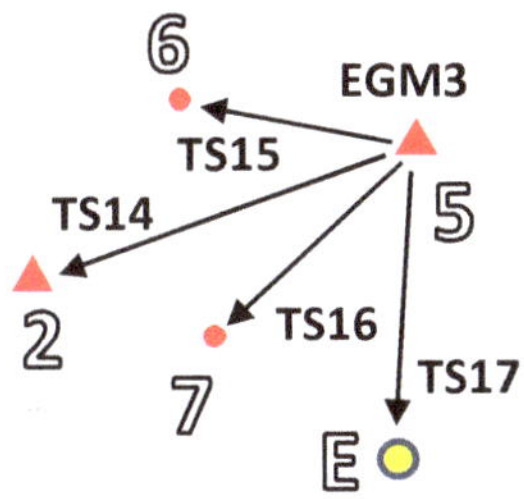

Figure 5. Surveying geometry at the total station position 5. The numbered points marked as TS relate to total station surveys, and EGM stands for the Earth Gravitational Model.

$$\begin{bmatrix} \varepsilon_{TS14} \\ \varepsilon_{TS15} \\ \varepsilon_{TS16} \\ \varepsilon_{TS17} \\ \hline \varepsilon_{EGM3} \end{bmatrix} = \mathbf{X} \begin{bmatrix} \mathbf{d}_E \\ \hline d\xi_3 \\ d\eta_3 \\ d\Sigma_3 \end{bmatrix} + \mathbf{U} \begin{bmatrix} \mathbf{d}_1 \\ \mathbf{d}_2 \\ \mathbf{d}_3 \\ \mathbf{d}_4 \\ \mathbf{d}_A \\ \mathbf{d}_B \\ \hline d\xi_1 \\ d\eta_1 \\ d\Sigma_1 \\ \hline \mathbf{d}_5 \\ \mathbf{d}_6 \\ \mathbf{d}_7 \\ \mathbf{d}_C \\ \mathbf{d}_D \\ \hline d\xi_2 \\ d\eta_2 \\ d\Sigma_2 \end{bmatrix} - \mathbf{y} \tag{44}$$

where

- $\gamma = \begin{bmatrix} \mathbf{d}_1^T & \mathbf{d}_2^T & \mathbf{d}_3^T & \mathbf{d}_4^T & \mathbf{d}_A^T & \mathbf{d}_B^T & | & d\xi_1 & d\eta_1 & d\Sigma_1 & \vdots & \mathbf{d}_5^T & \mathbf{d}_6^T & \mathbf{d}_7^T & \mathbf{d}_C^T & \mathbf{d}_D^T & | & d\xi_2 & d\eta_2 & d\Sigma_2 \end{bmatrix}^T$ is the vector of unknown corrections to all parameters adjusted on the second total station position, with zero expected value and a known covariance matrix:

$$\boldsymbol{\Sigma}_\gamma = \begin{bmatrix} \boldsymbol{\Sigma}_{\hat{\beta}} & \boldsymbol{\Sigma}_{\hat{\beta}\hat{\gamma}} \\ \boldsymbol{\Sigma}_{\hat{\beta}\hat{\gamma}}^T & \boldsymbol{\Sigma}_{\hat{\gamma}} \end{bmatrix} \tag{45}$$

- $\beta = \begin{bmatrix} \mathbf{d}_E^T & | & d\xi_3 & d\eta_3 & d\Sigma_3 \end{bmatrix}^T$ is the vector of corrections dX, dY, dZ to the approximate geocentric coordinates X, Y, Z of the new point E included in the model at the third total station position, and corrections $d\xi_3, d\eta_3, d\Sigma_3$ to the total station orientation angles;
- $\varepsilon = \begin{bmatrix} \varepsilon_{TS14}^T & \varepsilon_{TS15}^T & \varepsilon_{TS16}^T & \varepsilon_{TS17}^T & | & \varepsilon_{EGM3}^T \end{bmatrix}^T$ is the vector of total station measurement random errors $\varepsilon_x, \varepsilon_y, \varepsilon_z$ of the x, y, z coordinates of the points 2, 6, 7, E, respectively; and random errors $\varepsilon_\xi, \varepsilon_\eta$ of the deflection of the vertical components ξ_3, η_3 at the third total station position (at point 5): $\varepsilon_{EGM3}^T = \begin{bmatrix} \varepsilon_{\xi_3} & \varepsilon_{\eta_3} \end{bmatrix}$;
- $\mathbf{X}, \mathbf{y}$ is the known design matrix and free terms vector.

The solution of the observational–mixed model at the third total station position $\varepsilon = \mathbf{X}\beta + \mathbf{U}\gamma - \mathbf{y}$ is given by Equations (33)–(40).

The results of the sequential online adjustments of the total station positions were compared to those of the aforementioned classical offline adjustment of the total station traverse, including the merging points, as presented in the paper [28]. The obtained extremal coordinates' differences are only 10^{-6} m. That technically means that the proposed method and algorithm of the sequential online adjustment of the total station positions work correctly, and give the same results as the classical offline adjustment of the traverse, including the merging points, as expected.

2.4.3. Sequential Online Adjustments of the Total Station Observations

At the first total station position, the coordinate vectors of the station 1-point $\mathbf{GPS1} = \begin{bmatrix} X_1 & Y_1 & Z_1 \end{bmatrix}^T$ and the target 2-point $\mathbf{GPS2} = \begin{bmatrix} X_2 & Y_2 & Z_2 \end{bmatrix}^T$ have to be determined via a GNSS receiver. These position vectors and the vertical deflection components vector $\mathbf{EGM1} = \begin{bmatrix} \xi_1 & \eta_1 \end{bmatrix}^T$ are used for computation of the approximate value of the horizontal angle Σ_1. The observational mixed model $\boldsymbol{\varepsilon} = \mathbf{X}\boldsymbol{\beta} + \mathbf{U}\boldsymbol{\gamma} - \mathbf{y}$ for the first total station observation $\mathbf{TS1} = \begin{bmatrix} x_1 & y_1 & z_1 \end{bmatrix}^T$, including the GPS1, GPS2 and EGM1 data, is given by

$$\begin{bmatrix} \varepsilon_{\mathbf{TS1}} \\ \varepsilon_{\mathbf{GPS1}} \\ \varepsilon_{\mathbf{GPS2}} \\ \varepsilon_{\mathbf{EGM1}} \end{bmatrix} = \mathbf{X}[d\Sigma_1] + \mathbf{U}\begin{bmatrix} \mathbf{d}_1 \\ \mathbf{d}_2 \\ \hline d\xi_1 \\ d\eta_1 \end{bmatrix} - \mathbf{y} \tag{46}$$

where the covariance matrices $\boldsymbol{\Sigma}_{\varepsilon}, \boldsymbol{\Sigma}_{\gamma}$ of the vectors $\boldsymbol{\varepsilon}, \boldsymbol{\gamma}$ are known a priori.

The solution of the mixed model is given by following Equations (33)–(40).

For the second total station observation TS2, the mixed model is defined as

$$\varepsilon_{\mathbf{TS2}} = \mathbf{X}\mathbf{d}_3 + \mathbf{U}\begin{bmatrix} \mathbf{d}_1 \\ \mathbf{d}_2 \\ \hline d\xi_1 \\ d\eta_1 \\ d\Sigma_1 \end{bmatrix} - \mathbf{y} \tag{47}$$

where $\boldsymbol{\gamma} = \begin{bmatrix} \mathbf{d}_1^T & \mathbf{d}_2^T & | & d\xi_1 & d\eta_1 & d\Sigma_1 \end{bmatrix}^T$ is the vector of unknown corrections to all parameters adjusted previously, with zero expected value and a known covariance matrix:

$$\boldsymbol{\Sigma}_{\gamma} = \begin{bmatrix} \boldsymbol{\Sigma}_{\hat{\beta}} & \boldsymbol{\Sigma}_{\hat{\beta}\hat{\gamma}} \\ \boldsymbol{\Sigma}_{\hat{\beta}\hat{\gamma}}{}^T & \boldsymbol{\Sigma}_{\hat{\gamma}} \end{bmatrix} \tag{48}$$

Its solution is given by Equations (33)–(40).

The next total station observation vectors TS4, TS5, ... are included in the model and adjusted in the same way. The GPS3, GPS4, ... vectors can be included and adjusted after measurements at any time. For example, if the GPS3 vector is included after adjustment of the second total station observation TS2, the mixed model does not contain the new parameters' vector $\boldsymbol{\beta}$:

$$\varepsilon_{\mathbf{GPS3}} = \mathbf{U}\begin{bmatrix} \mathbf{d}_1 \\ \mathbf{d}_2 \\ \hline d\xi_1 \\ d\eta_1 \\ d\Sigma_1 \\ \hline \mathbf{d}_2 \end{bmatrix} - \mathbf{y} \tag{49}$$

where $\boldsymbol{\gamma} = \begin{bmatrix} \mathbf{d}_1^T & \mathbf{d}_2^T & \mathbf{d}_3^T & | & d\xi_1 & d\eta_1 & d\Sigma_1 \end{bmatrix}^T$ is the vector of unknown corrections to all parameters adjusted previously, with zero expected value and a known covariance matrix:

$$\boldsymbol{\Sigma}_{\gamma} = \begin{bmatrix} \boldsymbol{\Sigma}_{\hat{\beta}} & \boldsymbol{\Sigma}_{\hat{\beta}\hat{\gamma}} \\ \boldsymbol{\Sigma}_{\hat{\beta}\hat{\gamma}}^T & \boldsymbol{\Sigma}_{\hat{\gamma}} \end{bmatrix} \tag{50}$$

After measuring and adjusting all vectors at the first total station position, for example, in the sequence {TS1+ GPS1+ GPS2+ EGM1}, TS2, GPS3, TS3, TS4, TS5, GPS4, the adjusted sequential process is continued on the second position of the total station. Firstly, the previously adjusted position vectors GPS1, GPS2 and vector of the vertical deflection components $\mathbf{EGM2} = \begin{bmatrix} \xi_2 & \eta_2 \end{bmatrix}^T$ are used for computing the approximate value of the horizontal angle $\boldsymbol{\Sigma}_2$. The observational–mixed model $\boldsymbol{\varepsilon} = \mathbf{X}\boldsymbol{\beta} + \mathbf{U}\boldsymbol{\gamma} - \mathbf{y}$ for the first total station observation on the second position $\mathbf{TS6} = \begin{bmatrix} x_1 & y_1 & z_1 \end{bmatrix}^T$ is defined by

$$\begin{bmatrix} \varepsilon_{\mathbf{TS1}} \\ \varepsilon_{\mathbf{EGM1}} \end{bmatrix} = \mathbf{X}[d\Sigma_2] + \mathbf{U} \begin{bmatrix} \mathbf{d}_1 \\ \mathbf{d}_2 \\ \hline d\xi_1 \\ d\eta_1 \\ d\Sigma_1 \\ \hline \mathbf{d}_3 \\ \mathbf{d}_4 \\ \mathbf{d}_A \\ \mathbf{d}_B \\ \hline d\xi_2 \\ d\eta_2 \end{bmatrix} - \mathbf{y} \tag{51}$$

where the covariance matrices $\boldsymbol{\Sigma}_{\varepsilon}, \boldsymbol{\Sigma}_{\gamma}$ of the vectors are known a priori.

The solution of the mixed model is provided by solving Equations (33)–(40). Next, data obtained from the measurements at station 2 are included in the model and adjusted in the same way as the first total station position.

The adjusted sequential process is continued on the third position of the total station in the same way.

The results of the sequential online adjustments of the total station observations were compared to the classical offline adjustment of the total station traverse mentioned earlier, excluding the merging points presented in the paper [28]. The obtained extremal coordinates' differences are only in the order of 10^{-6} m. This technically means that the proposed method and algorithm of the sequential online adjustments of the total station observations work correctly, and give the same results as the classical offline adjustment of the traverse without merging points, as expected.

3. Discussion

In our studies, we have presented the algorithm for real-time adjustment of the integrated GNSS, total station, and EGM vertical direction data in the geocentric coordinate system. The proposed method has two modes: sequential online adjustment of the total station positions and online adjustment of the total station observations. Both modes have been tested on data from the spatial total station traverse measurements. In order to verify the correctness of the online algorithm, the total station traverse data were also adjusted using the offline algorithm. The offline adjustment was conducted with and without including the merging points.

Online and offline adjustment computation should be considered complementary approaches. They are an essential aspect of the assessment of surveying results, especially in issues related to displacement measurements and deformation monitoring. As demonstrated in the previous section, the accuracies offered by both procedures are comparable and utterly applicable to field measurements. Nevertheless, the main advantages of online adjustment are the ability to conduct reliable deformation monitoring and the ongoing fulfillment of the obtained measurement results for risk management purposes. In this context, attention should be paid to the problem of notifying users of risks arising from the construction and exploitation of the monitored object, as well as from various natural phenomena affecting changes in its structural geometry. Defining safety thresholds (so-called "limit checks") is not straightforward. This is primarily caused by the need to correctly identify a specific object and its surroundings, which in turn requires a reasonably long observation period and using multi-source data. Its integration in a standardized manner is one of the most significant theoretical challenges posed by modern geodesy.

Moreover, ongoing technological advancement influences the constant modification of once-accepted assumptions. As a rule, data sets representing surveying results captured continuously or quasi-continuously onsite must reliably identify outlier observations and determine the influence of other error sources, be they systematic, personal, or instrumental. To this end, complementary methods using estimation and optimization procedures are becoming increasingly important. As discussed in our paper, such a need to work online encourages the development of appropriate procedures and algorithms.

As demonstrated in the conducted studies, including information regarding the local distribution of plumb line parameters in total station surveying is expedient and desirable. To this end, total station and GNSS satellite measurements were concatenated within the EGM2008 geoid model. Such an approach unifies operating surveying instruments, driving them to be used precisely and universally. However, in the general assessment process, one should consider the specifics of the developed test network. After all, the object reveals specific features (urban, intensively built-up, and used areas) and geometrical features; the network is regular, set up following the art of measurement. We realize that depending on the altering conditions of surveys, the results may somewhat differ from those obtained in our experiment.

Nevertheless, as already shown in the literature review, similar work has been carried out in other conditions—for example, in the mountains—giving similar results. Additionally, incorporating further data sources into the observational system should not pose significant concerns. In such a case, the vectors of unknowns and the corresponding coefficient matrices will be modified, generally increasing the size of the corresponding arrays. Theoretically, this will cause increased demand for computer processing power; however, with the current parameters of available machines and computational optimization methods, this should be a manageable challenge.

4. Conclusions

In conclusion, we have formulated some closing statements and tracked the progress of further research in relevant fields. First, it should be noted that despite the many studies conducted worldwide on the integration and standardization of spatial data, the issue of effective data fusion still needs to be solved. This will mainly be achieved by the continuous development of measurement methods and surveying instruments, as well as by incorporating multi-source data into newly constructed surveying systems. Examples include numerous attempts to combine spatial (geometric) data with physical data. Secondly, scientific progress includes working on new geoid models and continuously improving existing ones. Such studies transform into continuous precision and accuracy validation of local plumb line parameters within the EGM framework. Undoubtedly, this requires ongoing test work using current data sources.

The results of our study can be summarized as follows:

- The total station traverse is very well spatially oriented by using the total station vertical axe deflection from normal to GRS 80 ellipsoid components obtained from EGM, and by measuring the coordinates of the total station standpoint using a GNSS receiver. There is no need to carry out GNSS measurements at any additional merging total station traverse points.

- The proposed methods and algorithms of the sequential online adjustment of the total station positions and the sequential adjustment of the single total station vector of observations, including the GNSS and EGM data in both cases, work correctly. They deliver the same results as the classical offline adjustment of the total station traverse.

The proposed algorithms allow for real-time surveying with the online adjustment of the observations in the geocentric coordinate system, providing current control of the adjustment results. Our online computation methodology, applied to the discussed surveys (total station/GNSS/EGM), can be used to directly model 3D objects in real time with high spatial accuracy in a geocentric coordinate system. It can also be applied to measure control points for further use in close-range photogrammetry (aerial and terrestrial). The procedure can be applied while defining reliable tie points for merging point clouds in laser scanning (both terrestrial and performed using UAVs). Such an approach is beneficial in BIM while constructing and updating relevant 3D data models.

As our survey network was near the GNSS reference station (at a distance not exceeding 250 m), we also acknowledge the need to test the solution for other-longer distances. Our future works also encompass testing and improving new, multi-source data algorithms for object diagnostic purposes.

Author Contributions: Conceptualization, K.K. and E.O.; methodology, E.O. and K.K.; validation, K.K. and Z.M.; formal analysis, E.O., K.K. and Z.M.; investigation, E.O., K.K. and Z.M.; writing—original draft preparation, K.K.; writing—review and editing, K.K. and E.O.; supervision, E.O.; project administration, K.K. All authors have read and agreed to the published version of the manuscript.

Funding: This research received no external funding.

Institutional Review Board Statement: Not applicable.

Informed Consent Statement: Not applicable.

Data Availability Statement: The data presented in this study are available upon request from the co-author Edward Osada: edward.osada@dsw.edu.pl.

Conflicts of Interest: The authors declare no conflict of interest.

References

1. Flah, M.; Nunez, I.; Ben Chaabene, W.; Nehdi, M.L. Machine Learning Algorithms in Civil Structural Health Monitoring: A systematic review. *Arch. Comput. Methods Eng.* **2021**, *28*, 2621–2643. [CrossRef]
2. Malekloo, A.; Ozer, E.; Al-Hamaydeh, M.; Girolami, M. Machine learning and structural health monitoring overview with emerging technology and high-dimensional data source highlights. *Struct. Health Monit.* **2022**, *21*, 1906–1955. [CrossRef]
3. Yuan, F.-G.; Zargar, S.A.; Chen, Q.; Wang, S. Machine learning for structural health monitoring: Challenges and opportunities, Sensors and Smart Structures Technologies for Civil, Mechanical, and Aerospace Systems. *Proc. SPIE* **2020**, *11379*, 1137903. [CrossRef]
4. Bao, Y.; Li, H. Machine learning paradigm for structural health monitoring. *Struct. Health Monit.* **2021**, *20*, 1353–1372. [CrossRef]
5. Luhmann, T.; Robson, S.; Kyle, S.; Boehm, J. *Close-Range Photogrammetry and 3D Imaging*; De Gruyter: Berlin, Germany; Boston, MA, USA, 2013; pp. 4–18, ISBN 978-3-11-060725-3.
6. Ullah, Z.; Al-Turjman, F.; Mostarda, L.; Gagliardi, R. Applications of Artificial Intelligence and Machine learning in smart cities. *Comput. Commun.* **2020**, *154*, 3123–3323. [CrossRef]
7. Ud Din, I.; Guizani, M.; Rodrigues, J.J.P.C.; Suhaidi, H.; Korotaev, V.V. Machine learning in the Internet of Things: Designed techniques for smart cities. *Future Gener. Comput. Syst.* **2019**, *100*, 826–843. [CrossRef]
8. Heunecke, O. Recent Impacts of Sensor Network Technology on Engineering Geodesy. In *The 1st International Workshop on the Quality of Geodetic Observation and Monitoring Systems (QuGOMS'11)*; Kutterer, H., Seitz, F., Alkhatib, H., Schmidt, M., Eds.; International Association of Geodesy Symposia; Springer, Science&Business Media LLC: New York, NY, USA, 2015; Volume 140, pp. 185–195. [CrossRef]
9. Karsznia, K.; Tarnowska, A. Proposition of an integrated geodetic monitoring system in the areas at risk of landslides. *Chall. Mod. Technol.* **2013**, *4*, 33–40.
10. Hein, G.W. Integrated geodesy state-of-the-art, 1986 reference text. In *Mathematical and Numerical Techniques in Physical Geodesy*; Sünkel, H., Ed.; Lecture Notes in Earth Sciences; Springer: Berlin/Heidelberg, Germany, 1986; Volume 7. [CrossRef]

11. Barzaghi, R.; Benciolini, B.; Betti, B. A numerical experiment of integrated geodesy. *Bull. Geod.* **1990**, *64*, 259–282. [CrossRef]
12. Karsznia, K. A concept of surveying and adjustment of spatial tacheometric traverses in the applications of integrated geodesy. *Acta Sci. Pol. Geod. Descr. Terr.* **2008**, *7*, 35–46.
13. Ranganathan, A. The Levenberg-Marquardt algorithm. *Tutoral LM Algorithm* **2004**, *11*, 101–110.
14. Wrześniak, A.; Giordan, D. Development of an algorithm for automatic elaboration, representation, and dissemination of landslide monitoring data. *Geomat. Nat. Hazards Risk* **2017**, *8*, 1898–1913. [CrossRef]
15. Dematteis, N.; Wrześniak, A.; Allasia, P.; Bertolo, D.; Giordan, D. Integration of robotic total station and digital image correlation to assess the three-dimensional surface kinematics of a landslide. *Eng. Geol.* **2022**, *303*, 106655. [CrossRef]
16. Yuwono, B.D.; Prasetyo, Y. Analysis Deformation Monitoring Techniques Using GNSS Survey and Terrestrial Survey (Case Studi: Diponegoro University Dam, Semarang, Indonesia). *IOP Conf. Ser. Earth Environ. Sci.* **2019**, *313*, 012045. [CrossRef]
17. Gargula, T. Adjustment of an Integrated Geodetic Network Composed of GNSS Vectors and Classical Terrestrial Linear Pseudo-Observations. *Appl. Sci.* **2021**, *11*, 4352. [CrossRef]
18. Qian, S.; Qian, S.; Xiaa, J.; Fosterb, J.; Falkmerc, T.; Hoe, L. Pursuing Precise Vehicle Movement Trajectory in Urban Residential Area Using Multi-GNSS RTK Tracking. *Transp. Res. Procedia* **2017**, *25*, 2356–2372.
19. Linder, W. *Digital Photogrammetry—Theory and Applications*; Springer: Berlin/Heidelberg, Germany, 2003; pp. 23–38, ISBN 978-3-662-06725-3.
20. Karsznia, K.; Osada, E. Photogrammetric Precise Surveying Based on the Adjusted 3D Control Linear Network Deployed on a Measured Object. *Appl. Sci.* **2022**, *12*, 4571. [CrossRef]
21. Condorelli, F.; Morena, S. Integration of 3D modelling with photogrammetry applied on historical images for cultural heritage. *VITRUVIO—Int. J. Archit. Technol. Sustain.* **2023**, *8*, 58–69. [CrossRef]
22. Robleda, P.G.; Ramos, P.A. Modeling and accuracy assessment for 3D-virtual reconstruction in cultural heritage using low-cost photogrammetry: Surveying of the Santa Maria Azogue church's front. *Int. Arch. Photogramm. Remote Sens. Spat. Inf. Sci.* **2015**, *XL-5/W4*, 263–270. [CrossRef]
23. Osada, E. *Geodesy*; Publishing House of the Wroclaw University of Technology: Wroclaw, Poland, 2002; pp. 161–209.
24. James, R. *Geodetic Calculations and Applications*; CreateSpace Independent Publishing Platform: Scotts Valley, CA, USA, 2017; pp. 89–91, ISBN 978-1547266562.
25. Osada, E.; Sergieieva, K.; Lishchuk, V. Improvement of the total station 3D adjustment by using precise geoid model. *Geod. Cartogr.* **2010**, *59*, 3–12.
26. Eshagh, M.; Ebadi, S. Geoid modelling based on EGM08 and recent Earth gravity models of GOCE. *Earth Sci. Inf.* **2013**, *6*, 113–125. [CrossRef]
27. Nocedal, J.; Wright, S.J. *Numerical Optimization*, 2nd ed.; Springer Science & Business Media LLC: New York, NY, USA, 2006; pp. 101–104, ISBN 978-0387303031.
28. Osada, E.; Owczarek-Wesołowska, M.; Ficner, M.; Kurpiński, G. TotalStation/GNSS/EGM integrated geocentric positioning method. *Surv. Rev.* **2017**, *49*, 206–211. [CrossRef]
29. Osada, E.; Sośnica, K.; Borkowski, A.; Owczarek-Wesołowska, M.; Gromczak, A. A direct georeferencing method for terrestrial laser scanning using GNSS data and the vertical deflection from global Earth gravity models. *Sensors* **2017**, *17*, 1489. [CrossRef]
30. Osada, E.; Owczarek-Wesołowska, M.; Sośnica, K. Gauss–Helmert Model for Total Station Positioning Directly in Geocentric Reference Frame Including GNSS Reference Points and Vertical Direction from Earth Gravity Model. *J. Surv. Eng.* **2019**, *145*, 04019013. [CrossRef]
31. Fahrmeir, L.; Kneib, T.; Lang, S.; Marx, B. *Regression: Models, Methods and Applications*; Springer: Berlin/Heidelberg, Germany, 2013; pp. 69–70.
32. Eshagh, M. Sequential Tikhonov Regularization: An alternative way for inverting satellite gradiometric data. *ZfV* **2011**, *136*, 113–121.
33. Osada, E. Adjustment of the total station data in real time. *Bolletino Di Geod. Et Sci. Affin.* **1996**, *55*, 121–130.
34. Jeudy, L.M.A. Generalyzed variance-covariance propagation law formulae and application to explicit least-squares adjustments. *Bull. Geod.* **1988**, *62*, 113–124. [CrossRef]
35. Rapp, R.H. *Geometric Geodesy Part II*; The Ohio State University, Department of Geodetic Science and Surveying: Columbus, OH, USA, 1993; p. 84.
36. *Leica Flex Line TS02 Plus, Data Sheet*; Leica Geosystems AG: Heerbrugg, Switzerland, 2013. Available online: https://www.leica-geosystems.com (accessed on 15 July 2023).
37. *Leica GS 10, Data Sheet*; Leica Geosystems AG: Heerbrugg, Switzerland, 2015. Available online: https://www.leica-geosystems.com (accessed on 15 July 2023).
38. Hirt, C. Prediction of vertical deflections from high-degree spherical harmonic synthesis and residual terrain model data. *J. Geod.* **2010**, *84*, 179–190. [CrossRef]
39. Osada, E.; Trojanowicz, M. Joint total-station and GPS positioning with the use of digital terrain and gravity models. *Geod. I Kartogr.* **1999**, *XLVIII*, 39–46.
40. Eshagh, M.; Berntsson, J. On quality of NKG2015 geoid model over the Nordic countries. *J. Geod. Sci.* **2019**, *9*, 97–110. [CrossRef]
41. Eshagh, M.; Zoghi, S. Local error calibration of EGM08 geoid using GNSS/levelling data. *J. Appl. Geophys.* **2016**, *130*, 209–217. [CrossRef]

42. Eshagh, M.; Ebadi, S. A strategy to calibrate errors of Earth gravity models. *J. Appl. Geophys.* **2014**, *103*, 215–220. [CrossRef]
43. Eshagh, M. Numerical aspects of EGM08-based geoid computations in Fennoscandia regarding the applied reference Surface and error propagation. *J. Appl. Geophys.* **2013**, *96*, 28–32. [CrossRef]
44. Eshagh, M. On the reliability and error calibration of some recent Earth's gravity models of GOCE with respect to EGM08. *Acta Geod. Geophys. Hung.* **2013**, *48*, 199–208. [CrossRef]
45. Eshagh, M. Error calibration of quasi-geoid, normal and ellipsoidal heights of Sweden using variance component estimation. *Contr. Geophys. Geod.* **2010**, *40*, 1–30. [CrossRef]
46. Kiamehr, R.; Eshagh, M. Estimation of variance components Ellipsoidal, Geoidal and orthometrical heights. *J. Earth Space Phys.* **2008**, *34*, 1000318.
47. Trojanowicz, M.; Osada, E.; Karsznia, K. Precise local quasigeoid modelling using GNSS/levelling height anomalies and gravity data. *Surv. Rev.* **2018**, *52*, 76–83. [CrossRef]

applied sciences

MDPI

Article

Experiment Study on Mechanical Evolution Characteristics of Coal and Rock under Three-Dimensional Triaxial Stress

Yabin Tao [1], Han Du [2,*], Ruixin Zhang [1], Jianzhao Feng [1] and Zhiyun Deng [2]

1 School of Energy and Mining Engineering, China University of Mining and Technology (Beijing), Beijing 100083, China; trippleh404@cumt.edu.cn (Y.T.); 47170039@lntu.edu.cn (R.Z.); f13313176404@163.com (J.F.)
2 State Key Laboratory of Hydroscience and Engineering, Department of Hydraulic Engineering, Tsinghua University, Beijing 100084, China; dengzhiyuncq@163.com
* Correspondence: mrkongfupanda@gmail.com; Tel.: +86-1824-181-0055

Abstract: The surrounding rock is in a complex stress environment and its mechanical behavior is also complex, especially after the excavation of the coal seam, the phenomenon of stress release of surrounding rock often occurs. The vertical stress and horizontal stress of the surrounding rock mass will have a series of complex changes. In underground engineering, rock mass is affected by dead weight pressure and tectonic stress. With coal mine production, the original stress of surrounding rock is demolished, and the destruction of surrounding rock is reflected in the loading and unloading failure of three-dimensional stress. Aiming at the phenomenon, this paper takes the Pingshuo East open-pit mine as the research background, and the experiments on physical and mechanical parameters of coal and rock mass was carried out, obtaining the coal and rock mechanics parameters, such as elastic modulus, Poisson's ratio, internal friction angle, cohesive force, etc. The stress strain curve was created based on the conventional triaxial experiment of coal and rock under different confining pressure conditions. According to the characteristics of these curves, we obtain underground engineering rock mass unloading stress–strain variation characteristics. Through establishing a stress–strain equation based on confining pressure, we finally describe the mechanical failure characteristics of rock under triaxial stress.

Keywords: complex stress environment; open-pit coal mine; failure mechanism; coal petrography; coal petrography; surrounding rock

Citation: Tao, Y.; Du, H.; Zhang, R.; Feng, J.; Deng, Z. Experiment Study on Mechanical Evolution Characteristics of Coal and Rock under Three-Dimensional Triaxial Stress. *Appl. Sci.* **2022**, *12*, 2445. https://doi.org/10.3390/app12052445

Academic Editor: Alexei Gvishiani

Received: 25 January 2022
Accepted: 22 February 2022
Published: 26 February 2022

1. Introduction

In recent years, rapid urbanization in developing countries is spurring the demand for coal resources to an unprecedented growth [1]. Therefore, shallow coal mine resources are decreasing constantly [2], coal mining has gradually transferred to deep mining [3], and many domestic coal mines are focusing on open-pit and underground mining models from exhaustion of coal mine shallow resources [4]. Along with the instance of the Pingzhuang West open-pit mine [5], there are also the Antaibao open-pit mine [6], China's Anjialing open-pit mine [7], and other open-pit and non-ferrous metal mines. When the two mining exploitation methods are used in the same area, the co-occurring processes of underground mining and open-pit influence each other [8]. In one aspect, the synchronization of the two working conditions will significantly change the strata structure, stress status, and formation of the mining of space [9,10]. On the other hand, more attention should be paid to the goafs at the bottom of underground mines, because the crown pillars are thin and a failure of the goaf close to the bottom may trigger a landslide disaster in the upper slope [11,12]. The interaction effects are predominantly evident in goaf stability analysis. The contributing factors of goaf stability are related to the changes of roadway lining distribution, the anchor, and surrounding rock. In particular, the surrounding rock coal mining brakes the equilibrium state of in situ stress field. The surrounding rock experiences deformation, destruction, caving, and stability under the coupled action of mining stress field and supporting stress field. Consequently, the stability of the surrounding rock greatly impacts the mine operations and personnel safety, and simultaneously reflects the mechanical evolution characteristics of coal petrography in a complex stress environment.

The significance of the destruction of original stress of the surrounding rock on existing roadways is a multidisciplinary problem and involves many aspects, such as the dynamic response of rock mechanics, the exploitation of the underground construction, the explosion vibration, the loose zone of the surrounding rock, and the propagation of the stress wave. Research methods have involved theoretical analyses, numerical simulations [13], field tests [14,15], and synthesis methods. Huang et al. [16] discussed the corresponding methods of controlling the surrounding rock stability, comparing general methods, and provided an important theoretical foundation for application in deep coal mines. Salmi et al. [17] dedicated considerable efforts to explore and summarize surrounding rock masses classification methods through empirical equations. Unlike the empirical techniques, numerical methods [18] can be used for evaluating deformation development progress of surrounding rock masses in mining geological and geotechnical conditions [19]. Li et al. [20] presupposed single monitoring information (e.g., resistivity profiles and source position messages) and presented multiparameter dynamic monitoring methods. These methods were validated using numerical simulation results. Jing et al. [21] took an anonymous coal mine of China as an example, explicating the numerical simulation results of the stress evolution, displacement field, overall distribution, and failure characteristics of the anchorage structure of surrounding rock with different rock bolt spacing. The results showed that when the maximum displacement between the roof and floor of the west wing track roadway reach a certain value, it can ensure the stability and safety of the excavated roadway. Finally, in association with field tests, Zhang et al. [22] conducted a research item in four different Chinese coal mining sites and evaluated the related response of influencing circumstances and factors of surrounding rock such as mining depth, support strength, and area of gob-side hanging roof. Yu et al. [23] developed a shaft and auxiliary shaft system for the stability and control technology of surrounding rock by the chamber and roadway ground of 850 m level in a mining area. He et al. [24] established a stability classification and order-arranging model of surrounding rock based on certainty measure theory. Chen et al. [25] employed X-ray diffraction technology to analyze the repair control technology of the surrounding rock in mining areas which contain clay minerals.

In the meantime, the accidents due to underground coal mining, such as roadways surrounding rock large deformation, account for more than 40% of all the accidents in mine construction and production. In this situation, global scholars and enterprise decision-makers have spent more effort on fundamentally recognizing the deformation mechanism and proposing reasonable control measures for roadways in deep underground areas. Most researchers have conducted numerous substantive research on deformation, failure mechanism, and control measures for surrounding rock in underground mining. He et al. [26] and Sun et al. [27] presented a design theory of nonlinear large deformation of surrounding rock deep underground and built up a bolt with constant resistance and large deformation. Kang and Liu et al. [28] analyzed the supporting difficulties and mechanism of soft fragmentized surrounding rock roadway with high stress, and they proposed a new kind of combined supporting system including a new kind of hollow grouting cable. Li and Wang [29] studied deformation characteristics of the surrounding rock in deep roadway with top coal, and they designed the pressure relief anchor box beam system. Singh et al. [30] studied the trigger factors and deformation mechanisms of failure of surrounding rock used at high stress levels from two different Indian open-pit and underground coupling coal mines. Sasaoka et al. [31] discussed the technologies for developing roadways in deep German coal mines under difficult ground conditions.

Although many previous studies have been proposed and conducted to date on this topic using theoretical analysis [32], numerical simulations [33,34], and field tests [35,36], research on the effects of seismic waves on existing roadways due to blasting in open-pit and underground coupled mining is lacking. In addition, because field measurements and laboratory mechanical experiment are scarce in open-pit coupled underground mining circumstances, most of the research results can only be used as reference data for surround rock in underground construction. Moreover, many studies are restricted due to geographical environmental, engineering background, and construction conditions, and the data are not suitable to settle this problem.

Considering the characteristics of surrounding rock in high-stress environments, this paper carries out work in the following aspects: According to the geological survey and geophysical characteristics of the East open-pit mine, regional geology, hydrogeology, and geophysical characteristics were analyzed [37,38], and the ore body and surrounding rock were sampled, respectively. The specific work carried out is as follows: (i) Physical properties of rock (natural bulk density, saturated water absorption, saturated bulk density) experiment; (ii) Rock deformation parameter experiment; (iii) Rock uniaxial compression experiment; (iv) Rock triaxial compression experiment; (v) Shear test of rock with variable angle mode. By carrying out indoor coal and rock mechanics experiments under triaxial stress, this paper provides a basis for roof mechanical model analysis and goaf stability analysis.

2. Materials and Methods

2.1. Engineering Background and Geological Setting

Pingshuo East open-pit mine (as shown in Figure 1) is subordinate to China coal Pingshuo group co., LTD. It is 4.42~5.47 km wide from east to west and 6.53~10.3 km long from north to south. The mining area is 48.73 km^2, and the mining depth of No. 11 coal floor is about 200 m on average. The main coal seams are No. 4, 9, and 11. The average thickness of coal seams in the whole area is 34 m, and the reserves are 1848.92 Mt. The approved production capacity is 20.0 Mt/a. A typical borehole histogram is shown in Figure 2.

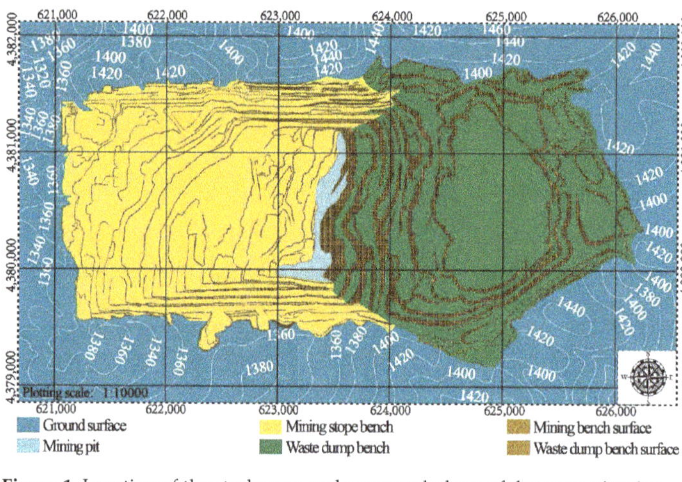

Figure 1. Location of the study area and geomorphology of the open-pit mine.

The stripping adopts the single-bucket truck discontinuous process, and the main stripping material is yellow sand soil, red clay, limestone, sandstone, etc. The bucket with capacity of 25 m^3, 32 m^3, and 55 m^3 is used for loading, and it is transported to the dump site by the supporting dump truck with load capacity of 200 t and 300 t. The coal mining adopts the semicontinuous process of single bucket—truck—semi-fixed crushing station—belt conveyor—coal preparation plant. The electric shovel is used in conjunction with the coal mining of the former machine, and the coal is transported to the coal crushing station by 200 t class truck.

Lithology	Depth(m)	Thickness(m)	Description
pebbly sandstone	102.50	12.10	
siltstone	107.94	5.44	
mudstone	117.48	9.54	
siltstone	118.40	0.92	
sandy mudstone	128.80	10.40	
kern stone	133.60	4.80	
sandy mudstone	135.30	1.70	
coal seam no.4	153.85	18.55	
sandy mudstone	159.46	5.61	
post stone	166.44	6.98	
sandy mudstone	174.35	8.06	
siltstone	179.25	4.90	
post stone	182.12	2.87	
sandy mudstone	185.61	3.49	
coal seam no.9	196.60	10.99	
sandy mudstone	199.93	3.33	
coal seam no.11	205.30	5.37	
siltstone	208.28	2.98	

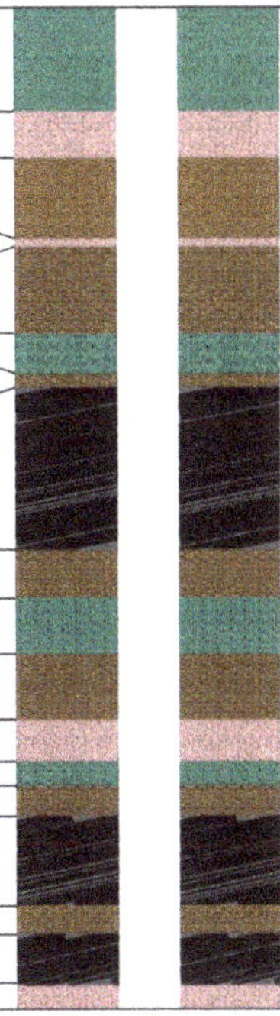

Figure 2. Lithology of the typical borehole.

2.2. Experimental Equipment and Experimental Scheme

The rock part of the experiment is carried out in accordance with the industry standards of the People's Republic of China, Rock Experiment Regulations for Water Conservancy and Hydropower Engineering (SL264-2001), Rock Experiment Regulations for Highway Engineering (JTG E41-2005) and Rock Experiment Method Standard for Engineering (GB/T50266-2013). The uniaxial compressive strength test specimens are generally cylindrical columns with a diameter of 5 cm (48~54 mm) and a height of 10 cm (the ratio of specimen height to diameter is generally 2.0–2.5); under the same water bearing state, no less than three specimens are prepared for each group. Meanwhile, the specimen size selected in this paper meets the standard requirements.

There were 19 groups of laboratory experiments, including five groups of rock physical properties experiment, six groups of uniaxial compression and deformation parameter experiment, four groups of triaxial compression experiment, and four groups of variable angle shear experiment.

According to conventional uniaxial and triaxial loading tests, the minimum uniaxial compressive strength of rock mass is coal rock with a compressive strength of about 20 MPa. In order to simulate the real stress of surrounding rock to the maximum extent, the confining pressures of rock mass were set at 0, 4, 8, 12, and 16 MPa during the triaxial test of surrounding rock. Meanwhile, the triaxial test can reflect the mechanical characteristics of rock mass in the actual stress field. According to the sampling situation, the triaxial test of rock mass is carried out and the loading rate is set as 0.1 mm/min. The relevant physical properties of rock samples tested through experiments are shown in Table 1.

Table 1. Density and water absorption calculation table.

Serial Number	Lithology	Specimen Size (Width × Length) mm × mm		Natural Quality g	Feed Water Quality g	Drying Quality g	Soil Natural Density g/cm³	Full Water Density g/cm³	Dry Density g/cm³	Bibulous Rate %
1-A-1-1	Coal	50.2	49.96	214.1	216.8	209.2	2.14	2.18	2.06	4.20%
1-A-1-2		50.4	50.51	218	217.9	208.4	2.05	2.04	2.1	4.87%
			mean value				2.1	2.11	2.08	4.54%
2-B-1-1	Mudstone	100.03	50.44	474.5	481.8	466.7	2.38	2.41	2.34	3.24%
2-B-1-2		99.6	50.5	495.9	501.9	490.3	2.49	2.52	2.46	2.37%
2-B-1-3		99.63	50.2	459.7	470.1	448.4	2.33	2.39	2.28	4.84%
2-B-1-4		99.31	50.46	480.3	490.7	468.7	2.42	2.47	2.36	4.69%
			mean value				2.4	2.45	2.36	3.78%
1-C-1-1	Fine sand-stone	99.39	50.42	493.7	505.6	491.6	2.49	2.55	2.48	2.85%
1-C-1-2		100.17	50.5	487.9	502.3	486.6	2.43	2.5	2.43	3.23%
1-C-1-3		101.25	50.61	467.5	480.3	465	2.3	2.36	2.28	3.29%
1-C-2-4		98.1	48.84	454.66	473.58	452.68	2.36	2.55	2.37	3.14%
			mean value				2.39	2.49	2.39	3.22%
1-D-1-1	Medium sand-stone	99.36	48.7	455.32	467.1	447.9	2.71	2.74	2.65	4.61%
1-D-1-2		98.92	48.74	457.86	469.6	450.7	2.55	2.56	2.44	4.19%
1-D-1-3		98.9	48.9	444.78	456.2	436.5	2.39	2.46	2.35	4.50%
1-D-1-4		100.1	48.72	451.48	463.56	442.68	2.42	2.43	2.37	4.70%
			mean value				2.42	2.43	2.37	4.70%

2.3. *Uniaxial Compressive Strength Test of Rock Mass*

The uniaxial compressive strength, elastic modulus, and Poisson's ratio of rock mass were obtained through indoor tests of uniaxial compressive strength, elastic modulus, and Poisson's ratio of rock mass. The tests were carried out in the Laboratory of Structural Mechanics, University of Science and Technology Beijing, China, and a WGE-600 microcomputer was used to control the universal screen test machine (as shown in Figure 3). The elastic modulus and Poisson's ratio of rock mass can be obtained according to relevant specifications and algorithms.

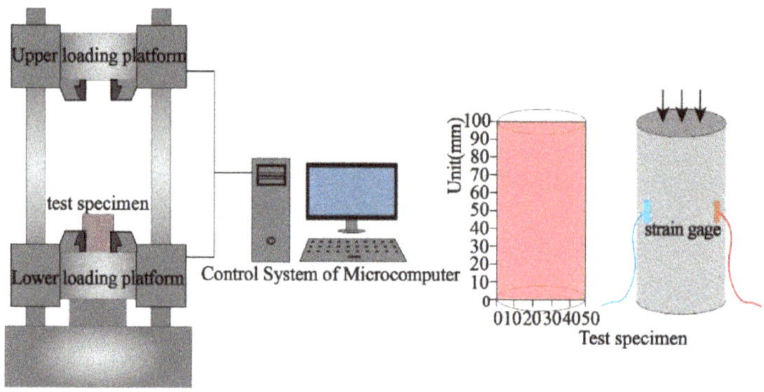

Figure 3. WGE-600 Experiment and investigation equipment with microcomputer control displaying of WGE-600.

2.4. *Shear Test of Rock with Variable Angle Mode*

Rock shear test is an experiment to determine the friction angle and cohesion of rock. The method is simple and feasible. Variable angle shear experiment usually adopts the mold change angle of 20°, 30°, and 40°. Then, the internal friction angle and cohesion of the rock mass can be obtained by drawing shear stress and normal stress curves of the failure surface.

The WGE-600 screen display universal experiment machine was also used in this experiment (as shown in Figure 3). Three sets of fixtures (20°, 30°, 40°) were used for variable angle shear specimen, as shown in Figure 4. Among them were the first group of four samples (mudstone), the second group of four samples (coal rock), the third group of three samples (fine sandstone). For the four groups of four samples (medium sandstone), shear strength calculation formula is as follows in Equation (1):

$$\sigma = P\sin\alpha/A \quad \tau = P\cos\alpha/A \tag{1}$$

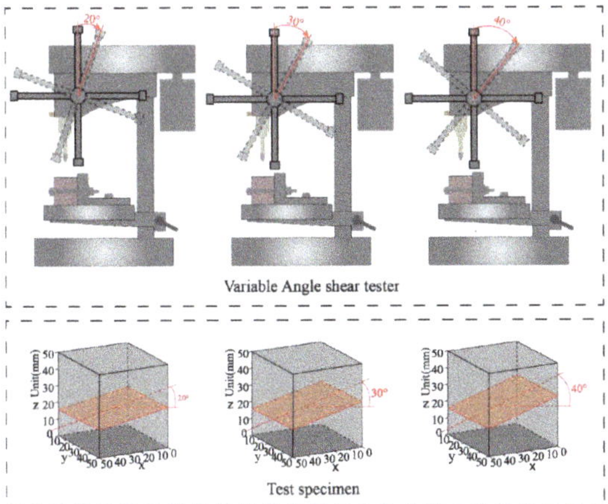

Figure 4. Schematic diagram of sample rock samples with different dip angles.

3. Results

3.1. Analysis of Uniaxial Compressive Strength of Rock Mass

Figure 5 shows the stress–strain curve characteristics of the specimen under uniaxial experiment. According to the experiment, under uniaxial pressure, the strength of medium sandstone is higher, followed by fine sandstone; mudstone is slightly lower than fine sandstone, and coal rock has the lowest strength. According to the experimental data, the uniaxial compressive strength of medium sandstone, fine sandstone, mudstone, and coal rock can reach 80 MPa, 50 MPa, 40 MPa, and 14 MPa, respectively.

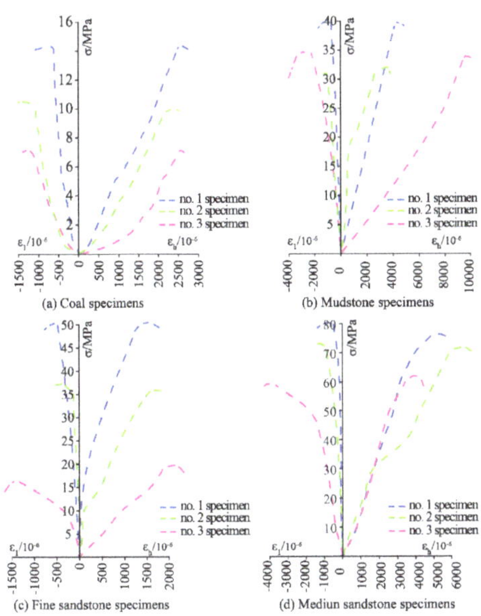

Figure 5. Stress–strain curves of rock specimens.

Figure 6 is the failure condition of rock sample according to the failure surface of the rock sample. It can be observed that the middle sandstone and fine sandstone mainly present shear failure under uniaxial compression. Figure 6(d-2,c-3) are typical representatives of shear failure, respectively. Some specimens also show tensile failure, but from the point of view of the failure strength of rock, rock mainly bears shear failure, and shear failure strength is also high.

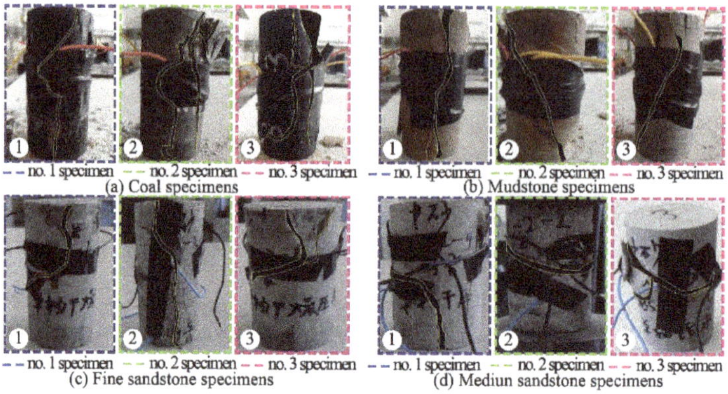

Figure 6. Failure state of rock specimens and samples.

Under uniaxial compression, mudstone and coal rock mainly exhibit splitting failure. Figure 6(b-1,a-1) are the typical representatives of splitting failure, respectively. There are also specimens showing shear failure, but in terms of the failure strength of rock, the rock is mainly subjected to splitting failure, and the shear failure strength is relatively low.

3.2. Analysis of Shear Test Results of Rock with Variable Angle Mode

According to the experimental data (Figures 7 and 8), the relationship between normal stress and shear stress is linearly fitted, and the fitting results are as follows in Equation (2).

$$\begin{cases} \tau = 0.96\sigma + 7.51(a) \\ \tau = 0.93\sigma + 3.42(b) \\ \tau = 0.91\sigma + 10.79(c) \\ \tau = 0.92\sigma + 9.24(d) \end{cases} \quad (2)$$

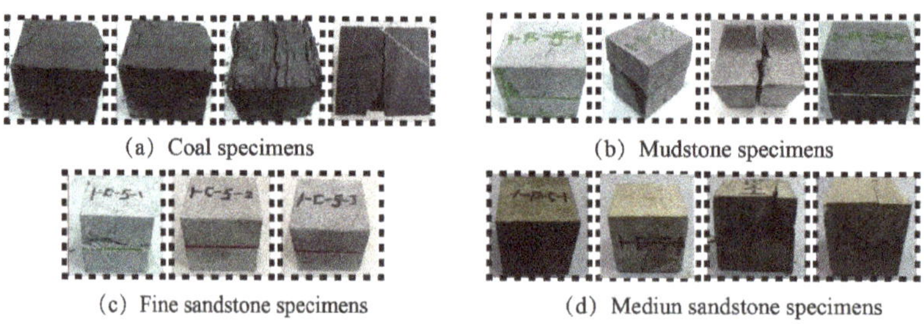

Figure 7. Coal rock variable angle shear specimen.

According to the experimental results and the fitted equation, it can be concluded that the cohesion of mudstone is 7.51 MPa and the internal friction angle is 43.3°, the cohesion of coal rock is 3.42 MPa and the internal friction angle is 40.6°, the cohesion of fine sandstone is 10.8 MPa and the internal friction angle is 30.73°, and the cohesion of medium sandstone is 9.24 MPa. The internal friction angle is 38.74°.

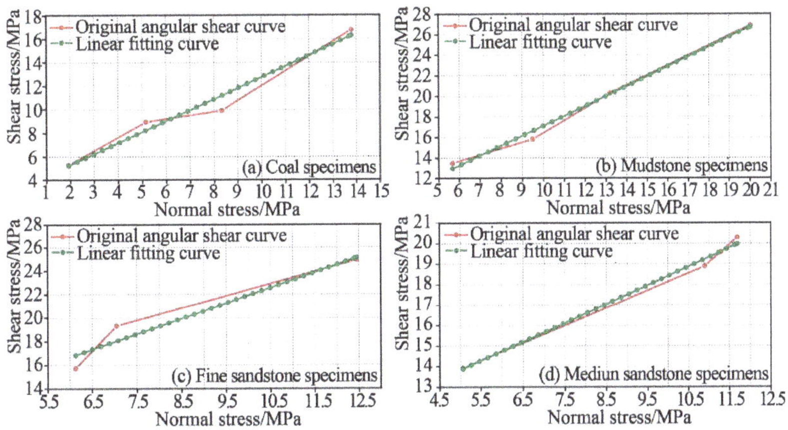

Figure 8. Relation curve for rock shear stress and normal stress.

3.3. Rock Mass Mechanical Response under Conventional Triaxial Loading Test

According to the experimental scheme, medium sandstone, mudstone, and fine sandstone were selected for rock triaxial experiment, designed as 0, 4, 8, 12, and 16 MPa, respectively.

According to the triaxial test, the cohesion and internal friction angle of rock mass can be obtained, and the variable angle shear test can also obtain the cohesion and internal friction angle of rock mass. In addition to obtaining some physical and mechanical parameters of rock mass, the triaxial test of rock mass was conducted to analyze the stress–strain curve characteristics of rock mass under triaxial stress environment (as shown in Figure 9). By sorting out the triaxial test results, the relationship between confining pressure and rock compressive strength can be obtained, and linear fitting can be carried out. The fitting results and the axial compressive strength of rock mass are as follows in Equation (3):

$$\sigma_1 = P_{\max}/A \tag{3}$$

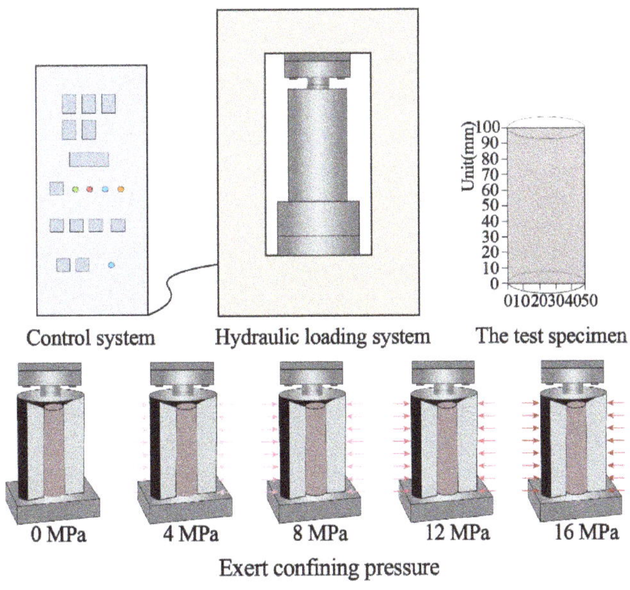

Figure 9. Triaxial experimental equipment.

The relationship between axial pressure and confining pressure can be expressed as follows in Equation (4):

$$\sigma_1 = \sigma_0 + k\sigma_3 \tag{4}$$

Cohesion and internal friction angle can be expressed as in Equations (5) and (6), as follows:

$$c = \frac{\sigma c(1-\sin\varphi)}{2\cos\varphi}, \varphi = \arcsin\frac{k-1}{k+1} \tag{5}$$

$$\begin{cases} \sigma_1 = 4.6\sigma_3 + 45.6(R^2 = 0.86)(a) \\ \sigma_1 = 2.5\sigma_3 + 101.7(R^2 = 0.98)(b) \\ \sigma_1 = 8.3\sigma_3 + 23.9(R^2 = 0.99)(c) \\ \sigma_1 = 5.85\sigma_3 + 17(R^2 = 0.90)(d) \end{cases} \tag{6}$$

3.4. Analysis of Rock Mass Fracture Characteristics

When rocks deal with complex stress conditions, their failure mainly presents as shear failure. In this study, the uniaxial test and triaxial test of rock mass were compared and analyzed. In the case of uniaxial compression, rock mass is generally shear failure, and there is a certain tensile failure phenomenon. However, in the case of confining pressure, the failure of rock mass is generally shear failure, and tensile failure rarely occurs. This is because the confining pressure constrains the transverse deformation of rock mass objectively.

The failure form or trace of rock mass is an important method to analyze the confining pressure failure mechanism. Taking the rock mass failure sample of fine sandstone (Figure 10) and stress–strain characteristic curve (Figure 11) as an example, the fracture trace of rock mass under uniaxial compression and triaxial confining pressure is analyzed and the fracture state of rock mass is analyzed.

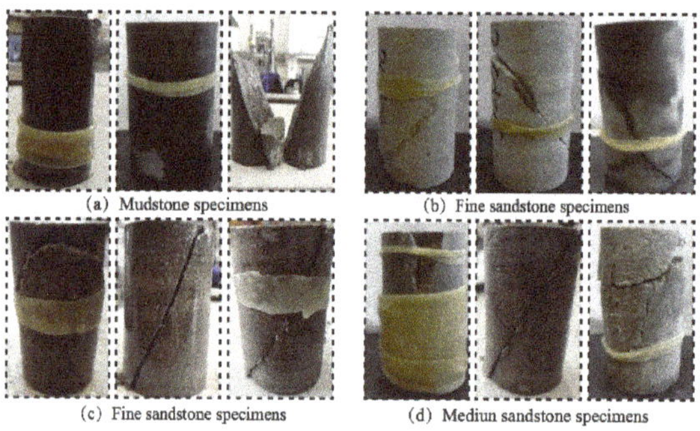

Figure 10. Prepare rock triaxial test samples.

On the other hand, it can be seen from Figure 12 that, under confining pressure, fine sandstone develops from multiple fractures at the beginning to a single fracture surface, which is manifested as compression shear failure of a single fracture surface. In general, with the increase of confining pressure, the angle of the fracture surface gradually becomes larger. When the confining pressure is 12 MPa, the angle of the fracture surface is about 80°, which significantly increases. From the analysis of the characteristics of the section, the fault surface of the fine sandstone with low confining pressure is relatively smooth. With the increase of confining pressure, the fault surface is more and more coarse, and the local expansion phenomenon is serious, indicating that under the condition of confining pressure, the confining pressure seriously restricts the transverse deformation of the rock, but, in general, the axial deformation of the fine sandstone is more serious.

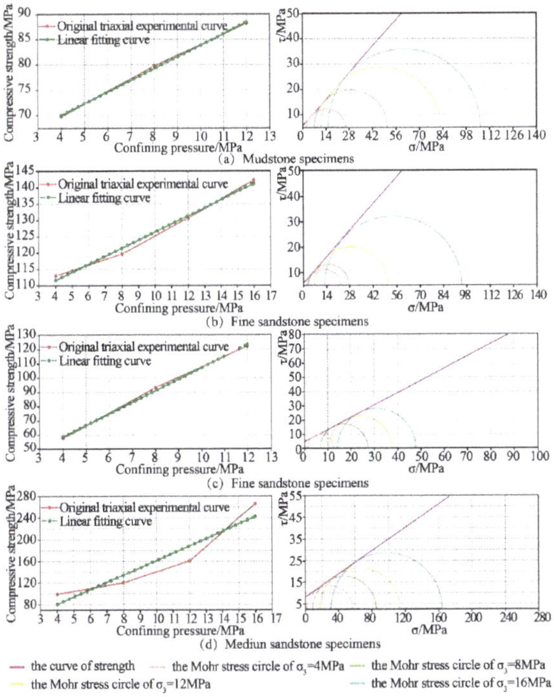

Figure 11. Triaxial test curve of mudstone, triaxial test curve of fine sandstone, triaxial test curve of fine sandstone, and triaxial test curve of medium sandstone.

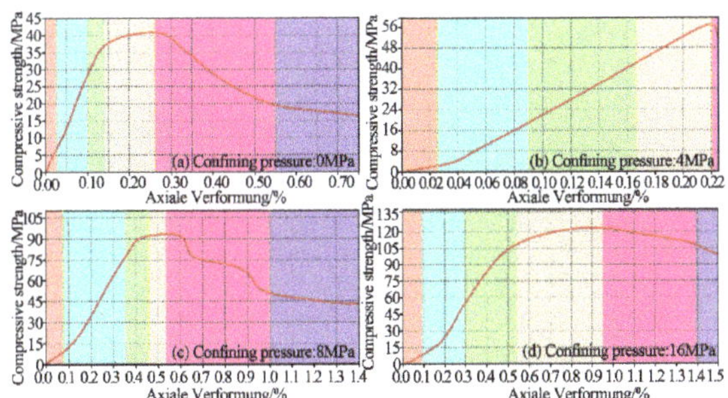

Figure 12. Total stress–strain curves of fine sandstone (group #3).

3.5. Stress and Strain Evolution of Rock Mass under Triaxial Stress

According to the indoor triaxial experiment, the characteristics of the total stress–strain curves of fine sandstone under confining pressure can be obtained. We aim to summarize the evolution law of these curves and better reflect the characteristics of rock stress and strain under actual conditions. In this study, statistical damage mechanics were used to quantitatively describe the damage and failure of fine sandstone. In the experimental process, it is assumed that the strength of each micro-element in fine sandstone follows Weibull distribution, and the damage and failure process is continuous. The distribution density function of micro-element strength in fine sand rock mass can be obtained as follows in Equations (7) and (8):

$$(\varepsilon) = \frac{m}{\varepsilon_0}\left(\frac{\varepsilon}{\varepsilon_0}\right)^{m-1} \exp\left[-\left(\frac{\varepsilon}{\varepsilon_0}\right)^m\right] \tag{7}$$

$$D = \frac{S}{S_m} = \int_0^{\varepsilon} (\varepsilon)d\varepsilon = 1 - \exp\left[-\left(\frac{\varepsilon}{\varepsilon_0}\right)^m\right] \tag{8}$$

Then, at the elastic stage of fine sandstone, the stress–strain curve of fine sandstone can be expressed by the following, Equation (9):

$$\begin{cases} \sigma_1 = (\lambda + 2G)\varepsilon'_1 + \lambda\varepsilon'_2 + \lambda\varepsilon'_3 \\ \sigma_2 = \lambda\varepsilon'_1 + (\lambda + 2G)\varepsilon'_2 + \lambda\varepsilon'_3 \\ \sigma_3 = \lambda\varepsilon'_1 + \lambda\varepsilon'_2 + (\lambda + 2G)\varepsilon'_3 \end{cases} \tag{9}$$

where σ_1 is the axial compressive strength, σ_2 and σ_3 are the confining pressures, where $\sigma_2 = \sigma_3$, λ and G are the Ramet constants, and ε'_1, ε'_2, and ε'_3 are the principal strains.

In the loading process of fine sandstone, the influence of hydrostatic pressure is fully considered; that is, the hydrostatic pressure is equal in the three principal stress directions during the loading process, and it is assumed that the deformation caused by hydrostatic pressure is ε_s; then, the expression of stress and strain of fine sandstone is as follows in Equation (10):

$$\begin{cases} \sigma_1 = E(\varepsilon_1 + \varepsilon_s) + v(\sigma_2 + \sigma_3) & \varepsilon'_1 = \varepsilon_1 + \varepsilon_s \\ \sigma_2 = E(\varepsilon_2 + \varepsilon_s) + v(\sigma_1 + \sigma_3) & \varepsilon'_2 = \varepsilon_2 + \varepsilon_s \\ \sigma_3 = E(\varepsilon_3 + \varepsilon_s) + v(\sigma_1 + \sigma_2) & \varepsilon'_3 = \varepsilon_3 + \varepsilon_s \end{cases} \tag{10}$$

In the triaxial experiment of fine sandstone, it is assumed that fine sandstone is continuous, medium, and isotropic. Therefore, for fine sandstone with isotropic anisotropic damage, the related computation is as the following Equation (11):

$$\varepsilon_1 + \varepsilon_s = \frac{1}{E}\left[\frac{\sigma_1}{1-D} - v\left(\frac{\sigma_2}{1-D} + \frac{\sigma_3}{1-D}\right)\right] \tag{11}$$

In the triaxial experiment, when the oil pressure is not loaded at the beginning of injection, the stresses in the three principal stress directions are equal. Therefore, the deformation ε_s caused by hydrostatic pressure can be obtained in the following Equation (12):

$$\sigma_3 = E\varepsilon_s + 2v\sigma_3 \quad \varepsilon_s = \frac{\sigma_3 - 2v\sigma_3}{E} \tag{12}$$

At the beginning of the triaxial experiment, the deformation caused by hydrostatic pressure is usually cleared before the experiment, and the data read during the experiment are ε_1, ε_2, and ε_3. In this study, the deformation caused by hydrostatic pressure is fully considered in the establishment of the mechanical model. Then, the expressions of axial pressure and confining pressure received by fine sandstone can be written as follows, in Equation (13):

$$\sigma_1 = E(\varepsilon_s + \varepsilon_1) + 2v\sigma_3 \quad \sigma_1 - \sigma_3 = E\varepsilon_1 \tag{13}$$

The failure of rock mass can be expressed by a function, and the general failure formula of rock mass is used as the formula of Equation (14), where K is the parameter related to the physical mechanics of fine sandstone, including the cohesion and internal friction angle of fine sandstone.

$$f(\sigma) - K = 0 \tag{14}$$

According to the existing research results, it can be seen that the Mohr Coulomb theory is used to deal with this kind of problem. The expression between principal stresses of the Mohr Coulomb theory can be written in the following Equation (15), and the general formula of the equation can be written as the following Equation (16).

$$\sigma_1 - \sigma_3 \frac{1+\sin\varphi}{1-\sin\varphi} = \frac{2C\cos\varphi}{1-\sin\varphi} \quad \varepsilon = (\varepsilon_1 + \varepsilon_s) - \frac{\frac{1+\sin\varphi}{1-\sin\varphi} - 2v}{E}\sigma_3 \tag{15}$$

$$D = \left[- \left(\frac{(\varepsilon_1 + \varepsilon_s) - \left(\frac{1+\sin\varphi}{1-\sin\varphi} - 2v\right)\frac{\sigma_s}{E}}{\varepsilon_0} \right)^m \right] \tag{16}$$

When fine sandstone is in the elastic stage, it can be considered that fine sandstone has no damage and $D = 0$:

$$D = 0, \varepsilon_1 + \varepsilon_s \le \left(\frac{1+\sin\varphi}{1-\sin\varphi} - 2v\right)\frac{\sigma_3}{E} \tag{17}$$

$$D = 1 - \exp\left[- \left(\frac{(\varepsilon_1 + \varepsilon_s)E - \left(\frac{1+\sin\varphi}{1-\sin\varphi} - 2v\right)\sigma_3}{E\varepsilon_0} \right)^m \right] \tag{18}$$

When the failure of fine sandstone occurs, $D \neq 0$, the stress–strain equation can be written into the following equation to express the failure degree of fine sandstone, and the relationship between stress and strain can also be obtained by Equation (23).

$$\varepsilon_1 + \varepsilon_{0s} \ge \left(\frac{1+\sin\varphi}{1-\sin\varphi} - 2v\right)\frac{\sigma_3}{E} \tag{19}$$

$$\sigma_1 - \sigma_3 = E\varepsilon_1 \exp\left[- \left(\frac{(\varepsilon_1 + \varepsilon_s)E - \left(\frac{1+\sin\varphi}{1-\sin\varphi} - 2v\right)\sigma_3}{E\varepsilon_0} \right)^m \right] dajia \tag{20}$$

$$\sigma_1 = E\varepsilon_1 \exp\left[- \left(\frac{(\varepsilon_1 + \varepsilon_s)E - \left(\frac{1+\sin\varphi}{1-\sin\varphi} - 2v\right)\sigma_3}{E\varepsilon_0} \right)^m \right] + \sigma_3 \tag{21}$$

$$\varepsilon_u = \left(\frac{1+\sin\varphi}{1-\sin\varphi} - 2v\right)\frac{\sigma_3}{E} \tag{22}$$

$$\sigma_1 = E\varepsilon_1 \exp\left[- \left(\frac{(\varepsilon_1 + \frac{\sigma_3 - 2v\sigma_3}{E})E - \left(\frac{1+\sin\varphi}{1-\sin\varphi} - 2v\right)\sigma_3}{E\varepsilon_0} \right)^m \right] + \sigma_3 \tag{23}$$

The above equation can be abbreviated as follows:

$$\sigma_1 = E\varepsilon_1 \exp\left[- \left(\frac{\varepsilon_1 + \varepsilon_s - \varepsilon_u}{\varepsilon_0} \right)^m \right] + \sigma_3 \tag{24}$$

The residual strength in fine sandstone can be considered to be approximately a straight line, as follows in Equation (25):

$$\sigma_1 - \sigma_3 = \sigma_p \tag{25}$$

The above equation is the process of establishing the stress–strain equation of fine sandstone under complex stress conditions. For the equation, the solution of the unknowns ε_0 and m has become the key problem of this equation.

For the solution of these unknowns, there are now enough means. For the solution of unknowns, origin fitting method is adopted, and finally the linear law between ε_0 and the confining pressure σ_3, and m and the confining pressure σ_3 is found. According to the experimental results, the elastic modulus of fine sandstone is 20.85 GPa, Poisson's ratio is 0.172, cohesion is 7.8 MPa, and internal friction angle is 48.5°. Finally, these parameters are brought into the equations, and the stress–strain relationship equation of fine sandstone under confining pressure can be established uniformly (as shown in Figures 13 and 14).

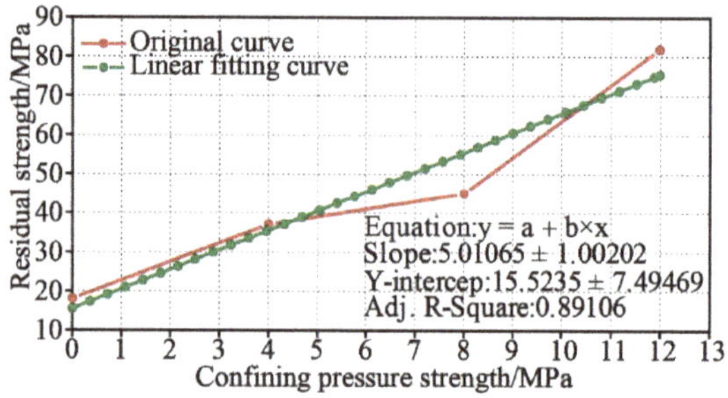

Figure 13. The relations between confining pressure strength and residual strength.

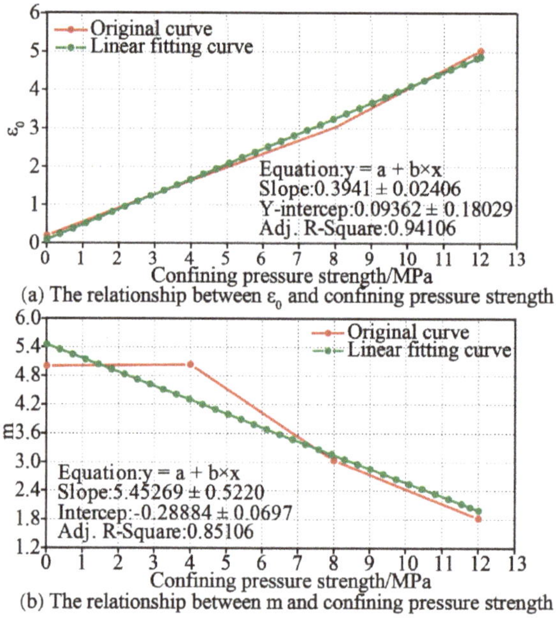

Figure 14. The relations between unknown parameters and confining pressure strength.

4. Discussion and Conclusions

In this study, physical property parameters of rock mass, such as density and water absorption, were obtained. Physical and mechanical experiments of rock mass were carried out in the laboratory, and mechanical parameters of rock mass were obtained, such as elastic modulus, Poisson's ratio, internal friction angle, and cohesion. The main conclusions of this study are as follows:

1. By carrying out physical and mechanical experiments of rock, the elastic modulus, Poisson's ratio, internal friction angle, and cohesion of coal and rock are obtained. The experimental results show that the compressive strength of the sample is in the order of medium sandstone > fine sandstone > mudstone.

2. The mechanical equation of fine sandstone under triaxial stress is established. On the basis of statistical damage mechanics, this chapter fully considers the influence of hydrostatic pressure in the experimental process, takes hydrostatic pressure as a part of the equation, and establishes a stress–strain equation based on confining pressure, which can describe the mechanical failure characteristics of rock under triaxial stress.
3. There is little difference in the cohesion and internal friction angle of the three rocks, among which mudstone has the lowest cohesion value under triaxial compression, indicating that the triaxial compression shear strength of the three rocks is similar. The cohesion of coal rock is the lowest, indicating that coal rock is easy to be destroyed by shear.

Author Contributions: Methodology, Y.T., H.D., R.Z., J.F. and Z.D.; writing—original draft, H.D.; writing—review and editing, R.Z., J.F. and Z.D. All authors have read and agreed to the published version of the manuscript.

Funding: This work was financially supported by the National Key Research and Development Plan (Grant No. 2018YFC1504902), the National Natural Science Foundation of China (Grant No. 52079068,52090081), the Key Research and Development Plan of Ningxia Hui Autonomous Region (Grant No. 2018BCG01003), the State Key Laboratory of Hydroscience and Engineering (Grant No. 2021-KY-04), and National Natural Science Foundation of China (No. 12102230).

Institutional Review Board Statement: Not applicable.

Informed Consent Statement: Not applicable.

Data Availability Statement: The data used to support the findings of this study are available from the corresponding author upon request.

Acknowledgments: The authors would like to thank the editor and the reviewers for their contributions.

Conflicts of Interest: The authors declare that they have no conflict of interest.

References

1. Katz, H.A.; Daniels, J.M.; Ryan, S. Slope-area thresholds of road-induced gully erosion and consequent hillslope-channel interactions. *Earth Surf. Process. Landforms* **2013**, *39*, 285–295. [CrossRef]
2. Zhang, J.; Li, M.; Taheri, A.; Zhang, W.; Wu, Z.; Song, W. Properties and Application of Backfill Materials in Coal Mines in China. *Minerals* **2019**, *9*, 53. [CrossRef]
3. Ye, Q.; Wang, G.; Jia, Z.; Zheng, C.; Wang, W. Similarity simulation of mining-crack-evolution characteristics of overburden strata in deep coal mining with large dip. *J. Pet. Sci. Eng.* **2018**, *165*, 477–487. [CrossRef]
4. Dong, L.; Sun, D.; Shu, W.; Li, X. Exploration: Safe and clean mining on Earth and asteroids. *J. Clean. Prod.* **2020**, *257*, 120899. [CrossRef]
5. Duan, B.; Xia, H.; Yang, X. Impacts of bench blasting vibration on the stability of the surrounding rock masses of roadways. *Tunn. Undergr. Space Technol.* **2018**, *71*, 605–622. [CrossRef]
6. He, M.; Feng, J.; Sun, X. Stability evaluation and optimal excavated design of rock slope at Antaibao open pit coal mine, China. *Int. J. Rock Mech. Min. Sci.* **2007**, *45*, 289–302. [CrossRef]
7. Du, S.; Feng, G.; Wang, J.; Feng, S.; Malekian, R.; Li, Z. A New Machine-Learning Prediction Model for Slope Deformation of an Open-Pit Mine: An Evaluation of Field Data. *Energies* **2019**, *12*, 1288. [CrossRef]
8. Yu, Y.; Wang, E.; Zhong, J.; Liu, X.; Li, P.; Shi, M.; Zhang, Z. Stability analysis of abutment slopes based on long-term monitoring and numerical simulation. *Eng. Geol.* **2014**, *183*, 159–169. [CrossRef]
9. Xu, C.; Liu, X.; Wang, E.; Zheng, Y.; Wang, S. Rockburst prediction and classification based on the ideal-point method of information theory. *Tunn. Undergr. Space Technol.* **2018**, *81*, 382–390. [CrossRef]
10. Liu, X.; Wang, S. Mine water inrush forecasting during the mining under waters. *Disaster Adv.* **2012**, *5*, 876–881.
11. Lin, P.; Liu, X.; Hu, Y.; Xu, W.; Li, Q. Engineering. Deformation stability analysis of Xiluodu arch dam under stress-seepage coupling condition. *Chin. J. Rock Mech. Eng.* **2013**, *32*, 1145–1156.
12. Lin, P.; Liu, X.; Hu, S.; Li, P. Large Deformation Analysis of a High Steep Slope Relating to the Laxiwa Reservoir, China. *Rock Mech. Rock Eng.* **2016**, *49*, 2253–2276. [CrossRef]
13. Liu, X.; Han, G.; Wang, E.; Wang, S.; Nawnit, K. Multiscale hierarchical analysis of rock mass and prediction of its mechanical and hydraulic properties. *J. Rock Mech. Geotech. Eng.* **2018**, *10*, 694–702. [CrossRef]
14. Hu, B.; Zhao, H.-B.; Wang, S.-J.; Liu, H.-N.; Peng, Y.-D.; Liu, X.-L.; Zhang, X.-P. Pull-out model test for tunnel anchorage and numerical analysis. *Rock Soil Mech.* **2009**, *30*, 1575–1582.
15. Liu, C.; Liu, X.; Peng, X.; Wang, E.; Wang, S. Application of 3D-DDA integrated with unmanned aerial vehicle–laser scanner (UAV-LS) photogrammetry for stability analysis of a blocky rock mass slope. *Landslides* **2019**, *16*, 1645–1661. [CrossRef]

16. Huang, X.; Liu, Q.; Shi, K.; Pan, Y.; Liu, J. Application and prospect of hard rock TBM for deep roadway construction in coal mines. *Tunn. Undergr. Space Technol.* **2018**, *73*, 105–126. [CrossRef]
17. Salmi, E.; Karakus, M.; Nazem, M. Assessing the effects of rock mass gradual deterioration on the long-term stability of abandoned mine workings and the mechanisms of post-mining subsidence–A case study of Castle Fields mine. *Tunn. Undergr. Space Technol.* **2019**, *88*, 169–185. [CrossRef]
18. Liu, X.; Wang, S.; Wang, S.; Wang, E.J. Fluid-driven fractures in granular materials. *Bull. Eng. Geol. Environ.* **2015**, *74*, 621–636. [CrossRef]
19. Lv, Q.; Wang, E.; Liu, X.; Wang, S. Determining the intrinsic permeability of tight porous media based on bivelocity hydrodynetics. *Microfluid. Nanofluid.* **2014**, *16*, 841–848. [CrossRef]
20. Li, J.; Hu, M.; Ding, E.; Kong, W.; Pan, D.; Chen, S. Multi-parameter numerical simulation of dynamic monitoring of rock deformation in deep mining. *Int. J. Min. Sci. Technol.* **2016**, *26*, 851–855. [CrossRef]
21. Jing, H.; Wu, J.; Yin, Q.; Wang, K. Deformation and failure characteristics of anchorage structure of surrounding rock in deep roadway. *Int. J. Min. Sci. Technol.* **2020**, *30*, 593–604. [CrossRef]
22. Zhang, N.; Yuan, L.; Han, C.; Xue, J.; Kan, J. Stability and deformation of surrounding rock in pillarless gob-side entry retaining. *Saf. Sci.* **2012**, *50*, 593–599. [CrossRef]
23. Yu, W.; Liu, F.J. Stability of close chambers surrounding rock in deep and comprehensive control technology. *Adv. Civ. Eng.* **2018**, *2018*, 40. [CrossRef]
24. He, H.; Yan, Y.; Qu, C.; Fan, Y. Study and Application on Stability Classification of Tunnel Surrounding Rock Based on Uncertainty Measure Theory. *Math. Probl. Eng.* **2014**, *2014*, 1–5. [CrossRef]
25. Chen, S.; Wu, A.; Wang, Y.; Chen, X.; Yan, R.; Ma, H. Study on repair control technology of soft surrounding rock roadway and its application. *Eng. Fail. Anal.* **2018**, *92*, 443–455. [CrossRef]
26. He, M.; Gong, W.; Wang, J.; Qi, P.; Tao, Z.; Du, S.; Peng, Y. Development of a novel energy-absorbing bolt with extraordinarily large elongation and constant resistance. *Int. J. Rock Mech. Min. Sci.* **2014**, *67*, 29–42. [CrossRef]
27. Sun, X.-M.; He, M.-C. Numerical simulation research on coupling support theory of roadway within soft rock at depth. *J. China Univ. Min. Technol.* **2005**, *34*, 166–169.
28. Kang, Y.; Liu, Q.; Gong, G.; Wang, H. Application of a combined support system to the weak floor reinforcement in deep underground coal mine. *Int. J. Rock Mech. Min. Sci.* **2014**, *71*, 143–150. [CrossRef]
29. Li, S.; Wang, Q.; Wang, H.; Jiang, B.; Wang, D.; Zhang, B.; Li, Y.; Ruan, G. Model test study on surrounding rock deformation and failure mechanisms of deep roadways with thick top coal. *Tunn. Undergr. Space Technol.* **2015**, *47*, 52–63. [CrossRef]
30. Singh, R.; Mallick, M.; Verma, M. Studies on failure behaviour of wire rope used in underground coal mines. *Eng. Fail. Anal.* **2016**, *70*, 290–304. [CrossRef]
31. Sasaoka, T.; Shimada, H.; Lin, N.Z.; Takamoto, H.; Matsui, K.; Kramadibrata, S.; Sulistianto, B. Geotechnical issues in the application of rock bolting technology for the development of underground coal mines in Indonesia. *Int. J. Min. Reclam. Environ.* **2013**, *28*, 150–172. [CrossRef]
32. Du, Q.; Liu, X.; Wang, E.; Wang, S. Strength Reduction of Coal Pillar after CO2 Sequestration in Abandoned Coal Mines. *Minerals* **2017**, *7*, 26. [CrossRef]
33. Liu, X.; Wang, S.; Wang, E. A study on the uplift mechanism of Tongjiezi dam using a coupled hydro-mechanical model. *Eng. Geol.* **2011**, *117*, 134–150. [CrossRef]
34. Dai, S.; Liu, X.; Nawnit, K. Experimental Study on the Fracture Process Zone Characteristics in Concrete Utilizing DIC and AE Methods. *Appl. Sci.* **2019**, *9*, 1346. [CrossRef]
35. Sun, H.; Liu, X.; Ye, Z.; Wang, E. Experimental investigation of the nonlinear evolution from pipe flow to fissure flow during carbonate rock failures. *Bull. Eng. Geol. Environ.* **2021**, *80*, 4459–4470. [CrossRef]
36. Sun, W.; Hao, J.; Liu, Y. A new proposed method for observing fluid in rock fractures using the enhanced X-ray image of digital radiography. *Geomech. Geophys. Geo-Energy Geo-Resour.* **2021**, *8*, 10. [CrossRef]
37. Liu, H.; Long, Q.; Liu, X.; Ye, Z.; Wang, E.; Du, W. Opencast Mining Technology j. Analysis of hydrogeological characteristics and water control measures of Pingshuo East open-pit mine. *Opencast Min. Technol.* **2021**, *31*, 78–91.
38. Jia, S.; Liu, T.; Yan, Y. Technology Innovation and Application j. Analysis of geological environment of pingshuo Group East open-pit Mine. *J. Clean. Prod.* **2014**, *15*, 126–127.

applied sciences

MDPI

Article

GIS-Based Planning and Web/3D Web GIS Applications for the Analysis and Management of MV/LV Electrical Networks (A Case Study in Tunisia)

Mohamed Hafedh Hamza [1,2,*] and Mohamed Chmit [2]

1 Department of Geography and Geographic Information Systems, Faculty of Arts and Humanities, King Abdulaziz University, Jeddah 21589, Saudi Arabia
2 Geomatics Section, Department of Geology, Faculty of Science of Tunis, University of Tunis El Manar, Tunis 1068, Tunisia; chmitmohamed@gmail.com
* Correspondence: mhhamza@kau.edu.sa or hafedh.hamza@fst.rnu.tn

Abstract: Geographic Information Systems (GISs) have an essential part to play in the management and planning of electricity distribution. Since the management of electricity network data was previously conducted in Tunisia based on paper maps and plans, the purpose of this study is to present a case for the planning of an MV/LV (Medium Voltage/Low Voltage) electrical network in the region of Medjez El Bab (North-West of Tunisia), based on GIS, Web, and 3D Web GIS, to create an intelligent electricity network, which will be a decision-making tool. Analyses of vehicle transport and pedestrian accessibility between installations and a generation of Origin-Destination cost matrix to calculate the average transport distances between the service points were conducted. Moreover, an analysis of the network's impedance allowed carrying out different scenarios to optimize performance and could obtain more efficient routes. The different analyses carried out were crucial for the maintenance of the electrical network and for future urban planning. A 3D virtual city has been developed to visualize graphical and attribute data for the study area. Web and 3D Web GIS applications that allow the publication of interactive maps on the Web as well as database information have been developed to offer users the possibility of consulting produced products by using the internet. A website related to the study was equally developed to gather the different obtained results.

Keywords: MV/LV network; GIS-based planning; spatial/network analysis; 3D virtual city; Web/3D Web GIS

Citation: Hamza, M.H.; Chmit, M. GIS-Based Planning and Web/3D Web GIS Applications for the Analysis and Management of MV/LV Electrical Networks (A Case Study in Tunisia). *Appl. Sci.* **2022**, *12*, 2554. https://doi.org/10.3390/app12052554

Academic Editors: Alexei Gvishiani and Boris Dzeboev

Received: 11 February 2022
Accepted: 24 February 2022
Published: 28 February 2022

1. Introduction

The main objective of planning electricity distribution networks is to determine the necessary conditions from a topological, structural, and physical point of view [1]. This goal has become a major issue in recent years for distribution planning services due to various changes occurring in the electricity sector [2,3]. The construction of new electrical facilities such as substations, transformer, and distribution lines depends on the available areas in the study area [4]. Determining the geographical location of electrical installations and identifying the elements of the electrical network constitute the great challenges that arise in the planning of the distribution network.

Geographic Information Systems (GISs) are computational tools that allow georeferenced databases to be associated with digital maps to facilitate the visualization of the geographical characteristics of a study area [5]. GISs help render data analyses easier to handle and provide high capabilities in dealing with large spatial data [6]. GISs are composed by data analysis and optimization tools to solve problems in various fields of engineering [7]. Effective deployment of GIS in power distribution would go a long way in the power industry in terms of being efficient and financially viable in addition to enhancing the consumer's satisfaction due to uninterrupted quality power supply [8]. The spatial and network analysis tools available in the GIS have been used in various planning studies of the electricity distribution network around the world [9–12]. Different GIS software could be used, whether commercial, such as ArcGIS, MapInfo Professional, or Global Mapper, or free, such as QGIS or GRASS GIS.

In addition to the basic functionalities of GIS, the WebGIS technique often integrates additional functionality by using data from different sources. WebGIS, a combination of Web and GIS, has become a growing discipline since its beginning in 1993. In 1993, Xerox Corporation's research center in Palo Alto (Santa Clara, CA, USA) developed a Web-based map viewer, marking the origin of Web GIS. This experiment made it possible to retrieve interactive information from the Web, rather than providing access to purely static files [13]. GIS became a key Internet application that allowed the public to take advantage of the benefits of the Web [14].

The data management of the electricity network was previously conducted in Tunisia with a classic method, by using paper maps and plans; the idea of this work was to create an intelligent electricity network and, thus, allow making appropriate decisions and advances [15]. It consists of carrying out GIS applications for the analysis and management of the MV/LV electricity network of the national Tunisian electricity and gas company STEG in the region of Medjez El Bab located in the governorate of Beja (worthwest of Tunisia) and seting up a Web GIS application related to it. The electrical network is made up of pylons, transformer stations, and power lines. MV lines are generally of the overhead type (sometimes underground near urban sites), and among these MV lines, there are bypasses that supply LV substations serving to supply low voltage to customers of the network [16]. The steps that will be followed in this study consist of creating a very large-scale database to set up the MV/LV network, defining and correcting topology rules (duplicates, inclusions, etc.), determining spatial relationships between objects, displaying the two MV/LV databases at several scales, and practicing several spatial and network analyses by enabling an intelligent study of the transport network and an efficient organization and coordination of vehicles and pedestrian technicians, and this will enable making the right decisions by developing strategic routing plans.

The use of GIS as a tool for managing electricity distribution will facilitate the use of electricity network data and the rapid resolution of technical problems with a double gain in time and material and will also make communication easier between the managers of this resource and the corresponding consumers. The planning of the distribution network is also an important element of urban planning and needs to be assessed comprehensively in order to improve the level of decision making and investment benefits as demonstrated in the study of Zheng et al. in China in 2012 [17].

Many other studies related to GIS-based electricity network management were established around the world as the study of Mentis et al. in Nigeria in 2015 [18], which discussed GIS-based electrification planning with a least cost approach. The study of Candelisea and Westacottc in the United Kingdom in 2017 [19] analyzed using GIS with the integration of photovoltaics within the UK electricity network. The study of Ashkezari et al. in Oman in 2018 [20] related to the implementation of a GIS performing real-time monitoring of the electrical network and producing its spatial representation, making it possible to control, search, and analyze related information. The study of Taye et al. in [21] established rural electrification planning using GIS in Ethiopia. All these studies demonstrated the importance of GIS as a powerful management tool for electrical networks.

Web GIS and 3D Web applications allowing the publication of interactive maps on the Web as well as the information of the database will be developed in this study to offer users the possibility of consulting the resulting products on the Internet. In addition, a website gathering the results of the study will be developed.

In the present study, we will try to demonstrate the importance of the use of GIS for electrical networks and we show the advantages of the use of Web GIS and 3D Web GIS, enabling the consultation of cartographic products via the Internet; we further show how the integration of GIS data into web platforms will significantly improve the use of both data and applications.

This paper is structured in four sections: Introduction; Methodology; Results and discussion; and Conclusion.

2. Methodoloy

2.1. Study Area

The study area (Figure 1) is located in the Northwest of Tunisia in the eastern part of the governorate of Beja. It lies between 36°45′/36°33′ N and 9°23′/9°45′ E. Its total area is 634 km^2. The study area is characterized by a heterogeneous and highly varied topographical surface: plains and hills set between mountain chains (up to 528 m high) that dominate from the northwest to the south. However, plains and hills are the dominant topographical feature of the area (almost 75%). The hills that bound the region in its northwest and southeast parts do not exceed 300 m. The plains dominate the northeastern part of the area with elevations ranging from 10 to 50 m. The vegetation cover is generally much degraded on the mountains where foothills are cleared. The mountains are, therefore, areas of concentration of water with a strong or medium erosive potential, while the foothills are little or moderately eroded because of their very low slope.

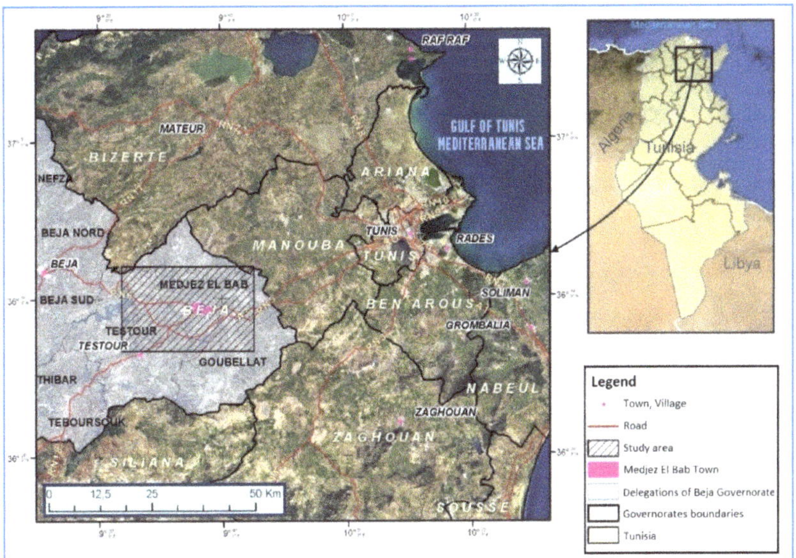

Figure 1. Study area.

2.2. Planning of the Electrical Networks Using GIS Tools

The first step is to process cadastral data covering the city of Medjez El Bab and integrate them into a spatial database at a municipal level. The second one consists of processing physical data around the city of Medjez El Bab (Grand Medjez El Bab area) to create another spatial database at a local scale that is able to be implemented afterwards with two geodatabases at different scales and uses. The completion of that part of the study is articulated in the following phases: data acquisition, database design, the establishment of some important thematic maps, and the elaboration of a 3D virtual city.

2.2.1. Data Acquisition

Nine official topographic maps produced by the national office of topography and cadaster OTC between 1982 and 1988 at a scale of 1: 25,000 [22] were georeferenced and digitized, and topology rules were used to minimize digitizing errors. Ten old topographic plans at a scale of 1:1000 produced in dwg format were also used [23]. These plans were processed: Manual and automatic topological corrections were performed on them. Maxar 2020 background image related to our study area with a 1 m resolution was extracted from the "World Imagery" service and was used to update data related to topographic maps and plans data. Maxar is a product of DigitalGlobe, an industry leader in satellite imagery and geospatial intelligence [24]. Moreover, a GPS Leica GS08plus was used for the purpose of updating the electrical network. The different steps of data acquisition are presented in Figure 2.

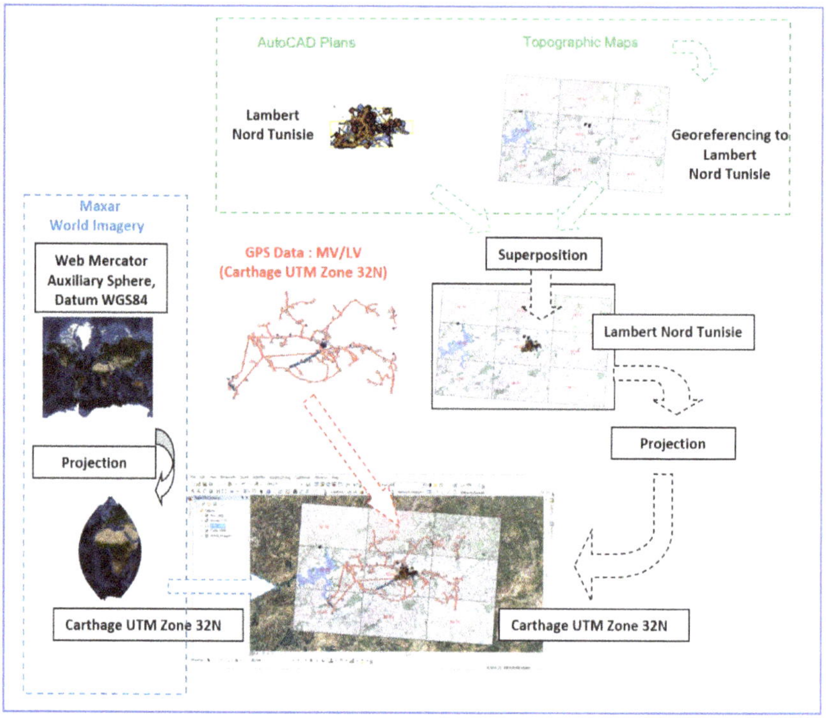

Figure 2. Steps followed for data acquisition.

2.2.2. Database Design

The database design was accomplished by applying MERISE methodology. MERISE is an Information System Design and Development methodology first introduced in the early 1980s, and it was widely used in France [25]. MERISE proceeds to separate treatment of data and processes, where the data-oriented view is modeled in three stages, from conceptual to logical to physical [26].

A *Conceptual Data Model (CDM)* helps analyze the conceptual structure of an information system and identifies the principal entities to be represented, their attributes, and the relationships between them [27]. The conceptual data model consists of tables, each of which must include a field that uniquely identifies each record (primary keys). The choice of tables was made in our study at two regional levels: a first level relating to the delegation (Grand Medjez El Bab which represents the entire project area) and a second level relating to the municipality of Medjez El Bab. The *Conceptual Data Model (CDM)* was then transited into a Physical Data Model (PDM). The architecture and structural details of the GeoDatabase are shown in Figure 3.

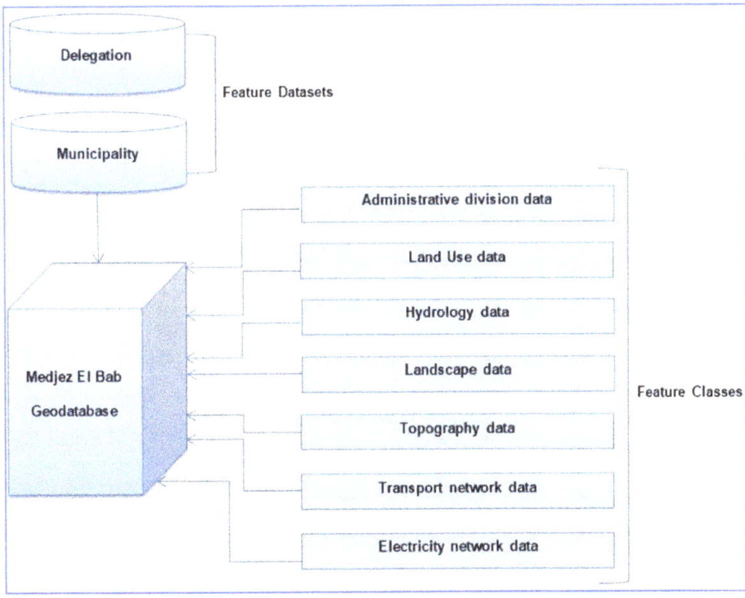

Figure 3. Geodatabase architecture and structural details.

The next phase of the process was to structure the data in a multi-representation Geodatabase. In this phase, it is necessary to proceed with cartographic generalization. Cartographic generalization is the process of controlling the amount of detailed information portrayed in a map. This also includes using the appropriate scale, purpose, and medium of the map. This form of generalization commonly consists of reducing the visual detail of data by reducing the map scale when the map purpose suggests the need for a simpler design. As analytical approaches to geography arose in the 1950s and 1960s, generalization, especially line simplification and raster smoothing, was a target of study [28,29]. The generalization of maps has become necessary due to automatic production of maps on the web and the increased amount of detailed GIS data available [30]. A multi-representation of multiple Geodatabases at different scales is shown in Figure 4. The last step was to determine spatial relationships between features. Relationship classes in the geodatabase manage the associations between objects in one class (feature class or table) and objects in another. Objects at either end of the relationship can be features with geometry or records in a table.

2.2.3. Elaboration of Thematic Maps

After data extraction and topological corrections, many thematic maps were established using different GIS tools. The extracted Digital Elevation Map (DEM) was based on an interpolation method specifically designed for the creation of hydrologically correct digital elevation models, and the ANUDEM method was developed by Michael Hutchinson [31,32]. That map showed that the plains and hills represent the dominant topographic aspect in the area (nearly 75%). The slope map extracted (Figure 5) showed that 62% of the study area had a slope less than 6%, 28% of the land had slopes between 6 and 18%, 5% had relatively steep slopes of 18 to 30%, and about 5% of the land had steep slopes greater than 30%. It follows that about 10% of the total study area has a slope greater than 18%, representing lands threatened by moderate to severe erosion. The steeply sloping areas are located in the transition zones between reliefs and plains and are occupied by flourishing agriculture. These areas can, however, be the site of severe erosion even with relatively low slopes. This erosion is accentuated by torrential runoff in the autumn and winter, which can together directly cause flooding and stripping of the fertile land, which will be sites of severe erosion even with relatively low slopes. This erosion is accentuated by torrential runoff in autumn and winter, which can together directly cause flooding and stripping of the fertile cover of the plains surrounding the landforms. An updated land use map was extracted based on different topographic maps and plans, the Maxar 2020 image, and GPS data (Figure 6). Many other maps related to MV/LV electrical networks were established as the classification map of the MV electrical network, the map of the MV voltage domains, and the map of MV/LV transformer stations (Figure 7).

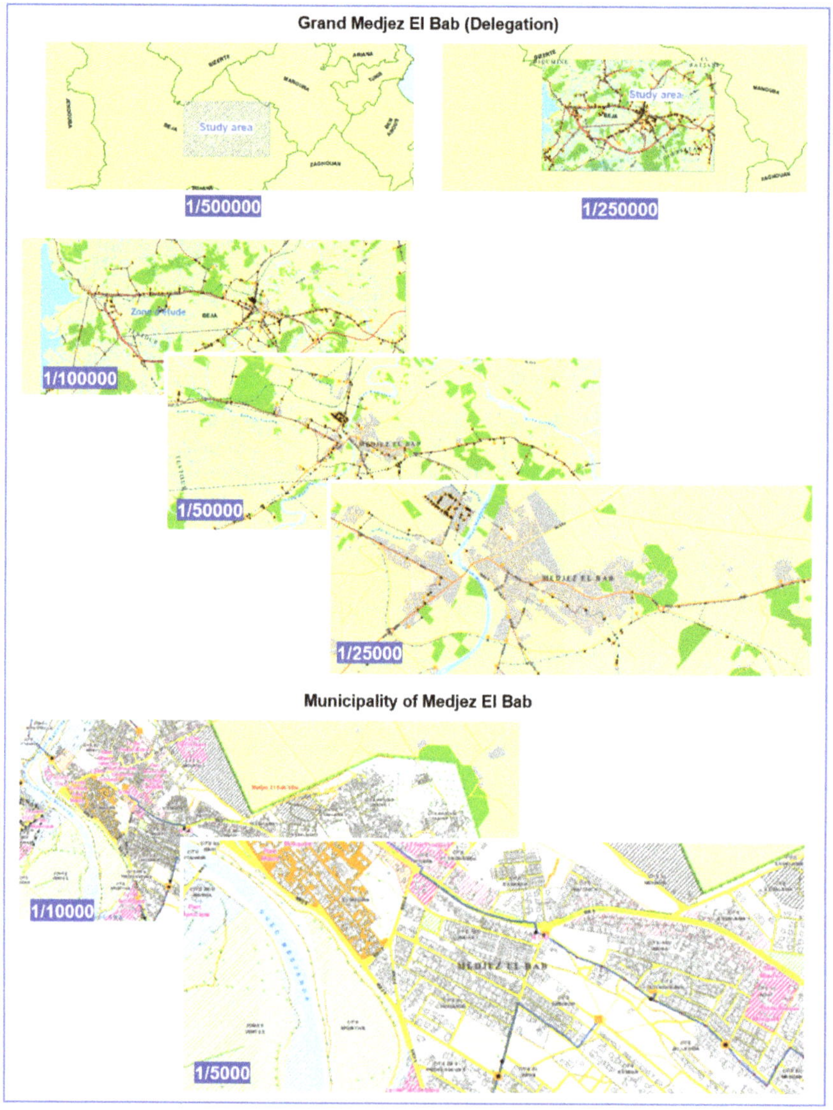

Figure 4. Multi-representation schema of multiple Geodatabases at different scales.

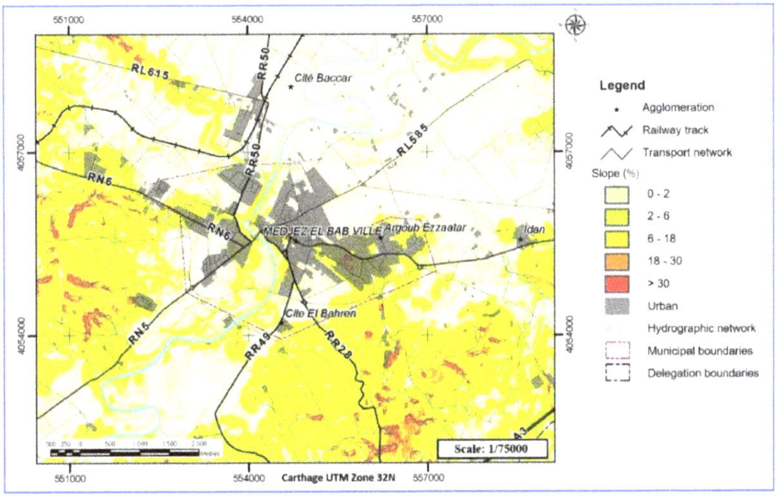

Figure 5. Slope map.

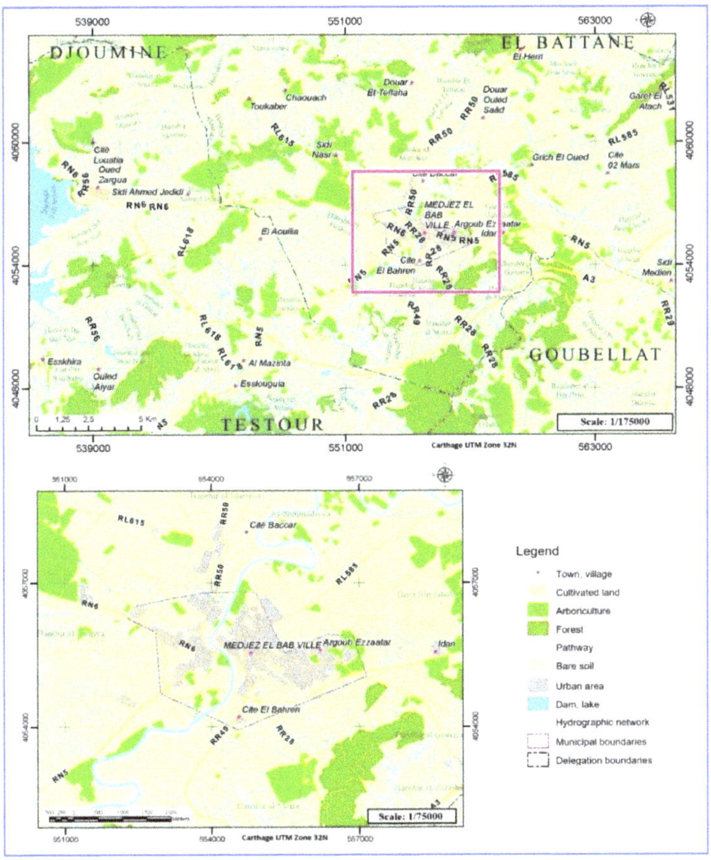

Figure 6. Land use map.

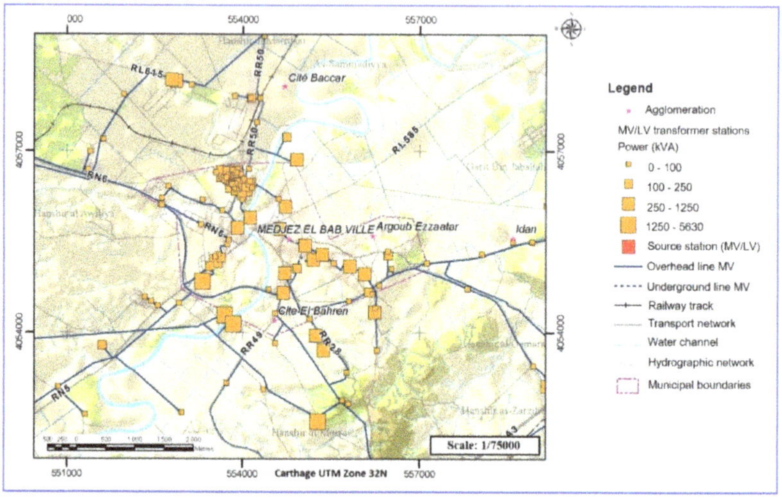

Figure 7. MV/LV transformer stations.

2.2.4. Elaboration of A 3D Virtual City

A three-dimensional city model is a digital representation of the Earth's surface and its related objects such as buildings, trees, vegetation, and some manmade features belonging to the urban area [33]. This realistic visualization represented in our study is a prototype that can help decision makers in future projects. Future projects may relate to the sites of works or other electrical network extensions, while taking into account what has been performed. In order to allow different levels of modeling, several levels of details (LOD) have been defined: LOD 0 (Regional model): this is a digital terrain model (2.5D) showing the entire landscape and for which its data come mainly from airborne laser; LOD 1 (Urban model): "block model" in which the buildings are schematized in the form of blocks without roof structures, thus giving an idea of the distribution of the height of the buildings; LOD 2 (Urban model): same as the above, but with textures for facades and roofs (oblique photogrammetry); LOD 3 (Urban model): same model as the above but with a more detailed architectural level; LOD 4 (Interior model): truly architectural "walkable" model, that is to say that it has models of the interior of buildings [34].

The CityGML standard was used in our study, which is an XML-based encoding for 3D representation, allowing the sharing and storage of virtual models of cities and landscapes. It is executed as an open data model deployed as an application schema for Geography Markup Language 3, the extensible international standard for spatial data exchange released by the Open Geospatial Consortium (OGC). CityGML offers a standard model to describe 3D objects in terms of their geometry, topology, semantics, and appearance. It also contains generalization hierarchies between classes, aggregations, relationships between objects, and spatial properties [35,36].

We proceeded to model the geographic space, which delaminates a region of 10 km around the city of Medjez El Bab. The first step is to display the digital terrain model in a 3D planimetric view. We then modeled various elements that make up MV/LV networks based on the drawing commands (geometry, coloring, and modification) and photos taken in the field. The modeling of the land use components of our study area was performed by adding the different data layers (urban component, roads, hydrography, and electrical network) in the scene and superimposing them with the relief and then performing an extrusion of the buildings to create realistic 3D objects (extrude according to the heights and the number of floors). The 3D model of the study area obtained is shown in Figure 8.

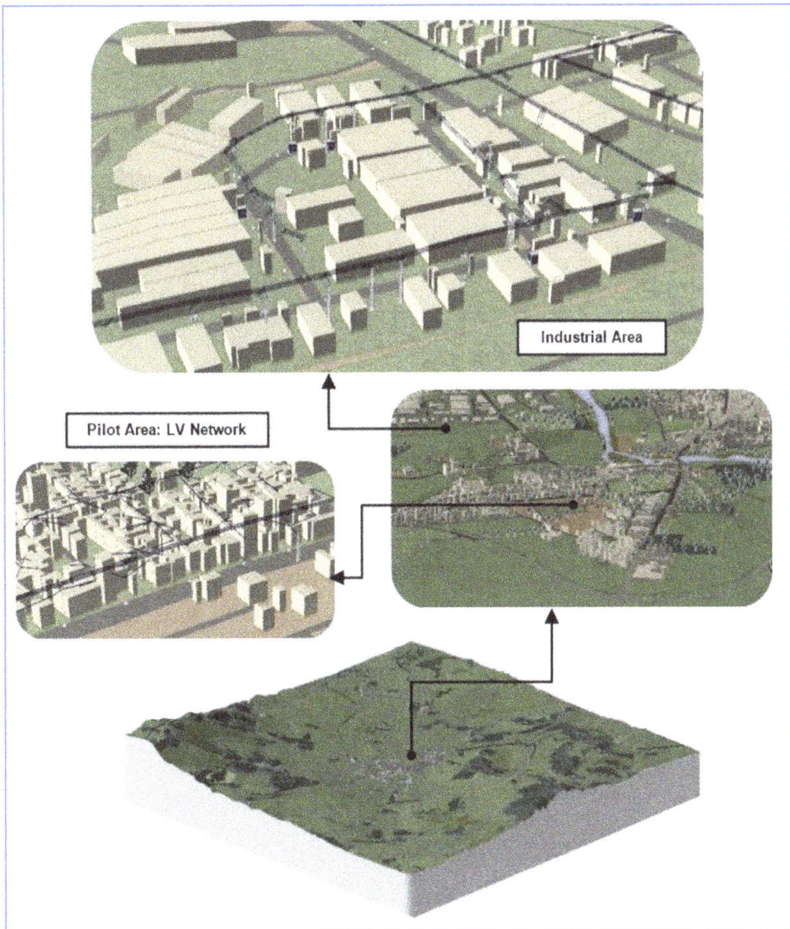

Figure 8. 3D model of the study area.

2.3. Web and 3D Web GIS Applications and Website Development

2.3.1. Web and 3D Web GIS Applications

This phase is related to the realization of a Web GIS in order to offer users the possibility of consulting the produced products by using the internet. The world had 4.79 billion Internet users in 2020 or 5.5% more than in 2019. This is 166% more users that existed 10 years ago [37]. The Internet and, in particular, the World Wide Web enable easy access to spatial data and applications that can offer fast and simple solutions to the spatial needs of people, such as public services and business operations [38]. Web-based GIS applications have gained popularity due to their ease of use and simplicity [39].

In the framework of our study related to an electrical network, the client side is a spatial data visualization interface (Map Viewer) based on GeoExt. The latter is a web application developed under OpenLayers to create and publish maps. This interface allows adding WMS (Web Map Service) layers from GeoServer and from an API (Application Programming Interface) such as Google Maps. Indeed, we have realized two cartographic interfaces: One for a global visualization of data using the database of the Greater Medjez El Bab (where the MV network is located); and a second one to display municipal data (where the LV network is located). Figure 9 for example shows, for example, the municipality cartographic interface.

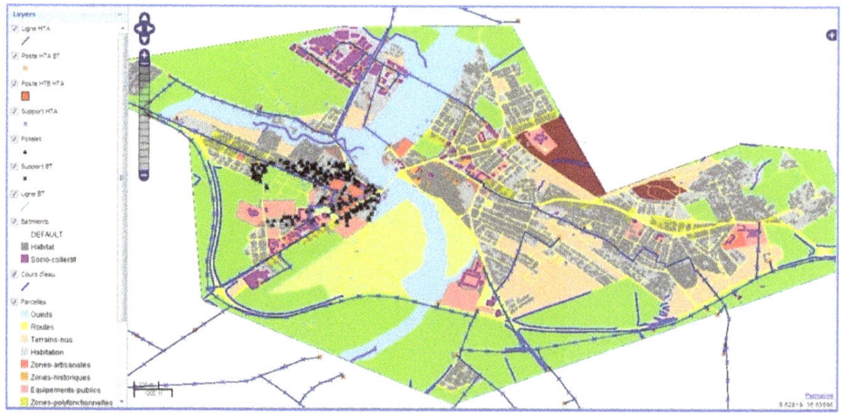

Figure 9. Medjez El Bab Municipality web cartographic interface.

On the other hand, 3D Web GIS has become a real requirement in several fields of study thanks to its devices for analyzing 3D models linked to spatially referenced data [40]. It is important to note that standardization initiatives over the last decade have facilitated the implementation of 3D Web GIS. Thanks to the Google Earth plugin and JavaScript API, we were able to integrate the specificities of Google Earth and its 3D renderings into web pages. In fact, we used API to create points and lines, projected images on a relief, added 3D models or import KML files, and we also developed advanced 3D mapping applications. The client side is a spatial data visualization interface based on the Ext.ux.GEarth code. This interface allows adding KML layers from Google Earth Community and an API as Google Maps. After modeling "3D virtual city" data for later storage in the server, KML files were created in Google Earth by adding feature classes that were previously modeled. Using Google Earth, we modeled KML data in 3D according to the graphic symbols. The data layers were then organized in a directory to update relative information in order to obtain a 3D virtual model. To place the resulting file on the server, we shared and sent the content to the Google Earth Community in order to obtain a network link to our model. The map interface is realized by using Google Earth API. After exporting the data to Google Earth Community (Online Map Server) and writing the Java Script file using the KML network link, a web map interface was created where the map background and the different layers can be visualized. From this interface, one can activate or deactivate the layers, view or un-view the legend, and zoom in or out. We are also able to move through the layers, change the scale, identify the different types of layers, and print and display the ruler (Figure 10).

This created interface offers a variety of dynamic mapping features from the side toolbar, including the following: display the different map layers, add KML/KMZ layers, search for a location, display the status bar (coordinates, elevation, etc.), display different options (grid, planisphere map, scale, and atmosphere), enable or disable navigation (mouse), zoom control (zoom in and zoom off), display information related to selected objects (attributes), display Google Earth layers (borders, roads, 3D buildings, relief, etc.), add sunlight, and view the map in real time.

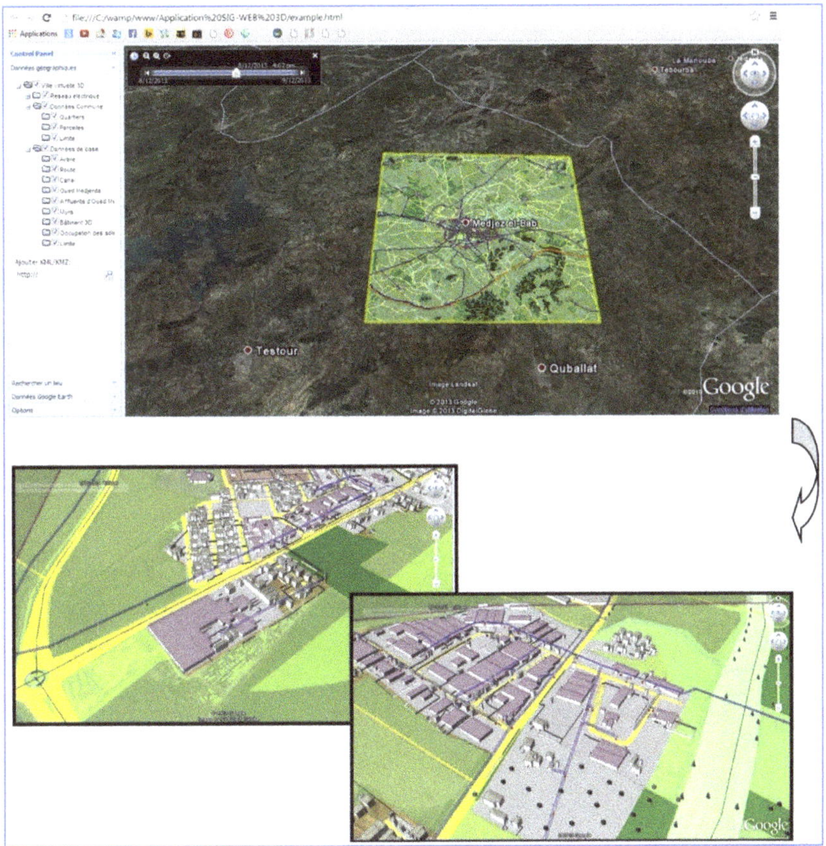

Figure 10. 3D web cartographic interface.

2.3.2. Website Development

The structuring of the website consists in building the global architecture of the website by organizing different information. The first step was to make an inventory of the contents of the website and to group them by theme in order to constitute headings (rubricing). Our website was structured as shown in Figure 11. The used editor was WordPress and WampServer as a web development platform (comprising Apache and MySQL servers). The programming languages were CSS, PHP, XML, and JavaScript.

The produced website included multiple interfaces: a Home Welcome window where the slider, the presentation of the project, and the different organizations involved in the project are displayed; an "Introduction" window that presents the scope of the study and a general idea about the city of Medjez El Bab; a window named "Thematics" on which there is a presentation of Geomatics, GIS, used tools, and details of the electric network; a "Results and Analysis" window that presents the results of thematic data, spatial analysis, network analysis scenarios, and 3D modeling; a "Media Library" window where the user can view photos (city photos, photo gallery taken in the field), videos, maps, and articles; and an "Interactive applications" window. From that window, three interactive applications can be displayed: "Grand Medjez El Bab area," "Municipality of Medjez El Bab area," and "3D virtual city"; an "About" window with general information; and, finally, a "Contact" window with location map and phone and email contacts.

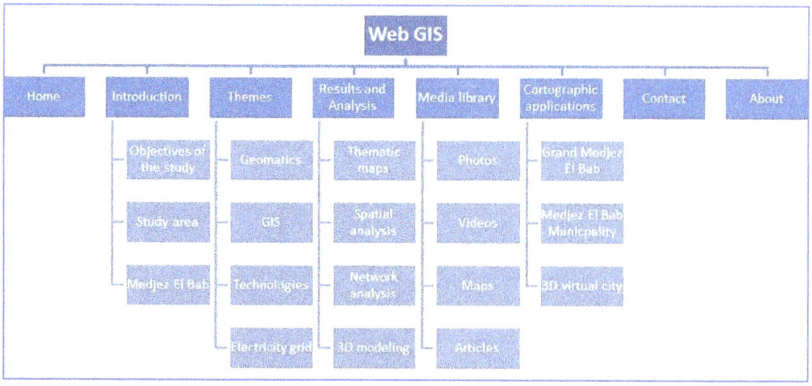

Figure 11. Website structuring.

2.4. Methodology Graphical Summary

The following figure (Figure 12) shows a summary of the different steps of the followed methodology.

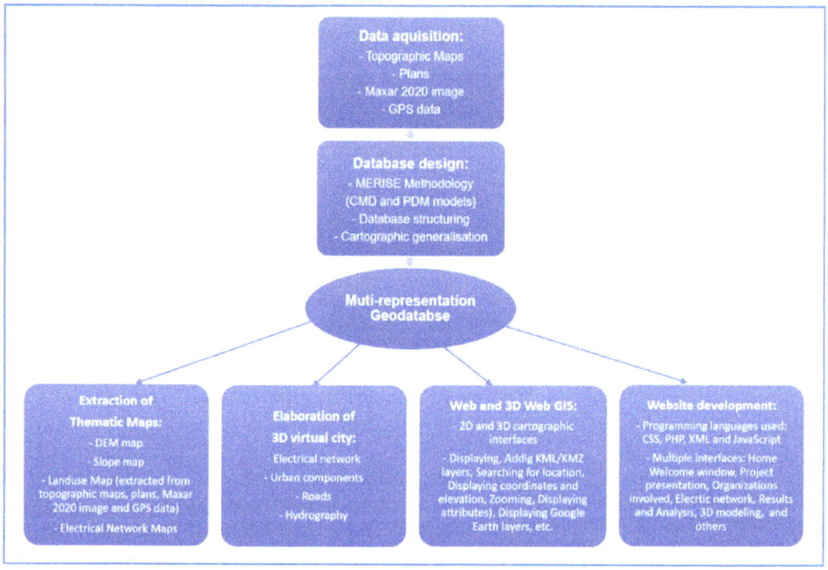

Figure 12. Graphical summary showing different methodology steps.

3. Results and Discussion

The geometric accuracy in such studies is very important and depends on the accuracy of the used data. In our study, we used topographic maps of 1: 25,000 dating from 1982 and 1988 and plans of 1:1000 dating from 2006. As the data generated from these maps and plans are not up to date, we were obliged to update it. We relied, in our case, on free Maxar 2020 imagery with a 1 m resolution. Moreover, we established a GPS campaign to accurately map the components of the electricity network: GPS Leica GS08plus was used in moving mode with centimetric accuracy (Horizontal: 1 cm, Vertical: 2 cm) for that purpose. In Figure 13, we can see, at different scales, the obtained cartographic results and we can see, in part D of the same figure, that the accuracy of the results is good: We can see the position of the MV electrical network (presented as a blue line) with high geometric accuracy (this electrical network is located in the right side of the road RN6 in that case). The establishment of GIS-based planning in the management of electrical networks depends clearly on the availability and accuracy of cartographic data that are not always available, especially in developing countries such as Tunisia. The problem of a lack of data can currently be solved by using free good-resolution satellite images that have become available in recent years, such as Maxar data. Moreover, establishing GPS campaigns for network mapping can be an important tool in such studies, especially with highly accurate GPS receivers such as the one used in our study.

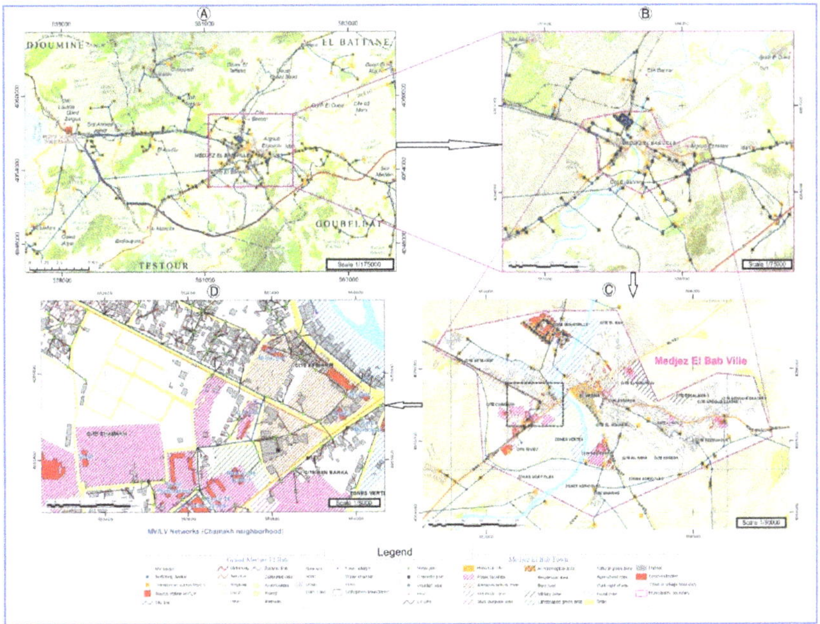

Figure 13. Presentation at different scales of MV/LV electrical networks.

GIS-based planning will be very useful for the national Tunisian company of electricity and gas STEG, which previously used paper cartographic data and which organized all its troubleshooting, restoration, or extension operations in a traditional manner. It will also provide the possibility to analyze the accessibility of vehicles and pedestrian technicians between electrical installations. It will also permit the generation of an Origin–Destination cost matrix in order to calculate the average transport distances between the service points and different localities. Moreover, it will allow the analysis of network impedance, which in itself will result in different possible scenarios to optimize performance and obtain more realistic routes. These various analyses will be crucial for the maintenance of the electricity network as well as for future urban planning.

Many thematic maps have been produced in the study area as DEM and slope maps extracted using GIS spatial analysis operations, the land use map updated based on a Maxar 2020 free image and GPS accurate data, and many other maps related to MV/LV electrical networks. On the other hand, GIS network analysis operations have enabled the efficient organization and coordination of vehicles and pedestrians and an intelligent analysis of the transportation network. Other network analyses have also helped to make the right decisions by developing strategic routing plans. The extracted slope map showed that 62% of the study area had low slopes, 28% had medium slopes, 5% had relatively steep slopes, and about 5% of the land had high steep slopes. This map was used to analyze an STEG technician's pedestrian accessibility. International standards say that the average speed of walking is between 3.2 and 5.2 km/h on flat ground and that average varies according to slope direction and values [41]. The analysis was based on making, by using GIS tools, an intersection between the slope map and the electrical network. The extracted slope data were then analyzed, thus allowing the verification of the variation of speed and time taken as a function of slope value and direction. This last operation proved, by calculating speed and travel times in an automatic manner, that the greater the slope in the direction of the topographic descent, the greater the speed and the lower the travel time, and the greater the slope in the direction of the ascent, the lower the speed and the greater the travel time. Speed varied from 5.5 to 6 km/h in the downhill direction and from 3.8 to 4.6 km/h in the uphill direction. Moreover, two topographic profiles were established to show the importance of topographic roughness in pedestrian accessibility. These profiles were analyzed using GIS network analysis tools to check the effects of slope on speed and travel time: Example 1: Distance = 2.2 km, time = 36 min, average speed = 3.66 km/h; Example 2: Distance = 2.2 km, time = 22 min, average speed = 6 km/h. The obtained results seem to be logical; in fact, the speed in example 1 (route with rugged topography) is lower than the speed in example 2 (route with less rugged topography); it shows that the slope of the path influences the speed (Figure 14).

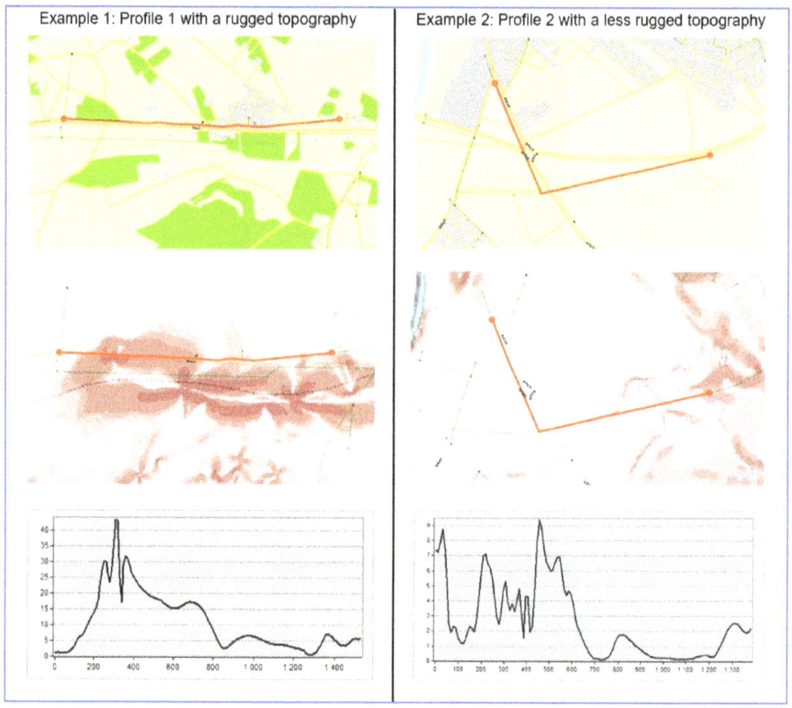

Figure 14. The two profiles used as examples to show the variation of time and speed according to slope.

Many studies about pedestrian crossing were established around the world using GIS as the study of Lassarre et al. in 2012 [42], who presented a GIS-based methodology for identifying pedestriancrossing patterns in order to store and integrate information on pedestrian trips and crossings made during trips with other geographical information, including road network, traffic, and facilities, in the region of Villeneuve d'Ascq-Lille in France. We can also cite the study of Aldemar et al. in 2020 [43] who conducted a GIS and microsimulation-based multicriteria decision analysis for the evaluation of pedestrian crossings in Turkey.

Concerning accessibility in using vehicle transport, seven categories of roads with different reference speeds have been considered: Motorways, Motorway junction, Primary roads (National roads), Collector roads, Secondary roads (Local roads), Tertiary roads (Unclassified roads), and Quaternary roads (Agricultural tracks) with the following respective maximum speeds: 100, 60, 80, 70, 60, 40, and 30 km/h. GIS tools were carried out on the transport network using several criteria: speed, time, and hierarchy. Many network analyses were established in our study, such as the determination of the shortest path between one of the MV/LV transformer stations and the source station located in the region of "Oued Zarga." In this case, the extracted length of the shortest path is 19.2 km with a travel time of 24 min (Figure 15). Assuming, for example, that the RN5 national road is cut due to road works, the shortest path extracted will be different with a length of 25.5 km and a travel time of 38 min. The route is shorter in the first scenario, and this can be explained by different factors such as the speed limit in the type of path crossed, which differs from one type of path to another. Other analyses of the calculation of routes according to a chosen impedance (best route with distance impedance; best route with time impedance) were also established.

Figure 15. Analysis of the shortest path between the source station and the transformer station.

Many studies have been conducted around the world about road network analysis, such as the study of Das et al. in India (2019) [44]. In that study, the authors made a road network map of Guwahati city to find the shortest route to solve traffic problems. We can also cite the study established by Shafabakhsh et al. (2017) [45] on which GIS was used as a system for managing and analyzing urban accidents on the road network by applying a combination of spatial and statistical methods.

An Origin Destination Cost Matrix was also established by using GIS to calculate average transport distances between the service points at different localities (cities and villages) and the MV/LV transformer stations. The same analysis was performed using time impedance. The following data must considered to calculate the transport cost: type of vehicle, average consumption, and fuel price.

A network service area is a region that encompasses all accessible roads (roads within a specified impedance). Service areas related to the city of Medjez El Bab indicating path accessibility according to time at 5, 10, 15, 20, and 30 min were extracted (Figure 16) and that has enabled the organization of service missions according to their duration.

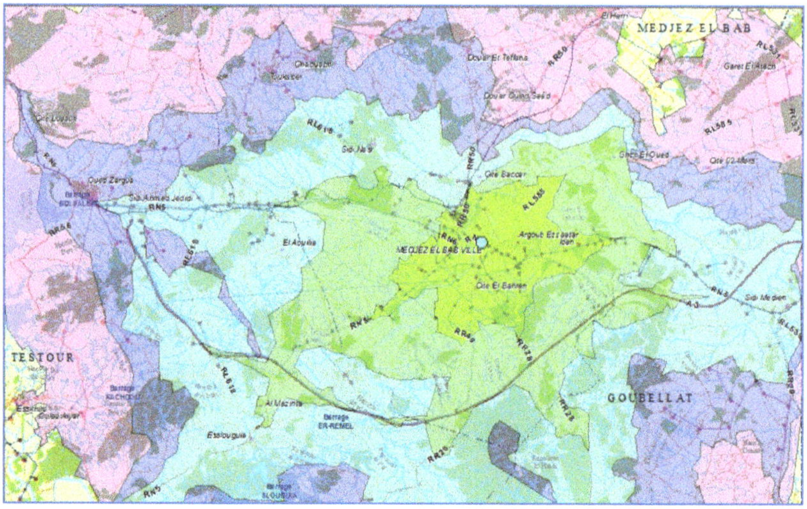

Figure 16. Polygons representing service areas of 5, 10, 15, 20, and 30 min from Medjez El Bab City.

It is quite clear that the analyses of electrical networks and the exploitation of their data, as well as the management of electrical distribution, are significantly better with the use of GIS spatial and network analysis tools. Moreover, the obtained results will be crucial in the case of a power failure or in the case of maintenance work and ensure the rapid resolution of technical problems and can also be used in future urban planning. Similar procedures will produce time, material, and organizational gains compared to conventional ones.

Currently, the electrical GIS system mainly adopts a two-dimensional electronic map to show the electrical network and related information. However, it is difficult to meet the requirements of the panoramic display due to the lack of a vertical viewing angle. A 3D visualization will be beneficial to realize a panoramic visualization of data in a 3D virtual environment [46]. A three-dimensional representation of the study area as well as the electrical networks was also carried out, providing a realistic visualization of our study area.

Web and 3D Web GIS applications offer users the possibility to consult the made products by the internet. The client side is a spatial data visualization interface (Map Viewer) based on GeoExt. This interface allows adding WMS layers from GeoServer and an API such as Google Maps. In our study, we were able to integrate Google Earth and its 3D renders into the web pages thanks to the Google Earth plugin and JavaScript API. We were able to create points and lines, to perform the projection of images, to add 3D models, and to import of KML files. We have also been able to develop advanced 3D mapping applications. The client side is a spatial data visualization interface based on the Ext.ux.GEarth code. This interface allows us to add KML layers from the Google Earth community and from an API such as Google Maps. Through the elaborated website, we were also able to create a link between Google Maps and the photos of the power poles with an explanatory video sequence to collect the maximum amount of information about the electrical network. on the interactive windows of the website, it is possible to establish a display of different scales: a display at the level of the delegation with a scale of 1:25,000 and a display at the level of the municipality with a scale of 1:5000.

Web GIS and 3D Web GIS demonstrated many advantages over traditional GIS, enabling the possibility of consulting cartographic products via the Internet. According to Jianhui and Xiangnan (2021) [47], traditional GIS services cannot meet the needs of professional energy applications. The integration of GIS data into web platforms will significantly improve the use of the data and, therefore, the use of the applications. Müller et al. (2016) [48]

pointed out that information gaps restrict the dissemination of sustainable electrification technologies and that interactive WebGIS tool could address these gaps. The developed database was ultimately used to set up Web GIS and 3D Web GIS applications and a website with multiple interfaces. This can be used to include thematic data, spatial analysis, network analysis scenarios, 3D modeling, and many others interactive applications in the study area. With the developed website, we were also able to link Google Maps and the photos of the power poles with an explanatory video sequence to gather as much information as possible about the electricity network. Moreover, on the interactive windows of the website, it was possible to set up a display at different scales.

The product developed in our study can be improved by using more accurate data, such as commercial Maxar satellite images with high accuracies ranging from 0.3 to 0.03 m. The present study did not consider factors as natural hazard factors and anthropogenic or human impact factors such as urban sprawl and urban density. The consideration of different natural factors that could influence the electrical network such as rains, winds, storms, lightning, and landslide could significantly improve such a study. A study established by Entriken and Lordan in 2012 [49] discussed, among the impacts of extreme weather events, events related to transmission and distribution systems including electrical networks, as well as damage recovery from winter storms, windstorms, and floods. Human factors such as urban density were considered in the study of Antoniucci et al. in 2021 [50], in which an analysis of urban density and household electricity consumption was carried out in Italy. Urban and industrial sprawls could equally be considered in similar studies as they can affect the capacity of the electrical network that must, therefore, be reviewed in light of new needs. These sprawls could affect the capacity of the network, hence the need to establish a study on the different sprawl scenarios in the future. Urban sprawl was considered, for example, for electrical consumption in the Spanish case in the study of Lasarte et al. in 2018 [51]. Agricultural activities could also be an important human factor, as many of them are connected to the electricity grid as providing irrigation and livestock water, as it was clarified in the study of Che-Castaldo et al. in 2021 [52].

4. Conclusions

In this paper, we discussed the contribution of GIS in the field of planning of the electric networks with the aim to improve the capacities and the output of the national company of electricity and gas in Tunisia (STEG).

A complete database related to the study area gathering all useful data was created. Different topographic maps and plans related to the study area were used, such as 1:25,000 topographic maps, 1:5000 plans, and Maxar 2020 background free image with a 1-m resolution. The use of similar free good-resolution satellite images can be a solution to the lack of data or non-updated data; however, the use of commercial Maxar images with a resolution ranging from 0.3 to 0.03 m will extend data accuracy. Moreover, GPS centimetric data were used to update electrical networks and to also verify the accuracy of the obtained data.

Many thematic maps such as DEM and slope maps, land use map, and electrical networks maps were established. Based on GIS network and spatial analysis operations integrating slope and MV/LV network data, it was possible to conclude that the accessibility of each category of the network depends on slope value and direction. Different categories of roads were considered in the analysis of the vehicle transport network using the following criteria: speed, time, and hierarchy. The objective was the analysis of the shortest path, the calculation of routes according to a chosen impedance (best route with distance impedance; best route with time impedance), the generation of the origin–destination cost matrix to calculate average transport distances between service points or different localities (towns and villages) and the MV/LV transformer stations. We also performed the same type of analysis by using temporal impedance. The calculation of the service areas related to the accessibility of the paths according to different times was also established to organize service missions. The study and analysis of pedestrian technicians' movement network also enabled the development strategic routing plans.

This study demonstrates, similarly to many studies related to the management of electrical networks based on GIS established in recent years around the world, the importance of GIS as a powerful tool for managing electrical networks. The planning of the electricity network will be important in urban planning and needs to be comprehensively evaluated in order to improve the quality of decision making and, consequently, improve the benefits of the investment.

Panoramic display requires 3D visualization, which will be beneficial for realizing more realistic data visualization; for this reason, a 3D representation of the study area as well as the electrical networks has been created, providing a realistic visualization of the study area.

The established Web GIS and 3D Web applications offer users the possibility to consult the made products via the internet. We were also able to integrate Google Earth and its 3D renders into the web pages and to add KML layers from the Google Earth community and from an API such as Google Maps. We have also been able to develop advanced 3D mapping applications.

By using the elaborated website, we were able to create a link between Google Maps and the photos of the power poles with an explanatory video sequence to collect the maximum amount of information about the electrical network, and it was possible display different cartographic products at different scales.

The present study did not consider factors such natural hazard factors, including weather and landslides. It does not consider human impact factors such as urban sprawl and urban density and agricultural and industrial impact. Taking these into account, all these factors in future works will significantly improve similar studies.

This study opens wide perspectives; in fact, it would be possible to reproduce similar work established in that pilot zone in other zones related to the national Tunisian electricity and gas company STEG across the country. This will help decision makers to better manage electrical networks and urban planning in the future.

Author Contributions: Conceptualization, M.H.H. and M.C.; Data curation, M.H.H. and M.C.; Formal analysis, M.H.H. and M.C.; Investigation, M.C.; Methodology, M.H.H.; Project administration, M.H.H.; Resources, M.H.H. and M.C.; Software, M.H.H.; Supervision, M.H.H.; Validation, M.H.H.; Visualization, M.H.H. and M.C.; Writing—original draft, M.H.H.; Writing—review & editing, M.H.H. All authors have read and agreed to the published version of the manuscript.

Funding: This research received no external funding.

Institutional Review Board Statement: Not applicable.

Informed Consent Statement: Informed consent was obtained from all subjects involved in the study.

Data Availability Statement: Not applicable.

Conflicts of Interest: The authors declare no conflict of interest.

References

1. Gonen, T. *Electrical Power Transmission System Engineering, Analysis and Design*, 3rd ed.; CRC Press, Taylor and Francis: New York, NY, USA, 2014. [CrossRef]
2. Wei, W.; Liu, F.; Mei, S. Charging strategies of EV aggregator under renewable generation and congestion: A normalized nash equilibrium approach. *IEEE Trans. Smart Grid* **2016**, *7*, 1630–1641. [CrossRef]
3. Luo, Z.; Hu, Z.; Song, Y.; Xu, Z.; Lu, H. Optimal coordination of plug-in electric vehicles in power grids with cost-benefit analysis—Part II: A case study in China. *IEEE Trans. Power Syst.* **2013**, *28*, 3556–3565. [CrossRef]
4. Mejía-Alzate, M.L. Análisis Interorganizacional en la Gobernanza Turística de la Ciudad de Colombia. *Rev. Lat.-Am. Tur.* **2018**, *4*, 8–22. [CrossRef]
5. Câmara, A.M.M.; Gilberto, S.D.F.; Marília, S.C. *Spatial Analysis and GIS: A Primer*; Image Processing Division, National Institute for Space Research (INPE): Rio de Janeiro, Brazil, 2004.

6. Hamza, M.H.; Added, A.; Rodriguez, R.; Abdeljaoued, S.; Ben Mammou, A. A GIS-based DRASTIC vulnerability and net recharge reassessment in an aquifer of a semi-arid region (Metline-Ras Jebel-Raf Rafaquifer, Northern Tunisia). *J. Environ. Manag.* **2007**, *84*, 12–19. [CrossRef] [PubMed]
7. Gutiérrez Pueblo, J.; Gould, M. *SIG: Sistemas de Información Geográfica*; Síntesis: Madrid, Spain, 1994.
8. Steede-Terry, K. *Integrating GIS and the Global Positioning System*; ESRI: Redlands, CA, USA, 2000.
9. Mejia Alzate, M.A.; Melo Trujillo, J.D.; Zambrano-Asanza, S.; Padilha-Feltri, A. Spatial-temporal growth model to estimate the adoption of new end-use electric technologies encouraged by energy-efficiency programs. *Energy* **2020**, *191*, 116531. [CrossRef]
10. Mejia Alzate, M.A.; Melo Trujillo, J.D.; Padilha Feltrin, A.; Sánchez Zuleta, C.C.; Fernández Gutiérrez, J.P. Geographical information systems as a Tool to assist the electricity distribution Networks planning. *Revista EIA* **2018**, *15*, 71–85. [CrossRef]
11. Nawaz-ul-Huda, S.; Burke, F.; Azam, M.; Naz, S. GIS for power distribution network: A case study of Karachi, Pakistan. *GEOGRAFIA Malays. J. Soc. Space* **2012**, *8*, 60–68.
12. Kaijuka, E. GIS and rural electricity planning in Uganda. *J. Clean. Prod.* **2007**, *15*, 203–217. [CrossRef]
13. Putz, S. Interactive information services using World Wide Web Hypertext. In Proceedings of the First International Conference on World Wide Web, Geneva, Switzerland, 25–27 May 1994.
14. Longley, P.A.; Goodchild, M.F.; Maguire, D.J.; Rhind, D.W. *Geographic Information Systems and Science*, 2nd ed.; Wiley: Chichester, UK, 2005.
15. Datta, A.; Mohanty, P. Enterprise GIS and Smart Electric Grid for India's power sector. In Proceedings of the IEEE PES Innovative Smart Grid Technologies Conference (ISGT), Washington, DC, USA, 24–27 February 2013; pp. 1–7. [CrossRef]
16. Chmit, M. Conception et réalisation d'une application SIG-WEB pour l'analyse et la gestion des réseaux électriques HTA/BT de la ville de Medjez El Bab. Master's Thesis, Université de Tunis El Manar, Tunis, Tunisia, January 2014.
17. Zheng, F.M.; Ma, L.; Liu, N.; Chen, J. Assessment for Distribution Network Planning Schemes of Urban Electric Power System. *Energy Procedia* **2012**, *14*, 1067–1074. [CrossRef]
18. Mentis, D.; Welsch, M.; Nerini, F.F.; Broad, O.; Howells, M.; Bazilian, M.; Rogner, H. A GIS-based approach for electrification planning—A case study on Nigeria. *Energy Sustain. Dev.* **2015**, *29*, 142–150. [CrossRef]
19. Candelise, C.; Westacott, P. Can integration of PV within UK electricity network be improved? A GIS based assessment of storage. *Energy Policy* **2017**, *109*, 694–703. [CrossRef]
20. Ashkezari, A.D.; Hosseinzadeh, N.; Chebli, A.; Albadi, M. Development of an enterprise Geographic Information System (GIS) integrated with smart grid. *Sustain. Energy Grids Netw.* **2018**, *14*, 25–34. [CrossRef]
21. Taye, B.; Workineh, T.; Nebey, A.; Kefale, H.A. Rural electrification planning using Geographic Information System (GIS). *Cogent Eng.* **2020**, *7*, 1836730. [CrossRef]
22. Office de la Topographie et du Cadastre OTC Topographic Maps 1/25000: Beja S.E (Feuille N° 18), Tebourba S.E (Feuille N° 19), Tebourba S.O (Feuille N° 19), Oued Ezzarga (Feuille N° 26), Oued Ezzarga (Feuille N° 26), Mjaz al Bab NE (Feuille N° 27), Mjaz al Bab NO (Feuille N° 27), Mjaz al Bab SE (Feuille N° 27), Mjaz al Bab SO (Feuille N° 27), 1982 to 1988. Available online: https://www.otc.nat.tn/ (accessed on 10 February 2022).
23. Ben Lassoued, T. *Topographic Plans of Medjez El Bab, Scale 1:1000*; BETBEL Technical Consultant Office: Tunis, Tunisia, 2006.
24. Maxar Technologies. Maxar Technologies Awarded Four-Year Global EGD Contract by the U.S. Government for On Demand Access to Mission-Ready Satellite Imagery. Sensor and Systems. 28 August 2019. Available online: https://www.maxar.com (accessed on 10 February 2022).
25. Rochfeld, A.; Tardieu, H. MERISE: An information system design and development methodology. *Inf. Manag.* **1983**, *6*, 143–159. [CrossRef]
26. Avison, D. MERISE: A European Methodology for Developing Information Systems. *Eur. J. Inf. Syst.* **1991**, *1*, 183–191. [CrossRef]
27. Sherman, R. Chapter 8: Foundational Data Modeling. In *Business Intelligence Guidebook from Data Integration to Analytics*; Elsvier: Amsterdam, The Netherlands, 2015; pp. 173–195.
28. Perkal, J. *An Attempt at Objective Generalization: Discussion Papers of The Michigan Inter-University Community of Mathematical Geographers*; Proba obiektywnej generalizacji, Geodezja i Karografia; (1958) English translation; 1965; Volume VII, pp. 130–142. Available online: http://www-personal.umich.edu/~{}copyrght/image/micmg/perkal2/perkal2.pdf (accessed on 18 June 2021).
29. Tobler, W.R. *Numerical Map Generalization: And, Notes on the Analysis of Geographical Distributions*; Department of Geography, University of Michigan: Ann Arbor, MI, USA, 1966; Volume 3, pp. 10–16. [CrossRef]
30. Nyangweso, D. Cartographic Generalization in Multi-scale Environment: Case study of Lamu County, Kenya. *Int. J. Sci. Res. (IJSR)* **2013**, *5*, 804–813. [CrossRef]
31. Hutchinson, M.F. Calculation of hydrologically sound digital elevation models. In Proceedings of the Third International Symposium on Spatial Data Handling, Sydney, Australia, 17–19 August 1988.
32. Hutchinson, M.F. A new procedure for gridding elevation and stream line data with automatic removal of spurious pits. *J. Hydrol.* **1989**, *106*, 211–232. [CrossRef]
33. Singh, S.P.; Jain, K.; Mandla, V.R. Virtual 3D city modeling: Techniques and applications. *ISPRS-Int. Arch. Photogramm. Remote Sens. Spat. Inf. Sci.* **2013**, *XL-2/W2*, 73–91. [CrossRef]
34. Biljecki, F.; Ledoux, H.; Stoter, J.; Vosselman, G. The variants of an LOD of a 3D building model and their influence on spatial analyses. *ISPRS J. Photogramm. Remote Sens.* **2016**, *116*, 42–54. [CrossRef]

35. Mignard, C.; Nicolle, C. Merging BIM and GIS using ontologies application to urban facility management in ACTIVe3D. *Comput. Ind.* **2014**, *65*, 1276–1290. [CrossRef]
36. Tekdal-Emniyeti, E.; Haefele, K.; Isele, J.; Celik, R.N. 3D documentation of historical sites and buildings for interdisciplinary works; Proc. SPIE 8085. In *Videometrics, Range Imaging, and Applications XI*; International Society for Optics and Photonics: Bellingham, WA, USA, 2011; p. 80850V. [CrossRef]
37. Internet World Stats. Usage and Population Statistics. 2020. Available online: http://www.internetworldstats.com/stats.htm (accessed on 18 June 2021).
38. Auer, M.; Zipf, A. 3D WebGIS: From Visualization to Analysis. An Efficient Browser-Based 3D Line-of-Sight Analysis. *ISPRS Int. J. Geo-Inf.* **2018**, *7*, 279. [CrossRef]
39. Siddiqueea, M.Z.H.; Strzalkab, A.; Eicker, U. Publication of energy consumption data of Scharnhauser park via Web GIS. In Proceedings of the Conference of the Applied Geoinformatics for Society and Environment, Stuttgart, Germany, 12–17 July 2009; Volume 103.
40. Von Schwerin, J.; Richards-Rissetto, H.; Remondino, F.; Agugiaro, G.; Girardi, G. The MayaArch3D project: A 3D WebGIS for analyzing ancient architecture and landscapes. *Lit. Linguist. Comput.* **2013**, *28*, 736–753. [CrossRef]
41. Victor, N. Modélisation de l'accessibilité piétonne à Luxembourg-Ville, Rapport de Stage, CEPS/INSTEAD, Master SIG et Gestion de l'Espace, Université Jean Monnet, Saint-Etienne, 54 p. (+ Annexes et Livret Cartographique). 2010. Available online: https://www.academia.edu/12162636/Mod%C3%A9lisation_et_visualisation_de_l_accessibilit%C3%A9_pi%C3%A9tonne_%C3%A0_Luxembourg-ville (accessed on 18 June 2021).
42. Lassarre, S.; Bonnet, E.; Bodin, F.; Papadimitriou, E.; Yannis, G.; Golias, J. A GIS-based methodology for identifying pedestrians' crossing patterns. *Comput. Environ. Urban Syst.* **2012**, *36*, 321–330. [CrossRef]
43. Alemdar, K.D.; Tortum, A.; Kaya, Ö.; Atalay, A. Interdisciplinary evaluation of intersection performances-A microsimulationbased MCDA. *Sustainability* **2021**, *13*, 1859. [CrossRef]
44. Das, D.; Ojha, A.K.; Kramsapi, H.; Baruah, P.P.; Dutta, M.K. Road network analysis of Guwahati city using GIS. *SN Appl. Sci.* **2019**, *1*, 906. [CrossRef]
45. Shafabakhsh, G.A.; Famili, A.; Bahadori, M.S. GIS-based spatial analysis of urban traffic accidents: Case study in Mashhad, Iran. *J. Traffic Transp. Eng.* **2017**, *4*, 290–299. [CrossRef]
46. He, B.; Mo, W.X.; Hu, J.X.; Yang, G.; Lu, G.J.; Liu, Y.Q. Development of power grid Web3D GIS based on Cesium. In Proceedings of the IEEE PES Asia-Pacific Power and Energy Engineering Conference (APPEEC), Xi'an, China, 25–28 October 2016; pp. 2465–2469. [CrossRef]
47. Jianhui, W.; Xiangnan, M. Design and Research on the Management of the Electricity System of WebGIS; Lecture Notes in Electrical Engineering. In *International Conference on Frontier Computing. FC 2020*; Singapore, 10–13 July 2020, Chang, J.W., Yen, N., Hung, J.C., Eds.; Springer: Berlin, Germany, 2020; Volume 747. [CrossRef]
48. Müller, M.F.; Thompson, S.E.; Kelly, M.N. Bridging the Information Gap: A WebGIS Tool for Rural Electrification in Data-scarce Regions. *Appl. Energy* **2016**, *171*, 277–286. [CrossRef]
49. Entriken, R.; Lordan, R. Impacts of extreme events on transmission and distribution systems. In Proceedings of the IEEE Power and Energy Society General Meeting, San Diego, CA, USA, 22–26 July 2012. [CrossRef]
50. Antoniucci, V.; Bisello, A.; Marella, G. Urban Density and Household-Electricity Consumption: An Analysis of the Italian Residential Building Stock. In *Smart and Sustainable Planning for Cities and Regions: SSPCR 2019*; Green Energy and Technology; Bisello, A., Vettorato, D., Ludlow, D., Baranzelli, C., Eds.; Springer: Cham, Switzerland, 2021. [CrossRef]
51. Lasarte, E.; Rubiera, F.; Cuartas, B.M. Energy consumption and urban sprawl: Evidence for the Spanish case. *J. Clean. Prod.* **2018**, *172*, 3479–3486. [CrossRef]
52. Che-Castaldo, J.P.; Cousin, R.; Daryanto, S.; Deng, G.; Feng ML, E.; Gupta, R.K.; Hong, D.; McGranaghan, R.M.; Owolabi, O.O.; Qu, T.; et al. Critical Risk Indicators (CRIs) for the electric power grid: A survey and discussion of interconnected effects. *Environ. Syst. Decis.* **2021**, *41*, 594–615. [CrossRef]

applied sciences

MDPI

Article

Deep Contrast Learning Approach for Address Semantic Matching

Jian Chen, Jianpeng Chen, Xiangrong She, Jian Mao and Gang Chen *

Institute of Smart City Research (Wuhu), University of Science and Technology of China, Wuhu 241000, China; chenj@ustc.win (J.C.); chenjp@ustc.win (J.C.); shexr@ustc.win (X.S.); nh146@163.com (J.M.)
* Correspondence: cheng@ustc.win; Tel.: +86-173-5294-4872

Abstract: Address is a structured description used to identify a specific place or point of interest, and it provides an effective way to locate people or objects. The standardization of Chinese place name and address occupies an important position in the construction of a smart city. Traditional address specification technology often adopts methods based on text similarity or rule bases, which cannot handle complex, missing, and redundant address information well. This paper transforms the task of address standardization into calculating the similarity of address pairs, and proposes a contrast learning address matching model based on the attention-Bi-LSTM-CNN network (ABLC). First of all, ABLC use the Trie syntax tree algorithm to extract Chinese address elements. Next, based on the basic idea of contrast learning, a hybrid neural network is applied to learn the semantic information in the address. Finally, Manhattan distance is calculated as the similarity of the two addresses. Experiments on the self-constructed dataset with data augmentation demonstrate that the proposed model has better stability and performance compared with other baselines.

Keywords: address matching; smart city; contrast learning; neural networks; data augmentation

Citation: Chen, J.; Chen, J.; She, X.; Mao, J.; Chen, G. Deep Contrast Learning Approach for Address Semantic Matching. *Appl. Sci.* **2021**, *11*, 7608. https://doi.org/10.3390/app11167608

Academic Editors: Alexei Gvishiani and Boris Dzeboev

Received: 21 July 2021
Accepted: 17 August 2021
Published: 19 August 2021

1. Introduction

Geographical addresses are the most important basic data resources in the construction of a smart city. How to dig out potential associations between address texts and use the result to serve for standardization construction is a key issue that directly affects the level of smart city construction.

The early research methods on address mainly focused on text similarity. The literal similarity between the two geographical addresses was calculated from a certain measurement dimension and the threshold was manually set [1]. Specifically, the edit distance [2–4] is a traditional way which defines the similarity as the minimum number of character editing operations required to convert one string to another string, which is very easy to be applied in real work. Subsequently, Jaccard [5] brought up a new way which obtains a more accurate effect on short address by calculating the local similarity of two addresses, but it does not work well for long addresses. Afterwards, the N-gram approach based on vector space was proposed [6], which converts addresses to vector representations in the same vector space, and then calculates the similarity using mathematical methods for example cosine similarity [7]. Compared with previous methods, the N-gram approach improves the effect and obtains a better performance. Nowadays, all the traditional methods mentioned above are still inadequate.

More recently, with the diversification of addresses and the higher requirement to process a large number of addresses than before, traditional address matching methods obviously cannot meet the requirements. A new method based on address structure and address element extraction is proposed, which uses the hierarchical syntax tree to identify address and then do further address matching work [8]. Basically, the way of acquiring address element is mainly by dividing into word segmentation with dictionaries, probability distributions, such as conditional random fields, hidden Markov models [9,10], or natural language word segmentation tools (Jieba, THULAC, etc.). Some scholars have put forward corresponding solutions to the extraction of address elements: some rule-based and fuzzy Chinese address coding methods are raised to establish a standard address database and generate matching rules [11–13]. However, this method relies on the completeness of address database, and it is difficult to formulate all the rules as the growth of database. Tian proposed an optimized Chinese address matching model and provided a coding service with this model [14]. Zhang used Bert as pretrained embedding model and applied CRF algorithm to extract address element and semantic features [15]. Comber embed word2vec into CRF to convert address elements into a fixed dimension semantic representation vector [16].

Nevertheless, semantic information cannot be obtained effectively when dealing with longer address records and the distribution of information density is uneven. To address this kind of problem, follow-up research begins to apply a neural network to do this task, such as using CNN or RNN [17]. Santos proposed a multi-model fusion approach which apply RNN and GRU as address semantic information modeling [18]. This model has achieved a good performance improvement compared to single similarity measurement-based models and some supervised learning methods. Next, Lai combined the advantages of RNN and CNN models, proposed the RCNN model, which uses a bidirectional structure and embeds BiRNN structure into convolutional layer [19]. This kind of structure effectively reduces the network noise and maximizes the ability to extract the context information of addresses. In the field of smart cities, there are also many researchers trying to introduce deep learning to cope with city development issues, such as the cities expansion and personalized POI recommendation [20,21]. Karimzadeh proposed a geographic analysis system for NER recognition, which can efficiently sort out geographic analysis problems [22]. After that, the deep learning methods are verified to solve the spatial data and the urban geographic problems, and prove that deep learning ways are the most effective way and have broad application prospects [23]. Grekousis have analyzed more than 140 articles about using artificial neural networks to settle the urban geographic problems. In summary, artificial neural networks have obvious advantages over traditional methods [24].

The semantic representation of addresses based on deep learning is essentially an NLP problem [25,26]. In the field of NLP, contrastive learning algorithms have recently been widely proposed [27,28]. Contrastive learning algorithm is a subset of deep learning. Its goal is to bring the enhanced new samples as close as possible in the embedding space and make different samples as far away as possible. How to construct examples is an important issue in contrastive learning. For the translation task, Yang changed the number of omitted words, word frequency, and part of speech according to the actual translation, designed different types of negative examples to realize data augmentation [29]. Wu and Meng proposed to use word deletion, reordering, and substitution to achieve it [30,31]. However, due to the inherent discrete characteristics of Chinese addresses, it is extremely difficult to implement data augmentation simply through text processing.

In summary, the approaches are difficult to cope with addresses that contain complex structures or contain redundant information. The reason is that these methods are lacking in terms of understanding the semantics of the address, and they also cannot extract the semantic features of the address well. At the same time, these models often focus on dataset built with specific conditions that do not provide effectively help on enhancing generalization ability, which cannot fundamentally improve the model performance.

To address these problems, this paper proposes a contrast learning address matching algorithm. First, ABLC use the Trie syntax tree structure to construct a standard address tree to extract address elements, then uses Bi-LSTM and CNN models to embed the address into vectors with semantic information. Following that, we introduce the attention mechanism to get position-aware information from the context, so as to further improve the accuracy of semantic representation. In the end, the corresponding Manhattan distance is calculated between two semantic vectors of address pair, which can be considered as the similarity of two addresses. Furthermore, we introduced a data augmentation method to extend the existing address dataset, and then the model has been significantly improved in the stability and the ability to perceive similar addresses is competitive.

The contributions of this paper are as follows: (1) Propose a contrast learning address matching algorithm that captures similarities and differences between the input address pairs so as to achieve the judgment of similarity and dissimilarity of address pairs. (2) Propose a semantic-based address representation model with a hybrid neural network that incorporates an attention mechanism. The model extracts local and global features of the input data in addition to giving higher weights to important information in the address, so as to more effectively capture key information from addresses. (3) Propose an address data augmentation method to improve the performance of the model. By constructing an address enhancement dataset based on the uniqueness of addresses and combining the dropout strategy to achieve data enhancement, the overall performance of the model is improved with better generalization capability.

2. Materials and Methods

In this section, we propose a semantic-based address matching framework according to the characteristics of the address. We first use the Trie syntax tree to build a standard address model and apply it to extract address elements. Additionally, then we create a contrast learning model which is based on a hybrid neural network, to perform semantic representation of the address. Finally, the similarity between address pair is obtained by calculating the Manhattan distance. Furthermore, the data augmentation method is introduced to construct address datasets, which improves the accuracy of address matching and the performance of the model. The address matching framework is referred to as the ABLC model and the algorithm description is as below Algorithm 1 shown.

Algorithm 1 The ABLC algorithm

Input: address set $\{A\}$, address text pair $(a_i \in A, a_j \in A)$
Output : similarity of two address text $sim(a_i, a_j)$

```
Initialize sepResult with null
divisionTree ← BuildTree(A)
for ele in [a_i, a_j] do
  for node in divisionTree do
    if headof(ele, len(node)) == node:
      sepList ← node
      ele.delete(node)
    if node == LastNode(A):
        sepList ← ele
    sepResult ← set_List
similarity ← ABLC(sepResult [0], sepResult [1])
sim(a_i, a_j) ← similarity
```

2.1. Problem Definition

We define address matching in this paper according to the below description: assume D_{sa} containing N address datasets $D_{sa} = \{sa_1, sa_2, \ldots sa_n\}$, for a certain element sa_i from D_{sa}, the task goal of this paper is to find an address pair $\{sa_i, sa_j\}$ and satisfy: $similarity(sa_i, sa_j) \geq \eta$,where $sa_i \in D_{sa}, sa_j \in D_{sa}$ and $sa_i \neq sa_j$, η is the set threshold.

2.2. Address Model

The particularity of the Chinese language leads to the particularity of Chinese addresses, which is mainly reflected in the following aspects: (1) Multiple: An address contains multiple place names; (2) Hierarchical: Address description is usually in sequence from large area to small area; (3) Detailed: The standard address contains the place name of each level. The Chinese address is composed of multiple address elements and a valid address element should include one of different level address names, such as the administrative division, the street, the neighborhood, the door, the landmark, and the point of interest. Several address description patterns commonly used now are: administrative division + street (road or lane) + house number; administrative division + community (natural village) + house number; administrative division + (street, road and lane) + point of interest (marker). Administrative divisions can be divided into provinces, cities, districts (counties), streets (towns), and communities (administrative villages).

The Trie syntax tree is a kind of hash-tree structure. Generally, it is used to store and sort a large number of strings. Unlike a binary tree, the key point of Trie syntax tree is that the string is not directly stored in the node, but is determined by the position of the node in the tree. Its advantage is to minimize unnecessary string comparisons and improve the query efficiency. All descendants of node have the same prefix, which is the string corresponding to this node. Additionally, the root node corresponds to an empty string. Basically, not all nodes have corresponding values, only the leaf nodes and some internal nodes have relevant values. This paper constructs the Trie syntax address tree, as shown in Figure 1, which is used to extract address elements.

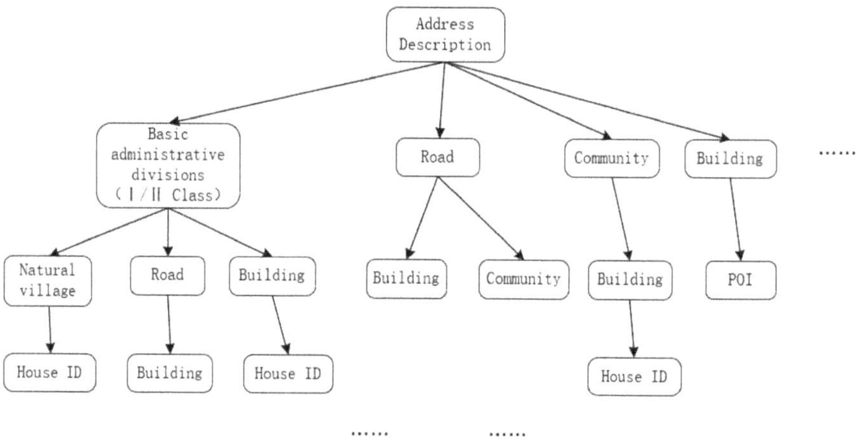

Figure 1. Trie syntax address tree.

2.3. *Address Semantic Contrast Learning Model*

This section introduces a semantic-based address contrast learning model which is fused with attention mechanism, Bi-LSTM and CNN network. The model is established based on the characteristics of the Chinese address and advantages of each sub-network in the hybrid neural network model. It accepts the input of the address pair, and, respectively, generates the semantic vector representation of the address, and finally determines whether the address pair is similar by calculating the Manhattan distance. The overall structure of the model is shown in Figure 2. The contrast learning model contains embedding stage, Bi-LSTM stage, CNN stage, attention stage, and semantic distance calculation stage. The specific details of each stage are explained as below.

2.3.1. Embedding

The embedding stage mainly focuses on converting the Chinese address into vectors, that is, maps the input address into a fixed m × n matrix. Chinese address is actually a special language description which the words have no formal delimiters, such as blank space. Therefore, the address needs to be segmented before word embedding and we should pay more attention to dividing the place name address into various address elements. Each address element is equivalent to a word in Chinese. This paper adopts Jieba's word segmentation algorithm and loads a custom word segmentation database to split address.

The construction of the custom database is based on the particularity of city place names and addresses to supplement the correct segmentation of unidentified names by Jieba.

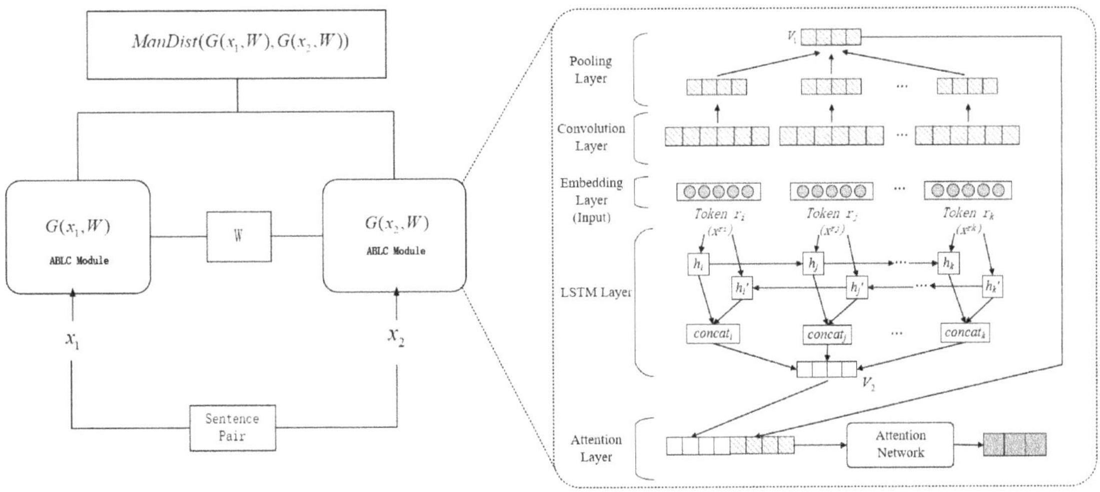

(**a**) (**b**)

Figure 2. (**a**) Overall framework of the contrast learning model. (**b**) The structure of ABLC model.

Suppose the address A is composed of N words, namely $A = \{a_1, a_2, \ldots, a_N\}$. For each word in address A, you can use the word vector dictionary $D^w \in R^{d^w|V|}$. Where V is the number of the vocabulary and d^w is the dimension of the vocabulary. The word vector dictionary D^w is obtained through learning, and the dimension of the word vector d^w is set according to requirements. Therefore, the vector of words a_i in address A is:

$$e_i = D^w V^i \tag{1}$$

where V^i is a vector of length $|V|$, and its value is 1 at e_i and 0 at the rest position. In this way, the vector of address A can be expressed as $e = \{e_1, e_2, \ldots, e_T\}$.

This paper limits the maximum length N = 20 after word segmentation for each address A. The size of the vocabulary is 10 W, and the dimension of the word vector is 300, that is, each address is mapped into a 20 × 300 vector after the embedding layer, which is used as the input of the subsequent stage.

2.3.2. Bi-LSTM

LSTM is a kind of RNN, mainly to solve the problem of gradient disappearance and gradient explosion in the training process. LSTM has better performance in long sequences [32]. The LSTM neural network uses three gate structures: input gate, forget gate and output gate to maintain and update the increase and decrease in information in the cell. However, a one-way LSTM can only process information in one direction, and cannot process information in another direction. The bidirectional LSTM is a further extension to solve the defects of LSTM. This paper uses bidirectional LSTM to extract feature information to learn address features fully. Specifically, two different LSTM neural network layers are used to traverse from the front and the back of the Chinese address, respectively, so that the address information of the two directions can be saved. Compared with the one-way LSTM, Bi-LSTM cannot only save the previous context address information, but also consider the future context address information. Therefore, the semantic representation is extracted more completely.

First, the forget gate generates a value f_t between 0 and 1 based on the output h_{t-1} from the previous memory unit and input data x_t, to determine how much information is lost in the last long-term state. h_{t-1} and x_t through the input gate to determine the update information to i_t, and in addition, through a tanh layer to get the new candidate memory unit information C_t'. Additionally, the last long-term status C_{t-1} is updated to C_t through the operation of the forget gate and the input gate. Finally, the judgment is obtained from the output gate, to multiply the value o_t between −1 and 1. The multiply result h_t is used to determine which state characteristics of the current memory cell are output. As shown in the following formula:

$$f_t = \sigma(W_f \cdot [h_{t-1}, x_t] + b_f) \tag{2}$$

$$i_t = \sigma(W_i \cdot [h_{t-1}, x_t] + b_i) \tag{3}$$

$$C_t' = \tanh(W_C \cdot [h_{t-1}, x_t] + b_C) \tag{4}$$

$$C_t = f_t * C_{t-1} + i_t * C_t' \tag{5}$$

$$o_t = \sigma(W_o \cdot [h_{t-1}, x_t] + b_o) \tag{6}$$

$$h_t = o_t * \tanh(C_t) \tag{7}$$

This model uses LSTM to solve long-term dependence, and combines the complementary information of the positive and negative directions of Bi-LSTM to fully learn the address text characteristics as shown in Figure 3. In this experiment, the number of hidden neurons is 100, and the dropout parameter is set to 0.5.

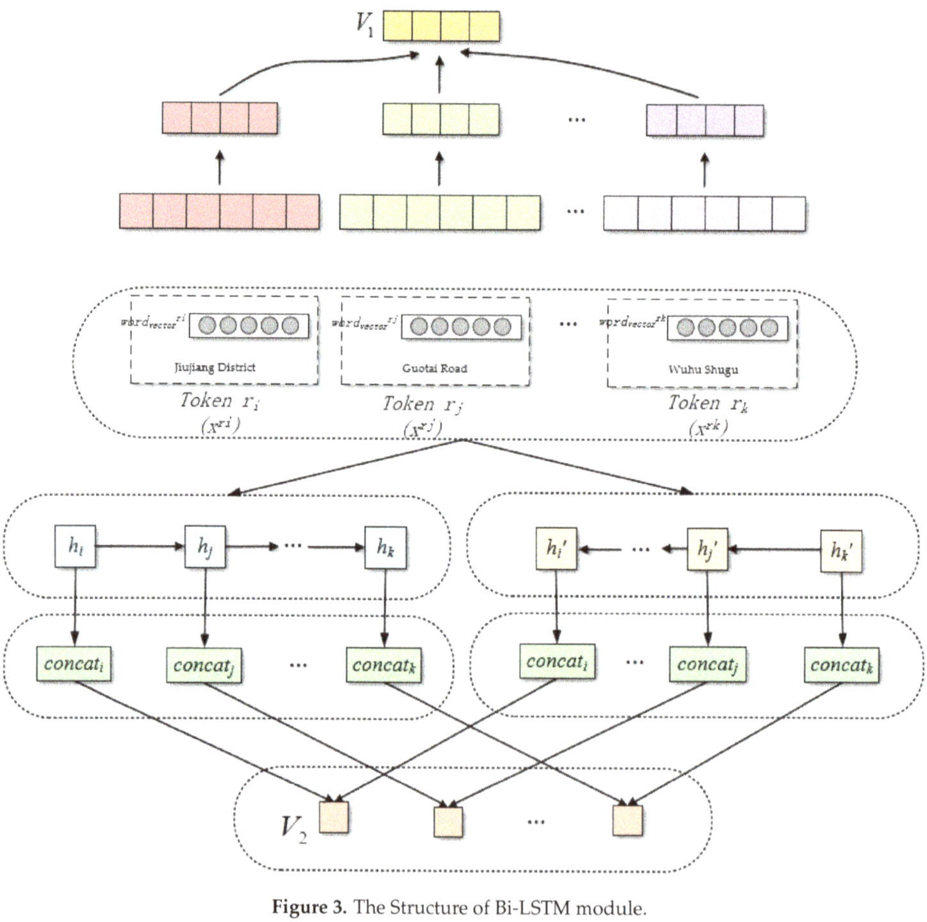

Figure 3. The Structure of Bi-LSTM module.

2.3.3. CNN

Convolutional neural network CNN has achieved good results in the field of computer vision [33], and the convolution kernel pooling is actually a process of feature extraction. The idea of CNN is to localize the overall data, use the convolution kernel function to extract the features in each local data, and then reconstruct all the fragmented features. Finally, the extraction of the overall information is realized under the guidance of the objective function.

Address text has multi-name and hierarchical property, that is, it is a text composed of a series of geographical entities, such as "Wuhu Shugu A Block 6 (POI), Guotai Road No. 2 (jieluxiang), Jiujiang District (District/County), Wuhu City (City), Anhui Province (Province)". The changes in the different levels of the Chinese address are consistent with the application scenarios of the CNN window. Based on this, the core convolution form based on CNN is used to extract the features of the address-level data. This paper uses 1-dimensional Convolution1D for convolution. The specific convolution structure is shown in the Figure 4: First, ZeroPadding1D is used to fill the edges of the input word vector matrix with zero values, and then 100 filters with a length of 5 convolution kernels are used for convolution. It is equivalent to using a $100 \times 5 \times 300$ convolution kernel to perform a convolution operation on the output matrix of the embedding layer. After the convolution operation, the extractable size is $20 \times 5 \times 300$. Then, select MaxPooling1D with pool_size of 2 to sample the convolved features, that is, take the maximum value of the convolved local area, and finally the output dimension is 20×100, as the input of the next stage.

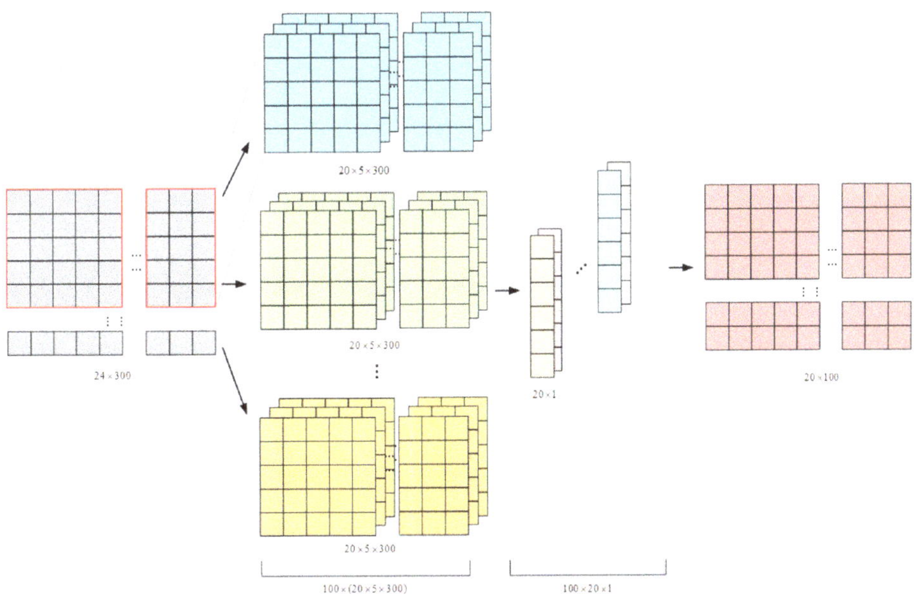

Figure 4. The structure of CNN module.

2.3.4. Attention

The human visual perception to the external world is not the full range, but focuses on a specific part according to the purpose [34]. In the field of NLP, self-attention simulates this learning process of humans. For a specific character, a certain weight is assigned to the character based on the whole text, and then integrates all the weights to determine the semantic representation of the character. According to the habit of describing addresses in Chinese, it is customary to put meaningful words or words of specific addresses in front of the expression, so different weights should be assigned to each word. For example: "1st village,1st village group", "No. 6, 1st floor, 1st community", "No. 1, building 11, district 4, 1st community", "No. 5–6, Zone E, 1st mall", "1 Building No. 11 Facade, 1st road, 1st community ". In this part, we propose to use the attention mechanism to represent the semantic information of the address, so that the semantic vector can express richer semantic information by assigning different weights.

The definition: H is the input vector containing $[h_1, h_2, \ldots, h_T]$, where T is the length of the sentence. The input vector at this stage is derived from the weighted output of the CNN and Bi-LSTM. The related formulas are described as follows:

$$A' = \tanh(H) \tag{8}$$

$$\alpha = \text{softmax}(\mathrm{W}^{\mathrm{T}} A') \tag{9}$$

$$A'' = H\alpha^T \tag{10}$$

where $H \in \mathrm{R}^{d^w \times T}$, d^W is the dimension of the word vector, W is obtained through training, and W^{T} is transposition, A'' is the vector representation after the attention stage.

Then, the final representation of each address vector is:

$$A = \sum_{axis=1,i=1}^{d^w} A'' \tag{11}$$

Among them, each row vector of the matrix is added to obtain the final vector.

2.3.5. Manhattan Distance

This paper applies Manhattan distance to calculate the similarity between a pair of addresses. The definition $A^{left} = (A^l{}_1, A^l{}_2, \ldots, A^l{}_n)$ and $A^{right} = (A^r{}_1, A^r{}_2, \ldots, A^r{}_n)$ vectors are, respectively, semantic representation of the address pair after attention stage, then the Manhattan distance of A^{left} and A^{right} can be expressed as:

$$Md = \sum_{i=1}^{n} \left| A^{left}{}_i - A^{right}{}_i \right| \tag{12}$$

Use the sigmoid function to predict the final similarity y value:

$$y = sigmoid(Md) \tag{13}$$

3. Results

3.1. Dataset

In order to evaluate the stability of the model proposed in this paper, we leverage a standard address library to construct an address data sets containing 195,405 pairs of address, and then employs manual marking to mark whether the two addresses are similar or not. An example of address pair is shown in Table 1. From the address pair dataset, we select 10% of the address pairs as the test sets, which contains 13,027 pairs of similar and 6513 pairs of non-similar. The ratio of positive and negative samples is around 2:1. For the remaining address dataset, we use a ten-fold cross-validation strategy for training and verification. In the data preprocessing stage, we use the third-party tool Jieba to segment the addresses. Considering that the address, as a short text with a special structure, may contain a large number of unique vocabularies of place names, we used a custom stop vocabulary list when segmenting words:

Table 1. Address pairs on custom dataset.

Address 1	Address 2	Similarity
Dormitory of Xinhua Bookstore, Chaowu Road, Jinhe Community, Wuwei City	Interior of Xinhua Bookstore, Chaowulu, Jinhe Community, Wucheng Town, Wuwei County, Anhui Province	1
No. 1, Wuteng Village, Xinwu Economic Development Zone, Wuhu County, Wuhu	Xiaocun Nature Village, Zhongyao Village Villagers Committee, Liulang Town, Wuhu County, Anhui Province	0

3.2. Data Augmentation

Data augmentation is an effective way to expand the sample sets by way of changing the training data. The larger the data size and the better the quality, the trained model is able to get better predictive and generalization capabilities. Address pairs on data augmentation are constructed as Table 2. It is different from the image field through the introduction of noise or cropping to achieve data enhancement [35–37]. In the field of NLP, small changes in string may lead to huge deviations in meaning, so it is hard to perform simple transformations on data. In text classification tasks, scholars have proposed several text enhancements based on noise which is synonym substitution, random insertion, random exchange, and random deletion [38]. Aiming at the particularity of the address, this paper adopts a data enhancement method based on the dropout strategy. Essentially, data enhancement is implemented by two ways. One way is to concatenate address with itself as a positive sample, or with a random sample from the rest addresses as a negative sample. The other way is to send same sample to dropout structure twice [39]. Specifically, assuming that the address A is input with the dropout semantic representation model, the vector obtained is $h^{(0)}$, and then the same address A is input into the semantic representation model (in this case, another random dropout) to obtain the vector $h^{(1)}$. We treat $h^{(0)}$, $h^{(1)}$ as a pair positive example.

Table 2. Address pairs on data augmentation.

Address 1	Address 2	Similarity
Xinhua Bookstore Dormitory of Xinhua Bookstore, Chaowu Road, Jinhe Community, Wuwei City, Interior of Xinhua Bookstore, Xinhua Bookstore, Chaowu Road, Jinhe Community, Wucheng Town, Wuwei County, Anhui Province	The interior of Xinhua Bookstore, Chaowulu Xinhua Bookstore, Jinhe Community, Wucheng Town, Wuwei County, Anhui Province, Dormitory of Xinhua Bookstore, Chaowu Road, Jinhe Community, Wuwei City	1
No. 1, Wuhu Wuteng Village, Xinwu Economic Development Zone, Wuhu County, Wuhu, Xiaocun Nature Village, Zhongyao Village Villagers Committee, Liulang Town, Wuhu County, Anhui Province	Xiaocun Nature Village, Zhongyao Village Villagers Committee, Liulang Town, Wuhu County, Anhui Province, No. 1, Wuhu Wuteng Village, Xinwu Economic Development Zone, Wuhu County, Wuhu	1
Dormitory of Xinhua Bookstore, Chaowu Road, Jinhe Community, Wuwei City, Interior of Xinhua Bookstore, Chaowu Road, Jinhe Community, Wucheng Town, Wuwei County, Anhui Province	No. 1, Wuhu Wuteng Village, Xinwu Economic Development Zone, Wuhu County, Wuhu, Xiaocun Nature Village, Zhongyao Village Villagers Committee, Liulang Town, Wuhu County, Anhui Province	0

3.3. Experiment

In this study, the word2vec model is used as the semantic representation model. After the address pairs are indexed as a predefined vocabulary list, the sentences are embedded as a list of word indexes. Lists that less than 20-dimensional are padded with 0 to 20-dimensional coding. As for the setting of hyperparameters, considering the possible length of the address, the output dimension of each word in the semantic embedding layer is set to 768 dimensions, and the overall semantic representation dimensions of each address in address pair are both set to 100. After the semantic representation, two semantic vectors are obtained separately and taken as the input of the next network layer.

Considering the size of the dataset, during the network training process, the batch size of training set is adjusted to 1024. The model also used a two-layer Bi-LSTM network and CNN layer to obtain global context information and local context information. To enhance the difference of two address, ABLC used a dropout structure and probability is set to 0.5. The output is fused into a $X \in \mathbb{R}^{25\times 100}$ feature matrix and sent to the self-attention network to get more position-aware information in address descriptions. Finally, two 100-dimensional representation vectors are used as output of the semantic representation to calculate the Manhattan distance. After four layers of full connection compression, the output of last layer is seen as the similarity of the two addresses.

In order to judge the prediction result of the model, we select accuracy, precision, recall, and F1 score as evaluation indicators. The accuracy reflects the model accurate to judge of "similar/dissimilar" and the F1 score reflects the overall performance of the model.

3.3.1. Parameter Experiment Analysis

The relevant parameters in this work are shown in Table 3. In order to verify the stability of the model parameters used in this paper, we have constructed a number of comparative experiments to prove it. The experiment contains multiple models with different batch sizes and learning rates. The model design is shown in the Table 4:

Table 3. Parameter name and corresponding value.

Parameter Name	Parameter Value
epoch	25
batch_size	1024
optimizer	Adam
learning_rate	0.01
dropout	0.5

Table 4. Parameter comparative experiment.

Model No.	Model Setting
1	learning_rate = 0.1
2	learning_rate = 0.001
3	learning_rate = 0.0001
4	batch_size = 512, learning_rate = 0.001
5	batch_size = 1500, learning_rate = 0.1

The experiment is carried out under the same training set, and the comparison results of the training indicators are shown in the Table 5:

Table 5. Parameter experiment comparison results.

Model	F1 Score	Accuracy	Recall	Precision
ABLC	0.9504	0.9563	0.9460	0.9552
1	0.9362	0.9439	0.9315	0.9413
2	0.9234	0.9343	0.911	0.9402
3	0.8926	0.8435	0.9798	0.8197
4	0.9263	0.9362	0.9137	0.9436
5	0.9381	0.9458	0.9356	0.9407

The specific analysis of the influence of each parameter element on the model prediction results is as follows:

As shown in Figure 5, when the learning rate is set to 0.01, the model converges and achieves a good convergence effect at the same time after 25 epochs. After the learning rate is set to a lower number like 0.0001, the overall average F1 score drops about 15%. The explanation of this experimental result can be expressed as that high learning rate will make the parameter update amplitude large in each iteration of the model. So that the model fails to converge and misses the extreme value during the iteration process. If the learning rate is too small, the convergence rate will be low. Moreover, the minimum point may not be reached and the convergence quality also will be poor.

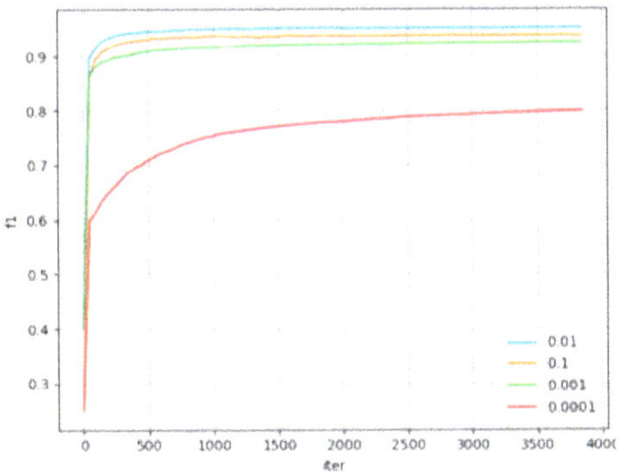

(a)

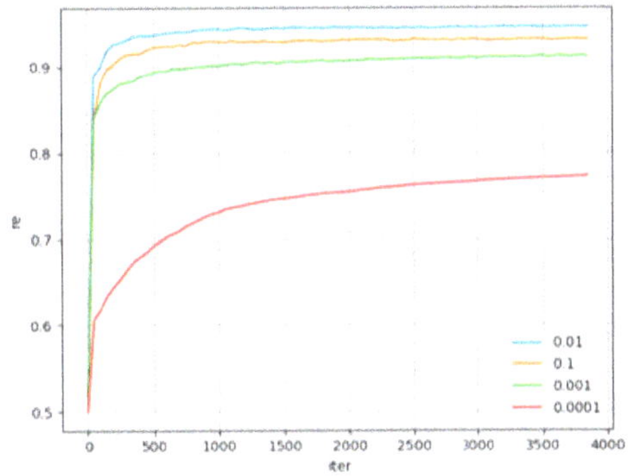

(b)

Figure 5. (**a**) F1 score at different learning rates on the training set. (**b**) Recall at different learning rates on the training set.

The experimental results from Figure 6 show that large batches could enable the model to obtain potential information in the datasets more quickly, but the overall gradient update times will be reduced accordingly. Because of that, the model often fails to reach the minimum value, and a small batch will give the model more opportunities to update parameters. Additionally, the result also shows that the adjustment of learning rate can profoundly affect the results of model prediction, and the gap can hardly be closed by changing the batch size.

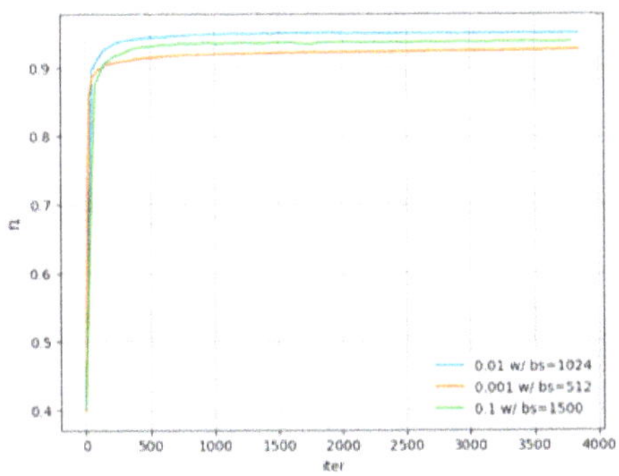

(a)

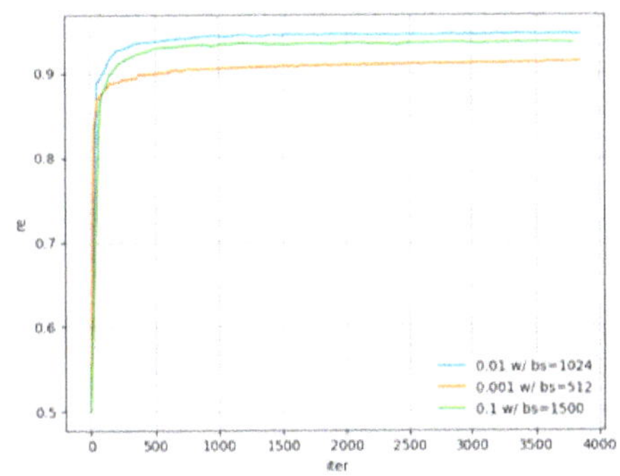

(b)

Figure 6. (**a**) F1 score at different learning rate and batch size on the training set. (**b**) Recall at different learning rate and batch size on the training set.

A too large or too small epoch number cannot lead this model to optimal results. When the training rounds are insufficient, the model cannot obtain enough information, and the performance of the trained model is poor. However, at the same time, too high training rounds will cause two problems. First, the model tends to overfit and the results between the training set and the test set are quite different. Secondly, the model may learn a large number of non-representative features and lead the prediction results to a worse direction. According to these conclusions, we select 25 as the best epoch number.

3.3.2. Analysis of Ablation Experiments

In order to verify the stability of the proposed module that integrates context and location information in this study, we designed a number of models with removing partial model structure for comparison experiments. The specific model design and the experimental results of the three models are shown in Table 6. Taking every 50 training samples as a round of iteration, the recall rate and F1 score are recorded. The changes of these two metrics with the training process are shown in Figure 7.

Table 6. Experiment results on ablation analysis.

Model Name	F1	Accuracy	Recall	Precision
ABLC	0.9504	0.9563	0.9460	0.9552
ABLC (BiLSTM + attention)	0.9448	0.9512	0.9428	0.9468
ABLC (CNN + attention)	0.9178	0.9297	0.9020	0.9413

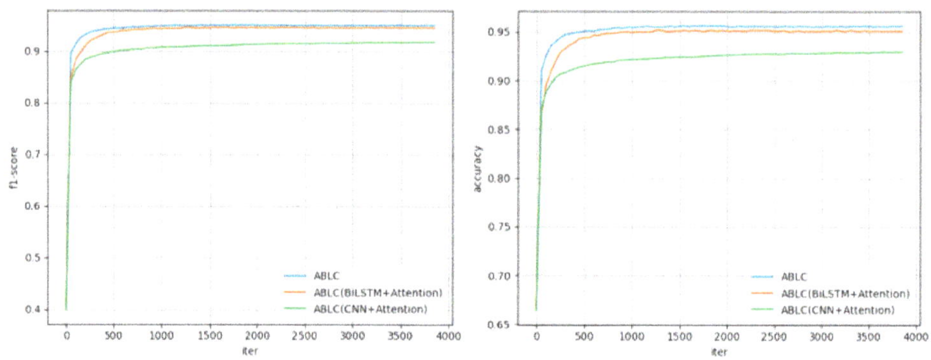

(a) (b)

Figure 7. (**a**) F1 score of different models on ablation analysis. (**b**) Accuracy of different models on ablation analysis.

As shown in Figure 7, in terms of and F1 score and accuracy, the model proposed in this paper has the best overall performance and stable performance. Compared with the model performance after ablation, the F1 score is improved by about 3–10%. This result proves the overall performance of the model decline when only considering the global context information obtained by Bi-LSTM or the local information obtained by CNN. The decline indicates that the model cannot effectively capture part of the key information in the address. At the same time, the prediction effect of the model has been effectively improved after combining the contextual global information and the local information related to the location. In addition, the F1 score proves that the accuracy of the model is not affected by the ratio of positive and negative examples in the dataset, but the learning ability of the model is indeed enhanced.

3.3.3. Comparative Experiment Analysis

In order to prove that the model proposed in this paper can achieve better results, we select some baseline models as the reference for performance comparison. Considering that the address similarity calculation problem is simplified into a judging problem of "similar" and "non-similar", it can be regarded as a disguised binary classification problem. We compare the approach proposed in this paper with multiple mainstream text classification approaches, including deep learning methods and machine learning methods. We utilized a random forest and SVM as comparison baseline models (2011) [40]. Additionally, we compared ESIM proposed by Kang (2020) for address semantic matching based on deep learning [41]. In addition, we introduce FastText (2016) and TextRCNN (2015) algorithms as comparison approaches [42,43]. The prediction result is shown in the following table, and the comparison of the change trend of some indicators during the training process is shown in the figure.

From the score shows in the Table 7, it can be concluded that the ABLC model has better improvement from several dimensions compared to other baseline models. From the semantic information point of view, the accuracy improvement of the ABLC model is round 4–10% than other models. The improvement proves that our model does have certain advantages in the classification results.

Table 7. Experiment results on comparison experiment.

Model Name	F1 Score	Accuracy	Recall	Precision
ABLC	0.9504	0.9563	0.9460	0.9552
ESIM	0.8992	0.9146	0.9051	0.9020
SVM	0.7267	0.7782	0.7125	0.7662
FastText	0.6763	0.812	0.6132	0.7569
TextRCNN	0.8062	0.8774	0.7733	0.8424
ABLC-1(Xlnet)	0.8142	0.7515	0.8348	0.7947

4. Discussion

Further detailed analysis, in terms of the convergence speed during the training process, FastText has a relatively simple and clear structure, so the convergence is more effective and faster than other models. Due to the fact that TextRCNN uses bidirectional RNN structure to obtain context information, it has certain advantages over FastText on the overall information acquisition, so the result shows TextRCNN has obvious performance strengths compare to FastText. This result presents two conclusions. First of all, the semantic extraction approach that using single sentence has certain advantages over using whole sentence pair to obtain context information as input of the network. Although bidirectional RNN can make up for the deficiencies on calculating distribution of the important words, it will ignore the comparative information from the sentence pair, so it is inferior to the ABLC model in performance. Secondly, the effect difference between the TextRCNN and FastText is not very large, indicating that the additional position-wise information introduced by attention mechanism has a certain improvement, but the effect is relatively limited. This conclusion can be explained as that the address is a special text information based on certain rules. Additionally, then the distribution of semantic information related to its position is often more fixed. Therefore, even though the position information is referred, the semantic gap between addresses is small, and the overall performance improvement of model is not very obvious.

Compared with ABLC model, ABLC-XLNet does not give an effective promotion, and has a certain amount of decline, as shown in Figure 8. As a possible explanation to this, address is a special branch of Chinese string, which contains proper nouns that refer to different places. XLNet has the ability to improve the performance on many downstream tasks, but for Chinese address descriptions, the model may lack of responsiveness to proper nouns of place. Because of this, the embedding performance cannot get a significant enhancement, and the prediction performance metrics of the model have a certain range of oscillations [44,45].

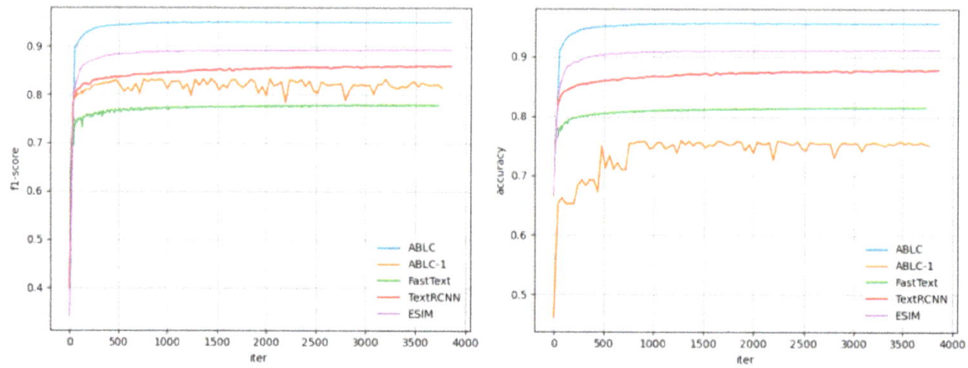

(a) (b)

Figure 8. (**a**) F1 score of different models compare with ABLC. (**b**) Accuracy of different models compare with ABLC.

Except the ABLC model proposed in this paper, the ESIM model has the best overall performance because ESIM uses local inference and inference synthesis techniques to achieve information extraction and can capture both local and global features of information. However, compared with the ABLC model proposed in this paper, its performance is slightly weaker, probably because the ABLC contrast learning model gives higher weight to the key information in addition to acquiring local and global features of information, and is good at capturing the similarities and differences between inputs via using contrast learning algorithm, thus having better classification capabilities.

A special case shown in the Table 8, even though this paper uses the semantic similarity to do the task of matching address, the ABLC model determines that the relationship between address 1 and address 2 is "not similar", whereas, the subject in address 1 and the subject in address 2 refer to the same building. One possible explanation is the training set does not contain such related information, or the lack of relevant external knowledge to supplement, so that the model cannot find out some subjects with related relationships.

Table 8. A special case of failure determination.

Address 1	Address 2	Similarity
No. 51, Changjiang Middle Road, Fanluoshan Street, Jinghu District, Wuhu City	Human Resources Security Bureau, Jinghu District, Wuhu City, Anhui Province	0

5. Conclusions

Aiming at the current problem of unrecognizable redundant information in Chinese addresses, this paper proposes a contrast learning address matching model based on attention-Bi-LSTM-CNN network. The model first extracts the address elements using Trie syntax tree according to the characteristics of Chinese addresses, followed by using Bi-LSTM to obtain the sentence-level information of addresses, as well as using CNN to obtain the word-level information in addresses, and combining with the attention mechanism to focus on the key information in addresses and assign higher weights. After the complete extraction of semantic information of the addresses, the final comparison of address similarity is achieved using the Manhattan distance. In addition, data augmentation is applied to construct the address augmentation dataset, which is combined with the dropout strategy to achieve data augmentation. The comparison with various benchmark models shows that our proposed model has better performance. For the next step we will consider that for one thing to study the association between addresses and geographic entities, for another thing to try to introduce information such as geographic information maps to enhance the accuracy of recognition. In addition, the generalization ability to unknown address needs to do further research.

Author Contributions: J.C. (Jian Chen) and G.C. proposed the original idea and conducted the organization of the content. J.C. (Jianpeng Chen) carried out the experiments and analysis the results. J.C. (Jian Chen) wrote this paper. X.S. and J.M. put forward some suggestions on the modifications of this paper. All authors have read and agreed to the published version of the manuscript.

Funding: This research was funded by Key Research and Development Plan of Anhui Province in 2021(202104a05020071).

Institutional Review Board Statement: Not applicable.

Informed Consent Statement: Not applicable.

Data Availability Statement: Not applicable.

Conflicts of Interest: The authors declare no conflict of interest.

References

1. Lee, B.H.Y.; Waddell, P.; Wang, L.; Pendyala, R.M. Reexamining the influence of work and nonwork accessibility on residential location choices with a microanalytic framework. *Environ. Plan. A* **2010**, *42*, 913–930. [CrossRef]
2. Zhang, Z.; Hadjieleftheriou, M.; Ooi, B.C.; Srivastava, D. Bed-tree: An all-purpose index structure for string similarity search based on edit distance. In Proceedings of the 2010 ACM SIGMOD International Conference on Management of Data, Indianapolis, IN, USA, 6–10 June 2010; pp. 915–926.
3. Levenshtein, V.I. Binary Codes Capable of Correcting Deletions, Insertions and Reversals. *Soviet Phys. Doklady* **1966**, *10*, 707.
4. Bilenko, M.; Mooney, R.J. Adaptive Duplicate Detection Using Learnable String Similarity Measures. In Proceedings of the 9th ACM SIGKDD International Conference on Knowledge Discovery and Data Mining, Washington, DC, USA, 24–27 August 2003; Association for Computing Machinery: New York, NY, USA, 2003; pp. 39–48.
5. Jaccard, P. Nouvelles Recherches Sur la Distribution Florale. *Bull. Soc. Vaudoise Sci. Nat.* **1908**, *44*, 223–270.
6. Banerjee, S.; Pedersen, T. *The Design, Implementation, and Use of the Ngram Statistics Package*; Springer: Berlin/Heidelberg, Germany, 2003.
7. Li, B.; Han, L. Distance weighted cosine similarity measure for text classification. In Proceedings of the International Conference on Intelligent Data Engineering and Automated Learning, Salamanca, Spain, 10–12 September 2013; Springer: Berlin/Heidelberg, Germany, 2013; pp. 611–618.
8. Kang, M.; Du, Q.; Wang, M. A New Method of Chinese Address Extraction Based on Address Tree Model. *Acta Geod. Cartogr. Sin.* **2015**, *44*, 99–107.
9. Laferty, J.D.; McCallum, A.; Pereira, F.C.N. Conditional Random Fields: Probabilistic Models for Segmenting and Labeling Sequence Data. In Proceedings of the 18th International Conference on Machine Learning, San Francisco, CA, USA, 18–24 July 2001; Morgan Kaufmann Publishers Inc.: Burlington, MA, USA, 2001; pp. 282–289.
10. Rabiner, L.; Juang, B. An introduction to hidden Markov models. *IEEE ASSP Mag.* **1986**, *3*, 4–16. [CrossRef]
11. Sun, Z.; Qiu, A.G.; Zhao, J.; Zhang, F.; Zhao, Y.; Wang, L. Technology of fuzzy Chinese-geocoding method. In Proceedings of the 2013 International Conference on Information Science and Cloud Computing, Guangzhou, China, 7–8 December 2013; IEEE Computer Society: Washington, DC, USA, 2013; pp. 7–12.
12. Xueying, Z.; Guonian, L.; Boqiu, L.; Wenjun, C. Rule-based approach to semantic resolution of Chinese addresses. *J. Geo-Inf. Sci.* **2010**, *12*, 9–16.
13. Cangxiu, C.; Bin, Y. A rule-based segmenting and matching method for fuzzy Chinese addresses. *Geogr. Geo-Inf. Sci.* **2011**, *27*, 26–29.
14. Tian, Q.; Ren, F.; Hu, T.; Liu, J.; Li, R.; Du, Q. Using an optimized Chinese address matching method to develop a geocoding service: A case study of Shenzhen, China. *ISPRS Int. J. Geo-Inf.* **2016**, *5*, 65. [CrossRef]
15. Zhang, H.; Ren, F.; Li, H.; Yang, R.; Zhang, S.; Du, Q. Recognition Method of New Address Elements in Chinese Address Matching Based on Deep Learning. *ISPRS Int. J. Geo-Inf.* **2020**, *9*, 745. [CrossRef]
16. Comber, S.; Arribas-Bel, D. Machine learning innovations in address matching: A practical comparison of word2vec and CRFs. *Trans. GIS* **2019**, *23*, 334–348. [CrossRef]
17. Kim, Y. Convolutional Neural Networks for Sentence Classification. In Proceedings of the 2014 Conference on Empirical Methods in Natural Language Processing (EMNLP), Doha, Qatar, 25–29 October 2014; Association for Computational Linguistics: Stroudsburg, PA, USA, 2014; pp. 1746–1751.
18. Santos, R.; Murrieta-Flores, P.; Calado, P.; Martins, B. Toponym matching through deep neural networks. *Int. J. Geogr. Inf. Sci.* **2018**, *32*, 324–348. [CrossRef]
19. Lai, S.; Xu, L.; Liu, K.; Zhao, J. Recurrent convolutional neural networks for text classification. In Proceedings of the AAAI Conference on Artificial Intelligence, Austin, TX, USA, 25–30 January 2015; p. 29.

20. He, J.; Li, X.; Yao, Y.; Hong, Y.; Jinbao, Z. Mining transition rules of cellular automata for simulating urban expansion by using the deep learning techniques. *Int. J. Geogr. Inf. Sci.* **2018**, *32*, 2076–2097. [CrossRef]
21. Ding, R.; Chen, Z. RecNet: A deep neural network for personalized POI recommendation in location-based social networks. *Int. J. Geogr. Inf. Sci.* **2018**, *32*, 1631–1648. [CrossRef]
22. Karimzadeh, M.; Pezanowski, S.; MacEachren, A.M.; Wallgrün, J.O. GeoTxt: A scalable geoparsing system for unstructured text geolocation. *Trans. GIS* **2019**, *23*, 118–136. [CrossRef]
23. Du, P.; Bai, X.; Tan, K.; Xue, Z.; Samat, A.; Xia, J.; Li, E.; Su, H.; Liu, W. Advances of four machine learning methods for spatial data handling: A review. *JGSA* **2020**, *4*, 1–25.
24. Grekousis, G. Artificial neural networks and deep learning in urban geography: A systematic review and meta-analysis. *Computers Environ. Urban Syst.* **2019**, *74*, 244–256. [CrossRef]
25. Young, T.; Hazarika, D.; Poria, S.; Cambria, E. Recent trends in deep learning based natural language processing. *IEEE Comput. Intell. Mag.* **2018**, *13*, 55–75. [CrossRef]
26. Zhang, L.; Wang, S.; Liu, B. Deep learning for sentiment analysis: A survey. *Wiley Interdiscip. Rev. Data Mining Knowl. Discov.* **2018**, *8*, e1253. [CrossRef]
27. Rumelhart, D.E.; Hintont, G.E.; Williams, R.J. Learning representations by back-propagating errors. *Nature* **1986**, *323*, 533–536. [CrossRef]
28. Klein, T.; Nabi, M. Contrastive self-supervised learning for commonsense reasoning. *arXiv* **2020**, arXiv:2005.00669.
29. Yang, Z.; Cheng, Y.; Liu, Y.; Sun, M. Reducing word omission errors in neural machine translation: A contrastive learning approach. *Proc. ACL* **2019**, 6191–6196.
30. Meng, Y.; Xiong, C.; Bajaj, P.; Tiwary, S.; Bennett, P.; Han, J.; Song, X. Coco-lm: Correcting and contrasting text sequences for language model pretraining. *arXiv* **2021**, arXiv:2102.08473.
31. Wu, Z.; Wang, S.; Gu, J.; Khabsa, M.; Sun, F.; Ma, H. Clear: Contrastive learning for sentence representation. *arXiv* **2020**, arXiv:2012.15466.
32. Karim, F.; Majumdar, S.; Darabi, H.; Chen, S. LSTM fully convolutional networks for time series classification. *IEEE Access* **2017**, *6*, 1662–1669. [CrossRef]
33. Khan, S.; Rahmani, H.; Shah, S.A.A.; Bennamoun, M. A guide to convolutional neural networks for computer vision. *Synth. Lect. Computer Vision* **2018**, *8*, 1–207. [CrossRef]
34. Shaw, P.; Uszkoreit, J.; Vaswani, A. Self-attention with relative position representations. *arXiv* **2018**, arXiv:1803.02155.
35. Shijie, J.; Ping, W.; Peiyi, J.; Siping, H. Research on data augmentation for image classification based on convolution neural networks. In Proceedings of the 2017 Chinese Automation Congress (CAC), Jinan, China, 20–22 January 2017; IEEE: New York, NY, USA, 2017; pp. 4165–4170.
36. Shorten, C.; Khoshgoftaar, T.M. A survey on image data augmentation for deep learning. *J. Big Data* **2019**, *6*, 1–48. [CrossRef]
37. Sakamoto, T.; Yokozawa, M.; Toritani, H.; Shibayama, M.; Ishitsuka, N.; Ohno, H. A crop phenology detection method using time-series MODIS data. *Remote Sens. Environ.* **2005**, *96*, 366–374. [CrossRef]
38. Wei, J.; Zou, K. Eda: Easy data augmentation techniques for boosting performance on text classification tasks. *arXiv* **2019**, arXiv:1901.11196.
39. Bouthillier, X.; Konda, K.; Vincent, P.; Memisevic, R. Dropout as data augmentation. *arXiv* **2015**, arXiv:1506.08700.
40. Chang, C.C.; Lin, C.J. LIBSVM: A library for support vector machines. *ACM Trans. Intell. Syst. Technol.* **2011**, *2*, 1–27. [CrossRef]
41. Lin, Y.; Kang, M.; Wu, Y.; Du, Q.; Liu, T. A deep learning architecture for semantic address matching. *Int. J. Geogr. Inf. Sci.* **2020**, *34*, 559–576. [CrossRef]
42. Joulin, A.; Grave, E.; Bojanowski, P.; Douze, M.; Jégou, H.; Mikolov, T. Fasttext. zip: Compressing text classification models. *arXiv* **2016**, arXiv:1612.03651.
43. Zhou, X.; Chen, X.; Song, J.; Zhao, G.; Wu, J. Team Cat-Garfield at TREC 2018 Precision Medicine Track. In Proceedings of the TREC, Gaithersburg, MD, USA, 14–16 November 2018.
44. Yang, Z.; Dai, Z.; Yang, Y.; Carbonell, J.; Salakhutdinov, R.; Le, Q.X. Xlnet: Generalized autoregressive pretraining for language understanding. *Adv. Neural Inf. Process. Syst.* **2019**, *32*, 5754–5764.
45. Devlin, J.; Chang, M.W.; Lee, K.; Toutanova, K. Bert: Pre-training of deep bidirectional transformers for language understanding. *arXiv* **2018**, arXiv:1810.04805.

applied sciences

MDPI

Article

Use of 2SFCA Method to Identify and Analyze Spatial Access Disparities to Healthcare in Jeddah, Saudi Arabia

Bandar Fuad Khashoggi and Abdulkader Murad *

Faculty of Architecture and Planning, Department of Urban and Regional Planning, King Abdulaziz University, Jeddah 80210, Saudi Arabia; bkasoghie@stu.kau.edu.sa
* Correspondence: amurad@kau.edu.sa

Abstract: The issue of reducing spatial disparities in access to healthcare is one of the most important healthcare planning issues that policy makers and planners investigate and consider as a key focus until present time. A healthcare system that meets the requirements of availability and affordability will be useless if the spatial accessibility to healthcare is not provided to all equally. Therefore, this study aims to identify and analyze spatial disparities in access to healthcare centers in Jeddah, Saudi Arabia. The two-step floating catchment area (2SFCA) method was used to measure spatial accessibility of healthcare centers based on the travel time threshold (i.e., 30-min drive time in this study). The GIS technology was used to execute the 2SFCA method. A geodatabase, which includes the population districts, locations of healthcare centers, and road network, was created. Some procedures were performed within the road network database to set the travel time that is considered as an essential step to compute the origin–destination (OD) cost matrix. The OD matrix was later used as the source for calculating provider-to-population ratios and the spatial accessibility scores for population districts. The results of the study revealed spatial disparities in access to healthcare centers in Jeddah city. The majority of the Jeddah population (i.e., 97.51%) have accessibility to healthcare centers, but with disparate levels. The central districts have a higher access score compared to the rest of the city's districts. Most districts that do not have accessibility to healthcare centers are concentrated in the southeast of the city. The results can help local health planners improve spatial equity in access to healthcare centers through giving the less-served districts a priority when allocating future healthcare centers in Jeddah city.

Keywords: healthcare centers; 2SFCA method; spatial accessibility; travel time threshold; GIS; OD cost matrix

Citation: Khashoggi, B.F.; Murad, A. Use of 2SFCA Method to Identify and Analyze Spatial Access Disparities to Healthcare in Jeddah, Saudi Arabia. *Appl. Sci.* **2021**, *11*, 9537. https://doi.org/10.3390/app11209537

Academic Editor: Alexei Gvishiani

Received: 7 September 2021
Accepted: 6 October 2021
Published: 14 October 2021

1. Introduction

Health is one of the most important basic requirements for a human life [1]. Improving public health for all people by ensuring equal access to healthcare is considered one of the primary Sustainable Development Goals (SDGs) [2,3]. Therefore, healthcare is one of the main pillars that governments seek to provide, finance, and manage in order to preserve communities' public health. Furthermore, the healthcare system is considered an important indicator reflecting the level of evolution and urbanization of urban communities. Therefore, studying this field has become a priority for many researchers, especially with the evolution of computer technologies that are able to study and analyze a reality of healthcare systems in populated areas) [4–6]. The healthcare system consists of organizations and institutional resources that seek to improve public health for all by providing a variety of services meeting their current and future needs, ensuring their access to these services equally with the least money, time, and effort [4]. Nevertheless, half of world's population has difficulty accessing healthcare [7]. Thus, these communities will not be able to perform their functions with high efficiency, which negatively affects social and economic development over time. Decision makers, especially urban planners in developing and developed countries, have been seeking to overcome this challenge by providing an equitable and efficient healthcare system [5,8–10].

However, the issue of reducing spatial access disparities to healthcare is one of the most important issues of healthcare planning that planners and policy makers investigate and consider as a key focus until the present time. The challenge is how to ensure that spatial accessibility to healthcare is given to all individuals equitably [6,11,12]. The spatial access disparities to healthcare in urban contexts resulted from several factors. First, the rapid and continuous increase in the population and urban growth of cities. Second, the urbanization of poverty, where people living under conditions of poverty suffer from health burdens that should not be underestimated. The third factor is the lack of government expenditure on healthcare provision. Governments try to provide healthcare services to all equally, but, in contrast, they face difficulties spending on healthcare sector development, which leads to an inability of healthcare systems to meet the growing demand for healthcare [4,10]. The fourth factor is structural barriers, such as a limited availability of transportation options, a lack of road network, and poor road conditions. These barriers can prevent or limit access to healthcare providers, which may contribute to creating more spatial disparities [13,14]. Overall, availability of preceding factors contributes to creating spatial inequality in access to healthcare, thus causing the emergence of some communities spatially facing a difficulty of accessing healthcare in urban areas compared to other communities that have perfect access to healthcare [9,10].

Spatial disparities in access to healthcare can be identified and analyzed by measuring the spatial accessibility to healthcare that researchers have extensively investigated recently [15–18]. Measurement of spatial accessibility estimates the movement opportunity between two different destinations and calculates the degree of relative difficulty that individuals face in accessing healthcare [19]. There are several spatial factors (e.g., supply and demand locations, distance or travel time) and non-spatial factors (e.g., socioeconomic variables of population) that inevitably affect how people access and utilize healthcare. Many studies have used and combined such factors to obtain more accurate and realistic results of measuring spatial accessibility to healthcare [20–25]. Measuring spatial accessibility to healthcare helps in formulating planning scenarios that indicate where the healthcare services should be allocated to improve poor spatial accessibility to services. That eventually contributes to minimizing spatial disparities in access to healthcare [26,27]. Several studies have identified and analyzed spatial disparities in access to healthcare by measuring spatial accessibility. For example, Horner and Mascarenhas [28] analyzed a location-based spatial accessibility to identify regional disparities in access to dental services in the State of Ohio. The results revealed disparities in distribution of dental services. For reducing spatial disparity in access to healthcare, Jin et.al. [29] measured and analyzed the spatial accessibility to healthcare services using the travel time technique (i.e., drive time) to assess an inequality of spatial access to healthcare in Deqing County, China. Furthermore, Tseng and Wu [30] integrated spatial and nonspatial factors (i.e., socioeconomic status) to assess spatial disparities in access to community care resources for providing a comprehensive consideration of equal allocation.

The two-step floating catchment area (2SFCA) method is one of the most recent and common methods used to measure a spatial accessibility to healthcare. It is considered a special case of the gravity model. At first, Radke and Mu [31] proposed the 2SFCA method in the form of a spatial deconstruction method, and then Luo and Wang [32,33] developed it to be an appropriate method for measuring the spatial accessibility, taking into consideration supply and demand together. Studies have proven that the 2SFCA method is easier to use and can calculate scores of spatial accessibility compared to the gravity model [18,34–36]. Subsequently, many improvements were made to the original 2SFCA method to better deal with the effect of distance decay within the catchments of demand and supply (i.e., considering disparities in spatial accessibility within catchments as a function of distance). Improvements were started by Luo and Qi [37] who introduced the enhanced 2SFCA (E2SFCA), and then Dai and Wang [38] who provided the kernel density 2SFCA (KD2SFCA), and so on [39]. Nevertheless, many studies have used the 2SFCA method to measure the spatial accessibility to healthcare [22,36,40–43].

The geographical information systems (GIS) play an important role in spatial decision-making related to public health. Furthermore, it can effectively deal with healthcare planning issues by applying statistical and analytical methods. The GIS could be used to enhance the outputs of measuring the spatial accessibility to healthcare. This could be performed using different GIS-based measures and models depending on spatial and nonspatial factors to assess spatial inequalities, identify shortage areas of healthcare, and determine the optimal allocation of health resources [6,32,33,44] However, the GIS technology has been used in many studies concerned with measuring spatial accessibility to healthcare using the 2SFCA method. For example, Wang and Luo [22] used the GIS technology to execute the 2SFCA method to measure spatial accessibility based on travel time, which was calculated by building the (O–D) cost matrix between the physician location and the population location. Areas with poor accessibility to primary healthcare were defined by integrating spatial and nonspatial factors. In another example, [36] executed the 2SFCA method using the GIS platform to identify and analyze spatial disparities in access to healthcare in Alwar district of Rajasthan. The results of study will help decision makers improve the road network and allocate future healthcare centers in the less-served areas in this district.

Jeddah city has been witnessing a continuous increase in its population and urban growth. Globally, it has become one of the fastest growing urban areas due to its high population growth rate, which is about 3.5% annually [45]. According to the report of "Global Metro Monitor 2011", Jeddah is among the 200 fastest growing metropolitan areas, surpassed only by Shanghai [46]. The rapid population growth in Jeddah will naturally lead to an increase in the demand for various services and creating disparities in the access to such services [47]. One of the most important strategic goals of Saudi Vision 2030 is to facilitate access to healthcare services at the right time and place [48,49]. Consequently, providing healthcare with equal access for all individuals is considered an important issue for the Saudi Ministry of Health and relevant decision makers. Hence, this study aims to identify and analyze spatial disparities in access to healthcare centers run by the Ministry of Health (MOH) in Jeddah city, Saudi Arabia. However, based on previous studies, this is the first study that applies the 2SFCA method to measure GIS-based spatial accessibility of the MOH healthcare centers in Jeddah depending on the travel time (i.e., drive-time threshold). This will help identify and analyze the spatial access disparities to healthcare centers in Jeddah. The results of this study could be helpful to the local health planners in Jeddah through giving the less-served districts a priority when allocating future healthcare centers. That will hopefully improve equity of healthcare service access, thus improving the quality of life for the Jeddah population.

2. Materials and Methods

2.1. Study Area

The study area is the City of Jeddah, which is located in the west of Saudi Arabia, and it is an important urban and commercial center, which has a great economic weight. Moreover, Jeddah is the second largest city after the capital of Saudi Arabia, Riyadh. It is one of the most important seaports on the Red Sea coast [50]. Jeddah is the first gateway that receives millions of people who annually come to Saudi Arabia to perform Hajj and Umrah rituals [51]. Jeddah urban area is approximately 1253.26 km^2 [52]. Jeddah has witnessed a quantum leap in all areas of development, which has contributed to a massive increase in its population and urban growth over time. The total population of Jeddah is about 4.1 million people [53]. The rate of population growth in Jeddah is about 3.5% annually [45]. Consequently, Jeddah city is considered as an ideal case that deserves extensive research of spatial accessibility of healthcare services.

The 2SFCA can measure the spatial accessibility at different census levels for a population by using any possible spatial distance variable as an input value within the network analysis [34]. Since the data of population are available at the level of urban Jeddah districts, the 2SFCA method will be applied to the level of districts. This level can provide an accurate representation of healthcare provider shortages and reflect the level of accessibility for the population related to each demand location (i.e., the district) in Jeddah. However, Jeddah has a good structure of healthcare services run by the Ministry of Health (MOH) or other government agencies, as well as the private sector. For example, there are five general hospitals and 47 healthcare centers run by the MOH in Jeddah. Furthermore, Jeddah has a wide range of private hospitals and clinics that are well distributed throughout the city. For this study, only the healthcare centers run by the MOH will be covered.

2.2. Requirements, Collection, and Preparation of Data

As mentioned, this study aims to identify and analyze spatial disparities in the access to the MOH healthcare centers in Jeddah city by measuring spatial accessibility of such services. The original 2SFCA method executed by the GIS technology was used to calculate spatial accessibility scores by considering the catchment area based on the travel time threshold. Spatial factors (i.e., locations of population, locations of healthcare centers, and travel time) were only used to measure spatial accessibility using the 2SFCA method. We did not include nonspatial factors in this study (i.e., socioeconomic variables and demographic characteristics of the population) due to a lack of this type of data at the districts level of Jeddah city. However, to achieve the purpose of this study, we captured three GIS coverages, adding their nonspatial data (Table 1). These coverages were (1) healthcare center locations, (2) population districts, and (3) the road network linking between the population threshold and the healthcare centers. As shown in Table 1, some characteristics of roads (i.e., length and speed limit of roads) were added to the attribute table of road network to estimate a travel time for the car-based transportation between the population threshold and the healthcare centers, where the transportation by buses, bikes, and walking is remarkably low in Jeddah city. All of the preceding data were processed in the ArcGIS Software and used to measure the spatial accessibility to healthcare centers in order to identify and analyze disparities of spatial access to such services in Jeddah.

Table 1. A summary of data requirements.

Dataset	Data Type	Description	Data Format
Population	Spatial	Urban district boundaries	Polygon
	Attribute	Census data at the level of urban districts: numbers and density of population, etc.	Excel table
MOH healthcare centers	Spatial	Locations of healthcare centers	Point
	Attribute	Name and location address, etc.	Excel table
Road network	Spatial	Road centerline	Line
	Attribute	Road ID, name, type, length, and speed limit of roads, etc.	Excel table

The above data were collected from various sources in paper format (data were not digital), then entered into the GIS through the digitization method. First, spatial boundaries of districts and related census data were collected from the report of Jeddah Urban Indicators Production issued by the Jeddah Urban Observatory (JUO) in 2015. To build the database within the ArcGIS Software, the census data were converted from paper format to Excel table by the digitization method. Furthermore, the spatial boundaries of districts were represented as a polygon format within the ArcGIS Software, where each polygon represents one district that has a unique ID number (record) within the census attribute table. Second, the addresses of the MOH healthcare centers in Jeddah city were identified through the interactive map on the MOH website (https://www.moh.gov.sa/Pages/Default.aspx) [accessed on 27 June 2021]. Those were later geo-coded to be the locations of facilities that were represented in a points format within the ArcGIS Software. Each point represents one location that has a unique ID number (record) within the related attribute table. The attribute data of the locations were collected from various references, such as the report of Jeddah Urban Indicators Production issued by the Jeddah Urban Observatory (JUO) in 2015, and the General Census of Population, Housing, and Facilities reports issued by the General Authority for Statistics in Saudi Arabia. The attribute data were then converted from paper format to Excel table by the digitization method to build the database within the ArcGIS Software. Third, the roads network and related attribute data were collected from the report of Jeddah Local Plan issued by Jeddah Municipality in 2015 [54], and Jeddah Explorer website (http://maps.jeddah.gov.sa/) [accessed on 21 May 2021]. Centerlines of roads were identified and represented as a lines format within the ArcGIS Software. Each line represents one road that has a unique ID number (record) within the related attribute table. The attribute data were also converted from paper format to Excel table by the digitization method to build the database within the ArcGIS Software.

2.3. Network Analysis

An approach of network analysis is an appropriate approach to deal with routing and transport issues by providing accurate and quick solutions [55]. Several inquiries related to linear networks can be answered by network analysis, especially with advances in the GIS technology recently. This type of analysis requires an availability of network data to find distance between two points within the network. The network data are represented in linear features, such as streets, roads, rivers, etc. A set of applications could be created using the network analysis, the most common of which are: (1) finding the best route or the closest facility by travel time or distance, (2) calculating service areas to determine areas within a specific travel threshold, (3) creating the OD cost matrix to find the cost to access the closest destinations from each origin, and (4) identifying the location of the new facility that provides the best response times for the community, etc. [50,56,57].

The ArcGIS Network Analyst Extension provides the network-based analysis that can spatially solve various routing problems. This analysis could be performed using many spatial tools, such as new route, new closest facility, new service area, new OD cost matrix, and new location-allocation analysis [50,55]. Users use the Network Analyst to solve many problems and inquiries of their network requirements. In other words, the Network Analyst provides several features that enable users to find the quickest or shortest route, plan routes, allocate new facilities, estimate drive-time service area, and so on. The network could also include some cost values (e.g., slope, distance, time, or other flow attributes) [50]. Furthermore, users could utilize the Network Analyst to dynamically produce a network based on realistic conditions, such as speed limits of roads, turn restrictions, and traffic characteristics. To run tools of Network Analyst Extension, the network dataset under ArcCatalog module should firstly be created based on the road network that includes the travel time value for each road (line segment) within the study area [55].

To perform the network analysis task, all data, such as population districts' boundaries and healthcare centers, should be consistent with the road network dataset, as well as in a format that a user could spatially map and analyze (e.g., population districts' boundaries that are presented as polygons should be converted to point shapefile) [34]. The point shapefile can be produced by using the "Feature to Point" function within the ArcGIS Software. In other words, this function creates centroids that are used to represent the districts of Jeddah. The use of this feature allows snapping of centroids along the road lines at the closest point to the network for the analysis. By this process, network distance measurements from the facility location can be determined. The district centroid represents the center of demand point for a population that a district polygon includes. Such a demand point is then used, along with the facility location point, to calculate the network distance that identifies which populations are within the drive-time catchment area [35]. Overall, using this feature is useful, especially given that some districts of Jeddah include limited main and secondary roads. However, the tool of "Origin–Destination (OD) Cost Matrix", which falls under the ArcGIS Network Analyst Extension, is used in this study to execute the 2SFCA method. The OD matrix is used as a source for calculating provider-to-population ratios, and scores of spatial accessibility for population districts.

2.3.1. Setting Travel Time

As mentioned, the 2SFCA method is used to measure the spatial accessibility of the MOH healthcare centers in Jeddah based on a travel time (i.e., drive-time threshold in this study). Therefore, some procedures should firstly be performed within the road network database to set the travel time that is also considered as an essential step to create the OD cost matrix later. According to the available data, four types of roads were used for this study, which are expressways, main, secondary, and local. The Ministry of Municipal and Rural Affairs (MOMRA) in Saudi Arabia issued the Engineering Design Manual for Roads in 2019. In this manual, speed limits are identified according to types of roads, where the desired speed of expressways is 100 km/h, the desired speed of main roads is 80 km/h, the desired speed of secondary roads is 60 km/h, while the desired speed of local roads is 50 km/h. To create a network capable of calculating the travel time, three main fields were added to the road network database within the GIS environment. First, the field of road length was added to calculate the length of each road (line segment) in meters. Assigning the length value for each road was performed by using the "Calculate Geometry Tool". Second, the field of speed limit was also added to assign a speed limit of travel for each road type in kilometer/hour. Third, the field of travel time was also added to assign the travel time value in minutes it would take to travel within each road type based on its speed limit. Assigning the travel time value for each road type was performed by using the "Field Calculator Tool". After calculating the travel time for each road type (line segment), some procedures were performed to identify potential problems of the road network shapefile, before producing a network dataset under ArcCatalog module. The problem of duplicated segments resulting from the existence of multiple names and features was solved by applying the "Must Not Overlap" rule within the ArcMap, where the road network does not consider travel or turn restrictions. Furthermore, the (Integrate) tool was used to ensure that all segments were connected, where many segments had no connectivity; thus, the Network Analyst Tools could not be perfectly run.

2.3.2. Creating the Origin–Destination (OD) Cost Matrix

After estimating the travel time and performing a set of procedures within the road network database, a network dataset was produced under the ArcCatalog module in preparation for creating the OD cost matrix that is used as a source to execute the 2SFCA method. Nevertheless, this study used the maximum travel time, which was the 30-min drive time, as a reference to calculate the accessibility score of the MOH healthcare centers in Jeddah. This value of time was determined according to many sources indicating that the 30-min drive time is the rational time to access the healthcare service. For example, Nichols et al. [58] mentioned that a 30-min drive time is a reasonable value for accessing healthcare facilities in Mississippi, USA. One of the most important results of the Project of Ontario CR Pilot was that 66% of sick people could access the healthcare within a specific travel time, estimated at 30 min [59]. Furthermore, the Health Resources and Services Administration (HRSA) has considered populations traveling more than 30 min to access healthcare are at risk for inadequate healthcare [60]. In other words, these populations reside in areas that have a shortage of physicians or facilities.

However, the tool of "OD Cost Matrix" is performed within the GIS environment to calculate scores of spatial accessibility within the drive-time threshold. This tool creates a dataset built by capturing all district centroids within a 30-min drive time (catchment threshold) from each healthcare center. It begins at the location of the first record of healthcare center by searching for all records of locations of district centroids that are located within the cut-off limit of 30-min drive time. This process is repeated through all records of locations of healthcare centers [34,35]. The result of this process is a table containing all the origin–destination pairs. This table shows origins first and then destinations that are sorted from closest to farthest based on travel time. Overall, this process is an important step for calculating scores of spatial accessibility using the 2SFCA method.

2.4. Measuring Spatial Accessibility of the MOH Healthcare Centers Using 2SFCA Method

During the past decade, the 2SFCA method has been used widely to study and analyze the spatial interaction between healthcare providers (supply) and populations (demand) by measuring and assessing spatial accessibility to healthcare. Researchers have preferred to use the 2SFCA method to evaluate healthcare accessibility due to several factors, the most important of which are (1) the ease and flexibility of data requirements, (2) the possibility of representing the capability of a population to travel over boundaries, (3) the unrestricted utilization of all locations within a catchment threshold with a possibility to deal with overlapping catchments, thus giving more realistic modeling results, as well as, (4) the possibility of using the travel-time threshold to overcome problems of distance impedance within catchment areas. Therefore, the 2SFCA method was identified as the appropriate method to achieve the purpose of this study, which is to identify and analyze spatial access disparities to the MOH healthcare centers in Jeddah.

The 2SFCA method could catch an area twice according to demand (i.e., population) and supply (i.e., healthcare providers). The 2SFCA method represents a model of healthcare provider-to-population ratio [36]. It is a method for creating an index that includes a set of combined information to compare among different locations. Measuring spatial accessibility depends on three main factors, which are supply (healthcare providers), demand (population), and travel time between them. Using the 2SFCA method, a spatial accessibility could be measured as a ratio between demand and supply in two steps with consideration of the impedance measure (i.e., travel time). Practically, such two steps evaluate the availability at the locations of healthcare providers as a ratio to the population who fall within the travel time catchments, and then sum up the ratios resulting from the first step for each population location [35].

The 2SFCA method uses the dichotomous distance decay function to create a map of spatial accessibility scores to be classified as accessible–inaccessible by considering the travel time catchment areas. Practically, this function assigns accessible value = 1 within the travel time catchments and inaccessible value = 0 outside of catchments. In other words, equal weights of 1 are given to all population falling within the catchments in both steps (i.e., full accessibility score); thus, they have accessibility to healthcare services. In contrast, equal weights of 0 are given to all population falling outside the catchments (i.e., accessibility score is zero); thus, they have no accessibility to healthcare services, and this indicates that the population of such areas should travel beyond the catchment threshold to access healthcare service [32,34,35].

The results are indexed scores of spatial accessibility that reflect the level of accessibility for population related to each demand location (i.e., district centroid). A spatial accessibility to healthcare services is evaluated according to a scale that includes a lowest score and the highest possible score. The lowest score is zero, which means no accessibility to healthcare, while, the higher the accessibility score, the greater the access to healthcare. Thus, if the supply is greater than demand, it is expected that the score of accessibility will be greater for the population residing close to healthcare services. On the other hand, the score of accessibility will be low if the demand is highly greater than the supply, even if the distance between the demand and supply is small [35]. Nevertheless, the 2SFCA method could be represented by the following two steps [32]:

Step 1. Calculating the provider-to-population ratio (R) for each healthcare provider; thus, for each provider (j), search all population locations (k) that are within a threshold travel time (d_0) from location (j) (that is, catchment area j), and calculate the provider-to-population ratio, (R_j), within the catchment area:

$$R_j = \frac{S_j}{\sum_{k \in \{d_{kj} \leq d_o\}} P_k}, \tag{1}$$

where (P_k) is the population of district (k) whose centroid falls within the catchment ($d_{kj} < d_0$), (S_j) is the number of providers at location (j); and (d_{kj}) is the travel time between (k) and (j). Shortly, this defined the provider-to-population ratio (R_j) within a catchment area. This represents a potential demand for the healthcare provider.

Step 2. Calculating the accessibility score (A) for each population district; thus, for each population location (i), search all provider locations (j) that are within the threshold travel time (d_0) from location (i) (that is, catchment area i), and sum up the provider-to-population ratios, (R_j), at these locations:

$$A_i^F = \sum_{j \in \{d_{ij} \leq d_o\}} R_j = \sum_{j \in \{d_{ij} \leq d_o\}} \frac{S_j}{\sum_{k \in \{d_{kj} \leq d_o\}} P_k},$$

where (A_i^F) represents the accessibility at population location (i) based on the (2SFCA) method, (R_j) is the provider-to-population ratio at the provider location (j) whose centroid falls within the catchment centered at (i) (i.e., $d_{ij} < d_0$), and (d_{ij}) is the travel time between (i) and (j). A greater value of (A_i^F) means a better accessibility at a location. Shortly, the second step allocates provider-to-population ratios to the population.

To perform this process, the ID fields were added to attribute tables of healthcare centers and population districts to create the origin–destination (OD) cost matrix. These reference fields actually represent the OriginID and DestinationID in the matrix table. After creating the matrix table, functions such as "Join" and "Sum" were used to assign the provider-to-population ratios and calculate the accessibility score for each population district. Use of these functions created a series of tables that actually represent the procedures of execution of the 2SFCA method. However, the results of 2SFCA will show the level of spatial accessibility to healthcare centers within a catchment threshold. In other words, the provider-to-population ratios will be calculated and summed up to identify and analyze spatial access disparities to healthcare services within the catchment threshold.

3. Results

3.1. Spatial Distribution of the MOH Healthcare Centers and Analysis

The first step after creating the geo-database was to understand the current condition about the healthcare centers and initially analyze their spatial distribution in Jeddah city. Healthcare centers are distributed in various parts of Jeddah city. They provide a primary healthcare to the population of Jeddah. A closer look at Figures 1 and 2 indicates that there is imbalanced spatial distribution of healthcare centers in the city. Healthcare centers highly cover most of the central districts; in contrast, the northern and southern districts are less covered by healthcare centers. In other words, the primary healthcare delivery system is not responding to continuous spatial expansion of Jeddah. It is also clear that healthcare centers are concentrated in the high population density areas, which are often concentrated in the city core, where they can serve a larger population, while the spatial concentration of these centers decreases in medium to low population density areas. Furthermore, most of the healthcare centers are located near to main roads, which means that they—in principle—can be accessed by public and private transport.

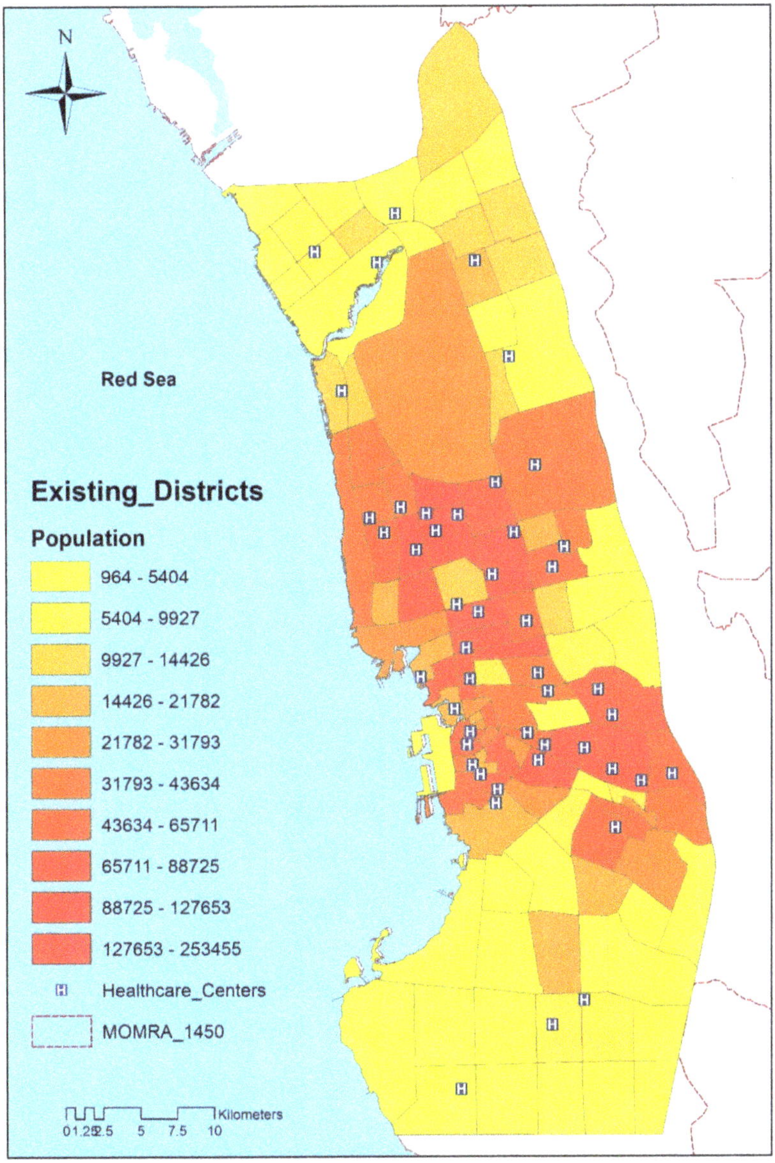

Figure 1. Spatial distribution of population districts and healthcare centers. Note: districts classified by population using the Natural Breaks (Jenks) within the ArcGIS Software.

For further investigation, the tool of standard deviational ellipses (SDE) was used to analyze the spatial distribution of healthcare centers and defining its relationship to the population concentration in Jeddah. Nevertheless, this spatial statistic function can analyze and compare spatial distribution of features and determine their patterns, orientations, and potential spatial directions. This tool is useful for identifying point patterns that take a directional orientation. The standard deviational ellipse could be calculated using point locations or weights to various points based on their attributes. Using this tool, the directional distribution and tendency for a group of features (e.g., areas or points) can be measured by computing the standard distance in directions of x, y, and z separately. Practically, the standard deviation of the x co-ordinates and the y co-ordinates is calculated from the mean center for identifying the ellipse axes. A new feature class containing an elliptical polygon centered on the mean center for all point features will be created. The attribute values for this output ellipse polygon include two standard distances represented in the long and short axes, and the orientation of the ellipse. The orientation represents the rotation of the long axis measured clockwise from noon. The GIS could provide a sense of directional orientation through a set of features drawn on the map; in contrast, calculation of the standard deviational ellipse helps make the trend more obvious. This tool can be helpful to several GIS applications, for instance, comparing the distributions of categories of health conditions, identifying ellipses for the spread of disease with passage of time, defining the directional distribution for a series of crimes, and detecting distributional trends of travel behavior [53].

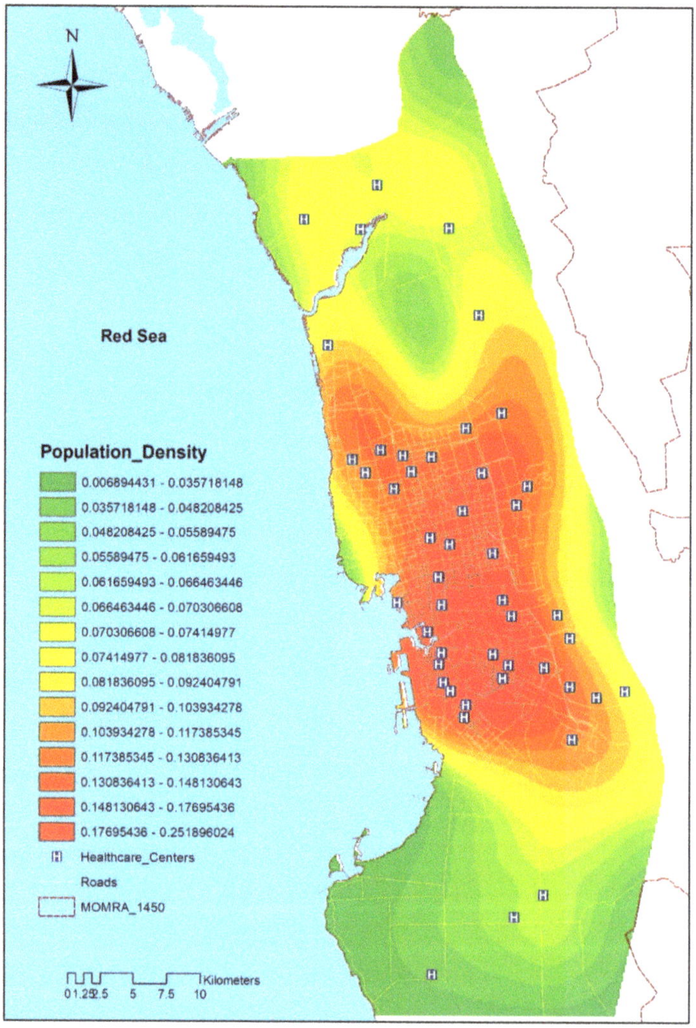

Figure 2. Spatial distribution of healthcare centers and population density. Note: population density classified by Kernel Density Model within the ArcGIS Software.

However, the SDE was selected based on the healthcare center location (i.e., point features) in this study. Figure 3 shows the output of SDE for the spatial distribution of the healthcare centers in Jeddah, which took the clustered pattern. It is clear that the directional orientation for healthcare centers is in line with the population concentration in Jeddah, where most of the centers are more concentrated and spread most widely over the central part of the city, while this concentration for centers decreases towards the north, south, and east (i.e., peripheral districts). This significant concentration of centers in the central part of the city may be due to the availability of many districts with a small area and high population density in this part, where these centers can serve a larger population.

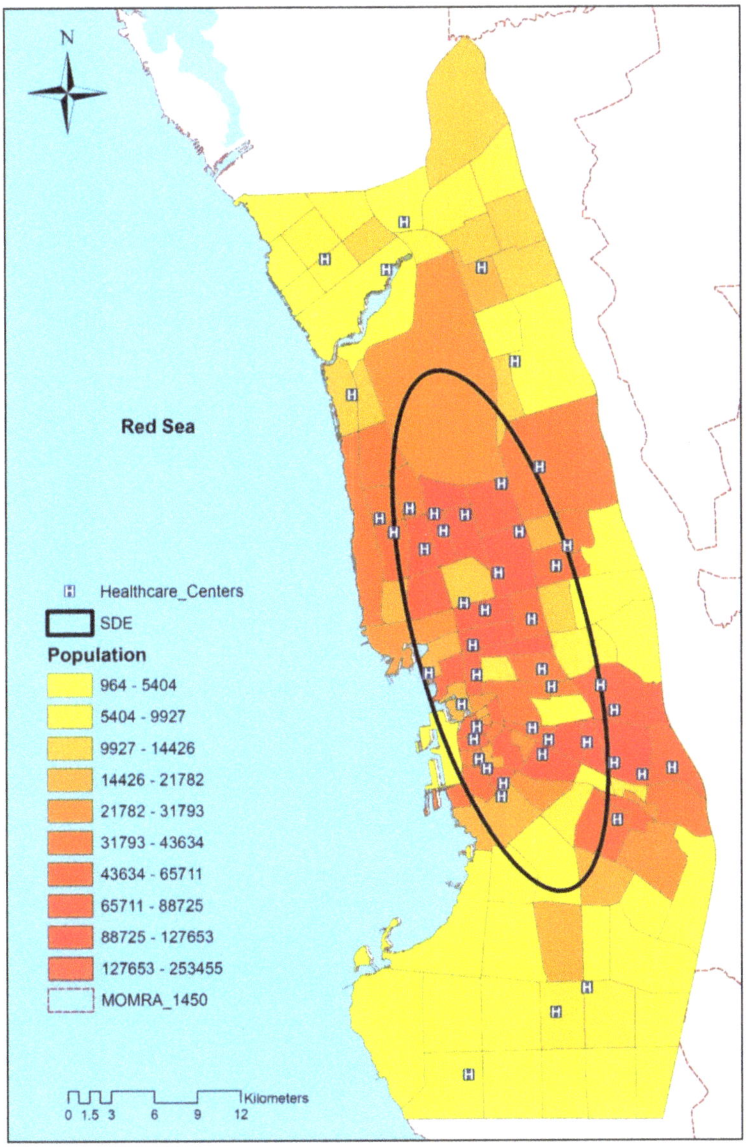

Figure 3. Standard deviational ellipse (SDE) for the spatial distribution of healthcare centers using ArcGIS Software.

3.2. Spatial Access Disparities to the MOH Healthcare Centers: Analysis of 2SFCA Results

The primary analysis in the previous section indicated that there is a disparity in the spatial distribution of the MOH healthcare centers in Jeddah, where it turns out that the central districts are well covered by centers compared to the peripheral districts that are less served by centers. However, the map of accessibility score (Figure 4) was created using the function of dichotomous distance decay (weight stands at 1 within a 30-min drive-time catchment area and 0 outside). The results of 2SFCA show remarkable disparities in spatial accessibility to healthcare centers within a catchment. Naturally, the difference in the number of healthcare centers available within the catchments contributed to creating the disparities in access to such centers. As shown in Figure 4, the results show that each possible district has an indexed accessibility score based on population. Scores of spatial accessibility were classified by Natural Breaks (Jenks) within the GIS environment. The scores do not reflect acceptable accessibility to healthcare centers, while they provide a relative measure by which accessibility can be evaluated and spatial disparities in access to services can be identified within the 30-min drive-time catchment area. This measure was created by comparing the spatial relationships between the supply and demand across space. The lowest score was zero, which means no accessibility to services, while, the higher the score of spatial accessibility, the greater the opportunity to access services. Thus, the score of spatial accessibility increases if the supply was greater than demand; in contrast, the score decreases if the demand was highly greater than the supply, even if the distance between the demand and supply was small. Nevertheless, the results obtained for urban Jeddah districts showed that the maximum accessibility score was 0.000013, which was obtained by 20 districts concentrated in the central part of the city, while the minimum accessibility score was 0, which was obtained by 14 districts concentrated in the peripheral parts of the city. The mean was 0.000011 and the standard deviation was 0.000002.

As shown in Figure 4, districts can be divided into two main parts (red part and green part). First, for the red part with a score of zero, it represents districts that have no accessibility because they are beyond a 30-min travel time to access healthcare centers. These districts represent 12.39% of the total number of urban Jeddah districts. Despite the fact that the area of these districts represents about a quarter of the total area of urban Jeddah districts, they include about 2.49% of the total Jeddah population. Residents of these districts should travel a more than 30-min drive time to access the healthcare centers. It was expected that such peripheral districts would have a score of zero (i.e., no accessibility), where they are less served by the healthcare centers compared to other districts in the city. Furthermore, they have a limited road network that does not sufficiently connect them with other parts of the city. Second, for the green part, it represents the districts that have accessibility to healthcare centers within the catchment threshold, but with disparate levels of accessibility. These districts represent 87.61% of the total number of urban Jeddah districts and include 97.51% of the total Jeddah population. The darker the green districts, the greater the accessibility to services. Central districts have a higher score of accessibility compared to the rest of the Jeddah districts that have a low accessibility score, especially in the north and south of the city. Nevertheless, districts with the greatest accessibility have a vast majority of the Jeddah population that are concentrated in the city center, while districts with the lowest accessibility have smaller numbers of the population. The largest part of the central districts is located in the highest indexed accessibility score, reflecting the higher provider-to-population ratios, where this part contains a large number of districts with a small area and high population density. In contrast, the lowest classification within the indexed accessibility score reflects the low provider-to-population ratios because the districts associated with such a classification are large in area compared to central districts. Consequently, there were fewer districts within the catchment of healthcare centers. Overall, a vast majority of the Jeddah population had a spatial accessibility to the healthcare centers. Table 2 summarizes the relevant results.

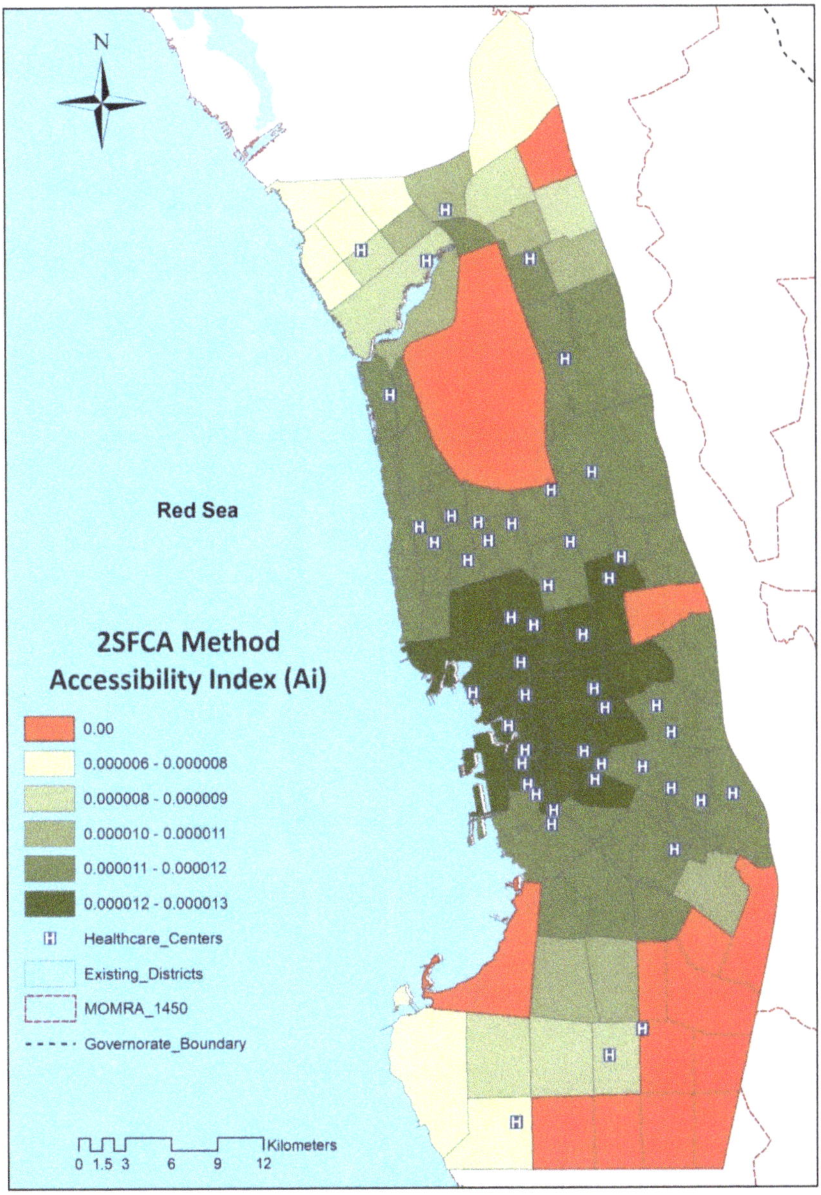

Figure 4. Results of spatial accessibility score of healthcare centers using 2SFCA method at the districts level within the ArcGIS Software.

It is clear from (Figure 5) that there is a relationship between a high accessibility and the spatial concentration for healthcare centers, as well as the road network. Districts with the greatest accessibility (e.g., central districts) have many healthcare centers and contain a good and highly efficient road network. Although there are the large number of central districts with a large population concentration, and although most of the central districts' roads have low speed limits and witness high traffic density that may increase the travel time between origin and service, the score of accessibility of these districts was high due to the spatial concentration of a large number of healthcare centers in such districts, where a large number of the population falls within their catchments. In addition, it is also clear from Figure 5 the proximity of healthcare centers to the expressways and main roads that cross many central districts and taken by the population to access healthcare centers. These levels of roads also have high traffic density within certain segments of them, but, nevertheless, they contribute to reducing travel time between origin and service due to their high speed limits. In contrast, the score of accessibility of peripheral districts decreases and may reach zero, despite the proximity of healthcare centers to the population of these districts. This is attributed to the high population number compared to the number of healthcare centers available within the catchments. Consequently, the population of these districts should compete more for healthcare close to them or should take a longer travel time that may exceed the catchment threshold to access services. This is because these districts have fewer healthcare centers than the central districts; in addition, such healthcare centers are located far from the expressways and main roads. The spatial accessibility of the population of these districts can be improved by increasing the healthcare center-to-population ratios. This can be achieved through multiple approaches, such as allocating additional healthcare centers in the less-served areas (i.e., the southern districts), or optimizing locations of some centers to achieve a more equitable distribution. In addition, it is important to improve the infrastructure of the road network in the southern or even northern districts, along with completing the construction of some roads located in these districts. Furthermore, new roads should be built in the southern districts to enhance their connectivity with other parts of the city, especially given that these districts basically have a limited road network. This will greatly contribute to improving spatial equity in accessibility by reducing travel times taken between the population residing in these districts and healthcare facilities.

Table 2. Districts and population with spatial accessibility using 2SFCA.

Statement	Total	(%) of Total
Districts with accessibility	99	87.61%
Districts without accessibility	14	12.39%
Population with accessibility	3,941,169	97.51%
Population without accessibility	100,715	2.49%
Area with accessibility (km^2)	949.55	75.74%
Area without accessibility (km^2)	303.71	24.22%

Figure 6 shows the 2SFCA results histogram of the healthcare centers at the districts level. The histogram shows that results depart from the normal distribution; however, the distribution of accessibility scores was not random, and the results were due to the interaction between supply, demand, and the travel-time threshold. The analysis indicates that the data distribution is greatly skewed negatively with a high kurtosis. According to DeCarlo [61] and Becker [35], this high kurtosis is a significant indication that there is a profusion of outliers derived from multivariate data within the large dataset, where the data have a greater peak than the normal distribution, and the distribution has more values in the tails compared to the normal distribution. Thus, the result was a high frequency of incidences around the mean, with a long left tail containing sets of incidences away from the mean. This high frequency of incidences, especially on the right side of the mean, is attributed to the presence of a large portion of central districts in the highest indexed accessibility score that reflects the higher provider-to-population ratios, where this part contains many districts with a small area and high population density. On the other hand, the low frequency of incidences away from the mean, especially on the left side of the mean, is attributed to the low provider-to-population ratios within the indexed score, reflecting the presence of districts that are large in area and less in number and population density compared to the central districts. Therefore, this confirms that the distribution does not follow a normal distribution of data.

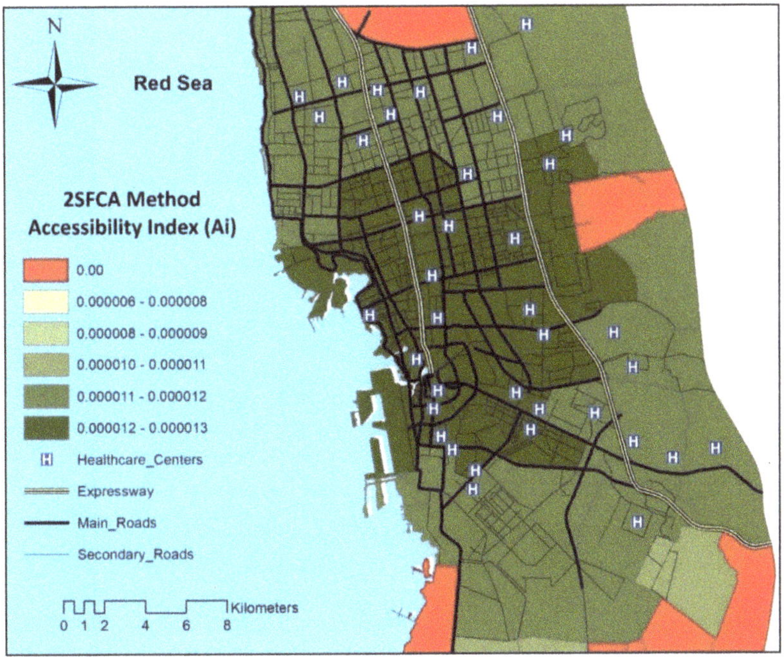

Figure 5. Districts with the greatest accessibility and their relationship to healthcare centers and road ne work. Note: The local road level was excluded from this map.

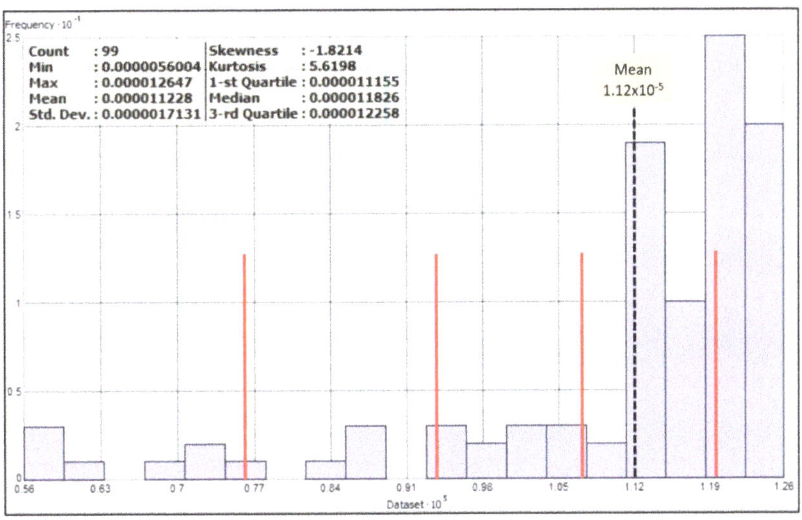

Figure 6. 2SFCA results histogram of MOH healthcare centers. Note: (1) districts with a score of 0 were excluded from the results, and (2) the red lines are the classification breaks.

4. Discussion

Governments seek to provide an equitable and efficient healthcare system that can improve the public health for all and ensure equal access to healthcare. Nevertheless, half of the world's population has a difficulty of accessing healthcare; thus, communities will not be able to perform their functions efficiently, which negatively affects social and economic development over time. Therefore, studying the reality of the healthcare system and how to improve its performance in populated areas has attracted the attention of many researchers and city planners recently, especially with the development of computer technologies (e.g., GIS technology). One of the most important healthcare planning issues is how to reduce the spatial disparities in access to healthcare. This issue results from the unbalanced spatial distribution of health resources, which contributes to the emergence of communities having poor spatial access to healthcare compared to others. Therefore, it is important to identify and analyze the spatial disparities in access to healthcare within the populated areas to give answer to where the healthcare services should be allocated. This will eventually minimize such disparities and ensure that spatial access to healthcare is given to all individuals equitably. Hence the importance of this study, which aimed to identify and analyze spatial disparities in access to the MOH healthcare centers in Jeddah city. Using GIS, the 2SFCA method was used to measure spatial accessibility to healthcare centers based on a 30-min drive time.

There is an extensive use of travel time in studies concerned with measuring spatial accessibility to healthcare than a travel distance, where the travel time can provide a better indication of spatial barriers to healthcare. Using GIS, the travel time can be estimated for each line segment within the road network based on (1) the speed limit specified for each segment and (2) the physical barriers [62]. In this study, the road network of Jeddah was used to estimate the travel time (impedance measure) in preparation for using a car-based drive-time threshold to measure the spatial accessibility to healthcare centers in Jeddah. The private car is the dominant mode of transportation, which is used more than buses, bikes, or even walking in Jeddah. It represents 95% of all trips within the city, and there is a lack of public transportation [52]. Accordingly, car ownership rates have been witnessing a huge increase in Jeddah, where about 90% of households own one or more cars, while the other 10% rely heavily on local taxis and buses for transportation purposes [63]. The use of different modes of transportation is not preferred in Jeddah due to (1) bad condition of buses, deficient infrastructure, and low accessibility [64], which leads to taking long times to access the desired destination, especially with the availability of high traffic congestion resulting from the full reliance on using private cars; and (2) districts are not suitable for walking or bicycling; thus, residents travel long distances by private cars in order to access required services.

Since private car is the only convenient way to travel in Jeddah, what about those who do not have a private car and reside, for example, in areas away from healthcare facilities? How can they access healthcare at the right time, especially with the lack of public transportation in Jeddah? In fact, the use of drive time as a measure of spatial accessibility to healthcare may not reflect the experience of those who do not have a private car. According to Jordan et.al. [65], spatial access to healthcare needs to select an appropriate measure, where there are several measures of varying complexity and specificity. However, better measures, which integrate public and private transportation, are needed to accurately reflect the experience of those who are without their own transport. This will contribute to mitigating spatial inequity in access to healthcare. There are several studies such as Mao and Nekorchuk [66] and Zhou et al. [67] that used multiple transportation modes to measure spatial accessibility to healthcare. Accordingly, decision makers should formulate intervention priorities aimed at improving the public transportation system in Jeddah, which contributes to facilitating access to healthcare services at the right time, especially for those who do not have cars. Furthermore, a future study that integrates multiple modes of transportation is needed to accurately assess spatial access disparities to healthcare services in Jeddah.

However, the results of study indicate that the MOH healthcare centers are insufficient in Jeddah city, where the outputs of the 2SFCA method revealed spatial disparities in access to healthcare centers. Spatial accessibility to healthcare centers is unevenly distributed and concentrated, where it is remarkably higher in central districts than peripheral districts. The disparity in access is largely a result of the unbalanced spatial distribution of healthcare centers, where most of which are concentrated in central districts and their concentration decreases towards the edge of the city. Furthermore, peripheral districts have a limited road network that does not sufficiently connect them with other parts of the city, which negatively affected the accessibility of the population residing in these districts. About 97.51% of the Jeddah population have access to healthcare centers, but with disparate levels of accessibility. In contrast, the area of districts that has no accessibility is approximately a quarter of the total area of urban Jeddah districts, and it includes about 2.49% of the total population of Jeddah. The percentage of the population that has no accessibility is, in fact, very small compared to the total population in the study area. However, identifying the areas in which these populations live is a necessary procedure for formulating planning interventions that can reduce spatial disparities in access to health resources, where a healthcare system that meets the requirements of availability and affordability will be useless if the spatial accessibility to healthcare is not provided to all equally [68].

It seems logical that government institutions or even the private sector focus on allocating healthcare services in densely populated areas in order to provide access to the majority of the population. For instance, it was found that the population of the central districts in Al-Madinah, Saudi Arabia can easily access health services with the least distance and fastest time because of the high spatial concentration of these services at the level of central districts [69]. Moreover, according to Luo et al. [18], the central urban areas of Wuhan have a higher or the highest level of medical accessibility that is attributed to the spatial concentration for most population and medical services in these areas. In the case of Jeddah, it is clear that the Ministry of Health has given a priority to densely populated districts (i.e., central districts) when allocating healthcare centers in order to serve a larger population. Therefore, we find that the central districts had a higher or the highest spatial accessibility scores because of the spatial concentration for the population and healthcare centers in these districts. However, one of the most important strategic goals of Saudi Vision 2030 is to facilitate access to healthcare services at the right time and place. Therefore, the Ministry of Health has recently focused on how to provide healthcare with equal access for all populations of Saudi cities. In Jeddah, for example, a number of new healthcare centers have recently been operated at the level of peripheral districts, especially in the northern and southern districts of the city, in order to reduce spatial disparities in access to healthcare in Jeddah. This is confirmed by the results of this study, which indicated that the peripheral districts, especially those located in the southern part of the city, are less-served areas by healthcare centers.

This study used the 30-min drive-time threshold to calculate spatial accessibility scores, similar to some previous studies that considered that the acceptable travel time to access healthcare should not exceed a 30-min drive time [58–60]. However, it might come to mind, if the travel time threshold was increased by more than 30 min or reduced by less than 30 min, would this affect spatial accessibility scores? Furthermore, would this change affect the magnitude of spatial disparities, especially given that 5–10 min in travel time is a reasonable fluctuation in travel time for daily commutes in and out of the city? In fact, several studies [16,18,70] indicated that the use of different travel-time thresholds can affect the analysis and evaluation of healthcare accessibility. For example, in the study of Luo et al. [18], the accessibility of medical services for elderly people in Wuhan, China was measured using the E2SFCA method based on different travel-time thresholds (i.e., 10 min and 60 min). The results of the study indicated that accessibility scores were affected when using two time thresholds, which led to the change in the magnitude of spatial disparities. When using the 10-min threshold, the accessibility scores presented the pattern of cluster spatial distribution and there were 84.01 million elderly people who had higher medical accessibility. In contrast, the scores increased when using the 60-min threshold, and there were more than 128.31 million elderly people who had higher medical accessibility.

In the case of Jeddah city, we expect that the use of different time thresholds will affect the results of the accessibility scores. Increasing the threshold by more than 30 min may improve the overall level of spatial accessibility, where healthcare centers further away from the population centroid will become accessible. Spatial accessibility scores may excessively increase at the level of central districts, with a slight increase for peripheral districts due to the existence of a limited road network in these districts. In general, increasing the threshold may contribute to reducing the spatial disparities in access to healthcare centers, especially given that some districts may be reclassified as districts with access to healthcare centers. On the other hand, reducing the threshold by less than 30 min may contribute to creating more spatial disparities, especially with the shortage of healthcare centers and the existence of a limited road network in certain districts without others. However, further research is needed to determine the effect of using different time thresholds on the magnitude of spatial disparities in access to healthcare centers in Jeddah. Therefore, our future work could investigate this issue by applying the E2SFCA method to measure spatial accessibility based on different time thresholds.

Although the study successfully demonstrated that remarkable spatial disparities in access to healthcare centers exist within urban Jeddah districts, it has certain limitations: (1) the presented study was based on population data that exceeded five years, and, therefore, more recent data are required to obtain more accurate results; (2) the study only covered the healthcare centers run by the Ministry of Health without considering private healthcare centers due to the lack of a spatial database for these centers. In contrast, identifying and geo-coding the addresses of these centers will cost researchers great time and effort that is not consistent with the time schedule given to complete this study. We expect that including private healthcare centers will improve the overall level of spatial accessibility, even with the limited road network in the peripheral districts, where these centers are well distributed over most parts of the city. However, more research is needed on the extent to which the inclusion of these centers contributes to reducing spatial disparities in access to healthcare in Jeddah. (3) Our initial research orientation was to integrate spatial and nonspatial factors (i.e., social and economic variables and demographic characteristics of the population) to accurately assess spatial disparities in access to healthcare centers in Jeddah, but the lack of data related to nonspatial factors at the districts level has led to the use of only the spatial factors in this study.

Researchers have become aware of the importance of integrating spatial and nonspatial factors to be an effective method for assessing access to healthcare [22]. Nonspatial factors, such as poverty, income, age, language, disability status, or even availability and cost of transportation, can affect access to healthcare facilities [71]. For instance, Law et.al. [72] found that the utilization of physicians and meeting healthcare needs differ from one neighborhood to another in Hamilton, Ontario. They revealed that residents of the lowest income neighborhoods have lower utilization of physicians, as well as their health needs not being met [73]. Furthermore, entire district populations aged 65 and over have poor access to primary health services in Naples, Italy, especially in the city suburbs [74]. Accordingly, it is important that future research investigates the assessment of spatial disparities in access to healthcare services in Jeddah by integrating spatial and nonspatial factors, using improved versions of the original 2SFCA method, where these improvements can better deal with the effect of distance decay within the supply and demand catchments compared to the original method that is limited to using the dichotomous distance decay function.

5. Conclusions

Providing equal spatial access to healthcare has become crucial, especially with increasing population growth in urban contexts. In other words, a healthcare system that meets the requirements of availability and affordability will be useless if the spatial accessibility to healthcare is not provided to all equally. Hence, this study sought to identify and analyze spatial disparities in access to the MOH healthcare centers in Jeddah city. Using

GIS, spatial accessibility was measured using the original 2SFCA method based on the 30-min drive-time threshold.

Overall, the results showed that 97.51% of the Jeddah population have access to healthcare centers, but with disparate levels of accessibility. The majority of those are concentrated in the central districts of the city. In contrast, 2.49% of the population is classified as unserved populations, where they should travel more than 30 min to access a healthcare center. Most of these districts are concentrated in the southeast of the city. Nevertheless, spatial disparities in access to healthcare centers can be reduced in Jeddah by applying several approaches, such as allocating new centers in less-served areas or optimizing locations of some centers at the level of Jeddah districts. Moreover, it is necessary to enhance the spatial connectivity between districts with poor access and other parts of the city. This can be achieved by building or completing road construction in some districts that basically have a limited road network, such as the southern districts. In addition, strengthening the existing collective transport network at the level of peripheral districts may improve spatial access to healthcare centers. Consequently, implementing such recommendations could reduce travel time taken between populations residing in peripheral districts and healthcare facilities by less than 30 min; thus, some of these districts may be reclassified as districts with access to healthcare centers. However, there is a need for further investigation on how much the spatial access to healthcare would be improved if road and public transport networks were developed in Jeddah, especially at the level of peripheral districts.

This study only covered the MOH healthcare centers in Jeddah according to the available data and the time schedule given for preparing this study; however, this is not enough to take a comprehensive conception of the reality of the healthcare system in Jeddah. Therefore, this study can be considered as a basis for further investigations, which is important to focus on measuring spatial accessibility to healthcare providers (governmental and private) in Jeddah or even other Saudi cities using the methodology of the 2SFCA method or its enhanced versions. Furthermore, it is also essential that these investigations consider integrating spatial and nonspatial factors to identify and assess spatial disparities in access to healthcare more accurately. However, in contrast, the lack of spatial data on healthcare providers (i.e., facility location address) and nonspatial data (e.g., socioeconomic variables at the level of urban population districts) may constitute an obstacle to conducting such investigations. It is possible to conduct surveys to collect this data at the level of urban population districts, but it may require more time, effort, and resources to accomplish this task.

However, the results of this study can be considered as an assistant reference for the MOH decision makers when developing the primary healthcare sector in Jeddah by giving the less-served districts defined by this study a priority when allocating future healthcare centers. This will contribute to improve spatial equity in access to healthcare centers.

Author Contributions: Conceptualization, A.M.; methodology, A.M. and B.F.K.; software, B.F.K.; validation, A.M. and B.F.K.; formal analysis, A.M. and B.F.K.; investigation, A.M. and B.F.K.; resources, A.M. and B.F.K.; data creation, B.F.K.; writing—original draft preparation, B.F.K.; writing—review and editing, A.M. and B.F.K.; visualization, B.F.K.; supervision, A.M. All authors have read and agreed to the published version of the manuscript.

Funding: This research received no external funding.

Institutional Review Board Statement: Not applicable.

Informed Consent Statement: Not applicable.

Data Availability Statement: Not applicable.

Conflicts of Interest: Authors declare no conflict of interest.

References

1. CESCR—Committee on Economic, Social and Cultural Rights. The Right to the Highest Attainable Standard of Health. United Nations Human Rights. 2020. Available online: https://www.refworld.org/pdfid/4538838d0.pdf (accessed on 26 May 2020).
2. Griggs, D.; Stafford-Smith, M.; Gaffney, O.; Rockstrom, J.; Ohman, M.C.; Shyamsundar, P.; Steffen, W.; Glaser, G.; Kanie, N.; Noble, I. Policy: Sustainable development goals for people and planet. *Nature* **2013**, *495*, 305–307. [CrossRef] [PubMed]
3. United Nations. Transforming our world: The 2030 Agenda for Sustainable Development. 2021. Available online: https://sdgs.un.org/2030agenda (accessed on 8 July 2021).
4. Sherif, A. Towards Spatial Justice in Urban Health Services Planning. Ph.D. Thesis, University of Utrecht, Utrecht, The Netherlands, 2007.
5. Drissy, A. The Evolution of Health Expenditure in Algeria and its Effectiveness in the Reform of the Health System during the Period (2013–2004). *Algerian J. Glob. Econ. Policies* **2015**, *6*, 137–145.
6. Khashoggi, B.F.; Murad, A. Issues of Healthcare Planning and GIS: A Review. *ISPRS Int. J. Geo Inf.* **2020**, *9*, 352. [CrossRef]
7. World Health Organization. World Bank and WHO: Half the World Lacks Access to Essential Health Services, 100 Million Still Pushed into Extreme Poverty Because of Health Expenses. 2017. Available online: https://www.who.int/news-room/detail/13-12-2017-world-bank-and-who-half-the-world-lacks-access-to-essential-health-services-100-million-still-pushed-into-extreme-poverty-because-of-health-expenses (accessed on 21 December 2019).
8. Geertman, S.C.M.; Van Eck, J.R.R. GIS and models of accessibility potential: An application in planning. *Int. J. Geogr. Inf. Sci.* **1995**, *9*, 67–80. [CrossRef]
9. Curtis, S. *Health and Inequalities: Geographical Perspectives*; Sage: London, UK, 2004.
10. Mokgalaka, H. GIS-Based Analysis of Spatial Accessibility: An Approach to Determine Public Primary Healthcare Demand in Metropolitan Areas. Master's Thesis, University of Cape Town, Cape Town, South Africa, 2015.
11. McGrail, M.; Humphreys, J. Spatial access disparities to primary health care in rural and remote Australia. *Geospat. Health* **2015**, *10*, 358. [CrossRef] [PubMed]
12. Oladimeji, A. Measurement of Spatial Accessibility and Disparities to Pharmacies in Lucas County and Multnomah County. Master's Thesis, The University of Toledo, Toledo, OH, USA, 2018.
13. Mandal, A.; Disparities in Access to Health Care. News Medical Life Sciences. 2019. Available online: https://www.news-medical.net/health/Disparities-in-Access-to-Health-Care.aspx (accessed on 26 September 2021).
14. Varela, C.; Young, S.; Mkandawire, N.; Groen, R.; Banza, L.; Viste, A. Transportation Barriers to Access Health Care for Surgical Conditions in Malawi a cross sectional nationwide household survey. *BMC Public Health* **2019**, *19*, 1–8. [CrossRef]
15. Shrestha, J. Evaluation of Access to Primary Healthcare: A Case Study of Yogyakarta, Indonesia. Master's Thesis, University of Twente, Enschede, The Netherlands, 2010.
16. Ye, H.; Hyun, K. Measuring Spatial Health Disparity Using a Network-Based Accessibility Index Method in a GIS Environment: A Case Study of Hillsborough County, Florida. *Int. J. Geospat. Environ. Res.* **2014**, *1*, 2.
17. Tao, Z.; Cheng, Y.; Zheng, Q.; Li, G. Measuring spatial accessibility to healthcare services with constraint of administrative boundary: A case study of Yanqing District, Beijing, China. *Int. J. Equity Health* **2018**, *17*, 7. [CrossRef]
18. Luo, J.; Chen, G.; Li, C.; Xia, B.; Sun, X.; Chen, S. Use of an E2SFCA Method to Measure and Analyse Spatial Accessibility to Medical Services for Elderly People in Wuhan, China. *Int. J. Environ. Res. Public Health* **2018**, *15*, 1503. [CrossRef]
19. Long, J. Modelling Accessibility. In *Geographic Information Science & Technology Body of Knowledge*; 3rd Quarter 2019 ed.; Wilson, P.J., Ed.; 2017. [CrossRef]
20. Mohan, J.; Joseph, A.E.; Phillips, D.R. Accessibility and Utilization: Geographical Perspectives on Health Care Delivery. *Trans. Inst. Br. Geogr.* **1986**, *11*, 121. [CrossRef]
21. Khan, A.A.; Bhardwaj, S.M. Access to Healthcare: A Conceptual framework and its relevance to Health Care Planning. *Eval. Health Prof.* **1994**, *17*, 60–76. [CrossRef] [PubMed]
22. Wang, F.; Luo, W. Assessing spatial and nonspatial factors for healthcare access: Towards an integrated approach to defining health professional shortage areas. *Health Place* **2005**, *11*, 131–146. [CrossRef] [PubMed]
23. World Health Organization. *Barriers and Facilitating Factors in Access to Health Services in the Republic of Moldova. Health Policy Paper*; WHO: Geneva, Switzerland, 2012; pp. 4–22. Available online: http://www.euro.who.int/__data/assets/pdf_file/0018/183510/e96775-final.pdf (accessed on 9 July 2021).
24. Zhang, L. Measuring Primary Health Care Accessibility in Mississippi State Using an Extended Kernel Density 2SFCA Method. Master's Thesis, Louisiana State University, Baton Rouge, LA, USA, 2015.
25. Bauer, J.; Klingelhöfer, D.; Maier, W.; Schwettmann, L.; Groneberg, D.A. Spatial accessibility of general inpatient care in Germany: An analysis of surgery, internal medicine and neurology. *Sci. Rep.* **2020**, *10*, 19157. [CrossRef] [PubMed]
26. Murad, A. Using GIS for Planning Public General Hospitals at Jeddah City. *JKAU* **2005**, *3*, 3–22. [CrossRef]
27. Fu, C. Planning Towards Equal Spatial Accessibility of NCI Cancer Centers Across Geographic Areas and Demographic Groups in the U.S. Ph.D. Thesis, Louisiana State University, Baton Rouge, LA, USA, 2015.
28. Horner, M.W.; Mascarenhas, A.K. Analyzing location-based accessibility to dental services: An Ohio case study. *J. Public Health Dent.* **2007**, *67*, 113–118. [CrossRef]
29. Jin, C.; Cheng, J.; Lu, Y.; Huang, Z.; Cao, F. Spatial inequity in access to healthcare facilities at a county level in a developing country: A case study of Deqing County, Zhejiang, China. *Int. J. Equity Health* **2015**, *14*, 67. [CrossRef]

30. Tseng, M.-H.; Wu, H.-C. Integrating Socioeconomic Status and Spatial Factors to Improve the Accessibility of Community Care Resources Using Maximum-Equity Optimization of Supply Capacity Allocation. *Int. J. Environ. Res. Public Health* **2021**, *18*, 5437. [CrossRef]
31. Radke, J.; Mu, L. Spatial decomposition, modeling and mapping service regions to predict access to social programs. *Geogr. Inf. Sci.* **2000**, *6*, 105–112. [CrossRef]
32. Luo, W.; Wang, F. Spatial accessibility to primary care and physician shortage area designation: A case study in Illinois with GIS approaches. In *Geographic Information Systems and Health Applications*; Skinner, R., Khan, O., Eds.; Idea Group Publishing: Hershey, PA, USA, 2003; pp. 260–278.
33. Luo, W.; Wang, F. Measures of spatial accessibility to health care in a GIS environment: Synthesis and a case study in the Chicago region. *Environ. Plan. B Plan. Des.* **2003**, *30*, 865–884. [CrossRef] [PubMed]
34. Almudaris, S. Measuring Accessibility to Primary Care Physicians in The Nashville Metropolitan Statistical Area. Master's Thesis, Western Kentucky University, Bowling Green, KY, USA, 2011.
35. Becker, C. A Spatial Analysis of Veteran Healthcare Accessibility. Master's Thesis, University of Southern California, Los Angeles, CA, USA, 2016.
36. Kanuganti, S.; Sarkar, A.; Singh, A. Quantifying Accessibility to Health Care Using Two-step Floating Catchment Area Method (2SFCA): A Case Study in Rajasthan. *Transp. Res. Procedia* **2016**, *17*, 391–399. [CrossRef]
37. Luo, W.; Qi, Y. An enhanced two-step floating catchment area (E2SFCA) method for measuring spatial accessibility to primary care physicians. *Health Place* **2009**, *15*, 1100–1107. [CrossRef]
38. Dai, D.; Wang, F. Geographic disparities in accessibility to food stores in southwest Mississippi. *Environ. Plan. B Plan. Des.* **2011**, *38*, 659–677. [CrossRef]
39. James, B., Jr.; Delamater, P.L. Examination of spatial accessibility at micro- and macro-levels using the enhanced two-step floating catchment area (E2SFCA) method. *Ann. GIS* **2018**, *25*, 219–229. [CrossRef]
40. Guagliardo, M.F. Spatial accessibility of primary care: Concepts, methods and challenges. *Int. J. Health Geogr.* **2004**, *3*, 3. [CrossRef] [PubMed]
41. Yang, D.H.; Goerge, R.; Mullner, R. Comparing GIS-Based Methods of Measuring Spatial Accessibility to Health Services. *J. Med. Syst.* **2006**, *30*, 23–32. [CrossRef]
42. McGrail, M.R. Spatial accessibility of primary health care utilising the two step floating catchment area method: An assessment of recent improvements. *Int. J. Health Geogr.* **2012**, *11*, 50. [CrossRef] [PubMed]
43. Putra, M.; Utami, N. Measuring Spatial Healthcare Facilities Accessibility Using Two-Step Floating Catchment Analysis (2SFCA) (Case Study: Cianjur Regency, Indonesia). *Semin. Nas. Geomatika* **2021**, *11*, 11–20.
44. Fradelos, E.; Papathanasiou, I.; Mitsi, D.; Tsaras, K.; Kleisiaris, C.; Kourkouta, L. Health Based Geographic Information Systems (GIS) and their Applications. *Acta Inform. Med.* **2014**, *22*, 402–405. [CrossRef]
45. Jeddah Municipality. The Importance of Jeddah City. 2020. Available online: https://www.jeddah.gov.sa/Jeddah/About/index.php (accessed on 3 November 2020).
46. Brookings Institution. Global Metromonitor 2011: Volatility, Growth, and Recovery. Global MetroMonitor. Brookings Institution 2011. p. 11. Available online: https://www.brookings.edu/wp-content/uploads/2016/06/0118_global_metro_monitor.pdf (accessed on 24 September 2020).
47. Al-Enezi, N. Analyzing Urban Growth and its Directions in Jeddah City of Using Remote Sensing Techniques and Geographic Information Systems (GIS). Ph.D. Thesis, Umm Al-Qura University, Al-Qura, Saudi Arabia, 2019.
48. MOH. *Annual Report of the Ministry of Health for the Financial Year 2018*; Annual Reports; MOH: Riyadh, Saudi Arabia, 2018; pp. 180–182. Available online: https://www.moh.gov.sa/Ministry/About/Documents/MOH-Report-2018.pdf (accessed on 13 October 2020).
49. Vision 2030. National Transformation Program. 2020. Available online: https://vision2030.gov.sa/ar/programs/NTP (accessed on 16 October 2020).
50. Murad, A. Using GIS for Determining Variations in Health Access in Jeddah City, Saudi Arabia. *ISPRS Int. J. Geo-Inf.* **2018**, *7*, 254. [CrossRef]
51. Helmi, M. The Ability of the Local Planning Authority to Implement Zoning Regulations: A Case Study of Jeddah, Saudi Arabia. Ph.D. Thesis, University of Newcastle upon Tyne, Newcastle, UK, 2015.
52. Jeddah Urban Observatory. *Report of Urban Indicators Production for Jeddah City*; Issue of 2015; Jeddah Municipality: Jeddah City, Saudi Arabia, 2015.
53. Murad, A.; Khashoggi, B.F. Using GIS for Disease Mapping and Clustering in Jeddah, Saudi Arabia. *ISPRS Int. J. Geo-Inf.* **2020**, *9*, 328. [CrossRef]
54. Jeddah Municipality. *Jeddah Local Plan. Jeddah Plans*; Jeddah Municipality: Jeddah City, Saudi Arabia, 2015; p. 47.
55. Akay, A.; Aziz, B. GIS-Based Forest Road Network Model for Forest Protection Purposes. In Proceedings of the 38th Annual COFE Meeting—Engineering Solutions for Non-Industrial Private Forest Operations, Kentucky, KY, USA, 19–22 July 2015; p. 267. Available online: https://cofe.org/pdfs/COFE_2015.pdf (accessed on 18 July 2021).
56. Comber, A.; Brunsdon, C.; Radburn, R. A spatial analysis of variations in health access: Linking geography, socio-economic status and access perceptions. *Int. J. Health Geogr.* **2011**, *10*, 44. [CrossRef]

57. Esri. Types of Network Analysis Layers. 2021. Available online: https://desktop.arcgis.com/en/arcmap/latest/extensions/network-analyst/types-of-network-analyses.htm#ESRI_SECTION1_DEAE22E63F944F6C958668B8C4AA96DA (accessed on 19 August 2021).
58. Nichols, E.N.; Bradley, D.L.; Zhang, X.; Faruque, F.; Duhé, R.J. The geographic distribution of mammography resources in Mississippi. *Online J. Public Health Inform.* **2014**, *5*, 226. [CrossRef]
59. Brual, J.; Gravely-Witte, S.; Suskin, N.; Stewart, D.; Macpherson, A.; Grace, S. Drive time to cardiac rehabilitation: At what point does it affect utilization? *Int. J. Health Geogr.* **2010**, *9*, 27. [CrossRef]
60. Allred, J.; Duffrin, C.; Brinkley, J.; Jones, J. High-risk travel distance and number of primary care visits in a North Carolina Medicaid population. *Front. Public Health Serv. Sys. Res.* **2015**, *4*, 55–60. [CrossRef]
61. DeCarlo, L.T. On the Meaning and Use of Kurtosis. *Psychol. Methods* **1997**, *2*, 292–307. [CrossRef]
62. Murad, A. Using geographical information systems for defining the accessibility to health care facilities in Jeddah City, Saudi Arabia. *Geospat. Health* **2014**, *8*, 661. [CrossRef] [PubMed]
63. Qurnfulah, E. The Negative Impacts of Subdivision Regulation on The Residential Built Environment: Jeddah'S Experience. Ph.D. Thesis, University of Newcastle, Newcastle, UK, 2015.
64. Aljoufie, M. Exploring the Determinants of Public Transport System Planning in Car-dependent Cities. *Procedia Soc. Behav. Sci.* **2016**, *216*, 535–544. [CrossRef]
65. Jordan, H.; Roderick, P.; Martin, D.; Barnett, S. Distance, rurality and the need for care: Access to health services in South West England. *Int. J. Health Geogr.* **2004**, *3*, 21. [CrossRef] [PubMed]
66. Mao, L.; Nekorchuk, D. Measuring spatial accessibility to healthcare for populations with multiple transportation modes. *Health Place* **2013**, *24*, 115–122. [CrossRef]
67. Zhou, X.; Yu, Z.; Yuan, L.; Wang, L.; Wu, C. Measuring Accessibility of Healthcare Facilities for Populations with Multiple Transportation Modes Considering Residential Transportation Mode Choice. *ISPRS Int. J. Geo Inf.* **2020**, *9*, 394. [CrossRef]
68. Li, Y.; Vo, A.; Randhawa, M.; Fick, G. Designing utilization-based spatial healthcare accessibility decision support systems: A case of a regional health plan. *Decis. Support Syst.* **2017**, *99*, 51–63. [CrossRef]
69. Abdelkarim, A. Integration of Location-Allocation and Accessibility Models in GIS to Improve Urban Planning for Health Services in Al-Madinah Al-Munawwarah, Saudi Arabia. *J. Geogr. Inf. Syst.* **2019**, *11*, 633–662. [CrossRef]
70. Zhu, L.; Zhong, S.; Tu, W.; Zheng, J.; He, S.; Bao, J.; Huang, C. Assessing Spatial Accessibility to Medical Resources at the Community Level in Shenzhen, China. *Int. J. Environ. Res. Public Health* **2019**, *16*, 242. [CrossRef]
71. NCBI. Health-Care Utilization as a Proxy in Disability Determination: 2. Factors That Affect Health-Care Utilization. 2021. Available online: https://www.ncbi.nlm.nih.gov/books/NBK500097/ (accessed on 30 September 2021).
72. Law, M.; Wilson, K.; Eyles, J.; Elliott, S.; Jerrett, M.; Moffat, T.; Luginaah, I. Meeting health need, accessing health care: The role of neighbourhood. *Health Place* **2005**, *11*, 367–377. [CrossRef] [PubMed]
73. Shah, T.; Bell, S.; Wilson, K. Spatial Accessibility to Health Care Services: Identifying under-Serviced Neighbourhoods in Canadian Urban Areas. *PLoS ONE* **2016**, *11*, e0168208.
74. Carpentieri, G.; Guida, C.; Masoumi, H. Multimodal Accessibility to Primary Health Services for the Elderly: A Case Study of Naples, Italy. *Sustainability* **2020**, *12*, 781. [CrossRef]

applied sciences

Article

A Spatial Fuzzy Co-Location Pattern Mining Method Based on Interval Type-2 Fuzzy Sets

Jinyu Guo and Lizhen Wang *

School of Information Science and Engineering, Yunnan University, Kunming 650500, China; jyguo@mail.ynu.edu.cn
* Correspondence: lzhwang@ynu.edu.cn

Abstract: The goal of spatial co-location pattern mining is to find subsets of spatial features whose instances are often neighbors in a geographical space. In many practical cases, instances of spatial features contain not only spatial location information but also attribute information. Although there have been several studies that use type-1 fuzzy membership functions to mine spatial fuzzy co-location patterns, there is great uncertainty associated with such membership functions. To address this problem, we propose a spatial fuzzy co-location pattern mining method based on interval type-2 fuzzy sets. First, we collect the interval evaluation values of the interval data of attribute information from experts to form granular data. Next, the original type-1 fuzzy membership function is extended to a granular type-2 fuzzy membership function based on elliptic curves. We use a gradual method to adjust the parameters of the fuzzy membership function so that its footprint of uncertainty satisfies both the connectivity and the given confidence. Based on this granular type-2 fuzzy membership function, we fuzzify the attribute information of instances and define the concepts of fuzzy features and fuzzy co-location patterns. A fuzzy co-location pattern mining algorithm based on spatial cliques is then proposed, termed the FCPM-Clique algorithm. In order to improve the efficiency of the algorithm, we propose two pruning strategies. In addition, we extend two classical spatial pattern mining algorithms, the Join-based algorithm and the Joinless algorithm, to mine fuzzy co-location patterns based on interval type-2 fuzzy sets. Many experiments on synthetic and real-world datasets are conducted, the performance of the three algorithms is compared, and the effectiveness and efficiency of our proposed FCPM-Clique algorithm is demonstrated.

Keywords: spatial data mining; fuzzy co-location pattern; interval type-2 fuzzy set; clique

Citation: Guo, J.; Wang, L. A Spatial Fuzzy Co-Location Pattern Mining Method Based on Interval Type-2 Fuzzy Sets. *Appl. Sci.* **2022**, *12*, 6259. https://doi.org/10.3390/app12126259

Academic Editors: Alexei Gvishiani and Boris Dzeboev

Received: 29 May 2022
Accepted: 16 June 2022
Published: 20 June 2022

1. Introduction

Spatial co-location pattern mining is an important branch of spatial data mining. A spatial co-location pattern is a subset of spatial features whose instances are often neighbors in space. For example, {hotel, restaurant} may be a prevalent co-location pattern because hotels and restaurants often neighbor each other in a city center. Spatial co-location pattern mining is widely used in the fields of botany [1], geographic information science [2–4], geology [5], urban facilities distribution, etc. For example, geologists may be interested in neighboring minerals in the same area [4].

In recent years, many algorithms for mining spatial co-location patterns have been developed. These algorithms can be divided into two types. Apriori-like algorithms [1,5–8] are one type. This algorithm first generates the candidate co-location patterns, then spends a lot of time generating row-instances of the candidate co-location patterns and checking their prevalence. Another algorithm is the algorithm based on prefix trees [2,3,9–14]. This algorithm generates maximal cliques from spatial datasets to discover prevalent spatial co-location patterns, which is generally more efficient than an Apriori-like algorithm. We have noticed that spatial data are often accompanied by attribute information expressed by specific values, such as the content of heavy metals in a certain geographical location [4].

However, existing spatial co-location pattern mining algorithms mainly focus on the location information of spatial data but ignore their attribute information.

In recent years, some researchers have used type-1 fuzzy sets to process the location information of spatial data and mine fuzzy co-location patterns (FCPs). However, type-1 fuzzy sets have a large amount of uncertainty. Type-2 fuzzy sets can model both intra-individual uncertainty and inter-individual uncertainty, while type-1 fuzzy sets can only model intra-individual uncertainty. Every expert can model a type-1 fuzzy set (intra-individual uncertainty), and the type-1 fuzzy sets of different experts are often different. Type-2 fuzzy sets can be used to fuse these different type-1 fuzzy sets, so as to consider the uncertainty between individuals.

In recent years, people have made great progress in the study of type-2 fuzzy sets [15–20]. Many researchers have applied fuzzy set theory to association rule mining. Anuradha et al. [21] used type-2 fuzzy sets to mine multi-level association rules. Lin et al. [22] used type-2 fuzzy sets to discover fuzzy frequent item sets efficiently. Kalia et al. [23] used type-2 fuzzy sets and genetic algorithms to mine fuzzy rules in high-dimensional classification tasks. Wang et al. [24] and Yang et al. [25] introduced fuzzy set theory into spatial co-location pattern mining and proposed a new metric method. Molina et al. [26] used a hierarchy to represent fuzzy association rules. Lin et al. [27] quickly discovered fuzzy frequent item sets from quantitative database based on type-2 fuzzy sets. Zhang et al. [28] mined the relationship between different types of crime rate based on fuzzy association rules. Wang et al. [29] and Anari et al. [30] combined a fuzzy set and rough set and proposed a new method of mining fuzzy association rules. In addition, fuzzy set theory is also widely used in attribute data mining. Gupta et al. [31] and Sun et al. [32] introduced fuzzy weight into fuzzy multi-attribute mining. Meng et al. [33] used interval type-2 fuzzy sets to solve multi-attribute decision-making in sponge city construction. Tao et al. [34] and Farhadinia et al. [35] applied ordered fuzzy sets to multi-attribute decision-making systems.

In recent years, many type-2 fuzzy membership functions have been proposed, such as ladder type, triangular type, Gaussian type and S type [36,37], π type [38], etc. However, these functions have a common feature: there are many parameters that determine the confidence and the width of the FOU, and the coupling between these parameters is very high, which makes it very difficult to select the appropriate parameters, for example, a type-2 fuzzy membership function based on an extended π function [38]. However, this is not the case for type-2 fuzzy membership functions based on elliptic curves [39,40] as there is only one parameter. Therefore, we generate an interval type-2 fuzzy membership function based on an elliptic curve.

We have noticed that spatial data are often accompanied by attribute information expressed by specific values, such as the content of heavy metals in a certain geographical location [4]. People are usually not sensitive to a specific value of the attribute information. However, when the attribute value changes fundamentally—that is, when it changes from a certain range to a new range—people will notice this and attach importance to the result. In this paper, in order to solve the large deviation caused by the method of mining fuzzy co-location patterns based on type-1 fuzzy sets [4], a method of mining fuzzy co-location patterns based on interval type-2 fuzzy sets and cliques from spatial datasets with attribute information is proposed.

The purpose of this paper is first to construct a special kind of interval type-2 fuzzy membership function—a granular type-2 fuzzy membership function based on interval type-2 fuzzy sets, which is used to improve the original type-1 fuzzy membership function. This reduces the deviation caused by the subjective evaluation of experts as far as possible, that is, to reduce the deviation caused by the uncertainty of a type-1 fuzzy membership function. Furthermore, we would like to find the influence of the uncertainty of fuzzy membership functions on the prevalence degree of fuzzy co-location patterns (FCPs) and measure the prevalence degree.

2. Innovation

Our paper makes the following contributions to the field of fuzzy co-location pattern mining:

(1) In the fuzzy processing of corresponding attribute information based on the original type-1 fuzzy membership function, a granular type-2 fuzzy membership function based on an elliptic curve is generated from the evaluation data. We adopt a gradual method to adjust the parameters of the function so that the type-2 membership function not only meets the connectivity [11] but also makes its confidence reach the given threshold (set to 85%).

(2) Based on the abovementioned granular type-2 fuzzy membership function, we define the concepts of fuzzy membership interval, upper bound participation ratio, lower bound participation ratio of fuzzy features, upper bound participation index, and lower bound participation index for FCP mining. Further, to smooth the "sharp boundary" of pattern prevalence and reduce the deviation caused by the subjective understanding of the prevalence degree of FCPs, we propose the concepts of absolutely prevalent FCPs, FCPs with a prevalence tendency degree, and absolutely non-prevalent FCPs.

(3) Based on spatial cliques, a method of mining FCPs is proposed. First, an algorithm for generating spatial cliques from spatial datasets is presented, and then an Apriori-like algorithm and two pruning strategies are proposed to discover the absolutely prevalent FCPs and FCPs with a prevalence tendency degree. In addition, we apply interval type-2 fuzzy sets to traditional co-location pattern mining algorithms and form an FCPs mining method based on interval type-2 fuzzy sets and the traditional Join-based algorithm, as well as another FCPs mining algorithm based on interval type-2 fuzzy sets and the traditional Joinless algorithm.

3. Related Definitions and Theorems

3.1. Definitions

Traditional co-location patterns can reflect the spatial association of instances but ignore the influence of other information—besides location information—on the mining results. In order to articulate our method, we give the following definitions:

Definition 1 (Data fuzzification). *The most important step of attribute data fuzzification is to determine the fuzzy membership function. According to the fuzzy membership function, the attribute data is expressed as a fuzzy membership.*

Supposing the fuzzy set over the domain is F, then the fuzzy membership function is U_F. We use membership function U_F to describe the fuzzy set F, that is, U_F: $x \rightarrow [0, 1]$.

In this paper, we use the upper boundary and lower boundary of the footprint of uncertainty (FOU) to describe the type-2 fuzzy membership function. That is, we introduce the upper bound membership function $\overline{U}_F(x)$ and the lower bound membership function $U_F(x)$.

Definition 2 (Granular evaluation data). *We use the expert evaluation method to evaluate the interval membership degrees of the interval values under a fuzzy set to form granular evaluation data, so as to reduce the error when the specific membership degree of a specific value under a fuzzy set is evaluated.*

Definition 3 (Fuzzy feature). *Fuzzy features refer to different types of things with fuzzy attributes in a space. A fuzzy feature set refers to the set of fuzzy features, expressed as* **Fuzz_Fea** = {$Fuzz\_f_1$, $Fuzz\_f_2$, ... , $Fuzz\_f_n$}. *A fuzzy feature is represented as* $Fuzz\_f_i$ $(1 \leq i \leq n)$.

The instances of a fuzzy feature are the sampling points of the fuzzy feature in the space and the fuzzy membership degree for the fuzzy feature when it is greater than 0.

Definition 4 (Neighborhood relationship). *A spatial dataset refers to the set of instances with type (feature), location information, and attribute information in a space. Suppose* **Fuzz_Fea** *is a set of features and* **S** *is the set of their instances, for any two instances, s_1 and s_2, if the Euclidean distance between the two instances is less than the threshold min_dist given by the user—that is, distance(s_1, s_2) $\leq$ min_dist—then the two instances satisfy the neighborhood relationship, which is represented as R(s_1, s_2).*

Definition 5 (Fuzzy co-location pattern). *A fuzzy co-location pattern (FCP)* **Fuzz_c** *is a subset of the set of fuzzy features, that is,* **Fuzz_c** $\subseteq$ **Fuzz_Fea**. *The size of an FCP is the number of fuzzy features of the pattern, which is expressed as |* **Fuzz_c** *|.*

Definition 6 (Row-instance and table-instance). *Suppose there is a subset of spatial instances* **S** *={s_1, s_2, ... , s_k} where any two instances are neighbors. If* **S** *contains all the fuzzy features of an FCP* **Fuzz_c** *and there is no subset of* **S** *that can contain all the fuzzy features of* **Fuzz_c**, *then* **S** *is called a row-instance of* **Fuzz_c**. *All row-instances of* **Fuzz_c** *constitute the table-instance of* **Fuzz_c**.

Definition 7 (Absolute row-instance and absolute table-instance). *Suppose there is a spatial instances$_i$ (i∈{1, 2, ... k}). Ifs$_i$ is an instance of the fuzzy feature Fuzz_f_1 and also an instance of the fuzzy feature Fuzz_f_2, thens$_i$ is called an absolute row-instance of a fuzzy co-location pattern {Fuzz_f_1, Fuzz_f_2}. For a fuzzy co-location pattern, all its absolute row-instances constitute its absolute table-instance.*

Definition 8 (Upper bound participation ratio and lower bound participation ratio). *Suppose there is an FCP* **Fuzz_c** *= {Fuzz_f_1, Fuzz_f_2, ... , Fuzz_f_k}, the lower bound participation ratio of the fuzzy feature Fuzz_f_i(i∈{1, 2, ... k}) is expressed as $\underline{PR}$(***Fuzz_c***, Fuzz_f_i), which is defined as the ratio of the sum of the lower bound membership degrees of the instances that appear non-repeatedly in the table-instance of the FCP* **Fuzz_c** *to the sum of the lower bound membership degrees of all the instances of Fuzz_f_i. The upper participation ratio is expressed as $\overline{PR}$ (***Fuzz_c***, Fuzz_f_i), which is defined as the ratio of the sum of the upper bound membership degrees of the instances that appear non-repeatedly in the table-instance of the FCP* **Fuzz_c** *to the sum of the upper bound membership degrees of all the instances of Fuzz_f_i, namely:*

$$\underline{PR}\left(\boldsymbol{Fuzz\_c}, Fuzz\_f_i\right) = \sum \underline{u}\left(\pi_{Fuzz_{f_i}}\ table\_instance(Fuzz\_c)\right) / \sum \underline{u}(table\_instance(\{Fuzz\_f_i\}));$$

$$\overline{PR}(\boldsymbol{Fuzz\_c}, Fuzz\_f_i) = \sum \overline{u}\left(\pi_{Fuzz_{f_i}}\ table\_instance(Fuzz\_c)\right) / \sum \overline{u}(table\_instance(\{Fuzz\_f_i\}))$$

where π is the relational projection operation, u(x)denotes the lower bound membership of x, and $\overline{u}$(x)denotes the upper bound membership of x.

Definition 9 (Upper bound participation index and lower bound participation index). *The upper bound participation index of FCP* **Fuzz_c** *is expressed as $\overline{PI}$(***Fuzz_c***), which is defined as the minimum of the upper bound participation ratio of all fuzzy features Fuzz_f_i in* **Fuzz_c**, *which is:*

$$\overline{PI}(\boldsymbol{Fuzz\_c}) = min_{i=1}^{k}\{\overline{PR}\,(\boldsymbol{Fuzz\_c}, \mathrm{Fuzz}f_i)\}.$$

The lower bound participation index of FCP ***Fuzz_c*** is expressed as $\underline{\mathrm{PI}}$(***Fuzz_c***), which is defined as the minimum of the lower bound participation ratio of all fuzzy features *Fuzz_f_i* in ***Fuzz_c***, which is:

$$\underline{\mathrm{PI}}(\boldsymbol{Fuzz\_c}) = min_{i=1}^{k}\{\underline{\mathrm{PR}}\,(\boldsymbol{Fuzz\_c}, \mathrm{Fuzz}f_i)\}.$$

Definition 10 (Prevalent fuzzy co-location pattern). *According to Definition 9, an FCP* ***Fuzz_c*** *has an upper bound participation index and a lower bound participation index. For an FCP* ***Fuzz_c*** *when given a prevalence threshold min_prev, if* $\overline{PI}(\textbf{Fuzz\_c}) < min\_prev$, ***Fuzz_c*** *is called an absolutely non-prevalent pattern. If* $\underline{PI}(\textbf{Fuzz\_c}) \geq min\_prev$, ***Fuzz_c*** *is called an absolutely prevalent pattern. If* $\underline{PI}(\textbf{Fuzz\_c}) < min\_prev \leq \overline{PI}(\textbf{Fuzz\_c})$, ***Fuzz_c*** *is called an FCP with prevalence tendency degree, and the prevalence tendency degree of the fuzzy pattern is:*

$$\theta = \left(\overline{PI}\left(\textbf{\textit{Fuz\_c}}\right) - \text{minprev}\right) / \left(\overline{PI}(\textbf{\textit{Fuz\_c}}) - \underline{PI}(\textbf{\textit{Fuz\_c}})\right)$$

We call the absolutely prevalent FCPs and FCPs with a prevalence tendency degree prevalent FCPs.

Example 1. We take the content of heavy metals, copper and zinc, in the topsoil of an area as an example. We divide the areas into the agricultural area (functional area A), residential area (functional area B), industrial area (functional area C), and commercial area (functional area D). Then, we sample the surface soil of each functional area and then measure the heavy metal content at each sampling point. The number of sampling points of functional areas, spatial location of sampling point, copper content, and zinc content are shown in Table 1. The spatial position of the sampling points are shown in Figure 1.

Table 1. Attribute information of sampling points.

Functional Area	The Number of Sampling Points	The Content of Copper	The Content of Zinc
A	1	62	226
A	2	31	105
A	3	46	136
B	1	86	183
B	2	29	112
B	3	45	150
B	4	57	124
C	1	63	210
D	1	51	155

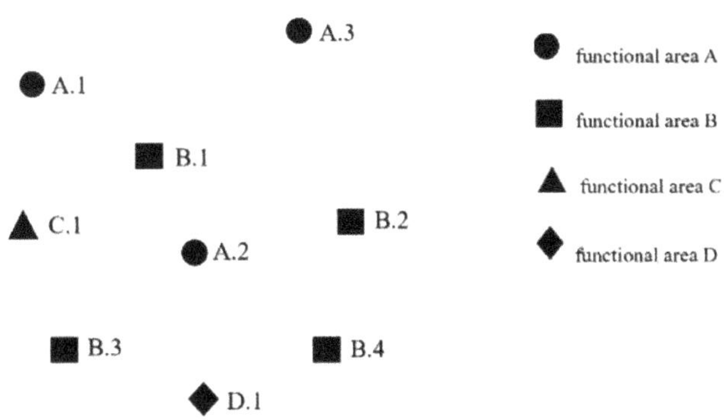

Figure 1. Spatial distribution of sampling points.

Suppose there is an FCP ***Fuzz_c*** = {A.Cu(M), B.Cu(H), C.Cu(M)}. Here, the fuzzy feature B.Cu(H) represents the B functional area with a high copper content. The size of the FCP is 3. According to the granular type-2 fuzzy membership function constructed in Section 4.1, we obtain the interval membership of the attributes of the instances in Table 1, as shown in Table 2. As can be seen from Table 2, the instances of fuzzy feature A.Cu(M) are A.1, A.2, and A.3; the instance of fuzzy feature B.Cu(H) is B.1; and the instance of fuzzy feature C.Cu(M) is C.1. If A.2, B.1, and C.1 are neighbors, {A.2, B.1, C.1} is a row-instance of the FCP ***Fuzz_c***. According to Definition 8, the lower boundary participation ratio of these three fuzzy features is $\underline{PR}$(***Fuzz_c***, A.Cu(M)) = 0.2256, $\underline{PR}$(***Fuzz_c***, B.Cu(H)) = 1, $\underline{PR}$(***Fuzz_c***, C.Cu(M)) = 1 and the upper bound participation ratio of these three fuzzy features are $\overline{PR}$(***Fuzz_c***, A.Cu(M)) = 0.2866, $\overline{PR}$(***Fuzz_c***, B.Cu(H)) = 1, $\overline{PR}$(***Fuzz_c***, C.Cu(M)) = 1, so the lower participation index is $\underline{PI}$(***Fuzz_c***) = 0.2256 and the upper bound participation index is $\overline{PI}$(***Fuzz_c***) = 0.2866. If the prevalence threshold *min_prev* = 0.2, then $\underline{PI}$(***Fuzz_c***) > *min_prev*, that is, ***Fuzz_c*** is absolutely prevalent. If the prevalence threshold *min_prev* = 0.3, then $\overline{PI}$(***Fuzz_c***) < *minprev and* ***Fuzz_c*** is absolutely non-prevalent. If the prevalence threshold *min_prev* = 0.25, then $\overline{PI}$(***Fuzz_c***) > *minprev* > $\underline{PI}$(***Fuzz_c***) and the prevalence tendency degree $\boldsymbol{\theta}$ of ***Fuzz_c*** is $\boldsymbol{\theta}$ = 0.6.

Table 2. The interval membership of the fuzzy attribute of the sampling point.

Functional Area	Number of Sampling Points	Cu(L)	Cu(M)	Cu(H)	Zn(L)	Zn(M)	Zn(H)
A	1	0	[0.4403, 0.7456]	[0.0104, 0.0831]	0	0	1
A	2	0	[0.3614, 0.6848]	0	0	[0.5414, 0.8298]	0
A	3	0	[0.8, 0.9591]	0	0	[0.5191, 0.8146]	0
B	1	0	0	[0.283, 0.5931]	0	0	[0.3905, 0.7023]
B	2	0	[0.3152, 0.6386]	0	0	[0.7138, 0.9252]	0
B	3	0	[0.7633, 0.9455]	0	0	[0.2495, 0.5656]	0
B	4	0	[0.5786, 0.8991]	0	0	[0.8315, 0.9695]	0
C	1	0	[0.4069, 0.717]	[0.0174, 0.1141]	0	0	1
D	1	0	[0.9169, 0.9896]	0	0	[0.1702, 0.4586]	[0.0335, 0.1698]

3.2. Lemma and Theorem

Lemma 1. *The upper bound participation index (or lower bound participation index) of an FCP and the upper bound participation ratio (or lower bound participation ratio) of fuzzy features in a pattern decrease monotonically with the increase of the size of the pattern.*

Proof. If instance s_1 of fuzzy feature $Fuzz_f_1$s is included in the row-instances of fuzzy pattern ***Fuzz_c***, then when fuzzy pattern ***Fuzz_c′*** ⊆ ***Fuzz_c***, instance s_1***Fuzz_c′*** must also be included in the row-instances of ***Fuzz_c′***. According to Definitions 8 and 9, the denominator of the upper bound participation ratio (or lower bound participation ratio) of the same fuzzy feature in the pattern and pattern ***Fuzz_c*** is the same, and the numerator decreases with the increase of the size of the pattern. Therefore, the upper bound participation ratio (or lower bound participation ratio) of the fuzzy features decreases monotonically. According to the definition of the upper bound participation index and lower bound participation index, the upper bound participation index (or lower bound participation index) is the minimum of the upper bound participation ratio (or lower bound participation ratio) of all fuzzy features in an FCP. Therefore, the upper bound participation index (or lower bound participation index) decreases monotonically with the increase of the size of the pattern. □

Theorem 1. *If a k-1 size pattern is absolutely non-prevalent, then any k size patterns that it contains are also absolutely non-prevalent.*

Proof. For a k-1 size pattern ***Fuzz_c***, if a k size pattern ***Fuzz_c′*** $\supseteq$ ***Fuzz_c***, according to **Lemma 1**, $\overline{\mathrm{PI}}(\boldsymbol{Fuzz\_c'}) \leq \overline{\mathrm{PI}}(\boldsymbol{Fuzz\_c})$. If this k-1 size pattern ***Fuzz_c*** is absolutely non-prevalent, then $\overline{\mathrm{PI}}(\boldsymbol{Fuz\_c}) < min\_prev$. Therefore, if $\overline{\mathrm{PI}}(\boldsymbol{Fuz\_c'}) \leq \overline{\mathrm{PI}}(\boldsymbol{Fuz\_c}) < min\_prev$, then k size pattern ***Fuzz_c′*** is also absolutely non-prevalent. □

4. Methods

4.1. A Mining Method Based on Spatial Cliques

We propose a method based on spatial cliques to mine all prevalent FCPs. Our method first uses a method based on a prefix tree to generate spatial cliques from spatial datasets; then an Apriori-like algorithm is used to generate candidate FCPs, find row-instances of candidates from the generated cliques, and then filter prevalent FCPs.

In this section, we discuss this approach in six subsections: generating granular type-2 fuzzy membership functions from type-1 fuzzy membership functions based on granular data (Section 4.1.1), materializing neighborhood relationships (Section 4.1.2), enumerating cliques (Section 4.1.3), generating candidate FCPs (Section 4.1.4), finding table-instances of candidates from cliques (Section 4.1.5), and filtering prevalent FCPs (Section 4.1.6).

4.1.1. Generating Granular Type-2 Fuzzy Membership Functions from Type-1 Fuzzy Membership Functions Based on Granular Data

When the specific membership degree of a specific value under a fuzzy set is evaluated, the deviation is often large. However, when the interval membership degree of an interval value in a fuzzy set is evaluated, the deviation can be greatly reduced. Therefore, we use the expert evaluation method to evaluate the interval membership degree of interval values under a fuzzy set to form granular data. Furthermore, according to the original type-1 fuzzy membership function, we generate a granular type-2 fuzzy membership function based on elliptic curves and gradually adjust the parameter of the granular type-2 fuzzy membership function by means of equal expansion.

(1) Count the granular evaluation data.

In this process, we need to choose the appropriate length of the interval membership degree and the appropriate length of the evaluation interval data so that experts can make accurate judgments. We take the interval membership degree belonging to the fuzzy set Cu(M) as an example when the copper content is between 10 and 50. The interval length of the content we chose to evaluate is 5 and the length of interval membership is 0.1. If the length of the evaluated data is small, we set the interval length to 2—for example, when the copper content is between 6–20. We have counted the data of 1000 geologists who evaluated the interval value of copper content under the fuzzy set Cu.(M). The number in the particle indicates the number of evaluators who consider that an interval value corresponds to the interval membership. The darker the color is, the more people are evaluated in this interval. Figure 2a shows the number of evaluators of interval memberships.

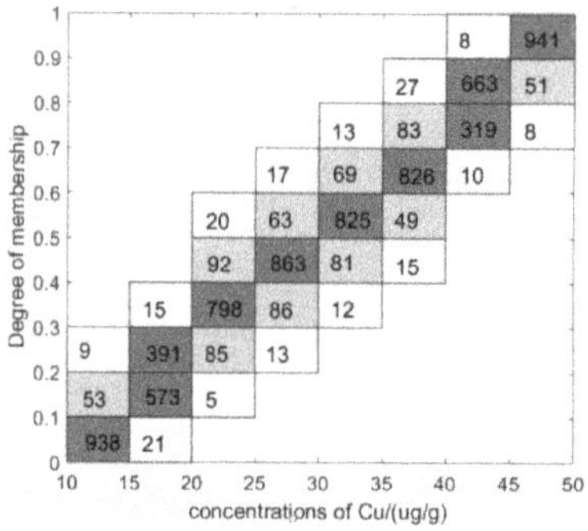

(a)

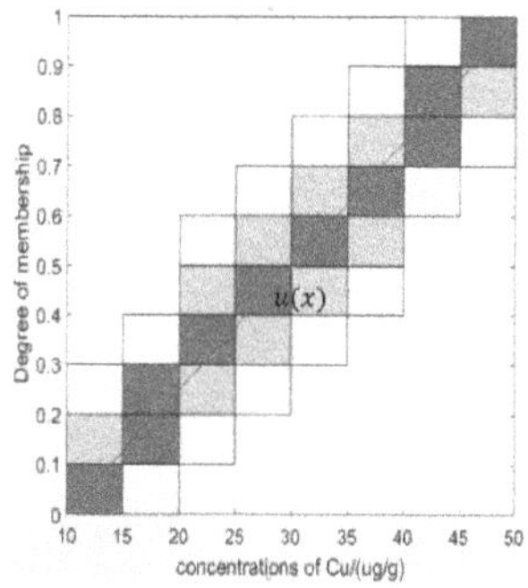

(b)

Figure 2. (**a**) Number of evaluators in interval data; (**b**) Type-1 membership function.

(2) The original type-1 fuzzy membership function is drawn based on the granular evaluation data.

In this example, according to the construction method in [4], we obtain the type-1 fuzzy membership function of heavy metal copper. We draw the type-1 fuzzy membership function, which is expressed as $y = \frac{x-a}{b-a} = \frac{x-10}{40}$, as shown in Figure 2b. Here, a is the lowest value that experts believe the heavy metal content tends to belong to the middle content, and b is the value that experts believe the copper content must belong to the middle content.

(3) Constructing a type-2 membership function based on the original type-1 membership function.

The granular type-2 fuzzy membership function we proposed is represented by the upper bound fuzzy membership function and the lower bound fuzzy membership function. When the original type-1 membership function is a decreasing function, the formula is expressed as:

Upper bound function $\overline{u}(x) = \sqrt[g]{1 - \left(\frac{x-a}{b-a}\right)^g}$.

Lower bound function $\underline{u}(x) = 1 - \sqrt[g]{1 - \left(\frac{b-x}{b-a}\right)^g}$ $(a \leq x \leq b)$

When the original type-1 membership function is an increasing function, the formula is expressed as:

Upper bound function $\overline{u}(x) = \sqrt[g]{1 - \left(\frac{b-x}{b-a}\right)^g}$

Lower bound function $\underline{u}(x) = 1 - \sqrt[g]{1 - \left(\frac{x-a}{b-a}\right)^g}$ $(a \leq x \leq b)$ where a is the lowest value that experts start to believe has a tendency to belong to the fuzzy set and b is the value that experts believe must belong to the fuzzy set.

The area covered by the upper bound function and the lower bound function is expressed as the FOU. Here, g determines the degree of expansion and contraction of the elliptic curve. The larger the value of g, the larger the FOU; the smaller the value of g, the smaller the FOU. In this example, when $g = 1.1$, our granular type-2 fuzzy membership function is shown in Figure 3a.

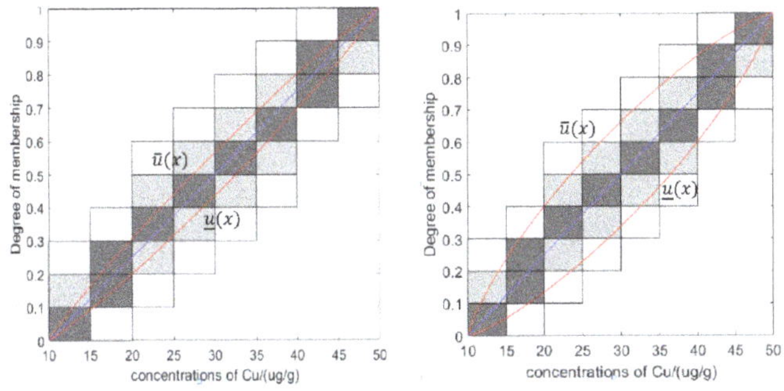

Figure 3. (a) Type-2 membership function when g = 1.1; (b) Type-2 membership function when g = 1.283.

A type-2 fuzzy membership function needs to satisfy two conditions: ① The FOU between the upper bound fuzzy membership function and the lower bound fuzzy membership function must be connected; and ② the proportion of data contained in the FOU to all data must reach a given threshold, that is, the confidence must reach a given threshold, which we set at 85% in this article. The granular data represents that, for any specific value in interval data, the possibility of taking any specific membership value in the interval membership is the same. Therefore, for interval data, the confidence of the type-2 membership function is $\sum_{i=1}^{n} S_i * x$. Here, n is the number of granules in the interval data, S_i is the ratio of the area of the ith granule in the FOU to the area of the ith granule, and x are the number of evaluators in this granule. Within a range, the average of the confidence of all interval data is the confidence of the granular type-2 fuzzy membership function in this range.

(4) Determine the parameters of the granular type-2 fuzzy membership function.

We use a gradual approach to determine the unique parameter g in the granular type-2 fuzzy membership function. First, we set an increment $\alpha = 0.1$, when $g = 1 + k_1\alpha$ ($k_1 = 1, 2, \ldots, n$). Then, we observe whether the FOU is continuous. If there is ψ ($\psi \in \{1, 2, \ldots, n\}$) when $g = 1 + \psi\alpha$, the FOU is continuous; when $g = 1 + (\psi + 1)\alpha$, the FOU is not continuous. Then, set the gradual value to one-tenth of the original (i.e., $\beta = 0.1\alpha$), $g = 1 + \psi\alpha + k_2\beta$ ($k_2 \in \{1, 2, \ldots, 10\}$), and so on, until the accuracy reaches the user's requirement and the FOU is connected. This process is shown in Table 3.

Table 3. Connectivity of the membership function with different values of g.

g	1.1	1.2	1.3	1.4	1.5	1.6	1.7	1.8	1.9	2.0
Is it connected?	Yes	Yes	No	No	No	No	No	No	No	No
g	1.21	1.22	1.23	1.24	1.25	1.26	1.27	1.28	1.29	1.3
Is it connected?	Yes	Yes	Yes	Yes	Yes	Yes	Yes	Yes	No	No
g	1.281	1.282	1.283	1.284	1.285	1.286	1.287	1.288	1.289	1.29
Is it connected?	Yes	Yes	Yes	No	No	No	No	No	No	No

When g = 1.283, the FOU of the granular type-2 fuzzy membership function is connected. At this time, the data in the FOU accounts for 85.13% of the total evaluation data, that is, the confidence of FOU is 0.8513, which reaches the default threshold. This is shown in Figure 3b.

The type-2 membership function we finally obtain from the granular data is shown in Figures 4 and 5.

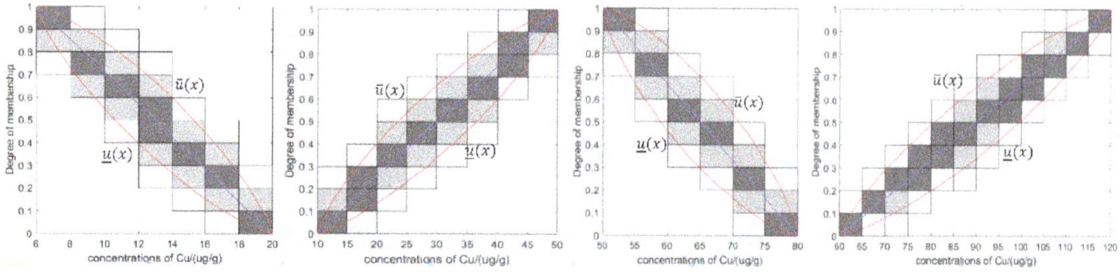

(a) Low content (b) Middle content (c) Middle content (d) High content

Figure 4. The granular type-2 fuzzy membership function for copper.

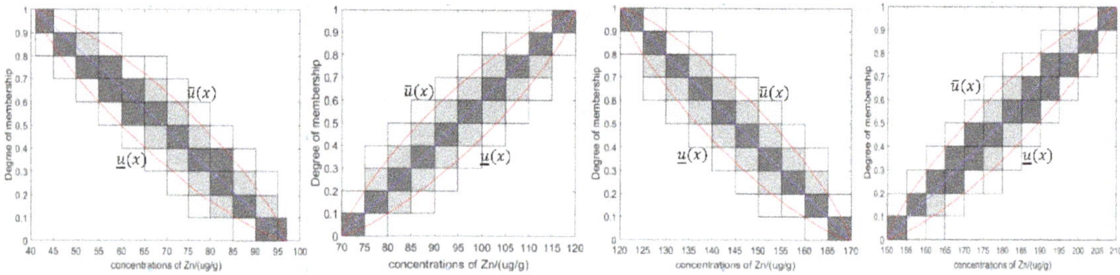

(a) Low content (b) Middle content (c) Middle content (d) High content

Figure 5. The granular type-2 fuzzy membership function for zinc.

The upper bound membership function and the lower bound membership function of the granular type-2 fuzzy membership function are shown in Figures 4 and 5. The values of g of the granular type-2 fuzzy membership function of copper are 1.296, 1.283, 1.272, and 1.272, respectively. The values of g of the granular type-2 fuzzy membership function of zinc are 1.28, 1.284, 1.284, and 1.272, respectively. The confidences of the granular type-2 fuzzy membership function of Cu are 0.8502, 0.8513, 0.8172, and 0.8953, respectively. The confidences of the granular type-2 fuzzy membership function of Zn are 0.8605, 0.8904, 0.8910, and 0.9022, respectively. The confidence of the granular type-2 fuzzy membership function in Figure 5c is 0.8172. If the confidence needs to be above 0.85, we can continue to select the appropriate step value and improve parameter g by the gradual method.

(5) Removing unconnected areas in the FOU.

After removing unconnected areas in the FOU, we can construct the segmented granular type-2 fuzzy membership function. Note that if the step value is too small, it may not reach the threshold, and if it is too large, it will lead to too many unconnected regions.

In this example, we set the step value as 0.1, then g = 1.372, and the constructed type-2 fuzzy membership function is shown in Figure 6a. After removing the disconnected area, the piecewise function is shown in Figure 6b and the confidence is 0.8721, reaching the given threshold. In this case, the fuzzy membership function is expressed as:

$$\overline{u}(x)=\begin{cases} \sqrt[1.372]{1-\left(\frac{x-50}{30}\right)^{1.372}} & \\ 0.8(60<x<61.3582) & \\ 0.7(65<x<65.0204) & (50\leq x\leq 80) \\ 0.5(70<x<71.018) & \\ 0.3(75<x<75.6912) & \end{cases}$$

$$\underline{u}(x) = \begin{cases} 1 - \sqrt[1.372]{1 - \left(\frac{80-x}{30}\right)^{1.372}} \\ 0.7(54.3088 < x < 55) \\ 0.5(58.982 < x < 60) \\ 0.3(64.9796 < x < 65) \\ 0.2(68.6415 < x < 70) \end{cases} \quad (50 \leq x \leq 80)$$

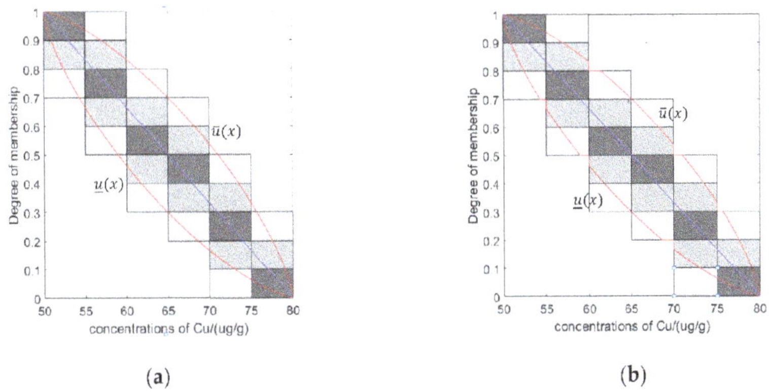

Figure 6. Granular type-2 fuzzy membership function when g = 1.372.

From the above process, it can be seen that, based on the elliptic curve, the interval type-2 fuzzy membership function can be quickly constructed so that its FOU meets the connectivity and the confidence reaches the given threshold.

4.1.2. Generating FCP Mining Cliques (FCPM-Cliques)

According to the spatial neighborhood relation R, a spatial dataset can be expressed as a neighbor graph. We sort all of the features in the lexicographical order of their names. Instances of different features are sorted by their features, and instances of the same feature are sorted by their index values. We provide the following definitions:

Definition 11 (Small neighbor instance set). *For instance, $s \in S$, with F_i is the feature of instance s. A set of instances $SN(s) = \{s' \in S \mid F_j < F_i \cap R(s, s')\}$ is defined as a small neighbor instance set of s, where F_j is the feature of s'. SNs(s) is a set of all the instances that are smaller than s and have spatial neighbor relationships with s.*

Definition 12 (Big neighbor instance set). *For instance, $s \in S$, with F_i is the feature of instance s. A set of instances $BN(s) = \{s' \in S \mid F_j > F_i \cap R(s, s')\}$ is defined as a big neighbor instance set of s, where F_j is the feature of s'. BNs(s) is a set of all the instances that are bigger than s and have spatial neighbor relationships with s.*

In order to quickly find the neighborhood relationship among all instances, we use a method based on grids [6]. After finding the neighborhood relationship among all instances, we get the neighborhood table of all instances, as shown in Table 4.

Table 4. Neighborhood table of instances.

SNs	Instance	BNs
-	A.1	{B.1}
-	A.2	{B.2}
-	A.3	{B.3, B.4, C.1, D.1}
{A.1}	B.1	-
{A.2}	B.2	-
{A.3}	B.3	{C.1}
{A.3}	B.4	{C.1, D.1}
{A.3, B.3, B.4}	C.1	{D.1}
{A.3, B.3, B.4, C.1}	D.1	-

Suppose there is a set of spatial instances $S = \{s_1, s_2, \ldots, s_k\}$, if there is $\{R(s_i, s_j) \mid 1 \le i \le k, 1 \le j \le k\}$, then S is a spatial clique—where a spatial clique is a fully connected subgraph in the instances distribution diagram. After obtaining the neighbor relationship between all the instances, we build a tree structure and adopt an effective method, called the Neighborhood Driven Method (NDM), to generate all spatial cliques.

For a spatial clique, the first instance of a clique is called the head of the clique and is represented by $H_{clique}.B_{clique}$ [OBJ]. Last, all the cliques with instance H_s as the head are represented by $clique_{A.1}$-clique. For example, for a clique $H_{clique_{A.1}}$ = {A.1, B.2, C.2, D.3}, [OBJ] $B_{clique_{A.1}}$ = A.1, [OBJ] $clique_{A.1}$ = {B.2, C.2, D.3}, $H_{A.1}$ is a $H_{A.1}$-clique.

After getting the neighborhood relation of all instances, the NDM aims to find these cliques headed by s_i for each instance s_i ($1 \le i \le n$, where n is the total number of instances).

Definition 13 (NDM-Based clique tree (N-tree for short)). *An N-tree is a tree structure; its root node is labeled as "root". In an N-tree, each node except "root" represents only one instance. Every branch of an N-tree is a spatial clique, which is called an N-clique. We use four strategies to construct an N-tree in the order of instances in the neighborhood table. Figure 7 shows an example of constructing an N-tree.*

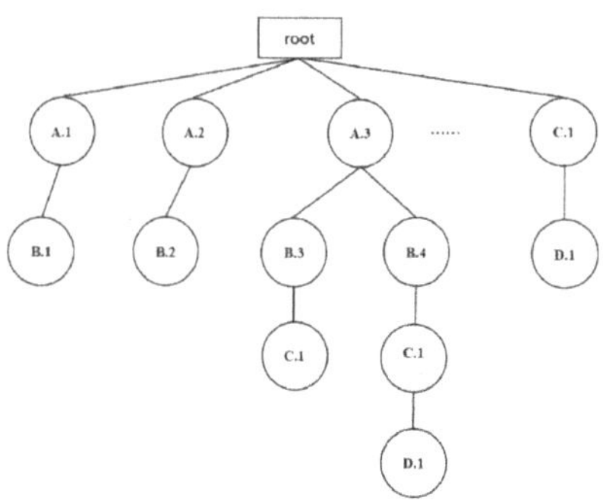

Figure 7. An example of an N-tree constructed by the NDM.

In a clique $cli = \{s_1, s_2, \ldots, s_k\}$, for any instance for s_i of cli, $\{s_{i+1}, s_{i+2}, \ldots, s_k\}$ is contained in BNs(s_i) and $\{s_1, s_2, \ldots, s_{i-1}\}$ is contained in SNs(s_i). Therefore, we check whether $\{s_2, s_3, \ldots, s_n\} \subseteq$ SNs(s)∩BNs(s_1) when we want to check whether an instance s and cli can form a new clique, which means that s is a neighbor to every instance in the clique and therefore a new clique can be formed. Four strategies are proposed according to the different situations of SNs(s)∩BNs(s_1).

Suppose there is a clique $cli = \{s_1, s_2, \ldots, s_k\}$ and an instance $s \in$ BNs(s_1).

Strategy 1: If the size of the clique |*cli*| = 1, SNs(*s*)∩BNs(s_1) = ∅, *s* and *cli* can form a new clique because $s \in$ BNs(s_1) and $s_1 \in$ SNs(*s*).

Strategy 2: If the size of the clique |*cli*| > 1, SNs(*s*)∩BNs(s_1) ⊇ {$s_2, s_3, \ldots, s_n$}, s and *cli* can form a new clique because $\forall s_i \in cli$ and there is R(s, s_i).

Strategy 3: If the size of the clique |*cli*| > 1, *cli'* = SNs(*s*)∩BNs(s_1) ⊆ {$s_2, s_3, \ldots, s_n$}, *s* and s_1∪ *cli'* can form a new clique.

Strategy 4: If the size of the clique |*cli*| > 1, SNs(*s*)∩BNs(s_1) ⊄ {$s_2, s_3, \ldots, s_n$} and {$s_2, s_3, \ldots, s_n$}⊄ SNs(*s*)∩BNs(s_1), but *cli'* = SNs(*s*)∩BNs(s_1) ∩ {$s_2, s_3, \ldots, s_n$}≠ ∅, *s* and *cli'* ∪s_1 can form a new clique.

The NDM uses four strategies to construct an N-tree in the order of instances in the neighborhood table. For instance *s* in the neighborhood table, we can get SNs(*s*). According to the four different situations of SNs(*s*) ∩ BNs(s_1), we adopt four strategies to build the N-tree. After traversing each instance in the neighborhood table, an N-tree is completely established. The pseudocode for the NDM is as follows (Algorithm 1):

Algorithm 1 : Neighborhood Driven Method.

```
Input:
nbsl: neighborhood relationship list (including SNs and BNs of each in
S: a set of instances
Output:
list of N-cliques
Steps:
nTree = Initialize N-tree;
For each instance s In nbsl.Instances Do
    queue = SNs(s);
    While NotEmpty(queue) Do
        head = nTree.AddHeadNode(queue.Out);
        bodies = GetCurrentBodies(head);
        relation = queue∩BNs(head);
        If bodies == null || relation == null Then
            head.AddNode(s)
        Else
            For Each list l in bodies do
                If l == relation Then
                    s and cli can form a new clique;
                Else If relation ⊇ l Then
                    s and cli can form a new clique;
                Else If relation ⊆ l Then
                    s and relation ∪ s1 can form a new clique;
                Else If NotEmpty(l ∩ relation) Then
                    s and relation ∩ l ∪ s1 can form a new clique;
                End If
            End For
        End If
    End While
End For
```

We use an example to illustrate the process of constructing an N-tree using the NDM. Figure 7 shows an example of constructing an N-tree according to the order of the instances in Table 4. In order to generate a clique with A.3 as its head (that is $H_{A.3}$-clique), we start from A.3—there is BNs(A.3) = {B.3, B.4, C.1, D.1}. If we start from B.3, there is SNs(A.3) ∩ BNs(B.3) = ∅, and, according to Strategy 1, A.3 and B.3 can form a clique. In the same way, there is SNs(A.3) ∩ BNs(B.4) =∅, where A.3 and B.4 can form a clique. Then, SNs(A.3) ∩ BNs(C.1) = {B.3, B.4} according to Strategy 2, C.1 and {A.3, B.3} can form a new clique and C.1 and {A.3, B.4} can also form a new clique. Finally, SNs(A.3) ∩ BNs(D.1) = {B.4, C.1} according to Strategy 3, D.1 and {A.3, B.4, C.1} can form a new clique.

Definition 14 (Complete clique set). *Given a set of cliques scl of a spatial dataset materialized by spatial neighborhood relationships and the set containing all cliques of the spatial dataset is represented as ascl, if the following formula is satisfied, scl is a set of complete cliques:*

$\forall$ cl $\in$ ascl, $\exists$ cl′ $\in$ scl $\Rightarrow$ cl $\in$ cl′.

Lemma 3. *N-cliques are complete.*

Proof. Given a clique $cl = \{s_1, s_2, \ldots, s_k\}$ where s_i is an instance, if *cl* is a maximal clique, $\forall s \notin cl$, *s*, and *cl* cannot form a new clique. It is not difficult to see that the NDM can generate all the maximal cliques. According to Definition 14, the set containing all maximal cliques is complete. Thus, N-cliques are complete. □

The four strategies (Strategy 1–Strategy 4) can make decisions concerning which node can be added to a clique to form a new N-clique, and, according to **Lemma 3**, we know that N-cliques are complete. Therefore, the cliques generated by the NDM are correct and complete.

In the process of mining spatial co-location patterns, we calculate the participation ratio of features according to the number of instances in a spatial clique. However, in the process of mining fuzzy spatial co-location patterns, we calculate the participation ratio of fuzzy features according to the membership degree of the attributes of instances in a spatial clique. Therefore, we pay more attention to the association between attributes of instances in the spatial clique, and we propose a kind of FCPM-clique and make the following definition: In a spatial clique, if there is an association between fuzzy attributes of instances, the fuzzy attributes of these instances can form FCPM-cliques.

We find that, in a clique, there is an association between fuzzy attributes of different instances, so they can form FCPM-cliques. There is also an association between fuzzy attributes with different attributes in an instance, so these can also form FCPM-cliques.

For example, for two neighbor instances A.1 and B.1, the fuzzy attribute Cu(M) of A.1 and the fuzzy attribute Cu(H) of B.1 can form an FCPM-clique {A.1.Cu (M), B.1.Cu (H)}, which can be used as a row-instance of FCP {A.Cu(M), B.Cu (H)}—as shown in Figure 8. For instance A.2, its fuzzy attributes Cu (H) and Zn (H) can form an FCPM-clique {A.2.Cu (H), A.2.Zn (H)}, which can be used as a row-instance of FCP {A.Cu(H), A.Zn (H)}—as shown in Figure 9.

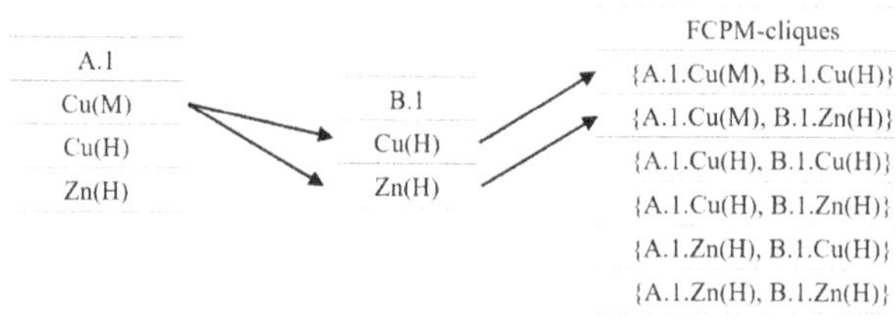

Figure 8. An example of constituting FCPM-cliques.

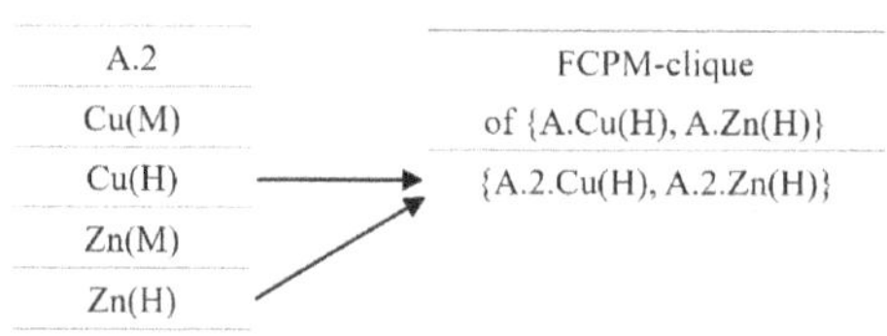

Figure 9. An example of constituting FCPM-cliques with only one instance.

For a candidate FCP, first we find the co-location pattern consisting of all the non-recurring features in the fuzzy candidate pattern. Then, we find the row-instances that contain all the fuzzy features of the fuzzy candidate pattern from the table-instance of the co-location pattern. These row-instances are the row-instances of the candidate fuzzy pattern. Through the FCPM-clique, we can find these spatial cliques containing all fuzzy features of the candidate FCPs.

4.1.3. Generating Candidate FCPs

In the process of generating candidate FCPs, we find that two fuzzy features with different features can form a candidate FCP; two fuzzy features with the same feature and different attributes can also form a candidate FCP. However, the FCP composed of fuzzy features with the same features and the same attributes has no practical significance, such as {A.Cu(L), A.Cu(M), A.Cu(H)}. In order to avoid producing such a meaningless candidate FCP, we propose an Apriori-like algorithm and bucket to generate complete candidate FCPs. The steps of the algorithm (Algorithm 2) are as follows:

Algorithm 2: The algorithm for generating candidate FCPs based on buckets.

Input: (1) Fuzzy feature set ***Fuzz_Fea*** = {Fuzzf_1, Fuzzf_2, ..., Fuzzf_n}
(2) Feature set ***Fea_List*** = {Fea_1, Fea_2, ..., Fea_n}
(3) Instance set of fuzzy features **S** = {s_1, s_2, ..., s_n}
(4) Spatial proximity relation R
(5) Minimum distance threshold *min_dist*
(6) Minimum prevalence threshold *min_prev*
Output: C_{k+1}: k + 1 size candidate FCPs
Step:

1.if (k == 1)
2.   for each ***Fuzz_f*** in ***Fuzz_Fea*** do
3.   SetId(***Fuzz_f***)
4.   Initialize_bucket(***Fuzz_f***)
5.else if (k == 2)
6.   for each ***Fuzz_f_1*** in one feature
7.   for each ***Fuzz_f_2*** in other feature
8.   Generate a candidate pattern from ***Fuzz_f*_1** and ***Fuzz_f*_2**
9.   for each ***Fea*** in ***Fea_List***
10.   for each ***Fuzz_f_1*** in the first attribute
11.   for each ***Fuzz_f_2*** in the second attribute
12.   Generate a candidate pattern from ***Fuzz_f*_1** and ***Fuzz_f*_2**
13.   Put fuzzy patterns with the same features and the same attributes into the same bucket
14.else
15.   for(i = 1; i < *bucket_count*; i++)
16.   for each fuzzy pattern in *bucket*[i]
17.   for(j = i + 1; j < *bucket_count*; j++)
18.   for each fuzzy pattern in *bucket*[j]
19.   if the first k – 1 fuzzy features of the two k-size fuzzy patterns are the same
20.   Generate a fuzzy pattern c_{k+1}
21.   if (check(k − 1, c_{k+1}, bucket))
22. c_{k+1} is a new candidate fuzzy pattern
23.   Put fuzzy patterns with the same features and the same attributes into the same bucket

Steps 2–4: we put all fuzzy features with the same feature and the same attribute into the same bucket. Steps 6–8: size-2 candidate FCPs are generated by connecting the fuzzy features contained in any two different features. Steps 9–12: since the number of attributes is 2, all fuzzy features contained in a feature can be divided into two sets of fuzzy features with the same features and different attributes by connecting any two fuzzy features in the two sets of fuzzy features, and a size-2 candidate FCP is generated. Step 13: put the fuzzy patterns with the same features and the same attributes into the same bucket. Steps 15–22: generate k + 1 size candidates from k size prevalent fuzzy patterns; for any two prevalent fuzzy patterns, check whether the first k − 1 fuzzy features of the two fuzzy patterns are the same, and if they are same, connect them to generate a k + 1 size candidate fuzzy pattern.

Finally, we check whether each sub-pattern of the newly generated $k+1$ size candidate fuzzy pattern is prevalent. Step 23: put the fuzzy patterns with the same features and the same attributes into the same bucket.

4.1.4. Find the Row-Instances of the Candidate FCPs by the Column-Filter Method

In order to improve the efficiency of finding row-instances of candidate FCPs, we propose a column-filter method to find the row-instances of the candidate FCPs from the row-instances of co-location patterns. First, we find the co-location pattern consisting of all the non-recurring features in the candidate FCP. Suppose the number of fuzzy features contained in the FCP is n, we obtain the row-instance of the FCP through filtering n times. In the first filtering, for all row-instances of the co-location pattern, we filter all row-instances with the first fuzzy feature. In the second filtering, for the row-instances obtained by the first filtering, we filter the row-instances with the second fuzzy feature, and so on. In the nth filtering, for the row-instances obtained by the nth filtering, we filter the row-instances with the nth fuzzy feature. The row-instances obtained by the nth filtering have all the fuzzy features of the FCP, so these are the row-instances of the FCP.

For the fuzzy features contained in each feature, we calculate the ratio of the number of instances with the fuzzy feature to the number of instances with the feature, that is, $\alpha_1 = N_{Fuzz\_f_1}/N_{f_1}, \alpha_2 = N_{Fuzz\_f_2}/N_{f_2}, \ldots, \alpha_n = N_{Fuzz\_f_n}/N_{f_n}$. Thus, for a fuzzy feature $Fuzz\_f_i$, it has a weight α_i. Then, we sort the fuzzy features of the candidate FCP in order according to the value of α_i from smallest to largest. Through the method, in the table-instance of the co-location pattern, the pruning rate of pruning the row-instances without the first fuzzy feature is the highest in the first filter. This greatly reduces the number of row-instances that need to be filtered in the second filter, and so on, until the row-instances that contain all the fuzzy features of the candidate FCP are filtered. In this way, we can greatly reduce the time cost to judge these row-instances and the efficiency is improved to a great extent. The pseudocode of this algorithm (Algorithm 3) is as follows:

Algorithm 3: The column-filter method.

Input:
A candidate FCP ***Fuzz_c*** = {Fuzzf_1, Fuzzf_2, …, Fuzzf_n}
Table-instance ***Table_c*** of co-location pattern that contains the candidate FCP
Output:
Table-instances of candidate FCP **Table[*n*]**
Step:
1. ***Table*** = ***Table_c***;
2. for each fuzzy feature Fuzzf_i in the candidate FCP ***Fuzz_c***
3.   for each *row-instance* in ***Table***
4.   if (the *row-instance* doesn't contain *Fuzz_f_i*)
5.   ***Table***.Delete(the *row-instance*);
6.   end for
7.   **Table[*i*]** = ***Table***;
8. end for
9. if *i* == *n*
10.   **Table[*i*]** is the table-instance of the candidate FCP

Step 1: The filtering operation begins with ***Table_c***, a table-instance of the co-location pattern that consists of the features of the candidate FCP. Steps 2–8: For all row-instances of the table-instance in ***Table_c***, we judge whether they have the first fuzzy feature of the candidate FCP. For any one of the row-instances, if it does not have the first fuzzy feature, we delete it; in this way, we filter out all the row-instances with the first fuzzy feature. Then, we judge whether these row-instances have the second fuzzy feature and filter out the row-instances that have both the first fuzzy feature and the second fuzzy feature, and so on. Steps 9–10: If the selected row-instance has all the fuzzy features of the candidate

FCP, then the row-instance is a row-instance of the candidate FCP and all row-instances of the candidate FCP make up its table-instance.

We compare the number of judgment operations between the column-filter method proposed in this section and the row-filter method proposed in Section 5.1 on average. For a candidate FCP of size p, assume that the number of row-instances of the co-location pattern composed of its features is q.

According to the row-filter method in Section 5.1, for each row-instance, we need to judge whether each instance has corresponding fuzzy features in turn. When there is at least one instance in a row-instance that does not have corresponding fuzzy features, the row-instance cannot be used as a row-instance of the candidate fuzzy co-location pattern. On average, each instance in a row-instance has the same probability of not having corresponding fuzzy features, so the expected number of judgment operations of a row-instance is $\frac{1+2+3+\ldots+p}{p} = \frac{\frac{p(p+1)}{2}}{p} = \frac{p+1}{2}$. We need to judge each row-instance in turn, so the expectation of the total number of judgment operations is $\frac{q(p+1)}{2}$.

According to the column-filter method in this section, we set the number of fuzzy sets to three. In the average case, each fuzzy set contains the same number of instances, and each filtering operation can filter 1/3 of the row-instances. Therefore, the expectation of the number of judgment operations of the column-filter method is $q + \frac{q}{3} + \frac{q}{3^2} + \ldots + \frac{q}{3^{p-1}} = \frac{3q}{2}\left(1 - \frac{1}{3^p}\right)$. When q is a constant value and p is large, $\frac{3q}{2}\left(1 - \frac{1}{3^p}\right) < \frac{q(p+1)}{2}$. Therefore, the column-filter method is more efficient than the row-filter method.

4.1.5. Filtering Prevalent FCPs

For a candidate FCP, after obtaining all its row-instances (these row-instances make up its table-instance) according to **Definitions 8** and **9**, we can calculate the upper bound participation index and lower bound participation index of the FCP. Then, according to the upper bound participation index, the lower bound participation index, and the prevalence threshold, we can filter prevalent FCPs. Therefore, we propose two pruning strategies to filter complete and correct prevalent FCPs from candidate FCPs.

Pruning strategy 1 (Fuzzy feature pruning): *In a candidate FCP, for any of its fuzzy features $Fuzz\_f_i$, if the upper bound participation ratio of the fuzzy feature is less than the given prevalence threshold min_prev, the candidate fuzzy pattern is absolutely non-prevalent.*

Proof. For a candidate fuzzy co-location pattern $\boldsymbol{c}$, its upper bound participation index $\overline{\mathrm{PI}}(\boldsymbol{c}) = min_{i=1}^{k}\{\overline{\mathrm{PR}}(\boldsymbol{c}, Fuzz\_f_i)\}$ and $\overline{\mathrm{PR}}(\boldsymbol{c}, Fuzz\_f_i)$ is the upper bound participation ratio of the ith fuzzy feature in $\boldsymbol{c}$. If the upper bound participation ratio of the fuzzy feature $Fuzz\_f_i$ is less than the given prevalence threshold, that is, $\overline{\mathrm{PI}}(\boldsymbol{c}) = min_{i=1}^{k}\{\overline{\mathrm{PR}}(\boldsymbol{c}, Fuzz\_f_i\boldsymbol{c})\} < min\_prev$, the upper bound participation index of is less than the given prevalence threshold *min_prev*. In this case, according to Definition 10, is absolutely non-prevalent. □

Pruning strategy 2 (Absolute table-instance pruning): *For a candidate FCP* **c**, *if the lower bound participation index of its absolute table-instance $\underline{\mathrm{API}}(\boldsymbol{c})$ is greater than or equal to the given prevalent threshold min_prev, then the FCP is an absolutely prevalent FCP.*

Proof. For a candidate FCP $\boldsymbol{c}$, its absolute table-instance is included in its table instance, so the lower bound participation index of pattern $\boldsymbol{c}$ is greater than that of its absolute table-instance $\underline{\mathrm{API}}(\boldsymbol{c})$, that is, $\underline{\mathrm{PI}}(\boldsymbol{c}) \geq (\boldsymbol{c})\ \underline{\mathrm{API}}(\boldsymbol{c})$. If $\underline{\mathrm{API}}(\boldsymbol{c}) \geq min\_prev$, then $\underline{\mathrm{PI}}(\boldsymbol{c}) \geq \underline{\mathrm{API}}(\boldsymbol{c}) \geq min\_prev$, according to Definition 10, c is absolutely prevalent. □

4.1.6. Time Performance Analysis

The time cost of our proposed method is mainly divided into three parts: generating candidate FCPs, generating FCPM-cliques, and finding table-instances of the candidates. In the process of generating candidate FCPs, suppose the number of features is N_{fea}, the number of attributes is N_{att}, the number of fuzzy sets is N_{fuzzy_set}, and let $N_{f\_a} = N_{fea}*N_{att}$, where N_{f_a} is the number of buckets (we put all fuzzy features with the same feature and the same attribute into the same bucket). The computational complexity of generating candidate patterns is $O(\sum_{i=2}^{N_{f\_a}}(C_{N_{f\_a}}^{i} * N_{fuzzy\_set}^{i}))$, where $C_{N_{f\_a}}^{i}$ refers to selecting i buckets from N_{f_a} buckets arbitrarily; $N_{fuzzy\_set}^{i}$ is the i power of N_{fuzzy_set}, which refers to selecting a fuzzy feature from i buckets, respectively, and combines these fuzzy features into a candidate fuzzy co-location pattern. N_{ins} [OBJ], and for any instance s of the neighborhood table, the number of small neighbor instances is |SN(s)| on average. There is EN_{ins} [OBJ] * |SN(s)|, where E is the number of edges in the neighborhood relation of the spatial dataset, so the computational complexity of constructing N-cliques is $O(E)$. In the NDM algorithm, the main time cost is to compute *queue* ∩ BNs(*head*). In the process of finding the table-instances for the candidates, the time complexity is $O(T)$, where T is the number of all row-instances found from the cliques for all candidates N_{row} [OBJ] on average (that N_{row} [OBJ] on average), there is $O(T) = ON_{row}$[OBJ] $\sum_{i=2}^{N_{f\_a}}(C_{N_{f\_a}}^{i} * N_{fuzzy\_set}^{i}$[OBJ])).

4.2. Extending Traditional Algorithms to Discover Prevalent FCPs

The traditional Join-based and Joinless algorithms cannot discover prevalent FCPs. Therefore, we extend the two algorithms to enable them to mine prevalent FCPs.

4.2.1. An Extended Join-Based Algorithm for Mining Prevalent FCPs

We use the method in Section 4.1.3 to generate candidate FCPs and then use the traditional Join-based algorithm to generate co-location patterns and their row-instances. Finally, we use the row-filter method to generate table-instances of the candidate FCPs. For each row-instance, we judge whether each instance has the corresponding fuzzy features in turn. When there is at least one instance in a row-instance that does not have corresponding fuzzy features, the row-instance cannot be used as a row-instance of a candidate fuzzy co-location pattern. The specific steps for the row-filter method are as follows:

The pseudocode for the row-filter method (Algorithm 4) is as follows:

Steps 1–2: show that for any candidate FCP cp_i we find the non-repeated feature set sf contained in all fuzzy features of this candidate FCP. Steps 4–6: show that if the number of all the features in sf is 1 (at this time, there is only one feature in sf, represented as f_1), the instances in f_1 that have all the fuzzy features in the candidate FCP are selected as the row-instances of cp_i. Steps 7–9: show that if the number of all the features in sf is greater than 1, the row-instance that has all the fuzzy features in the candidate FCP is selected as the row-instances of cp_i.

4.2.2. An Extended Joinless Algorithm for Mining Prevalent FCPs

For a circle with a distance threshold of the neighborhood relationship as its radius, instances in the circle have a star neighborhood relationship. The traditional Joinless algorithm uses the star partition model to materialize the neighborhood relationship of spatial datasets; so, in the process of generating row-instances of co-location patterns, it uses the instance lookup operations instead of a large number of Join operations.

Algorithm 4: The row-filter method.

Input:
cps: the set of candidate FCPs
T_{ins}: the table-instance of the co-location pattern consisting of non-recurring features in the candidate FCP
Output:
all row-instances of candidate FCPs
Steps:
1. For each candidate FCP cp_i In ***cps*** Do
2. suppose $cp_i = \{\text{Fuzz}f_1, \text{Fuzz}f_2, \ldots, \text{Fuzz}f_j\}$; here, Fuzz$f_m$ $(1 \leq m \leq j)$ is a fuzzy feature, the set of the non − recurring features of all fuzzy features in cp_i is sf $= \{f_1, f_2, \ldots, f_k\}$ $(1 \leq k \leq j)$;
3. For each row − instance R_{ins} In T_{ins} Do
4. if | *sf* | == 1
5. $f_1\_satisfy$.instance = filter(instances of f_1 that have all the fuzzy features in the candidate fuzzy pattern simultaneously)
6. cp_i_table_instance.Add($f_1\_satisfy$.instance)
7. else
8. $R_{ins}\_satisfy$= filter(R_{ins} that have all the fuzzy features in the candidate fuzzy pattern simultaneously)
9. cp_i_table_instance.Add($R_{ins}\_satisfy$)
10. End For
11. End For

We improve the traditional Joinless algorithm and propose an extended Joinless algorithm (Algorithm 5) to mine prevalent FCPs:

Step 1: spatial data sets are materialized into disjoint star neighbor sets. Step 4: $k + 1$ size candidate FCPs are generated from k size prevalent FCPs. Steps 5–9: star instances of candidate FCPs are generated from star neighbor sets, which requires that the first feature of candidate FCPs is the same as the feature type of the central object of the star neighbor sets. For example, the star instances of FCP {A.Cu(M), B.Zn(M), C.Cu(H)} are found from the neighbor set of feature A. Steps 10–12: show that if the number of features that do not repeatedly appear in the candidate FCP is 1, the instances of the unique feature that have all fuzzy features in the candidate FCP simultaneously are selected as the row-instances of the candidate FCP. Steps 13–15: show that if the number of features that do not repeatedly appear in the candidate FCP is 2, the star instances with size-2 that have all the fuzzy features of the candidate FCP simultaneously are selected as the row-instances of the candidate FCP. Since the spatial proximity relation is symmetrical, a star instance with size-2 is a clique instance. Steps 16–20: show that if the number of features that do not repeatedly appear when a candidate FCP is n ($n \geq 3$), the star instances with size-n that have all fuzzy features of the candidate FCP are temporarily selected as the row-instances of the candidate fuzzy co-location patterns. Then, we check whether the upper bound participation index of the candidate FCP is lower than the given prevalence threshold *min_prev*; if so, the candidate FCP is non-prevalent. If the upper bound participation index of the candidate FCP is greater than *min_prev*, we check whether the star instances of the FCP are clique instances, that is, whether each instance of a star instance are neighbors to each other; if so, the star instance is a clique instance, and if not, the star instance is discarded. Step 21: we calculate the upper bound participation index and lower bound participation index of candidate FCPs and compare this with the given prevalence threshold *min_prev*. We then obtain the absolutely prevalent FCPs and the FCPs with prevalence tendency degrees.

Algorithm 5: The extended Joinless algorithm.

Input: (1) Fuzzy feature set ***Fuzz_Fea*** $= \{\text{Fuzz\_}f_1, \text{Fuzz}f_2, \ldots, \text{Fuzz}f_n\}$
(2) Instance set of fuzzy features **S** = $\{s_1, s_2, \ldots, s_n\}$
(3) A set of star neighbors of the feature Fuzz_f_i: $\{ sn_{f_1}, sn_{f_2}, \ldots, sn_{f_n}\}$
(4) Spatial proximity relation R
(5) Minimum distance threshold *min_dist*
(6) Minimum prevalence threshold *min_prev*
Output: Prevalent FCP set ***Fre_P***
Variable:
k: The size of FCPs
C_k: candidate k size FCPs
SI_k: Star instance set of k size candidate patterns
CI_k: Group instance set of k size candidate patterns
PI_k: The set of star instances have all fuzzy features in candidate patterns simultaneously
P_k: prevalent k-size FCPs
MemSum_up[k]: The sum of upper bound membership degrees of all instances of all fuzzy features in k size FCP.
MemSum_down[k]: The sum of lower bound membership degrees of all instances of all fuzzy features in k size FCP.
Step:
1. SN = gen_star_neighbor(***Fea***,**S**,R);
2. P_1 = ***Fuzz_Fea***; k = 1;
3. while($P_k \neq \varnothing$){
4. C_{k+1} = gen_candidate(P_k);
5. for i in 1 to n
6. for x where $f_i = cf_1$, cf_1 is the first feature of C_{k+1}
7. SI_{k+1} = gen_star_instance(C_{k+1}, x);
8. end for
9. end for
10. if t = 1 (t is the number of non recurring features in a candidate fuzzy pattern)
11. PI_{k+1} = The instances of the only feature that have all fuzzy feature of the candidate fuzzy co-location pattern;
12. $CI_{k+1} = PI_{k+1}$;
13. if t = 2
14. PI_{k+1}= The row − instances in SI_{k+1} which have all fuzzy features of the candidate fuzzy co-location pattern;
15. $CI_{k+1} = PI_{k+1}$;
16. else do
17. PI_{k+1}= The star instances SI_t that have all fuzzy features of the candidate fuzzy co-location pattern;
18. C_{k+1} = filter_coarse_prevalent_colocations(C_{k+1}, PI_{k+1}, *minprev*);
19. CI_{k+1} = gen_clique_instances(C_{k+1}, PI_{k+1});
20. end do
21. P_{k+1}= filter_prevalent_colocations(C_{k+1}, CI_{k+1}, *min_prev*);
22. $k = k + 1$;
23. ***Fre_P*** = ***Fre_P*** $\cup P_{k+1}$;
24. }

Our expansion includes: For a candidate FCP, first, we find the pattern consisting of all the non-recurring features in the candidate FCP; then, we find the star instances that have all the fuzzy features of the candidate FCP from the star instances of the co-location pattern. These star instances are the star instances of the FCP. Finally, we find the clique instances of the FCP from the star instances (at this time, the clique instances are the row-instances of the FCP) (Steps 17–19). When the number of features in a candidate FCP is 1, the instances of the only feature that have all fuzzy features in the candidate FCP simultaneously are selected as the row-instances of the candidate FCP (Steps 10–12). When the number of features in a candidate FCP is 2, the size-2 star instances that have all the fuzzy features in the candidate are selected as the size-2 row-instances of the candidate (Steps 13–15). For example, suppose there is a candidate FCP {A.Cu(M), B.Cu(M)}, feature A is the first feature of the candidate FCP, according to Steps 13–16. From this, we obtain all the star-instances of feature A (Phase 1). Then, we get all the row-instances of co-location pattern {A, B} from the star-instances of feature A (Phase 2). Then, from these row-instances, we filter the row-instances with fuzzy features A.Cu(M) and B.Cu(M), which are the row-instances of the candidate FCP. According to Steps 10–12, we filter the instances that have fuzzy features A.Cu(M) and A.Zn(M) from the instances of feature A; these instances are the row-instances of the candidate {A.Cu(M), A.Zn(M)} (Phase 3), as can be seen in Table 5.

Table 5. An example of the extended Joinless algorithm.

Phase 1		**Phase 2**	**Phase 3**	
Star Instances of Feature A	A	{A, B}	{A.Cu(M), B.Cu(M)}	{A.Cu(M), A.Zn(M)}
A.1, B.2, C.2 A.2, B.3, C.1 A.3, B.5, C.2 A.4, C.3	A.1 A.2 A.3 A.4	A.1, B.2 A.2, B.3 A.3, B.5	A.1, B.2 A.2, B.3	A.1

5. Results

In this section, we evaluate our methods using synthetic and real datasets, and a wide range of experiments have been carried out. We used a method similar to Ref. [2] to generate synthetic datasets and compared the performance of the three algorithms in terms of the number of features, the number of instances, the prevalence thresholds, the distance thresholds, and the size of the fuzzy sets. We used two real datasets to evaluate our methods: one is the heavy metal content of the topsoil sampling points in Quanzhou City, and the other is the heavy metal content of the topsoil sampling points near the industrial area of Jiayuguan. All our experiments are carried out using C#. The Intel PC we use has the Windows 10 operating system, Intel Core i5-4258@2.40 GHz, and 8 GB memory.

5.1. Results for Synthetic Datasets

A method similar to Ref. [6] is used to randomly generate synthetic data. First, the size of the region of our instances distribution is set to $D \times D$, and the whole region is divided into several grids with the size *min_dist* × *min_dist*, where *min_dist* is the distance threshold of the instance neighbors. After determining the number of features and the total number of instances, we determine the number of instances owned by each feature according to the Poisson distribution.

In a grid of the region, the data density (number of edges) of all neighbor grids is controlled by a value called *clump*; we set the value of *clump* to 1, in accordance with the general situation described in Ref. [6]. The number of edges in all instances is approximately equal to $E \times clumpy$. When *clump* = 1, the number of edges of all instances is E, where E is generated randomly.

(1) The effect of distance thresholds on the results

We investigate the influence of different distance thresholds on the execution time by changing the distance threshold of the neighborhood relationship between instances, as shown in Figure 10a. We find that with the increase of distance threshold, the execution time of the three methods increases because the number of instance neighbors will increase (that is, the number of edges will increase) and the average number of row-instances of each FCP will also increase. We also find that with the increase of distance threshold, the growth rate of the execution time of our method is much slower than that of Join-based and Joinless methods. This is mainly because our method searches cliques with all fuzzy features in candidate FCPs and uses them as row-instances of candidate FCPs instead of generating row-instances through connections. In the experiment, the Join-based FCP mining algorithm is represented by "Join-based", the Joinless FCP mining algorithm is represented by "Joinless", and the FCP mining algorithm based on interval type-2 fuzzy sets and cliques is represented by "NDM".

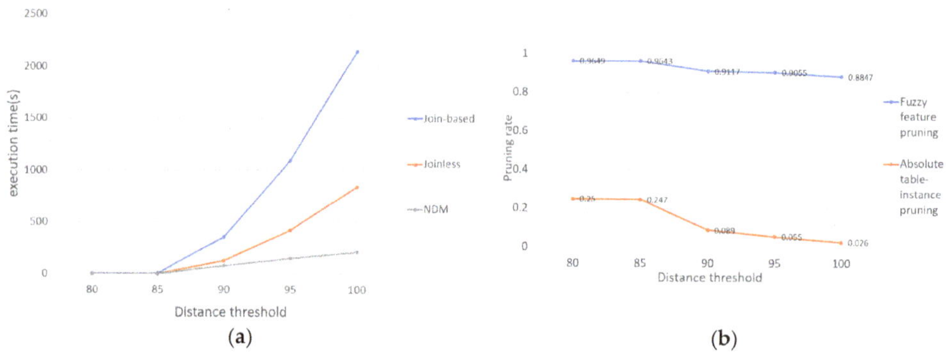

Figure 10. (**a**) Effect of the neighbor distance thresholds; (**b**) Pruning degree of the two strategies.

With the increase of distance threshold, the trend of the pruning degree of the two pruning strategies is shown in Figure 10b. For a spatial dataset, the number of absolute table-instances is certain and will not change with the change of distance threshold. However, the average number of row-instances of FCPs will increase with the increase of distance threshold, so the pruning degree of the absolute table-instances pruning will decrease. In addition, with the increase of distance threshold, the average number of row-instances of each FCP will also increase. Therefore, in a candidate FCP, the number of fuzzy features whose upper bound participation ratio is below the prevalence threshold will decrease, the number of prevalent FCPs also increases, and the pruning rate of fuzzy features will also decrease. In the process of mining fuzzy co-location patterns, the combination of features, attributes, and fuzzy sets will generate a lot of fuzzy features and a large amount of candidate fuzzy co-location patterns. As the total number of row-instances of co-location patterns is certain, many candidate fuzzy co-location patterns have few row-instances, so the pruning degree of the fuzzy feature pruning strategy is very high. On average, for a candidate fuzzy co-location pattern, the number of absolute row-instances it contains is small, so the pruning degree of the absolute table-instance pruning strategy is low.

(2) The effect of prevalence threshold on the results

In this subsection, we study the influence of changing the prevalence threshold over a certain range on the execution time. As shown in Figure 11a, we find that the execution time of these three methods will decrease with the increase of the prevalence threshold. With higher prevalence thresholds, the Join-based (or Joinless) FCP mining algorithm can prune a lot of candidates, so when the prevalence threshold increases, the time cost will be reduced rapidly. For our method based on interval type-2 fuzzy sets and cliques, the prevalence threshold only participates in filtering the absolutely prevalent FCPs and FCPs with prevalent tendency degrees, so the execution time will decrease slowly with the increase of the prevalence threshold. When the prevalence threshold is higher, the time cost of the Join-based (or Joinless) FCP mining algorithm is less than our method. This is because the number of prevalent FCPs with higher $\overline{PI}$ and $\underline{PI}$ is far less than that with lower $\overline{PI}$ and $\underline{PI}$. In other words, when the prevalence threshold changes within a certain range, the time cost of the method based on interval type-2 fuzzy sets and cliques is more stable, which is more conducive to finding prevalent FCPs.

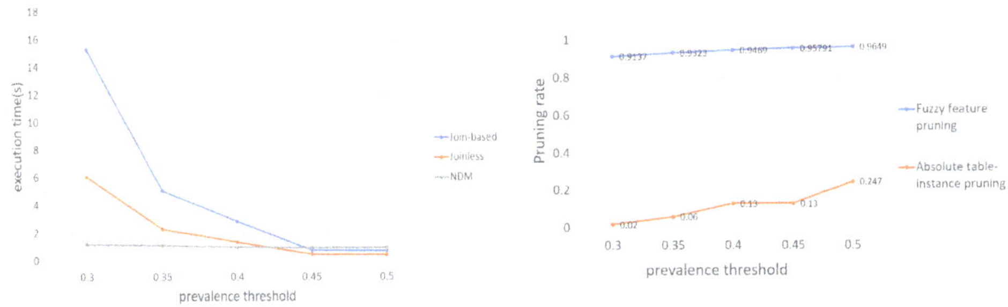

(a) (b)

Figure 11. (**a**)Effect of the prevalence threshold; (**b**) Pruning degree for the two strategies.

The effect of increasing the prevalence threshold on the pruning degree of the two pruning strategies is shown in Figure 11b. With the increase of the prevalence threshold, the number of absolutely non-prevalent patterns also increases, so the pruning rate of the fuzzy feature pruning increases. When the prevalence threshold increases, the number of absolute table-instances is unchanged, but the number of absolutely prevalent patterns and patterns with a prevalent tendency degree will decrease, and the number of pruned candidates for the absolute table-instance pruning strategy will also decrease. When the speed of the number of pruned candidates for the absolute table-instance pruning strategy is less than the speed of the reduction of the number of prevalent patterns, the pruning degree of absolute table-instances will increase.

(3) The effect of the number of fuzzy features on the results

In this subsection, we examine the effect of increasing the number of fuzzy features on the execution time. The number of fuzzy features is the number of features multiplied by the number of fuzzy attributes multiplied by the number of fuzzy sets. In this experiment, the number of fuzzy attributes is set to 2, the number of fuzzy sets is set to 3, and the number of features we select are 3, 6, 9, 12, and 15, respectively. As shown in Figure 12a, we find that the execution time of the three methods first increases and then decreases because when the number of features increases from 3 to 6, the number of candidate FCPs also increases rapidly; the number of operations that generate or examine row-instances of candidate FCPs also increases. When the number of fuzzy features continues to increase, the average number of row-instances of candidates will decrease because the three methods satisfy the downward closure property, so the pruning strategy in this paper can be used to effectively prune absolutely non-prevalent candidates in the process of generating prevalent FCPs. In addition, for the method based on interval type-2 fuzzy sets and cliques, the number of edges also affects its execution time. When the number of fuzzy features increases from 18 to 36, the number of edges increases. When the number of fuzzy features increases from 36 to 90, the number of edges decreases. When the number of fuzzy features is 72, the efficiency of the Joinless FCP mining method is higher than the other two methods; this is because our pruning strategies effectively prune many candidates. In other words, when the number of fuzzy features is small, our method based on interval type-2 fuzzy sets and cliques is more efficient than the other two methods.

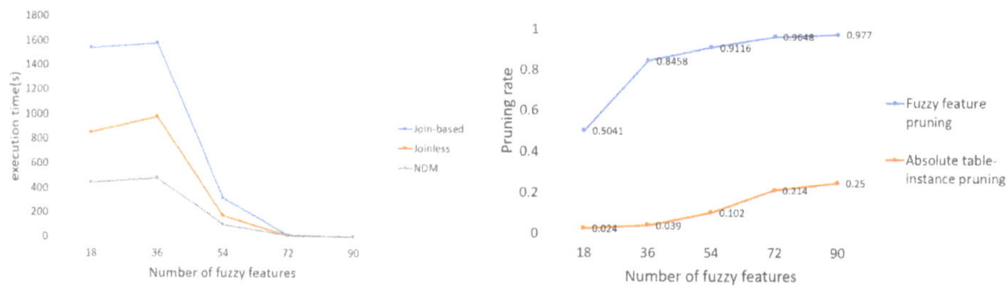

(a) (b)

Figure 12. (**a**) Effect of the number of features; (**b**) Pruning degree of the two strategies.

The effect of increasing the prevalence threshold on the trend of the pruning degree of the two pruning strategies is shown in Figure 12b. With the increase in the number of features in the candidate FCPs, the number of fuzzy features whose upper bound participation degree $\overline{PR}$ is less than the prevalence threshold will also increase. Therefore, the pruning rate of the fuzzy feature pruning strategy also increases. When the number of features increases, the number of prevalent FCPs decreases, but the number of FCPs pruned by the absolute table-instances pruning strategy decreases more slowly, so the pruning degree of absolute table-instances continues to rise.

(4) The effect of the number of instances on the results

In this subsection, we examine the effect of increasing the number of instances on execution time. As shown in Figure 13a, we find that with the increase of the number of instances, the execution time of the three methods increases. When the number of instances is 50,000, the efficiency of the FCP mining algorithm based on interval type-2 fuzzy sets and cliques is much faster than the Join-based (or Joinless) approach. This is because when the data in an area becomes denser, the average number of row-instances of FCPs will be more, and the Join-based (or Joinless) method needs to spend a lot of time to connect and generate (or filter) row-instances. The method based on interval type-2 fuzzy sets and cliques first generates cliques and then searches the row-instances of candidates from cliques, which saves a lot of time cost caused by connection or filtering operations. Therefore, when the instances in the region are dense, the efficiency of the method based on interval type-2 fuzzy sets and cliques is obviously higher than that of the Join-based (or Joinless) approach.

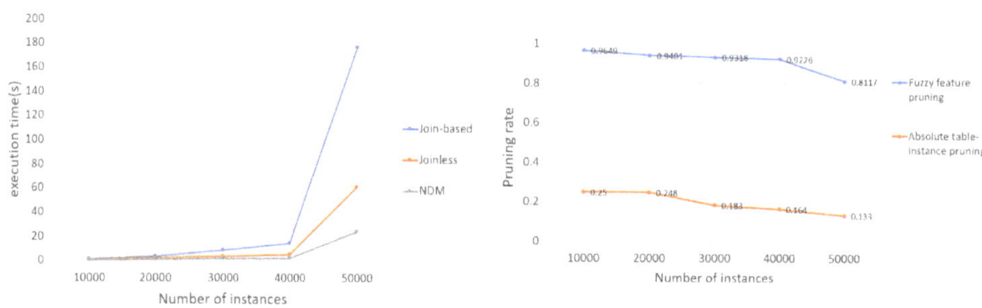

(a) (b)

Figure 13. (**a**) Effect of the number of instances; (**b**) Pruning degree of the two strategies.

With the increase of the number of instances, the trend of the pruning degree of the two proposed pruning strategies is shown in Figure 13b. As the number of instances increases, the average number of row-instances of each FCP will also increase. Therefore, the pruning rate of the fuzzy feature pruning strategy will decrease. The number of FCPs pruned by the absolute table-instance pruning strategy increases, but the number of prevalent

FCPs increase faster, so the pruning degree of the absolute table-instance pruning strategy decreases.

(5) The effect of the number of attributes on the results

In this paper, we take the content of the heavy metals, copper and zinc, in topsoil as an example and construct the granular type-2 fuzzy membership function for copper and zinc, respectively, so there are two attributes. Here, we only keep the zinc attribute of the instances and examine the influence of the number of attributes on the execution time, as shown in Figure 14a. We find that when the number of attributes decreases, the number of fuzzy features will also decrease, and then the generated candidate FCPs and prevalent FCPs will also decrease a lot—and the execution time greatly reduces.

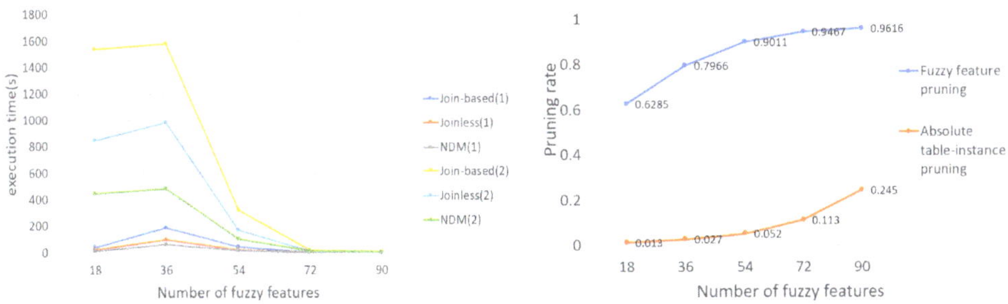

(a) (b)

Figure 14. (**a**) Effect of the number of attributes; (**b**) Pruning degree of the two strategies.

The trend of the pruning degree for the two proposed pruning strategies is shown in Figure 14b when there is one attribute. This is similar to the trend of the pruning rate when there are two attributes. With one attribute, with the increase of the number of features, the number of fuzzy features whose upper bound participation ratio $\overline{PR}$ is less than the prevalence threshold will also increase, so the pruning rate of the fuzzy feature pruning strategy will also increase. When the number of features increases, the number of prevalent FCPs decrease, but the number of FCPs pruned by the absolute table-instance strategy decrease more slowly, so the pruning degree of the absolute table-instance pruning strategy continues to rise.

5.2. Results for Real Datasets

In this section, we examine the validity of our proposed approach through the use of two real datasets. The first real dataset is the spatial distribution data for heavy metals in the topsoil of Jiayuguan City, and is shown in Figure 15a. It is divided into four functional areas: industrial area, agricultural area, living area, and Gobi area. According to the proportion of the region, these data are projected on to a space of $850 * 670$. The second real dataset is the spatial distribution data for heavy metals in the urban topsoil of Quanzhou City, and is shown in Figure 15b. It is divided into five functional areas: traffic area, commercial area, urban green area, residential area, and industrial area, and we project these data on to a space of $700 * 500$ according to the proportion of the region.

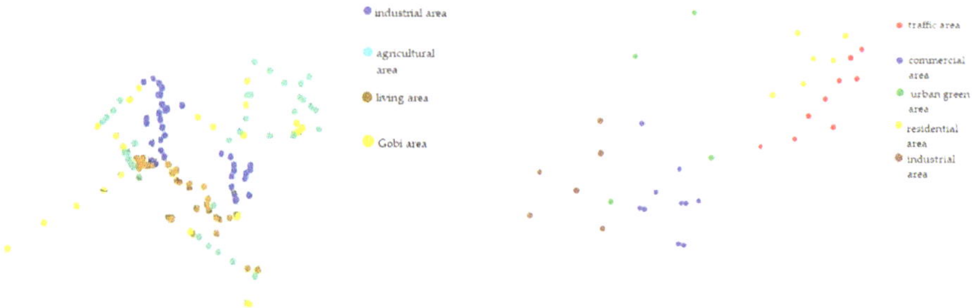

(a) (b)

Figure 15. (**a**) Real dataset 1; (**b**) Real dataset 2.

For dataset 1, we set the distance threshold *min_dist* of the neighborhood relationship to 50, prevalence threshold to 0.3, and mined 112 prevalent FCPs. For dataset 2, we set the distance threshold *min_dist* of the neighborhood relationship to 50, prevalence threshold to 0.3, and mined 46 prevalent FCPs.

Table 6 shows part of the mining results for dataset 1 and Table 7 shows a part of the mining results for dataset 2. In Table 6, A, B, C, and D, respectively, represent the industrial area, agricultural area, residential area, and Gobi area. For example, an FCP {A.Cu(L), C.Cu(L)} indicates that an industrial area with a low copper content and a residential area with a low copper content often tend to be located together, and the degree of this trend is 0.6125. An FCP {B.Cu(M), B.Zn(L), C.Cu(M), C.Zn(L)} indicates that agricultural areas with a low zinc content and middle copper content, and residential areas with a low zinc content and middle copper content often tend to be located together—and the degree of this tendency is 1. In Table 7, A, B, C, D, and E, respectively, represent the traffic area, commercial area, urban green area, residential area, and industrial area. For example, an FCP {A.Zn(H), D.Zn(H)} indicates that traffic areas with a high zinc content and residential areas with a high zinc content often tend to be located together, and the degree of this tendency is 1. An FCP {A.Cu(M), A.Zn(H), D.Cu(M)} indicates that traffic areas with a middle copper content and high zinc content, and residential areas with a middle copper content often tend to be located together, and the degree of this tendency is 0.9047.

Table 6. Part of the mining results for real dataset 1.

Prevalent Fuzzy Co-Location Pattern	Prevalence Tendency Degree	Lower Participation Index and Upper Participation Index	Participation Index for the Traditional Method
{A.Cu(L), C.Cu(L)}	0.6125	0.0709, 0.6621	0.2312
{A.Cu(L), D.Cu(L)}	0.2328	0.036, 0.3801	0.1169
{A.Cu(M), C.Cu(M)}	0.2636	0.1567, 0.3513	0.2384
{D.Cu(M), D.Zn(H)}	1	0.4781, 0.9135	0.6599
{D.Cu(L), D.Zn(M)}	0.9096	0.2414, 0.8896	0.6681
{A.Zn(L), A.Cu(M), C.Zn(L)}	0.2636	0.1567, 0.3513	0.2384
{A.Zn(L), C.Zn(L), D.Cu(M)}	0.2167	0.1749, 0.3346	0.2492
{B.Cu(L), B.Zn(L), C.Zn(L)}	0.8586	0.2597, 0.5448	0.4342
{B.Cu(M), B.Zn(L), C.Cu(M), C.Zn(M)}	0.7039	0.1804, 0.5844	0.3358
{B.Cu(M), B.Zn(L), C.Cu(M), C.Zn(L)}	1	0.3933, 0.738	0.5726

Table 7. Part of the mining results for real dataset 2.

Prevalent Fuzzy Co-Location Pattern	Prevalence Tendency Degree	Lower Participation Index and Upper Participation Index	Participation Index for the Traditional Method
{A.Cu(M), A.Zn(M)}	0.7965	0.2438, 0.5199	0.3661
{A.Zn(H), D.Zn(H)}	1	0.4114, 0.966	0.8528
{A.Cu(M), D.Zn(H)}	0.8208	0.2438, 0.5575	0.3661
{E.Cu(L), E.Zn(L)}	1	0.3657, 0.4952	0.4425
{A.Cu(M), A.Zn(M), D.Cu(M)}	0.7965	0.2438, 0.5199	0.3661
{A.Cu(M), A.Zn(H), D.Zn(M)}	0.9047	0.2828, 0.4633	0.3853
{A.Cu(M), A.Zn(M), D.Cu(M), D.Zn(H)}	0.7965	0.2438, 0.5199	0.3661
{A.Cu(M), A.Zn(H), D.Cu(M), D.Zn(H)}	0.9047	0.2828, 0.4633	0.3853

For the real dataset 1, 112 prevalent FCPs (26 absolutely prevalent FCPs and 86 FCPs with a prevalence tendency degree) were mined according to our measurement method, and 63 prevalent FCPs were mined according to the traditional measurement method. For the real dataset 2, 46 prevalent FCPs (11 absolutely prevalent FCPs and 35 FCPs with a prevalence tendency degree) were mined according to our measurement method, and 37 prevalent FCPs were mined according to the traditional measurement method. We find that our method finds more prevalent FCPs (including absolutely prevalent FCP and FCPs with a prevalence tendency degree) than the method of mining spatial FCPs based on type-1 fuzzy sets.

In the method of mining FCPs based on type-1 fuzzy sets, the membership degree of the attribute of the spatial instance is a specific value because of the type-1 fuzzy membership function. Therefore, the participation index of the mined FCPs is also a specific value. We compare the participation index of FCPs with the prevalence threshold and we can divide these FCPs into prevalent FCPs and non-prevalent FCPs. For example, we apply the method of mining FCPs based on type-1 fuzzy sets to dataset 1. The prevalence threshold is set to 0.3 and the participation index of FCP {A.Cu(L), C.Cu(L)} is 0.2312, which is non-prevalent. The participation index of {B.Cu(L), B.Zn(L), C.Zn(L)} is 0.4342, which is prevalent.

6. Discussion

The membership of attributes of spatial instances is determined by the fuzzy membership function. However, the disadvantage of using type-1 fuzzy membership functions is that the type-1 fuzzy membership function itself has a large amount of uncertainty. The application of type-1 fuzzy sets in mining FCPs will result in large deviations. Therefore, we apply interval type-2 fuzzy sets to the process of mining FCPs in order to reduce the deviation caused by the uncertainty of type-1 fuzzy membership functions and find the influence of the uncertainty of the fuzzy membership function on the prevalence degree of FCPs. Moreover, our method is more efficient than the traditional FCPs mining method based on type-1 fuzzy sets [4].

Compared with the method of mining FCPs based on type-1 fuzzy sets, our method has the following advantages:

(1) By constructing granular type-2 fuzzy membership functions, we reduce the deviation caused by the uncertainty of type-1 fuzzy membership functions. In this paper, the membership degrees of the attributes of spatial instances are expressed by interval values, and then the mined FCPs have an upper bound participation index and a lower bound participation index. We propose a method to measure the prevalence degree of FCPs when there is uncertainty in the fuzzy membership function, which

provides a basis for judging the influence of the uncertainty of the fuzzy membership function on the results of mining FCPs.

(2) Due to the uncertainty of fuzzy membership functions, we find FCPs whose lower bound participation index is less than the prevalence threshold but the upper bound participation index is higher than the threshold. This shows that in all the possible values of the participation index of these FCPs, some values are higher than the prevalence threshold and some values are lower. That is to say, such an FCP has a certain tendency to be prevalent, but also has a certain tendency to be non-prevalent; we regard this as a kind of potential prevalent FCP. For example, in the real dataset 1, the FCP {A.Cu(M), C.Cu(M)} has a lower bound membership of 0.1567, an upper bound membership of 0.3513, and a prevalence tendency degree of 0.2636. We can find that when there is uncertainty in the fuzzy membership function, {A.Cu(M), C.Cu(M)} has a 26.36% tendency degree to be prevalent. However, in the traditional method based on type-1 fuzzy sets, the participation index of the FCP is 0.2384 when the prevalence threshold is 0.3; we can only find that it is non-prevalent and cannot find the influence of the uncertainty of the fuzzy membership function on its participation index and prevalence degree. When using type-1 fuzzy sets, we cannot find the potential association between the middle concentration of copper in the industrial area and the middle concentration of copper in the living area, even if the association is very significant.

(3) We use the prevalence tendency degree to measure the prevalence degree of FCPs. This allows us to find the FCPs whose lowest possible value of participation index is higher than the prevalence threshold. These FCPs are still prevalent when there is uncertainty in the fuzzy membership function, so we define them as absolutely prevalent FCPs and we regard them as a kind of stable and prevalent FCP. However, in the traditional method based on type-1 fuzzy sets, it is difficult to find this kind of FCP according to the participation index and the prevalence threshold. For example, in the real dataset 1, an absolutely prevalent FCP {B.Cu(M), B.Zn(L), C.Cu(M), C.Zn(L)} has a lower bound membership of 0.3933 and an upper bound membership of 0.738. We can easily find that the pattern is stably prevalent and we can find the influence of the uncertainty of the fuzzy membership function on the prevalence degree of FCPs. However, in the traditional method based on type-1 fuzzy sets, the participation index of the FCP is 0.5726 and it is difficult to judge whether this pattern is prevalent when there is uncertainty in the fuzzy membership function.

7. Conclusions

In the work of traditional FCPs mining, the generation of table-instances of an FCP takes up most of the execution time. In this paper, we propose a method based on interval type-2 fuzzy sets and cliques. Firstly, cliques are generated from spatial datasets, and then we look for row-instances of the candidate FCP from the cliques, thus saving the time of generating row-instances of candidates through a large number of connection operations.

Compared with type-1 fuzzy membership function, a type-2 fuzzy membership function more truly reflects the situation when there is uncertainty in the membership degree of fuzzy attributes. In this case, our method effectively measures the prevalence degree of fuzzy co-location patterns, and finds stable and prevalent fuzzy co-location patterns and fuzzy co-location patterns with a prevalence degree. It provides a reference for people to find the influence of the uncertainty of membership functions on the prevalence degree of fuzzy co-location patterns. At the same time, our method has some limitations, such as the time cost of the gradual method of adjusting the parameter of interval type-2 fuzzy membership function, which is high, and the efficiency is low when the number of instances or features is very large.

For future work, we will consider using distributed and parallel mining to mine fuzzy co-location patterns. The combination of features, attributes, and fuzzy sets will generate a lot of fuzzy features and a large amount of candidate fuzzy co-location patterns, which will lead to a tremendous time cost. Distributed and parallel mining can greatly improve the efficiency of mining fuzzy co-location patterns. In addition, we consider using interval type-3 fuzzy sets to deal with higher-level uncertainty.

Author Contributions: Conceptualization, J.G. and L.W.; methodology, J.G.; software, J.G.; validation, J.G.; formal analysis, J.G.; investigation, J.G.; resources, J.G.; data curation, J.G.; writing—original draft preparation, J.G.; writing—review and editing, J.G. and L.W.; visualization, J.G.; supervision, L.W. All authors have read and agreed to the published version of the manuscript.

Funding: This work is supported by the National Natural Science Foundation of China (61966036, 61662086, 62062066), and the Project of Innovative Research Team of Yunnan Province (2018HC019).

Institutional Review Board Statement: Not applicable.

Informed Consent Statement: Not applicable.

Data Availability Statement: Datasets used to support this study are available at https://kns.cnki.net/kcms/detail/detail.aspx?dbcode=CJFD&dbname=CJFDLAST2019&filename=CJXB201905011&uniplatform=NZKPT&v=2Dyga9ZPvUiLfRITaVsaHLvqJzi6SIk_IdPyJhHbblobEZ2c9SSMLXdtsP51wdS3.

Acknowledgments: The authors are very grateful to the reviewers for their valuable suggestion in improving this paper. The authors would like to thank Peizhong Yang, Pingping Wu for the insightful comments on the early drafts of the manuscript.

Conflicts of Interest: The authors declare no conflict of interest.

References

1. Wang, L.; Wu, P.; Chen, H. Finding Probabilistic Prevalent Colocations in Spatially Uncertain Data Sets. *IEEE Trans. Knowl. Data Eng.* **2011**, *25*, 790–804. [CrossRef]
2. Bao, X.; Wang, L. A clique-based approach for co-location pattern mining. *Inf. Sci.* **2019**, *490*, 244–264. [CrossRef]
3. Bao, X.; Lu, J.; Gu, T.; Chang, L.; Xu, Z.; Wang, L. Mining Non-Redundant Co-Location Patterns. *IEEE Trans. Neural Netw. Learn. Syst.* **2021**, *99*, 1–14. [CrossRef] [PubMed]
4. Wu, P.; Wang, L.; Zhou, Y. Research on spatial co-location pattern mining with fuzzy attributes. *Comput. Sci. Explor.* **2013**, *7*, 348–358. [CrossRef]
5. Ouyang, Z.; Wang, L.; Wu, P. Spatial Co-Location Pattern Discovery from Fuzzy Objects. *Int. J. Artif. Intell. Tools* **2017**, *26*, 1750003. [CrossRef]
6. Yoo, J.S.; Shekhar, S.; Smith, J.; Kumquat, J.P. A partial join approach for mining co-location patterns. In Proceedings of the 12th Annual ACM International Workshop on Geographic Information Systems—GIS '04, Washington, DC, USA, 12–13 November 2004; pp. 241–249. [CrossRef]
7. Yoo, J.S.; Shekhar, S. A join-less approach for co-location pattern mining. *IEEE Trans. Knowl. Data Eng.* **2006**, *18*, 1323–1337. [CrossRef]
8. Huang, Y.; Shekhar, S.; Xiong, H. Discovering co-location patterns from spatial datasets: A general approach. *IEEE Trans. Knowl. Data Eng.* **2004**, *16*, 1472–1485. [CrossRef]
9. Berry, A.; Pogorelcnik, R. A simple algorithm to generate the minimal separators and the maximal cliques of a chordal graph. *Inf. Process. Lett.* **2011**, *111*, 508–511. [CrossRef]
10. Dasari, N.S.; Desh, R.; Zubair, M. pbitMCE: A bit-based approach for maximal clique enumeration on multicore processors. In Proceedings of the 20th IEEE International Conference on Parallel and Distributed Systems (ICPADS), Hsinchu, Taiwan, 16–19 December 2014; pp. 478–485. [CrossRef]
11. Wang, L.; Bao, Y.; Lu, Z. Efficient Discovery of Spatial co-location Patterns Using the iCPI-tree. *Open Inf. Syst. J.* **2009**, *3*, 69–80. [CrossRef]
12. Wang, L.; Zhou, L.; Lu, J.; Yip, J. An Order-clique-based Approach for Mining Maximal Co-locations. *Inf. Sci.* **2009**, *179*, 3370–3382. [CrossRef]
13. Yang, P.; Wang, L.; Wang, X. A MapReduce approach for spatial co-location pattern mining via ordered-clique growth. *Distrib. Parallel Databases* **2020**, *38*, 531–560. [CrossRef]
14. Tran, V.; Wang, L.; Chen, H.; Xiao, Q. MCHT: A maximal clique and hash table-based maximal prevalent co-location pattern mining algorithm. *Expert Syst. Appl.* **2021**, *175*, 114830. [CrossRef]
15. Wang, F.; Mo, H. Some fundamental issues on type-2 fuzzy sets. *Acta Autom. Sin.* **2017**, *43*, 1141. [CrossRef]
16. Torshizi, A.D.; Zarandi, M.H.F. A new cluster validity measure based on general type-2 fuzzy sets: Application in gene expression data clustering. *Knowl. Based Syst.* **2014**, *64*, 81–93. [CrossRef]
17. Tabakov, M.; Chlopowiec, A.; Chlopowiec, A.; Dlubak, A. Classification with Fuzzification Optimization Combining Fuzzy Information Systems and Type-2 Fuzzy Inference. *Appl. Sci.* **2021**, *11*, 3484. [CrossRef]

18. Lin, C.-J.; Jeng, S.-Y.; Lin, H.-Y.; Shiou-Yun, J. Design and Verification of an Interval Type-2 Fuzzy Neural Network Based on Improved Particle Swarm Optimization. *Appl. Sci.* **2020**, *10*, 3041. [CrossRef]
19. Son, L.H. Enhancing clustering quality of geo-demographic analysis using context fuzzy clustering type-2 and particle swarm optimization. *Appl. Soft Comput.* **2014**, *22*, 566–584. [CrossRef]
20. Heidarzade, A.; Mahdavi, I.; Mahdavi-Amiri, N. Supplier selection using a clustering method based on a new distance for interval type-2 fuzzy sets: A case study. *Appl. Soft Comput.* **2016**, *38*, 213–231. [CrossRef]
21. Anuradha, R.; Rajkumar, N. A Novel Approach in Mining Specialized Coherent Rules in a Level-Crossing Hierarchy. *Int. J. Fuzzy Syst.* **2017**, *19*, 1782–1792. [CrossRef]
22. Lin, J.C.-W.; Zhang, Y.; Fournier-Viger, P.; Hong, T.-P. Efficiently Updating the Discovered Multiple Fuzzy Frequent Itemsets with Transaction Insertion. *Int. J. Fuzzy Syst.* **2018**, *20*, 2440–2457. [CrossRef]
23. Kalia, H.; Dehuri, S.; Ghosh, A.; Cho, S.-B. Surrogate-Assisted Multi-objective Genetic Algorithms for Fuzzy Rule-Based Classification. *Int. J. Fuzzy Syst.* **2018**, *20*, 1938–1955. [CrossRef]
24. Wang, X.; Lei, L.; Wang, L.; Yang, P.; Chen, H. Spatial Co-location Pattern Discovery Incorporating Fuzzy Theory. *IEEE Trans. Fuzzy Syst.* **2021**, *30*, 2055–2072. [CrossRef]
25. Yang, P.; Wang, L.; Wang, X.; Zhou, L. SCPM-CR: A Novel Method for Spatial Co-location Pattern Mining with Coupling Relation Consideration. *IEEE Trans. Knowl. Data Eng.* **2021**, *99*, 1. [CrossRef]
26. Molina, C.; Ruiz, M.D.; Serrano, J.M. Representation by levels: An alternative to fuzzy sets for fuzzy data mining. *Fuzzy Sets Syst.* **2019**, *401*, 113–132. [CrossRef]
27. Lin, J.C.-W.; Ahmed, U.; Srivastava, G.; Wu, J.M.-T.; Hong, T.-P.; Djenouri, Y. Linguistic frequent pattern mining using a compressed structure. *Appl. Intell.* **2021**, *51*, 4806–4823. [CrossRef]
28. Zhang, Z.; Huang, J.; Hao, J.; Gong, J.; Chen, H. Extracting relations of crime rates through fuzzy association rules mining. *Appl. Intell.* **2019**, *50*, 448–467. [CrossRef]
29. Wang, T.; Zhou, M. Zhou. Integrating rough set theory with customer satisfaction to construct a novel approach for mining product design rules. *J. Intell. Fuzzy Syst.* **2021**, *41*, 331–353. [CrossRef]
30. Anari, Z.; Hatamlou, A.; Anari, B. Finding Suitable Membership Functions for Mining Fuzzy Association Rules in Web Data Using Learning Automata. *Int. J. Pattern Recognit. Artif. Intell.* **2021**, *35*, 2159026. [CrossRef]
31. Gupta, P.; Mehlawat, M.K.; Grover, N. A Generalized TOPSIS Method for Intuitionistic Fuzzy Multiple Attribute Group Decision Making Considering Different Scenarios of Attributes Weight Information. *Int. J. Fuzzy Syst.* **2019**, *21*, 369–387. [CrossRef]
32. Sun, P.; Chen, S.; Zhi, Y. Multiple Attribute Variable Weight Fuzzy Decision-Making Based on Optimistic Coefficient Method. *Int. J. Fuzzy Syst.* **2021**, *23*, 573–583. [CrossRef]
33. Meng, F.; Li, S. A new multiple attribute decision making method for selecting design schemes in sponge city construction with trapezoidal interval type-2 fuzzy information. *Appl. Intell.* **2020**, *50*, 2252–2279. [CrossRef]
34. Yan, T.; Han, C. A Novel Approach of Rough Conditional Entropy-Based Attribute Selection for Incomplete Decision System. *Math. Probl. Eng.* **2014**, *2014*, 1–15. [CrossRef]
35. Farhadinia, B. Hesitant fuzzy set lexicographical ordering and its application to multi-attribute decision making. *Inf. Sci.* **2016**, *327*, 233–245. [CrossRef]
36. Choi, B.-I.; Rhee, F.C.-H. Interval type-2 fuzzy membership function generation methods for pattern recognition. *Inf. Sci.* **2009**, *179*, 2102–2122. [CrossRef]
37. Lee, C.-H.; Pan, H.-Y. Performance enhancement for neural fuzzy systems using asymmetric membership functions. *Fuzzy Sets Syst.* **2009**, *160*, 949–971. [CrossRef]
38. Liao, T. A Procedure for the Generation of Interval Type-2 Membership Functions from Data. *Appl. Soft Comput.* **2017**, *52*, 925–936. [CrossRef]
39. Kayacan, E.; Sarabakha, A.; Coupland, S.; John, R.; Khanesar, M.A. Type-2 fuzzy elliptic membership functions for modeling uncertainty. *Eng. Appl. Artif. Intell.* **2018**, *70*, 170–183. [CrossRef]
40. Kayacan, E.; Maslim, R. Type-2 Fuzzy Logic Trajectory Tracking Control of Quadrotor VTOL Aircraft with Elliptic Membership Functions. *IEEE/ASME Trans. Mechatron.* **2017**, *22*, 339–348. [CrossRef]

applied sciences

MDPI

Article

Deep Transfer Learning Model for Semantic Address Matching

Liuchang Xu [1,2,3,4], Ruichen Mao [5], Chengkun Zhang [6], Yuanyuan Wang [7], Xinyu Zheng [1,3,4], Xingyu Xue [1,3,4] and Fang Xia [1,8,*]

1 College of Mathematics and Computer Science, Zhejiang A&F University, Hangzhou 311300, China
2 College of Computer Science and Technology, Zhejiang University, Hangzhou 310063, China
3 Key Laboratory of Forestry Intelligent Monitoring and Information Technology of Zhejiang Province, Zhejiang A&F University, Hangzhou 311300, China
4 Key Laboratory of State Forestry and Grassland Administration on Forestry Sensing Technology and Intelligent Equipment, Zhejiang A&F University, Hangzhou 311300, China
5 Zhejiang Laboratory, Hangzhou 311121, China
6 School of Earth Sciences, Zhejiang University, Hangzhou 310027, China
7 Ocean Academy, Zhejiang University, Zhoushan 316021, China
8 Institute of Digital Village, Zhejiang A&F University, Hangzhou 311300, China
* Correspondence: xiaf@zafu.edu.cn

Abstract: Address matching, which aims to match an input descriptive address with a standard address in an address database, is a key technology for achieving data spatialization. The construction of today's smart cities depends heavily on the precise matching of Chinese addresses. Existing methods that rely on rules or text similarity struggle when dealing with nonstandard address data. Deep-learning-based methods often require extracting address semantics for embedded representation, which not only complicates the matching process, but also affects the understanding of address semantics. Inspired by deep transfer learning, we introduce an address matching approach based on a pretraining fine-tuning model to identify semantic similarities between various addresses. We first pretrain the address corpus to enable the address semantic model (abbreviated as ASM) to learn address contexts unsupervised. We then build a labelled address matching dataset using an address-specific geographical feature, allowing the matching problem to be converted into a binary classification prediction problem. Finally, we fine-tune the ASM using the address matching dataset and compare the output with several popular address matching methods. The results demonstrate that our model achieves the best performance, with precision, recall, and an F1 score above 0.98.

Keywords: semantic address matching; deep transfer learning; pretraining model; fine tuning

Citation: Xu, L.; Mao, R.; Zhang, C.; Wang, Y.; Zheng, X.; Xue, X.; Xia, F. Deep Transfer Learning Model for Semantic Address Matching. *Appl. Sci.* **2022**, *12*, 10110. https://doi.org/10.3390/app121910110

Academic Editors: Alexei Gvishiani and Boris Dzeboev

Received: 7 August 2022
Accepted: 5 October 2022
Published: 8 October 2022

1. Introduction

Addresses are used to describe a unique spatial location on Earth and are usually expressed in the form of an addressing system [1]. In recent years, with the rapid development of location services, massive amounts of industry data based on addresses as spatial information have started to emerge. Address matching is a crucial application in address services, which compares addresses with the same location in different address databases to obtain the best match with the search address and to determine position on a map [2]. Traditional address matching technology is challenged by the prevalence of high-precision address matching in urban industries, such as logistics and online taxi services. Therefore, an effective address matching method is required to facilitate the provision of accurate and efficient intelligent spatial location services and to promote the development of smart cities.

The pattern of arrangement of address elements varies from country to country. For instance, the US address pattern is "room number + street + state + country", and it performs well in creating a national geodatabase [3]. Japanese addresses, on the other hand, are coded based on location and geographic relativity, with the overall order being the opposite of the address pattern used in the US, and generally without "streets" [4]. In general, the address patterns of the above countries are nested and relatively standardized. However, Chinese addresses are relatively more difficult to match due to their complex context and rules, mainly due to the following reasons: (1) Chinese addresses are written without separators; (2) Chinese addresses often contain landmarks or POI and topology (e.g., road intersections); (3) different government departments manage addresses, leading to confusion; and (4) address assignment and updating lags behind rapid urban renewal [5]. The data objects of this study are Chinese addresses.

Address matching is generally divided into matching based on rule-based or statistical methods and semantic similarity matching based on machine learning and deep learning. Character-based approaches match addresses by calculating their string similarity metrics and then manually establishing a threshold or a particular classifier to identify a match [6,7]. String similarity metrics include edit distance and its variants [8–10], Jaccard similarity metric [11], and Jaro distance and its variants [12,13]. Among them, Santos et al. compared 13 different string similarity metrics for place name matching and found that adjusting the similarity threshold was the key to achieving good performance [14]. In addition, the calculation of cosine similarity between embeddings based on N-grams is also a common method [15,16]. This method has better performance compared with traditional metrics. Recently, Yong et al. proposed a normalization method based on the Euclidean distance between the address to be processed and the address in the standard library, but it is only applicable to some specific datasets [17]. Another type of methods is address element based, which segments out address elements by rules or statistical methods and then compares the address elements and their hierarchy to determine whether they match [18–20]. Lin et al. point out that the degree of matching of address elements depends on whether they can be extracted correctly [21]. In general, dictionary queries [22], probability statistics such as CRF [23] and HMM [24], and creating matching rules [25,26] are the basic ways for retrieving address elements. Another common method is to construct a decision tree consisting of matching rules, each corresponding to a path in the tree. Kang et al. proposed an address matching tree model based on the analysis of the spatial constraint relationship between address elements; this requirement makes the address model more complex [27]. Focusing on the wrong word separation problem, Luo and Huang suggested a method based on a trie tree and finite-state machine [28]. The aforementioned techniques, however, frequently struggle when dealing with nonstandardized (missing address elements or represented by POI) and complexly structured addresses (such as the Chinese address feature aforementioned).

In recent years, the area of artificial intelligence has seen tremendous progress in natural language processing (NLP), most of which is attributable to deep learning's enhanced performance. Word2vec [29], ELMo [30], GPT [31], BERT [32], XLNet [33], ERNIE [34], and ELECTRA [35] are a few of these classical language models. Since addresses as special textual descriptions, more and more studies in address matching has also introduced natural language models based on deep learning [36]. Cruz et al. analyzed 41 papers on address matching published between 2002 and 2021 and discovered that most of the relevant studies have used deep learning methods. Among them, consistent with the above in this paper, due to the complexity of Chinese addresses, Chinese address matching accounted for half of the studies [37].

Comber et al. used CRF and word2vec for address matching to extract the semantics of addresses without designing complex rules [38]; Zhang et al. provides a convolutional neural network (W-TextCNN) for Chinese address pattern classification [39]. With the popularity of gating mechanism neural networks, address matching and normalizing based on LSTM and GRU have been carried out by an increasing number of researchers [40–43]. Santos et al. used a deep neural network based on bidirectional GRUs for place name matching [44]; Shan enriched the address context by collecting address data on the Internet and trained an address representation model with two LSTMs and attention mechanisms to extract address vectors [45]. While Li et al. incorporated the hierarchical relationship between address elements into a neural network and proposed a BiLSTM-based multitask learning method [46], Chen et al. proposed a contrast learning address matching model based on attention-Bi-LSTM-CNN networks (ABLC) [47]. Subsequently, more and more researchers have used the attention mechanism in their address matching models [48–50]. With the popularity of pretrained language models, Lin et al. used the classical enhanced sequence inference model (ESIM) [51] for address record pair modelling [21], whereas Xu et al. and Qian et al. used the BERT model. Xu et al. proposed a BERT-based model for extracting address semantic representations to achieve the fusion of address semantics and geospatial information [36]; Qian et al. combined BERT and LSTM, and proposed a hierarchical region-based approach for geolocation of Chinese addresses [52]. However, all of the aforementioned methods require the extraction of address semantic features to embedding, and this can affect the effectiveness of address semantic understanding, as has been demonstrated in the field of NLP [32].

In summary, when dealing with nonstandardized addresses with complicated structures, the aforementioned approaches still lack a level of comprehension of address semantics, which negatively impacts the accuracy of address matching.

To address the above problems, we use a deep transfer learning approach. First, we pretrain an addresses corpus so that our address semantic model (abbreviated as ASM) can learn unsupervised address contexts to better understand address semantics. Then, we use the address-specific geospatial property to build a labelled address matching dataset, allowing the matching problem to be converted into a binary classification prediction problem. Finally, fine-tuning the ASM with the address matching dataset allows the model to improve its performance significantly.

The contributions of this paper are as follows: (1) A neural network based on a multihead self-attention mechanism and a permutation-based target task is used to train the ASM for a large-scale corpus in an unsupervised automated manner. The ASM can learn address semantics better. (2) A deep transfer learning approach is used to achieve semantic address matching by fine-tuning the ASM, which improves the matching accuracy. (3) A semantic address matching dataset construction method is proposed to convert address matching into a classification prediction task. The method constructs an address matching dataset with labels using location information as the inference condition. (4) Results demonstrate that with the transfer learning approach, a better-performing downstream task such as address matching can also be achieved with microsupervision.

The remainder of this paper is organized in four sections. Section 2 introduces the materials used in our study, as well as the data processing procedures. The methodology adopted is also demonstrated in Section 2, including the pretraining and fine-tuning based on XLNet. The results of our experiments are analyzed in Section 3. Section 4 presents our conclusions and the future work of this study.

2. Materials and Methods

In this section, we introduce a deep transfer learning approach in NLP and propose a semantic address matching framework. First, we tokenize all address data to be used as model input in the pretraining phase. Then, we use the XLNet model [33] to pretrain the address corpus and make the model understand the address semantics by learning contextual information. Finally, we construct a supervised dataset for semantic address matching, fine-tune the pre-trained ASM for address matching, and compare it with multiple models to evaluate the accuracy of the ASM.

2.1. Dataset

Address records for the raw data were manually collected in 2019 from various government departments. The geographical area to which this data refers is Shangcheng District, Hangzhou, Zhejiang Province, China. The address dataset contains a variety of location description types, including standard addresses, nonstandard addresses, POIs, road intersections, place name abbreviations, and so on. The preprocessed address dataset amounted to 1,552,532, consisting of three fields of address records, longitude, and latitude, which served as the address corpus for the pretraining phase.

To use the address data for semantic address matching, we created a dataset of address pairs with labels based on the address corpus. Based on the set of addresses filtered with the same coordinates, we performed manual matching using a standard address database. In addition, to give the model better prediction performance and generalization capabilities, we augmented the dataset with easy data augmentation [51] methods for text classification tasks, mainly using synonym replacement, address element deletion, and address element insertion. To improve the robustness of the model, we constructed mismatched address pairs in the set of address pairs with Jaccard similarity coefficients [11] greater than zero. We finally obtained a dataset of 64,358 address pairs and corresponding labels, a sample of which is shown in Table 1. The statistical features of the dataset used for semantic address matching are shown in the Table 2, where we used the difference in the number of characters, Levenshtein distance [8], and Jaccard similarity coefficient [11] to show the similarity of address pairs in the dataset. Unmatched address pairs will perform worse in terms of text similarity, in line with our common sense.

Table 1. Examples of some data in the labelled address dataset.

Address S_a	Address S_b	Label
Block 3, No. 15 Haiyue Road, Hangzhou City (杭州市海月路15号3幢)	Block 3, Haiyue Garden Residential Unit, Shangcheng District, Hangzhou (杭州上城区海月花园3幢)	1
Block 3, No. 15 Haiyue Road, Hangzhou City (杭州市海月路15号3幢)	Hangzhou Haiyue Bathing Centre (杭州海月洗浴中心)	0

Table 2. Statistical characteristics of the labelled address dataset.

Statistical Characteristics	Value
Total number of address pairs	64,358
Number of matching address pairs	32,179
Number of unmatched address pairs	32,179
Average length difference for all the address pairs	4.76
Average length difference for matching address pairs	3.73
Average length difference for unmatched address pairs	5.79
Average Levenshtein distance for all the address pairs	12.16
Average Levenshtein distance for matching address pairs	7.28
Average Levenshtein distance for unmatched address pairs	17.05
Average Jaccard similarity coefficient for all the address pairs	0.46
Average Jaccard similarity coefficient for matching address pairs	0.68
Average Jaccard similarity coefficient for unmatched address pairs	0.24

2.2. Semantic Address Matching Definition

In this paper, we study the address matching in the absence of a standard address database, referred to as the semantic address matching task. The following description defines semantic address matching:

Given the address dataset: $D = \{add_1, add_2, \ldots, add_n\}$, the goal of semantic address matching is to find each address pair: (add_i, add_j), satisfying $add_i \doteq add_j$, where $add_i \in D$, $add_j \in D$, $i \neq j$, and $\doteq$ represent the comparison operator. The operation objects on either side of the comparison operator refer to the same real-world object with the same coordinates. It is important to note that no information other than the string itself and its corresponding geospatial information is utilized in this study to calculate the similarity of two addresses. Therefore, the task addressed in this study focuses on the problem of matching addresses with the same location instead of address disambiguation. In addition, due to the many different representations of the same location, we believe that it is not possible to achieve a correct match without processing from a natural language understanding perspective. Therefore, in our study, "address semantic understanding" refers to the textual understanding of the address corpus, while "address semantic reasoning" used for address matching is based on the spatial relationship reasoning of addresses.

2.3. Pretraining Phase Using the Address Corpus Based on XLNet

This section presents a transfer learning-based pretraining model for address semantics: address semantic model (ASM). The ASM is based on the characteristics of Chinese addresses, combined with the advantages of semantic understanding in deep learning natural language models. The model takes as input a single character of a Chinese address that has been tokenized, and uses a multihead self-attention-based semantic extraction module to help the model understand the semantics of the address with the objective of permutation unknown character prediction. For the practical training problem resulting from the prediction objective, a two-stream self-attention structure for target position representations is used. The overall structure of the ASM is shown in Figure 1.

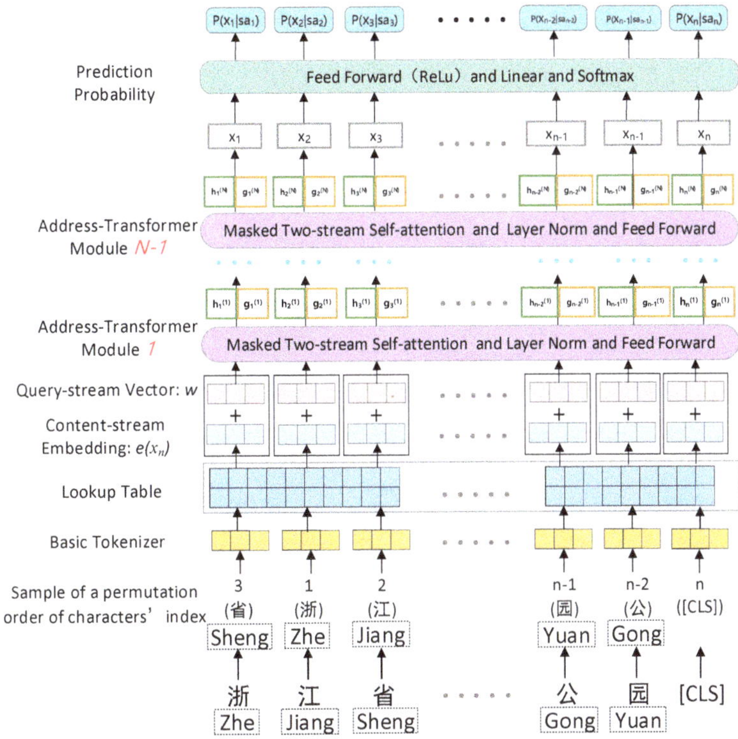

Figure 1. The overall structure of the ASM.

2.3.1. Tokenization of Address Characters

The conversion of Chinese addresses into input that can be received by the ASM is the basis for training. Since Chinese addresses are not like alphabetic forms of languages, such as English, they do not have delimiters. Therefore, most Chinese address studies start with the segmentation of address elements. Due to the unique hierarchy of addresses, partitioning addresses into various address elements is already a problem worth studying. Our study, however, aims to convert the complex address matching into a classification problem that can be automated for computer computation. Although the commonly used SentencePiece method [53] in NLP can automate the segmentation of Chinese addresses by counting high-frequency co-occurring characters combined into subword units and constructing dictionaries, the subwords obtained by its segmentation are too long, and some of the segmented words do not conform to the common sense of Chinese addresses, which will affect the semantic understanding during pretraining.

We therefore use the Basic Tokenizer, which tokenizes a character as a unit. It separates words and symbols according to spaces. We first add blank characters before and after each character of the address. Then the characters are matrix-transformed according to the lookup table to become the input of the one-hot encoding, and the activated dimensions in the one-hot encoding are the index number corresponding to the character in the dictionary key–value pair. In this study, two dictionaries—one with non-Chinese characters and the other with solely Chinese characters—are created once the individual characters from each address have been obtained. These dictionaries have 9425 and 3491 characters, respectively.

2.3.2. Objective of Permutation Unknown Character Prediction and Two-Stream Self-Attention Structure

The objective of permutation language modeling is derived from the XLNet model [30]. Without altering the character order of the original text, the target employs rearrangement to sabotage the index order of text descriptions. This training target not only preserves the high-order and long-range dependencies present in the text context, but also improves on the disadvantages of past autoregressive language modeling's targets that could only exploit unidirectional contexts (forward or backward), enabling a pretrained model to utilize deep, bidirectional contextual information more effectively. Addresses, as special natural languages incorporating geospatial information and hierarchy, need to fully utilize the bidirectional contextual information, so we use a permutation language model objective for pretraining the address corpus.

Specifically, we assume that given an address record X of length T, there are a total of T! sequences of permutations. If all permutations are traversed and the parameters of the model are shared, then the model must be able to learn the context of all positions. We take a simplified address record, for example, "Hangzhou Underwater World" ("Hang Zhou Hai Di Shi Jie" in Chinese pinyin), and predict the third character "Hai" in a different order, as shown in Figure 2. In Figure 2b, for instance, the address permutation is disordered as 3→2→4→1→6→5, so when predicting "Hai", there is no address context character, and the prediction can only be made based on the previous hidden state. For Figure 2f, the "Hai" (3) character learns all five context characters except itself.

The objective function of XLNet is to maximize the log-likelihood function of the target subsequence conditional on the nontarget subsequence:

$$\max_{\theta} \mathrm{E}_{z \sim Z_T}[\log p_\theta(X_{z>c} | X_{z \leq c})] = \mathrm{E}_{z \sim Z_T}\left[\sum_{t=c+1}^{|z|} \log p_\theta(x_{z_t} | X_{Z<t})\right] \quad (1)$$

where Z_T denotes the set of all permutations of the index of an address record of length T; $z \in Z_T$ is one of the sequences of indexed permutations, where z_t denotes the t-th element of the sequence of indexed permutations, and $z_{<t}$ denotes the first t-1 elements of z; $\mathrm{E}_{z \sim Z_T}$ denotes the maximum Expectation; and p_θ denotes the predicted probability. In addition, XLNet used the partial prediction optimization. It slices a permutation z into two subsequences, $z_{\leq c}$ and $z_{>c}$, where c is the slice point that slices the two subsequences into a nontarget sequence and a target sequence, respectively.

While the above objective of permutation unknown character prediction works well for understanding address semantic by removing ambiguity from the target prediction, it creates the problem that the model does not know the position of the character to be predicted in the original address record. Therefore XLNet [33] introduces a two-stream self-attention structure to let the model know where the character to be predicted is located explicitly, which consists of two sets of hidden representations instead of one. The two streams of representations are updated with a shared set of parameters as follows:

$$g_{z_t}^{(m)} \leftarrow Attention(Q = g_{z_t}^{(m-1)}, KV = h_{z<t}^{(m-1)}; \theta), (query\ stream: use\ z\ but\ cannot\ see\ x_{z_t}) \tag{2}$$

$$h_{z_t}^{(m)} \leftarrow Attention(Q = h_{z_t}^{(m-1)}, KV = h_{z\leq t}^{(m-1)}; \theta), (content\ stream: use\ both\ z_t\ and\ x_{z_t}) \tag{3}$$

where Q, K, V denote the query, key, and value in an attention operation [54]; h_{z_t} denotes the content representation, which serves a similar role to the standard hidden states in Transformer, and g_{z_t} denotes the query representation, which only has access to the contextual information $X_{z<t}$ and the position X_{z_t}, but not the content X_{z_t}.

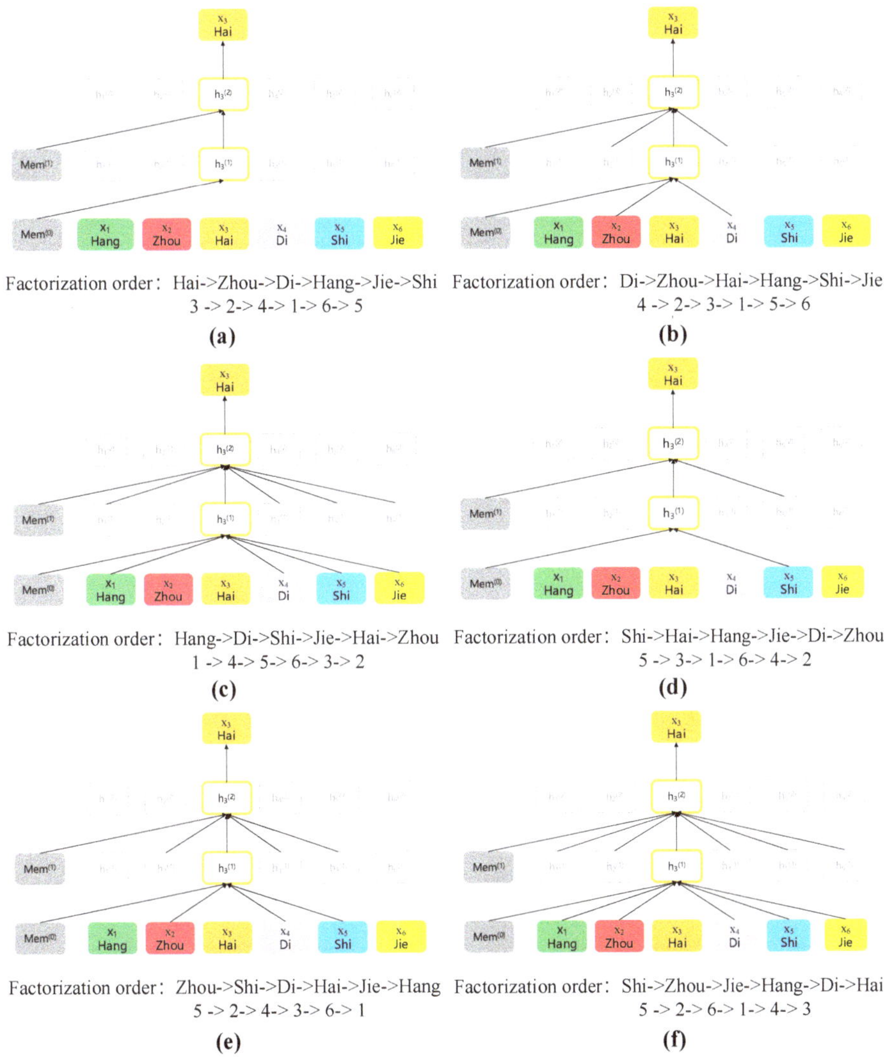

Figure 2. Illustration of the permutation language modeling objective for predicting "Hai" given the same simplified address record but with different factorization orders.

In addition, we employ Transformer-XL with a multihead self-attention mechanism as an address semantic feature extractor [55]. Transformer-XL integrates two important techniques, namely, the relative positional encoding scheme and the segment recurrence mechanism. This allows for better adaptation to the two-stream attention permutation language model. As the number of semantic feature extraction structures affects the performance of the model in subsequent experiments, each layer of the Transformer-XL module is tentatively defined in this section as the address-transformer module.

2.4. Fine-Tuning for Semantic Address Matching

Fine-tuning is an implementation of deep transfer learning, which refers to adding task-relevant structures and parameters to an already-trained model, and then retraining on a task-relevant corpus [56]. We therefore used a newly constructed labelled address matching corpus for the semantic address matching, adding a new neural network structure for a fine-tuned learning model and training framework based on the classification task. The network structure is first superimposed with a layer of fully connected feedforward neural networks for nonlinear transformation, with an activation function of *tanh*, which is mathematically formulated as follows:

$$t(x) = \frac{e^x - e^{-x}}{e^x + e^{-x}} \tag{4}$$

After obtaining the probability distribution features using the fully connected neural network, we then connected the fully connected neural network without the activation function for linear transformation. Since semantic address matching is a binary classification task of whether to match, the output of this layer is two-dimensional. Finally, we passed the output probability distribution score of this layer into the SoftMax normalization function to predict the probability of matching or not matching the address pair, respectively. We designed the deep semantic address matching model (abbreviated as DSAMM) with the following objective function. Here, given that the size of the number of address string pairs per batch iteration is *batch_size*, the predicted probability output is *prob(batch_size,2)*, and the true label sequence is *label(batch_size)*, the true label probability for each address pair is as follows:

$$true\_prob(batch\_size) = gather(prob, Label) \tag{5}$$

The final objective function is obtained by taking logarithmic values of the probabilities and then summing them (i.e., log transformation) and averaging them. The objective function is specified below:

$$\frac{\sum_{i=1}^{batch\_size} \log(true\_prob[i])}{batch\_size} \to 0 \tag{6}$$

The accuracy metrics used in this study include precision, recall, and F1 score [57]. Precision calculates the proportion of true positive samples out of those predicted to be positive; recall reflects the rate at which positive examples in this are predicted to be accurate and, in semantic address matching, refers to the percentage of correctly matched pairs out of all address pairs that should be correctly matched; and the F1 score is the harmonic mean of precision and recall.

3. Results and Discussion

3.1. Address Semantic Model Pretraining

We examine the semantic understanding effectiveness of the ASM by examining the prediction accuracy of address characters for permutation language objective. In the experimental design, we refer to the influencing factors of pretraining in a study by Xu et al. and use the number of address-transformer modules and whether the numbers in the address records are replaced with uniform identifiers as the independent variables for the analysis of the pretraining hyperparameters [36]. The purpose of the experiments in this section is to validate the effectiveness of the ASM without testing or predicting it, so only the training set and the verification set are required for pretraining the model. Due to the large address corpus, we set the proportions of training set and verification set to be approximately 99% (1,537,532) and 1% (15,000), respectively.

The optimal values for the relevant hyperparameters were determined by drawing on previous studies of pretrained language models and previous experiments. It is shown in Table 3. In terms of the number of address-transformer modules, we set the number of modules to 6, 8, 10, and 12 to observe the performance of the target task in different situations. As shown in Figure 3, the training loss gradient for different numbers of modules decreases rapidly until about 40k steps, and then keeps decreasing slowly and gently in the following iterations, and basically levels off in the last 40k steps, indicating that the ASM instances have been adequately trained. Additionally, the comparison of each ASM instance reveals that the positions of the four curves overlap more, which indicates that the number of address-transformer modules has little influence on prediction performance. As shown in Table 4, the accuracy of the model validation under the above four comparisons ranged from 90.5% to 91.5%, and the target accuracy increased slightly with the number of modules, indicating that the more the number of address-transformer modules, the better understanding of address semantic. However, due to the tiny increase in accuracy and the large increase in training time, building six layers of modules was the most cost-effective option. The findings of this study are consistent with those of Xu et al. [36].

Table 3. Setting values for each hyperparameter of the ASM.

Hyperparameters	Are Numbers Replaced with "CODE I"	
	Replaced	**Not Replaced**
num_address_transformer_module	12	6, 8, 10, 12
batch_size	32	32
hidden_size	768	768
num_head	4	4
dim_head	8	8
FFN_hidden_size	2048	2048
Dropout	0.1	0.1
K	3	3
sequence_len	32	32
mem_len	24	24
η	0.00005	0.00005
epoch	4	4
voc_size	3491	9425

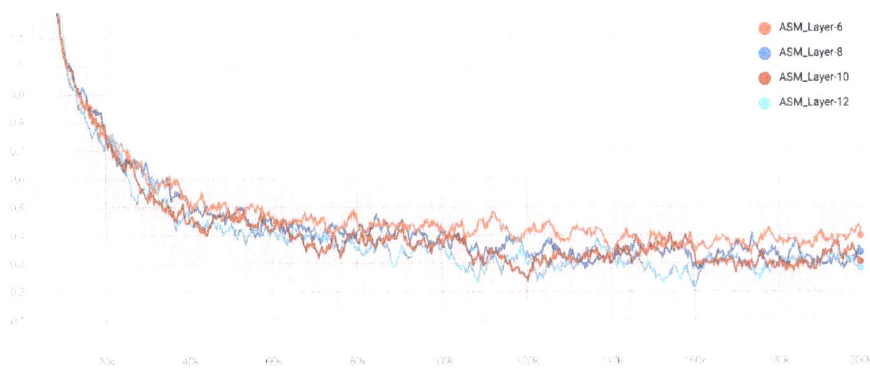

Figure 3. Comparison of training loss gradient curves for the ASM with a different number of address-transformer modules, where the curve smoothness is set to 0.9, and the opaque curve is after the smoothing setting and the translucent curve is the original loss gradient curve.

Table 4. Values for the ASM indicator for a different number of address-transformer modules.

Number of Address-Transformer Modules	Training Time	Training Loss	Mean Evaluation Loss	Accuracy
6	6 h 19 m 36 s	0.3864	0.3615	90.62%
8	8 h 21 m 54 s	0.3395	0.3258	90.98%
10	10 h 23 m 5 s	0.3016	0.3003	91.21%
12	12 h 32 m 45 s	0.2831	0.2793	91.47%

Considering that most of the numbers in the address records have only geospatial property and no semantic information (e.g., "No. 116 Tianmushan Road" and "No. 226 Tianmushan Road", there is no difference in their contexts other than numbers, so the model cannot make accurate predictions at all. Therefore, we replaced all Arabic numerals in the address corpus with a uniform identifier: "CODE". Since Xu et al. [36] used BERT [32] to construct the ALM for pretraining the address corpus, we used their method for comparison. Figure 4 shows a comparison of the training loss gradient curves of the replaced and unreplaced "CODE" corpus under the ASM. The training loss curve of the model with the "CODE" replacement is always below the original address corpus, demonstrating that the numbers confound the predictive target of the model and reduce the predictive power of the model. As can be seen from Table 5, the prediction accuracy of the ASM increased by approximately 7 percentage points after the replacement with "CODE". In addition, when compared with the ALM, our model's prediction accuracies all improved, indicating that the ASM performs better in understanding address semantics.

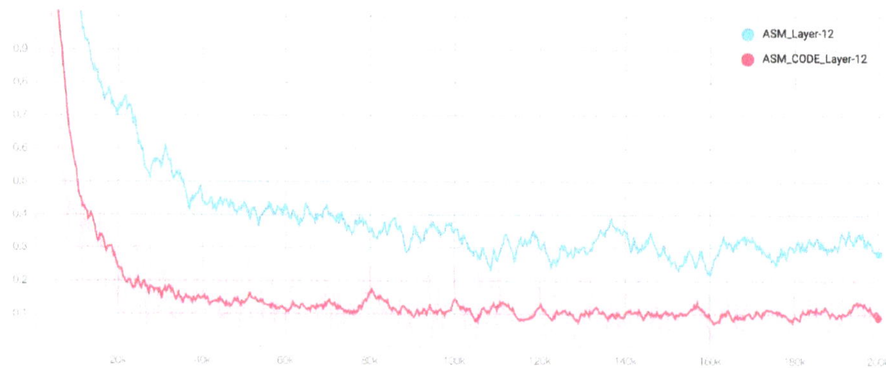

Figure 4. Comparison of training loss gradient curves for the ASM with Arabic numeric code replacement and without replacement, where the curve smoothness is set to 0.7, and the opaque curve is after the smoothing setting and the translucent curve is the original loss gradient curve.

Table 5. The prediction accuracy of the ASM and the ALM in the validation datasets.

Prediction Accuracy	Without "CODE" Replacement	"CODE" Replacement
ALM [40]	90.31%	97.22%
ASM	91.47%	98.53%

3.2. Fine-Tuning for Semantic Address Matching

To explore whether there is overfitting, we tested the validation dataset by fine-tuning it while also spacing the number of iterations by 500, and determined the set of hyperparameters that worked best. In this study, 80% of the data were randomly taken as the training dataset and the remaining 20% as the validation dataset. The loss gradient curves for training and validation are shown in Figure 5.

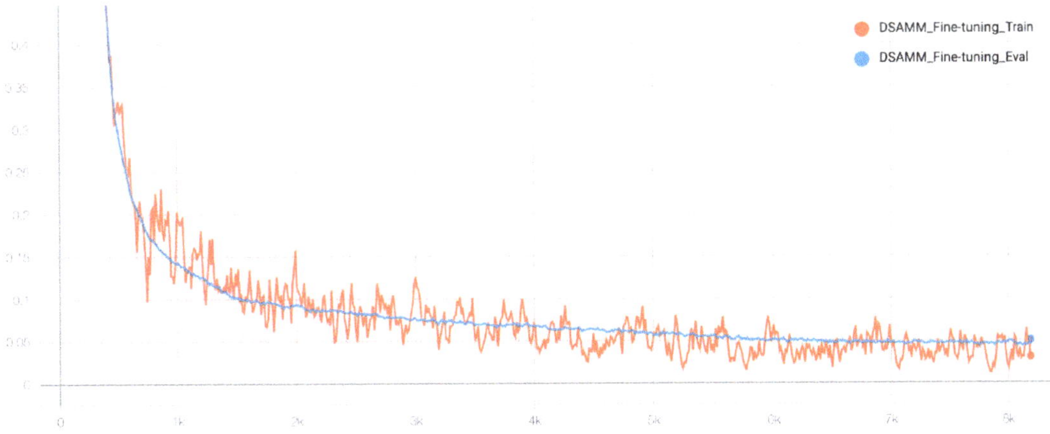

Figure 5. Training loss-gradient curve overlaid with validation loss-gradient curve of the DSAMM, where the curve smoothness is set to 0.7, and the opaque curve is after the smoothing setting and the translucent curve is the original loss gradient curve.

Figure 5 shows that the training loss value of the DSAMM decreases rapidly before the steps of training iterations are 1k, indicating the model's excellent learning ability and ability to make a good assessment of whether address pairs match even before the dataset is entirely learned. This has inspired us to investigate whether better matching can be achieved even with microsupervision, that is, with a small amount of supervised training data.

Between 2k and 6k training iteration steps, the training loss value decreases more gently and steadily, indicating that the model is still learning the task objectives for the supervised data. Between the final 6k and 8k, the training loss values have largely flattened out, demonstrating that the DSAMM instance has been sufficiently learned by that number of iterations to warrant further training. The final training loss values ranged from 0.01 to 0.05, indicating that the fine-tuning training was effective, and the exact metric values will be discussed further in the subsequent comparative experimental analysis. In addition, the trend of the loss values is like that of the training loss values, with a "high rate of decline—slow decline—gradual levelling off". The final loss values for the validation datasets range from 0.04 to 0.05, indicating that there is no overfitting. We selected the optimal iteration step of the validation datasets as the DSAMM instance after training 8 epochs.

We set up comparative experiments with various gradients of the proportion of training sets to examine whether the DSAMM only needs a limited number of labelled training datasets to attain high matching precision. Figure 6 displays the matching prediction precision with 1%, 5%, 10%, 15%, and 20% of the original training datasets. When employing only 15% of the initial training set of labelled address pairings, the DSAMM achieves a matching precision of above 0.80, and 0.84 when using 20%. It demonstrates that the DSAMM can perform well in weakly supervised learning, most likely due to the transfer learning employed in our study. Our model has fully understood the address semantics using self-supervised learning in the pretraining phase, and when then fine-tuned for task-based learning using partially supervised data, it is able to combine the advantages of the two phases mentioned above, allowing the model to perform at a high level on the task with less supervised training data. This also coincides with the research that first proposed the idea of pre-training [58].

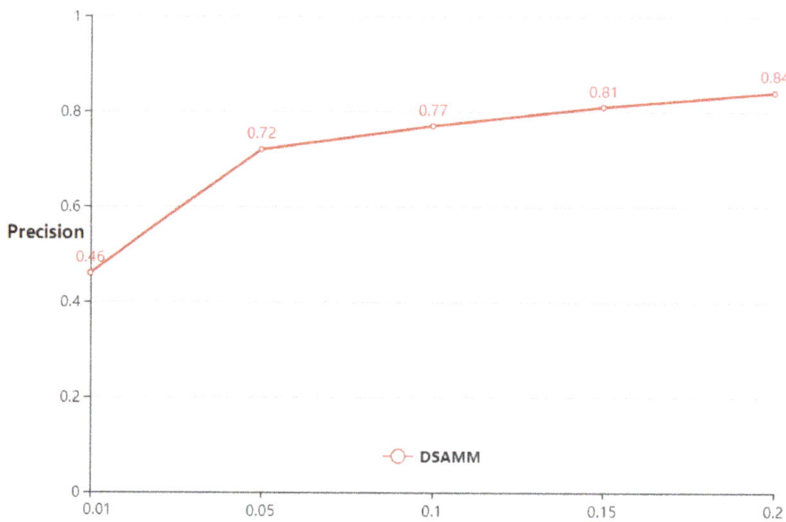

Figure 6. Comparison of the precision of the validation set of the DSAMM instances with different scaled training sets with labels.

3.3. *Comparative Experiment Analysis of the Address Matching*

To evaluate the semantic address matching performance of the DSAMM we proposed, several baseline models were selected for comparison. The methods compared include character-based matching methods, machine-learning-based matching methods, and deep-learning-based matching methods, where the character-based matching methods in this paper use Levenshtein distance [8], Jaccard similarity coefficient [11], and Jaro similarity [12] to measure string correlation, followed by a random forest (RF) classifier [59] and a support vector machine (SVM) classifier [60] to determine whether the address pairs match. In terms of comparing machine-learning-based matching methods, we compared with the method proposed by Comber et al. [38]. It uses CRF to label the address elements, then uses word2vec for embedding, and applies RF and SVM for classification prediction. In terms of deep-learning-based matching methods, we compared with the method proposed by Lin et al. [21]. It uses a two-stage model, with the first stage using word2vec to obtain embeddings of address pairs for use as input to the next stage of the deep neural network, and the second stage using a typical deep learning model for interaction-based text matching, ESIM [47], which is directly for address matching.

The results of the comparison of the three metrics for each method are shown in Table 6. As the F1 score is the most representative and important evaluation metric, we present the F1 scores for each method in the form of a bar chart for a more visual comparison. As shown in Figure 7, the Jaccard similarity coefficient stands out when using a string-similarity-based approach for the semantic address matching. The classifier employing RF as the task regularly outperforms SVM, and this is also true for the other two groups using machine learning techniques, according to performance comparisons.

When comparing the machine-learning-based matching methods, the CRF method performs worse, probably because the pretraining corpus is more susceptible to influence. In addition, the performance of the machine-learning-based matching methods did not outperform the string-based matching methods, which may be since Comber et al. used relatively generic English addresses, whereas Chinese addresses have their own uniqueness and, therefore, lead to different conclusions [38].

Table 6. Comparative evaluations of different address matching methods in precision, recall, and F1 score.

Comparison Methods	Predictive Results		
	Precision	Recall	F1 Score
Levenshtein distance + SVM	0.849	0.795	0.822
Levenshtein distance + RF	0.875	0.819	0.847
Jaccard similarity coefficient + SVM	0.890	0.881	0.885
Jaccard similarity coefficient + RF	0.914	0.911	0.912
Jaro similarity + SVM	0.902	0.807	0.854
Jaro similarity + RF	0.891	0.858	0.874
word2vec + SVM	0.832	0.762	0.797
word2vec + RF	0.910	0.879	0.895
CRF + word2vec + SVM	0.826	0.751	0.788
CRF + word2vec + RF	0.895	0.846	0.871
word2vec + ESIM	0.969	0.961	0.965
DSAMM	0.987	0.982	0.984

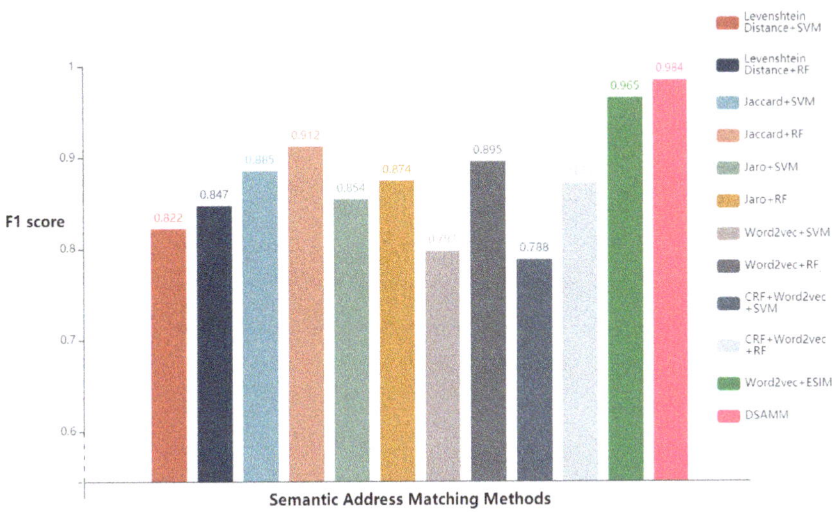

Figure 7. Comparative evaluations of different address matching methods in F1 score.

The deep learning method of "word2vec + ESIM" outperformed the string-similarity-based and machine-learning-based methods in a comparison of results, with all the three evaluation metrics having values above 0.96. This shows how the deep learning framework can significantly increase the accuracy of the semantic address matching. All the final metric values attained by training the DSAMM instances in our study were above 0.98, with the F1 value coming in at 0.984, which represents a notable improvement in prediction evaluation values and the best metrics. This suggests that the migration learning model used in this work can effectively increase the accuracy of semantic address matching. Additionally, we constructed the model without segmenting the address text for elements, which significantly increased the effectiveness of address matching.

4. Conclusions

In this study, we built the ASM with strong address semantic understanding using a pretraining approach to semantic modelling of a vast and complicated address corpus. We introduced a fine-tuning approach in deep transfer learning to achieve a high accuracy of semantic address matching. The main conclusions of this study are as follows:

1. The ASM was constructed using a self-supervised pretrained language model. The experimental results demonstrate that the ASM can achieve a high level of accuracy with the objective of predicting unknown characters.
2. The DSAMM was constructed using the fine-tuning approach in deep transfer learning. The results of the comparison experiments showed that the DSAMM performed the best, with all the metrics above 0.98.
3. It is shown that utilizing deep transfer learning, high address matching accuracy can be attained with only a few labelled training datasets.

However, this study has some limitations. (1) We treated each address record as a sentence, which resulted in the hierarchy of addresses being ignored; (2) our experimental area was limited to a city, which ignored the ambiguity of place names. Therefore, in a future work, we will consider, first, studying how to ensure the accuracy of address matching when there are semantic similarities or ambiguities in addresses at a larger regional scale, and, second, trying to introduce geographical information associated with addresses and learning external knowledge to assist in achieving better address matching.

Author Contributions: Conceptualization, L.X. and F.X.; methodology, L.X., C.Z. and R.M.; validation, X.X. and X.Z.; formal analysis, L.X., F.X. and R.M.; data curation, Y.W.; writing—original draft preparation, L.X.; writing—review and editing, L.X., F.X. and X.Z.; visualization, X.X.; supervision, F.X. All authors have read and agreed to the published version of the manuscript.

Funding: This research was funded by the National Natural Science Foundation of China Grant Nos. 42001354, 42050103; Natural Science Foundation of Zhejiang Province Grant Nos.LGG22D010001, LQ19D010011, LGN22D010003; Scientific Research Fund of Zhejiang Provincial Education Department Grant No. Y202147221; and Scientific Research and Development Foundation of Zhejiang A&F University Grant No. 2020FR060, 2020FR064.

Institutional Review Board Statement: Not applicable.

Informed Consent Statement: Not applicable.

Data Availability Statement: The data used to support the findings of this study are available from the corresponding author upon request.

Acknowledgments: The authors would like to thank the editor and the reviewers for their contributions.

Conflicts of Interest: The authors declare no conflict of interest. The funders had no role in the design of the study; in the collection, analysis, or interpretation of data; in the writing of the manuscript; or in the decision to publish the results.

References

1. Drummond, W.J. Address matching: GIS technology for mapping human activity patterns. *J. Am. Plan. Assoc.* **1995**, *61*, 240–251. [CrossRef]
2. Longley, P.A.; Goodchild, M.F.; Maguire, D.J.; Rhind, D.W. *Geographic Information Systems and Science*; John Wiley & Sons: Hoboken, NJ, USA, 2005.
3. Holtzheimer, P.E. Introduction to the GBF/DIME: A primer. *Comp. Environ. Urban Syst.* **1983**, *8*, 133–173. [CrossRef]
4. Harada, Y.; Shimada, T. Examining the impact of the precision of address geocoding on estimated density of crime locations. *Comput. GeoSci.* **2006**, *32*, 1096–1107. [CrossRef]
5. Chen, J.; Chen, J.; She, X.; Mao, J.; Chen, G. Deep Contrast Learning Approach for Address Semantic Matching. *Appl. Sci.* **2021**, *11*, 7608. [CrossRef]
6. Lee, B.H.; Waddell, P.; Wang, L.; Pendyala, R.M. Reexamining the influence of work and nonwork accessibility on residential location choices with a microanalytic framework. *Environ. Plan. A* **2010**, *42*, 913–930. [CrossRef]
7. Sun, Z.; Qiu, A.G.; Zhao, J.; Zhang, F.; Zhao, Y.; Wang, L. Technology of fuzzy Chinese-geocoding method. In Proceedings of the 2013 International Conference on Information Science and Cloud Computing, Guangzhou, China, 7–8 December 2013; 2013; pp. 7–12.
8. Levenshtein, V.I. Binary codes capable of correcting deletions, insertions, and reversals. *Sov. Phys. Dokl.* **1966**, *10*, 707–710.
9. Navarro, G. A guided tour to approximate string matching. *ACM Comput. Surv. CSUR* **2001**, *33*, 31–88. [CrossRef]
10. Zhang, Z.; Hadjieleftheriou, M.; Ooi, B.C.; Srivastava, D. Bed-tree: An all-purpose index structure for string similarity search based on edit distance. In Proceedings of the 2010 ACM SIGMOD International Conference on Management of data, Indianapolis, IN, USA, 6–10 June 2010; pp. 915–926.
11. Jaccard, P. Nouvelles recherches sur la distribution florale. *Bull. Soc. Vaud. Sci. Nat.* **1908**, *44*, 223–270.

12. Jaro, M.A. Advances in record-linkage methodology as applied to matching the 1985 census of Tampa, Florida. *J. Am. Stat. Assoc.* **1989**, *84*, 414–420. [CrossRef]
13. Winkler, W.E. String Comparator Metrics and Enhanced Decision Rules in the Fellegi-Sunter Model of Record Linkage. In Proceedings of the Section on Survey Research Methods, American Statistical Association, Anaheim, CA, USA, 6–9 August 1990; pp. 354–359.
14. Santos, R.; Murrieta-Flores, P.; Martins, B. Learning to combine multiple string similarity metrics for effective toponym matching. *Int. J. Digit. Earth* **2018**, *11*, 913–938. [CrossRef]
15. Banerjee, S.; Pedersen, T. The design, implementation, and use of the ngram statistics package. In *Proceedings of the International Conference on Intelligent Text Processing and Computational Linguistics, Mexico City, Mexico, 16–22 February 2003*; Springer: Berlin/Heidelberg, Germany; pp. 370–381.
16. Pedersen, T.; Banerjee, S.; McInnes, B.; Kohli, S.; Joshi, M.; Liu, Y. The Ngram statistics package (text:: nsp): A flexible tool for identifying ngrams, collocations, and word associations. In Proceedings of the workshop on multiword expressions: From parsing and generation to the real world, Oregon, Portland, 23 June 2011; pp. 131–133.
17. Wang, Y.; Liu, J.; Guo, Q.; Luo, A. The standardization method of address information for pois from internet based on positional relation. *Acta Geod. Cartogr. Sin.* **2016**, *45*, 623.
18. Xueying, Z.; Guonian, L.; Boqiu, L.; Wenjun, C. Rule-based approach to semantic resolution of Chinese addresses. *J. Geo-Inf. Sci.* **2010**, *12*, 9–16.
19. Cheng, C.; Yu, B. A rule-based segmenting and matching method for fuzzy Chinese addresses. *Geogr. Geo-Inf. Sci.* **2011**, 3.
20. Li, L.; Wang, W.; He, B.; Zhang, Y. A hybrid method for Chinese address segmentation. *Int. J. Geogr. Inf. Sci.* **2018**, *32*, 30–48. [CrossRef]
21. Lin, Y.; Kang, M.; Wu, Y.; Du, Q.; Liu, T. A deep learning architecture for semantic address matching. *Int. J. Geogr. Inf. Sci.* **2020**, *34*, 559–576. [CrossRef]
22. Wu, D.; Fung, P. Improving Chinese tokenization with linguistic filters on statistical lexical acquisition. In Proceedings of the Fourth Conference on Applied Natural Language Processing, Stuttgart, Germany, 13–15 October 1994; pp. 180–181.
23. Lafferty, J.; McCallum, A.; Pereira, F.C. Conditional Random Fields: Probabilistic Models for Segmenting and Labeling Sequence Data. 2001. Available online: https://repository.upenn.edu/cgi/viewcontent.cgi?article=1162&context=cis_papers (accessed on 6 August 2022).
24. Bokaei, M.H.; Sameti, H.; Bahrani, M.; Babaali, B. Segmental HMM-based part-of-speech tagger. In Proceedings of the 2010 International Conference on Audio, Language and Image Processing, Shanghai, China, 23–25 November 2010; pp. 52–56. [CrossRef]
25. Tian, Q.; Ren, F.; Hu, T.; Liu, J.; Li, R.; Du, Q. Using an optimized Chinese address matching method to develop a geocoding service: A case study of Shenzhen, China. *ISPRS Int. J. Geo-Inf.* **2016**, *5*, 65. [CrossRef]
26. Koumarelas, I.; Kroschk, A.; Mosley, C.; Naumann, F. Experience: Enhancing address matching with geocoding and similarity measure selection. *J. Data Inf. Qual. JDIQ* **2018**, *10*, 1–16. [CrossRef]
27. Kang, M.; Du., Q.; Wang, M. A new method of Chinese address extraction based on address tree model. *Acta Geod. Cartogr. Sin.* **2015**, *44*, 99.
28. Luo, M.; Huang, H. New method of Chinese address standardization based on finite state machine theory. *Comput. Appl. Res.* **2016**, *33*, 3691–3695. (In Chinese)
29. Mikolov, T.; Chen, K.; Corrado, G.; Dean, J. Efficient estimation of word representations in vector space. *arXiv* **2013**, arXiv:1301.3781.
30. Peters, M.; Neumann, M.; Iyyer, M.; Gardner, M.; Zettlemoyer, L. Deep contextualized word representations. *arXiv* **2018**, arXiv:1802.05365.
31. Radford, A.; Narasimhan, K.; Salimans, T.; Sutskever, I. Improving Language Understanding by Generative Pre-Training. 2018. Available online: https://s3-us-west-2.amazonaws.com/openai-assets/research-covers/language-unsupervised/language_understanding_paper.pdf (accessed on 6 August 2022).
32. Devlin, J.; Chang, M.W.; Lee, K.; Toutanova, K. Bert: Pre-training of deep bidirectional transformers for language understanding. *arXiv* **2018**, arXiv:1810.04805.
33. Yang, Z.; Dai, Z.; Yang, Y.; Carbonell, J.; Salakhutdinov, R.R.; Le, Q.V. Xlnet: Generalized autoregressive pretraining for language understanding. *Adv. Neural Inf. Process. Syst.* **2019**, *32*, 1–11.
34. Zhang, Z.; Han, X.; Liu, Z.; Jiang, X.; Sun, M.; Liu, Q. ERNIE: Enhanced language representation with informative entities. *arXiv* **2019**, arXiv:1905.07129.
35. Clark, K.; Luong, M.T.; Le, Q.V.; Manning, C.D. Electra: Pre-training text encoders as discriminators rather than generators. *arXiv* **2020**, arXiv:2003.10555.
36. Xu, L.; Du, Z.; Mao, R.; Zhang, F.; Liu, R. GSAM: A deep neural network model for extracting computational representations of Chinese addresses fused with geospatial feature. *Comput. Environ. Urban Syst.* **2020**, *81*, 101473. [CrossRef]
37. Cruz, P.; Vanneschi, L.; Painho, M.; Rita, P. Automatic Identification of Addresses: A Systematic Literature Review. *ISPRS Int. J. Geo-Inf.* **2021**, *11*, 11. [CrossRef]
38. Comber, S.; Arribas-Bel, D. Machine learning innovations in address matching: A practical comparison of word2vec and CRFs. *Trans. GIS* **2019**, *23*, 334–348. [CrossRef]

39. Zhang, C.; Guo, R.; Ma, X.; Kuai, X.; He, B. W-TextCNN: A TextCNN model with weighted word embeddings for Chinese address pattern classification. *Comput. Environ. Urban Syst.* **2022**, *95*, 101819. [CrossRef]
40. Luo, A.; Liu, J.; Li, P.; Wang, Y.; Xu, S. Chinese address standardisation of POIs based on GRU and spatial correlation and applied in multi-source emergency events fusion. *Int. J. Image Data Fusion* **2021**, *12*, 319–334. [CrossRef]
41. Liu, J.; Wang, J.; Zhang, C.; Yang, X.; Deng, J.; Zhu, R.; Nan, X.; Chen, Q. Chinese Address Similarity Calculation Based on Auto Geological Level Tagging. In Proceedings of the International Symposium on Neural Networks, Moscow, Russia, 10–12 July 2019; Springer: Cham, Switzerland; pp. 431–438.
42. Li, H.; Lu, W.; Xie, P.; Li, L. Neural Chinese address parsing. In Proceedings of the 2019 Conference of the North American Chapter of the Association for Computational Linguistics: Human Language Technologies, Minneapolis, MN, USA, 2–7 June 2019; Volume 1, pp. 3421–3431.
43. Li, P.; Luo, A.; Liu, J.; Wang, Y.; Zhu, J.; Deng, Y.; Zhang, J. Bidirectional gated recurrent unit neural network for Chinese address element segmentation. *ISPRS Int. J. Geo-Inf.* **2020**, *9*, 635. [CrossRef]
44. Santos, R.; Murrieta-Flores, P.; Calado, P.; Martins, B. Toponym matching through deep neural networks. *Int. J. Geogr. Inf. Sci.* **2018**, *32*, 324–348. [CrossRef]
45. Shan, S.; Li, Z.; Yang, Q.; Liu, A.; Zhao, L.; Liu, G.; Chen, Z. Geographical address representation learning for address matching. *World Wide Web* **2020**, *23*, 2005–2022. [CrossRef]
46. Li, F.; Lu, Y.; Mao, X.; Duan, J.; Liu, X. Multi-task deep learning model based on hierarchical relations of address elements for semantic address matching. *Neural. Comput. Appl.* **2022**, *34*, 8919–8931. [CrossRef]
47. Chen, Q.; Zhu, X.; Ling, Z.; Wei, S.; Jiang, H.; Inkpen, D. Enhanced LSTM for natural language inference. *arXiv* **2016**, arXiv:1609.06038.
48. Shan, S.; Li, Z.; Qiang, Y.; Liu, A.; Xu, J.; Chen, Z. DeepAM: Deep Semantic Address Representation for Address Matching. In *Proceedings of the Asia-Pacific Web (APWeb) and Web-Age Information Management (WAIM) Joint International Conference on Web and Big Data, Chengdu, China, 1 August 2019*; Springer: Cham, Switzerland; pp. 45–60.
49. Zhang, H.; Ren, F.; Li, H.; Yang, R.; Zhang, S.; Du, Q. Recognition method of new address elements in Chinese address matching based on deep learning. *ISPRS Int. J. Geo-Inf.* **2020**, *9*, 745. [CrossRef]
50. Gupta, V.; Gupta, M.; Garg, J.; Garg, N. Improvement in Semantic Address Matching using Natural Language Processing. In Proceedings of the 2021 2nd International Conference for Emerging Technology (INCET), Belagavi, India, 21–23 May 2021; pp. 1–5.
51. Wei, J.; Zou, K. Eda: Easy data augmentation techniques for boosting performance on text classification tasks. *arXiv* **2019**, arXiv:1901.11196.
52. Qian, C.; Yi, C.; Cheng, C.; Pu, G.; Liu, J. A coarse-to-fine model for geolocating Chinese addresses. *ISPRS Int. J. Geo-Inf.* **2020**, *9*, 698. [CrossRef]
53. Kudo, T.; Richardson, J. Sentencepiece: A simple and language independent subword tokenizer and detokenizer for neural text processing. *arXiv* **2018**, arXiv:1808.06226.
54. Vaswani, A.; Shazeer, N.; Parmar, N.; Uszkoreit, J.; Jones, L.; Gomez, A.N.; Kaiser, L.; Polosukhin, I. Attention is all you need. In Proceedings of the Advances in Neural Information Processing Systems 30 (NIPS 2017), Long Beach, CA, USA, 4–9 December 2017; pp. 6000–6010.
55. Dai, Z.; Yang, Z.; Yang, Y.; Carbonell, J.; Le, Q.V.; Salakhutdinov, R. Transformer-xl: Attentive language models beyond a fixed-length context. *arXiv* **2019**, arXiv:1901.02860.
56. Dai, A.M.; Le, Q.V. Semi-supervised sequence learning. In Proceedings of the Advances in Neural Information Processing Systems 28 (NIPS 2015), Montreal, QC, Canada, 7–12 December 2015; pp. 3079–3087.
57. Powers, D.M. Evaluation: From precision, recall and F-measure to ROC, informedness, markedness and correlation. *arXiv* **2020**, arXiv:2010.16061.
58. Hinton, G.E.; Salakhutdinov, R.R. Reducing the dimensionality of data with neural networks. *Science* **2006**, *313*, 504–507. [CrossRef]
59. Breiman, L. Random forests. *Mach. Learn.* **2001**, *45*, 5–32. [CrossRef]
60. Chang, C.C.; Lin, C.J. LIBSVM: A library for support vector machines. *ACM Trans. Intell. Syst. Technol. TIST* **2011**, *2*, 1–27. [CrossRef]

MDPI AG
Grosspeteranlage 5
4052 Basel
Switzerland
Tel.: +41 61 683 77 34

Applied Sciences Editorial Office
E-mail: applsci@mdpi.com
www.mdpi.com/journal/applsci

www.ingramcontent.com/pod-product-compliance
Lightning Source LLC
Chambersburg PA
CBHW041936260926
51850CB00018B/195